T0336065

Leaders in Animal Behavior
The Second Generation

Animal behavior, as a discipline, has undergone several key transitions over the last 25 years, growing in both depth and breadth. Key advances have been made in behavioral ecology and sociobiology; in the development of studies integrating proximate and ultimate causation; in the integration of laboratory and field work; and in advances in theoretical work in areas such as sexual selection, foraging and life-history traits. Thus it is appropriate to relate the individual stories of those who have had significant impacts on the field as we know it today. *Leaders in Animal Behavior: The Second Generation* is a collection of autobiographies from 21 individuals who have been peer selected, and have provided unique and important contributions to the field in the past 25 years.

LEE DRICKAMER received a Ph.D. in zoology from Michigan State University and is Regents' Professor Emeritus at Northern Arizona University where he teaches introductory biology and classes in animal behavior, ornithology, and behavioral ecology. He is currently the Animal Behavior Society historian and his research interests include population and conservation biology and the behavioral ecology and reproductive traits of rodents and prairie dogs.

DONALD DEWSBURY is Professor Emeritus of Psychology at the University of Florida. His primary interests are in the history of psychology; primarily comparative and experimental psychology and related parts of the biological sciences. He serves as historian and archivist for Divisions 1 and 26 of the American Psychological Association, and is past president of Divisions 1 (Society of General Psychology) and 26 (History of Psychology) and of the Animal Behavior Society. For much of his early career he worked in reproductive and social behavior in various animal species.

LEADERS IN ANIMAL BEHAVIOR

The Second Generation

Edited by

Lee C. Drickamer
Northern Arizona University

Donald A. Dewsbury
University of Florida

CAMBRIDGE
UNIVERSITY PRESS

CAMBRIDGE
UNIVERSITY PRESS

University Printing House, Cambridge CB2 8BS, United Kingdom

One Liberty Plaza, 20th Floor, New York, NY 10006, USA

477 Williamstown Road, Port Melbourne, VIC 3207, Australia

314-321, 3rd Floor, Plot 3, Splendor Forum, Jasola District Centre, New Delhi - 110025, India

103 Penang Road, #05-06/07, Visioncrest Commercial, Singapore 238467

Cambridge University Press is part of the University of Cambridge.

It furthers the University's mission by disseminating knowledge in the pursuit of
education, learning and research at the highest international levels of excellence.

www.cambridge.org
Information on this title: www.cambridge.org/9780521517584

First published 2010
First paperback edition 2022

A catalogue record for this publication is available from the British Library

Library of Congress Cataloging in Publication data
Leaders in animal behaviour : the second generation / edited by Lee Drickamer,
Donald Dewsbury.
p. cm.
Summary: "Animal behavior, as a discipline, has undergone several key transitions over the last 25 years,
growing in both depth and breadth. Key advances have been made in behavioural ecology and socio-biology,
in the development of studies integrating proximate and ultimate causation, in the integration of laboratory and
field work, and in advances in theoretical work in areas such as sexual selection, foraging and life-history traits.
Thus it is appropriate to relate the individual stories of those who have had significant impacts on the field
as we know it today. Leaders in Animal Behavior: The Second Generation is a collection of autobiographies
from 21 individuals that have been peer selected, and have provided unique and important contributions
to the field in the past 25 years" – Provided by publisher
ISBN 978-0-521-51758-4 (hardback)
1. Ethologists – Biography. 2. Animal behavior. I. Drickamer, Lee C.
II. Dewsbury, Donald A., 1939– III. Title.
QL26.L426 2009
591.5092′2–dc22
2009038902

ISBN 978-0-521-51758-4 Hardback
ISBN 978-0-521-74129-3 Paperback

Contents

Preface

The original volume, *Leaders in the Study of Animal Behavior*, was published in 1985. Since then, a number of significant developments have occurred in animal behavior. As the field grows and matures, several threads have emerged. The biggest of these developments has been behavioral ecology, along with sociobiology, which began in earnest in the 1970s. Another thread involves studies of animal cognition and culture; a third thread includes the use of new technologies and methods to explore problems in animal behavior. A fourth thread is the development of evolutionary psychology, and a final thread is the re-emergence of integrating studies of proximate and ultimate causation of behavior. In the summer of 2006, one of us (LCD) approached the other (DAD), who had edited the first volume, with the notion that a second volume, covering these developments in animal behavior over a 20–25 year period, was warranted. We agreed to work together to produce this volume of autobiographies of the second generation of leading individuals in animal behavior.

The initial question we faced, as was the case for the earlier volume, was the matter of selecting potential participants. We discussed this together and with colleagues to develop a thorough procedure that would provide the opportunity for wide consideration of individuals who have contributed to animal behavior during a period extending from the 1960s to the present time. The time period includes a portion of the chronology that was used for the first book; we did this because the contributions of many individuals to our field of study were not recognized by the early 1980s. Their work has become quite significant during the ensuing decades and they thus deserved consideration.

Our selection process involved three steps. First, we developed a panel of 15 judges. We attempted to gather a wide diversity of panel members, including individuals from 6 countries and representing all subfields of modern animal behavior. Second, each judge was asked to nominate up to 10 individuals, without ranking them. We compiled the list of nominees and elected to place the top 43 names on the ballot. Third, we sent ballots to all members of the panel and asked them to rate their top 10 choices in rank order; all individuals who were nominees could not vote for or rank themselves. When that information was compiled, a list of 25 names emerged as the top choices. We contacted all of these individuals to ask them to write an autobiography. Several declined the opportunity to write an autobiography, either for time reasons, or because they had previously published a similar essay. These included John Garcia, Jane Goodall, Robert Trivers, and Fernando Nottebohm.

The present volume includes autobiographical essays from 21 leading figures in the second generation of those who led animal behavior down some new avenues and advanced our knowledge of and methods for obtaining information about all aspects of animal behavior. We reviewed the initial submissions and made minor content and editorial suggestions. We elected not to be heavy-handed in terms of editing, believing that the individual styles of the various authors should remain, as they are representative of different approaches to writing and organization of materials pertaining to their own scientific lives.

We believe that this new book, covering a second generation of those who have led the field of animal behavior, should serve several useful functions. First, it will make an excellent book for seminar discussions pertaining not just to the history of our discipline, but also as a set of windows on how different individuals approached the study of animals. Examining the various influences that affected these scientists and exploring the progression of their individual careers can provide significant insight for students and, indeed, those of us already working in animal behavior. Second, the book serves as a reference and record of what occurred in animal behavior during the last quarter of the twentieth century and early years of the twenty-first century. As a discipline, animal behavior is currently undergoing significant changes, both in terms of the coverage of different topics, as in the renewed focus on integrating proximate and ultimate factors, and also with respect to using new technologies to explore problems in ways that had heretofore not been possible. This latter includes such methods as analyzing hormones and other body chemistry from urine or feces collected in the field. It also includes the use of DNA technology to define relationships between individuals in animal societies and the ability to dissect the developmental processes underlying particular behavior traits.

We appreciate the good work of our panel of experts that aided in the selection process. This group included: Chris Barnard; Michael Beecher; Gordon Burghardt; Hugh Drummond; Jeff Galef; Patricia Gowaty; Felicity Huntingford; Jeff Lucas; Douglas Mock; Chuck Snowdon; Judy Stamps; Michael Taborsky; and Meredith West. We thank Martin Griffiths at Cambridge University Press for genial, enthusiastic, and rapid handling of all phases of the book's production. We thank Patrick Bateson for aiding in making the connection to Cambridge University Press as a potential publisher. Sarah Hrdy deserves our appreciation for providing the title for this volume.

Dewsbury, D. A. (Ed.). (1985). *Leaders in the Study of Animal Behavior: Autobiographical Perspectives*. Lewisburg, PA: Bucknell University Press and Dewsbury, D. A. (Ed.). (1989). *Studying Animal Behavior: Autobiographies of the Founders* Chicago: University of Chicago Press (Paperback edition).

1

Understanding ourselves

RICHARD D. ALEXANDER

Biographies, as generally written, are not only misleading but false. The author makes a wonderful hero of his subjects; he magnifies his perfections, if he has any, and suppresses his imperfections. History is not history unless it is the truth.

Abraham Lincoln

Myth does not mean an absence of truth but a concentration of truths.
Doris Lessing

Would that all of our autobiographical myths could be concentrations of truths!

Autobiographies are trickier than biographies. Stanley Elkin (1993) has suggested that everyone has a worthwhile life story to tell about herself or himself, but no one is likely

Leaders in Animal Behavior: The Second Generation, ed. L. C. Drickamer & D. A. Dewsbury. Published by Cambridge University Press. © Cambridge University Press 2010.

to tell it completely and accurately. Part of the reason for these failures, said Elkin, is that we never reveal everything in the darkest corners of the basements of our lives (his actual phrase was the "nasty hoard" in the "secret cellar"). Another part is that, sometimes, we simply cannot recall and interpret accurately, even if honestly, what really happened, when, and why.

When autobiographies are requested in scientific contexts, the accounts are probably expected to center around the scientific work of the author. This may or may not be the author's first choice in autobiographic materials. But such a focus surely eases the first reason for imperfection, though not eliminating it entirely. All scientists are likely to have at least a few too-dark secrets in the basements of their professional performances, especially in the sociality and ancillary responsibilities of their science. The focus on science also retains difficulties with regard to recalling and interpreting honestly, because all of us so-called scientists have gradually but certainly adjusted and re-adjusted our views of the tendrils of understanding and influence that were generated during the earliest stages of our budding careers, and that have contributed to our becoming what we are. We cling to our interpretations of the steps in our development and performance because they seem to make sense to us, and they are likely to exalt us more than the alternatives; and also because, as is surely healthy in moderation, we tend to like our own versions of ourselves. The saving grace is that, if the connecting aspects of our grown-up version of how we, as we might think, "came to be famous" were put together early enough in our careers, they may actually have influenced significant portions of our life's itinerary.

There is likely a parallel to all of this in the evident reluctance of people in general, including at least most biologists, to accept ourselves – meaning all humanity – as having evolved through a process of differential reproduction, an acceptance that necessarily calls for submitting to a thorough revealing of how we have evolved and what we are evolved to be, and to do. Even universal darkest corners of basements – and *a fortiori* those who publicize them carelessly – can be difficult to tolerate.

When I was invited to provide an autobiography for the first volume, I rejected the idea because, at age 52, I still held the fond belief that I had scarcely begun. Now, on the verge of age 80, thinking otherwise seems a little easier. As might be expected, I have sometimes favored information and activities not represented in my published work, or obscurely represented there. Because the different topics that held my attention across the past 70 years or so did not appear in a simple non-overlapping sequence, the reader will find me returning to earlier dates each time the subject changes. My professional attention to human behavior and evolution did not develop until the mid-1960s, but its origins, I realized belatedly, were older than any of my other academic interests.

Early life

I was born, and lived during the first 16 years of my life, in a modest farmhouse in Sangamon Township, Piatt County, Illinois. Our farm was not far from the north branch of the Sangamon River, and the extensive wooded areas along the river became my principal boyhood haunt. My family made its living from a 151-acre general-purpose farm operated "on the shares" with the landlord, a high school classmate of my parents. We grew corn, oats, clover, alfalfa, and pastures, all as livestock feed – no cash crops. We bred, raised, and marketed hogs and beef calves, and sold cream from several milk cows and eggs from a large flock of hens. We relied heavily on chickens for our own meat because a chicken was the appropriate size for one meal so that there was little or no need for an icebox. Virtually all of the meat, eggs, and milk products that we consumed came from our own animals, and most of our vegetables and fruit came from our two large gardens. We separated cream and skim milk with a hand-cranked separator kept in the kitchen, and made butter with a small hand-cranked churn. Skimmed milk (today's "no fat" milk) was fed to the hogs. My mother preserved meat, vegetables, and fruit enough to last the winter, nearly all that we needed – at first in Mason jars, later with a hand-cranked home-canning machine. Her cook stove was fueled initially with wood, later with coal. She used a hand-operated washing machine and wringer at first, then acquired a used machine operated by a step-start gasoline engine. Our house was heated by two free-standing oil stoves, one in the living room and one in the dining room.

My home environment those first 16 years was a rich one, full of hard work and the incessant demands of a complicated livestock operation. All of our farm work was done with horses until I was 13, the same year that electric lines reached our farm. The independent play and exploration of farm kids in those less complicated days, when there were virtually no "No Trespassing" signs and a higher proportion of unsupervised and unrestricted activities, now seem to me to have been unusually conducive to development of a creative and imaginative approach to life (cf. Alexander 1991b, 2001a, 2004, 2005b, 2006b, mss. 1–3, 5)

I have two siblings: an older sister, Nell Beadles (dentist's wife and homemaker), and a younger brother, Noel (farmer). Both of my parents, Archie Dale Alexander and Katherine Elizabeth Heath Alexander, attended college briefly, and each taught in a one-room country grade school for a few years before changing to farming. I attended a one-room country school for seven years, starting at age five. The second year I was boosted by my teacher, Mrs. Edna Williams, to third grade, the same grade as my older sister (this same sequence occurred for my mother, my father, and my aunt Ruth; other than my father, Ruth was the only one of her eight siblings to complete high school). My school was elegantly spare, with no library other than an 8-volume set of *Compton's Pictured Encyclopedia*. When I was in seventh grade, an 8-volume set of *Book Trails* was added. In seventh and eighth grades, we were required at the end of the school year to spend a day at the county seat, taking written examinations constructed by Charles MacIntosh, the county superintendent of schools, to verify that we were qualified to proceed into high school.

My high school graduated 46 students in my class of 1946. I had spent six years in 4-H and four in the Future Farmers of America. I was chosen by the local Rotary Club as the "outstanding boy" in my high school class, but I graduated fifth or so, behind a slate of scholarly girls. I never won an athletic letter, a failure that has always bothered me. I did win "outstanding student" awards in art and agriculture, and a blue ribbon in a saxophone quartet at a state band contest; and I achieved State Farmer status in the Future Farmers of America, only the second instance in my high school (the other my oldest male first cousin). I sang in the school chorus and in a male quartet, and played on "scrub" teams (nowadays called "junior varsity") in football and basketball. I also tried once to run the mile in track competition and came in last.

My favorite course in high school was English, with an emphasis on poetry. The use of language, in all of its aspects, never stopped being a passion, and I always regarded it as my second choice in teaching. It occurred to me long ago that even such a seemingly simple source as a dictionary can be used to extract broadly interesting and deeply significant information about ourselves as members of the human species.

I remember a serious discussion with our English teacher, Miss Katharine Turner, about Robert Frost's intended meaning for his poem *Mending Wall*. Miss Turner suggested to the class that the poem showed that farmers are generally too set in their ways to change in accord with newer times. Sensitive about the sometimes tense interactions of my father and his neighbor, who each held responsibility for half of the livestock fence between their two farms, I argued to her that the poem reflected Frost's belief that for neighbors to cooperate in repairing that no longer essential stone wall between their properties helped maintain their friendship and cordiality. I liked Frost's, "Good fences make good neighbors." But I haven't forgotten his question to someone who asked him to explain one of his poems. He is reported to have sat in silence for a moment, before he said, "What do you want me to do, say it in a worser way?"

In 2001, I was inducted into the Monticello Community High School Hall of Fame – a delightful experience that included riding with Lorrie, my best friend now of 61 years, and wife of 58 years, and with some of the grandchildren of my brother (our own four grandchildren were attending school in California). On the afternoon of Homecoming day we traveled slowly in a broad circle through the town of Monticello, in the lead car of the parade, with a huge sign on the door with my name on it. During the entire ride we called back and forth with friends, relatives, and classmates among the crowds that lined the parade route, some of whom I hadn't seen in more than 60 years.

In the fall of 1945, Helen Burgoyne, a teacher of Latin, whom I had not even known was my high school advisor, walked past my girlfriend and me in the hallway and paused to ask if I intended to go to college in the fall. I said I didn't know, and wondered aloud why she had asked. She looked at me for a moment and said, "Come with me." She took me to the school office and showed me my scores on college entrance examinations and other tests, all of previously unknown significance to me. I was dumbfounded, by her interest, by the existence of the tests, and by the scores she showed me. When I explained that my family had no funds to send two offspring to college at the same time, she told me about Blackburn College, a southern Illinois school with a student work plan and a total expense of US$250 for the entire year, including room and board. That was, of course, a time when a farm hand's pay was approximately a dollar a day, and room and board; and when a new Ford V-8 sedan cost a little over US$700.

After five semesters at Blackburn (described in detail in Alexander 2001b), I transferred to the Illinois State Normal College (now Illinois State University), where I majored in education in biology, graduating in August 1950. I had decided I needed courses not given at Blackburn, and also realized I could work more hours outside classes there, earning my way completely by attending classes and retaining jobs 12 months of the year (Blackburn restricted students to 15 hours of work each week). Because of this schedule, and my indecision about professions, I accumulated over 150 semester hours of coursework. Without my being conscious of the reasons then, most of my optional courses were human-oriented, on such topics as introductory and educational psychology, abnormal sociology, philosophy, history, economic geography, and others.

During undergraduate years my interests progressed through a sequence of perhaps predictable changes in anticipated and desired professions: dairy farmer, veterinarian, county farm advisor, chemist, teacher, biologist, and finally entomologist. Two of these dreams failed for specific reasons. My necessarily out-of-state application to veterinary school, when I was a sophomore with a single course in biology, was rejected. I also discovered that the severe headaches that frustrated me in chemistry laboratories were a reaction to nitric acid fumes.

My "off-the-farm" work began at age 14 or 15, when I detassled seed corn. A year or so later, as a result of the World War II shortage of metal for repairing combines, I "ran a rack" (loaded and drove a wagon transporting bundles of oats to the thresher) with a team of horses, across 2–3 weeks and many miles, on the last old-time threshing run in Piatt County, Illinois (Alexander ms 2). At ages 15 and 16, I worked parts of two summers on a neighboring farmer's commercial hay baler (Alexander ms 2), and spent much of one summer mowing fencerows with a hand scythe on a neighboring farm – a half mile a day. During two summers, 17 and 18, I worked in a gravel quarry, first as clay-picker on a rock-crusher, and later as grease monkey on the gravel washing plant. For a short time I worked 60 feet above the ground helping construct overhead bins for sand and gravel, a stint that earned me the union wages of a high steel worker. I also went through the initial steps of learning to weld and to operate a dragline crane and a bulldozer (Alexander ms 2). At Blackburn I fired furnaces, tended the lawn, felled trees, hauled garbage, painted interiors and exteriors of buildings (dorms, offices, and a chapel), janitored classrooms, and helped tear out and remove an old railroad. At Normal I worked at several jobs simultaneously: handyman for multiple families, washing dishes and glassware in a restaurant and a chemistry department, and serving as biology department teaching assistant and stockroom clerk. As a graduate student at Ohio State University I was of course a teaching assistant. During the first summer break I bottled milk at the university dairy and later made illustrations for an entomology textbook being authored by two professors: my eventual doctoral advisor, Donald J. Borror, and Dwight M. Delong.

My roommate, across nearly all four years of undergraduate work, and ultimately my 60-year closest and most influential male friend, was the late Dr Carl Walter Campbell. He was, like me, an Illinois farm kid. As a University of Florida professor, he came to be regarded as the first-ranked tropical fruit horticulturist in the world and one of the three most important of all time. As students, Carl and I worked at the same jobs, took almost all the same courses, sang together in choirs, glee clubs, and quartets, and for nearly two years cooked most of our meals in basement kitchens. Among other things we learned how to make a meal of bread and milk-and-flour gravy that included half a pound of pig liver, the last item costing seven cents. Carl and I thrived on thinking of ourselves as deep and profound thinkers, and serious critics of all that surrounded us. No topic was outside our limits. While undergraduates, we generated a strong interest in American folk music and occasionally performed together. As a result I learned to play, rudimentarily and left-handed, guitar, fiddle, harmonica, and five-string banjo, and I searched out and saved a few nine-teenth century local songs, known only to one or two of the then oldest people in the farming community of my childhood (Alexander ms 2). Carl, who died in 2006, was an extraordi-nary intellectual influence across my entire adult life, but especially during our under-graduate years. In 2006, he and I received honorary doctorates in science from Blackburn College at the same podium on the same day without the then Blackburn officials knowing beforehand of our relationship. As well, only two years apart, we received the inaugural lifetime achievement awards of our most appropriate but quite different scientific societies. We each had a hand in starting one of those now international societies, dealing, respec-tively, with tropical fruit horticulture and human behavior and evolution.

Although I rejected a career in agriculture after only one year in college, I never lost my attachment to farming as a way of life. As a result, Lorrie and I have operated an 80-acre farm continuously for 35 years. We have raised mostly hay, and maintained several pastures; continuously bred and raised horses and cattle; and started and trained riding horses. Our major operation has been maintaining breeding herds of horses to support my effort to understand and write about horse social behavior and the horse–human interaction (Alexander 2001a, 2006b, ms 3). Included among our crops were our hard-working, hard-thinking daughters, Susan and Nancy, now school teachers in California, and, more recently, four diverse, exceptional, and dear grandchildren who continue to visit us regularly: Alex and Lydia Turner, Morgan Johnson, and Winona Johnson-Alexander.

I always liked school, but I sometimes had trouble staying with its specific projects. If I had been in elementary school more recently, I might have been diagnosed as a victim of one of the now widely discussed short-attention-span afflictions. As a freshman in college, having to study hard for the first time in my life, I realized that, compared with fellow students in my study hall, I had difficulty continuing an assignment until it was finished. I contemplated this problem seriously, until I generated the conscious strategy of continuing an intellectual activity only until my interest began to lag, then changing immediately to the most attractive alternative and repeating, returning to the temporarily abandoned projects one by one – over and over if necessary – until all were completed. I think I eventually learned to follow this procedure without conscious effort. It has made me an insufferably frustrating co-author and caused my working spaces to be cluttered to the point of being intolerable to others. But, even though I may have taken the strategy to extremes, it seems to have worked reasonably well. Several years ago a publisher's representative sat down with me at a meeting of the Human Behavior and Evolution Society, noted that her company had recently published the autobiography of a certain prominent evolutionary biologist, and suggested that I might also write an autobiography for them. I replied that I hoped to write an autobiography someday, but it was not an immediate goal. I said I had a number of partly finished book manuscripts on my computer that I regarded as more pressing. After a moment she asked how many book manuscripts I had on my computer. I answered immediately and truthfully, "Fifty-five." After a significant period of silence (or a period of significant silence!), she rose and said, "Thank you," as she departed without looking back. I still have about 40 of those manuscripts to go, so I have little hope of reversing her opinion that I am stark raving mad.

In the fall of 1949, as a college senior, I applied for a high school science teaching job in southern Illinois, was interviewed, and accepted the position. While the contract was in the mail, two professors, Ernest M. R. Lamkey (bacteriologist) and Donald T. Ries (entomologist), heard about my prospective job and phoned me repeatedly, urging me instead to consider graduate school. I told them I knew nothing about graduate school, and, because of my father's serious (soon to be terminal) illness, neither my parents nor I had funds for such a purpose. As a result, each of these two thoughtful professors secured a teaching fellowship for me, by communicating with old friends, Ries with Howard Evans at Cornell and Lamkey with Alvah Peterson at Ohio State. I chose Ohio State because it was closer to my home in Illinois (thus, to my ill father), and entomology over botany and zoology, because of how well I had liked Ries's undergraduate entomology course. Otherwise, according to Howard Evans, I would have been his first doctoral student at Cornell. I likely would have studied wasp behavior because I was already attracted to it. Instead I became the first doctoral student of Donald J. Borror and studied, across the first phase of my professional life, speciation, acoustical behavior, sexual behavior, and aggression in the singing insects, and the evolution of life history patterns in crickets. My Master's advisor at OSU was Lamkey's old college roommate, Alvah Peterson.

A week after I obtained the B.Sc., Lorraine Kearnes, whom I met at Blackburn, and I were married and left immediately for Columbus, Ohio. On my first day at OSU, before classes had commenced, I talked with Dr Peterson about courses. One of them, we decided, would be an independent research course on dragonflies, an insect group that had fascinated me while I was amassing an insect collection for an undergraduate entomology course. I thought of dragonflies as the dinosaurs of the insect world. I had first paid close attention to them in an abandoned and partly flooded gravel pit in Normal, Illinois. When I left Peterson's office that afternoon I went directly to the Olentangy River, carrying an aquatic net and jars, and began to collect dragonfly and damselfly juveniles. By the end of the term I had distinguished the juveniles of all of the local species, and matched all but one of them to the known adults in the area. That one may have been a new species with cryptic adults. I never found out. But I had become permanently imprinted on the Odonata, and I have continued across many years an effort to explain their unique – perhaps uniquely bizarre – copulatory behavior, in conjunction with my interest in the origin of insect wings and the phylogeny of mating behavior in the Arthropoda (Alexander 1961, 1967a, ms. 4; Alexander *et al.* 1997; Carle 1982).

After the first term at OSU, I nearly dropped out of graduate school because of a B− grade in Donald J. Borror's course on insect systematics, which I regarded as my most important course. I studied alone that first term, knew virtually nothing about graduate school or the department, and believed (erroneously) that Borror couldn't possibly expect us to learn "all that stuff." Lorrie earnestly talked me into continuing one more term, during which I managed to elevate my almost defunct self-respect by obtaining the highest grade in the second term of Borror's (to me, and to many others) all-important course.

In August 1951, I completed a master's degree with a thesis on the biology of arthropods living inside shelf fungi. I collected fungi everywhere I could within about 100 miles of the University, placing them in paper bags and watering the fungi at intervals, saving some immatures for illustration, and allowing the rest to become adults so that I could identify them. The entomology graduate students at Ohio State then were a wonderful group. Regardless of the nature of their thesis work, nearly every one had chosen a particular group of insects to study, biologically and systematically. Carloads of entomology graduate students traveled to unglaciated southeastern Ohio on weekends, often sleeping overnight in the woods. It was an exhilarating and rewarding time that I believe had a strong influence on my later scientific activities.

During the last part of that summer I completed 50 or 60 drawings for Borror's entomology text (including a drawing of a male ant used on the cover spine) before entering the Army in September. Most of my drawings are still in that remarkable textbook, which may have set a record by being continuously in print, and in classroom use, for 54 years; it is currently a text in the introductory entomology course at the University of Michigan. Imagine my pleasure when, for the most recent (7th) edition of what is now called *Borror's Biology of Insects*, I received a request from Charles A. Triplehorn, one of the current authors, and a close friend and graduate student colleague, 1950–51, for use of two of my more recent drawings.

In the fall of 1951, I was drafted into the Army – probably because my conscience, from having experienced the extreme patriotism of WWII, prevented me from completing the examination that determined whether or not a graduate student could be drafted. It was later explained to me in front of the Monticello, Illinois, courthouse and all the other departing inductees, that I was drafted out of graduate school because the fellow who otherwise would have gone had to stay home and help his father harvest corn. I accepted this decision with a faint humor but without resentment. In Chicago I was invited to apply for a direct commission. When I realized I would have to serve three years instead of two, plus the six months before the commission could be approved, I discarded the application, enabling me to return to Ohio State in the fall of 1953. Luckily for me, the war was formally ended less than two weeks before I was scheduled to be shipped to Korea as an infantry rifleman, one of the most dangerous roles in an extremely dangerous war. After basic training I became an entomologist in the Medical Service Corps at Fort Knox, Kentucky, in the winter inspecting mess halls and in the summer locating potential malaria-carrying mosquito breeding sites for the pest control unit. I was assigned to the latter job because of the concern that returning soldiers might bring Korean malaria back with them. During that time I also explored the caves of Fort Knox with a fellow soldier, became interested in speciation in the blind cave beetles there, and hoped for a while to do my doctoral research on that problem at the

University of Chicago, with Allee and Emerson, until I learned that both had retired. In 1960, Emerson served on the committee that awarded me the AAAS Prize.

Career stage 1: the singing insects

While in the Army I successfully sought to work with Donald J. Borror when I returned to OSU. When I walked into his office he handed me a new Magnemite battery-operated tape recorder with a hand-cranked spring-powered motor, and D cell batteries powering the recording apparatus. It was one of the first battery-operated tape recorders made following American acquisition of tape recording technology from the Germans after winning the war. Borror simply said to me, "Why don't you take this out in the field and see what you can learn about the singing insects." I did that, and by late summer I had managed to tape-record, collect, and identify all of the species of crickets, katydids, and cicadas in the vicinity of Columbus, Ohio. Included were many undescribed and misidentified species. Dr Edward S. Thomas, an astute and kind lawyer and natural historian curating insects in the Ohio State Museum, who knew most of the singing insects of Ohio well, took me under his wing. I also wrote for advice to Bentley B. Fulton, then a professor at North Carolina State. He sent me a wonderful letter welcoming me to the study of singing insects, and telling me exactly how he would proceed in my situation. Later I visited him, and we became good friends. When he died, I was invited to honor him by lecturing to the general session of the North Carolina Academy of Science. I titled my talk *Dr Bentley B. Fulton and the Singing Insects.*

Among other things, Fulton was the first to hybridize different species of crickets and show, by using ingenious devices he constructed himself, that the songs of the hybrids were distinctive, and never heard in nature despite micro-geographic intermixing of individuals of the species involved. Though he didn't mention it, his (1933) experiments also showed that species differences, as well as "complex organs" within species, are due to a history of what Darwin called "numerous, slight, successive modifications" (see also Alexander 1968b, 1978, 1979, ms 1). Fulton thus demonstrated that species differences can be explained by micro-evolution, and that, at least in this case, so-called macro-evolution is simply micro-evolution extended. The relevant one of Darwin's several remarkable challenges was: "If it could be demonstrated that any complex organ existed, which could not possibly have been formed by numerous, successive, slight modifications, my theory would absolutely break down." (Darwin 1859, p. 189; for other challenges, see Alexander 1979, ms. 1). Needless to say, not this challenge, nor perhaps any of Darwin's other challenges, has been met. This particular challenge of Darwin's, and Fulton's extension of it to species differences, nullifies a major argument of those who think "intelligent design" is necessary to explain species differences (see also Alexander 1978, 1979, 2008, ms 1).

My first contribution to science was a paper delivered in 1954 at the Ohio Academy of Science meetings in Athens, Ohio. The title was *Songs of the House Cricket.* It was an effort to analyze and explain functionally the array of stridulatory signals made by the introduced European House Cricket, *Acheta domesticus* Linnaeus, using the tape recorders acquired by Borror, and the only OSU audiospectrograph, made available by Joseph Hynek (later director of the UFO project) in the astronomy observatory. The talk was something of a disaster. I had ambitiously – and recklessly – planned to synchronize slides of audiospectrographs of songs with tape recordings of the same songs. Perhaps this had never been attempted before, at least publicly. Unknown to me, the OSU professor, Carl Reese, who agreed to show the slides for me, had opened his car door into the street during the noon hour, just before my talk, and a passing motorist had slammed the car door around against the side of his car. The poor man came in late, still traumatized, and immediately got my slides irreversibly out of synchrony with the tapes. As a result the talk was hopelessly confusing and ran well over its time. In charge of the session was Alvah Peterson, a rigid taskmaster with respect to time-keeping in scientific meetings. He was known for responding to pleas for a little more time by calling for a vote from the audience on whether the speaker should be allowed to continue. No one wanted to run that gauntlet, and Peterson was not lenient just because I had been his master's student. Somehow I struggled through that first talk at a scientific meeting, but at the end there was considerably more laughter than clapping.

In 1955, I delivered two talks on my research at the Cincinnati meetings of the Entomological Society of America. One of the sessions, a symposium on systematics, was hosted by Professor Theodore H. Hubbell, then Director of the Museum of Zoology at the University of Michigan. Hubbell and I had a mild disagreement in the discussion following my talk, and I uncompromisingly expressed a strong view. He thought I was describing species without having located any morphological differences between them, although I hadn't actually done that – yet. He ducked his head, in what I would later learn was his politely humble manner, and asked, "What about the poor curator who cannot distinguish specimens of the different species?" I replied – I'm not sure how much defensively and how much arrogantly – "Well – if I were a curator I would rather label a specimen *Gryllus* sp., and know it was correct, than label it *Gryllus assimilis* and know it was wrong." At that time I "knew" I would never be a curator because I had declared more than once that I would not take such a job; nor did I know that Hubbell had labeled virtually all U.S. *Gryllus* specimens in the University of Michigan Museum of Zoology as *Gryllus assimilis*, and of course added on each label: "det. T. H. Hubbell."

Some time later, I learned that Hubbell and Delong had subsequently been on an ESA committee together, and had discussed me as a possible job candidate at the University of Michigan. As a result, even though the Museum of Zoology had hired a second entomologist in 1956, an unprecedented third position was established, and I was invited to give a job seminar. After my seminar, in a five-minute encounter in the darkness of the back porch of Hubbell's home, the chair of the Zoology Department, Dugald E. S. Brown, a molecular biologist, hired me half-time in each of the two units with a verbal offer of US$5200 for a 12-month position. There were no other candidates. This whole event is one of several I have described in this essay that probably can no longer occur "legitimately" in academia in America.

In August 1957, Lorrie and I moved our family to the University of Michigan, and I began as Instructor in Zoology and Curator of Insects. Almost 44 years later, following a five-year period as Director of the Museum of Zoology, I retired, delighted to be the Theodore H. Hubbell Distinguished University Professor Emeritus of Evolutionary Biology and Curator Emeritus of Insects. Some time before the retirement I asked Dr Hubbell, then in his late 80s, why he had hired me after what I remembered as my smart alec performance in our long-ago interaction in Cincinnati. True to his "Prince of a Man" nature (as he was designated by his friend and colleague, Dr Irving J. Cantrall), Dr. Hubbell replied simply, "I respected your opinion."

I have always regarded my student life as a series of accidents and last-minute assists from perceptive people, mainly teachers. If Helen Burgoyne had not paused in a hallway in the fall of 1945, and asked me if I was going to college, I may not have gone. If Professors Lamkey and Ries had not invested considerable time and effort in encouraging and enabling me to start graduate school, I almost certainly would never have done it. At the end of my first semester as a graduate student, if Lorrie had not pleaded for me to try one more term, I definitely would have dropped out. If Dwight DeLong had not secured a truly last-minute 15-month Rockefeller Foundation postdoctoral position for me when I failed to obtain a suitable position after graduating, I likely would not have remained in academia. If Donald Borror had not passed on to me his own invitation to speak on singing insects at that 1955 Entomological Society of America symposium on systematics, if Theodore H. Hubbell had not been the chair of that symposium, and if Hubbell had not subsequently been on that ESA committee with DeLong, a position would not have been created for me at the University of Michigan. It is difficult to imagine what might have happened. I had been offered a temporary position by the chair of the entomology department at the University of Illinois, but I have always believed that I would have gone back to Illinois as a farmer.

Despite the spur-of-the-moment nature of my choice of entomology for graduate work (rather than zoology or botany), I never regretted it, largely because of three truly outstanding entomology professors at the Ohio State University, whose memories I revere: Alvah E. Peterson, Donald J. Borror, and Dwight M. Delong. They and Theodore H. Hubbell primed the pumps of whatever success I have realized. Borror started me on a wonderful set of problems when he handed me one of his Magnemite tape recorders. He also turned out to be the kind of advisor who demonstrates continually how to get important things done and lets his students conduct their own projects. There could not have been a better advisor for me.

By the time I had completed my doctoral thesis I had shown that nearly a third of the species of singing insects in central and southeastern Ohio were as yet unrecognized, and others were recognized but wrongly classified, wrongly understood, or wrongly named. I was astonished that almost half of the singing insects on the OSU campus were undescribed. I had become fascinated by several special problems in crickets, including mating behavior, male aggression and territoriality (Alexander 1961), the evolution of life history variations (Alexander 1968a), and a new method of speciation that Robert S. Bigelow and I later called allochronic speciation (Alexander and Bigelow 1960). A more appropriate adjective would have been allohoric – referring to seasonal separation of adults – but we didn't like its sound, or the possible confusion that might have been engendered, involving gradual sympatric seasonal divergence in the seasonal mating period, which is not likely to result in speciation.

In 1959 Bigelow and I presented the speciation paper at the Darwin Centennial meeting of the Society for the Study of Evolution at the University of Chicago, and later published it in *Evolution* (Alexander and Bigelow 1960). Neither of us doubted that we were correct, despite significant skepticism and subsequent molecular information convincing others that it could not have happened as we said (Mayr 1963; Harrison 1979). We knew that species which overwinter in two different life stages, so widely separated on the life cycle that they breed non-overlappingly in their brief adult seasons, are subject to speciation via accidental and more or less sudden seasonal separation of adults (Alexander 1968a). We knew that there are several pairs of sister species of Orthoptera with this particular difference. A third species that we regarded from the first as having some of the relevant attributes of the ancestor of the two allochronic but phenotypically, ecologically, and geographically almost identical species turned out to be able to produce both life cycles among the offspring of a single female (Walker 1980; Masaki and Walker 1987; R. D. Alexander, unpublished). We also realized that the strongly divergent selection resulting from the dramatic shift between the two life cycles could possibly result in genetic happenstances that could confuse the effort to place the speciation event temporally, or on a phylogenetic tree of *Gryllus* species. The only questions seem to be how anciently and how often such speciation has taken place. We came to understand the difficulty in verifying precisely how even "ordinary" instances of speciation take place, difficulties that restrict biologists to understanding speciation mainly from broadly comparative study – including representations of all the stages of speciation by combining multiple instances of speciation, rather than being able to view all stages of speciation in single cases. These difficulties cause biologists to withdraw from unusual or unique forms of speciation, in particular when such cases require learning obscure or unfamiliar details of the biology of the organisms involved (Alexander 1967c).

My work with cricket biology eventually resulted in an effort to compare the evolution of mating behavior in all arthropods (Alexander 1967a) and generated the theory, still unfalsified, I think, that insect wings were courtship devices prior to their becoming flying organs (Alexander and Brown 1963), whether or not they actually began with still another function (cf. Kukalova-Peck 1978, 1983). This suggestion correlated with the discovery that primitively wingless hexapods do not copulate directly, and that the earliest copulations were almost certainly luring acts in which the female mounted the male. As with modern crickets and some others that copulate female-above, the female is stimulated to mount and remain mounted by a wide variety of glands and other stimuli on the male's back, usually exposed by the male lifting and vibrating the wings, and sometimes including the female eating the male's wings (Alexander 1967a; Alexander and Otte 1967).

The area I was covering in my field work on singing insects kept expanding outside Ohio, and because of a Rockefeller Foundation postdoctoral grant I was able to work in most of the eastern United States. After moving to Michigan in 1957, I began to include all of the U.S., and Mexico down to Cuernavaca. Eventually I was able to distinguish the songs of nearly 1000 species of mostly North American crickets, katydids, and cicadas. Once a song has been learned, the geographic and ecological distribution, abundance, seasonality, and diurnal activity of that species can be learned perhaps more quickly than can be accomplished with any other kind of organism; and locating and collecting individuals is also much quicker and easier. The reason is that among sympatric and synchronic species, each species has a unique song, and the songs can not only be used to locate and approach individuals, but also are distinguishable while one is driving, sometimes at speeds of 30–40 mph. Some insect songs can be recorded continuously with a microphone in a parabola held in the window of a slow-moving vehicle, thereby capturing the songs of all singing males in a transect along the road, demonstrating continuous changes in songs, for example, across hybrid zones (for extensive data from a long and complicated hybrid zone, see Alexander 1968b). On many occasions my students and I have been able to capture the only individual heard singing across hundreds of miles, even though it was often singing in a hidden location, or in the top of a tall tree. I learned that the field crickets (*Gryllus* spp.), said by James A. G. Rehn and Morgan Hebard in 1919 to be a single species all over the New World (a conclusion repeatedly maintained by Rehn), included a large number of different species. Three decades ago, as my primary interests were changing, I turned over all of my *Gryllus* specimens and records, including scatterdiagrams of songs and morphology, to Dr David Weismann, who presently estimates there are at least 50 U.S. species in the genus (personal communication). I also made all of my singing insect materials available to Dr Thomas J. Walker.

One of the discoveries that puzzled me, in studying the singing insects, was the absence of character displacement (Alexander 1969). I could not convince myself that this paradox had been resolved by anyone. Eventually I decided it was important to work on a different continent and compare the two faunas with regard to this and other questions. Daniel Otte, my third doctoral student (following Kenneth C. Shaw and Mary Jane West), and I took our families and flew to Australia in 1968, using funds from National Science Foundation and Guggenheim grants, and from the University of Michigan, which allowed me to take a six month sabbatical leave followed by nine months of off-campus duty. Otte and I traveled 46,000 miles in 12 months in Australia, in a Land Rover, and discovered approximately 376 new cricket species, and 492 species in all (Otte and Alexander 1983). We also came away with recorded songs of approximately 35 cricket species that we were never able to collect, chiefly because they lived in the tops of tall trees in jungles where the trees could not be felled because of density and vines.

As in the U.S., we found no obvious cases of character displacement. Using songs, we were able to develop better maps of distribution of even the most commonplace species than could have been accomplished by using all of the Australian specimens in all of the collections of the world. On our first trip north in Australia we predicted, and then were able to demonstrate, that two species of *Teleogryllus* – being used in a large experiment in Melbourne, and expected to cause field hybridization that would produce sterile or out-of-season hybrids and interfere with rangeland depredation by field crickets – were living in sympatry without hybridizing. When we reported this to the experimenters, from a region none of them had inspected, they checked what we had said, and abandoned their experiment. We also discovered that what was regarded as two or three cricket species depredating the outback rangelands was actually many times that, thus an enormously more difficult proposition. I have always believed that these discoveries alone more than justified the financial cost of our work in Australia, scoffed at by the media (e.g. Paul Harvey), as well as Indiana's U.S. Representative Richard Roudebush. The eventual result was a confrontation between the Director of the National Science Foundation and the Congress regarding two grants, mine and that of Gordon Orians for biological work on red-winged blackbirds.

Since our monograph on Australian crickets was published (Otte and Alexander 1983), Dan Otte has gone on to become the world's all-time outstanding cricket systematist, doubling the known world species of crickets while also expanding his thesis work on grasshoppers in comprehensive and superbly illustrated monographs of the North American fauna. When I once asked him why he turned to crickets rather than continuing solely with the grasshoppers with which he had worked as a graduate student, he said simply, "Australia."

One of the most amazing groups of insects, studied by Thomas E. Moore and me primarily in the 1950s and 1960s, comprises the 17- and 13-year cicadas. Among other things we discovered that, rather than the one or two species generally accepted, this group included three 13-year species and three 17-year species, each species with a sister species having the other life cycle. Subsequently, two of my students, John Cooley and David Marshall, in 1997, discovered a seventh species, evidently derived from a 17-year population in the area where the sets of species with the two life cycles meet. This peripheral population had evidently changed its life cycle from 17 years to 13 years and eventually spread well into the 13-year species' geographic range, and as a result was still undergoing character displacement in song (Cooley *et al.* 2001; Marshall & Cooley 2000). This ongoing case of character displacement, in species with such long life cycles, and the necessarily infrequent encounters between species, fuels the speculation that one reason character displacement is so rarely observed is that it typically quickly reaches the stage at which selection is no longer favoring divergence in the relevant traits, and that condition spreads rapidly throughout its range. The details of the life histories, speciation, and acoustical communication of the periodical cicadas have made it one of the insect wonders of the world (Alexander and Moore 1962; Alexander 1967c, 1968b; Cooley *et al.* 2001).

The first time I participated in a discussion of how my lifetime interest in biology might have been generated was at a faculty lunch with several biologists at the University of Michigan 50 years ago. I was surprised to learn there that many of my colleagues had been youthful natural historians, or insect collectors, some preparing scientific papers while they were still in high school. As with Sewell Wright, some of them spoke of reading Darwin at early ages. Excepting my rambling explorations in the local woods and along the Sangamon River, and my efforts to make pets of a large number of local mammals and birds (raccoons, possums, a gray fox, 13-lined ground squirrels, white-footed mice, a hawk, a crow, pigeons, starlings, and betsy beetles), I had little experience with such things. I did have an older first cousin (one among 34) who, while in high school, netted butterflies, ran strings lengthwise through their bodies, and used them to decorate his room. On this account I regarded him as perverse, and still do; perhaps he was, because he became a daredevil pilot rather than an entomologist, eventually dying in a crop-dusting accident in Georgia, caused by a wing-dipping greeting to a Georgia farmer, which unfortunately contacted a power line.

During approximately half of my time concentrating on the singing insects, evolutionary biologists were largely restricted to studies of evolutionary pattern: fossils, and comparative studies derived from phylogenies of species. Attention to adaptations was so limited and general as to be largely trivial. It was a time when evolution was often defined, perhaps timorously, as change across time. Population genetics seemed to be flourishing, but, owing to the manner in which fitness was represented in equations, it was for decades typically practiced in such a way that the effect of differential reproduction was seen as maximizing the fitness of populations. Eventually the fallacious argument was made that rapid evolution constitutes a threat to the survival of a population because rare new beneficial mutations automatically lower the fitness of the abundant alleles already present (Brues (1964) explained the error; cf. Alexander 1988). Serious studies of evolutionary processes, especially efforts to understand evolutionary adaptation in detail, began to flourish following George Williams's 1966 book *Adaptation and Natural Selection*.

Career stage 2: evolutionary adaptation, social behavior, and the evolution of eusociality

In 1955, while I was a postdoc, and program chairman of the Columbus Entomological Society, Donald R. Meyer (an OSU psychologist) and I cooperated to bring Theodore C. Schneirla to the OSU campus to discuss European ethology, and to speak to the entomologists about army ants and to the psychologists on the ontogeny of emotions in higher mammals. As a result Schneirla became a good friend who stopped in to visit me at Michigan whenever he came to see Norman R. F. Maier, his friend and co-author on one of the earliest American texts on animal behavior. Maier, a behavioral and later industrial psychologist, was at that time the only U-M person who had received what was then referred to as the AAAS Prize (now known as the Newcomb Cleveland Prize and given for a different kind of contribution). These two men happened to be among the first people I told in 1961 – proudly, if I remember correctly(!) – that I had just won that prize. The three of us were having lunch together in downtown Ann Arbor.

When I decided to dedicate *Darwinism and Human Affairs* to the great teachers in my life, I realized with a shock that I would have to include Don Meyer. His course discussing European ethology, using Tinbergen's *The Study of Instinct* as the text (criticized severely in the course), frustrated me so that I spent most of the time between sessions figuring out how to demonstrate to him what I regarded as the error of his ways. I wasn't entirely enthusiastic about European ethology, mostly because of some of its theoretical assumptions; but I had soft spots for several of its suggestions, such as seeking to work out the entire behavioral repertoires (ethograms) of species in evolutionary terms. I also regarded Meyer as terribly lazy because he consistently came to class saying he had lost his notes, or was unprepared. He would walk into his classroom with nothing in his hands, sit on his desk facing us, tell us he had lost his notes – or something else of the same nature – and ask us what we would like to discuss. Long afterward I realized that because such behavior had kept me stirred up and angry, he had (accidentally, I thought then) taught me an immense amount. I spent most of my time between his class periods trying to find ways to thwart his arguments. Forty years after taking Meyer's course, and 17 years after including him in the dedication of my book, at a used book sale in Ypsilanti, Michigan, I found a symposium volume that included a paper of his on rhesus monkeys. In it I read with astonishment his contention that he had deliberately conducted his classes in the way that he did because he regarded formal lectures as an imperfect way to teach. There is no doubt that Donald R. Meyer and Theodore C. Schneirla influenced my life significantly, even if in directions neither of them would necessarily favor.

I think it is accurate to say that the second phase of my career as an evolutionary biologist was evidenced by a growing interest in social behavior (Alexander 1974), and that the experiences I have just described were instrumental. This interest actually began in 1954, not only from my studies of crickets, but from wondering how what is now called eusociality (societies with queens and sterile workers and soldiers) could have evolved.

Between 1954 and 1958, I was impressed by the work of Charles Michener and Howard Evans, who, in the analytical methods of that day, were approaching the problem of eusociality by matching the phylogenies of bees and wasps, respectively, with comparative studies of their behaviors that are associated with eusocial life. But I was more interested in termites than in the Hymenoptera, and eventually more interested in the effects of the process of evolution than in tracing long-term patterns. I generated the idea then that the main precursor of eusociality might be the tending of offspring until they reached either adulthood or a juvenile stage at which they could assist in the rearing of younger offspring. I searched for instances in which offspring were indeed reared to adulthood while remaining in direct association with parent(s). I was at first surprised that such a situation seemed extremely rare, except in eusocial forms, causing me to wonder whether that situation led to most or all cases of the evolution of eusociality. This question remained in my mind across several decades, as a manuscript was slowly generating, starting in the mid-1970s, in which I sought to develop this idea. I realized that if offspring were reared to adulthood, the rearing situation would have to be relatively safe. Termites were my model, and I understood that the dead logs in which many termites live are expansible fortresses made of food. It seemed to me that a newly adult offspring reared in such a safe place, particularly where successive generations of offspring could be reared successfully, might well gain genetically some-times, by remaining in the nest and tending younger offspring, as juvenile termites do. The alternative of setting out to start a new nest would sometimes be less successful than helping in the existing nest. I had not developed the idea quite this far when I first encountered Bill Hamilton's (1964) intimidating papers quantifying the benefits of assisting close relatives. As a result, the paper that had been generating for so long, titled *The evolution of eusociality*, was not completed until 1991, as the first chapter in *The Biology of Naked Mole Rats*, and co-authored by two of my doctoral students, Katherine M. Noonan and Bernie Crespi (Alexander, Noonan, and Crespi, 1991). Noonan worked on eusocial paper wasps, and

Crespi, who was Bill Hamilton's student until Bill returned from the U-M to Oxford as a Royal Society Research Professor, was a student of Thysanoptera, in which he had discovered eusocial forms (Crespi *et al.* 2004).

My efforts to understand eusociality were greatly enhanced by my long-time association with a beetle in the family Passalidae that lived in a forest near my family's farm. As a boy I carried these "betsy beetles" in my pocket in penny matchboxes, and I made wagons and other vehicles such that I could hitch the beetles up in teams, by tying threads of "harness" to their anterior horns. I carried them to school, and to church, for times when the proceedings seemed too absolutely boring. In church I toyed at length with the temptation to allow a betsy beetle to climb the dress of some girl – any girl – who happened to be sitting in front of me. Luckily, I never was bold enough to allow a climbing beetle to reach bare skin.

I have since admired passalids, and I watched their behavior in terraria in my office across decades. In the 1970s I wrote and published a laboratory exercise for introductory zoology, and used them in that course, which was enrolling *c.* 525 U-M students each term, for the several years that I taught the course (Alexander 1967b). The passalids I used have a rich repertoire of stridulations, including acoustical exchanges with their larvae, which evolved their stridulatory device independently of that used by their parents. A tongue and groove fastener keeps the adult's thick, tough, black, polished forewings together (passalids are also called "patent leather" beetles). When the forewings are pried apart, and their complexly folded yellowish translucent underwings are extended, the dorsal surface of the abdomen is revealed to be as clear as Cellophane, so that the movements and structural features of the internal organs of the beetles can be watched as long as is desired. When the beetles are eventually released, they carefully fold their underwings, snap together the elytra, and walk off undamaged, ready to repeat everything for the next student.

As an added attraction, these passalids carry up to 30 or more species of mites on their external surface, many of them specific to particular locations on the beetle. But I eventually realized that using them in large zoology classes has the potential for extinguishing the North American species. They are already gone from two of the habitat niches in which I initially found them, one in Illinois, the other in Michigan. Luckily, those who followed me in teaching the behavior labs in introductory biology at Michigan were not as entranced with passalids as I was, and dropped that exercise when I stopped teaching the introductory course. They also dropped my exercise on cricket behavior (Alexander 1972a) and another of my design that consisted of bringing in dozens of fragments of "habitats" in appropriate containers so that the 525 students (in laboratory groups of 25–35) could have two-hour indoor "field trips" in which they could sort and distinguish large numbers of species (and contemplate the concept as well as the diversity). They could also see various life stages in metamorphosis, and directly observe predation, parasitism, mutualism, and other biological phenomena, involving members of nearly all animal phyla, alive in fragments of their "natural" habitats (Alexander 1972b).

My fondness for passalids, coupled with a 1947 *Scientific Monthly* paper by Clarence Hamilton Kennedy (then an OSU professor emeritus), titled *Child labor of the termite society versus adult labor of the ant society*, were instrumental in generating the questions

I raised about eusociality in the early 1950s, leading me eventually to argue that Hamilton may not have been correct in his rather tentative suggestion that haplodiploid sex determination, and the resulting closer genetic relationship of full sister workers in the Hymenoptera (3/4 rather than 1/2 genetic overlap), might explain why the Hymenoptera evolved eusociality 13 or so times and the rest of the insects only once (termites). Noonan, Crespi, and I argued instead that the chief precursor of eusociality was life in a relatively safe location coupled with parental care up to either adulthood in the offspring, or to some earlier life stage that enabled the offspring to take on duties within the nest that benefited younger siblings, or aided the reproductive adults in other ways (Crespi has since labeled this, "Alexander's Factory-Fortress Model" for the evolution of eusociality. We argued that the degree of risk in starting a new family, weighed against the safety of the parental nest, the possibility of aiding close relatives, and the potential for expansion of the nest (allowing sizable increases in family or colony size), could be the appropriate variables. Beyond the mere plausibility of this argument, we used what we thought to be the best comparisons available, the abundance of extensive parental care among extant non-eusocial species closely related to the eusocial Hymenoptera (extensive, but not lasting until the offspring achieved adulthood while actually associating with their mother), and the rarity of continuous direct parental care to adulthood in the absence of eusociality (meaning that hymenopteran females that merely stock with food items and then close the nest and leave it are not as likely to give rise to eusocial descendants as females that feed offspring directly to adulthood and perhaps as well continue to defend the nest until the offspring are mature).

From the ancestors of probably tens of thousands of non-eusocial extensively parental Hymenoptera there had evolved 13 or so cases of eusociality, compared (when we began these considerations) to one case in all other insects (termites). Among all other insects, there were then known only a handful of cases of parental care lasting to or near adulthood of the offspring, and a few hundred with somewhat less extensive parental care. We concluded that the termites, perhaps more than the Hymenoptera, had crucial special advantages. First, they live in expansible fortresses made of their food (dead logs and trees), enabling them to be relatively safe at all times – even sufficiently as to favor neoteny – and to evolve step-by-step increasingly effective defensive structures and behavior. Second, insects with gradual rather than complete metamorphosis not only can produce opportunities for older juvenile offspring to aid younger juveniles, but can also allow juvenile stages to evolve gradually, to become increasingly better helpers as they move toward adulthood. We noted that entomologists had long referred to termite workers as "permanent juveniles" that have evolved to senesce and die without assuming the usual attributes of adults. Hymenoptera could become significant helpers only as adults, after a (typically) non-helping, almost entirely dependent larval stage followed by a non-helping pupal stage. Moreover, neither hymenopteran juvenile stage is of a nature that – as with gradual metamorphosis – facilitates a step-by-step improvement in helpership features as the juvenile grows and matures.

Long after I had generated most of these arguments, I realized that my hypothesis was vulnerable. A great many mammals and birds rear offspring to suitable stages for offspring helping, and have gradual metamorphosis, but at that time none were described as being eusocial. I could have argued that mammals and birds haven't been around, or parental, long enough, but that seemed a lame proposition. In the expansible food-filled fortresses of termite-like insects, new adults or late juveniles might not have as much impetus to leave the parents' nest immediately. But most birds and mammals don't live in such long-term safe and expansible hideaways. Indeed, the altriciality of songbirds has been attributed to the importance of rapid growth and development, facilitating earlier departure from the increasingly unsafe nest (Ricklefs 1983; cf. Alexander 1990b).

I decided that the best way to explain my argument was to generate an elaborate description of a hypothetical vertebrate, which, if it existed, would be eusocial. Starting around 1976, I did this, and presented the argument, including 15 or 20 predicted attributes of the hypothetical species, in a series of six lectures given in North Carolina, Michigan, Colorado, and Arizona (3 lectures). At Northern Arizona University, Terry Vaughn, a mammalogist who had recently been on sabbatical leave in Africa, introduced me to naked mole rats (NMRs) after telling me enthusiastically that the hypothetical eusocial vertebrate I had just described in my lecture was "a perfect description of the naked mole rat of Kenya!" At that time no one seemed to be entertaining notions of eusociality in any vertebrate. I had never heard of naked mole rats. I asked Terry Vaughn what was known about them, and he said nothing much except that they live underground in groups. I asked him if anyone was working on them and he told me about Jennifer Jarvis, a professor at Capetown University in South Africa, previously at the University in Nairobi, Kenya. He said that, along with some others, she was interested in their physiology because of their virtual hairlessness and their lack of control of body temperature. I wrote to Jenny, explained my hypothesis, and asked a series of questions designed to tell me whether or not NMRs are eusocial. Her prompt replies virtually demonstrated that they are. At the time I was completing *Darwinism and Human Affairs*, and I decided I could not go to Africa until that book was in press. I asked my former doctoral student, John Hoogland, if he and his wife, Judy, would like to go with Lorrie and me to visit Jenny and get her to go with us to Kenya to study and capture some NMRs. John decided not to go because he wanted to remain concentrated on his studies of prairie dogs. I approved of that decision, and the world can see from John's magnificent work on prairie dogs that it was a good one (e.g. Hoogland 1995). I then asked two other former students, Paul Sherman and Cynthia Kagarise Sherman, if they would like to participate in the study. They were enthusiastic, and the four of us went to South Africa in December 1979. When we arrived, Jenny presented me with a manuscript arguing that NMRs are eusocial. I cannot deny that I felt a twinge of proprietorship about the seminal idea, but fortunately I elected simply to help her with the paper, and spent most of that first night sitting on the cover of the toilet seat in the bathroom of Lorrie's and my hotel room, working on it while Lorrie slept. I suggested a more direct title, and worked over the entire paper, handing it back to her the next morning. We asked her to come to Kenya with us and offered to cover her travel expenses and food. She could not come with us immediately, so we went on ahead and waited in Nairobi for her to arrive. There she arranged for us to rent a Land Rover. The five of us went to Mtito Andei, between the game parks Tsavo East and Tsavo West. There we engaged the services of a local Kikuyu man, Anthony Simon N'Dalinga Chondo, who had previously helped Jenny capture NMRs. With Jenny's and N'Dalinga's help we collected about 120 individuals, placing them in two sizable flat tin containers, and carried them by plane back to the University of Michigan and Cornell University.

Jenny brought with her to Kenya her manuscript, which she had reworked to reflect my comments. Paul and I both read it this time and commented critically. Some time after we had returned to the U.S., Jenny sent me the final version of the paper, telling me that *Science* had rejected it and asking what I thought she should do. I worked through the paper again, reducing it substantively without, I believed, removing anything crucial, and advised her to resubmit it, behaving as if the editor had invited her to improve it for possible publication. Then I contacted the *Science* editor, arguing that the paper was the most important mammal discovery of the twentieth century, that it certainly would be published, and that it should be published in *Science*. This time it was accepted (Jarvis 1981). It was surely no accident that the paper was initially rejected. It was in need of further editing, but, more importantly, as with perhaps most biologists then, the reviewers must have found incredible the notion of a mammal with queens and workers. They could well have supposed that Jenny had been influenced by unnecessarily exuberant or even mentally aberrant entomologists! Across the next few years, the team of NMR enthusiasts agreed to work on their separate projects, meeting for discussion at least once in Ann Arbor, cooperating in wonderful good humor, and holding off separate publications until the volume on the biology of naked mole rats was finally produced (Sherman *et al.* 1991).

My former doctoral student, Stanton Braude, who has now devoted more than a quarter of a century to the field study of NMRs in Kenya, published a list of my predictions about vertebrate eusociality (Braude 1998). Although I (sadly!) did not predict either hairlessness or heterothermy, the set of predictions I used in my lectures in the 1970s almost certainly matched no species in the world except the naked mole rat. I was delighted when, concerning those eusociality predictions, the Austrian mathematical ecologist Karl Sigmund, in his 1993 book on mathematical ecology, wrote (p. 118) that, "This splendid feat of theoretical biology ranks with the prediction of the planet Neptune by astronomers." Later I transmitted to him, via one of his colleagues, my contention that predicting eusociality in one animal species among at least tens of millions is surely incredibly more difficult than predicting one more planet in a "puny" group of eight. I cautioned the messenger to explain that I was grinning mischievously when I gave him the message. He assured me with a smile of his own that, "Karl will understand completely!"

Passalid beetles are not yet absent from the still developing theater concerning the evolution of eusociality. More than 30 years ago, the late botanist Laura Berkeley, and subsequently her husband, Jack Schuster, carried out with me undergraduate research projects on the North American passalid species I had been watching in terraria in my office. When Jack asked my advice about continuing to work on passalids in graduate school, I recommended that he work in Florida with my Ohio State graduate student colleague, Thomas J. Walker, a superb scholar studying mainly singing insects. Jack did that, and among other things became one of the foremost students of the world fauna of

Passalidae. In March 2007 he came from Guatemala City to the University of Michigan to examine Passalidae in the Museum of Zoology collection. There I peppered him with questions regarding whether any of the 600 or so world species of Passalidae tend their offspring in such a fashion as to fit my model of incipient eusociality, and might have become fully eusocial. To be fair, searching for eusociality has not been a central theme in Jack's life work as a passalid systematist (but see Schuster and Schuster 1997). It is also possible that, as the variety of eusocial species continues to grow, we will have to generalize and perhaps simplify or expand the definition of eusociality. Jack and Laura Schuster have reported that most Passalidae live in family groups, including both parents, larvae, pupae, and teneral mature offspring. Young passalids "must eat the feces of the mature adults" (inoculated with bacteria and fungi from the digestive tract of adults). "Larvae and adults cooperate in pupal case construction and teneral adults repair pupal cases of siblings." These kinds of interactions qualify as "complex subsociality" or incipient eusociality and seem to support the "factory-fortress" hypothesis (see above). I continue to regard passalid beetles as an important frontier yet to be crossed in the study of eusociality. This transition may, however, be much more unlikely for Passalidae than for termites, because Passalidae suffer from the handicaps of all holometabolous insects with respect to generating helpers among their juveniles.

Career stage 3: evolution and human behavior

Until I entered graduate school I had never been exposed to even the possibility of taking a course in evolution. I have no memory of any undergraduate or graduate course teaching me anything significant about the evolutionary process. Unlike my colleagues at Michigan I had not read Darwin, or any other documents that caused me to think about evolution. As a doctoral student I tried to understand Sir Ronald A. Fisher's 1930 book, but I was studying species and started with that chapter. I too hastily decided that Fisher didn't understand species and speciation, and neglected the rest of his book until I co-taught evolutionary ecology at Michigan with Don Tinkle. Beginning with that course, Fisher's essays on topics such as allelic dominance, sex ratio selection, heroism, runaway sexual selection, and the first quantification of kin selection – involving bright color in poisonous caterpillars moving in sibling groups – began the process of making him one of my biological heroes for his contributions to adaptive behavior. It is telling that modern biologists took so long to understand Darwin's various challenges, and as well to understand Fisher's adaptive hypotheses.

Although there was one course in evolution taught at Ohio State in the 1950s, I don't think I ever knew a graduate student who had taken it. Instead we went *en masse* to the geology department and fulfilled the evolution requirement by taking a course in paleontology. We sacrificed process for pattern because we knew, in a vague way, that evolution courses based on process tended then to be fuzzy and often simply wrong. Even population genetics was generally regarded as frustrating; and of course we were correct in being skeptical because there was little thought of assessing levels at which selection operated most strongly, therefore no way to predict or understand directions of selection. Williams's arguments provided the basic understanding that allowed biologists and others to begin to examine organisms in light of the evolutionary process confidently and effectively.

Unfortunately, the human-oriented social sciences, medical sciences, political sciences, and some other formal activities had already generated answers to their various questions and established research languages and strategies in the late nineteenth century and early twentieth century. Religion, of course, had done so long before. Along with negative views about the ethics or fairness of some human adaptations, and too-frequent oversimplifications of their developmental backgrounds – both made more evident by increased attention to how the evolutionary process works – this situation created a general tendency to doubt, avoid, and derogate virtually all "adaptationist" approaches. It did not help that some senior and distinguished evolutionary biologists, who did not wish to go back and straighten out their now revealed errors, as well as others with political and ideological stances, also took up the cudgels against Williams's arguments. This widespread negative attitude toward efforts to explain human adaptations has by no means disappeared, but it has certainly waned, at least in some circles.

Because biologists did not soon enough provide a way for budding social and medical scientists, and others, to incorporate the evolutionary process into their newly organizing societies, attitudes toward accepting such a ponderous and seemingly intrusive paradigm would probably have been skeptical from the start. Biologists have been just as reluctant to incorporate studies of humans into their own departments. As a result, biology and all the other disciplines that were trying to understand the human species have tended to remain on separate paths. This situation has surely been a significant force in delaying the study of evolutionary adaptations of the human species and for that reason, perhaps, delayed as well our ability to act on the acute global problems that are beginning to be frightening to thoughtful people. They are unprecedented both in their topics and in that they require global agreement to be solved (Alexander ms 1).

In 1949, when I was waiting to take my turn as a new 19-year old student teacher, I was appalled at the difficulty the student teacher who preceded me was having with the class. I had not been sufficiently aware of the possibility of disciplinary problems. I began to search frantically for a way to begin that would enable me instantly to capture the interest of the class. In a newspaper I read about a species of kittiwake that unlike related kittiwake species laid its eggs on cliff ledges. Its eggs, the article said, were more triangular than those laid in nests on the ground. I told this story to the class the first day and asked them how they thought this had come about. Two hands shot into the air, one of a boy from a nearby orphanage, whose goal was to become a veterinarian, and one of a girl whose father was a professor (and who later informed me that a B+ grade I had just given her was the only grade other than an A she had ever received). I allowed the boy to speak first and he said (paraphrased), "As the egg is being laid, the female kittiwake squeezes the reproductive tract while the shell is being formed, in a way that causes the egg to take a triangular form." Without commenting I turned to the girl. She exclaimed, "That's not it at all! The eggs of ledge-nesting kittiwakes at first varied in how triangular they were, and the less triangular ones rolled off the cliff more often. This eventually caused all the surviving eggs of the cliff-nesters to be triangular." I explained how both answers could be correct, and not in conflict, though only the latter one explained why the cliff-nesting kittiwake alone laid triangular eggs. Not until much later did I realize that, in the very first student comments in my very first class meeting, two high school kids had described what came to be widely understood as the proximate mechanism and ultimate function by which adaptive traits can be described and understood. I wished I had been astute enough to compliment them sufficiently, and explain fully the beauty of their two answers combined, as well as explore the possibility of alternative explanations. Teaching that high school course became my introduction to the evolutionary process and hypothesizing about adaptation.

How and why did I make the transition, initiated around 1967, from studying the singing insects as a systematist and behaviorist to eventually writing some 50 articles and two books about how evolution applies to humans and, in particular, human behavior? I decided in 1954, the day after I passed the written and oral preliminary examinations for the doctorate, that I wanted to be an *evolutionary* biologist. A few years later I realized I would like to think of myself (grandly!) as trying to falsify the hypothesis that everything about life is a result of evolution, in the successions of environments that have occurred across the history of life (evolution, after all, is a simple process; it is the accumulated effects of the almost endless changing successions of environments, large and small in their influence, external and internal to the organism, that have caused life, and the developmental and other processes across the lifetime, to be diverse and almost unbelievably complicated – and, indeed, somewhat jerry-rigged.)

The approach I set for myself would require that I proceed eventually to the most difficult of all traits (behavior) and the most difficult of all species (humans). I was delighted with this thought because I had realized by then that I had long been interested in where humans had come from, what they are like, and how they had gotten to be the way they are. I believed by then that such interests were the reasons I had taken a number of special optional courses about humans and their activities. I also understood that, if I followed my intent, I would eventually have to try to understand, in evolutionary terms, the human behaviors most difficult to understand – for example, social behavior, culture, morality, religion, humor, music, and the arts. If I could not solve those problems I could not generate a strong likelihood that the implication in my falsification effort was reasonable. I also realized that to fail completely with any human topic would be, at least for my grand goal, a disaster. This is why I have published on all of the above topics, even when I knew that I had not yet achieved unique and unassailable theory (Alexander 1974–1991a, 1993, 1998, 2005a, 2006a, 2008). I felt that revealing even incomplete and unsatisfactory results might assist others in continuing to more satisfactory conclusions.

Some time after I had decided I would become an evolutionary biologist, I realized I knew of no individual in evolutionary biology that I could accept as a bona fide, complete, and satisfactory hero. For some reason I felt then that everyone should have a hero in his own field. Several evolutionary biologists had qualities that I appreciated very much, but none completely filled the bill. To me, then, Charles Darwin seemed a remote figure, as did Fisher, although I eventually recognized both as true intellectual giants (Darwin in particular – cf. Alexander ms 1). So I decided, quite consciously, to create for myself a composite hero from contemporary biologists. The people who made up that composite hero were Howard Evans (because of his strong and effective devotion to field work), Ernst Mayr (because of his determination, his persistence, and his enthusiasm for eschewing wishy-washy-ness and continually giving the students of the world something concrete to shoot at), Donald J. Borror (because of his constant devotion to his work and his unpretentious passion for getting things straight), Alvah Peterson (because of his common sense, his almost regal kindness, and his scholarly approach to everything), and George Gaylord Simpson (because he seemed to me one of the most brilliant, serious, and unapologetic students of the evolutionary process).

As an instructor, my initial duties at Michigan were teaching laboratories in introductory biology. After a year or two I offered a course in insect behavior, which was taken by only two students, both undergraduates. Both became academic biologists. Mary Jane West-Eberhard is a member of the National Academy of Sciences, who became a student of social wasps and a close friend and co-author with Howard Evans. Bruce Goldman is a professor at the University of Connecticut, who has studied the physiology and behavior of naked mole rats.

Around 1960 I asked the biology department head, the molecular biologist Dugald E. S. Brown, if I could initiate a new course for upper undergraduates and graduate students, which I would title *Animal Behavior and Evolution*. I was skeptical of my chances, and prepared to leave Michigan if the request was denied. At the time there was no course on animal behavior and no course on evolution in the U-M biology department. Some evolution was taught in an ecology course, and it was discussed minimally in courses such as entomology and vertebrate natural history. But the large introductory course in zoology was based on an almost ridiculous caricature of evolution. My questions about evolution and behavior, submitted for the written doctoral examinations in biology, were often treated by colleagues as jokes (e.g. "Why female coyness?"), and almost never accepted on the examinations. To my surprise, however, Brown acted as though he was wondering why I hadn't asked for a new course before. He proceeded to give me good advice and strong support. I went on to teach some version of animal behavior and evolution to classes, eventually of 100–200 graduates and undergraduates, for almost 40 years. For a few years I co-taught a new course, evolutionary ecology, with Donald Tinkle. I also taught seminars on animal and human behavior and evolution nearly every term for the 44 years of my tenure at Michigan. It would be folly for me to try to list all of the distinguished ecologists, systematists, behaviorists, physicians, anthropologists, psychologists, and others who happened to be in Tinkle's and my courses. Two of them have autobiographies in this volume. Several are members of the National Academy of Sciences.

Approximately 12 years before my retirement in 2001, I changed the title of my course to *Evolution and Human Behavior*. I took up the human species the same way I took up the study of other species – as a systematist, and in the particular way that systematics was seen in the middle of the twentieth century. That is, I began to explore the biology of the human species when, as I said in *The Biology of Moral Systems*, "biology" means everything about the life and natural history of the group or species in question, through the eyes of an evolutionary biologist (Alexander 1987). After my retirement, the anthropology course taught by my former student Beverly Strassmann, currently U-M Associate Professor of Anthropology, became the U-M's central course in human behavior and evolution.

In early 1949, when I transferred to Normal as a second semester junior, I saw in the catalogue a course called philosophy. I read the abstract and was exhilarated. I believed I had finally found the course that I had always wanted to take. I have no memory of ever having known there were courses called anthropology, in any school I attended, until I took the job at Michigan in 1957. Philosophy at Illinois State Normal proved a huge disappointment. At the beginning of the course we were assigned to read a set of philosophical "classics." I was exuberant about this task when I began it. Every night at eight when my evening work washing dishes in a downtown restaurant was finished, I would run several hundred yards to the library reading room and read until the library closed at eleven. But I quickly began to feel that I was reading material within which I could find statements that I liked, or that intrigued me, and others that made no sense, or had no basis. There was no scientific backing, and thus no real distinction between the two categories. Because I could find no way to establish one view over another, or to detect errors, I felt as though I was being cheated, left – indeed, encouraged – to choose the statements I thought were reasonable and then treat them as factual. I never finished reading a single one of those assigned classics. Eventually I simply gave up. On the last night that I walked out into the darkness and away from that assignment, as the library was closing, I muttered out loud to myself, "Philosophers are no better off than theologians: they have no decent theory." Today that seems a surprising statement, certainly for a 19-year old version of myself; but it is exactly what I said, under those exact circumstances. I held no ill feeling toward the philosophy professor, an excellent teacher named Francis Belcher. I thought enough of him that almost a year after I had left for graduate school I sent him an essay titled *Biology for Survival*, and explained in an accompanying letter that I felt I had done a poor job of answering when he deliberately challenged me in class with the excellent question, "Why would *anyone ever* want to become a *biologist!*"

As my interest in pursuing questions about evolution and human behavior grew, I began to recall the effect during my childhood of my family's faithful attendance at the United Methodist Church in a tiny ghost town in Piatt County, Illinois. I remembered being puzzled because so much of what was discussed in that church involved causes and effects – events – that I never had experienced in everyday life outside the church. I recalled sitting in the church on Easter mornings and wondering what it meant for the preacher to say that a man had risen from the dead and ascended into Heaven to sit at the right hand of God the Father Almighty. I developed a habit of sitting in the back of the church so that I could watch the backs of the necks of the farmers I knew well. I wanted to know if they squirmed in their seats, or if their necks became a little redder, or if anything else indicated discomfort with assertions and claims that seemed to me to be unlikely, or even outlandish. I never perceived any such response. This caused me eventually to realize that I was in some sense on the outside of this local community, looking in. I never forgot that. But years later, when I was contemplating studying the human species in evolutionary terms, my undergraduate pre-occupation with courses in the social sciences and history and the arts and philosophy seemed finally to make sense. I knew I would have to start with simple questions about humans but continue on a gradual course toward tackling the most difficult questions imaginable, at least about living creatures. I believed that such questions would be the most severe tests of possible falsification of evolution as the explanation of all life.

At some time during the later stages of my formal schooling I became aware that I had known, since high school days, that the only place the questions I wanted to answer about humans were raised was in that little country church. In retrospect I believe that I have wished to have this truly important set of questions answered since I was a child. Probably, I was interested because I found it difficult or impossible to accept much of what was said in the church; perhaps the principal reason was a certain difficulty in thinking metaphori-cally. Nowhere outside that church do I recall receiving even a smidgeon of an answer to my questions about humans: not in school, from elementary school all through undergraduate school, not from my parents, not from any teacher, not in any publication. Eventually I asked my mother why we had never discussed the questions of where humans came from, what they are really like, and how they got to be that way. She did not answer for a while, and then, without looking at me, said, "We didn't have time." I objected, noting that in our family we discussed all kinds of other questions. I pointed out that there is always time on a farm to discuss such things because so many of our tasks could be carried out while thinking hard about something entirely different. She did not respond, and in a brief bout of sympathy I asked if she thought it might be because we had no reasonable theory. After another period of silence she said in a subdued voice, "That might have been part of it." My mother was devoted to her church, even if in a social way more than a narrowly religious one. I understood I would not receive additional help from her with that question.

In the mid-1960s Professor Emmet T. Hooper asked me if I would take over from him the job of reviewing Robert Ardrey's *The Territorial Imperative*. I declined. Undaunted, he somehow entered my office when I was not there and left the materials involved on my desk. I kept looking at them and finally decided, after all, to review Ardrey's book, and to include in the review Konrad Lorenz's *On Aggression*. Eventually I asked Don Tinkle to co-author the essay (Alexander and Tinkle 1968a).

In early 1968, the Alexander and Otte families – four adults, two teenage girls, and two 1½ year old twin girls – left together for Australia to monograph the crickets of that continent. In Melbourne, Professor Murray Littlejohn asked me to speak to the Royal Society of Victoria on human behavior and evolution. I declined, saying I knew nothing except what I had written in the Ardrey–Lorenz review. He suggested I talk on that topic, but I didn't want to do that. The cricket expedition left shortly afterward, in a new Land Rover towing a two-wheeled trailer, to spend several months in northern Queensland. When we returned to Melbourne, Murray had scheduled the talk he had asked me to give. Neither he nor I remembers whether I eventually agreed to do it, or if he simply put me down for it. In any case, everything was in good humor, and I gave the talk, titled *The Search for an*

Evolutionary Philosophy of Man. During parts of the lecture the audience was deathly quiet. As Murray and I walked down the street afterward, he said thoughtfully, "The clerical robes and collars were not there; but their ghosts were!" I decided to turn the lecture into an essay and spent many hours on the project in the University of Melbourne library (Alexander 1971).

By the time I returned from Australia I had discovered a statement by the University of Chicago anthropologist David Schneider that the asymmetrical treatment of cousins in many pre-technological societies could not be explained by "biology." That use of the term "biology," I had come to realize, was mostly restricted to social scientists and philosophers, and roughly meant "genetics and physiology." I felt that it was a way of referring to what many people call "hard-wired," "innate," or "instinctive" traits, thinking that such adjectives contribute to an understanding of how humans acquire their traits – falsely, I believe, or at least clumsily and imperfectly. The events between the zygote and the organism, during which all of the genetic elements in the genome are necessarily selected in the direction of cooperating completely with their fellow units within the genome (even if they never achieve it), are too numerous and complex to be so labeled (Alexander 1990a, 1991a). In my opinion, the "genetic and physiological" definition of "biology" has had a broadly significant negative effect, hindering progress in the study of not only evolutionary adaptation, particularly as applied to humans, but our understanding of development (ontogeny), learning, and all related biological problems. It happened because the human-oriented disciplines developed before serious study of human adaptation was possible, and because ontogeny is so complex that we are reduced to manufacturing terms to use when we are really too ignorant to have much of an idea about what is actually involved (West-Eberhard 2003).

It seemed to me that if evolution can explain all aspects of all life – including learning and culture – society-wide patterns of culture should not be frequently or significantly contrary to a history of natural selection, even if they are learned (Alexander 1979; Flinn and Alexander 1982); otherwise the capacity for the cumulative learning of culture could not continue to evolve. I set out to find as many statements like that of Schneider as I could, so that I could see whether they actually were inconsistent with a history of natural selection. I supposed that such patterns were learned, but I felt that the relevant learning capabilities and tendencies had surely been patterned by natural selection, such that Schneider's kind of statement would at least be misleading. I was willing to work from the assumption that our main problem is to find out more about how learning has evolved, and how it works, rather than accept that culture is independent of changes saved as a result of differential reproduction for reasons other than massive and rapid alterations of the human environment.

It was not easy to locate statements such as Schneider's. I went through the journals article by article, sometimes page by page. Most articles in anthropology journals then gave little clue of such topics or conclusions in their titles or abstracts. I did find two other examples that seemed to me important. (1) The avunculate, or mother's brother phenomenon, which had also been examined in detail by Schneider and his co-author Kathleen Gough (1961). These terms refer to situations in which men show a great deal of attention to their sister's children, sometimes more attention than to their spouses' children. (2) Asymmetry in cousin marriages. I set out to discover whether these different phenomena had features consistent with a history of natural selection. Every case of that sort that I eventually discovered did indeed turn out to have an evolutionarily consistent solution. Eventually I brought most of the solutions together in *Darwinism and Human Affairs* (1979; see also Alexander 1998, 2006a).

In the mid-1960s I was examining a paper that seemed to show that the human brain cavity continued to expand well after it had exceeded the brain cavities of related primates, and may even have increased its rate of expansion afterward. I pondered for a long time what hostile forces of nature could be responsible for the continuing growth of an organ as calorically expensive as the human brain. Eventually I decided that the only qualifier has to be humans themselves. It seemed to me immediately that this could only happen if humans lived in groups and had as their most important "enemies" other groups of humans. I pursued this idea and eventually argued that humans had become sufficiently ecologically dominant that all of Darwin's "Hostile forces of nature" – predators, parasites, diseases, food shortage, climate, and weather – had been reduced, or changed in relative importance, in a fashion that caused human groups to use a greater proportion of their calories in being directly and acutely competitive with other human groups for certain special resources, including mates. Thus, internecine battles could become efforts not only to displace but to destroy other humans (typically men and boys) but also to gain other humans as resources (typically women and girls). I argued that this is the reason for the continuing prevalence of different forms and magnitudes of warfare (and its relatives), and that the alternation of intense and organized cooperation within groups and adversarial relations between groups sets up a perpetual favoring of genes that improve the ability and tendency of humans to further the subtlety and complexity of their sociality and all its mechanisms, and to change quickly to meet the immediate circumstances. I referred to the result of all this as an endless Balance-of-Power Race. Some authors have suggested that it should have been termed an imbalance-of-power race, but maintenance of multiple groups can occur only if groups continue to remain balanced in ways that prevent wholesale annihilation. Weak groups may disappear, but stronger groups must continue to fill in the spaces if multiple competitive groups are to persist. I believe that the most intense and frequent efforts would typically be on the part of slightly weakened groups that needed to continually re-establish balances that would lower the likelihood of all-out contests. Because the members of a single species behaving this way are bound to exchange genes continually (or repeatedly), evolutionary selection can continue indefinitely to favor those who are able to form and live in groups that can protect themselves. It is difficult to imagine the kinds of environmental and situational changes that could halt the ensuing evolutionary races in brain size and function and their consequences.

I have argued (Alexander 1990–2008) that (1) the combination of the uniquely large and complex human (primate) brain, and its middle and late life social (kin group) functioning across much of the adult lifetime, is responsible for (2) the doubling of the human lifetime compared to all our extant ape relatives, and by extension, (3) the long juvenile life and learning period, and the altriciality of the human baby, as preparation and enhancement of later learning and use of knowledge, facilitated by (4) the unique ability of humans to learn the relative genetic overlap among virtually their entire kin groups, owing to (5) concealment of ovulation, which allowed accurate knowledge of genetic relationships of different kin in even multi-male groups and thereby enabling virtually all of the uniquely social human traits (Alexander 2008), and (6) menopause, as transfer of effort from offspring production to support of up to the entire kin group midway through the adult lifetime of women.

How have humans accomplished this? They do not grow continually. They are not armored or surrounded by individuals sufficiently devoted to protecting them. We need only consider a set of human traits that menopause also demonstrates: living in kin groups under conditions that have spawned the huge brain that causes the human baby's head to be the limiting factor in successful birth, and to become the most calorically expensive – and the most remarkable – organ of the human body. … The human brain and all of its correlates in learning, cognition, consciousness, extensive and elaborate scenario-building, memory, and other intellectual features enable human individuals to increase their reproductive output via kin help and make evolution of longer lifetimes adaptive. The collection of mental adaptations in humans – and their continued elaboration late in life – can be so significant that individuals that have become seriously senescent in physical attributes can nevertheless remain important, or even essential, to the survival of families and kin groups. … It is obviously possible to evolve a longer adult life if the rate of reproduction can be sufficiently increased as to delay or reverse senescent trends caused by late-acting deleterious pleiotropic effects of genes that have their advantageous effects primarily in earlier life (Williams 1957). … It is thus not surprising that age is so often venerated, and that some kinds of leaders are desired or required to be as old as, or older than, the usual age of death in non-human primates. Although there have been suggestions that the altriciality of the human baby is responsible for the extension of the adult human lifetime, and its intelligence (e.g., Hutchinson 1965, Williams 1992), the reverse – or some kind of successive reinforcement – also seems possible: that selection favoring the collection of mental capabilities of adults may have caused those capabilities to be further enhanced by beginning and increasing their elaborateness and the earliness of [learning and] development in the human juvenile

(Alexander 2008).

I consider the evolution of the combination of Ecological Dominance and the related Balance-of-Power Race among human groups to be the most general and most important adaptation of the human species (Alexander 2008). By this I mean that virtually all of the unusual or unique traits of humans are molded around this combination of traits. The resulting amity–enmity axis – the extreme cooperativeness within the local unit of competition and its correlated ability to marshal against other local units of competition – is currently the world's greatest problem, and this has some possibility of being true indefinitely. Because of this unique set of adaptations, the traits of humans are often decidedly different from, or even opposite to, those of related species – or, in some cases, opposite to all such traits in all species. To me it seemed that, in attempting to analyze adaptations, one has to consider the human species as an *N* of one. As a result I developed an approach to the analysis of humans that I called the Jigsaw Puzzle Method (Alexander 1990b). I considered different traits as if they were pieces of a jigsaw puzzle that can only be completed by discovering how all of the pieces (traits) are related functionally. Some time after I generated this idea, I realized that theoretical physicists must similarly examine the only physical universe available to them. My thought that this approach resembled that of the theoretical physicists trying to assemble all of the functions of the universe was confirmed by Professor Gordon Kane, a U-M Professor of Physics, who assured me, "You are doing exactly what we are doing."

I have had great difficulty seeking to solve what I see as the four most difficult aspects of human behavior to understand thoroughly: music, humor, the arts, and religion. Twice I have literally "copped out" while giving lectures on the arts to large numbers of people. The first time I was delivering a special lecture sponsored by the U-M College of Literature, Science, and the Arts. The second involved an international symposium on human behavior and evolution in Stockholm, Sweden. In the first of these two lectures I stopped short, at the end of a 50-minute talk, because I suddenly realized that the theory I was developing was not worthy. In Stockholm, on the night before my lecture, I could not convince myself that I could justify lecturing as though I could describe and support a satisfactory theory of the arts. So I gave a general lecture on human evolutionary adaptation instead. I rationalized to the audience that no such lecture had been given in the series, and that one was needed. At the end of my lecture I delivered a ten-minute condensed version of my original topic, the arts. The best published version I have yet been able to muster as an explanation of the arts, including humor, music, and religion, appear in *Darwinism and Human Affairs* (Alexander 1979), and in my chapter in the volume *Darwinism and Philosophy* (Alexander 2005a). I had already published a long paper on humor in *Ethology and Sociobiology* (1986a) and an important addendum on humor concerning social-intellectual play in *The evolution of the human psyche* (Alexander 1989). My hypotheses about music were summarized in an abstract of a talk delivered to the Human Behavior and Evolution Society 14 March 1988. The abstract was published in the program of the meeting. A more elaborate discussion of the evolutionary background of music appeared in a course text, titled *Understanding Humanity*, printed and distributed in Ann Arbor, 1997–98, for my course in Evolution and Human Behavior.

I think there is sometimes confusion about the different approaches of researchers, and what they are trying to accomplish, in adopting an evolutionary approach to understanding the human species. Paleontologists and archaeologists are tracing patterns of change through evolutionary time, as by examining and dating fossils, tools, weapons, and other relicts. Such investigators are primarily seeking to explain where humans have been, when they were there, their long-term movements, and what they did at different times and in different places. Others are using molecular information, sometimes taken from fossils, sometimes from comparative study of extant humans around the globe, to reconstruct approximately the same aspects of long-term human history. My own primary effort, since 1967, has been to use knowledge of how the evolutionary process – differential reproduction – works, and how it has through its cumulative effects resulted in humans being what they are today. Obviously, in the end, all of these different approaches to patterns of change and the cumulative processes of change must necessarily be combined and synthesized repeatedly to continue generating the most accurate overall picture of human nature and history. We are surely in a position to understand ourselves profoundly, as we will need to be if we are to begin to solve, for the first time, the succession of global problems that are descending upon us.

My graduate students and postdoctoral associates

Anything and everything that I may have accomplished as a biologist after moving to Michigan has been facilitated by the 33 doctoral students whose committees I have chaired between 1958 and the present, and almost as many postdoctoral fellows and other doctoral students who have worked more or less directly with me. There is no way to measure the importance of the hours we sat in Rooms 2009, 2080, and the Faculty Conference Room in the University of Michigan Museum of Zoology, arguing, lunching, and laughing over all of the topics mentioned in this essay, and many more. Those former students and postdocs are all splendid people, and keen observers and analysts of the world and its inhabitants.

I was fortunate to become enamored with the evolutionary study of social behavior at a time when students were beginning to show the same interest, and at a time at Michigan when biology graduate students were admitted on the basis of their application qualifications, as determined by a department-wide committee, rather than to fill slots in particular fields or in the laboratories of particular faculty members. They were assured of financial support for five years via teaching assistantships, and they were given a little over a year to complete a preliminary qualifying examination and select a sub-discipline and a doctoral chair. These arrangements made the biology doctoral students at Michigan financially and otherwise more independent than is often the case. This is why so little of the student research described in this essay was done, or published, in a formally cooperative way with me. Our interactions were virtually all informally cooperative, and my students typically wrote doctoral theses based entirely on their own independent research, more often than not funded by their own research grants.

Starting in 1966 (Williams 1966), my students and I recognized that it is parsimonious to assume that evolutionary selection is typically (but not always, of course: cf. Alexander 1974, 1993; Alexander and Borgia 1978) most effective at the lowest levels of organization of life – meaning it is the best hypothesis to entertain at the start. As a result of their beginning with this hypothesis, my doctoral students were among the first investigators to realize and demonstrate the importance of recognizing individuals in field studies of adaptiveness in the social behavior of organisms such as milkweeds, damselflies, aphids, social wasps, acacia ants, field crickets, grasshoppers, true katydids, meadow katydids, thrips, dung flies, leopard frogs, Sierra toads, bullfrogs, bank swallows, cockatoos, mountain bluebirds, prairie dogs, ground squirrels, dolphins, cicadas, field mice, naked mole rats, white-tailed deer, and humans.

Concluding reflections

In January 2001, I retired with the intent of writing children's books, books about horses and horse people, poems, songs, and articles and books of local historical interest. I have done these things, and am doing them still, to the best of my ability; and I continue to enjoy every minute. The beginnings of some of these writings took place more than 60 years ago, but my more than a half century as a professional biologist seems to have trained me in precisely the wrong directions for significant contributions to literature or the arts. Perhaps my efforts to reverse such training are somehow part of my continuing striving to understand in evolutionary terms the arts and such related human activities as music and humor. Perhaps I am simply not talented in the appropriate directions. I continue, however, to feel strongly that those of us trained in evolutionary biology – indeed, in systematics, and in accounts and syntheses of all the life attributes of individual species – have a special kind of knowledge and skill, which includes a responsibility to build toward a profoundly more penetrating understanding of ourselves, as individuals and as part of the cooperating and competing collective of humanity. In the end, regardless of how reductionistically we begin, we must understand ourselves, and indeed all life, as whole organisms. Despite the most dramatic and valuable starts with the genetic materials, the almost unbelievable complexity of development, owing to the nearly complete cooperation of genes in genomes (Alexander 1993, 2005a; West-Eberhard 2003; Burt and Trivers 2006), will prevent a fully ontogenetic exposing of ourselves from being possible for a very long time. Theodosius Dobzhansky (1961) summarized the basic reasons almost a half century ago:

Heredity is particulate, but development is unitary. Everything in the organism is the result of the interactions of all genes, subject to the environment to which they are exposed. What genes determine is not characters, but rather the ways in which the developing organism responds to the environment it encounters.

Recently, an evolutionary biologist was quoted as saying, "Science doesn't answer questions about the meaning of life." The same person also said, "To say that just because something is terribly complex it needs a supernatural explanation is to give up on the scientific enterprise." Taken together, these two statements, from the same person in the same article, highlight the reluctance of biologists to speak plainly about evolution as an explanation of human life and human traits, as well as all life and its traits, including religion. Assuming that a supernatural is not to be invoked, what approach other than science exists to take over when "terribly complex" questions such as the meaning of life are contemplated? On what basis can an evolutionary biologist argue that science cannot analyze something, such as the meaning of life, necessarily derived in some fashion, however indirectly, from the evolutionary past of the human species? All of the attributes we engage to even pose questions such as the meaning of life are somehow results of the evolutionary process. The science of biology does not hesitate to seek the answers to all questions about other living creatures, including the "meaning" of all aspects of nonhuman life (their evolved significance as well as their personal meanings for humans). I regard it as unfortunate that we humans withdraw from at least hypothesizing that every aspect, every trait, every result of change in life has to be an outcome of the cumulative process of evolution as it has occurred in the dauntingly extensive and intricate successions of environments across history.

At the risk of being judged a hopeless megalomaniac, I will say here that it is my greatest regret, so late in my lifetime of thought and research, that I have been inadequate in my attempts to discover and explain how people everywhere can understand themselves sufficiently better from knowledge of evolution as to change the sociality of global humanity in a positive way. Regardless of the pace of technological and other scientific advances, understanding of ourselves in evolutionary terms – understanding sufficiently profound that it requires at least a temporary ability to withdraw slightly and judge ourselves as if we were aliens, or members of a different species – may always be necessary if we are to recognize and accept the most important sources and reasons for change in the social life of humans. I regret my inability to identify confidently even the first steps of a solution to the long-standing central problem of humanity that derives from the prevalence, throughout our history, of uniquely ferocious and frequent inter-group competitions within our own species.

A hydrogen bomb is an example of mankind's enormous capacity for friendly cooperation. Its construction requires an intricate network of human teams, all working with single-minded devotion toward a common goal. Let us pause and savor the glow of self-congratulation we deserve for belonging to such an intelligent and sociable species.

(Robert S. Bigelow, 1969. The Dawn Warriors*)*

References

Alexander, R. D. (1961). Aggressiveness, territoriality, and sexual behavior in field crickets (Orthoptera: Gryllidae). *Behaviour* **17**: 130–223. Reprinted (1974, in part) in *Territoriality: Landmark Papers in Animal Behavior*, ed. A. W. Stokes, p. 364. New York: Dowden, Hutchinson, and Ross.

Alexander, R. D. (1967a). The evolution of mating behavior in arthropods. *Symp. R. Ent. Soc. Lond.* **2**: 78–94.

Alexander, R. D. (1967b). The [passalid] beetle. Introductory Zoology Laboratory Exercise. Ann Arbor, MI: Edwards Brothers Printers.

Alexander, R. D. (1967c). *Singing Insects: Four Case Histories in the Study of Animal Species.* Pattern of Life Series. Chicago, IL: Rand McNally.

Alexander, R. D. (1968a). Life cycle origins, speciation, and related phenomena in crickets. *Q. Rev. Biol.* **43**(1): 1–42.

Alexander, R. D. (1968b). Arthropods. In *Animal Communication: Techniques of Study and Results of Research.*, ed. T. Sebeot, pp. 167–216. Bloomington, IN: Indiana University Press.

Alexander, R. D. (1969). Comparative animal behavior and systematics. *Proc. Int. Conf. Systematics* (Ann Arbor, Michigan, July 1967), National Academy of Science Pub. 1962, pp. 494–517.

Alexander, R. D. (1971). The search for an evolutionary philosophy of man. *Proc. R. Soc. Victoria* **84**: 99–120.

Alexander, R. D. (1972a). Behavior associated with reproduction in field crickets. In *A Laboratory Manual: Introduction to Biology*, ed. D. G. Shappirio, B. E. Frye, & P. M. Ray. Ann Arbor, MI: Campus Publishers.

Alexander, R. D. (1972b). Diversity of organisms. In *A Laboratory Manual: Introduction to Biology*, ed. D. G. Shappirio, B. E. Frye, & P. M. Ray. Ann Arbor, MI: Campus Publishers.

Alexander, R. D. (1974). The evolution of social behavior. *A. Rev. Ecol. Syst.* **5**: 325–83.

Alexander, R. D. (1978). Evolution, creation, and biology teaching. *Am. Biol. Teacher* **40**: 91–107.

Alexander, R. D. (1979). *Darwinism and Human Affairs*. Seattle, WA: University of Washington Press.

Alexander, R. D. (1986a). Ostracism and indirect reciprocity: the reproductive significance of humor. *Ethol. Sociobiol.* **7**: 253-270R. [Also published in German as *Ostrazismus und indirekte Reziprozität die reproduktive Bedeutung des Humors*, pp. 79–99. Berlin: Duncker & Humbolt.]

Alexander, R. D. (1987). *The Biology of Moral Systems*. Hawthorne, NY: Aldine de Gruyter.

Alexander, R. D. (1988). The evolutionary approach to human behavior: what does the future hold? In: *Human Reproductive Behavior: A Darwinian Perspective*, ed. L. L. Betzig, M. Borgerhoff Mulder, & P. W. Turke, pp. 317–41. London: Cambridge University Press.

Alexander, R. D. (1989). The evolution of the human psyche. In *The Human Revolution*, ed. C. Stringer and P. Mellars, pp. 455–513. Edinburgh: University of Edinburgh Press.

Alexander, R. D. (1990a). Epigenetic rules and Darwinian algorithms: the adaptive study of learning and development. *Ethol. Sociobiol.* **11**: 241–303.

Alexander, R. D. (1990b). How did humans evolve? Reflections on the uniquely unique species. *Univ. Mich. Zool. Spec. Publ.* **1**: 1–38. (http://humannature.com/ep/reviews/ep04132.html)

Alexander, R. D. (1991a). Social learning and kin recognition – an addendum. *Ethol. Sociobiol.* **12**: 387–99.

Alexander, R. D. (ed.) (1991b). *Mom's Story. The Life of Katharine Elizabeth Heath Alexander Stutzenstein*. Manchester, MI: Woodlane Farm Books.

Alexander, R. D. (1993). Biological considerations in the analysis of morality. In *Evolutionary Ethics*, ed. M. H. Nitecki and D. V. Nitecki, pp. 163–96. Albany, NY: State University of New York Press.

Alexander, R. D. (1998). *Understanding Humanity. The Human Species in Evolutionary Perspective*. [Printed by Dollar Bill Copying, Ann Arbor, MI, as 1997–1998 text in a course in evolution and human behavior, currently in preparation for publication.]

Alexander, R. D. (2001a). *Teaching Yourself to Train Your Horse. Simplicity, Consistency, and Common Sense from Foal to Comfortable Riding Horse*. Manchester, MI: Woodlane Farm Books.

Alexander, R. D. (2001b). *Club 48. A Personal Account of Blackburn College and the Men of Butler Basement. 1946–48.* Manchester, MI: Woodlane Farm Books.

Alexander, R. D. (2004). *The Red Fox and Johnny Valentine's Blue-Speckled Hound*. Manchester, MI: Woodlane Farm Books.

Alexander, R. D. (2005a). Evolutionary selection and the nature of humanity. Chapter 15. In *Darwinism and Philosophy*, ed. V. Hosle and Ch. Illies, pp. 301–48. University of Notre Dame Press.

Alexander, R. D. (2005b). *Pop's Story: A Midwestern Farm Boy's Memories of Times with his Father*. Manchester, MI: Woodlane Farm Books.

Alexander, R. D. (2006a). The challenge of human social behavior. A review essay stimulated by Hammerstein, Peter (ed). 2003. Genetic and Cultural Evolution of Cooperation. *Evol. Psychol.* **4**(2):1–28. (http://human nature.com/ep/reviews/ep04132.html)

Alexander, R. D. (2006b). *Playin' Cowboy: The Coontail Blue and Other Horse Tales.* Manchester, MI: Woodlane Farm Books.

Alexander, R. D. (2008). Evolution and human society, understanding the human species and its immediate ancestors, and HBES talk handout from keynote presentation at Kyoto HBES Meeting. *Human Behav. Evol. Newsl.* August 2008.

Alexander, R. D. (ms. 1). Darwin's challenges and the future of human society. In *Predictions: Breakthroughs in Science, Markets, and Politics,* ed. F. Wayman, X. Williamson & X. Buenode Mesquita. Ann Arbor, MI: University of Michigan Press. (In press.)

Alexander, R. D. (ms. 2). *Stealing Watermelons: Tales from Sangamon Township.* Manchester, MI: Woodlane Farm Books.

Alexander, R. D. (ms. 3). *The Social Behavior of Horses and Horse People.* Manchester, MI: Woodlane Farm Books.

Alexander, R. D. (ms. 4). Insect mating, wings, and phylogeny.

Alexander, R. D. (ms. 5). *Thumping on Trees.* Manchester, MI: Woodlane Farm Books.

Alexander, R. D. & Bigelow, R. S. (1960). Allochronic speciation in field crickets, and a new species *Acheta [Gryllus] veletis. Evolution* **14**: 334–6. [Reprinted in *Papers on Evolution,* ed. Ehrlich, Holm, & Raven. 1969. Little, Brown Publ.]

Alexander, R. D. & Borgia, G. (1978). Group selection, altruism, and the levels of organization of life. *A. Rev. Ecol. Syst.* **9**: 449–74.

Alexander, R. D. & Brown, W. L. (1963). Mating behavior and the origin of insect wings. *Univ. Mich. Occas. Pap.* **628**: 1–19.

Alexander, R. D. & Moore, T. E. (1962). The evolutionary relationship of 17-year and 13-year cicadas with three new species (Homoptera: Cicadidae: *Magicicada*). *Univ. Mich. Mus. Zool. Misc. Pub.* **121**: 1–59.

Alexander, R. D. & Otte, D. (1967). The evolution of genitalia and mating behavior in crickets (Gryllidae) and other Orthoptera. *Univ. Mich. Mus. Zool. Misc. Pub.* **133**: 1–62.

Alexander, R. D. and Tinkle, D. W. (1968a). A comparative review (of *On Aggression* and *The Territorial Imperative*). *BioScience* **18**(3): 245–8. [Reprinted (1970, in part) in *The Contemporary Scene: Readings on Human Nature, Race, Behavior, Society, and Environment,* ed. P. B. Weiss. New York: McGraw-Hill.

Alexander, R. D., Noonan, K. M., & Crespi, B. (1991). The evolution of eusociality. In *The Biology of the Naked Mole Rat,* ed. P. W. Sherman, J. Jarvis, & R. D. Alexander, pp. 3–44. Princeton, NJ: Princeton University Press.

Alexander, R. D., Marshall, D. C. & Cooley, J. (1997). Evolutionary perspectives on insect mating. In *The Evolution of Mating Systems in Insects and Arachnids,* ed. B. Crespi & J. Choe, pp. 4–31. Princeton, NJ: Princeton University Press.

Bigelow, R. S. (1969). *The Dawn Warriors: Man's Evolution toward Peace.* Boston, MA: Little, Brown.

Borror, D. J. and Delong, D. M. (1954). *An Introduction to the Study of Insects.* New York, NY: Rinehart and Company.

Braude, S. (1998). The predictive power of evolutionary biology and the discovery of eusociality in the naked mole-rat. *NCSE Reports* **17**(4): 12–15.

Brues, A. (1964). The cost of evolving vs. the cost of not evolving. *Evolution* **18**: 379–83.

Burt, A. and Trivers, R. (2006). *Genes in Conflict: The Biology of Selfish Gene Elements.* Cambridge, MA: Belknap Press of Harvard University Press.

Carle, F. L. (1982). Evolution of the odonate copulatory process. *Odontologica* **11**: 271–86.

Cooley, J. R., Simon, C., Marshall, D. C., Slon, K. & Ehrhardt, C. (2001). Allochronic speciation and reproductive character displacement in periodical cicadas supported by mitochondrial DNA, song pitch, and abdominal sternite coloration data. *Molecular Ecology* **10**: 661–72.

Crespi, B. J., Morrie, D. C. & Mound, L. A. (2004). *Evolution of Ecological and Behavioural Diversity. Australian Acacia Thrips as Model Organisms.* Canberra, Australia: Australian Biological Resources Study and Australian National Insect Collection, CSIRO.

Darwin, C. R. (1859). *On the Origin of Species*. London: Murray.

Darwin, C. R. (1871). *The Descent of Man and Selection in Relation to Sex.* 2 vols. New York, NY: Appleton.

Dobzhansky, T. (1961). In *Insect Polymorphism*, ed. J. S. Kennedy, p. 111. London: Royal Entomological Society.

Elkin, S. (1993). Out of one's tree. *Atlantic Monthly* (January): 69–77.

Evans, H. E. (1958). The evolution of social life in wasps. *Proc. 10th Int. Congr. Entomol.* **2**: 449–57.

Fisher, R. A. (1930) (1958). *The Genetical Theory of Natural Selection,* 2nd edn. New York, NY: Dover Press.

Flinn, M. & Alexander, R. D. (1982). Culture theory: the developing synthesis from biology. *Human Ecology* **10**: 383–400.

Fulton, B. B. (1933). Inheritance of song in hybrids of two subspecies of *Nemobius fasciatus* (Orthoptera). *A. Rev. Entomol. Soc. Am.* **26**: 368–76.

Hamilton, W. D. (1964). The genetical evolution of social behavior, I, II. *J. Theor. Biol.* **7**: 1–52.

Harrison, R. G. (1979). Speciation in North American field crickets: evidence from electrophoretic comparisons. *Evolution* **33**: 1009–23.

Hoogland, J. L. (1995). *The Black-Tailed Prairie Dog: Social Life of a Burrowing Mammal.* Chicago, IL: University of Chicago Press.

Hutchinson, G. E. (1965). *The Ecological Theater and the Evolutionary Play.* New Haven, CT: Yale University Press.

Jarvis, J. U. M. (1981). Eusociality in a mammal: cooperative breeding in naked mole-rat colonies. *Science* **212**: 571–3.

Kennedy, C. H. (1947). Child labor of the termite society versus adult labor of the ant society. *Scient. Mthly* **65**: 309–24.

Kukalova-Peck, J. (1978). Origin and evolution of insect wings and their relation to metamorphosis, as documented by the fossil record. *J. Morphol.* **156**: 53–126.

Kukalova-Peck, J. (1983). Origin of the insect wing and wing articulation from the arthropodan leg. *Can. J. Zool.* **61**: 1618–19.

Marshall, D. C. & Cooley, J. R. (2000). Reproductive character displacement and speciation in periodical cicadas, with description of a new species, 13-year *Magicicada neotredecim. Evolution* **54**: 1313–25.

Masaki, S. & Walker, T. J. (1987). Cricket life cycles. *Evol. Biol.* **21**: 349–423. (http://buzz.ifas.ufl.edu/k340lm87.pdf)

Mayr, E. (1963). *Animal Species and Evolution.* Cambridge, MA: Harvard University Press.

Michener, C. D. (1958). The evolution of social life in bees. *Proc. 10th Int. Congr. Entomol.* **14**: 299–342.

Otte, D. & Alexander, R. D. (1983). *The Australian Crickets (Orthoptera: Gryllidae).* Philadelphia, PA: Acad. Nat. Sci. Monograph 22.

Ricklefs, R. E. (1983). Avian postnatal development. In *Avian Biology*, vol. 7, ed. D. S. Farner, J. R. King, and K. C. Parkes, pp. 1–83. New York, NY: Academic Press.

Schuster, J. C. & Schuster, L. B. (1997). The evolution of social behavior in Passalidae (Coleoptera). In *The Evolution of Social Behavior in Insects and Arachnids*, ed. J. C. Choe & B. J. Crespi, pp. 260–9. Cambridge: Cambridge University Press.

Sherman, P. W., Jarvis, J. U. M. & Alexander, R. D. (eds). (1991). *The Biology of the Naked Mole-Rat*. Princeton, NJ: Princeton University Press.

Sigmund, K. (1993). *Games of Life*. Oxford: Oxford University Press.

Trivers, R. L. (1971). The evolution of reciprocal altruism. *Q. Rev. Biol.* **46**: 35–57.

Walker, T. J. (1980). Mixed oviposition in individual females of *Gryllus firmus*: graded proportions of fast-developing and diapause eggs. *Oecologia* **47**: 291–298. (http://buzz.ifas.ufl.edu/g464lwa80.pdf)

West-Eberhard, M. J. (2003). *Developmental Plasticity and Evolution*. Oxford: Oxford University Press.

Williams, G. C. (1957). Pleiotropy. Natural Selection, and the evolution of senescence. *Evolution* **11**: 398–411.

Williams, G. C. (1966). *Adaptation and Natural Selection*. Princeton, NJ: Princeton University Press.

Williams, G. C. (1992). *Natural Selection, Domains, Levels, and Challenges*. New York, NY: Oxford University Press.

2

Motherhood, methods, and monkeys: an intertwined professional and personal life

JEANNE ALTMANN

Ever since I was a little girl, I wanted to be a behavioral ecologist, or live in Africa and study monkeys, or... No, my sense of careers was much more limited, as were my experiences with wildlife.

Born in New York City in 1940, I was raised after infancy in suburban Maryland just outside Washington, D.C. in a sequence of small apartments/flats that were imbedded in stretches of such apartments, with scattered patches of grass lawn and occasional playgrounds populated with swings and slides and climbing bars. With almost no public transportation, no family car until my late preteen years and bicycle riding restricted to sidewalks, I had little exposure there to the world of animals that would become so important in my life, although I believe that I was always an observer of people. Ecological context was reinforced by family: my mother was highly allergic to virtually all mammals (except us), my father was extremely and increasingly overprotective, and both were New Yorkers to the core, by both background and temperament. No, I did not start from an attraction to nature or animals or animal behavior. I started from a love of reading and math and science, a love of puzzles, a thirst for playing with ideas and relationships and alternatives, all of which were more a part of my upbringing and inclinations than was the world of nature. The animals, behavior, the passion for African savannahs and North American woods would come later. Interestingly, my sister eventually became a farmer in rural West Virginia; what a puzzle our environmental choices were to our city and suburban-focused parents!

Leaders in Animal Behavior: The Second Generation, ed. L. C. Drickamer & D. A. Dewsbury. Published by Cambridge University Press. © Cambridge University Press 2010.

Academic excellence was highly valued by my parents both in principle and in practice, but the message was a mixed one for a girl: a college education was a must, but a must in the nineteenth century model, in order to be a better wife and mother, and also to have an employment skill 'in case' something happened to one's husband. At the same time, I later came to feel that, unlike my younger sister, I was treated as 'first son', the one who was expected to inherit parental values and traditions. Confused and dissatisfied and with little sense of the route to shaping my future, but in classic 1950s fashion, I lived an unquestioning and relatively unexamined life in many respects. Yet I loved math and science, was eager to travel, and so somewhere developed the idea, I think from an elementary school teacher who had previously taught at Army bases around the world, that I could earn a degree in math, and then travel the world by teaching math everywhere. Although my father had the not uncommon view that teaching was a lower calling than 'doing', I had no idea what else I might do with a math or science degree, especially if I also wanted to travel. My maternal grandmother somewhat reluctantly granted my radical high school graduation wish for a slide rule, agreeing to the purchase provided that I would buy a slide rule that was small enough to hide in a purse, a considerable concession for her I suspect, and one that conferred a great feeling of possibility to me.

University was affordable for my family if I went to a nearby public college where I could live at home, and so when my father was transferred to Los Angeles as I finished high school, my journey as math major started at UCLA after a summer job at the National Institutes of Health (NIH) in Bethesda, Maryland. I returned to that job after a first year at UCLA, meeting there a tall, handsome, and captivating biologist, the person who would become the love of my life and introduce me to a totally different world. Stuart Altmann was completing his Harvard PhD thesis in biology after spending two years studying rhesus monkey behavior on Cayo Santiago, an island off Puerto Rico, working for NIH. 'I went to the zoo with a zoologist' read my simple message home after our first date. When, during our subsequent short-term coast-to-coast 'courtship' we decided to marry, Stuart told my parents that I wouldn't have much money but would experience a lot of travel! They took a while to recover from their shock and dismay, not at Stuart but at such an early marriage. When we started a family two years later, my mother's wail was 'a baby before a BA?' To this day, however, I remain amazed that at 18 I made the best decision of my life, one I've never regretted. That I was able to raise two truly amazing children and then also have a wonderfully satisfying career, strikes me as such marvelous luck. It did take luck on top of commitment and passion and a few very important sources of encouragement, including Stuart, to surmount the primarily gender-related obstacles that I often found discouraging and demoralizing. Serendipity played a major role in my professional and personal development over the next 20 years.

I had enjoyed and was challenged by some of my UCLA courses and had a range of interesting part-time math tutoring jobs, but the University was not the right academic environment for me, particularly as the math department was open about not wanting women math majors as they would be a waste of time; I was ready for change. With little money and only a single academic year before Stuart would submit his thesis, I enrolled as a part-time math undergraduate at MIT, obtained a part-time job doing computer programming for Beatrice Whiting at Harvard, and helped Stuart with data analyses for his thesis (S. Altmann 1962), spending many a delightful brown-bag lunch among the ants with Stuart's advisor, Ed Wilson. Thus began my education outside the classroom, my exposure to a number of outstanding academics, and my experience of a number of the research threads that would ultimately be woven into the tapestry of my academic life. Beatrice Whiting and her husband John designed and directed an ambitious six-culture study of child social behavior (Whiting & Whiting 1975). In 1959, this project was radical in a number of ways that I only fully appreciated later. Six field teams spread around the world to collect data in standardized ways to address hypotheses about the process of socialization and the cultural contingencies involved. Such a level of hypothesis testing, systematic data protocols, planning, and coordination were unprecedented at a time and in a field in which individualism and exploration still predominated. The premise and execution of this distributed, simultaneous field work is all the more striking when one considers the status of international communication in the late 1950s. Moreover, taking advantage of the recent development of computers, data were going to be entered (onto computer punch cards!), verified and subjected to exploratory analysis as they came in from the field. This resulted in my introduction to computer programming and provided my first experience with this emerging technology several years before I would otherwise have had that exposure. Only half jokingly, the Whitings, who knew Stuart's thesis project, suggested that perhaps we should add the Cayo Santiago rhesus monkeys as a seventh culture to see what would be interesting in the comparison.

As a result of working with data from both the six-culture study and Stuart's rhesus monkey project, my first introduction to behavioral field research was one that treated as normal a focus on research design, data analysis, hypothesis-testing, and data comparability, not common at the time. I also developed from the beginning the idea that discipline boundaries were permeable, that good, useful ideas and techniques were to be found in different fields, and that humans were an interesting species to compare in rigorous ways to nonhumans. This was an essential mindset for me a decade later when I began the investigations of behavioral sampling methods that led to my observational methodology paper (Altmann 1974). Perhaps it was just as well, at least for me, that I was unaware that the highly respected woman who was the lead director of the six-culture study and who mentored me for that academic year by outstanding example did not have tenure, and would not receive it until she was almost ready to retire decades later. However, this oblivion, like that regarding school racial segregation until I was in secondary school, is shocking and embarrassing for me to imagine now, especially since during those same times many of us were aware of, and grew up experiencing, religious prejudice and segregation.

My exposure to computer programming was unusual at the time and was reinforced in the summer of 1960 by working at NIH again, this time in Wilfred Rall's group in the Office of Mathematical Research, which was exploring mathematical models of neural networks. There, I again encountered mind-stretching and stimulating science from mentors who encouraged me simply by inclusion, thereby communicating implicitly that I was doing what I was meant to do.

Once again I transferred universities, this time as Stuart assumed a faculty position in Zoology at the University of Alberta in Edmonton. The math department there was welcoming and challenging, and I was invited to join the honors program for my final year. However, I declined because by then I was expecting our first child and would need to attend school part-time, an option that I learned wasn't available for an honors degree student. Despite this disappointment, finishing my degree part-time in the evening and spending most of my time parenting was not all disadvantages. I cannot imagine, either emotionally or intellectually, having missed the intimate and intense involvement in raising my children, at once the most challenging and fulfilling experience of my life. At the same time, my intellectual yearnings required more. My enjoyment of mathematics was greatest in pure math, and my success and pleasure was greatest when I lost all sense of time or place for innumerable uninterrupted hours or even days, when I was in a state that has been termed 'Flow' (sensu Csíkszentmihályi 1990), something that at least for me was incompatible with being the primary child-care person at home. Although I was unaware of it at the time, my parenting experience was probably a significant fork in the road that eventually led me to focus on behavior rather than mathematics – yet it would be over a decade before my professional direction was really set.

A little over a year later, in 1963, our toddler Michael and I joined Stuart for what was to be a 15-month study of primate communication in a natural setting, and the plan was that as Stuart and a PhD student from another university conducted behavioral observations I would promptly tabulate and analyze data they collected, a radical idea at the time and still relatively uncommon. However, this was the explorer's era in tropical field research, even more so in studies of large mammals, and many scientists wanted 'their own species'. Because a previous 10-month field study had been conducted on baboons in a park on the outskirts of Nairobi in Kenya and another short study had been conducted in South Africa, the PhD student was concerned that baboons had 'been done'. The student dropped out of the project at the last minute, and I joined Stuart in observations. What an amazing, life-changing year that was in so many ways.

At 23 and with camping experience that was limited to a very happy summer week spent in Girl Scout camp in the mountains of Virginia during several successive middle school years, plus a camping honeymoon that involved two weeks of camping from Los Angeles to Cambridge, Massachusetts, I had no expectation, positive or negative, of what life would be like when we left Edmonton for East Africa, with stops particularly for Stuart to meet with Robert Hinde in Cambridge, Detlev Ploog and Konrad Lorenz in Munich, and Hans Kummer in Zurich. In each case, Stuart's stimulating science discussions were complemented by more personal ones as the wives and children welcomed Michael and me with great generosity. Vreni Kummer provided my first advice and hints of experiences to come 'living in the bush'. It's difficult to describe how unusual and welcome such warmth by strangers was, especially in Europe of 1963, a time when hoteliers were even surprised that we travelled with a child, particularly with no nanny for caretaking of him and to feed him out of sight rather than in the public dining room!

In Nairobi, we were again fortunate, in that Irv and Nancy DeVore were back in Kenya for a brief return to the site of their Nairobi Park baboon research. Before Irv's first trip to study baboons, he had visited Cayo Santiago to learn Stuart's behavioral data collection methods on the rhesus monkeys, and the DeVores visited us at Harvard on their way back from Kenya at the end of their study. Now, we visited Nairobi Park together, and they were generous with advice on everything from baboon observations to shopping. Although Irv and Nancy were confident that we would settle in Nairobi Park with all its advantages of the nearby city and habituated animals, they also suggested some other Protected areas we might visit in Kenya and northern Tanzania (then Tanganyika) in our search for field sites. These included Amboseli, in Kenya, where Irv had spent a few weeks during his field study. With advice from various local wildlife biologists as well, we developed a route and shopping list, then began preparing for two months of independent living and travel on the primarily unpaved roads. Our goal was to survey olive and yellow baboon populations of southern Kenya and northern Tanzania in search of a field site that had relatively undisturbed populations that would be sufficiently visible for detailed behavioral studies. We bought a used long-wheelbase Land Rover and outfitted it simply to store all we would need while being able to use it for toddler's playroom and adults' rooftop observation site during the day, sleeping quarters at night (Altmann & Altmann 1970). While vehicle modifications were being fabricated, and when transport was available, we began to develop an eye for baboons and baboon behavior through observations at Nairobi Park where the tourist-familiar baboons were readily observable.

Despite Nairobi's inviting climate and gardens full of native and exotic flora, we were eager to leave the city for our 'safari', and we began our travels shortly after Michael's second birthday, as soon as the vehicle was ready. First on our route was Amboseli Reserve, at that time a dusty seven hour drive to the base of Mt. Kilimanjaro. From mile-high, lush Nairobi, we descended across the Athi plain on the unpaved main Nairobi–Arusha road to the border town of Namanga, where we continued to descend, on a less traveled route, into the Amboseli basin, centered on the now-seasonal remains of the Pleistocene Lake Amboseli. Here in the Amboseli–Longido basin was my introduction to tropical ecosystems and for me, unlike for most tropical biologists, it was on the East African savannahs that I 'imprinted'.

Amboseli was home to a large population of the least disturbed and most observable baboons that we were to see anywhere over the next two months. As highly terrestrial, omnivorous, and highly social primates, baboons, even more than other cercopithicines, were, and remain today, ecologically as well as geographically widespread, frequently coming into contact and conflict with humans and their livelihoods, especially in agricultural regions, increasingly so as agriculture increasingly expands at the expense of other land use. As a result, baboon (sub)species throughout Africa often either flee humans or approach them, particularly near tourist or park ranger areas, where baboons come to see humans as sources of food. Then, as now, African wildlife in many areas are preserved and most unaffected by humans in the arid and semi-arid regions that are unsuited for agriculture and are inhabited instead by pastoralists such as the Maasai. UC-Berkeley PhD student Tom Struhsaker had surveyed areas in Kenya and Uganda for study sites for his vervet monkey research, and he, too, had settled in Amboseli for similar reasons (Struhsaker 1967). In 1963–4, Amboseli was a

44

Jeanne Altmann

Figure 2.1. Baboons foraging on the savannahs of Amboseli, Kenya, at the base of Tanzania's Mt. Kilimanjaro. The animals were already relatively tolerant of observers in or on the roof of a vehicle when we first went to Amboseli, but they initially fled from people on foot. Since the mid-1970s all our observations have been conducted on foot.

Figure 2.2. Baboons are one of the most highly social of primate species. Even in this semi-arid environment where as much as three-quarters of the daytime is spent foraging, time is always taken for social grooming and resting.

large Maasai Reserve, devoid of tourist accommodation except for a small tented camp, and open to visitors only in the dry season because the access road, which was transected by a seasonal river, unpredictably flooded during rains. We had the whole area almost to ourselves much of the year and had no communication or supply resources except for monthly trips to

Nairobi or Arusha, which we alternated with Tom, not very far from whose tent we pitched our own two tents. The isolation also meant that we were on our own for all tasks from water-hauling and diaper washing through cooking. For much of the year, we couldn't find any help that would do these tasks, no-one who would live in a tent where lions, elephants, buffalo, and leopard were regular visitors. All in all, this was a year of maximum multi-tasking as childcare and camp living activities were juggled along with searching for and observing baboons, good preparation for the rest of my life. By her own example, my mother had taught me that if a mother's employment wasn't essential for family survival, one would gain opportunities to pursue non-domestic activities by being faster and more efficient at the domestic ones, and it was a message I internalized early on.

That first year in Kenya was an immersion introduction to fieldwork, animal behavior, and baboon ecology for me, while baboons and the other wildlife were the focus of our family's life, because the year was also one in which we had more life-threatening medical crises than we were to have in total since then, many of my animal behavior memories from that period remain diffuse and fragmentary to this day. My field 'year' suddenly was cut short in April when our son suffered from a complete paralysis from the neck down. We barely succeeded in crossing the river to reach Nairobi, where he ultimately was kept near an iron lung for the next two months until he recovered enough for me to return with him to the Clinical Centre at the National Institutes of Health in the US. After four months of separation and with much relief, our family reunited when Stuart joined us from Kenya in August. Anyone who has experienced the thrill of a child's first steps can perhaps imagine the even greater thrill of a child's first steps after a life-threatening paralysis.

Figure 2.3. Baboon females form close, persistent ties, particularly with mothers, daughters, and other close relatives in this matrilocal species. Mothers' initial grooming of her infants of both sexes develops within a couple of years into reciprocal relationships with daughters though not with sons.

What was to be our last year in Canada was a busy one personally and professionally. I gave my first scientific presentation in December, on progression order in baboons, in a symposium Stuart organized on *Communication and Social Interactions in Primates*. I then spent much of my available work time that spring helping edit the resultant University of Chicago symposium volume (Altmann 1967), both of which were important 'firsts' for me. Reviewing and editing a diverse range of papers introduced me to the literature in a new way and helped me continue to develop my interest in methodology. The new year also saw Stuart hospitalized for a back injury and me for a mid-pregnancy miscarriage, finally bringing to a close our 'medical year' that had begun in Kenya. That summer, we prepared to leave Edmonton, headed to the suburbs of Atlanta, Georgia, and to Emory University's Yerkes Primate Research Center. We made the move in our sort of style – Stuart, Michael, and I, with camping supplies for a few weeks, piled into or on top of our royal blue Volkswagen 'beetle' complete with bright red aerodynamic wooden roof carrier that Stuart constructed for the journey. What a sight we must have been at the campsites along the way as we and our belongings spilled out in the evening and folded precisely in again, usually the next morning – we couldn't take too long on this journey as I was mid-pregnancy again, and the 'beetle' had no spare space.

We located a small wedge-shaped and steep wooded lot near Emory and began having a house built to nestle in the hillside. A few months after our daughter Rachel was born, we happily moved into the 'construction zone'. I initially joined Stuart part-time analyzing the data from our Kenya year. But within a year the initial promise that I could work with Stuart was rescinded as Emory extended their nepotism rules to the Primate Center. Stuart offered to fight this decision, but various prior exclusions left me demoralized and unwilling. Instead, this seemed an ideal time to explore my early interest in teaching mathematics beyond the tutoring that I had always done to varying degrees ever since I began university.

I soon learned that I was in the right place and time from the perspective of the nearby public school system as well. The racially segregated school system was finally about to integrate, under pressure locally from civil rights organizations and many African–American parents, and nationally from the federal government. One type of 'carrot' provided by the federal government came in the form of additional funds for math programs in the historically black schools. Because math teachers were already in very short supply in the state, temporary teaching certificates were available to those who had a math degree but not an education degree. I first did substitute teaching in the recently integrated high school. When I applied for a regular job and distressed the administrators by answering honestly their questions about religion and frequency of church attendance, they promptly decided that the way to deal with several 'problems' at once was to offer this Jewish math major a one-year job developing and implementing a remedial math program in the town's historically black elementary school. I enthusiastically jumped at the opportunity for social and professional reasons. I spent 1967–8 in one of the most stimulating, educationally satisfying, and exhausting jobs of my career as five of us joined long-standing black colleagues, replacing those of their colleagues who integrated the previously white schools during one of the most historic civil rights and political years of the 1960s. At the request of the parents of my poverty-level students I started an evening math class for parents so they could help their children with homework, and I taught school-time enrichment for the top math students in my son's school. I subsequently completed my Master of Arts in Teaching (MAT) degree in math, continued tutoring, and worked part-time supervising math student teachers. At every turn and task, though, I found myself observing behavior and crossing disciplines, as when I found that teaching or tutoring math often involved teaching language and logic skills, or noticing handwriting impediments to arithmetic performance in low-performance students, or noticing the extent to which students of all ages devised much more creative ways to solve math problems or even do basic arithmetic tasks than the single proscriptions taught in the schools. Neither I nor my students ever fit neatly into the boxes to which we were assigned, something that I didn't originally recognize or find comfortable but with which I was increasingly becoming comfortable. My slow and somewhat twisty path was perhaps the right and even necessary one for me.

I'm not sure what route my career might have taken if we had stayed in Georgia, but in 1969, Stuart received a research grant that enabled us to return to Kenya for a few months. Soon afterwards, he also accepted a faculty offer from the University of Chicago. At the same time, Stuart made a suggestion that ultimately helped tip my career direction back toward behavioral studies. He thought researchers would find useful a guide to ways of analyzing behavioral data that had been gathered in different ways. I was intrigued by a convergence of my interests. With an impending move, I was not in a position to take on employment beyond occasional tutoring, so I dove into the observational behavior literature to get a feel for the current state of the topic. Being an outsider to the field, this was something that I particularly needed. For the same reason, I could also bring a fresh perspective. By the time we settled in Chicago in autumn 1970, I was 'hooked'. We returned to Amboseli in July of 1971 to begin habituating a group of baboons, Alto's Group, one of those that we had censused and followed for some months in 1969. Alto's Group became the study group for Glenn Hausfater's thesis research in which he tested Stuart's earlier queueing model of male priority-of-access to mating opportunities (Hausfater 1975). Thus, what later became a long-term research program began with a focused short-term study, as has been the case for most if not all research projects that developed into long-term ones and that have provided models for recent ones that have been started with explicit long-term goals.

With our younger child, Rachel, now in school for a few morning hours each weekday, I was able to focus some time on reviewing the methods in observational studies that ranged across species from ants and wasps to humans, and from an array of disciplines including psychology, education, anthropology, and zoology. A few striking findings were soon evident. The first was that for only a small proportion of studies, and a particularly small proportion of field studies, were the descriptions of the behavioral sampling adequate to identify the key characteristics that one would need to make analysis decisions. Second was that well-described sampling schemes were virtually never justified in terms of their ability to answer the behavioral questions in which the authors were interested. And finally, the strengths and weaknesses in these respects were distributed across academic disciplines, taxa being studied, and researcher seniority. Consequently, the topic of my emerging project shifted from data analysis to the nature of relating behavioral questions to data collection, and my original cross-disciplinary tendencies became reinforced. Although writing has always been difficult for me, the project was such a perfect fit and so fascinating, and Stuart was so encouraging and willing to comment on successive drafts, that my earlier discouragement and fears gradually receded. When I had a draft of *Observational Study of Behavior: Sampling Methods* (J. Altmann 1974) ready to share more broadly, I sent it to many animal behaviorists, and was gratified when several generously gave me valuable feedback. Stuart declined my offer of co-authorship, years later telling me that he thought I would only receive adequate recognition if I was the sole author; with hindsight, I think he was probably right.

The sampling paper was in many ways the right paper at the right time. A need was there, and the paper's use spread rapidly even before publication. The approach was a very simple, first-principles one. It was not mathematical other than basic ideas of sampling spaces, estimation, and question formulation, but even that was unusual at the time, and unfortunately, sometimes missing or misunderstood even now. It was intended as a guide to thinking and planning and design, with the introduction as important as the subsequent evaluations of existing sampling methods. A recent small volume by Marian Stamp Dawkins provides a marvelous expansion on this theme (Dawkins 2007).

By the time I submitted the sampling paper to *Behaviour* in 1973, Rachel was in school 'full day', and to some extent thanks to my work on the sampling paper, I decided that I wanted to enter a PhD program, one that would lead to a career in behavioral ecology. I also chose to focus my research on baboon females and their young infants. This was a difficult decision, because in naturalistic studies, particularly of primates, this was considered primarily a women's topic, especially in an era that focused on male aggression and reproduction as the evolutionarily important topics in selection, in differential reproduction, and in survival. This perspective didn't make sense to me; rather, I thought it was clear that in natural environments the survival and reproductive challenges faced by female mammals were at least as great as those of males, especially so in species such as primates that experienced disproportionately long periods of gestation, lactation, and infant carrying during long daily travels while foraging. The period of investment in offspring constitutes a high-risk bottleneck period in terms of selection and one on which evolutionary biologists should be focusing rather than avoiding. Indeed, times were soon to change thanks to the increasing influence of Bob Trivers' work (Trivers 1972). None the less, research into behavioral aspects of evolution continued for decades to be hampered and misled by a focus on sexual vs. natural selection and by considering number of mates the key measure of differential reproduction. These misconceptions are only now being redressed, finally bringing studies together into an evolutionary framework.

The strongest biology program in the Chicago area was the one in which Stuart was a faculty member, and I wouldn't consider it both for that reason and because I wanted a return to school to further enhance the complementarity of our skills, not to increase the overlap. Most fortunately, the University of Chicago also has an explicitly interdisciplinary program in Human Development that had modeled in many ways on the one at Harvard in which

Figure 2.4. Individual mother–infant pairs have long been observed in close association with one of a group's adult males in a number of baboon populations. However, the functions of these associations, particularly for males, remained in dispute. Within the past decade, paternity analyses combined with behavioral data, demonstrated that the associated male is often the infant's father and that males in this polygynandrous species play an important supportive role in thir juvenile offspring's social interactions.

I worked with Beatrice Whiting over a decade earlier, one that sought to bring the perspectives of psychology, sociology, anthropology, and biology to bear on an understanding of the human lifecourse. With considerable enthusiasm on the part of several faculty members and puzzlement or skepticism on the part of others about the relevance of my goal to develop a thesis project on baboon ontogeny and parenting, I was admitted and joined the program in 1973. The faculty accepted my in-press sampling paper as meeting the requirement for a preliminary research project, making an exception regarding prior work for which I was grateful. When I graduated a little over five years later, I was also gratified that by then I had won over enough of the wonderfully diverse faculty that they nominated my thesis for the annual Social Science thesis prize, which I received.

Possibly more than for other students, the process of conducting my thesis research was a life-changing one, much more so than one might guess from an apparent continuity: I studied behavioral ecology of baboons before and continued to study them afterwards. It was in these years, though, that in many ways I found my niche and my 'voice', that I developed my focus on demography, on life histories in both the biological and the social science meanings of that phrase, on the integration of ecological context, family structure, and individual lives. It was during these years that I first traveled and spent time in the field without the family, albeit briefly. And though I was always respected intellectually by Stuart, I was understandably just

Jeanne Altmann

Figure 2.5. Although infants are primarily dependent on their mothers for nutrition, protection, and transport for most of a year, the pair is from the onset deeply imbedded in a highly social milieu.

starting to be recognized by others as a scientist, including being elected as editor from the American side of the journal *Animal Behaviour*. Like many students, I had trouble putting closure on my thesis; fortunately, taking over the journal editorship and my son Michael's upcoming high school graduation in 1979 provided two great incentives to finish. I wanted to complete my degree before Michael's graduation. In addition, I took over the journal at a stage when it would require a huge amount of effort to get out of a tough place. Editing the journal was another whole education in itself, in both science and sociology of science. It was easily the most time-consuming and rewarding professional service I've performed.

My study on the ecology of motherhood focused on the allocation challenges faced by simultaneous investment in survival, maintenance, and reproduction, particularly for non-seasonal breeders with high offspring investment in slowly maturing young. My research reinforced my original idea that studying mothers and infants was not the 'cute', soft female topic that it was often considered but rather the situation of particularly great opportunity for selection, especially so in challenging environments such as the arid African savannahs. I was intrigued by the extent to which the social environments of mothers benefitted or hindered their ability to succeed in negotiating this perilous period. Equally striking was the degree to which both parent and offspring were each dependent on the other for success. Although the interests of parent and offspring were not identical (Trivers 1972), their ability to find creative, mutually beneficial solutions to potential conflicts of interest seemed at the core of success in the primate lineage, perhaps a major factor in evolution of human ontogeny and sociality.

I faced a dilemma as I thought about publication of my thesis. I knew that each aspect of my study would attract more readership and a broader audience, especially from those not studying primates, if it was developed as one or more separate technical papers, and I felt that each was a broadly important topic. Yet, the intertwining of one with another, the contingencies between ecological and social context for example, cried out for an integrated approach, one that did not treat each aspect as an independent, neatly discrete box, lined up in a string of separate topical papers. I tried to do the separate papers but the farther along I went, the more I felt the pull for the integrated whole just too compelling. I've sometimes regretted that I didn't find a way to do both, perhaps in rapid succession, and if I were a speedier writer or more efficient or with a more exclusive research life, or if I hadn't been trying to maintain the long-term field project, perhaps I could have, but that was not to be. Life for scientists as well as our study subjects is often one of tradeoffs and imperfect decisions. In any case, the decision was made and the thesis was written as a book and quickly published under the title *Baboon Mothers and Infants* (J. Altmann 1980) (the publisher rejected 'ecology' in the title and wanted the addition of 'baboon'); it was broadly reviewed and well received. Although I had a baby before my BA, I did (barely) manage to finish my PhD before that 'baby' graduated from high school and before the second one entered.

Once again, I was at a crossroads. Editing *Animal Behaviour* provided a great challenge, especially so in an era of pre-electronic communication, but no salary, and I also needed to seek funding for the baboon research. Stuart, Glenn, and I had co-directed the Amboseli Baboon Project for the first decade as the Project gradually became one with long-term goals. Stuart's focus was primarily on feeding and foraging, Glenn's on dominance and reproduction in males, mine on demography, ontogeny, and female life-histories. We each had several other topics of interest as well as a number of shared ones. Most years, our family spent at least part of the year in Amboseli, as did Glenn, and each year one or two PhD students or post-doctoral fellows spent a year on individual short-term specific studies while maintaining demographic, reproductive, and agonistic data. After the first few years, we gradually introduced a growing set of common procedures and data sets that these participants were expected to collect in exchange for being able to draw on data from previous years and focus most of their time on specific independent projects. In 1982, the resultant Monitoring Guide was announced in society newsletters and made available to interested researchers on request, and in recent years the current version, authored by myself and Susan Alberts, has been available for downloading on our Amboseli Baboon Project website, www.princeton.edu/~baboon.

However, by the early 1980s, Glenn wanted to be independent of Baboon Project commitments, and Stuart wanted to focus on his book about infant feeding and nutrition rather than devote time to the fieldwork or data collection or long-term aspects of the Project maintenance or database management. My research questions and interests, on the other hand, had been and were increasingly longitudinal, lifespan, and intergenerational in nature.

When I completed my degree in Human Development, I could have tipped toward studies of either humans or baboons, but in the absence of funded local academic opportunities in human development and with the likelihood of more problems than advantages entailed by relocation for a faculty position, I leaned toward staying with the Amboseli baboons. I was already deeply invested in what had become one of the first, and is now one of the most

Figure 2.6. Amboseli observers endeavor to stay only as close to their subjects as needed for particular data collection, no closer than the most shy member of the group will tolerate. Here, Raphael Mututua successfully makes himself the most boring primate in the animals' world.

intensive and extensive, long-term field studies of any large mammal. My thesis research and initial demographic and developmental analyses only whetted my appetite for answering lifespan and intergenerational questions about the relationship of individual and group behavior to survival and reproduction. On another level, I realized that Mt Kilimanjaro, the African savannahs, and 'a feeling for the organism', had become too much a part of my being to leave them. Fortunately, Stuart and I had spent our adult lives living on one income, his, with my very modest contribution rarely covering more than flights to Kenya for the children, so we were great at economizing. We were satisfied continuing to manage that way, and as Michael entered university, we were aided greatly by the University of Chicago's generous offspring college tuition plan – Chicago and baboons it would be if I could obtain the research funding. The next year, I was very fortunate to receive a multi-year research grant, one that provided both research funds and some salary to follow up on findings and questions raised in *Baboon Mothers and Infants*; I was thrilled.

In addition to the early 1980s bringing a transition to sole responsibility for the baboon fieldwork and the growing data sets, it also marked a transition in primary field personnel that has now persisted for over two decades and has provided the field base for the next generation of our fieldwork. This transition involved intensive training of and participation by Kenyans at all levels and primary focus on multi-year personnel and projects. After a few false starts with other local assistants, in 1981 we began training Raphael Mututua, a recent O-level graduate, who 28 years later is a marvelous Project Manager as well as data gatherer. In 1989, Raphael was joined by Serah Sayiallel and in 1995 by Kinyua Warutere, the three constituting a 'dream team' with a continuity across years that has been essential to all our

Figure 2.7. Raphael Mututua, Serah Sayiallel, and Kinyua Warutere, the Amboseli field 'dream team'. Each follows a different study group most days. However, for several days at the end of each month, they spend a few hours together with each group (five groups since 1999 for example) to reinforce inter-observer agreement and to solve any emerging problems.

Amboseli research projects, but especially our ability to develop life-history studies of males that are now becoming as rich as those we had been conducting on females, especially rare in a male-dispersing species. The new efforts to study male life histories were also made possible by two American assistants, first Amy Samuels throughout the early and mid-1980s. By the mid 1980s, I also was able to attract the first of our Kenyan graduate students, Philip Muruthi, who received his MSc at the University of Nairobi, joined Princeton University for his PhD, and then returned to do conservation work in Kenya. Philip remains a valued colleague and a mentor to many young African wildlife scientists.

The 1980s were exciting times for the Amboseli project despite a number of significant threats to the fieldwork that ultimately were resolved. Even a single decade of continuous data on identifiable individuals was revealing insights into life-history plasticity and its sensitivity to both social and ecological environments, themes that became increasingly major ones in my thinking. The baboons also increasingly found themselves in a changing environment; as individuals and group, they exhibited interesting behavioral responses to those changes. In addition, partly in response to reviewer concerns, we made the demanding – and expensive – expansion of our studies from a single social group to multiple ones, starting by adding Hook's Group to intensive monitoring. This enabled us to study variability over space, foraging, and demographic contexts as well as over time.

Despite the early emphasis of primate field studies on feeding, fighting, and mating, by the late 1980s much more had been revealed about individual differences among females throughout their lives than about males. As with males of most species and with

Jeanne Altmann

Figure 2.8. Baboon males not only exhibit natal dispersal at about 7-8 years when they become fully adult; repeated, secondary dispersal is also common during adulthood. This poses a special challenge for them in the need to establish new social affiliative and agonistic relationships.

the dispersing sex of all species, theory and data remained basically cross-sectional, snapshots in time, rather than representing lifetime differences and age-specific data based on individuals over the life-course. This is a glaring gap to a great extent because individual differences in male life histories and reproductive success are much less stable across the lifespan than are those of females and, therefore, cannot be estimated well by short-term data. For boom-and-bust life courses such as those that characterize males of many species, including baboons, short-term studies overestimate male variance in fitness and overestimate the impact of factors such as dominance and number of mates obtained. The practical challenges of redressing this gap are great, however, especially in species with slow life histories and male dispersal. Our expansion to multiple study groups and the permanent trained presence of local assistants facilitated being able to monitor males for much more of their adulthood. This set the stage for considering in-depth studies of male life histories, exactly the topic on which a new PhD student wanted to focus.

In 1984, I joined the new Conservation Biology Department at Brookfield Zoo, where I was fully salaried for the first time. I also received modest but crucial field support for Amboseli, particularly for training Kenyan students. Shortly thereafter, I was offered a faculty position in Ecology and Evolution at the University of Chicago, so I arranged a split position between the Zoo and the University. Susan Alberts, who had spent a post-college year with me in Amboseli on a Watson Fellowship, was completing an MSc at UCLA, and was interested in tackling the 'male problem' for her PhD research. The topic, the time, and the person were right. Susan entered the Chicago PhD program soon thereafter. She proceeded to conduct a beautiful behavioral study of male maturation, dispersal, and mating, using both her own detailed several years of field data and our now 15-year longitudinal data set. Of even more long-term significance, Susan became committed to finding a way to obtain paternity information to combine with the male behavioral data, and we both began exploring a way to do that. At the same time, I was increasingly interested in elucidating the relationship between physiology and behavior without disturbance. Again, serendipity was on our side, and just as Susan was writing up her thesis, we established collaborations for physiology with Robert Sapolsky and for paternity determination with Bob Lacey at Brookfield Zoo, and Bob Wayne at the Zoological Society of London. The result was a launching in 1989 of morphological, physiological, and genetic sampling with Robert using the least invasive immobilization technique (blowpipe darting) that was available at the time. These collaborations produced a number of novel papers throughout the 1990s, including ones relating social factors to glucocorticoids, morphological comparisons of 'lodge' and wild-feeding baboons, and confirmation that at least under some conditions mating behavior distributions predict paternity distributions (see bibliography at www. princeton.edu/~baboon). These projects and testing of some longstanding hypotheses were made possible by a combination of our long-term data and methodological developments in field and laboratory.

These initial investigations provided compelling motivation to continue studies of hormones and behavior and of molecular ecology, as we had just scratched the surface. To answer the next set of questions we were interested in would require genetic samples from individuals whom we did not dart: younger individuals, more shy new immigrant males, and females throughout pregnancy. In terms of physiology, we also wanted to answer questions about more integrated, chronic hormone levels, which would require fully non-invasive sampling, such as urinary or fecal sampling. Fortunately, a few research groups around the world were starting to validate fecal steroid methods in primates, and a few others were starting to develop methods to use fecal sources of DNA for parentage studies. Meanwhile, Susan began her post-doctoral research in molecular population genetics to test a range of questions in behavioral ecology including ones for which the integration of intensive life-history and behavioral data with known parentage were the essential ingredients, and to pursue other questions in molecular ecology that extended use of genetic data beyond parentage. Our collaboration, rather than ending with her thesis in 1992, intensified, with both of us working on aspects of behavior, survival, and reproduction, and with each of us eventually developing a complementary laboratory component.

By the late 1990s, Michael and Rachel had established their own families and careers elsewhere, Stuart had transitioned to emeritus status, I accepted a faculty position at Princeton, and we made our first move in almost three decades. I felt very fortunate. I will probably always miss Chicago, some special colleagues there, and the University-wide and interinstitutional Committee on Evolutionary Biology that I had chaired for seven years. At the same time, I love where I am; what more can one ask for? In Princeton I established a fecal steroid hormone lab; Susan joined the faculty at Duke University, where she established a genetic lab. With 'startup' funds from both our universities followed by five-year support for multidisciplinary studies at the Directorate Level at the National Science Foundation, we were able to take this multipronged approach, something that is not possible through NSF individual program units that fund behavioral ecology.

About the same time, we started implementing a long-term plan, first for equal partnership and then for Susan's assuming the leadership in Amboseli. Long-term longitudinal field studies are in themselves rare, but I found none with a continuity and transition plan, yet the need, the obligation, felt compelling. In the decade since then, at least a few others have been developed or are under consideration, some explicitly stimulated by our effort.

This first decade of the twenty-first century has been at least as exciting as any previous one. I find it interesting and very satisfying that I am no longer asked when 'baboons will be done', such a frequent question in the first decades of the baboon project. If anything, the research potential of the Amboseli baboon population continues to increase rather than to decrease or even reach a plateau. This is partly due to capitalizing on emerging techniques and to emerging environmental challenges the baboons encounter. However, it is also due to the increasing value of the data that originally motivated the long-term project: systematically collected life-history, demographic data on known individuals throughout their lifespan, within the context of variable social and ecological environments. With cumulative years, we have learned and continue to identify what patterns and aspects of lives are relatively stable and exhibit little variance across time and space, even across individuals in some cases, and conversely what aspects are highly age-dependent, contingent on ecological and social factors, and now also genetic ones. Intensive and extensive research on the Amboseli baboon population has also enabled us to begin to identify physiological, behavioral, and life-history tradeoffs, conditions under which those tradeoffs occur, and the extent to which some individuals are relatively unconstrained in their options whereas others are much more constrained. In addition, we can now investigate patterns and variability late in life, during aging, at the opposite end of the lifespan where my studies started almost four decades ago. Another reason for increasing value and excitement is our ability to 'get under the skin' now, to answer some questions about physiology and genetics and to do so primarily while retaining hands-off methodologies.

No one biological system or population is ideal for all questions in behavioral ecology, nor does one provide findings that will hold over all species or even over space and time for that species. But I'm amazed and gratified by what the baboons have enabled us to do to answer a range of fundamental, exciting questions and by their potential to do so into the future. I've felt fortunate, and continue to be amazed at how much fun one can have planning research, observing behavior, analyzing data, and emerging with confirmation of ideas, or having them blown out of the water, or something complex in between!

The themes of survival, reproduction, the importance of ecological and social context, maternal effects, ontogeny, demography, and the relationship of relatedness and social structure to demographic structure, all were topics of study in the first decade of the baboon project. All were ones for which as many questions remained as were answered. Even as some completely new themes have been developed as well, most first publications became one end of a thread that has continued into later decades, often to the present, and that developed more complex colors and textures, ones increasingly interwoven with each other that continued or re-emerged in exciting directions with the help of new methodologies, diverse collaborations, and changing developments in the fields of ecology, evolution, and behavior (see the Amboseli Baboon Research Project website, www.princeton.edu/~baboon).

Why invest in a long-term study, one that spans multiple generations, several decades? Why stay with a single species, a single population, especially one with a slow life history? Even I have periodically asked myself that question through the years, especially when renewal of local research permits or grant funding has been at risk, but also when the weight of managing a full-time project and staff halfway around the world, plus the US components, becomes overwhelming. Yet I know at the core that not just my life but our understanding of ecology, evolution, and behavior would be much more limited if all rigorous research were conducted within a narrow range of perhaps easier possibilities.

References

Altmann, J. (1974). Observational study of behavior: sampling methods. *Behaviour* **49**: 227–67. [Reprinted (1996) in *Foundations of Animal Behavior*, ed. L. D. Houck & L. C. Drickamer. University of Chicago Press.

Altmann, J. (1980). *Baboon Mothers and Infants*. Cambridge, MA: Harvard University Press. [Reprinted (1990) with updated preface, by The University of Chicago Press.]

Altmann, S. A. (1962). A field study of the sociobiology of rhesus monkeys. *Macaca mulatta. Ann. NY Acad. Sci.* **102**: 338–435.

Altmann, S. A. (1967). *Social Communication Among Primates*. Chicago, IL: University of Chicago Press.

Altmann, S. A. (1998). *Foraging for Survival*. Chicago, IL: University of Chicago Press.

Altmann, S. A. & Altmann, J. (1970). *Baboon Ecology: African Field Research*. Chicago, IL: University of Chicago Press.

Csíkszentmihályi, M. (1990). *Flow: the Psychology of Optimal Experience*. New York, NY: Harper & Row.

Dawkins, M. S. (2007). *Observing Animal Behaviour: Design and Analysis of Quantitative Data*. Oxford: Oxford University Press.

Hausfater, G. (1975). *Dominance and Reproduction in Baboons: A Quantitative Analysis*. Basel: Karger.

Struhsaker, T. T. (1967). Ecology of vervet monkeys *(Cercopithecus aethiops)* in the Masai-Amboseli Game Reserve, Kenya. *Ecology* **48**(6): 892–904.

Trivers, R. L. (1974). Parent-offspring conflict. *American Zoologist* **14**: 249–64.

Whiting, B. & Whiting, J. (1975). *Children of Six Cultures: A Psycho-Cultural Analysis*. Cambridge, MA: Harvard University Press.

3

Coming together

PATRICK BATESON

Much of my research life has been spent collaborating with scientists in fields other than my own. I think that ethologists have been lucky in as much as the competitiveness that disfigures many other areas of science has been unusual in ours. The collaborations in which I have been involved may have touched the other, more back-stabbing areas of research from time to time, but my contacts have invariably been marked by deep friendships and led, not simply to close personal relations, but also to bridge-building between disciplines. That explains the title of this piece. Despite the trend for specialisation and the self-protecting priesthoods that often come with it, ethologists have played an important role in bringing

Leaders in Animal Behavior: The Second Generation, ed. L. C. Drickamer & D. A. Dewsbury. Published by Cambridge University Press. © Cambridge University Press 2010.

together people who speak different scientific languages. The subject has served to promote unity in biology. Understanding, as Niko Tinbergen did, that many different questions could be asked about biology and, as Robert Hinde did, that these questions could be asked at many different levels has been crucial in this process.

Early life

Several years before I was born, my parents bought a plot of land at the top of the Chiltern Hills half way between London and Oxford. They didn't have much money and my father designed the low-roomed and cheaply built house in which I was born. He specialised in timber-drying and had the house panelled with plywood. My parents' friends teased them that it would feel like living in a packing case. In fact the house had an extraordinary warmth about it and everybody who came there loved it, but I am sure that they were also attracted by my Norwegian mother's vivacity and my father's charm. I was born in this house on 31 March 1938. My brother, Jon, had been born five years before. He has had much more turmoil in his life than I ever had, but we have remained strongly attached to each other to this day.

From a very early age, I told anybody who asked that I wanted to be a "biologist" without having any clear idea what that might entail. The reason was that I had a kinsman who was a very eminent scientist. His name was William Bateson and, as one of the champions of Gregor Mendel, he coined the term "genetics". He was a cousin of my grandfather and had died 12 years before I was born, but the family was evidently very proud of him and often referred to him. Many years later in my life I wrote an assessment of him and discovered how far ahead of his time he had been (Bateson 2002).

My father volunteered for the British Expeditionary Force at the beginning of the Second World War. He was wounded at Dunkirk and captured, spending five years in a German prisoner of war camp. I had no memory of him until he returned, haggard, when I was seven. At our home during the War we had many refugee children staying with us, particularly from my mother's country, Norway. Some of them have been close friends ever since. Roaming among the Chiltern beech woods and along the escarpment as we did, I couldn't have had a happier childhood despite the prolonged absence of my father.

Some years after the war a well-known naturalist, Richard Fitter, brought his family to live in a neighbouring house. I learned a lot from him and developed an enthusiasm for ornithology that flowered in my teens. After returning from Germany, my father took a job in London and during the week the family lived in a house in Chelsea caring for Old Uncle Ned, as we called him, who was the younger brother of William Bateson. When I was 13 I went to Westminster School in London, and was greatly influenced by its liberal character, openness and intellectual style. In my second year there I spent time at a bird observatory on the Northumberland coast where my interest in ornithology properly took off. I was fascinated by catching birds, studying them in the hand and putting rings on them in the hopes that, if they were migrants, they would be recovered in a faraway place. I spent several holidays at this bird observatory on the Northumberland coast. A schoolmaster who was visiting the bird observatory on one occasion told me that, after getting a first degree at University, I could go on to do something called a PhD involving the study of birds in depth. I had very little idea of what that meant at that time but it sounded like heaven and I set my heart on doing research. A little later I paid a visit to the Wildfowl Trust at Slimbridge in Gloucestershire and the Scientific Director, Geoffrey Matthews, told me that the biological subject to aim for was animal behaviour.

As a schoolboy I was much taken by a double biography of Charles Darwin and his so-called bulldog, Thomas Henry Huxley (Irvine 1955). This was my first taste of how important Darwin had been to modern biology. At the same time as I was starting to spread my wings academically at Westminster, I was also rowing enthusiastically and spent two years in the School's top crew of eight oarsmen and a cox. My chemistry teacher was appalled by my sporting success and told me in no uncertain terms that it was impossible to be the servant of two masters. He thought that the hard physical training required for rowing seriously was incompatible with academic achievement, but he under-estimated the confidence-building effects of one on the other. My biology master, in contrast, was wiser and more tolerant. I continued to row. And, with his guidance, my enthusiasm for biology grew steadily. In due course, I proved my chemistry teacher wrong and my biology teacher right, and won a place at King's College, Cambridge. This College was to play a big role in my life in years to come.

After leaving Westminster in December 1956, I went to Norway for the rest of the academic year and spent the first part of it with my maternal grandfather, Paal Berg, who had been Chief Justice and led the resistance to the German occupation during the Second World War. A great man, he was remarkably self-effacing. Quietly he helped me and I spent several months in the Zoology Museum, learning how to skin birds, reading about systematics and studying differences in the morphology of a single species. This gave rise to my first paper on geographical variation in ringed plover.

In the summer of 1957, again thanks to my grandfather, I joined a Norsk Polarinstitut expedition to Svalbard, the archipelago very close to the North Pole governed by Norway. My job was to take readings off a depth sounder as we plotted the structure of Raudfjord, in the north of West Spitsbergen. However, I had lots of time for bird-watching and for exploring the uninhabited mountains on my own. It wasn't wise and on one occasion I nearly killed myself on a steep scree that plunged down to the sea, but overall the experience was wonderful.

Cambridge and science

In October 1957 I went up to Cambridge to read Natural Sciences. That winter I went to the Edward Grey Institute student conference organised each year at St Hugh's College, Oxford, by David Lack. I gave a talk about my summer's experiences in Spitsbergen. Another Cambridge undergraduate, Chris Plowright, was at the conference and we hatched a plan to take an expedition to Spitsbergen the following summer to study the rare Ivory Gull. Niko

Tinbergen gave a talk about gulls at the conference because he was engaged at that time in writing up his comparative study ("The Progress Report" that became in effect a final report on this phase of his work). Chris and I talked to Niko about our Ivory Gull plans. Niko, with fond memories of his year in Greenland as a young man, was very keen to join us. He was all the keener because Ivory Gulls often nest on cliffs and only a few years before Esther Cullen had described the special cliff-nesting adaptations of Kittiwakes. We recruited to the expedition two other Cambridge undergraduates, who were keen ornithologists and photographers: Robin Hitchcock and John Cutbill. Niko leading our expedition was the plan for many months and he spent considerable time with us, planning what we should do. However, he had an ulcer that got worse in the spring and he was ordered by his doctor to stay in England, much to our sorrow and his. However, he continued to be very supportive and the Spitsbergen expedition went ahead.

After various adventures we arrived on board a Norsk Polarinstitut vessel in the mouth of Wahlenbergfjord in the north-east of the Svalbard archipelago where the Ivory Gulls were supposedly nesting. However, the fjord was full of ice and after some discussion we were dropped with all our stores and two boats on some islands in the mouth of the fjord. In the next three weeks as the ice cleared we were able to explore further and further into the fjord. One cliff which had had an Ivory Gull colony on it in the 1930s was no longer occupied. However, to our great delight, we found another cliff, 1000 feet high, and there, circling in front of it, were our pure white gulls. We lugged all our equipment to the top of the cliff and set up camp for the next five weeks. We were able to obtain a detailed film record of the gull's reproductive behaviour and learned a lot about its breeding biology. They did not have the cliff-nesting adaptations of Kittiwakes.

I spent the next six months writing up our work in two papers, one in *British Birds* and one in *Ardea* (Bateson & Plowright 1959). (This was done at the expense of revision for my second year exams.) Niko wrote a commentary on the second paper. He was very complimentary about our work, but a little annoyed when I questioned his claim that the gulls had an "innate" fear of swimming in the electrifyingly cold Arctic water. I said that they might just as well learn it after one icy dip. Typically, Niko included the alternative explanation in his published comments.

After that experience with the "Maestro", I was set on doing research for a doctorate with him. In my final undergraduate year at Cambridge, Niko invited me to spend a week in the Spring vacation at the large black-headed gull colony at Ravenglass to get a taste of doing field work with him. My main memory of that week was sustained discussions about science, politics and much else with Niko, Dick Brown and Juan Delius in a caravan parked on the sand dunes. In discussions, he seemed disconcerted when others made points he hadn't thought of himself or when one of his assertions was contradicted convincingly. Part of Niko's charm was his vulnerability. When a good point was made against an argument he had mounted, his consternation was almost palpable. But the charm and his boyish enthusiasm were utterly endearing. During my stay at Ravenglass I spent long periods in a hide watching curlew feeding. I noticed that the birds repeatedly touched the surface with their beaks but when they drove their beaks deep into the mud, they almost always caught something. I knew that their beak tips were packed with sense organs so it seemed likely that they could sense prey accurately. I measured the frequencies of each touch of the surface, each deep thrust and each time the bird caught something. Niko liked this approach to quantification.

I enjoyed my last year as undergraduate at Cambridge enormously. After reading the spread of scientific subjects required at Cambridge for the first two years, I could now specialise in Zoology. Finally I was doing what had I wanted to do since I was a child. It was wonderful to be able to think critically and creatively about the subject. I hadn't done well in exams in my previous two years because I had been diverted by so many other interesting matters, including my discovery of the delights of becoming a field ethologist. I now worked hard and with enthusiasm; my supervisors started to take an interest in me. The physiologist John Pringle was one of them. He was a shy, socially awkward person and often seemed forbiddingly remote. He was often lost in his thoughts – so much so that he was said to have walked past his wife in the street without noticing her. (Astonishingly, he loved ballroom dancing.) He taught me insect physiology and molecular biology and we got on well. He wanted me to do a PhD with him, but I said that I had already set my heart on doing a doctorate with Niko Tinbergen. John Pringle said: "Don't do that, you can do behaviour in your holidays." His attempt at intellectual seduction didn't work.

The class had many outstanding zoologists in it, including Peter Grant who, later in his career and with his wife Rosemary, went on to demonstrate Darwinian evolution in Galapagos finches, for which they both received the Royal Society's Darwin medal. We enjoyed each other's company. To my surprise I came top of the class in the final exams. My mentor in King's, George Salt, said that if I stayed in Cambridge I was much more likely to get a Research Fellowship than at Oxford since, at that time, King's only gave such Fellowships to its own graduates. I talked to Robert Hinde who was prepared to supervise me as a graduate, but who fully realised the delicacy of the situation as far as Niko Tinbergen was concerned. Robert said: "I don't want to lose Niko as a friend over this." Anyway, Robert telephoned Niko and explained the Research Fellowship situation, and Niko capitulated. As things turned out, I'm not sure that a DPhil with Niko would have been a good career decision for me. If I had worked on gulls as I had intended earlier that year, I might have landed myself in an area of research in which Niko himself was already losing interest and such work wouldn't have obviously led anywhere.

Before I started my doctoral research, I spent the early autumn with three bird-watching friends in Greece, David Balance, Robin Hitchcock and Ian Nisbet. In Thrace, some policeman stumbled out into a marsh, where we were watching waders, to arrest us. What were four young men with binoculars doing so close to the Bulgarian border? In the police station, by showing them our bird books and notes, we eventually persuaded the Greek police that we were eccentric Englishmen and not Bulgarian spies. Later Ian Nisbet and I travelled south to the Peloponnese and Crete. It was round the period of the full moon and each night we gazed through binoculars at the moon's face. Migrant birds could be seen flying across it. By noting the time and the position on the moon where the migrant was seen to enter and leave, we could later calculate each bird's flight direction. They were flying due south, heading out across the Mediterranean to Africa; at the time it was a finding of sufficient interest to be worth publishing and we did so (Bateson & Nisbet 1961).

Doctoral work on imprinting

Behavioural research at Cambridge was carried out at Madingley, five miles outside the City. The Ornithological Field Station was directed by Bill Thorpe, a shy but determined and far-seeing man. He had started his famous work on song-learning in birds and had already attracted some remarkable people to the field station, where he installed Robert Hinde to run the place. By the time I arrived in 1960, Richard Andrew, John Crook, Janet Kear, Peter Klopfer, Hugh Fraser Rowell and Peter Marler had already passed through and I only got to know them well later. Thelma Rowell was starting post-doctoral work with Robert on the role of maternal behaviour in the development of rhesus monkeys. John Fentress arrived shortly after me and, in addition to his doctoral work on voles, he obtained from Whipsnade Zoo a wolf cub, which took over his life. Thorpe had raised funds to build a new laboratory, and the field station was given a new title: "The Sub-Department of Animal Behaviour". The new building was rising from the ground in my first year as a graduate student.

Robert Hinde wanted me to work on imprinting. (We now have to call it "behavioural imprinting" because of the carelessly labelled "genomic imprinting", a fascinating but completely unrelated area of research.) Robert was a superb supervisor. He took tremendous trouble over the written work of his research students and he taught us how to think. He was not much interested in what he regarded as woolly theory and demanded clear thought at all times. At the best of times and to the best of his friends he is a formidable critic with an alarmingly penetrating eye. Many were the speakers who gave seminars in Robert's rooms in St John's College who came away not quite sure what had hit them or why they had been subject to such ferocious scrutiny. He didn't like generalisations and frequently warned us that little progress is made in the end if superficial straightforwardness and clarity are illusions.

Robert Hinde exerted an extraordinary influence on ethology, primatology and, more recently, studies of human behavioural biology and development. He has an astonishing ability to assimilate material from a wide variety of fields. He told me long after he had retired that he loved to go into his College to write and at lunch to sit next to somebody from another discipline who would give him a new reference. He has always revelled in diversity, but none the less certain distinctive themes ran through his own writing. He frequently referred to crossing and re-crossing the boundaries between different levels of analysis and the need to study processes. He wanted his research to be of use to humanity and was the British chairman of Pugwash, the international organisation for promoting peace. He has a deep concern about the causes of aggression and the peculiarly human institution of war.

I became interested in what ends the sensitive period for imprinting and, along with others, argued that it was due to a process of competitive exclusion. Experience with one class of objects made it more difficult for the birds to form attachments to another class and the greater the stimulus value of the first class the more rapid would be the exclusion effect. Later I found that once birds had been imprinted, they could use this perceptual experience to help them in discovering new sources of food. I came to feel that the supposed peculiarities of imprinting were due to the special context in which it occurred rather than to any unique features of the learning process. I became more and more convinced that imprinting shared many features with perceptual learning and that Konrad Lorenz's claim that this was a special form of learning was wrong. I described my experiments at the International Ethological Conference at The Hague in 1963. Lorenz was sitting in the front row of the audience. He became increasingly angry. At the end of my talk he got up and said "I'm going to direct my remarks not to you, but to Robert Hinde." He berated Robert for allowing me to work on domestic chicks. Lorenz had a deep loathing for the effects of domestication on animals, which had an unpleasant manifestation in his writings about humans during the Second World War, when he swam with the Nazi tide (Burkhardt 2005). As the ethological giants battled it out to no resolution, I smoked a cigarette. (Not long after, I gave up smoking – but that was for other reasons.)

Subsequently I wrote a long review of the literature on imprinting (Bateson 1966). On the strength of that review I was asked to give a plenary lecture at the Ethology conference in Stockholm in 1967. Konrad Lorenz was in the chair and was studiously polite. Although my relations with him never fully recovered after my talk four years before, it was obvious to me that he was a man of enormous charm and charisma.

Measuring behaviour and thinking creatively

It is hard to believe that ethology would have achieved what it did without Konrad Lorenz. Nevertheless, the subject was moving into a much more rigorous phase than in its early days. Many of the classical examples that figured so strongly in the first text books on animal behaviour would not pass editorial scrutiny in the twenty-first century. Lorenz used to say: "If I have one good example, I don't give a fig for statistics". Small samples, non-independence of measurements (when measurements were made), naïve or improper use of statistics (when statistics were used), lack of adequate controls (when experiments were carried out), not conducting experiments blind and generally using poor experimental design were all flaws in ethological work of that early period. I wrote a chapter in a book edited by Larry Weiskrantz about ethological methods of observing behaviour (Bateson 1968). Some years later a brilliant former student of mine, Paul Martin, returned to a position at Madingley from a post-doc in the United States. Paul produced for the students a detailed guide to the measurement of behaviour. I said to him that this would make a superb little book. We agreed to collaborate and in 1986 the first edition of *Measuring Behaviour* was published (Martin & Bateson 1986). This was followed by a second edition in 1994 and a third one in 2007. I like the book and was particularly pleased by the comment of one colleague, who remarked that it was not just a book about methodology in behavioural biology, it was a book about how to do science.

Over the years I had spent some of my time editing other people's writing. Shortly after returning to Cambridge in 1965 I agreed to become the European editor of *Animal*

Behaviour. I found that time devoted to assimilating the ideas and correcting the grammar of others sapped the energy that I had for my own writing, and I gave up the job after four years. However, I did learn to discriminate between the sound and well-constructed elements of a paper and the genuinely original. The two did not always – or even often – go together.

Notwithstanding the draining aspects of editing, Peter Klopfer and I started a new series called *Perspectives in Ethology* which ran for ten volumes over 20 years. We noted in our first editorial that in the early days of ethology, most of the major developments had been in the realm of ideas and the evidence had been largely anecdotal. As the subject became increasingly rigorous (as represented in the advice given in *Measuring Behaviour*), the requirement to be quantitative meant sometimes that relatively trivial measures were chosen at the expense of representing the complex and patterned nature of a phenomenon. An overview was likely to be lost and original thinking was in danger of being ridiculed. Peter and I felt that our series should encourage new perspectives even though, in their inchoate form, they might be vulnerable to attack. We quickly discovered that we were faced with a problem of our own making when nearly incomprehensible chapters were submitted to us. What standards should we use? Coherence and consistency might seem to be good editorial criteria and yet to insist on normality in these respects might destroy the embryo forming in an author's argument. We compromised, and the resulting volumes were often uneven – but always interesting. As each volume reached its completion I would usually end up enjoyably at the house of Peter and his wife Martha outside Durham in North Carolina to complete a new editorial. Invariably I was left breathless by their energy and the enormous range of their commitments. While they were out training for a marathon or some other enormously draining activity, I would opt out and tap away on the keyboard of their computer. The last volume was co-edited by Nick Thompson, who was meant to continue the series. However, he found that his own commitment to good writing often ran counter to the eclecticism that Peter and I had encouraged. The tensions that had always been present in the series finally overwhelmed it.

California and the hunt for underlying mechanisms

Tinbergen (1951) had stated in *The Study of Instinct* that it was the ethologists' job to carry the analysis of behaviour down the level of the physiologist. Like many others I took this seriously, feeling that it was important to understand how behaviour works. What are the neural mechanisms that underlie what the whole animal does? After I received my doctorate, I spent two years at the Stanford Medical Centre in California as a Harkness Fellow, working with an eminent neurosurgeon and neuropsychologist Karl Pribram. He could not have been more different from Robert Hinde. *All* theories were interesting to him. His lack of scepticism was remarkably liberating for someone with my background. On one occasion, however, he astonished me by announcing that everybody should be trained like a behaviourist – or a Cambridge ethologist. I know what he meant. It is much easier to go from rigour to creativity than the other way about.

I did not learn to become a neuro-surgeon with Karl but, working with amygdalectomised monkeys as I did, I understood why lesioned animals can make you look at normal animals in a quite different way. Apart from the taming effects of removing the amygdala, the lesioned monkeys studied the displays in a discrimination learning apparatus much more intently than the sham-operated animals and learned one complex task more rapidly.

My wife, Dusha, and I had married just before we left for California. The Harkness Fellowship required that we spend the middle summer of the two years we were in the United States travelling to each region of the country. When the time came I was less than eager because I was in the middle of a long experiment and did not want to spend three months as a tourist. As things turned out, of course, it was an extraordinary trip in which we camped in incredibly beautiful places, stayed with numerous delightful people and obtained a clear sense of the varied geography and cultural diversity of the United States. We dined out on this trip for years afterwards.

The major experiment I completed at Stanford was on perceptual learning in rhesus monkeys. I constructed panels with illuminated letters that could be put in a monkey's cage. Later the monkeys were required to discriminate between letters in some automated apparatus in which they were rewarded with food if they pressed the correct panel. If one of the letters was familiar, they learned the discrimination much more rapidly than if neither was familiar. However, if both letters were familiar from their experience in the home cage the monkeys learned the discrimination more slowly. I argued that the two letters had been classified together (Bateson & Chantrey 1972). Later, when I returned to Cambridge, a student of mine, David Chantrey, repeated the general idea of my experiment on monkeys after imprinting chicks with two objects. He found that if exposure to the two imprinting objects had started within less than 30 seconds of each other, the chicks had difficulty in discriminating between them in order to receive a food reward. They had classified them together. If, on the other hand, the gap between the onset of exposure was lengthened to more than five minutes, then the chicks subsequently performed the discrimination with ease (Chantrey 1972). We argued that they had classified them apart. Many years later, Rob Honey and I used a different design and confirmed that when the presentation of the two imprinting objects was mixed and close together in time then the birds learned the discrimination between them more slowly than when the presentations were separated (Honey *et al.* 1993).

I had returned to Cambridge in 1965 with a research post at the Sub-Department of Animal Behaviour and a Research Fellowship at my old College, King's. The traditional Cambridge Colleges still operated somewhat like monasteries, as though they were filled with resident bachelors. Those of us who were married were expected to dine with other Fellows at least several evenings each week. Over the years, as more and more Fellows had young families and partners who were working, this practice has dwindled. Even so, meeting Fellows from other disciplines was rewarding and sometimes had major benefits. It certainly did for me. One evening, by chance, I sat next to a medically trained neuro-scientist. He had been working on attention and habituation but was very interested in the effects of learning on the nervous system. After my experience at Stanford I was keen to bring my behavioural knowledge to bear on mechanisms that underlie learning. Behavioural imprinting seemed like a very good candidate because any effects of experience should stand out more prominently in an animal that had come straight out of a dark incubator. We recognised our common interests and thus began my long collaboration with Gabriel Horn, for that was who my companion at dinner was, on the neural basis of imprinting. We cast around for various biochemical measures that we could use and, after a seminar in London, had an offer of help from Steven Rose. He was, first and foremost, a biochemist but he had great interest in the effects of visual experience on the brain. The collaboration started among the three of us.

The job at the Cambridge end was to run the experiments, remove and dissect the brains and send the coded samples packed in dry ice to Steven who, not knowing the code, would analyse them. We first looked at protein synthesis and then messenger RNA synthesis. Big effects showed up, but how could they be interpreted? So many different things were confounded with the laying down of a memory of the imprinted object. The imprinted chicks had been more active than those in the control group. They had received more visual experience. They had been more attentive. The list of confounded variables was endless. By degrees we devised a series of procedures, each of which ruled out a different subset of possible explanations. It was like triangulation – no single experiment was conclusive, but taken together a clear conclusion emerged. We could eventually be confident that the particular area in the forebrain activated by imprinting was intimately concerned with the laying down of the representation of the imprinting object. This collaborative work among the three of us culminated in a long review in *Science* (Horn *et al.* 1973).

In one of our key experiments, the amount of imprinting experience the chicks received on the first day after hatching was varied. Some chicks (the 'under-trained' group) received only a limited opportunity for learning about the imprinting object on day one, whereas others (the 'over-trained' group) were imprinted for so long that they could learn little more about the imprinting object. On the second day of the experiment, all the birds were exposed to the same object for the same amount of time. As the length of exposure on the first day increased, so the biochemical activity associated with changes in the forebrain roof decreased on the second day (Bateson *et al.* 1973). Subsequently Gabriel and I, collaborating by then with Brian McCabe, combined this approach with a brain mapping technique and identified the critical region associated with imprinting (the intermediate and medial part of the hyperstriatum ventrale or IMHV) (Horn *et al.* 1979). When this part of the brain was removed with a small lesion the birds were unable to form a preference for the imprinting object. When it was removed immediately after imprinting the birds behaved as though they had not been imprinted. These discoveries led on to much more detailed analyses of the cellular and molecular events involved in memory formation and how the stored information is used in the overall organisation of the bird's behaviour (Horn 1998). Many others collaborated in the imprinting work at both the neural and the behavioural levels, including Mark Johnson and Johan Bolhuis, who wrote a review of imprinting for *Biological Reviews* to come out a quarter of a century after I had published mine in the same journal.

The role of behaviour in development

In 1968 Ellen Reese arrived at Madingley for a sabbatical with her husband Tom. At Mount Holyoke she taught students about operant conditioning. Ellie and I decided to collaborate in order to see whether young domestic chicks would work to present themselves with an imprinting object. I had noticed that the chicks would move around actively if no such object was presented to them – as if they were searching for something to which they could become attached. We built some apparatus in which a pedal set into the floor would activate an object that we knew they would approach and learn about. This object was built from a flashing light used on the top of police cars and ambulances. When activated, a translucent box with two sides blacked out revolved around a bulb with a red filter placed over it. I knew that chicks would approach this stimulus and would come to prefer it over other artificial stimuli. We arranged things so that a day-old chick, apparently searching for an imprinting object, would activate the light when it stood on the pedal. The chick would then approach the light, which took it off the pedal, and the light would go out. It had to learn to return to the pedal in order to restart the imprinting stimulus. Domestic chicks and wild Mallard ducklings rapidly learned to do this. All the factors that were known to affect imprinting, such as the characteristics of the stimulus and the chicks' prior experience, affected how readily they would work to present themselves with the stimulus (Bateson & Reese 1969). It was such a robust phenomenon that we decided to make a film about it, given that Ellie already had a lot of experience of making educational films. The film grew into a more general film about imprinting. Much of it was shot at the Wildfowl Trust at Slimbridge thanks to the great help provided by Janet Kear.

An important feature of what Ellie and I had done was well illustrated in our film. Far from waiting passively to be stimulated, the birds played an active role in their own development. As Peter Marler once put it, they had a hunger for learning.

Behavioural development in the cat

Bob Prescott, who had received his PhD from work at Madingley, arrived back at the laboratory in the mid-1960s with a plan to work on domestic cats. I was in charge of administration of the laboratory and it was my job to set up a cat colony for him. In doing so, I discovered that it was possible to keep cats at quite high densities. However, they fared much better if they had an outside run attached to their pens. The only problem about that was that, without proper defences, feral cats could come into contact with the colony cats and infect them. So we built outside pens but with an additional protective fence that gave the cat colony the appearance of a concentration camp. The cats did well, and Bob went on to show how important in olfactory communication were the various types of rubbing which are so familiar to cat owners. Some years later Bob moved to St Andrews and I adopted the cat colony.

Prill Barrett came to work with me on a project to look at the development of play in cats. One of the major conclusions of our quantitative work was that social play developed several weeks before object play (Barrett & Bateson 1978). A sharp increase in object play occurs 2–3 weeks after kittens start to take solid food. Kittens that are separated artificially from their mothers started to play with objects significantly earlier and did so more frequently than did the control group. They also developed earlier if their mothers' lactation had been temporarily blocked with the drug bromocriptine. Finally, kittens developed earlier when their mothers were rationed than when they were given an *ad libitum* diet (Bateson *et al.* 1990). The results suggested that the kittens' mode of development is triggered by an environmental forecast provided by the mother. Kittens that would have to fend for themselves at an early age gathered through play the essential skills they would need at the stage in development when they would have to put these skills to serious use. Very similar findings have been obtained by my students and associates using rats, in which the age of weaning is triggered by the maternal state (Smith 1991: Gomendio *et al.* 1995).

These conditional responses to the environment have prompted examination of many other such cases where the offspring's phenotype is triggered by information provided by the mother, an issue to which I shall return later. Far from being a time of conflict, weaning in both the cat and the rat seemed to be a time for two-way communication between mother and offspring to their mutual benefit (Bateson 1994).

My wife, Dusha, and I have been breeding pedigree cats for many years, first Russian Blues and then Egyptian Maus. Our kittens have been extremely affectionate and we wondered why so many people believed that pedigree cats are stand-offish and do not make good pets. You can, of course, breed for friendliness to humans. In studies conducted in our cat colony at Madingley friendly studs were more likely to have friendly offspring than unfriendly studs even though the fathers never had social contact with their sons and daughters (Turner *et al.* 1986; McCune 1995). For all that, the efforts of sensible breeders are wasted if they do not make some effort to expose their new litters to humans early in the kittens' lives. I have come across breeders who keep their queens isolated in outside pens. Maybe they think that the queens will eat their newborn offspring if humans reach in and handle the kittens. Or maybe they have neither the time nor the inclination to stroke and talk to the young kittens. Either way they are making a big mistake.

Kittens handled by humans from an early age make strikingly better pets than those not treated in this way. The statement has been confirmed by numerous scientific studies and by the experience of countless breeders who made the effort. The period over which such exposure to humans is effective probably extends to well after the kittens start to take solid food at about four to five weeks after birth. The precise amount of contact is also not critical – a little and often from early on and, if that is not possible, much more contact from six weeks after birth. But leave the contact too late and the cat will not make a good pet. The practice of leaving a young cat with its mother and siblings until the time that pedigree cats are usually sold – round about three months after birth – and not giving it much contact with humans is a recipe for reinforcing the belief that pedigree cats are unfriendly.

When my wife and I have had large litters of kittens (6 or more), some of them have been very small – particularly the last ones to be born. These small members of the litter have difficulty in competing for a good nipple when faced with the scrabbling of their bigger siblings. In these cases we have given them artificial cat's milk through a syringe. The kittens soon became so adept at sucking milk from the syringe that we had no need to press the plunger. These kittens, when they became more active, would come out of the nesting box when we were nearby and cry for their extra feed. Unsurprisingly, they made wonderful pets later in their lives.

Many of the studies of cats carried out by others had not made their way into print. Dennis Turner organised a conference in Zurich, bringing many of those who had worked on cats together. Later he and I co-edited the proceedings of the conference and brought them together into a successful book (Turner & Bateson 1988), which went into a second edition twelve years later.

The work on play in cats raised the question of whether this aspect of their behaviour was a consequence of domestication. Tim Caro, by then a post-doc based at Madingley, was working on cheetah in the Serengeti National Park of Tanzania; his presence there provided me with an opportunity to look at play in cheetah, using his knowledge of where to find the animals. That was the excuse, but in reality it gave me a chance to visit the African continent for the first time. My daughter Melissa, then aged 14, came with me. We were both bowled over by the experience. Before reaching the Serengeti, Josh Ginsburg, who was prospecting for a project on Grevy's Zebra, took us up to Samburu and then down into the Masai Mara through the Aberdare Mountains. The big mammals are, of course, wonderful but I was completely overwhelmed by the beauty and the abundance of the birds. In the Serengeti the privilege of being with as experienced a field-worker as Tim was wonderful – and we did, indeed, see cheetah cubs playing as intensely as domestic cats! For Melissa the experience was a tipping point, I suspect, and I watched with astonishment and delight as her career path moved ever closer to mine and she became a very successful research worker in her own right. Nowadays, people ask me whether I am Melissa's father.

A golden age at Madingley

Beginning in the 1970s, an outstanding group of graduate students and postdocs had gathered at Madingley. Robert Hinde's interests in primate behaviour had led him to encourage Jane Goodall and later Dian Fossey to write up their field work for Cambridge PhDs. This stimulated many other outstanding people to work on primates in the field and to come back to Madingley to write up their work for their doctorates. These included Tim Clutton-Brock, Dorothy Cheney, Sandy Harcourt, Phyllis Lee, Robert Seyfarth, Kelly Stewart and Richard Wrangham. At the same time I had some superb graduate students. Those working on imprinting in chicks included Jeremy Cherfas, David Chantrey, Patrick Green and Patrick Jackson. Paul Martin and Mike Mendl worked on play in cats and Sandra McCune on their socialisation. Dafila Scott was using her ability, acquired as a child, to distinguish between individual Bewick's swans in the nearby Fens in order to look at the benefits of staying with parents. Also in the field but in Africa, Chris Magin worked on hyraxes, Clare FitzGibbon on cheetah and Barbara Maas on bat-eared foxes. Gadi Katzir began his distinguished career working on jackdaws at Madingley, and Georgia Mason started on her path to fame by working on the stereotypies of mink. My office had been next to that of Joan Stevenson Hinde ever since I had returned from California and I benefited enormously from her wisdom and friendship. More recently, Nick Humphrey had joined the laboratory as an Assistant Director of Research, adding his brilliance to an already remarkable group.

In 1975 we celebrated with a conference the 25th anniversary of the founding by Bill Thorpe of the Madingley field station. Robert Hinde and I edited the chapters resulting from the conference (Bateson & Hinde 1976). The most cited chapter in the book was Nick Humphrey's on the social function of intellect, which was largely responsible for spawning a new field of research (Humphrey 1976). It also led Nick and me to plan a new research group in our College, King's, devoted to behavioural ecology and sociobiology. The field was attracting the brightest people in behavioural biology. We wanted to bring some of them to Cambridge. The King's College Sociobiology Group succeeded in doing that. It consisted of Brian Bertram, Tim Clutton-Brock, Robin Dunbar, Dan Rubenstein, Richard Wrangham and many others who came for shorter periods of time. The group ran a wonderful seminar series and subsequently published the proceedings of a culminating conference. For my part, it reawakened my interest in the evolution and function of behaviour.

Functional and evolutionary approaches

Although my interests had become focused on ontogeny, a subject that Tinbergen knew was important but which never really excited him, like many ethologists of my generation I was much influenced by him in my subsequent research. I found myself asking again and again, what is the current function of this pattern of behaviour or process? It raised questions about development that I would have never considered otherwise. Why, for instance, does the sensitive period for sexual imprinting occur later in the life-cycle than the sensitive period for filial imprinting? At the Bielefeld Ethology Conference in 1977 I suggested an explanation, which was later published (Bateson 1978, 1979). Filial imprinting is necessary to enable the young bird to discriminate between its own mother and other females of the same species that might attack it. Sexual imprinting enables the bird to avoid inbreeding and also avoid too much outbreeding, both of which were disadvantageous. To benefit to the greatest extent from its prior experience, the bird should learn about its siblings when they had moulted into their adult plumage. Hence the sensitive period for sexual imprinting should occur later in development than the sensitive period for filial imprinting. The long-term effects of early experience on mate choice, I suggested, are that birds prefer members of the opposite sex which are a bit different but not too different from what they had seen before. Later I found that, in Japanese quail, the most preferred partner was found to be a first cousin when the birds had been reared with siblings (Bateson 1982b, 1983). The optimal outbreeding idea was the central focus of my chapter in a book that I edited on mate choice (Bateson 1983).

The quail experiments led on to further work with Carel ten Cate in which we painted black spots onto white quail and found that they preferred mates who had a greater number of spots than the individuals with which they had grown up (ten Cate & Bateson 1989). The idea for the experiment had come from a conjecture that sexual selection might only involve one genetic change, namely in the sex that was chosen by the other sex. We suggested that in each generation the choosing sex, by the prior developmental process of sexual imprinting, calibrated the standard from which its sexual preferences would be offset (ten Cate & Bateson 1988).

The King's Sociobiology project eventually led to a conference and then to a book. In my chapter in the book I started to explore the active role of behaviour in evolution (Bateson 1982a). This interest has grown over the years (Bateson 2006a). Among other benefits it led to Kevin Laland coming to Madingley to start his own programme on social learning, which flowered in many directions, including his co-authorship of a book on niche construction. Most recently these interests of mine have led to collaboration with two Russian colleagues, Kostya Anokhin and Misha Burtsev, developing models of how plasticity might have evolved and then driven evolution.

Provostship of King's

By 1987 I had been in Cambridge for a long time and, observing friends and colleagues who had been refreshed by changing their universities, felt that it was the moment to move somewhere else. At that stage I was looking for a post in the United States. However, I was asked if I would stand for Provostship of my Cambridge College, King's. The Provost is the Head of the College and in other Colleges in Cambridge is called Master, President or Principal. After considerable hesitation, I agreed. When I was elected, I was the first scientist to become a Provost of King's. So began a completely different and extraordinarily interesting life in the middle of a College of great beauty and source of wonderful music.

One of the pleasures of the job was that my wife, Dusha, and I could work together in ways that both of us found satisfying. Up to that point she had devoted herself to bringing up our two bright and beautiful daughters, Melissa and Anna. They undoubtedly benefited from Dusha's care but by now they were growing up and it was possible for us to work together.

Although much went on during the day, official life continued into the late evening. A substantial chunk of the job began after coming out of interminable committees and went on until after breakfast next day when the overnight guests finally departed. Apart from all the other things she did, this part of keeping the show on the road fell heavily on Dusha's shoulders.

Jack Plumb, who had been Master of Christ's College and enjoyed entertaining Princess Margaret, the Queen's sister, asked us if we would invite her to one of the Carol Services in King's College Chapel. After we had agreed and were firmly on the hook, he said: "She will be staying with you, of course." Our jaws dropped but we could hardly say "No". Thus began a series of visits and an exposure to a *demi monde* totally unfamiliar to us. Princess Margaret liked to be treated with appropriate formality and was, I suspect, slightly ill at ease in an academic environment. However, as we got to know her, we respected her intelligence and we became increasingly attached to her.

Patrick Bateson

Among our many guests we particularly enjoyed the visit of the Dalai Lama, who insisted on having dinner with us rather than with his two monks who travelled with him. We found him to be a man of enormous warmth, intellectual curiosity and, indeed, humour. When we asked him how he could eat what we were having for dinner when he normally abstains from food after mid-day, he said that he would say a prayer. At a question and answer session arranged for the students he was asked whom he admired most in the world. Without hesitation, he said: "Gorbachev." Later Gorbachev was due to stay with us and I was asked vehemently by a Russian dissident, who had suffered terribly during the Brezhnev era, how I could have such a man staying in the Lodge. I was able to say that if he was good enough for the Dalai Lama he was good enough for me.

The visits of the Dalai Lama and Gorbachev demanded tight security, as did that of the Queen when she came for dinner after an Advent Carol Service. But these arrangements paled into insignificance compared with what was done for Salman Rushdie when, for the first time after the *fatwah* had been placed upon him four years before, he stood up in the Chapel and gave a public address. After the service he came for lunch in the Lodge. Virtually every long curtain in the Lodge had an armed Special Branch policeman hiding behind it. Each shrub in the garden was investigated to see whether it concealed a bomb. A retired and distinguished bachelor Fellow, whose rooms looked down onto the garden, rang the Porters to say that: "Some rough looking types are in the Provost's garden – with dogs."

As Provost I had to participate in the great tableau events in the Chapel, including the famous Nine Lessons and Carols on Christmas Eve, which was broadcast to 100 million people around the world. I am not a believer and some of my more militant atheist friends accused me of hypocrisy. However, I loved the beauty of those occasions. On Christmas Eve I had to read the last lesson, St John's grandly poetic version of the Creation. It starts: "In the beginning was the Word, the Word was with God and the Word was God." I wondered whether this poetic vision was a mistranslation. The King James version of the Bible was translated from Greek, and the Greek word for Word is *logos*. Now *logos* can be translated as "idea" and so an alternative version of the Gospel according to St John, Chapter 1, might have started more plausibly: "In the beginning was the Idea, the Idea was with God and the Idea was God."

As I sat in the Provost's stall in the Chapel and gazed up in wonder at the wonderful vault above, I saw things that were probably never intended. Running down the middle of the vault is a series of great stone bosses. Each one was supposed to be solid and weigh half a ton but recent research suggests that they are hollow. They are certainly not key-stones since the forces keeping the vault in place are taken by the ring of stones into which each boss sits. All the bosses do is provide decoration and since the vault was built at the beginning of the sixteenth century, they celebrate the Tudors, with alternate portcullises and Tudor roses. As I looked at a rose which I could see from its side I was sure I could see an old man with moustache and beard looking rather severely at me. So much for having mischievous thoughts in a place of worship.

The prominence that derived from being head of a Cambridge College meant that I was invited onto various national bodies. I became a Trustee of the newly formed Institute for

Public Policy Research, which was formulating policies for an as yet to be elected Labour Government. A few years later I joined the Museums and Galleries Commission and subsequently became its Vice-Chairman until, ironically given my own political interests, the Commission was dissolved by Tony Blair's new administration in 1997. These experiences opened windows into worlds that had hitherto been unknown to me.

I did not give up science once I became Provost, however. I was supposed to spend at most half of my time in the College; the rest of the time I was still an academic, teaching and doing research. Rob Honey came to work with Gabriel Horn and me on imprinting. We re-examined the process by which two imprinting objects are classified together if they occur closely together in time or classified apart if their exposures are separated in time. We used this phenomenon to dissect out by lesions the various memory systems that are established after imprinting (Honey *et al.* 1995).

Throughout my work on imprinting I had been conscious of the many factors that are known to affect the process and the need for theoretical models to aid thought. I worked on various ways to formalise such models (Bateson, 1991a). Gabriel Horn and I produced designs for a neural net but did not publish anything initially. He became sceptical about the benefits of this approach and we allowed the project to drop for many years. Then Gabriel went to a conference at which neural net models of parallel distributed processing were all the rage. He came back to me, both excited and somewhat remorseful, saying: "We should have published our ideas years ago." I then set to in earnest to write computer programs that would implement our thinking. It was some of the toughest intellectual work I had done. At the end of it, though, we had a neural net model for imprinting which showed how, with processes that were plausible in terms of what is known currently about the nervous system, realistic and complicated behaviour could be generated by simple underlying rules (Bateson & Horn 1994).

The welfare of deer hunted with dogs

One day in 1995, a Professor Savage came to see me in my office in King's. He had chaired a working group for the National Trust on the conservation and management of red deer on the Quantocks and on Exmoor, including the various implications of a ban on hunting the deer with hounds – a practice peculiar to that part of West England. Its terms of reference did not ask for any investigation of the ethical or moral issues. A prominent Methodist minister, Lord Soper, proposed, therefore, that the National Trust set up a balanced Working Party on deer hunting to examine issues relating to cruelty and animal welfare not addressed in the Savage Report, and led by an independent Chairman. In 1994 a resolution to do that was agreed by the National Trust. Professor Savage wanted to know whether I would chair the working party. I said that such a study would need new research and that such work could not be carried out by a committee. I would do the research if I had a free hand in carrying it out. The Council of the National Trust pondered my offer and then asked me to conduct a scientific study. The terms of reference were as follows: "To study suffering as a welfare factor in the management of red deer on National Trust properties on Exmoor and the

Quantock Hills having regard, *inter alia*, to the scientific evidence on stress induced in deer by hunting with hounds and by other culling methods, animal welfare legislation, and the likely effects, in so far as they can be estimated, of a hunting ban on suffering among deer."

Thus began a saga which took me into new territory. It would have been new for anybody because no study of what happens to a mammal when it is chased by dogs had been carried out. I was, however, deeply interested in issues to do with animal welfare. As President of the Association for the Study of Animal Behaviour in the late 1970s, I established an ethical committee and played a central part in establishing an approach that is now enshrined in British law protecting animals used for research. The principle is one of maximising scientific and medical benefit while minimising pain and suffering in the animals. The problem of how pain and suffering should be assessed could not be side-stepped and I suggested how the process might be made transparent (Bateson 1991b). I also gave much thought to the ethical issues involved (Bateson 2005).

In my approach to the welfare issues to do with hunting, I decided to use the triangulation technique that we had employed in our approach to the neural basis of imprinting, namely to employ different approaches, each of which was associated with its own subset of ambiguities. The approaches included: assessments of the behavioural responses of the hunted deer; the physiological state of hunted deer in relation to that found in suffering humans; the physical damage to the animal at the end of the hunt; the deer's ability to cope with the challenge of a prolonged hunt by dogs; and departures during the hunt from the conditions to which the animal is well adapted.

I paid many visits to Exmoor and the Quantocks to plan and develop the study and to observe hunts. Discussions were held with the main parties supporting or opposed to the hunting of red deer with hounds. I also had many talks and exchanges of correspondence with others who had knowledge of hunting, stalking, deer management and the biology of the red deer. The study that followed would not have been possible without the assistance of Liz Bradshaw, whom I had appointed after a big trawl for an appropriate Research Associate. Liz was based on Exmoor from the end of August 1995 until the end of January 1997. She observed a complete hunting season and followed many hunts. She was also responsible for overseeing the collection of red deer blood and muscle samples needed for assays and collaborated closely with me in analysing the scientific material arising from the study. The study necessarily involved a wide range of scientific knowledge on the veterinary, welfare, physiological and behavioural sides of the project, as well as expertise in the management of deer, and knowledge of culling methods. I was fortunate in being able to bring together a distinguished scientific panel covering the areas of knowledge concerned.

I reported to the Council of the National Trust in 1997, concluding that the level of total suffering of red deer would be markedly reduced if hunting with hounds were ended. At the end of the study I was left with no doubt that the deer were forced into a prolonged struggle which in no way resembled being hunted by wolves – nor were they playing, as was sometimes suggested implausibly by the supporters of stag-hunting. I argued that hunting with hounds could no longer be justified on welfare grounds, given the standards applied in other fields such as the transit and slaughter of farm animals and the use of animals in research. On the strength of my report the National Trust immediately banned hunting of red deer with hounds on its land. In due course we published our scientific findings (Bateson & Bradshaw 1997; Bradshaw & Bateson 2000).

My study was the first of its kind but it was quickly followed by another study commissioned by the Stag Hunts and the Countryside Alliance, who were appalled by what we had found. The study was led by Professor Roger Harris and, to some consternation of the sponsors, obtained results gratifyingly similar to those obtained in my study. Later the Government set up an Inquiry into the hunting of wild mammals with dogs chaired by a former senior civil servant, Lord Burns. Jointly with Professor Harris, I was contracted to provide the Inquiry with a report on the scientific evidence for the effects of hunting with dogs in England and Wales on the welfare of deer, foxes, mink and hare. We concluded that nothing in the course of a deer's natural life resembles being hunted by hounds and that the known patterns of predation by wolves do not resemble stag-hunting as practised by the Stag Hunts of Devon and Somerset. Taken together with the physiological effects of hunting, it was clear that hunting with hounds would not be tolerated in other areas of animal husbandry. Lord Burns and his committee concluded that: "Most scientists agree that deer are likely to suffer in the final stages of hunting. The available evidence does not enable us to resolve the disagreement about the point at which, during the hunt, the welfare of the deer becomes seriously compromised."

My own view is that a welfare problem arises at a much earlier stage in a chase than was suggested by Professor Harris and his colleagues. Professor Harris and his team found that deer killed after chases lasting around an hour had rectal temperatures at the maximum on their clinical thermometer (at or in excess of 43 °C). Red deer are heavy-bodied animals and do not have sweat glands. Consequently, when they are exerting themselves to the limit, they become very hot. In longer hunts, when they move more slowly and the carbohydrates for powering their muscles are consumed, temperatures at death are not so high, but by that stage they would have had time to cool down after the sprinting in the early stages of hunting, which they would have been unable to sustain.

The principal area of disagreement between scientists was over the time elapsing between the hunted deer exhausting their carbohydrate resources for powering their muscles and death at the hands of the hunters. Some believed (without any evidence) that the time would be brief but most agreed that during this period the welfare of the deer would be compromised. I did have some evidence from my study relevant to this issue. At the point when the carbohydrate resources are consumed the deer look very tired. Their heads drop and they cannot jump high fences any longer. In two cases this state was observed in hunted deer 20 minutes before death. In two cases it was 80 minutes before death and in four cases it was more than 100 minutes before death at the hands of the Hunts.

The media coverage of this work on hunting was enormous and I was vilified by those for whom hunting with hounds was their ruling passion. Every effort was made by the supporters of stag-hunting to ruin my reputation. In the middle of this I was elected Biological Secretary of the Royal Society, which discomfited my critics. In this new capacity I became involved in other controversies, such as the genetic modification of plants and animals. By that stage I was battle-hardened and enthusiastically threw myself into an effort to raise the public profile of the Royal Society as an independent source of authoritative information about science.

Plasticity and alternative pathways in development

A sabbatical in 1997–8 meant that I could write a book about development, which I'd wanted to do for years. This space gave me a sense once again of what fun it is to be an unencumbered academic. I wrote the book with Paul Martin, with whom I had collaborated on *Measuring Behaviour*. This new book was about how behaviour develops and was called *Design for a Life* because we wanted to emphasise the adaptive and seemingly well-designed aspects of so much of what happens in development (Bateson & Martin 1999). In retrospect the title might have been a mistake because the pre-Darwinian ideas about intelligent design were being taken up by the Creationists in their attempts to dress up their beliefs as a form of science. Be that as it may, one of the chapters was about alternative routes in development, picking up on a long-standing interest of biologists in qualitative polyphenisms, like those found in the castes of social insects, and quantitative norms of reaction. The capacity of organisms with the same genome to develop in dramatically different ways, depending on the conditions in which they find themselves, is widespread among plants and animals. I thought that it might well be true in humans and was intrigued by David Barker's discovery that people who had been small at birth were much more likely to suffer from heart disease in adulthood if they developed in an affluent society. Perhaps the mother provides a forecast of the conditions that her unborn offspring will face in later life and, as a consequence, the baby develops the body size and metabolism appropriate to those conditions. Usually the offspring's response would be adaptive but at times of rapid increases in affluence, individuals could be faced with a mismatch between their phenotypes and their environments.

The idea bore fruit and in 2003 I brought together a group of epidemiologists, evolutionary biologists, behavioural ecologists and developmental biologists at a palace belonging to Paulo Roberto Imperiale in Cremona. The meeting was organised by an old friend and collaborator of mine, Bruno D'Udine, who had been instrumental in bringing me to many superb occasions in Italy in the past. The Cremona workshop was a remarkable meeting of minds and was altogether a great success. In due course it gave rise to a report, which has been much cited ever since (Bateson *et al.* 2004). It led me to carry out research with James Curley and Frances Champagne on the induction of alternative phenotypes in mice. It also led to sustained collaboration with Peter Gluckman and other members of the Liggins Institute in Auckland New Zealand (Gluckman *et al.* 2005).

The Cambridge principle

When I first became excited about animal behaviour I had not supposed that I would spend much of my career studying development in the laboratory. Sometimes when I talk about what has become my chosen subject, people say to me that it all seems so complicated. Indeed, Richard Dawkins played back on me an old joke of mine, suggesting that people like me who study the Great Nexus (his ironic phrase, not mine) are obscurantists. In the 1970s I had playfully enunciated three heterodox principles for the behavioural biologist. The first one was "Never use a simple explanation if a more complicated one will do instead"; the second was "Never use a causal explanation if a teleological one will do instead." The first became known self-mockingly as "the Cambridge principle" because Robert Hinde, the driving intellect at the Sub-Department of Animal Behaviour at Cambridge, used to say repeatedly "Behaviour *is* complicated." The second principle became known as "the Oxford principle" for reasons not unconnected with Richard Dawkins. The third principle, relevant to studies of cognition and welfare, was "Treat animals like humans until you have good reason to think otherwise." Notwithstanding the jokes, my general concern has been with how the undoubted complexities of development might be made more tractable by uncovering principles that make sense of that complexity (Bateson 2006b).

However, focusing on over-simplified causal factors in the name of clarity has caused great confusion. The same goes for dichotomies. I certainly would not agree with the assertion that the world is populated by two kinds of people: those who like dichotomies and those who do not. As a member of the second camp, how could I? I have been consistently sceptical of the idea derived from folk psychology that a useful distinction exists between behaviour patterns that are innate and those that are not innate. This scepticism is not based on the assumption that all behavioural and cognitive systems develop in the same way. I do not think that just two different kinds of developmental processes are found and in this respect I had been much influenced by the writings of Danny Lehrman. In clarifying my ideas in recent years, I much enjoyed working with a philosopher of biology, Matteo Mameli (Mameli and Bateson 2006; Bateson and Mameli 2007). We argued that the triggering mechanisms involved in producing alternative phenotypes are unlikely to be similar to the many processes of learning, which, in their turn, probably differ radically from each other. Furthermore, the processes involved in the development of some neurobiological systems are probably quite different from those that require plasticity and may be more akin to those involved in the development of an organ such as a kidney. Yet the outcome of development will often have involved interplay between these different forms of development. This conclusion does not denigrate the value of identifying important features, such as robustness in the face of variation in environmental conditions, nor attempts to discover how they have evolved. It merely asserts that features may be co-opted for use in different ways and in different combinations.

Over-used metaphors from engineering such as "hard-wiring" and "pre-programming", applied globally to the outcome of development, fail to capture the character of the processes and once again invite the mistaken view that they can be contrasted with their opposites. I believe that a thorough investigation of developmental processes has been hindered by indiscriminate use of the labels "innate" and "acquired". Their use encourages researchers to bundle together notions that should be distinguished from each other and to assume that important questions about development have already been fully answered, when in reality they have not.

Patrick Bateson

Return to research

I have done my fair share of administration over my career. For many years I ran the laboratory at Madingley. I was lucky to work with some technical assistants, Tom Goss, Les Barden, Paul Heavens and John Owen, who made a great difference to my own research and to the running of the place. I gave up the Directorship of the laboratory when I became Provost of King's, handing over to Barry Keverne. I was Head of the Department of Zoology after Gabriel Horn retired and before Malcolm Burrows was appointed. Later I was also Biological Secretary and Vice-President of the Royal Society. The Royal Society has a motto "nullius in verba" which is an extract from a line by the Latin poet Horace. The line may be translated as follows: "I say no master has the right to swear me to obedience blind." In effect, the fundamental mode of conduct was to trust the authority of evidence rather than the authority of somebody's say-so. It is an excellent principle but it was not always easy to make the point that we needed evidence without sounding plaintive and self-serving: "More research is needed." Work for the Royal Society was the most interesting part of the various administrative jobs that I have done. I was responsible for a large group of some 350 Research Fellows, who are the jewels in the crown of the Royal Society. I also had to oversee the election of the full Fellows on the Biological side. I am proud to have introduced a simple expedient that virtually eliminated tactical voting on the specialist committees. People will often vote in favour of their personally preferred candidate in preference to someone who had emerged as the stronger candidate. Worse, they will give greater weight to their preferred candidate by giving the stronger candidate the lowest possible ranking. They will not do this, however, if their rank-order is made public. So the solution was obvious. Make known to all members of the committee how all the others voted. The effect was miraculous.

In addition to these jobs for the Royal Society I was responsible for producing a number of reports on endocrine disruptors, the use of animals in research, the use of genetically modified animals, and science and the public interest. I also gave evidence to select committees of the House of Lords on behalf of the Royal Society and repeatedly spoke on radio and television.

I am probably too untidy in my habits to have been a natural administrator. Sadly, the administrative load in academic life has grown so much that many of our brightest students look askance at a life in research. When I was on a fund-raising trip for my College in the United States, I obtained the details of many trusts and foundations. The number of administrators employed by these organisations was strongly and positively correlated with the length of time the bodies had been in existence. The same inexorable trend is to be found in universities. A very good reason is always found for introducing another layer of bureaucracy, ensnaring the unfortunate academics who had become involved, and providing more work for those who had not.

When I retired from my administrative jobs I felt an enormous sense of release. I was able to return to the work that I enjoy most of all. It has been wonderful to do the things that I love once again and recover the intellectual interplay and companionship involved in research collaborations and encounters at conferences. These contacts generate great stimulation and great personal warmth. Reflecting back on my career, most of my closest friends are people that I worked with or met in the course of my research life. And we really did come together. For these things I am enormously grateful.

References

Barrett, P. & Bateson, P. (1978). The development of play in cats. *Behaviour* **66**: 106–20.
Bateson, P. P. G. (1966). The characteristics and context of imprinting. *Biol. Rev.* **41**: 177–220.
Bateson, P. P. G. (1968). Ethological methods of observing behaviour. *The Analysis of Behavioural Change*, ed. L. Weistrantz, pp. 389–99. New York: Harper & Row.
Bateson, P. (1978). Sexual imprinting and optimal outbreeding. *Nature* **273**: 659–60.
Bateson, P. (1979). How do sensitive periods arise and what are they for? *Anim. Behav.* **27**: 470–86.
Bateson, P. (1982a). Behavioural development and evolutionary processes. In *Current Problems in Sociobiology*, ed. King's College Sociobiology Group, pp. 133–57. Cambridge: Cambridge University Press.
Bateson, P. (1982b). Preferences for cousins in Japanese quail. *Nature* **295**: 236–7.
Bateson, P. (1983). Optimal outbreeding. In *Mate Choice*, ed. P. Bateson, pp. 257–77. Cambridge: Cambridge University Press.
Bateson, P. (1991a). Are there principles of behavioural development? In *The Development and Integration of Behaviour*, ed. P. Bateson, pp. 19–39. Cambridge, Cambridge University Press.
Bateson, P. (1991b). Assessment of pain in animals. *Anim. Behav.* **42**: 827–39.
Bateson, P. (1994). The dynamics of parent-offspring relationships in mammals. *Trends Ecol. Evol.* **9**: 399–403.
Bateson, P. (2002). William Bateson: a biologist ahead of his time. *J. Genet.* **81**: 49–58.
Bateson, P. (2005). Ethics and behavioral biology. *Adv. Stud. Behav.* **35**: 211–33.
Bateson, P. (2006a). The adaptability driver: links between behaviour and evolution. *Biol. Theor. Integr. Devel. Evol. Cogn.* **1**: 342–5.
Bateson, P. (2006b). The nest's tale: a reply to Richard Dawkins. *Biol. Philosophy* **21**: 553–8.
Bateson, P. & Bradshaw, E. L. (1997). Physiological effects of hunting red deer (*Cervus elaphus*). *Proc. R. Soc. Lond.* B**264**: 1707–14.
Bateson, P. P. G. & Chantrey, D. F. (1972). Discrimination learning: retardation in monkeys and chicks previously exposed to both stimuli. *Nature* **237**: 173–4.
Bateson, P. P. G. & Hinde, R. A. (eds.) (1976). *Growing Points in Ethology*. Cambridge: Cambridge University Press.
Bateson, P. & Horn, G. (1994). Imprinting and recognition memory: a neural net model. *Anim. Behav.* **48**: 695–715.
Bateson, P. & Mameli, M. (2007). The innate and the acquired: useful clusters or a residual distinction from folk biology? *Devel. Psychobiol.* **49**: 818–31.
Bateson, P. & Martin, P. (1999). *Design for a Life: How Behaviour Develops*. London: Cape.
Bateson, P. P. G. & Nisbet, I. C. T. (1961). Autumn migration in Greece. *Ibis* **l03a**: 503–16.
Bateson, P. P. G. & Plowright, R. C. (1959). Some aspects of the reproductive behaviour of the ivory gull. *Ardea* **47**: 157–76.

Bateson, P. P. G. & Reese, E. P. (1969). The reinforcing properties of conspicuous stimuli in the imprinting situation. *Anim. Behav.* **17**: 692–9.

Bateson, P. P. G., Rose, S. P. R. & Horn, G. (1973). Imprinting: lasting effects on uracil incorporation into chick brain. *Science* **181**: 576–8.

Bateson, P., Barker, D., Clutton-Brock, T. *et al.* (2004). Developmental plasticity and human health. *Nature* **430**: 419–21.

Bateson, P., Mendl, M. & Feaver, J. (1990). Play in the domestic cat is enhanced by rationing the mother during lactation. *Animal Behaviour* **40**: 514–25.

Bradshaw, E. L. & Bateson, P. (2000). Welfare implications of culling red deer (*Cervus elaphus*). *Anim. Welfare* **9**: 3–24.

Burkhardt, R. W. (2005). *Patterns of Behavior.* Chicago, IL: University of Chicago Press.

Chantrey, D. F. (1972). Enhancement and retardation of discrimination learning in chicks after exposure to the discriminanda. *J. Comp. Physiol. Psychol.* **81**: 256–61.

Gluckman, P. D., Hanson, M. A., Spencer, H. G. & Bateson, P. (2005). Environmental influences during development and their later consequences for health and disease: implications for the interpretation of empirical studies. *Proc. R. Soc. Lond.* B**272**: 671–7.

Gomendio, M., Cassinello, J., Smith, M. W. & Bateson, P. (1995). Maternal state affects intestinal changes of rat pups at weaning. *Behav. Ecol. Sociobiol.* **37**: 71–80.

Honey, R. C., Horn, G. & Bateson, P. (1993). Perceptual learning during filial imprinting: evidence from transfer of training studies. *Q. J. Exp. Psychol.* **46B**: 253–69.

Honey, R. C., Horn, G., Bateson, P. & Walpole, M. (1995). Functionally distinct memories for imprinting stimuli: behavioral and neural dissociations. *Behav. Neurosci.* **109**: 689–98.

Horn, G. (1998). Visual imprinting and the neural mechanisms of recognition memory. *Trends Neurosci.* **21**: 300–5.

Horn, G., Rose, S. P. R. & Bateson, P. P. G. (1973). Experience and plasticity in the central nervous system. *Science* **181**: 506–14.

Horn, G., McCabe, B. J. & Bateson, P. P. G. (1979). An autoradiographic study of the chick brain after imprinting. *Brain Res.* **168**: 361–73.

Humphrey, N. K. (1976). The social function of intellect. In *Growing Points in Ethology*, ed. P. P. G. Bateson & R. A. Hinde. pp. 303–17. Cambridge: Cambridge University Press.

Irvine, W. (1955). *Apes, Angels, and Victorians.* London: Weidenfeld & Nicolson.

Mameli, M. & Bateson, P. (2006). Innateness and the sciences. *Biol. Philosophy* **21**: 155–88.

Martin, P. & Bateson, P. (1986). *Measuring Behaviour.* Cambridge: Cambridge University Press. (2nd edn 1993; 3rd edn 2007.)

McCune, S. (1995). The impact of paternity and early socialisation on the development of cats behaviour to people and novel objects. *Appl. Anim. Behav. Sci.* **45**: 109–24.

Smith, E. F. S. (1991). Early social development in hooded rats (*Rattus norvegicus*): a link between weaning and play. *Anim. Behav.* **41**: 513–24.

ten Cate, C. & Bateson, P. (1988). Sexual selection: the evolution of conspicuous characteristics in birds by means of imprinting. *Evolution* **42**: 1355–8.

ten Cate, C. & Bateson, P. (1989). Sexual imprinting and a preference for supernormal japanese quail. *Anim. Behav.* **38**: 356–7.

Tinbergen, N. (1951). *The Study of Instinct.* Oxford: Clarendon Press.

Turner, D. C. & Bateson, P. (1988). *The Domestic Cat: The Biology of its Behaviour.* Cambridge: Cambridge University Press. (2nd edn 2000.)

Turner, D. C., Feaver, J., Mendl, M. & Bateson, P. (1986). Variation in domestic cat behaviour towards humans: a paternal effect. *Anim. Behav.* **34**: 1890–2.

4

My life and hard times

JERRAM L. BROWN

Natural history at Cornell, 1948–54

Of three siblings in my family I was the oldest and six years apart from the middle child, a situation known to encourage development of self reliance when young. Born in 1930 to a college-educated musician-mother and a mathematician-father, then raised in a small (about 9,000 then), college town, Amherst, Massachusetts, I was expected to go to college. Entering Cornell University in 1948, I chose a major in chemistry. After my freshman year, however, I realized that I was outclassed by my chemistry classmates, many of whom had greater motivation and ability – a blessing in disguise actually. Relinquishing chemistry, I became free to indulge other inclinations. Courses in introductory ecology and animal behavior were not offered to undergraduates then at Cornell (1948–52), but it was possible

Leaders in Animal Behavior: The Second Generation, ed. L. C. Drickamer & D. A. Dewsbury. Published by Cambridge University Press. © Cambridge University Press 2010.

to study natural history. In the classroom I took naturalistic subjects, such as taxonomy of vascular plants, entomology, ichthyology, mammalogy and ornithology, as well as genetics (pre-DNA), evolution and physiology (with Donald R. Griffin). Fraternities held no interest for me, and I never entered one. A nice bonus was a series of guest lectures by Karl von Frisch on his work with honeybees.

My main interest was to explore natural areas everywhere, and to see as many species of birds as I could, starting in the northeastern United States. Bird watching was my excuse to observe animals and to seek out romantic and beautiful natural places near and far. In spring and winter vacations my friends and I organized adventurous trips, including one to subarctic Newfoundland and another to tropical Mexico to participate in a Christmas bird census at Xilitla, San Luis Potosí (Davis 1952).

A bit of self reliance also helped. To see new species of birds one summer I took a job in California; and in another summer, in southeastern Alaska. Not having much money, to get to these places I hitch-hiked from my home in Amherst, Massachusetts, round-trip to Kings Canyon National Park, California, and then in a later year round-trip to Ketchikan, Alaska, going up the Inside Passage as a stevedore on a freighter and back to Washington State as crew on a small fishing boat. In California I took weekend hitch-hiking trips to San Diego and Monterrey for sea birds, to the high Sierras for alpine birds, and the Central Valley for California Condors and marsh birds. On these trips I learned quite a bit about birds, truck drivers, salesmen, farmers, a waterfowl manager, a race-car driver, a car thief, etc. I am told that the Annual Finger Lakes Midwinter Waterfowl Census, a cooperative effort of the ornithological community at Cornell which my friends and I initiated in 1951 for some winter fun, is still going today.

In Alaska, as a "stream guard," I was left alone with a boat, a tent, a slab of bacon and other supplies and told to arrest fishing boats that violated the sacrosanct area at the mouth of the salmon spawning stream that I guarded. Before fishing season opened and before the boats appeared, I would go for a week or more without seeing a human being or even an airplane or boat; it was an isolated and pristine wilderness. Later some of the fishermen became my friends, and I did not have to arrest any of them.

My region was old-growth rain forest on the seaward side of Prince of Wales Island and densely vegetated. I shared trails with the bears that made them; wolves could be heard at night. In the bays were whales that I could "drive" up to with my outboard motor boat and almost touch (I didn't dare), and hundreds of Marbled Murrelets – seabirds that flew mysteriously into the trackless, forested mountains at dusk every night, where seabirds were not supposed to go, calling as they went, an ethereal chorus I shall never forget. Their nests, in tall trees, had not been discovered at that time. I wanted their discovery to be one of my first scientific papers, but it was instead about a range extension of the Northern Mockingbird to Alaska (Brown 1954).

Academically, my interests were aroused by a book that had appeared shortly before I entered college, *Taxonomy and the Origin of Species* by Ernst Mayr (1942). His treatment of geographic variation in birds was the centerpiece of his approach to the problem of the origin of species. This seemed like something I could do. It was the time of what was then called "the modern synthesis"; so I was also influenced by Dobzhansky (1937).

One of my summer jobs while at college was at the American Museum of Natural History in New York City. I was assigned the task of sorting a collection of Asian birds. I had considerable freedom and could chat with the African explorer–ornithologist James Chapin, and wander through the office of Ernst Mayr, who was almost never there. On one occasion Mayr asked me what I wanted to do with my life. I told him I would like to become an ornithologist. His advice, because I did not have impressive grades, was not to do so. Two of my college friends later worked there, Lester Short and Walter Bock.

I learned much about the natural world in my college wanderings, which I valued more than grades and the memorizing they required. I paid a price, however; my grades suffered, and I was unable to gain admittance into the first graduate schools to which I applied. Fortunately, Cornell allowed me to stay on in a master's program. Ironically, I could not bring myself to work with the ornithological faculty there because of my interest in natural selection and geographic variation and their emphasis on popular rather than scientific aspects of photography and tape recording bird songs. So I chose to work under Edward Rainey, a systematist who studied geographic variation in freshwater fishes of the southeastern United States. He set me in a promising direction, studying the genus *Fundulus* (New World killifish) and helped me over the hurdles of publishing scientific papers. I owe my early success to him especially. At the time I did not realize this, and I have always been sorry that I did not thank him more before he died.

Graduate schools and the military, 1952–60

In my first semester as a graduate student I decided to take my course work more seriously and put birdwatching on the back burner. I also met a person who had a major long-term influence on my life, Esther Rosenbloom. We were married within a year, in 1953. Fifty-five years later, after 3 children and 8 grandchildren (three in college or graduated by 2008), we are still going through life as a couple and grateful to be together. Esther majored in botany and so had a solid background in science; and we both had strong interests in the arts. She became an important collaborator in the field and office. We both are a little uncomfortable with the culture of conformity.

After receiving my M. S. (February 1954), I wanted to continue in evolutionary systematics but switch to birds. Rainey offered me the chance to obtain a Ph.D. in ichthyology with one more year, since my research had already exceeded some Ph.D. theses at Cornell University; but I was more confident now and eager for more challenging things. So I enrolled in a Ph.D. program at University of Oklahoma under a famous ornithologist, George Miksch Sutton. The people at Oklahoma were all wonderful to us; but as soon as I left Cornell I became eligible for the draft, and my luck ran out. Even though the Korean War was over, after one semester in Oklahoma I was forced to serve 23 months in the US Army. It was a colossal waste of time, but it allowed me time to publish papers from my

Master's thesis; it helped to pay for my Ph.D.; and it allowed us to have our first baby – for a mere US$13.

At the end of my military service in 1956 I had enough scientific publications (Brown 1954, 1955, 1956a,b, 1957, 1958a,b), I hoped, to impress admissions committees in spite of my undergraduate grades. Arriving in Berkeley I found that I would be handicapped in pursuing my plan to become an avian systematist because all the student curatorships had already been awarded to other students for 4 years, at the end of which I would be finished. So I looked around for a new and exciting field of research. Niko Tinbergen had recently moved to Oxford, UK, from the Netherlands and published two influential books in ethology in English (1951, 1953). Most previous literature in this field had been in German. This developing field, ethology or animal behavior, seemed promising intellectually and for employment, so I changed my goals. Berkeley at that time, however, lacked any faculty in this subject. Fortunately Peter Marler arrived a year later. His students then included Fernando Nottebohm and Mark Konishi. By that time, however, I was comfortably ensconced in the Museum of Vertebrate Zoology among friends (including Gordon Orians and William J. Hamilton, III) so Marler was not technically my advisor.

The faculty member who became my formal advisor, Frank A. Pitelka, was focused on lemming cycles in the Arctic at the time. This left me largely to my own devices. The greatest intellectual influence on my education at Berkeley was David Lack of Oxford, UK, through his book, *The Natural Regulation of Animal Numbers* (1954). Lack was a spokesperson for individual selection vs. group selection. His views strongly influenced my later ecological publications, which favored individual selection (e.g. Brown 1964a, 1969a, d).

My Ph.D. thesis on communication in the Steller's Jay (Brown 1964b) attracted modest attention for its analysis of a graded visual signal, crest elevation (Marler & Hamilton 1966, p. 378, figs. 10–14); but its move toward more neurobiological framing of questions and data was criticized, as I expected (Hailman 1965). On the side, I was able to study geographic variation in an avian species (Brown 1963b), as I had originally wanted as an undergraduate. One year, during winter and spring vacations, with help from Gordon Orians and two faculty members at the University of Arizona, Paul Martin and Joe Marshall, I did a quick study of the Mexican Jay (*Aphelocoma ultramarina*) that would be useful to me later on (Brown 1963c). One of my students later ridiculed me for not studying helping behavior at this time. My reason for not doing so was simple: at the time of our spring vacation there no eggs hatched and no young were being fed. I could not stay to observe feeding of young because I was a teaching assistant and had to return to work. While at Berkeley I also had classes from Frank Beach (sexual behavior) and Eckhard H. Hess (imprinting).

The ethology of the late 1950s was largely concerned with motivational models that were derived from watching the animals. They eschewed neurobiology. A few invertebrate neurobiologists were interested in behavior, as were some physiological psychologists who worked with mammals, but a neurobiological basis for ethological concepts was not being attempted except by Erich von Holst (von Holst & von Saint Paul 1958, 1960). So I began to study neurobiology and applied for a post-doctoral fellowship to study brain stimulation in Europe with a view toward bridging the gap between ethology and neurobiology. I had wanted to work with von Holst, but by the time he accepted me I had of necessity already made other arrangements.

A post-doctoral fellowship in Europe and the beginnings of neuro-ethology, 1960–65

Our family of five settled into life in Zürich, Switzerland, in the fall of 1960. I was soon learning the techniques of neurosurgery with Robert W. Hunsperger, who had worked with Nobel Prize winner W. R. Hess, in a building where Einstein once worked. We used brain stimulation at threshold currents to activate specific areas and lesions to suppress activity. Because ethological models of the 1950s had focused on "unitary drives" for attack and escape as the motivation of displays, we concentrated on regions of the hypothalamus, amygdala and midbrain associated with these behaviors. Unlike the first work of this type on birds (von Holst 1957; von Holst & von Saint Paul 1958), our work on cats depended on knowing the exact location in the brain of our stimulations and lesions (Brown *et al*. 1964, 1969a, b). This required extensive histological analysis and slowed down our publication.

For an ethologist trained in the United States a two-year residence in Europe (1960–62) provided a fantastic opportunity to become acquainted with the most famous ethologists of the day, who were all European, and to become familiar with current European work. I had a friendly walk alone with Konrad Lorenz around the Zürich Zoo before joining Heini Hediger, the zoo's director and an influential author, for an informal discussion. Esther and I also enjoyed very much a warm visit to Lorenz's research station at Seewiesen, Bavaria, during Fasching. Robert Hinde entertained us cordially at Cambridge University in his vast, elegant office. My visits with Niko Tinbergen at Oxford University and especially at his main study area for the Black-headed Gull near Ravenglass were wonderful. I enjoyed the group's friendly scientific discussions in the tents among the sand dunes at evening. On the return to Zürich we stopped at Leiden to visit the van Iersal group, where we had a memorable dinner with the Sevensters, who had worked on Sandwich Terns. We then drove north to Groningen, where G. P. Baerends and his wife, J. M. Baerends-van Roon, entertained us warmly (April 1961). We shall always remember this trip to Holland – and the chocolate-laced breakfasts. A separate trip to Stockholm allowed me to enjoy the hospitality of Eric Fabricius and talk shop with a colleague who was doing brain stimulation in pigeons (Akerman 1965).

Toward the end of my post-doc I wrote a summary of our work that began to integrate our neurobiological research with ethology. People with a conservative inclination did not welcome this, but it was widely read at the time (Brown & Hunsperger 1963) and even reprinted in books of readings (McGill 1965, 1977; Fox 1973). According to the first textbook of neuro-ethology (Ewert 1980), Brown and Hunsperger "introduced the term neuroethology in connection with their studies on the activation of agonistic behavior..." (p. 11). Other terms, such as *Verhaltensphysiologie*, ethophysiology, and ethoanatomy, had been used previously, but neuro-ethology caught on and is still popular today.

Teaching neuro-ethology, 1965–77

Finding a job in ethology was not difficult then because demand for people to teach ethology in America exceeded supply. I even turned down a job offer at a good university in 1960 to take my post-doctoral fellowship. Finally in 1962 I had to find a job somewhere, and I was nervous. I was applying for jobs in the USA but was in Europe and not available for interviews. Fortunately, an army friend from 1955–56, a Ph.D. chemist, notified me of a job at the University of Rochester in Rochester, NY. I applied and was shocked and relieved when they took me sight unseen. This was a joint appointment. My laboratory was in the newly created Center for Brain Research; but my office and teaching responsibilities were in the Department of Biology. I began teaching general biology and animal behavior; and in 1965 I offered a new course in neuro-ethology and advertised for graduate students in that field in the bulletin of the Animal Behavior Society.

My students and I investigated evoked vocalizations from the avian midbrain (Brown 1965), self-stimulation of the avian brain (Goodman & Brown 1966), neuro-ethology of fishes (Demski & Knigge 1971), the auditory system of lizards (Kennedy 1974), the mid-brain auditory system of the Red-winged Blackbird (Newman 1970, 1972), and winter foraging in squirrels (Lewis 1980) and gulls (MacLean 1986). One inspiration for my neuro-biological research had been my attempts to relate observations on social behavior in Steller's Jays to neural mechanisms (Brown 1964b). I did not have a chance to do neuro-ethological work using Steller's Jay, a western species, until 1969 (Brown 1973). I also presented a general summary of neuro-ethological approaches to the study of emotional behavior and ethological concepts (Brown 1969c), a review of neural mechanisms of avian vocalization with emphasis on motivational systems (Brown 1969b) and a review of neural mechanisms of aggression (Brown 1970b).

Behavioral ecology on the rise, 1962–69

At Berkeley, 1956–60, I had been much influenced by the debate between David Lack (1954) and proponents of group selection by population regulation (then principally Wynne-Edwards 1962). "To understand this, one must know that the central problem of ecology in 1960 seemed to many of us to be the regulation of population numbers" (Maynard Smith 1985, p. 351). Ecology has changed greatly since then, shifting emphasis toward community ecology. The principal test case in this debate centered on territoriality and its influence on population regulation. Thus the issue had importance then that was far greater than can be appreciated today. Further, in retrospect, it marked the beginning of the rise in interest in behavioral ecology. Some of my work at Berkeley was relevant (Brown 1963a), and I had not lost interest in such matters. I was able to contribute conceptually while still maintaining my neuro-ethological research (Brown 1964a, 1969a, d). The 1964 paper was far more influential than I had expected. In contrast, the 1969 paper in *American Naturalist*, although it was the first to derive the famous "ideal free distribution" (Fretwell & Lucas 1970), was little noticed. My name for it, the "optimal mix," did not catch on. In this paper, in addition to deriving a simple model, I showed using published data that titmice were actually maximizing their reproductive success through density-dependent habitat choice, rather than limiting it as required in the group selection theory of Wynne-Edwards (1962). Ironically, one modern ecologist told me that he dismissed the paper because he thought (mistakenly) that it was "group selectionist". The "opposite" was actually true.

It was my habit as an assistant professor to visit the library every Saturday morning to keep up with recent literature. One day in 1963, I came across an interesting paper by W. D. Hamilton titled "The evolution of altruistic behaviour" (1963). Immediately I wanted to go out and test some of the predictions that this simple idea suggested, but the possibilities for field work on this problem with birds in Rochester, NY seemed slim. I knew from previous field work (Brown 1963c) that I could easily capture, band and observe Mexican Jays, and I knew from a visit in 1956 that the Southwestern Research Station in the Chiricahua Mountains of southeastern Arizona would be feasible as a base for a field study. Years later, when a sabbatical year became possible for me, our whole family moved to Tucson, Arizona, for a year in the fall of 1968, where I undertook two studies, one a neuro-ethological study of the Steller's Jay (Brown 1973), and the other, a descriptive study of helping behavior in the Mexican Jay (Brown 1970a). My plan after the sabbatical was to continue behavioral ecology in the spring and do neuro-ethology the rest of the year. To my surprise, the response to my work on the Mexican Jay was far beyond expectation while the response to my neuro-ethological work was low keyed at best.

Hamilton and the offspring rule, 1972–86

As interest in helping behavior grew, my thoughts turned toward application of Hamilton's rule to field data on birds. There was one problem. As written, the rule was for participants of the same generation, as in Haldane's (1932) parable of the drowning brothers. In helping, however, costs to the altruist and benefits to the recipient were not in the same generation. Offspring were helping their parents. In order to evaluate field data on helping in birds I devised a simple modification. Many years had passed since the publication of Hamilton's rule, and I wondered why this had not been done before. So I wrote to W. D. Hamilton (July 1972) to see if he or somebody else had already done it. He recommended that I publish it; so I did (Brown 1975, pp. 735–6). This method was soon used in reviews of helping behavior (Brown 1978; Emlen 1978). Later I developed it more pedagogically, naming it the "offspring rule"(Brown 1987).

I had good relations with Hamilton and even wrote two letters of recommendation for him – one for the Crafoord Prize, which he shared in 1993. He was a guest in our country home Oct. 29, 1981. An inveterate naturalist like myself, he admired our field of autumn wild-flowers, principally goldenrods and asters, noting perceptively, "Late flowers in this part of North America certainly are a striking and delightful contrast to Europe: must be due ultimately to relatively guaranteed late fine weather with its consequent encouragement to bee activity" (in litt., May 4, 1981). I have not seen any papers testing this hypothesis. Later, in his autobiographical work (Hamilton 1996, p. xi), he thanked me for "early serious attention" (principally Brown 1970a, 1974). I was not a one-sided advocate, however; my position always was that Hamilton's rule provided one of several alternative hypotheses, all of which deserved consideration, especially in combinations. My position from the outset, frequently misrepresented simply as "kin selection," was that ecological factors, mainly territorial defense of resources and females, "set the stage" for relatedness to matter (Brown 1969d, 1974).

A textbook including behavioral ecology and neuro-ethology, 1975

At the University of Rochester I wrote out most of my lectures in animal behavior and made many of the chapters available to the students, using their feedback over the years to improve their clarity. I assembled these into a textbook (Brown 1975). This was in the days when everything was done on typewriters – mine was manual. The study of animal behavior was developing rapidly then, and this book incorporated not just new facts but bold new perspectives (in my opinion anyway). As I stated, "This is a book about behavior with the central, unifying theme of biological evolution" (p. xv). More specifically, "...the central concepts of this book are concerned with populations," i.e. how behavior evolves by natural selection. This new way of presenting animal behavior fused animal behavior with population biology, as in my earlier behavior-ecological papers (Brown 1964a, 1969d, a, 1974). This approach was also incorporated, though not so explicitly, in two other textbooks that appeared the same year (Alcock 1975; Wilson 1975). This fusion was, in fact, the central message of what came to be called "sociobiology," a term that misses the most fundamental point, which was the role of population biology in viewing behavior as the product of natural selection. It was not that we had just started to study social behavior – Tinbergen (1953) and others had already done that. I have developed this point further in a history of the study of avian behavior (Brown 1994). The basic idea is that as fields grow they tend to split, then later to fuse as newly recognized, "interdisciplinary" specialties.

As mine was a textbook of behavior rather than behavioral ecology, I included some chapters that I particularly liked because they presented behavior not in the traditional way of ethologists at that time, ignoring the nervous system, but with specific reference to neural mechanisms. For example, I tried to elucidate the neural mechanisms relevant to traditional Lorenzian concepts such as "fixed action patterns" and "innate releasing mechanisms." This neuro-ethological part of the book was not appreciated by the traditionalist reviewers, who seemed to be primarily concerned with what later became known as behavioral ecology.

In retrospect, I should have published this book a year earlier, which I could have done with only a little inconvenience. Even though I had written it for the college sophomores who took my course and judged the level to be appropriate in class surveys, the book was pitched at a level too high to compete for sales to undergraduates in the rest of the country. It was not written for my graduate students but was used in some graduate courses elsewhere. The book sold well enough to pay for my research in Australia, but I preferred not to raise or lower the level of the book for mere financial gain and so decided not to revise it.

I looked for collaborators to revise it, but found nobody. I had my hands full moving into behavioral ecology and decided to concentrate on that.

Australia: a field experiment, 1976–7

My interest in helping behavior in birds in relation to altruism in the Hamiltonian sense stimulated me to become more involved in behavioral ecology (Brown 1974). The idea of altruism was emotionally resisted by some ornithologists. After presenting a somewhat provocative paper called "Communism in an American Jay" during the Cold War, I was publicly chastised at a meeting of the American Ornithologists' Union by one of its presidents, Charles Sibley – a famous ornithologist notorious for angry outbursts – for using the word "altruism" (1970, Buffalo, NY).

An early issue arose from the observation that reproductive success was correlated with group size, i.e. with the number of helpers, thus appearing to support the notion that helpers benefitted parents. But was this relationship caused by the greater number of helpers in larger groups or was it caused by the larger groups living in better territories, an uncertainty considered earlier by Lack (1968) and bolstered by our data (Brown & Balda 1977) and others'? I proposed to do an experimental helper-removal study in Australia to resolve the issue, but the proposal was turned down by the National Science Foundation with instructions to stay in Arizona and study Mexican Jays. Such studies were not feasible in Mexican Jays, however, because the large group size precluded studying a large number of groups in one year.

Using money from sales of my book, we went to Australia for 11 months anyway, leaving the jay study with trusted assistants. Together with Douglas Dow, from the University of Queensland at Brisbane, without whom the work could not have been done, my wife and daughter, Sheryl, and I had a most enjoyable year mostly in the outback town of Meandarra, Queensland, among the 'roos, koalas and death adders, in a house where there was no electricity (except rarely by gasoline generator), no telephone, and no running water except off the roof. The local graziers (sheep and cattle ranchers) were wonderful to us. It was a productive year for babbler papers (Brown *et al.* 1982a, b, 1983; Brown & Brown 1981b; Brown 1979; Johnson & Brown 1980; Brown *et al.* 1978). Our field experiment was quickly imitated by others (e.g. Mumme 1992), and controversy over the effect of the helpers was basically resolved. When controlling territory quality experimentally, the "helpers" helped not just by bringing food (Brown *et al.* 1978) but also by causing greater nesting success (Brown *et al.* 1982b).

"The grand tour" of 1977

In Europe of the 1700s the educated were expected to take "the grand tour" to learn first-hand about great cultural landmarks of European history. A more appropriate "grand tour" for me was a naturalistic tour of the great biological "landmarks" of the world. We had already done North America and western Europe – during the Cold War – and had stopped in Hawaii on the way to Australia. While based in Brisbane we undertook a side trip to New Guinea to see birds-of-paradise and bowerbirds as well as pitohuis, cassowaries and many others. Arriving in Port Moresby, we then flew inland in a single-engine plane March 16, 1977 to Wau, where there was a biological station. Flying in, we had to weave up the valleys with mountains high on each side. Below we could see slash-and-burn clearings in the forest. The plane had to land on a runway that was on a slope, so we landed in the uphill direction. The station was comfortable and in a civilized town; the staff was very helpful. The wife of the director played the Hammerklavier Sonata of Beethoven on her piano for us. Poinsettia *trees* grew in her yard, truly an outpost of European culture in a country recently emerged from the Stone Age where, we were cautioned, pigs were more valuable than women. Flying out from Mt. Hagen, where many men wore grass skirts, to Port Moresby in a small propeller plane we passed low over endless stretches of seemingly uninhabited virgin forest (March 26, 1977). I wonder what is left of it.

Continuing around the world from Brisbane, on the way back to Rochester, we went first to Bali, which was a memorable cultural experience in itself as well as a naturalistic paradise. We proceeded on to Borneo, retracing some of the footsteps of Alfred Russel Wallace, to see proboscis monkeys and monkey-cup carnivorous plants in Kampung Baco National Park. We saw and heard whooping gibbons from a tower in the rainforest of Malaysia. We then flew via Sri Lanka (then under martial law) to Kerala in southern India, where with Rauf Ali we saw various langurs and monkeys. Our final stop before leaving India was at Mumbai with its venerable Bombay Natural History Society and a memorable visit with India's reigning ornithologist, Salim Ali, in his home.

We took off from India August 1 and headed for Africa, but our plane lost the use of one engine over the Indian Ocean, forcing us to limp back to Mumbai. We finally arrived a day late in Nairobi, Kenya. In the taxi that took us from Nairobi to Nakuru we were amazed not by the giraffes, elephants, topi and other antelope that one could see from the car but by the nonchalance of the Africans sharing our cab, who simply took them for granted and read their newspapers. At Naivasha we were guided by Dave and Sandy Ligon, who showed us their study area for woodhoopoes; and we met Uli and Heidi Reyer, who showed us their tiny kingfishers but said not a word about their helping behavior. After a day of watching wildlife with Bob Hegner, ending at a colony of bee-eaters that he was studying with Stephen Emlen, we were off to Zambia to visit Dale Lewis, a former post-doctoral fellow.

Zambia was relatively untamed. Here, unlike Kenya, one could get out of one's car in the parks. We were stampeded by elephants while our open jeep would not start, charged by a herd of hippos while crossing the Luangwa River in a flimsy rowboat, face to rear with an unexpected elephant at a turn of the trail in the bush too close for comfort, and worried about the fate of a colleague who had been dragged into the bush by a lion – he recovered. One night we had dinner with a local chief; we were served water buffalo stomach – a great greasy honor – so we ate it. Finally we were on our way to Paris to visit family for a few days, then back to "reality" in Rochester.

This round-the-world tour in 1976–77 covered a lot, but still omitted some major areas. Later years helped to complete the "tour" with scientific visits to South America (1984), Japan (1986), South Africa (1998), Taiwan (2002) and Beijing, China (2002) and a personal visit to Asia Minor (the isle of Lesvos in 2007).

A Ph.D. program in behavioral ecology, 1978–2002

In Rochester, I decided it was time to look for a job where I could be part of a larger Ph.D. program and concentrate more on behavioral ecology. Being paid no better than a high school teacher while at Rochester in spite of being a Professor, I also sought a raise. We finally chose the State University of New York at Albany. There was a bias there against ecology; the old guard that ran the department was concerned mainly with cell biology, and the new chairman was a molecular biologist. Their first act shortly after we arrived and bought a house was to tell me that they were not going to give me the laboratory they had promised in my official job letter because a biochemist wanted to double his own laboratory space. He was the top money man in the department at the time, so it was no contest. It was true that I could get along without a laboratory because my research was done in the field at a research station; but it became difficult for some of my students in later years as my interests changed, requiring molecular work.

The first years at Albany were actually pretty good. We became friends with the chair and other influential people in the department. We had similar interests in culture and the arts. After a few ecologists left who did not get grants and had not gotten along with the cell-molecular group, we had opportunities to hire two behavioral ecologists, Ronald Pulliam (later President of the Ecological Society of America and advisor to the Clinton administration) and Jim Gilliam. This put us in the front ranks of graduate schools that offered a Ph.D. in behavioral ecology at that time. Of course, as time passed other schools with large endowments caught up and surpassed us; but we had our time in the sun. We attracted excellent graduate students from all over the world including the United States. Our success was demonstrated when the National Research Council conducted a review of graduate education in biology in the United States in 1995. Of the 4 groups in our department, only our group passed; and we were congratulated by the Vice President for Research. This was not appreciated by our cell and molecular biologists.

An unexpected new direction: the MHC, 1978–98

Soon after my arrival in Albany (summer 1978) I took a small commuter plane to Ithaca, NY, to give a seminar. Somehow a conversation developed with the woman across the aisle. She turned out to be a respected geneticist, Lorraine Flaherty, based in Albany with expertise in the major histocompatibility complex (MHC). She thought I might be interested in recent laboratory experiments in which male mice tended to mate with a female that had an MHC haplotype dissimilar to their own (Yamazaki *et al.* 1976). I brooded over this for a while and eventually had an opportunity to discuss its importance and speculate a bit (Brown 1983b).

The matter would probably not have gone further except for some good luck. Kathleen Egid, a student of Sarah Lenington, who had worked with mate choice and the "t-allele" in wild mice (Lenington & Egid 1985; Egid & Lenington 1985), joined me for a postdoc (1986–89) and together we followed up on Yamazaki's discovery. We were able to extend Yamazaki's finding to females using congenic lines (Egid & Brown 1989), and we did some further work with one of my Ph.D. students (Eklund 1997a; Eklund *et al.* 1992, 1991; Eklund 1999, 1997b). We also published a review (Brown & Eklund 1994). Lacking expertise in genetics and without proper laboratory facilities, we eventually had to abandon this line of research. Fortunately Wayne Potts was able to carry it further with considerable success (Potts *et al.* 1991 and later papers). Further progress is likely in this field, as the vomeronasal olfactory system is now known to be sensitive to MHC peptides (Leinders-Zufall *et al.* 2004) and a class of chemosensory receptors that seems to be associated with social signals has been found in the olfactory epithelium (Buck & Buck 2006) .

Origin of the International Society for Behavioral Ecology in Albany, 1986

Behavioral ecology had its roots in the 1960s when cost–benefit studies of social behavior became interesting (Brown 1964a; Brown & Orians 1970; Orians 1969) and when foraging theory (MacArthur & Pianka 1966) and kinship theory (Hamilton 1963, 1964) were introduced. It was at first largely expressed in studies of the population biology of social behavior, a trend that solidified with three books published in 1975 (Alcock 1975; Wilson 1975; Brown 1975), reviewed in historical perspective by Brown (1994). Later treatments unified the field by joining foraging behavior with social behavior under the general framework of behavioral ecology (Krebs & Davies 1978 and later editions in 1984, 1987, 1991, and 1997).

Despite the longstanding popularity of behavioral ecology, by the early 1980s there was no society that gave primary consideration to this field. There was no place for all people interested in behavioral ecology to come together to listen to each other's papers and talk shop. Consequently, the bird people went to ornithological meetings, the insect people to entomological meetings, and so on. The only societies that might have brought the different taxonomic varieties of behavioral ecology together were the Animal Behavior Society (ABS), the American Society of Naturalists, and the International Society for the Study of Evolution; but these societies represented many other interests, and they too were specialized. Theorists, geneticists and empiricists still tended to go to different meetings, so far as behavioral ecology was concerned.

I was for many years personally frustrated by the lack of a behavioral ecology society and by the inertia that prevented its formation. Such a society would benefit all behavioral ecologists, naturally including our own faculty and students. What could be done? There was no sign that anyone was taking steps to form such a group. Others were frustrated too. At a meeting of the Animal Behavior Society in 1985 several behavioral ecologists got together and commiserated about the lack of appreciation of their work at these meetings. One member of this disgruntled group was C. P. L. (Chris) Barkan, a graduate student of mine. According to his report, there was no plan by this group to start a behavioral ecology society. He then urged me to assume a leadership role in founding a new society. I checked with our relevant faculty and students but found mixed support. Nevertheless, I judged that there was enough enthusiasm to proceed, even though there was surprising reluctance by some ecology faculty.

I refused to proceed unless the goal of forming a society was explicit and primary. After ascertaining that the local manpower, mainly graduate students, would be sufficiently numerous and motivated, our next task was to see if the job could be done with the money likely to be available. This meeting was a real financial gamble for us because there was no pre-existing society to cover our losses or to provide guidance or logistical support. There was no pre-existing membership list or mailing list. There was no precedent by which to predict attendance.

We chose to have the meeting on a weekend during the fall semester, counting on universities in the Northeast for our primary attendance (Oct 16–19, 1986). We wanted to attract a wider audience, however; so we tried to assemble a list of speakers that would have high visibility, and we invited people from everywhere. The 367 registered participants came from Florida, California, England, Germany, and much of the United States and Canada (but not Ithaca, NY).

To bring this off we felt that we should offer to pay expenses of the speakers. Therefore, Brown and Caraco wrote a proposal to the National Science Foundation to cover expenses, requesting US$17,509 (Title: International Meeting for Behavioral Ecologists). We received US$9,000. (National Science Foundation: BNS 8609880). The meeting cost altogether US$24,617. Fortunately we were able to pay all our bills.

We invited people to be speakers who represented a broad spectrum. We wanted to bring together theorists with empiricists. We wanted primatologists, entomologists, mammalogists, ornithologists etc. to talk to each other and to theorists. We wanted ecologists, behaviorists and evolutionary biologists to communicate with each other. In short, we wanted to break down some of the barriers that separated these people from different constituencies but with a common interest in behavioral ecology. We wanted a *really* modern synthesis.

When the invitations to speakers went out we could not promise financial support for them because we had not yet heard from the National Science Foundation. It was also a rather long trip for a weekend from Europe or the West Coast of North America. Perhaps the invitation was viewed by some as insufficiently honorific. In any case, many of the more senior and famous personages politely declined our invitation, including most of the women. We did, nevertheless, attract many outstanding speakers, including some from United Kingdom, Germany and Canada.

There were two schools of thought about the society among those present in 1986. Some felt that we had enough societies and journals already. Others felt that we needed either a new society, a new journal, or both. The attendees voted to establish a society and to study the possibility of a new journal, plans for which were approved and formalized at the next meeting of the society. I believe all participants returned home from the Albany meeting with the feeling that the meeting was a success scientifically and that they had participated in an historic event, namely the birth of the International Society for Behavioral Ecology and the plans for a new journal, *Behavioral Ecology*. Our goal was achieved, and the new society has prospered to this day.

Is helping selected? 1991–98

Perhaps helping is not actually selected but is simply a response to changed environmental conditions, as suggested for insects (West-Eberhard 1987) and birds (Jamieson 1989; Jamieson & Craig 1987, "unselected hypothesis"). The scenario envisioned was that delayed departure of offspring placed them in a new "environment" in which begging young were encountered before the offspring had laid eggs themselves. Nonbreeders would then respond to the stimuli from begging nestlings by feeding them because they already had the "neural machinery" to do so by virtue of having been selected to care for their own offspring. This reasonable proposal stimulated a reaction from selectionists that led to a collection of position papers in *The American Naturalist*. It was argued that cost–benefit studies did prove that selection had occurred (Emlen *et al*. 1991) and – correctly in my opinion – that they did not (Jamieson 1991).

This controversy lay at an impasse for several years because it was nearly impossible in the case of helping to satisfy the formal conditions for demonstration that natural selection had affected the trait (Endler 1986). Cost–benefit studies did not do it. The argument based on Endler's conditions still remains; but an indirect approach provided insight. Carol Vleck and I reasoned that if natural selection had acted on the feeding of young by nonbreeders, then their physiology might also be affected. We tested this hypothesis by examining levels of hormones that were already known to be associated with the feeding of young, especially prolactin, which had been implicated with helping in a study of Harris's Hawk (Mays *et al*. 1991; Vleck *et al*. 1991). We tested two simple predictions: (1) levels of prolactin should be higher in a species with helping than a closely related, sympatric species without helping; (2) levels of prolactin should rise *before* the appearance of begging young, in contrast to the response hypothesis, which would predict a rise *after* the appearance of begging young.

We found that levels of prolactin were much higher in the species with helpers, the Mexican Jay, than in the species without helpers living in the same region, the Western Scrub-Jay (Brown & Vleck 1998). This was true even in the nonbreeding yearlings and in both sexes. This result could not have been predicted by the rival "flexible strategy" or "unselected" hypothesis as delineated by Jamieson and Craig for birds.

Tests of the second hypothesis also supported the selectionist theory. Levels of prolactin rose in both sexes well before the first young hatched, as in another species of jay (Schoech *et al*. 1996). This result allowed us to reject the begging-young hypothesis because the levels of prolactin rose in March and early April, well before the first young appeared (May and late April). Although it could be argued that one species is not enough for a test of the hypothesis (as did one critical reviewer), our paper seemed to settle the issue.

Dispersal, Inbreeding, Pedigrees, Extra-pair copulations
and Mate Choice, 1990–2004

After working on helping behavior for many years and summarizing the literature at various times (Brown 1978, 1983a, 1987), I decided to leave helping to others and move on to other kinds of behavior. Helping behavior had long been related to delayed dispersal, but the relationship of delayed dispersal to inbreeding, genealogies, and mate choice had not been vigorously pursued. Helpers in most birds are individuals that have delayed breeding for self-benefit, delayed dispersal for self-benefit, and found themselves in a situation where they have poor opportunities for breeding but an option to help their parents (Brown 1969d, 1974). I came to this conclusion on my own but then discovered that Selander (1964) had proposed similar ideas in a publication on the systematics of cactus wrens, a fact that might have been overlooked much longer had Selander not been a reviewer of my paper on territory and population regulation (Brown 1969d).

Delayed dispersal leads to retention of relatives in the same social groups and in extreme cases to complex pedigrees in which many individuals have a relative in the same group (Brown & Brown 1981a). This in turn raises the probability of inbreeding unless inbreeding is selected against. We, therefore, set out to determine whether inbreeding occurs commonly or not in the Mexican Jay; we found that 5% of pairs were inbred by a genealogical criterion (Brown & Brown 1998). Next we asked whether inbreeding was deleterious or not; it was (Brown & Brown 1998). A further step was to ask how inbreeding is avoided. Avoidance by dispersal is one option, but we found that offspring commonly breed in the same group territory in which they hatched (Brown & Brown 1984). Even if they leave the home group, they commonly go only one territory away. They even breed in the presence of their father or mother (author's unpublished data).

Work on the MHC by ourselves and others had shown the probability of a mechanism of mate choice that had nothing to do with prevailing ecological ideas about mate choice. Since Darwin (1859, 1871) the literature on mating systems (Emlen & Oring 1977) and mate choice (Andersson 1994) had centered on macho males and the fascination that females have for them and their resources. We discovered as soon as technology and other restraints would allow that Mexican Jays were not genetically monogamous (Bowen *et al.* 1995) in spite of appearing to be socially monogamous (Brown 1963c). They were, in fact, highly non-monogamous genetically (Li & Brown 2000).

Possibly a female's main interest in some species was not in picking the most macho male but in picking a male whose genes would allow her young to be genetically competitive. After reviewing the literature, I proposed that especially in sexually monomorphic species, such as the Mexican Jay, without obvious signals or "badges" of superiority, an important factor in a female's choice of a mate might be the male's genetic influence on her young (Brown 1997, 1999). Specifically, "I argue that a female's strategy should be to find the alleles that best complement her own in at least some of her offspring"(p. 70). I did not argue that a female should choose the most heterozygous male.

This is not an easy theory to test, but we and others have obtained some evidence that is consistent with it. Following the lead of a controversial paper that found a positive relationship between genetic similarity of social mates and frequency of extrapair fertilization in three species of shorebirds (Blomqvist *et al.* 2002), we discovered a similar pattern in the Mexican Jay (Eimes *et al.* 2004) but with stronger background. Striking support for the heterozygosity theory was recently provided in a study of Antarctic Fur Seals (*Arctocephalus gazella*), which showed that certain females tend to move out of their harem to find a more heterozygous partner, even moving away from males with large harems to do so (Hoffman *et al.* 2007). Analogy with extrapair copulation comes to mind. Further work is needed.

Global warming and breeding seasons, 1995–present

As the Earth warmed steadily, especially after about 1970, a few people became interested in the possible effects on plants and animals. I knew early that our long-term data would allow us to test for an advance in initial laying dates in the population of jays that I studied, but I did not want to begin the analysis until I had sampled enough years. Although there was a paper on vegetation in the Alps that showed a correlation between plant distribution and warming (Grabherr *et al.* 1994) and a paper reporting that British amphibians were breeding earlier (Beebee 1995), there was apparently no evidence of this sort for birds and other organisms anywhere in the world when we started our analyses. We wanted to be first for birds at least. I knew that I would have to do this analysis in depth to be convincing to the many skeptics. One global-warming skeptic did indeed review our first attempt to publish. So we started in a small way by testing for associations between laying dates and various climatic variables (Brown & Li 1996). We also analyzed relationships between climatic variables and breeding success (Li & Brown 1999).

As it turned out we were not first in the world. An excellent analysis of breeding records of many species of British birds presented a convincing case for causation of earlier breeding by a warming climate (Crick, H. P. Q. *et al.* 1997). It does appear, though, that our paper presented the first evidence outside the UK that birds are breeding earlier (Brown *et al.* 1999). Close on our heels was a study on Tree Swallows (Dunn & Winkler 1999).

Ornithologists had ignored this subject. So Crick and I organized the first international symposium and round table dealing with global warming and birds to bring it to the attention of ornithologists at the International Ornithological Congress in Beijing, 2002 (Crick *et al.* 2006; Crick & Brown 2006). Our session was packed to overflowing. An interesting coincidence is that, like me, Crick had also studied helping in birds (Crick & Fry 1986; Crick 1992).

The Mexican Jay, 1969–2005

Many people in recent years know me for one thing that they cannot comprehend; namely, that I studied the same bird for 37 years. Why couldn't I have answered the main questions in a single year or less? The answer, of course, is that new questions are always coming along.

One can use the same system to answer many different questions. I have expanded on this model-system theme elsewhere (Brown 2001) so will not do so here.

I would emphasize, however, that this system is not good for certain questions that many people expected me to address. These include tests of Hamilton's rule. The reason is that one needs to have a large sample of simple groups for such questions, and when group sizes are large, as in Mexican Jays, the number of groups studied must be small. Also one needs a simple system; the Mexican Jay system, with its plural breeding and complex genealogies, is anything but simple. These are some of the reasons that we did our studies on whether helpers really matter in Australia, on a species with small group sizes and a simpler social organization.

Turning points

Looking at my scientific career one sees six rather drastic changes of direction: (1) from chemistry to zoology as an undergraduate; (2) from fish systematics to avian ethology in graduate school; (3) from field ethology to neuro-ethology as a postdoc; (4) from laboratory neuro-ethology to field behavioral ecology later; (5) from a focus on helping behavior in birds to the MHC and mate choice in relation to inbreeding; (6) from mate choice to global warming and laying dates. Each of these was both a challenge and a gamble for me, but each was fascinating and rewarding. New directions are always necessary in science, so perhaps it is worth asking what general features in my life pushed me to change direction so drastically and so often.

As a college senior I knew I was interested in evolution and systematics, but I did not have the knowledge to choose a good research project for myself. Fortunately, I had an advisor who offered me a list of suitable projects from which to choose. As I learned more about the literature, I felt, perhaps unconsciously, that I had to look for the very latest and most exciting events and use these as cues for my own research. This led me first to ethology and then to neuro-ethology as new fields. I could have kept doing neuro-ethology for the rest of my life, but I was fascinated with behavioral ecology. When Hamilton's rule appeared, that proved to be my next turning point. This time I was not just changing fields; I had identified a particular model that would guide me. It was a similar specificity later with regard to my interest in the MHC and Yamazaki's discovery. One could also see that global warming would become important in ecology before the biological literature on it began to mush-room. This, I suggest, reflects a certain maturity that the beginner typically lacks – as I did. With experience and a habit of reading widely, I was able to recognize potential turning points as soon as the papers were out. If there is anything to learn from my career, this is probably it. Read widely; keep current; and watch for potential turning points.

Obviously, there is not one single type of personality that succeeds in science. But it helped me to be somewhat skeptical of established belief, to be consciously aware of the scientific method (cogently expressed by Chamberlin 1897) as a way to settle differences of opinion, and to strive for objectivity. I found Sunday School to be mainly memorization rather than analytical thinking. I questioned things that tactful people do not question, and openly debated contentious subjects. I tended to find new and minority positions intriguing. Science was probably better for me than sales or politics.

Students

During my life I have seen many positive changes in the way we live. In contrast, I have also witnessed the near-total destruction of the world's supposedly inexhaustible ocean fisheries, worldwide pollution even into the polar regions, the destruction of endless tracts of virgin forest, and an exponential increase in risk of extinction of all kinds of life. Change was rapid, but for a teacher one thing remains the same – the ages of the students. For me it was always refreshing to chat with my undergraduate advisees, and my warmest memories of teaching come from interactions with undergraduates. It was more than just teaching, however; many confided their personal problems to me, and I may have prevented one suicide by bringing that person into our family during a difficult period.

When I moved to Albany my teaching responsibilities brought more contact with graduate students. My door was always open to them. I also made every attempt to bring graduate students together with the many guest speakers we entertained in our rural home. I have served on the thesis committees of 50 graduate students who completed their theses (M.S. and Ph.D.). Graduate students will always be my friends and favorite colleagues. A similar category consists of the 40 or so research assistants I worked with in the Chiricahua Mts. Some of them were my students, but at least 12 achieved Ph.D.s at other universities (Humberto Alvarez, Carlos Andres Botero, Amy Gemmell, Ted Kennedy, John McCormack, Elizabeth Sandlin, David Siemens, Adrienne Smith, Deborah Smith, Scott Stoleson, Pepper Trail, Scott Wilson). I hope that students will find something useful to them in this personal account.

Hard times

Although my life turned out to be more fulfilling than I could have imagined, there were difficult times. In 1984 I went to a surgeon to have a bleeding mole removed from my forearm. After the histology was done he announced with an ashen face that it was a melanoma, an aggressive cancer, which could have already spread out of control. So he removed even more tissue from where the mole had been, leaving quite a hole in my arm, which he skillfully stitched together. Various doctors gravely told me what my odds of survival for 5 years were: a little better than 50%. I began the field season with my arm in a sling. Not long afterward, a slight swelling in the afflicted arm raised fear that melanoma had spread, causing worry and another operation, this time more complicated. They removed my left axillary lymph nodes, unnecessarily as it turned out.

More troubles were in store. As part of a routine physical exam in 1993, I was given my first PSA test for prostate cancer. I had an advanced case which required surgical removal of my prostate. On the day of the operation (2/25/93) there was a snow storm so bad that even the usual plows could not clear our driveway. Fortunately, a neighbor had some heavy equipment that allowed me to get to the hospital on time. There I was told that the blood I had "given" earlier to be used in this operation was in Syracuse, not Albany. I insisted they get it, and the operation was done successfully later in the day. The operation left me connected to a catheter for about 6 weeks, but I was able to do the field work that year. To be on the safe side I elected to follow the operation with radiation treatments.

In August 2002, at age 72 after 40 years of teaching, I decided to scale down my scientific writing, give up teaching and begin a new life. We continued the regular field work in the Chiricahua Mts through 2004, closing up finally in 2005. We increased our attendance at local but world class opera, ballet, orchestras and chamber music. In warm weather I expanded my gardening, greatly increasing our garden-fresh salad consumption. In the non-gardening seasons we explored France, Spain, Italy and Greece in search of birds, Roman and Greek ruins, Neolithic cave paintings, good food and wine. Vive l'ethologie.

Acknowledgments

Apologies and respects to James Thurber (deceased), humorist and dog lover, whose title I have borrowed (Thurber 1933). I am indebted to many reviewers for the National Science Foundation and other agencies for supplying continuous funding for my researches, 1960–2006, except for one year. For comments on this manuscript I thank Esther Brown, John McCormack and the editors.

References

Akerman, B. (1965). Behavioral effects of electrical stimulation in the forebrain of the pigeon. I. Reproductive behaviour. II. Protective behaviour. *Behaviour* **26**, 323–50.

Alcock, J. (1975). *Animal Behavior. An Evolutionary Approach*, 1st edn. Sunderland, MA: Sinauer Associates.

Andersson, M. (1994). *Sexual Selection*. Princeton, NJ: Princeton University Press.

Beebee, T. J. C. (1995). Amphibian breeding and climate. *Nature* **374**, 219–20.

Blomqvist, D., Andersson, M., Kupper, C., *et al.* (2002). Genetic similarity between mates and extra-pair parentage in three species of shorebirds. *Nature* **419**, 613–15.

Bowen, B., Koford, R. R. & Brown, J. L. (1995). Genetic evidence for hidden alleles and unexpected parentage in the Gray-breasted Jay. *Condor* **97**, 503–11.

Brown, J. L. (1954). Mockingbird (*Mimus polyglottos*) in southeastern Alaska. *Murrelet* **34**, 11.

Brown, J. L. (1955). Local variation and relationships of the cyprinodont fish, *Fundulus rathbuni* Jordan and Meek. *J. Elisha Mitchell Scient. Soc.* **71**, 207–13.

Brown, J. L. (1956a). Distinguishing characters of the cyprinodont fishes, *Fundulus cingulatus* and *Fundulus chrysotus*. *Copeia* **1956**, 251–5.

Brown, J. L. (1956b). Identification and geographical variation of the cyprinodont fishes *Fundulus olivaceus* (Storer) and *Fundulus notatus* (Rafinesque). *Tulane Stud. Zool.* **3**, 119–34.

Brown, J. L. (1957). A key to the species and subspecies of the cyprinodont genus *Fundulus* in the United States and Canada east of the continental divide. *J. Wash. Acad. Sci.* **47**, 69–77.

Brown, J. L. (1958a). Geographic variation in southeastern populations of the cyprinodont fish *Fundulus nottii* (Agassiz). *Am. Midl. Nat.* **59**, 447–88.

Brown, J. L. (1958b). A nesting record of the scissor-tailed flycatcher in Nuevo Leon, Mexico. *Condor* **60**, 193–4.

Brown, J. L. (1963a). Aggressiveness, dominance and social organization in the Steller jay. *Condor* **65**, 460–84.

Brown, J. L. (1963b). Ecogeographic variation and introgression in an avian visual signal: the crest of the Steller's jay, *Cyanocitta stelleri*. *Evolution* **17**, 23–39.

Brown, J. L. (1963c). Social organization and behavior of the Mexican jay. *Condor* **65**, 126–53.

Brown, J. L. (1964a). The evolution of diversity in avian territorial systems. *Wilson Bull.* **76**, 160–9.

Brown, J. L. (1964b). The integration of agonistic behavior in the Steller's jay *Cyanocitta stelleri* (Gmelin). *Univ. Calif. Publ. Zool.* **60**, 223–328.

Brown, J. L. (1965). Vocalization evoked from the optic lobe of a songbird. *Science* **149**, 1002–3.

Brown, J. L. (1969a). The buffer effect and productivity in tit populations. *Am. Nat.* **103**, 347–54.

Brown, J. L. (1969b). The control of avian vocalization by the central nervous system. In *Bird Vocalizations*, ed. R. A. Hinde, pp. 79–96. Cambridge: Cambridge University Press.

Brown, J. L. (1969c). Neuro-ethological approaches to the study of emotional behaviour: stereotypy and variability. *Ann. N.Y. Acad. Sci.* **159**, 1084–95.

Brown, J. L. (1969d). Territorial behavior and population regulation in birds. *Wilson Bull.* **81**, 293–329.

Brown, J. L. (1970a). Cooperative breeding and altruistic behavior in the Mexican jay, *Aphelocoma ultramarina*. *Anim. Behav.* **18**, 366–78.

Brown, J. L. (1970b). The neural control of aggression. In *Animal Aggression. Selected Readings*, ed. C. R. Southwick, pp. 164–86. New York: van Nostrand Reinhold Company.

Brown, J. L. (1973). Behavior elicited by electrical stimulation of the brain of the Steller's Jay. *Condor* **75**, 1–16.

Brown, J. L. (1974). Alternate routes to sociality in jays – with a theory for the evolution of altruism and communal breeding. *Am. Zool.* **14**, 63–80.

Brown, J. L. (1975). *The Evolution of Behavior.* New York: Norton.

Brown, J. L. (1978). Avian communal breeding systems. *A. Rev. Ecol. Syst.* **9**, 123–55.

Brown, J. L. (1979). Growth of nestling grey-crowned babblers, with notes on determination of age in juveniles. *Emu* **79**, 1–6.

Brown, J. L. (1983a). Cooperation – a biologist's dilemma. In *Advances in the Study of Behavior*, ed. J. S. Rosenblatt, R. A. Hinde, C. Beer & M. Busnel, Vol. 13, pp. 1–37. New York: Academic Press.

Brown, J. L. (1983b). Some paradoxical goals of cells and organisms: the role of the MHC. In *Ethical Questions in Brain and Behavior*, ed. D. W. Pfaff, pp. 111–24. New York: Springer-Verlag.

Brown, J. L. (1987). *Helping and Communal Breeding in Birds: Ecology and Evolution.* Princeton, NJ: Princeton University Press.

Brown, J. L. (1994). Historical patterns in the study of avian social behavior. *Condor* **96**, 232–43.

Brown, J. L. (1997). A theory of mate choice based on heterozygosity. *Behavioral Ecology* **8**, 60–5.

Brown, J. L. (1999). The new heterozygosity theory of mate choice and the MHC. *Genetica* **104**, 215–21.

Brown, J. L. (2001). The Mexican Jay as a model system for the study of large group size and its social correlates in a territorial bird. In *Model Systems in Behavioral Ecology: Integrating Empirical, Theoretical and Conceptual Approaches*, ed. L. A. Dugatkin, pp. 338–58. Princeton, NJ: Princeton University Press.

Brown, J. L. & Balda, R. P. (1977). The relationship of habitat quality to group size in Hall's babbler (*Pomatostomus halli*). *Condor* **79**, 312–20.

Brown, J. L. & Brown, E. R. (1981a). Extended family system in a communal bird. *Science* **211**, 959–60.

Brown, J. L. & Brown, E. R. (1981b). Kin selection and individual selection in babblers. In *Natural Selection and Social Behavior: Recent Results and New Theory*, ed. R. D. Alexander & D. W. Tinkle, pp. 244–56. New York: Chiron Press.

Brown, J. L. & Brown, E. R. (1984). Parental facilitation: parent-offspring relations in communally breeding birds. *Behav. Ecol. Sociobiol.* **14**, 203–9.

Brown, J. L. & Brown, E. R. (1998). Are inbred offspring less fit? Survival in a natural population of Mexican Jays. *Behav. Ecol.* **9**, 60–3.

Brown, J. L. & Eklund, A. (1994). Kin recognition and the major histocompatibility complex: an integrative review. *Am. Nat.* **143**, 170–96.

Brown, J. L. & Hunsperger, R. W. (1963). Neuroethology and the motivation of agonistic behaviour. *Anim. Behav.* **11**, 439–48.

Brown, J. L. & Li, Shou-H. (1996). Delayed effect of monsoon rains influences laying date of a passerine bird living in an arid environment. *Condor* **98**, 879–84.

Brown, J. L. & Orians, G. H. (1970). Spacing patterns in mobile animals. *A. Rev. Ecol. Syst.* **1**, 239–62.

Brown, J. L. & Vleck, C. M. (1998). Prolactin and helping in birds: has natural selection strengthened helping behavior? *Behav. Ecol.* **9**, 541–5.

Brown, J. L., Hunsperger, R. W. & Rosvold, H. E. (1964). Combined stimulation in areas governing threat and flight behavior in the brain stem of the cat. *Prog. Brain Res.* **6**, 191–6.

Brown, J. L., Hunsperger, R. W. & Rosvold, H. E. (1969a). Defense, attack, and flight elicited by electrical stimulation of the hypothalamus of the cat. *Exp. Brain Res.* **8**, 113–29.

Brown, J. L., Hunsperger, R. W. & Rosvold, H. E. (1969b). Interaction of defence and flight reactions produced by simultaneous stimulation at two points in the hypothalamus of the cat. *Exp. Brain Res.* **8**, 130–49.

Brown, J. L., Dow, D. D., Brown, E. R. & Brown, S. D. (1978). Effects of helpers on feeding of nestlings in the grey-crowned babbler (*Pomatostomus temporalis*). *Behav. Ecol. Sociobiol.* **4**, 43–59.

Brown, J. L., Brown, E. R. & Brown, S. D. (1982a). Morphological variation in a population of grey-crowned babblers: correlations with variables affecting social behavior. *Behavi. Ecol.* **10**, 281–7.

Brown, J. L., Brown, E. R., Brown, S. D. & Dow, D. D. (1982b). Helpers: effects of experimental removal on reproductive success. *Science* **215**, 421–2.

Brown, J. L., Dow, D. D., Brown, E. R. & Brown, S. D. (1983). Socio-ecology of the grey-crowned babbler: population structure, unit size and vegetation correlates. *Behav. Ecol. Sociobiol.* **13**, 115–24.

Brown, J. L., Li, S.-H. & Bhagabati, N. (1999). Long term trend toward earlier breeding in an American bird: a response to global warming? *Proc. Nat. Acad. Sci. USA* **96**, 5565–9.

Buck, S. D. & Buck, L. B. (2006). A second class of chemosensory receptors in the olfactory epithelium. *Nature* **442** 645–50.

Chamberlin, T. C. (1897). Studies for students. The method of multiple working hypotheses. *J. Geol.* **5**, 837–48.

Crick, H. Q. P. (1992). Load-lightening in cooperatively breeding birds and the cost of reproduction. *Ibis* **134**, 56–61.

Crick, H. Q. P. & Brown, J. L. (2006). Symposium 08 Effects of global climate change on birds: evidence and predictions. *Acta Zool. Sin.* **52**, 153.

Crick, H. Q. P. & Fry, C. H. (1986). Effects of helpers on parental condition in red-throated bee-eaters (Merops bullocki). *J. Anim. Ecol.* **55**, 893–905.

Crick, H. Q. P., Dudley, C., Glue, D. E. & Thomson, D. L. (1997). UK birds are laying eggs earlier. *Nature* **388**, 526.

Crick, H. Q. P., Brown, J. L. & Rehfisch, M. M. (2006). RTD 19 Climate change impacts: key issues and future research. *Acta Zool. Sin.* **52**, 56–7.

Darwin, C. (1859). *On the Origin of Species*. London: Murray.

Darwin, C. (1871). *The Descent of Man and Selection in Relation to Sex*. New York: Modern Library, Random House.

Davis, L. I. (1952). Winter bird census at Xilitla, San Luis Potosi, Mexico. *Condor* **54**, 345–355.

Demski, L. S. & Knigge, K. L. (1971). The telencephalon and hypothalamus of the bluegill (*Lepomis macrochirus*): evoked feeding, aggressive and reproductive behavior. *J. Comp. Neurol.* **143**, 1–16.

Dobzhansky, T. (1937). *Genetics and the Origin of Species*, 1st edn. New York: Columbia University Press.

Dunn, P. O. & Winkler, D. W. (1999). Climate change has affected breeding date of tree swallows throughout North America. *Proc. R. Soc. Lond.* B**266**, 2487–90.

Egid, K. & Brown, J. L. (1989). The major histocompatibility complex and female mating preferences in mice. *Anim. Behav.* **38**, 548–50.

Egid, K. & Lenington, S. (1985). Responses of male mice to odors of females: effects of T- and H-2-locus genotype. *Behav. Genet.* **15**, 287–95.

Eimes, J. A., Parker, P. G., Brown, J. L. & Brown, E. R. (2004). Extrapair fertilization and genetic similarity of social mates in the Mexican Jay. *Behav. Ecol.* **16**, 456–60.

Eklund, A. C. (1997a). The effect of early experience on MHC-based mate choice in two B10.w mouse strains (*Mus domesticus*). *Behav. Genet.* **27**, 223–9.

Eklund, A. (1997b). The major histocompatibility complex and mating preferences in wild house mice (*Mus domesticus*). *Behav. Ecol.* **8**, 630–4.

Eklund, A. C. (1999). Use of the MHC for mate choice in wild house mice (*Mus domesticus*). *Genetica* **104**, 245–8.

Eklund, A., Egid, K. & Brown, J. L. (1991). The major histocompatibility complex and mating preferences of male mice. *Anim. Behav.* **42**, 693–4.

Eklund, A., Egid, K. & Brown, J. L. (1992). Sex differences in the use of the major histocompatibility complex for mate selection in congenic strains of mice. In *Chemical*

Signals in Vertebrates VI, ed. R. L. Doty & D. Müller-Schwarze, pp. 213–17. New York: Plenum Press.

Emlen, S. T. (1978). Cooperative breeding. In *Behavioural Ecology: An Evolutionary Approach*, ed. J. R. Krebs & N. B. Davies, pp. 245–81. Oxford: Blackwell Scientific Publications.

Emlen, S. T. & Oring, L. W. (1977). Ecology, sexual selection, and the evolution of mating systems. *Science* **197**, 215–23.

Emlen, S. T., Reeve, H. K., Sherman, P. W. *et al.* (1991). Adaptive versus nonadaptive explanations of behavior: the case of alloparental helping. *Am. Nat.* **138**, 259–70.

Endler, J. A. (1986). *Natural Selection in the Wild.* Princeton, NJ: Princeton University Press.

Ewert, J.-P. (1980). *Neuro-ethology.* Berlin: Springer Verlag.

Fox, M. W., Ed. (1973). *Readings in Ethology and Comparative Psychology.* Belmont, CA: Wadsworth.

Fretwell, S. D. & Lucas, H. L. (1970). On territorial behaviour and other factors influencing habitat distribution in birds. *Acta Biotheor.* **19**, 16–36.

Goodman, I. J. & Brown, J. L. (1966). Stimulation of positively and negatively reinforcing sites in the avian brain. *Life Sci.* **5**, 693–704.

Grabherr, G., Gottfried, M. & Pauli, H. (1994). Climate effects on mountain plants. *Nature* **369**, 448.

Hailman, J. P. (1965). Book review. *Q. Rev. Biol.* **40**, 324–5.

Haldane, J. B. S. (1932). *The Causes of Evolution.* New York: Longman.

Hamilton, W. D. (1963). The evolution of altruistic behaviour. *Am. Nat.* **97**, 354–6.

Hamilton, W. D. (1964). The genetical evolution of social behaviour. I. and II. *J. Theor. Biol.* **7**, 1–52.

Hamilton, W. D. (1996). *Narrow Roads of Gene Land: The Collected Papers of W. D. Hamilton,* Vol. 1, *Evolution of Social Behaviour.* San Francisco: W. H. Freeman/Spektrum.

Hoffman, J. I., Forcada, J., Trathan, P. N. & Amos, W. (2007). Female fur seals show active choice for males that are heterozygous and unrelated. *Nature* **445**, 912–14.

Jamieson, I. (1989). Behavioral heterochrony and the evolution of birds' helping at the nest: an unselected consequence of communal breeding? *Am. Nat.* **133**, 394–406.

Jamieson, I. (1991). The unselected hypothesis for the evolution of helping behavior: too much or too little emphasis on natural selection? *Am. Nat.* **138**, 271–82.

Jamieson, I. & Craig, J. L. (1987). Critique of helping behaviour in birds: a departure from functional explanations. In *Perspectives in Ethology*, ed. P. P. G. Bateson & P. H. Klopfer, Vol. 7, pp. 79–98. New York: Plenum Press.

Johnson, M. S. & Brown, J. L. (1980). Genetic variation among trait groups and apparent absence of close inbreeding in Grey-crowned Babblers. *Behav. Ecol. Sociobiol.* **7**, 93–8.

Kennedy, M. C. (1974). Midbrain regions relevant to auditory communication in the Tokay gecko, *Gekko gekko* L. PhD thesis, University of Rochester.

Krebs, J. R. & Davies, N. B., Eds. (1978). *Behavioural Ecology. An Evolutionary Approach*, 1st edn. Sunderland, MA: Sinauer Associates.

Lack, D. (1954). *The Natural Regulation of Animal Numbers*, 1st edn. Oxford: Clarendon Press.

Lack, D. (1968). *Ecological Adaptations for Breeding in Birds.* London: Methuen.

Leinders-Zufall, T., Brennan, P., Widmayer, P. *et al.* 2004. MHC class I peptides as chemosensory signals in the vomeronasal organ. *Science* **306**, 1033–7.

Lenington, S. & Egid, K. (1985). Female discrimination of male odors correlated with male genotype at the T locus: a response to T-locus or H-2 locus variability? *Behav. Genet.* **15**, 53–67.

Lewis, A. R. (1980). Patch use by gray squirrels and optimal foraging. *Ecology* **61**, 1371–9.

Li, S.-H & Brown, J. L. (1999). Influence of climatic variables on reproductive success in the Mexican Jay. *Auk* **116**, 924–36.

Li, S.-H. & Brown, J. L. (2000). High frequency of extrapair fertilization in a highly social, plural breeding species, the Mexican Jay (*Aphelocoma ultramarina*) revealed by DNA microsatellite markers. *Anim. Behav.* **60**, 867–77.

MacArthur, R. & Pianka, E. (1966). On the optimal use of a patchy environment. *Am. Nat.* **100**, 603–9.

MacLean, A. A. E. (1986). Age-specific foraging ability and the evolution of deferred breeding in three species of gulls. *Wilson Bull.* **98**, 267–79.

Marler, P. & Hamilton, W. J., III 1966. *Mechanisms of Animal Behavior*. New York: John Wiley & Sons.

Maynard Smith, J. (1985). In Haldane's Footsteps. In *Leaders in the Study of Animal Behavior*, ed. D. A. Dewsbury, pp. 347–54. Lewisburg, PA: Bucknell University Press.

Mayr, E. (1942). *Systematics and the Origin of Species*, 1st edn. New York: Dover. (Reprinted 1964.)

Mays, N. A., Vleck, C. M. & Dawson, J. (1991). Plasma luteinizing hormone, steroid hormones, behavioral role, and nest stage in cooperatively breeding Harris' Hawks (*Parabuteo unicinctus*). *Auk* **108**, 619–37.

McGill, T. E., Ed. (1965). *Readings in Animal Behavior*, 1st edn. New York: Holt, Rinehart, Winston.

McGill, T. E., Ed. (1977). *Readings in Animal Behavior*, 3rd edn. New York: Hold, Rinehart and Winston.

Mumme, R. L. (1992). Do helpers increase reproductive success? An experimental analysis in the Florida scrub jay. *Behav. Ecol. Sociobiol.* **31**, 319–28.

Newman, J. D. (1970). Midbrain regions relevant to auditory communication. *Brain Res.* **22**, 259–61.

Newman, J. D. (1972). Midbrain control of vocalizations in redwinged blackbirds. *Brain Res.* **48**, 227–42.

Orians, G. H. (1969). On the evolution of mating systems in birds and mammals. *Am. Nat.* **103**, 589–603.

Potts, W. K., Manning, C. J. & Wakeland, E. K. (1991). Mating patterns in seminatural populations of mice influenced by MHC genotype. *Nature* **352**, 619–21.

Schoech, S. J., Mumme, R. L. & Wingfield, J. C. (1996). Prolactin and helping behaviour in the cooperatively breeding Florida scrub jay (*Aphelocoma coerulescens*). *Anim. Behav.* **52**, 445–56.

Selander, R. K. (1964). Speciation in wrens of the genus *Campylorhynchus*. *Univ. Calif. Publ. Zool.* **74**, 1–224.

Thurber, J. (1933). *My Life and Hard Times*. New York: Harper Brothers.

Tinbergen, N. (1951). *The Study of Instinct*. Oxford: Oxford University Press.

Tinbergen, N. (1953). *Social Behaviour in Animals*, 1st edn. London: Methuen.

Vleck, C. M., Mays, N. A., Dawson, J. W. & Goldsmith, A. (1991). Hormonal correlates of parental and helping behavior in cooperatively breeding Harris' Hawks (*Parabuteo unicinctus*). *Auk* **108**, 638–48.

von Holst, E. (1957). Die Auslosung von Stimmungen bei Wirbeltieren durch "punktformige" elektrische Erregung des Stammhirns. *Naturwissenschaften* **21**, 549–51.

von Holst, E. & von Saint Paul, U. (1958). Das Mischen von Trieben (Instinktbewegungen) durch mehrfache Stammhirnsreiziung beim Huhn. *Naturwissenschaften* **23**, 579.

von Holst, E. & von Saint Paul, U. (1960). Vom Wirkungsgefuge der Triebe. *Naturwissenschaften* **18**, 409–22.

West-Eberhard, M. J. (1987). Flexible strategy and social evolution. In *Animal Societies: Theories and Facts*, ed. Y. Ito, J. L. Brown & J. Kikkawa, pp. 35–51. Tokyo: Japan Scientific Society Press.

Wilson, E. O. (1975). *Sociobiology*. Cambridge, MA: Harvard University Press.

Wynne-Edwards, V. C. (1962). *Animal Dispersion in Relation to Social Behavior*. New York: Hafner.

Yamazaki, K., Boyse, E. A., Mike, V. *et al.* (1976). Control of mating preferences in mice by genes in the major histocompatibility complex. *J. Exp. Med.* **144**, 1324–35.

5

Individuals, societies and populations

TIM CLUTTON-BROCK

Early days

Asked to write an autobiographical sketch by Lee Drickamer, I floundered. What to write about? How on earth to make sense of a loosely connected web of diverging interests, a series of fits and starts? But, reading Niko Tinbergen's entry in the first volume, I was reassured by his doubts. If he had difficulty in thinking what to do, doubts were probably the right place to start. It was reassuring to know that he, too, had not followed organised paths, that one foray had led to another, that things had somehow developed. But there were still other questions to resolve. How to balance personal and professional matters? "As best I can tell," one of the editors wrote on my first draft, "you have no interests or life outside your work." Hmph – but is it really sensible to share my more arcane interests with my colleagues? Will they understand my love of vegetable carving or my fascination with tenth century Persian pornography? And what to call it? I asked one of my ex-students. "How I Wasted My Time and Other People's Money" he suggested? Possibly not.

My family comes from the West of England and the Welsh Marches. As for many other people in this field, it all began for me with bird-watching, which was less a hobby than part of a way of life. My father was the headmaster of a large boys' school in Cheltenham in the west of England and, like many other academics, I was an only child. We moved between

Leaders in Animal Behavior: The Second Generation, ed. L. C. Drickamer & D. A. Dewsbury. Published by Cambridge University Press. © Cambridge University Press 2010.

Cheltenham and Pembrokeshire in the far west of Wales, where we had an isolated house on the Cleddau estuary. I spent school holidays in Pembrokeshire with curlews calling on the mudflats outside my window and wigeon grazing at the bottom of the garden and became a keen birdwatcher and naturalist. I fished for bass and mackerel, set long lines and static nets, sailed in summer and shot ducks in winter. Life had to be organised around the tides and I was always amazed when visiting friends did not immediately recognise that if you failed to catch the tide you usually had to wait till next day.

Pembrokeshire is known for its seabird islands – Skomer and Skokholm and, further out, the smaller Grassholm, covered in gannets. They carry large populations of puffins, razor-bills, guillemots and Manx shearwaters and the distinctive seabird smell of guano hangs around them. During the school holidays I visited the islands to watch birds and, later on, to help run the daily boat. I read most of the books of Ronald Lockley, the controversial Pembrokeshire equivalent of Frank Fraser Darling, and all those of Gavin Maxwell, the aristocratic otter lover of the western Highlands, and enjoyed Konrad Lorenz's popular account of animal behaviour, *King Solomon's Ring*.

At boarding school in Rugby, I had to choose between Classics, Humanities and Sciences when I was fourteen. I had no interest in Latin and was better with words than numbers so chose History, French and English Literature, specialising in History. The curriculum focussed on the period between 1485 and 1840, the formative period for the British constitution. Economic and social interpretations were fashionable and provided an explan-atory framework that made sense of what was going on: a sort of ecology of ideas. I found it satisfying to understand the economic and political forces that drove the ideological rhetoric but, though I was interested in history, the weekly outings of the bird club were the high points of my week. I joined the Brathay Exploration Group and spent part of my last summer holidays ringing petrels and shags on Foula, the most westerly of the Shetland Islands.

When I was seventeen, I won a place to read History at Cambridge and decided to take a year off. I arranged to work for the Zambian Game Department and, in the autumn of 1965, booked a berth to Beira in Mozambique on a Union Castle boat. Six weeks later I arrived in the Luangwa Valley, a network of game reserves and controlled hunting areas in north-eastern Zambia centred on the Luangwa River, covering several thousand square miles. Tall groves of grey-trunked mopane trees covered much of the floor of the Valley, interspersed with large open dambos with coarse grass and oxbow lakes closely grazed by antelope. The bird life was spectacular, and the Valley (as it was known in Zambia) carried some of the densest populations of large mammals in Africa. I worked for the redoubtable Johnny Uys, the chief warden of the Valley. Parts of the area were still lawless and poaching was widespread. The whole area was irregularly patrolled by parties of armed game guards, sometimes led by a ranger, and game guards were occasionally murdered by poachers and vice versa. I spent much of my time on two-week foot patrols: we would set off before dawn and walk until three or four in the afternoon and then make camp; fires had to be kept to a minimum to disguise our presence and, at night, we slept in the open or under rigged groundsheets if it was wet. The first night I was terrified and sleepless; the second, I was terrified and exhausted; and the third night I was simply exhausted and from then on slept without a problem.

Johnny Uys was an outstanding all-round naturalist and I learned to identify the birds and trees of the Valley from him – but the main focus was on the mammals. Elephant, hippo, buffalo and lions were abundant and, during the wet season, much of the time was spent on foot so that close encounters were common. The rangers and game guards were very experienced and handled them coolly, but over longer periods, risks stack up. When I revisited the valley fifteen years later, three of the five people I knew best – including Johnny Uys – had been killed by animals. Between patrols, I catalogued ivory that had been handed in, sorted museum specimens, totted up data for animal counts and analysed records of the distribution of mammals, birds and trees. Using Johnny Uys' data, I produced my first scientific paper, on the breeding of the previously unstudied rufous bellied-heron, *Butorides rufiventris*.

At the end of the year I returned to England to read History at Cambridge and found that the History Tripos still concentrated on British constitutional history between 1485 and 1840. Fresh from Luangwa, I wanted something else. Please, might I read Zoology? No. Please, please – and throw me out if I don't keep up? Certainly not. As an alternative, my understanding tutor, the plant ecologist Peter Grubb, suggested Archaeology and Anthropology. I decided to give it a try. The first-year course covered archaeology, social anthropology and physical anthropology (which would now be called biological anthropology). To someone who had spent the previous five years learning about the details of constitutional change over 400 years in the UK, a time-span that extended from eighty million BP to the present and covered the whole planet looked like big horizons and I was quickly hooked. In the long vacations, I went on archaeological expeditions to Afghanistan (to search the Helmand Valley for Neolithic settlements), to Tunisia (to map Stone Age sites) and to Greenland (to survey Viking longhouses), and rapidly learned enough to know that I did not want to specialise in archaeology.

Social anthropology was in the throes of structuralism and the idea that human behaviour was adjusted to environments was regarded as overly simplistic. When I persisted in asking questions about why societies differed, I was firmly told that I would never be any good at the subject if I persisted with my "materialistic" bias. As a result, I decided to specialise in physical anthropology, which was an eclectic mixture of human physiology, hominid palaeontology and primate behaviour. In the late 1960s, the standard text book was John Buettner-Janusch's *The Origin of Man* but, not long after it came out, BJ was imprisoned – first for running an LSD factory in his lab and, later, for attempting to poison the judge that sentenced him. Interesting times. Research on primate behaviour was proliferating and was just starting to ask exactly the kind of 'materialistic' questions about relationships between the 'structure' and organisation of social groups that were discouraged in social anthropology. Irwin Devore's *Primate Behavior* had been published in 1965 and included studies of most of the species that had been worked on systematically. At Cambridge, Robert Hinde had supervised Jane Goodall's PhD on chimpanzees and had recently taken on a second student, Dian Fossey, to work on gorillas.

Physical anthropology was based in the Duckworth Laboratory and had very few students. There were rows of Hadza skulls and more than a whiff of Gormenghast about the place. The head of the laboratory was the portly and erratic Jack Trevor but, in practice, the lab was run by the demonstrator, David Pilbeam, who specialised in research on Miocene and Pliocene apes and later held chairs in anthropology at both Yale and Harvard. The students who elected to read physical anthropology were a small group of enthusiasts, some of whom later became well known in the field: Ian Tattersall, David Chivers and Alison Richard.

The relatively narrow focus of the courses in physical anthropology gave me time to read widely outside the immediate subject. I spent two weeks reading Ernst Mayr's *Animal Species and Evolution*, with its enormous scope and great depths of knowledge. I went for supervisions in animal behaviour with Pat Bateson and read most of the first edition of Robert Hinde's brown textbook on animal behaviour which preceded the larger, blue second edition – *The Great Blue Bible* as it was known to Hinde's students. I read David Lack's books on population dynamics and regulation in birds and found them fascinating – as well as a number of standard ecological texts like Odum's *Fundamentals of Ecology*, which I found less interesting.

I was in no doubt as to where the most exciting work on animal societies was going on. Working at the Edward Grey Institute at Oxford, John Crook had explored the social behaviour of African weaver birds and had shown that the size and structure of colonies was closely related to the habitats they occupied. Forest-dwelling species were mostly solitary or nested in small groups, species living in dry woodlands lived in intermediate size colonies, and the largest colonies were found in species living in arid savannahs. He had recently moved to the Department of Psychology at Bristol and was turning his attention to primates. With Steve Gartlan, he had recently (1966) published a paper that related the size and structure of primate groups to differences in the habitats they occupied, arguing that forest species lived in small groups whereas the largest groups were found in terrestrial species living in open savannahs and open areas. As in his previous work on weaver birds (and virtually all other comparative work at that time), he demonstrated relationships between species differences and ecological variation by drawing up lists of species allocated to different habitat groups, and his papers contained no statistical tests.

I had read most of the primate studies and knew enough to realize that the size and structure of primate groups was not as closely related to habitat type as that of weaver birds and that Crook and Gartlan's scheme was over-simplified. Quite a number of forest-living, arboreal species lived in large groups whereas some savannah-dwelling terrestrial species, like patas monkeys, lived in quite small ones. For my final year dissertation, I collated the information on group size and range size available for primates and explored relationships with a range of ecological variables, including timing of activity, diet type and habitat type. My supervisor, the population geneticist John Gibson, saw no reason not to use formal statistics to test the significance of differences in behaviour between groups of species, and patiently guided me through my first statistics tests. I think this was one of the first applications of formal statistics to interspecific data of this kind – but neither John nor I was aware of this at the time. Our analysis showed that there was no simple relationship between habitat type and the structure of primate societies: different aspects of primate societies were related to different ecological parameters and some – like the adult sex ratio – did not appear to be consistently related to any of the ecological parameters.

After finishing my degree, I cast around for what to do. I went to see Robert Hinde, who was starting to send students to work with Jane Goodall in the Gombe National Park in Tanzania. Jane had students working on baboons and chimpanzees and suggested that I might work on red-tail monkeys, *Cercopithecus ascanius*. Little was known about the forest-living guenons although the social behaviour of vervet monkeys (then *Cercopithecus aethiops*) had been studied in detail, so a study of red-tails offered an interesting comparison of an arboreal forest-living species with a congeneric terrestrial savannah-dwelling species. Robert managed to find me a studentship and I started to prepare for two years' fieldwork at Gombe. Robert encouraged me to investigate the effects of feeding behaviour on social behaviour in detail and gave me Ian Newton's papers on feeding behaviour in finches to read. I also spent several days working with his assistants observing the rhesus macques that Robert was using for studies of mother–offspring relationships. Quantitative data on the social interactions and activity of individuals were collected at regular intervals during timed watches. The recording techniques were carefully designed and highly organised and Robert encouraged me to adapt them for work in the field.

Primate societies

In the autumn, I flew to Nairobi and then on to Kigoma, arriving in the middle of a thunderstorm. Down to the harbour, where the blue hills of the Congo were clearly visible twenty miles away on the opposite side of Lake Tanganyika, then two hours up the lake by boat to the National Park. On both sides, the lake was edged with the hills of the Rift escarpment, which ran down to rocky pebble beaches. At intervals, there were clusters of thatched huts used by fishermen who netted small fish (dagaa) at night, using lanterns, and dried them on the shingle by day, doing their best to keep the baboons away. The hills were bisected by steep-sided valleys with streams that ran into the Lake. Outside the Park, the lower slopes of the hills had mostly been cleared and were covered in long fire-resistant grasses, but within it the forest still came down to the water's edge. Jane's camp consisted of the main hut and feeding station in a clearing in the forest, half a mile up from the lake, and a collection of tin huts built at intervals around the path up from the shore. There were a dozen students and assistants working on the chimps and baboons. The team consisted of biologists and anthropologists from the USA, several of them from California, and rather more bookish types (me among them) sent out by Robert Hinde from Cambridge. Living conditions were primitive, the camp was very isolated, and social relations between members of the group were strained.

I set about trying to find red-tail monkeys. Walking up the ridges between valleys, you could see over the forest and sometimes, very occasionally, you could see red-tail monkeys. They lived in groups of five to ten which appeared occasionally above the forest canopy and then dipped down and disappeared from view. If you left the slopes and tried to follow them you were trapped in thickets of thorny vines; if you tunnelled through the undergrowth, the monkeys quickly disappeared. It all looked like a dead loss. There were also larger groups of 20–80 red colobus monkeys, which were more visible than the red-tails. I abandoned my original plans and decided to work on the colobus, which had also not yet been studied. Like the red-tails, they ran away as I tunnelled through the thick undergrowth on the forest floor, but I eventually solved this by gridding the group's range with over twenty miles of paths cut through the undergrowth at hundred yard intervals, and the monkeys gradually became used to my presence.

The obvious question was why red colobus lived in large multi-male groups, unlike the common or black and white colobus found in other parts of east Africa, which lived in harem groups of 5–10. My guess was that this must have something to do with the distribution of their food, so I set about mapping their ranging behaviour and measuring what they fed on. I adapted Robert Hinde's check sheets to estimate time spent feeding on different types of food and used focal sampling to monitor social interactions. Quantitative studies of feeding behaviour were common for birds but, at that stage, there had been few attempts to quantify the feeding behaviour of primates. It seems strange now, when quantitative techniques are standard, but I believe that this was one of the first times that check sheets were used in primate field studies. For my last four months in Africa, I moved to Kibale Forest in western Uganda, where there were both red colobus and the commoner black and white colobus, and carried out a brief comparative study of the two species. As at Gombe, the red colobus lived in large, multi-male groups, fed primarily on flowers and fruit, used a wide variety of different trees and had a sizeable home range. In contrast, the black and white colobus lived in small groups of 5–10, in home ranges less than a quarter of the size of those of the red colobus, and fed almost exclusively on three tree species, relying on mature leaves when reproductive parts were not available.

In 1970, I returned to Cambridge and spent a laborious year-and-a-half analysing my data. Computers were not yet widely available and quantitative data had to be extracted by hand. Buy a double folio sheet of paper, and divide it into rows and columns, subdivide each row and column into two and use a clutch of different coloured pencils. Now it was possible to allocate records of the frequency with which animals of different age and sex categories were seen eating different parts of each food species at different heights and sites at different times of day. Statistical tests were calculated by hand, using slide rules or, if you were lucky, an electro-mechanical calculator. These had rows of numbers on a revolving drum that click-clacked round and round, generating a lot of heat and noise, and eventually came up with your answer. The first electronic calculators came out while I was writing and could add, subtract, multiply and divide and even do roots and logs. They measured 18″ by 18″ and cost the equivalent of a half-year's salary for a post-doc.

I started to draft chapters and gradually tightened them up under Robert Hinde's expert guidance. Robert was the leading ethologist at the time and, while he was always helpful and very generous with his time, his comments could be caustic. You had to learn to write sense – fast. For part of this time I shared a room with Dian Fossey, who was also supervised by Robert. After Dian had had a draft returned by Robert there was often much slamming of

(a)

(b)

Figure 5.1. (a) An unusually clear view of a male red colobus in the Gombe National Park. (b) The more usual view of a group of sixty monkeys.

doors and occasional floods of tears. Dian, who was over six foot and no lightweight, referred to everyone in diminutives. Unsurprisingly, I was Little Tim, Pat Bateson (more surprisingly) was Little Pat. All except Robert Hinde – who was definitely and consistently Big R. Dian only recognised one silverback.

Gradually I built up a quantitative description of the monkeys' use of different foods. I suggested that, for energetic reasons, the red colobus required a relatively high-quality diet of the reproductive parts of plants and that this caused them to feed on different tree species in different months. The need to maintain access to a wide range of tree species, in turn, required them to occupy sizeable home ranges – which could support a relatively large number of animals. There were likely to be advantages to living in large groups, (including improvements in the animals' ability to detect and defend themselves against predators) and few energetic costs, since all individuals required access to a large range. Black and white colobus, in contrast, could subsist on a smaller range of species, eating mature leaves when reproductive parts were unavailable – which probably explained why, unlike red colobus, the species was able to colonise dry woodlands that showed a high degree of phenological synchrony. Since individuals did not require large ranges to maintain access to a diverse diet, and increases in group size imposed energetic costs, the most efficient way of life was for them to live in small groups, in small territories which they could defend effectively.

I finished my thesis and was very aware of the need to get my work into print. What to do? Which journals publish rapidly? I knew nothing about choosing journals – but *Nature* came out each week so it seemed reasonable to suppose that it must be faster than *Animal Behaviour* or *Animal Ecology*. Naively, I put together the core of my work as a full-length article for *Nature* and sent it off. A couple of months later, I got an acceptance note and a list of minor required changes. It all seemed unremarkable: the article appeared in due course (Clutton-Brock 1974) and I thought little of it. It was only much later that I realised how enormously lucky I had been. Though my paper was critical of Crook and Gartlan's (1966) paper, I have always supposed that John Crook must have generously persuaded *Nature* to publish it.

Red deer and Rum

My time at Gombe had taught me – repeatedly and forcefully – the limitations of working on arboreal monkeys in tropical rain forests. It took months to habituate groups. At any one time only a fraction of the animals of the group were visible and, even then, they were fifty to a hundred feet up in the forest canopy. I was keen to investigate the ecological mechanisms limiting population density and the extent to which they were affected by the form of breeding systems. Research on population dynamics in terrestrial vertebrates had focussed principally on birds and, as yet, little was known about the population dynamics of large, long-lived mammals.

I decided to investigate the possibility of working on red deer, the largest terrestrial mammal in Britain, usually regarded as conspecific with American elk. There were thought to be between 150,000 and 200,000 red deer in the Highlands and many of the large estates were maintained principally for deer stalking, but there had been no extensive study of their behaviour since Frank Fraser Darling had carried out a famous but largely non-quantitative study in Wester Ross in the 1930s. After several hundred years of clearance, regular burning and overgrazing, Scottish red deer lived in a virtually treeless environment where visibility was good even if the weather was often frightful. Government scientists had recently started to document their feeding ecology and population demography and there was interest in the possibility of farming them. It looked like a good opportunity to explore the interactions between social behaviour and population dynamics.

After finishing my first degree at Cambridge, I had spent part of the summer making a 16mm film of the work of Roger Short and Gerald Lincoln on the physiology of reproduction in red deer stags on Rum, in the Inner Hebrides. Rum was a Natural Nature Reserve, run by the Nature Conservancy, who were keen to foster ecological research on the deer. I immediately thought of basing the study there. The island was around eight by five miles and held around twelve hundred deer. By 1972, Roger Short's work had changed direction and he was now working principally on human reproduction, though he was still supervising the physiology of reproduction in female deer. Fiona Guinness had joined him and had learned to recognise most of the female deer using a quarter of the island; she was monitoring their breeding cycles and reproductive success. I visited the Nature Conservancy in Edinburgh and Inverness and arranged with them to stop the annual cull in the quarter of the island where Roger's group had been working so that we could examine the effects of rising population density on demography and behaviour.

I wrote an application for a NERC post-doctoral fellowship – and got it, which gave me two years salary at £1,200 per year and some field expenses. I decided to take it up at Oxford and to join Hans Kruuk, who had recently completed his work on spotted hyenas and was now starting to work on badgers and foxes. Niko Tinbergen was still the head of the Animal Behaviour Group but was now working on autism in children and was seldom around. Richard Dawkins was the group lecturer – his work at that stage was mostly concerned with the details of behavioural sequences and he appeared at intervals wreathed in perforated computer tape. Mike Cullen, from Psychology, was closely associated with the group and was a continuous source of stimulation, challenging us to focus on testing hypotheses rather than on descriptions. There was also a remarkable array of students and post-docs, many of whom remained in the field – including Larry Schaffer, Tim Halliday, Linda Partridge, Felicity Huntingford and David Macdonald. I settled in and then headed for Rum, where Fiona had agreed to stay on to help with the work.

Rum lies a dozen miles off the west coast of Scotland. It is mountainous and, in winter, its treeless slopes suffer the full force of Atlantic gales. When I arrived, I found, to my horror, that Fiona had arranged to go on a year-long trip to Indonesia starting in two weeks' time and that I had thirteen days to learn to recognise two hundred (mostly unmarked) animals in the study area. In ferocious weather, we trekked around the hills of the north of Rum, stopping to identify the individuals in each group of deer we came to. In the half-dark of a Hebridean afternoon in a force eight gale, with driving rain plastering binoculars and telescopes, Fiona would point to the barely visible outline of one hind after another. The characteristics that she used to identify the hinds were fine differences in coat colour, ear shape and physical conformation. "That's Myrtle – she has a pale coat" (it was black with rain), "shapely ears" (they were laid back) "and a grim expression" (her head was down and she was feeding). "And that's Cleopatra and her yearling – they have dark eyes, hairy ears and long legs." Cleopatra lay down immediately, facing away from me. The hinds were continuously on the move and individuals would be quickly lost among the other ten or twenty animals in the group as I struggled to write down what Fiona had said. Panic – of a very sodden kind. However, we kept going and gradually I learned the cues that Fiona was using. By the time Fiona left, I was a nervous wreck – but I was able to recognise almost all the hinds that regularly used the 12 km^2 northern part of the island where we worked.

It was immediately obvious that it would be much easier to collect quantitative data on all aspects of ecology and behaviour than it had been at Gombe. It was possible to recognise almost all the animals in the study area so that all data could be collected on recognisable individuals. You could sit and watch individuals for as long as you wished, so that focal sampling was easy. It was possible to walk set routes on a regular basis and to record the location of individuals, the size of groups and what plant community each animal was feeding on. Hind groups had restricted ranges so that it was usually possible to locate fifteen to twenty groups and record their members. Finally, Fiona had found that it was possible to catch, weigh and mark most of the calves born in the study area and to find the carcasses of most individuals that died, so that it was feasible to track the complete life-histories of individuals and investigate the factors affecting breeding success and survival.

I set up regular sampling schedules to assess the extent to which the deer used different plant communities and check sheets to measure the frequency of interactions, systematising the collection of information on reproductive success and life-histories, and began to use the newly developed statistical package SPSS. Over the year, I alternated between Rum and Oxford, spending rather over half my time on the island. In summer, it was idyllic and in the

Figure 5.2. Red deer on Rum in the rut.

Figure 5.3. The author collecting ticks from a tame hind on Rum (1975).

evenings I fished for mackerel and trout and set pots for crabs and lobsters. Fiona joined me back on Rum the next autumn. Late in my first year on Rum, I heard that the University of Sussex was advertising a lectureship in animal behaviour. I was just settling in at Oxford and on Rum and the last thing I wanted to do was to move, but my fellowship was only for two years so I decided to apply. I was called for interview by Richard Andrew and offered the job, which I decided to accept. I was sorry to leave the way of life I had been developing on Rum and in Oxford but it was a lucky decision.

Sociobiology and interspecific comparisons

At Sussex, I was technically in Richard Andrew's group but most members of the group were working at the interface of physiology and behaviour and their interests were different to mine. However, Paul Harvey had been appointed at the same time to a post in John Maynard Smith's group and we quickly discovered that we had very similar interests. Craig Packer, Anthony Rylands, Paul Greenwood and Georgina Mace joined us as PhD students.

During the three years that I spent at Sussex a revolution was occurring in animal behaviour. As a research field, the primary links of animal behaviour had been to physiology and psychology, studies of causation, motivation and development had predominated and few people devoted much time to working on the function and evolution of behaviour. For example, Robert Hinde's major review of animal behaviour in 1970 had devoted less than 5% of its length to studies of evolution and function. Connections to population genetics and the formal study of evolution were poorly developed and many evolutionary explanations of animal behaviour relied either implicitly or explicitly on group selection.

Between 1965 and 1975 the situation changed radically. In 1962, VC Wynne Edwards, Regius Professor of Natural History at Aberdeen, had published an enormous review of animal social behaviour (*Animal Dispersion in Relation to Social Behaviour*) in which he claimed that animals commonly regulate their numbers in advance of resource limitation through a variety of social mechanisms in order to avoid over-exploiting their food supplies. Unlike many functional explanations at that time, Wynne-Edwards was clear about the evolutionary mechanisms he was invoking. Adaptive regulation of population density presumably occurred, he argued, because groups or species that failed to limit their numbers exhausted their resources and died out, to be replaced by groups that avoided over-exploiting their resources. Wynne-Edwards' theory irritated both ecologists like David Lack (who believed that animal numbers were usually limited directly by resource avail-ability) and evolutionary biologists like George Williams and John Maynard Smith (who pointed out that selection within groups would operate against individuals that restricted their reproductive output in advance of resource limits.)

Wynne-Edwards' book stimulated G. C. Williams to write a clear and critical review of theories of adaptation that pointed out the flaws in group selection explanations of social behaviour (Williams 1966). Two years earlier, Bill (W. D.) Hamilton had established the framework of modern interpretations of social behaviour with the description of inclusive fitness and kin selection (Hamilton 1963, 1964) and over the period that I was at Sussex (1973–1976) theoretical concepts from population genetics, ecology and animal behaviour were being combined to produce a new research field. Seminal papers by Bill Hamilton, John Maynard Smith and Geoff Parker proved a basis for understanding the evolution of strategies of competition and cooperation and the origins of sex differences in morphology and behaviour. It was clear to all of us that rapid changes were occurring in our under-standing of animal behaviour, and there was excitement in the air.

In 1975, E. O. Wilson's *Sociobiology* synthesised the new conceptual framework that was developing but made no systematic attempt to assess the extent to which it explained the distribution of behaviour across species. One of the most obvious ways of testing its predictions was to investigate whether it could account for interspecific differences in social and reproductive behaviour and related characteristics. I dusted off the quantitative analyses of primate social behaviour that I had done as an undergraduate, brought the data sets up to date, and started to examine the extent to which species differences in social behaviour were related to group size and body size. Paul Harvey was interested in the results, and together we set out to explore systematic relationships between species differences in social behaviour, ecology, breeding systems and sexual dimorphism.

Our first analyses of interspecific data examined the distribution of species differences in group size and range size, showing that variation in group size was related to the risk of predation while range size was affected by group size and body size as well as by resource availability (Clutton-Brock & Harvey 1977b, 1978c). We then went on to investigate whether sex differences in body size and the relative size of male weaponry were consis-tently related to the nature of breeding systems, and explored the ecological correlates of variation in body size and relative brain size. The use of quantitative interspecific data to test evolutionary hypotheses was novel and not everyone was enthusiastic. Solly (subsequently Lord) Zuckerman, doyen of primatology and (ex) Chief Scientific Adviser to HM Government, launched a bitter attack on our work in a postscript to the re-publication of his (1932) *Social Life of Monkeys, Apes and Man* (Zuckerman 1981), in which he also panned field studies of primates in general and the work of Jane Goodall and other female primatologists in particular. I was rather gratified when, a few years later, I won the Zoological Society of London Scientific Medal, which was presented by the President – Solly Zuckerman. The official photograph showed us shaking hands but obviously both gritting our teeth.

Though analysis of interspecific data was (and is) an important component of attempts to test the generality of evolutionary explanation, it has its problems. Values for closely related species may not be independent of each other and it is often necessary to control for the effects of body size. Exactly how this is done can affect the distribution of values and the conclusions of the analysis. As the statistical complexity of analyses increased and the assumptions involved became less obvious, my interest began to wane. While Paul continued to develop analytical methods for controlling for taxonomic effects (and later produced a seminal book on the analysis of interspecific data (Harvey & Pagel 1991), I concentrated more on the red deer study on Rum, which had now been running long enough to bear fruit.

Trying to combine running a major field study of deer on Rum with the duties of a university lecturer at Sussex was difficult and I hankered after the freedom that I had previously had to time my movements to the needs of fieldwork. At Cambridge, Pat Bateson and Nick Humphrey persuaded King's College, which funded research groups in its Research Centre, to set up a group in sociobiology. I applied and was elected as a Senior Research Fellow for four years in 1976, together with Brian Bertram, Dan Rubenstein, Adam Lominicki and Richard Wrangham. Geoff Parker and Robin Dunbar also joined us for part of the time. Freed from the demands of teaching and administration, I could focus on developing the deer studies. The group provided great stimulation; I remember this as one of the happiest periods of my life.

There were other reasons too. Shortly before I returned to Cambridge, I met Dafila Scott, who was finishing off a PhD on Bewick's swans. Dafila had been brought up at the Wildfowl Trust at Slimbridge and had learned to recognise over 500 individual swans by the pattern of black and yellow on their faces. Her research was exploring the effects of parents on the behaviour, development and survival of their offspring and her interests were very similar to mine. Dafila came to visit me on Rum and I visited her study site at Welney; after I moved to Cambridge, Dafila and I lived together in King's. We married in 1980 and bought a dilapidated sixteenth century house on the edge of the Fens, where we have lived ever since. Our daughter, who is now a vet, was born the following year and our son two years later. Looking after them made it hard for Dafila to continue research and she began a career as an artist, initially illustrating books and later specialising in pictures of wildlife and landscapes. When I'm involved in fieldwork, Dafila usually comes too and paints or draws. We are also keen divers and fish watchers and try to fit in one diving or snorkelling trip each year.

Figure 5.4. Dafila and I shortly after our marriage (1980).

Sex differences and selection on females and males

The red deer study provided an opportunity to collect unusually accurate records of growth, reproduction and survival in large numbers of individuals throughout their lifespan. Fiona Guinness rejoined me on Rum in 1974 and spent much of the year there for the next thirty years, while in 1975 Steve Albon joined me as a research assistant and subsequently stayed on as a PhD student and postdoc, playing a central role in the demographic analysis. Our earliest investigations of the ecology of red deer focussed on the social and ecological factors affecting the breeding success of females. We showed that the timing and weight of calves at birth exerted an important influence on their growth, survival and eventual reproductive success. We measured the relative fitness costs of gestation and lactation (Clutton-Brock *et al*. 1989a) and showed that, as mothers approached the end of their lives and their future chances of successful reproduction fell, they invested a higher proportion of their resources in their calves (Clutton-Brock 1984). We also explored the effects of group size on the breeding success of females and the survival of their offspring (Clutton-Brock *et al*. 1982a).

Red deer are an archetypal polygynous species and the deer study provided an ideal opportunity to compare the selection pressures operating on males and females. Less than half of all mature stags breed successfully during their lives and there is intense competition for harems of females. Stags often become involved in escalated fights and we showed that they suffer a high risk of temporary or permanent injury (Clutton-Brock *et al*. 1979) and that they minimise the frequency of fighting by using roaring contests to assess rivals (Clutton-Brock & Albon 1979), providing one of the first examples of 'honest advertisement'. Later on, the work on roaring was developed by Karen McComb, who showed that roaring also helps to induce oestrus in females and that females pay more attention to stags that roar frequently. The different components of our work on red deer showed how the contrasting selection pressures operating on females and males had led to the evolution of sex differences in adult body size, growth, timing, habitat size and diet which, in turn, affected the impact of food availability and population density on their survival. In 1982, we drew together our results in a monograph that directly compared the behaviour and ecology of females and males (Clutton-Brock *et al.* 1982b).

In the early 1980s, genetic techniques for identifying paternity were developing rapidly; in 1984, Josephine Pemberton joined the project and began to investigate variation in male success based on genetic measures of paternity using electrophoretic techniques and, later, using DNA 'fingerprinting'. We showed that, as predicted, variance in lifetime breeding success was larger in males than females. In 1988, I edited a book, which drew together similar studies in other animals and compared the selection pressures operating on females and males in other breeding systems (Clutton-Brock 1988).

Later on, Loeske Kruuk joined the deer project and we began to explore the effects of genetic variation on phenotype and breeding success and the heritability of different traits. We showed that there was heritable variation in most fitness components as well as in body and antler size, though countervailing selection was common and lifetime reproduction success showed lower heritability than specific components of fitness. There was interest in the possibility that asymmetry of phenotypic traits ('fluctuating asymmetry' or FA) was heritable and related to breeding success and that female deer might select their mating partners on the basis of the symmetry in their horns or antlers. We carried out a detailed analysis of the symmetry of antlers in the deer and showed that FA in antler size and shape was neither correlated with fitness nor heritable (Kruuk *et al.* 2003). When we submitted our paper, we narrowly resisted titling it 'FA shows FA'!

Parental investment

One day on Rum, Marion Hall, who had been studying maternal behaviour in female deer in an enclosure on Rum, came up to me. "You know, Tim," she said, "I think hinds are more likely to fail to breed after they have raised a son." I had a look at the data, which seemed to bear this out – but the sample was small. Was this the case in the whole population? It was. Moreover, hinds were more likely to die after rearing sons than after rearing daughters – especially if they were relatively low in dominance ranking (Clutton-Brock *et al.* 1981; Gomendio *et al.* 1990). We measured the sucking behaviour of males and found that male calves sucked almost twice as frequently as females. As in many other mammals where adult males are substantially larger than females, males grew faster than females and required more food. When Dafila and I produced a son in 1983, we feared the worst and Dafila kept a record of how many times she had to breastfeed him each night. It needed no statistical comparisons to show that, until we eventually ignored his demands, our son required feeding more than twice as often as our daughter had at the same age.

In the deer, large dominant females were less affected by producing sons than smaller subordinates, whose sons tended to be poorly developed and often died before reaching maturity or failed to breed successfully as adults. This had the effect that the sons of dominant mothers had higher breeding success than their daughters – while the reverse was true for subordinates. Ten years earlier, Bob Trivers and Dan Willard had published a paper predicting effects of this kind in polygynous species and suggesting that, where they occurred, superior females should bias the sex ratio of their offspring towards sons (Trivers & Willard 1973). So did the deer do this? Yes: dominant hinds produced 65–75% males at birth, subordinates 35–45 (Clutton-Brock *et al.* 1984). However, further work showed that several processes may affect the sex ratio at birth and not all of them are necessarily adaptive. When Loeske Kruuk re-examined sex ratio variation in the Rum red deer after deer numbers had reached carrying capacity and the population was resource-limited, she found that the (average) sex ratio of calves born each season varied with density and winter weather, presumably reflecting the extent of sex differences in embryonic mortality (Kruuk *et al.* 1999a) – an effect that was absent in the early years in the project when numbers were below ecological carrying capacity.

By 1978, the sociobiology group at King's was ending and its members were dispersing. I decided to stay on in Cambridge and won a BBSRC fellowship and then a Royal Society Senior Research Fellowship. In 1987, I joined the Department of Zoology and have been there ever since. Our research on parental investment in sons and daughters in the deer stimulated my interest in other aspects of the evolution of parental care and I produced a monograph reviewing evolutionary aspects of care, which was published in the Princeton series that John Krebs and I edited (Clutton-Brock 1991). Writing broad reviews often highlights questions that need answering and one important question that was still unanswered concerned the precise relationship between sex differences in parental care, mating competition and sexual selection.

In 1972, Bob Trivers had argued that sex differences in competition were a consequence of differences in parental investment but his definition of parental investment was difficult to apply and his theory was commonly taken to suggest that sex differences in competition were a consequence of the relative contribution of the two sexes to parental care. Since in many fish and some amphibians, males are responsible for all parental care and still compete more intensively for breeding opportunities than females, there was obviously something wrong. Listening to Ingrid Ahnesjo and Anders Berglund's description of their work on pipefish at a meeting in Stockholm, I realised that, in species where males cared for young in nests, they often cared for several clutches of eggs at the same time whereas where they cared for eggs in pouches, they were often limited to a single clutch or even part of the clutch. This suggested that where males care and compete they might be able to process clutches of eggs faster than females can produce them while in species where males care and females compete, females may be able to produce clutches faster than males can care for them. A review of the literature showed that this was indeed the case (Clutton-Brock & Vincent 1991) and, with Geoff Parker, I went on to argue that sex differences in potential rates of reproduction in the two sexes (which depend on relative levels of parental investment) determine operational sex ratio and the comparative intensity of mating competition in the two sexes (Clutton-Brock & Parker 1992). Several students came to work with me on species with unusual care systems: Amanda Vincent came to work on sea horses, Peter Brotherton on dik dik, Rebecca Kilner on canaries, Sigal Balshine-Earn on cichlid fish, Lisa Thomas on assassin bugs and Andrea Manica on sergeant-major fish.

Population dynamics and demography

While the Rum deer study provided unique opportunities to compare the operation of selection in males and females, its principal aim was to determine how population density was regulated in large, long lived polygynous animals that did not defend feeding territories, unlike the monogamous birds which had been the principal focus of previous work on population regulation in vertebrates. Following the termination of the annual cull in 1972, deer numbers in our study area rose until the early 1980s and then stabilised, giving us an opportunity to investigate how changes in population density affected fecundity, survival and emigration and to assess the contribution that different changes made to limiting population size. We showed that juvenile mortality was strongly density-dependent and was the principal factor limiting numbers. Whereas juvenile mortality increased with population density, birth weight and neonatal survival were more strongly influenced by average temperatures in late winter, which affected the onset of plant growth. A female's birth weight influenced the birth weight and survival of her offspring; Steve Albon was able to show that variation in spring temperature generated cohorts of successful and unsuccessful mothers (Clutton-Brock *et al*. 1987).

We also found that rising population density affected the survival of juvenile males more than that of juvenile females, increased the proportion of adolescent males that emigrated, and depressed the number immigrating. One consequence was that, as numbers rose, the sex ratio of adults became progressively biased towards females and there was evidence that a similar trend occurred in other deer populations. Most Scottish deer forests derive much of their income from shooting males and deer managers often argue that high female numbers are necessary to produce large numbers of stags that can be culled. Our results instead suggested that high hind numbers depress the production of males and that deer forests that allow female numbers to reach ecological carrying capacity are likely to be reducing their annual offtake of males (Clutton-Brock & Lonergan 1994; Clutton-Brock *et al*. 2002b).

As a result of our work on population regulation, we became involved in assessing the causes and consequences of variation in deer density throughout the Highlands. The Red Deer Commission (now the Deer Commission for Scotland) had been set up in the 1950s to monitor changes in deer numbers and resolve conflicts of interest over deer management between crofters and landowners. Their expert counting teams had covered almost all of the Highlands several times on foot, providing estimates of deer numbers that were used to set (and enforce) culling targets. Using the RDC's counts, Steve Albon and I mapped deer density across the Highlands and investigated the factors affecting it, producing a book that synthesised existing knowledge of red deer ecology and management in Scotland (Clutton-Brock & Albon 1989). Deer numbers were rising throughout the Highlands and the RDC attributed this to the failure of landowners to cull adequate numbers of females – but our analysis suggested that annual culls had not been high enough to limit population growth for many years. In many parts of the Highlands, deer compete for food with hill sheep and as, on average, there were six sheep for every deer in the Highlands, changes in sheep numbers looked as if they might play a crucial role in the dynamics of Highland deer populations. We were able to show that, where sheep numbers were reduced, deer numbers rose (Clutton-Brock & Albon 1992) and we concluded that declining numbers of sheep, combined with warmer winters, probably accounted for much of the increase in deer numbers throughout the Highlands (Clutton-Brock & Albon 1992; Clutton-Brock *et al.* 2004a; Hallett *et al.* 2004).

Cycling sheep

In contrast to red deer and many other long-lived mammals, populations of some rodents show regular or semi-regular cycles, during which population density can vary by more than two orders of magnitude. Explaining their cycles has fascinated mammalian ecologists for half a century but it is usually difficult to identify the relative contributions of changes in fecundity, survival and emigration to changes in population size. In the early 1970s, I went to a talk on vole population cycles by the doyen of rodent ecologists, Charles Krebs, and it immediately struck me that what was needed was a study of a cyclical population where it was possible to recognise and follow large numbers of individuals throughout the cycle. A study of this kind could also be used to investigate variation in the effects of changes in density on the costs and benefits of reproductive decisions as well as the demographic processes limiting population size.

Were there any mammals where this would be possible? A faint bell rang in my head. On St Kilda, the most remote of the island groups in the Hebrides, there was a naturally regulated population of sheep, introduced in the Bronze Age, which were thought to show semi-regular population cycles. The sheep varied in colour from buff to black and were mostly horned. They were confined to the main island, Hirta, and the smaller island of Soay. Peter Jewell, who was married to my cousin Juliet, had worked on them in the sixties and had edited a book on their ecology and demography (Jewell *et al.* 1974) which showed that there were semi-regular crashes every 3–5 years when up to two-thirds of all sheep in the population died in late winter. With Peter Jewell and Steve Albon, I visited St Kilda in May 1984 and returned in April the next year to tag lambs.

The islands of St Kilda lie to the west of the Outer Hebrides and are uninhabited apart from a military base (which tracks rockets fired from the Outer Hebrides), a summer warden and occasional work parties. The only way there was by the occasional helicopter or by monthly visits of the large flat-bottomed LCL (Landing Craft, Logistic) run by the army, which sailed slowly out from Benbecula, taking twenty hours to get to the island. It was made to take a squadron of tanks and their crews and accommodation was in tiered bunks between the inner and outer skins of the boat. If the weather was good, it beached on St Kilda to unload stores for the Army, but if it was bad, it often hove to in the shelter of the island or returned to Benbecula. The boat had a slow, lurching roll that could make even the hardiest sailor sick, and there was competition to avoid being allocated the bottom bunk.

In 1985, the sheep population was at peak and 300–400 sheep regularly used the Village Bay area. Each day we walked around the island looking for recently born lambs. When we saw them, we stalked them – and then ran them down and tagged them. On the flats behind the beach, it was possible to catch them if you ran flat out but even twelve-hour-old lambs could run up slopes faster than a man and the only way to catch lambs on the sides of the main bay was to get above them and then charge down at breakneck speed from above and grab them. In the course of a month, we tagged 90% of all the lambs born to the Village Bay population. Steve Albon and I set about organising a programme to investigate the causes of the population oscillations. We returned to St Kilda in April 1986 to tag the next season's lambs and found a scene of carnage. It was a crash year and over half of all the sheep on the island had died. In some of the stone enclosures that dot the lower slopes of Village Bay, the corpses were piled up four deep. A sickly stench lay over the whole bay. We counted the carcasses and removed jaws and skulls for cleaning. We also established a regular census of the sheep population that measured their use of different areas and plant communities.

It soon became clear that, while the sheep crashed every 3–5 years, crashes varied in magnitude and timing, and the "cycle" was not fully regular. Gradually, we worked out what was happening (Clutton-Brock *et al.* 1991, 1997; Clutton-Brock & Coulson 2002; Grenfell *et al.* 1992). Immediately after a crash, the survivors were in poor condition, relatively few lambs were born and neonatal survival was low but, by the autumn, they had regained condition, many females born that spring and almost all over a year old were pregnant and many had conceived twins. Few animals died in winter and lamb survival the following spring was high so that, by midsummer, numbers had nearly doubled. The situation was repeated the following year since, as a result of the early independence of lambs and the

Figure 5.5. Soay ewe and twins.

Figure 5.6. Sorting sheep heads after a population crash on St Kilda.

early cessation of lactation in their mothers, breeding females were able to regain lost condition during the summer months when food was superabundant. As a result, almost all individuals conceived each autumn and there was no decline in fecundity as numbers increased. By the autumn of the third year, the size of the population had often tripled, exceeding the number that could be supported by the winter food supply. By late winter, all animals were in poor condition, parasite levels were high and many individuals died in early spring. Mild winter weather could reduce the magnitude, allowing relatively large numbers of older animals to survive, who would be killed by the next bad winter (Grenfell *et al.* 1998; Coulson *et al.* 2001). Parasites played an important role and I collaborated with Frances Gulland, Bryan Grenfell and Ken Wilson to show that parasite populations increased with sheep density and contributed to variation in winter mortality.

The sheep study also offered unusual opportunities for studying selection. With Josephine Pemberton, we developed a technique for catching most of the sheep in the Village Bay population each year, involving a large team of volunteers and long nets that we staked out. This allowed us to weigh large samples of individuals each year. At all ages, light individuals are less likely to survive and the reproductive history is also important: female sheep are spectacularly fecund, commonly become pregnant when they are around six months old, and, as adults, often produce twins. In crash years, virtually all individuals died that were breeding at less than a year old or which were rearing twins (Clutton-Brock *et al.* 1996b). It seemed surprising that selection had not reduced fecundity and possible that their high fecundity was a consequence of human selection in the past. However, the costs of early breeding and producing twins disappeared in non-crash years and, when we modelled the consequences of different reproductive strategies, it turned out that the survival costs of high fecundity in crash years were more than offset by increases in breeding success during intervening years. After the effects of reproductive history had been allowed for, it was possible to examine selection pressures on morphological and genetic traits and we went on to explore selection on coat colour and on the size and conformation of horns ((Moorcroft *et al.* 1996; Clutton-Brock 2004) and to show that selection pressures were strongest in crash years.

Conservation and tropical biology

In the late 1980s I nearly left academia. Derek Ratcliffe, the Chief Scientist of the Nature Conservancy (the British government's conservation agency) retired and his job was advertised. I was interested in becoming directly involved in conservation issues and the only way to find out about the job was to apply for it – so I did, and was astonished when I was offered it. However, I eventually decided to stay on at Cambridge. I was glad I did because, within a year, the government had decided that the Nature Conservancy was exerting too strong an influence and split it into three national agencies.

Instead of moving out of academia, I started to spend part of my time on conservation-related activities. My involvement with field research in Africa had shown me how important it was that Western ecologists should have experience of research in tropical as well as temperate areas. In addition, there was an urgent need to transfer expertise in ecology and conservation biology to students from tropical countries, where teaching was often rudimentary. In North America, the Organisation for Tropical Studies ran regular courses for university students in Central America, providing an important source of training in tropical ecology – but there was nothing similar in Europe. I discussed the problem with Steve Stearns, then at Basel, and we agreed to establish a confederation of European biologists that would run regular training courses in ecology in the tropics. Steve put together a meeting of people who might be interested from across the European Union, which agreed to set up an organisation called the Tropical Biology Association, while I set about finding the necessary money. For two years I had no luck. Then the UK government established the Darwin Initiative to foster the conservation of biodiversity; I applied for a large grant in the first round and, with strong support from Bob May, got it. We decided that courses should be made up of around 50% of European students and 50% of students from Africa or other tropical countries and should be taught by the best teachers that we could find. Leon Bennun joined us as the first director of TBA and was later succeeded by Rosie Trevelyan.

The first courses were held in Kibale Forest in western Uganda and were an immediate success for both African and European students. Under Leon and Rosie's leadership, the TBA has gone from strength to strength, increasing the number of courses per year to four and running courses in Kenya, Tanzania, Madagascar and Malaysia as well as Uganda; it has now trained over 1000 students drawn from more than fifty countries. Many of its graduates are filling important positions and sending their own students and employees on TBA courses. The enthusiasm of the students from Africa and Europe has made my involvement with the TBA one of the most rewarding activities of my career.

I also became involved in the organisation of a research programme run by the Royal Society at Danum in Borneo, chairing the committee that ran it. Danum was one of the few large areas of unlogged rainforest remaining on the island and the diversity of animals and plants was astonishing. The Royal Society helped to run a large field station there, used by many plant and animal ecologists. However, while Danum was ideal for research on plants and insects, the density of vertebrates was low and visibility was poor, so I was never tempted to try to work there myself.

In Britain, the frequency of bovine tuberculosis (TB) was rising and badgers were implicated as the possible source of the disease. In many parts of the West Country, badger numbers had been rising and one view was that this was the main cause of the increase in TB in cattle. I joined a government committee chaired by John Krebs to look into the problem. A strong pro-badger lobby disputed the evidence that cows commonly caught TB from badgers and it was decided to set up a major experiment to compare the effects on TB rates of doing nothing, of controlling badgers in areas where TB infections occurred, and of proactive badger culling. I was asked by the Department for Environment, Food and Rural Affairs (DEFRA) to organise this but it was incompatible with my involvement in the long-term field studies so I declined.

Leks and the evolution of coercion

Through the 1980s and 1990s, I continued to work on mammalian breeding systems. I started to investigate the puzzling phenomena of leks. In a number of insects, some birds, a few bats and several ungulates, mature males defend small, clustered territories that hold virtually no resources. Receptive females come to these leks to mate, aggregating on a relatively small number of the available territories, and eventually mate with the males holding them. Studies of lek-breeding animals commonly assume that this unusual mating system is driven by strong female preferences for mating with superior partners – though it is not clear why female mating preferences should be particularly important in species that lek.

In four successive years, we worked on fallow deer leks in Petworth Park in Sussex (Clutton-Brock *et al.* 1988, 1989b) while my students, Andrew Balmford, James Deutsch and Rory Nefdt, investigated leks in Ugandan kob and Kafue lechwe. In all three species, it quickly became clear that females moving to leks showed strong preferences for territories that either contained females or had previously been heavily used by them. James and Rory hired gangs of labourers and exchanged the topsoil between preferred and little-used territories. This showed that adding soil from heavily used territories to territories that attracted few females substantially increased the mating success of the males defending them, suggesting that females preferred particular territories rather than particular males. However, when we forced male fallow deer to change territories to see whether the same individuals maintained their relative success at different sites (Clutton-Brock *et al.* 1989b) they did, suggesting that here, females might be attracted to particular males. It was still possible that the same males were successful on different territories because they were better than others at defending females rather than because females preferred them as mating partners (Clutton-Brock *et al.* 1992) so we set up an experimental arena in another deer park, induced female fallow deer into oestrus and tested their preferences for different categories of males. To our surprise, females showed no consistent preferences for particular males

Figure 5.7. A fallow deer lek in full swing in Petworth Park (Sussex).

although they showed a strong preference for joining males with other females – and when we offered them a choice of paddocks containing males and females or ones containing only females they showed no preference for the former.

Our experiments with fallow deer suggested that the tendency for females to collect in particular territories might simply be a by-product of a tendency for females to aggregate (Clutton-Brock & McComb 1993; McComb & Clutton-Brock 1994) and we started to explore the possibility that leks might not, after all, be driven primarily by female mating preferences. One possibility was that male harassment played an important role. All lek-breeding ungulates live in unstable mixed-sex herds where individual males cannot guard receptive females effectively and females approaching oestrus often leave their usual herds after they start to be harassed by males, joining territorial males that are able to protect them (Clutton-Brock *et al.* 1993). Perhaps clustered territories provide females with greater safety: James Deutsch showed that, in kob, females avoided leks where the grass was long enough to hide predators and that the relative popularity of territories was related to their distance from cover that could hide predators. Perhaps holding a territory close to the territories of other males might also have benefits to males because it allowed them to gain females that left their neighbours' territories. We modelled this process (Stillman *et al.* 1993) and showed that defending territories in clusters could, indeed, generate significant substantial benefits to males because females that had joined a cluster were likely to remain there until they mated. The more closely we examined the behaviour of lek-breeders, the less convinced we were that the whole system was driven by female preferences for mating with particular males and the clearer it became that many different factors could affect the movement of females and the mating success of males (Clutton-Brock *et al.* 1993, 1996a). However, to sort out what was going on required experiments of a kind that would not be feasible in ungulates and by 1995 I felt that we had made as much progress as we could.

Our work on male harassment of females threw up an interesting question. Was it really in the interests of males to harass females to a point where their survival was in jeopardy? Geoff Parker and I developed models of harassment and showed that male coercive strategies were often likely to evolve to a point where they had serious costs to females (Clutton-Brock & Parker 1995a, b) and we demonstrated that this was the case, using tsetse flies as models (Clutton-Brock & Langley 1997). Subsequently we went on to explore the evolution of other forms of coercion. In many social mammals, dominant individuals regularly attack subordinates that repeatedly infringe their interests. We suggested that this was best interpreted as a form of punishment and that punishing strategies were the mirror image of reciprocity (Clutton-Brock & Parker 1995a). Punishment may often provide a more effective way of controlling the behaviour of potential competitors than rewards since dominant individuals can often reduce the fitness of subordinates at little cost to themselves and even the threat of punishment is often sufficient to deter subordinates. In contrast, manipulating the behaviour of others through rewards is inevitably costly, and invites cheating. A quick dip into political philosophy showed that Machiavelli had been well aware of this contrast.

Cooperative societies

By the 1990s, I was beginning to feel that I had worked on polygynous animals for long enough and wanted a new research direction. Specialised cooperative societies posed an obvious challenge to evolutionary biologists, as Darwin himself had recognised. Why should many individuals spend most of their lives helping others to breed rather then breeding themselves? The usual answer was that helpers were close relatives of the breeders they assisted and that cooperative breeding was driven primarily by kin selection (see (Emlen 1991, 1995) but there was also evidence that, in many societies, some helpers were not closely related to breeders (Reyer 1980, 1984; Cockburn 1998). It was also necessary to explain why highly developed cooperative systems had evolved in some species but not in others. And then there was the problem of humans, where cooperation between unrelated individuals is normal and widespread. How had this developed?

There had been a number of excellent studies of cooperative birds but few detailed studies of cooperative mammals. It was not difficult to see why. Some – like the cooperative mole rats – spend their lives underground and so are difficult to observe in the wild. Others, like African wild dogs, have enormous ranges, so that detailed studies of more than one or two packs present problems. The diurnal mongooses looked a possibility – several are cooperative breeders and dwarf mongooses had already been studied in detail by Jon Rood and Anne Rasa. Around this time, I saw the famous BBC film of meerkats (*Meerkats United*) and realised that meerkats could be a serious proposition for a detailed study of cooperation. They lived in open country and evidently habituated closely to photographers, and they had recently been studied in the southern Kalahari by David Macdonald and his students. The cultural boycott of South Africa requested by the ANC had ended and there were several British research groups already working in the country.

I contacted John Skinner at the Mammal Research Institute in Pretoria and he arranged for one of his assistants to take me to the Gemsbok National Park (now Kgalagadi Transfrontier Park) in the South African part of the Kalahari to look at them. We drove across the western Transvaal to the Park and then north up the dry bed of the Nossob River, past grazing herds of springbok, wildebeest and gemsbok, to Nossob Camp. The following day, we went out to search for meerkats and eventually located a group that had been habituated by French photographers. They were entirely tame and foraged happily for insects around our feet, totally ignoring us. There was virtually no grass; they spent almost all the day in the open and could be followed without difficulty. They looked ideal. I returned to the UK and raised the funds necessary to start up the project.

The following year, I returned to South Africa, bought vehicles, hired assistants and set up one study site at Nossob and one on ranch-land 100 km to the south-east, close to Vanzyl's Rus. Progress in the Park was rapid – though most meerkat groups ran when approached, it was possible to follow them in a vehicle and gradually to reduce their tolerance distance. Within a year, we were able to walk with ten groups. On the ranch, they were far more scared of people and we were forced to rely on tracking groups and watching them at their sleeping burrows. Eventually, one group became so habituated that it allowed us to walk with it; as habituated animals dispersed to form other groups, the number of habituated groups increased until we had as many as in the Park. After they were used to our presence, the meerkats readily took crumbs of hard-boiled egg or water from a rabbit feeding bottle and we used this technique to weigh individuals repeatedly – providing detailed records of both their foraging success and growth. As time went on, the number of groups we worked with increased to six, ten and, eventually, fifteen.

Kalahari meerkats are specialised cooperative breeders. Groups vary in size from three animals to over thirty or more and one female in each group virtually monopolises

Figure 5.8. Dafila weighing meerkats in the Kalahari Gemsbok Park.

reproduction, breeding 2–4 times per year. One male defends access to her and fathers almost all her young. Subordinate females that remain in their natal group are regularly threatened and attacked by the dominant female. They show lower levels of sex hormones than dominants and seldom attempt to breed unless the group includes a male who is not closely related to them and the dominant female's capacity to control them is weak (Clutton-Brock *et al.* 2001b; Carlson *et al.* 2004; Kutsukake & Clutton-Brock 2006; Young *et al.* 2006). When subordinate females do breed, dominants commonly kill their pups if they are pregnant but spare them if they are not. After pups are born, they are cared for by all group members. Subordinate females often lactate to pups born to dominants and all adult group members apart from the dominant male contribute to guarding the breeding burrow, usually for a day at a time (Clutton-Brock *et al.* 2000; Clutton-Brock *et al.* 2002c). All individuals also help to feed pups during their first 2–3 months, adjusting the proportion of food that they find that they give to pups to the frequency with which they beg. As pups develop, helpers teach them to catch and handle dangerous prey, such as scorpions (Thornton & McAuliffe 2006). Group members take turns to go on sentry duty when the group is foraging, giving regular calls to let the rest of the group know there is a sentry on guard and graded alarm barks when predators are sighted (Clutton-Brock *et al.* 1999b; Manser *et al.* 2001). The group cooperates to attack predators, to rescue group members that are being attacked and to feed sick adults that remain at the sleeping burrow.

Contributions to cooperative activities affect weight gain, although individuals adjust the level of their contributions to their condition so that costs to fitness appear to be small (Clutton-Brock *et al.* 1999b, 2000, 2001a; Russell *et al.* 2002, 2003, 2004; Clutton-Brock *et al.* 2002a). So why do subordinates help rather than conserving their resources? Most group members are close relatives, so individuals that cannot breed themselves gain 'indirect' fitness benefits by helping to rear young produced by the dominant female. In addition, our work showed that the survival of all individuals increased with group size: members of small groups were unable to post sentries for most of the time the group was foraging, rates of predation were higher, and larger neighbouring groups raided deep into their territories, killing pups and babysitters when they came across them and gradually pushing them out of their territories. In addition, individuals are more likely to disperse successfully if they are part of a large splinter group – and only big groups can produce large splinters. It pays all individuals to maintain size of the groups they live in, and the combination of direct and indirect benefits is sufficient to ensure that all group members contribute willingly to cooperative activities (Kokko *et al.* 2001; Clutton-Brock *et al.* 2003, 2004b, 2005).

The cooperative behaviour of the meerkats affects all aspects of their reproductive strategy and ecology. Since all group members help to rear pups, dominant females are able to raise three or even four litters per year and can maintain their status for up to ten years, producing up to a hundred offspring in their lifetime – almost twice as many as have been fathered by the most successful male red deer that we have recorded on Rum! However, most females never gain dominant status and most die after they are evicted from their natal group. As a result, competition between female meerkats for the dominant position is intense and females have evolved a number of unusual characteristics: they are more frequently aggressive to each other than males, and dominant females can displace another group member. When females acquire the dominant position in their group, their testosterone levels increase and they show a secondary period of growth (Clutton-Brock *et al.* 2006). Several of these characteristics are found in females in other cooperative mammals, including marmosets and naked mole rats, suggesting that they may be a consequence of intense intrasexual competition between females (Clutton-Brock 2007b). I went on to examine the distribution of competition between females for breeding opportunities in other systems and to argue that its effects were very similar to the effects of mating competition between males in polygynous species (Clutton-Brock 2007b).

After six years' work on meerkats in the Gemsbok Park, the authorities had tired of hosting scientists working on a species that had no urgent conservation needs; we ended the research in the Park and concentrated all our efforts at the ranch. When one of the farmers that owned part of the land we worked on decided to sell up, I raised the funds to buy his ranch and so ensure the future of the work. We removed the domestic stock and bought antelope to replace them. To monitor breeding success and survival in multiple groups, we needed manpower, and the size of the resident team rapidly increased. Owning the study site allowed us to develop the facilities necessary to maintain a large project in an isolated area. We built and graded roads, made bricks and houses, dug sewerage systems, pumped our own water (from thirty meters below the sand) and killed and butchered our own meat. I'm regularly amused at the common perception that scientists are impractical dreamers living in the upper storeys of ivory towers.

Several students and post-docs joined me to take advantage of the unique opportunities offered by the meerkats. Marta Manser, and later Joah Madden, worked on auditory and

Figure 5.9. As the meerkats become tamer they often use us as shelter, or even as a convenient place to go on guard (2006).

olfactory communication; Justin O'Riain, Anne Carlson and Andy Young investigated the physiological basis of cooperative behaviour; Alex Thornton and Sinead English worked on aspects of development; and Andy Russell and Sarah Hodge worked on the reproductive strategies of females. With Rosie Woodroffe, Mike Cant, Jason Gilchrist, Sarah Hodge and Matt Bell I helped to develop a parallel long-term study of cooperation and reproduction in banded mongooses in Uganda and, with Mandy Ridley, studies of Arabian and pied babblers. In both cases, we were able to habituate individuals to observation from within a few feet and to weigh them whenever we wished. Habituated meerkat groups also provide marvellous opportunities for wildlife film-makers since the animals can be filmed from a few feet away. The *National Geographic* sent a team to shoot a half-hour television film, and the BBC did another. David Attenborough visited for footage for one of his mega-series and the obligatory shot with a meerkat on his shoulder. The Toyota Corporation filmed a commercial with us and gave us a Land Cruiser. Then one day my phone rang. It was Caroline Hawkins from Oxford Scientific Films (OSF). They were considering making a series on meerkats that followed the lives of individuals, an animal soap opera. Could I tell her where it could best be shot? I told Caroline that our study area was the ideal location and I agreed to collaborate on the understanding that the films would tell meerkat life as it really is – a mixture of extreme cooperation and ruthless conflict. Not too many cuddly images, a limit to the degree of anthropomorphic interpretation and no claims that meerkat society provides an ideal model for humans. But I was sceptical that it would be a success. Would the public really want to watch thirteen half-hour programmes featuring life in one meerkat group?

The OSF film crew arrived to start shooting and, over the next six months, obtained footage of the social lives of the meerkats that has few parallels in animal documentaries. They could set up cameras so close to individuals that the camera's reflection is sometimes visible in their pupils. They filmed eviction and infanticide and were able to follow individual meerkats after they had been evicted from the group. They documented raids, war dances and extended battles between groups as well as attacks by snakes, eagles and owls. The eventual series was called *Meerkat Manor* and rapidly established a much larger audience, in both the US and Europe, than anyone had expected. Three more series followed, each of thirteen programmes, and the meerkat phenomenon was eventually reviewed in *Time* magazine, where a perceptive reporter commented that it was like a human soap opera where a predator could stop by at any moment and eat one of the characters. In 2007, I produced a popular book that told the (true) life story of the central character in the series, and used this as a vehicle to explain the evolution of animal societies (Clutton-Brock 2007a).

I am still visiting the Kalahari twice each year and am also still actively involved in the red deer study where my work is focusing, suitably enough, on the ecology of ageing and the factors affecting longevity (Nussey *et al.* 2006; Clutton-Brock & Isvaran 2007) and on the effects of climate change. The three other long-term studies I have helped to maintain, on Soay sheep, banded mongooses and pied babblers, are still running, generating a regular flow of new discoveries. In total, over 100 PhD students and post-docs have used them to explore questions about ecology or evolution. Each year Dafila and I still spend part of our time at the house in Pembrokeshire where I lived as a child and I am well aware how strong an effect my time there has had on the course of my life. Our children have gone on to develop related interests and jobs: Amber is a vet, specialising in embryo transfer in horses, while Peter, our son, is working in a government team organising carbon credits.

Hindsight

At a meeting in Cambridge to celebrate the centennial of Darwin's death in 1987, Richard Lewontin gave a bleak assessment of developments in evolutionary biology since Darwin's time (Lewontin 1983). He argued that relatively little progress had been made and that evolutionary biologists had devoted too much of their time to studies of natural history. John Maynard Smith responded to him. Would he not concede that we now understood far more about the evolution of animal breeding systems and societies than Darwin had? Lewontin grudgingly acquiesced.

Maynard Smith was right in identifying the understanding of animal breeding systems and societies as one of the major advances of the past half-century. Since 1960, there has been a striking development in our understanding of the evolution of reproductive strategies, and the structure of animal populations which extends from microorganisms to the higher primates and man. Evolutionary theorists, population geneticists, ecologists and epidemiologists, reproductive physiologists, neurobiologists, primatologists, ethologists, sociologists and behavioural ecologists have all played an important part in this synthesis, and it has been extremely exciting and very rewarding to have been involved. The framework of theory that has developed plays a crucial role in helping to structure our interpretation of the evolution of variation in behaviour, physiology, and anatomy. In combination with empirical studies, it provides the perspective necessary to identify exceptions, paradoxes and unanswered questions, stimulating novel questions and original research.

There is much still to be done and many important areas of social evolution and ecology that have still received little attention. To understand the operation of selection in different species, we need to know much more about the contrasts in selection pressures between clonal systems, and societies where all individuals are capable of breeding and eusocial societies. Our understanding of the ecological mechanisms limiting density is still confined to a small number of species and little is known of the effects of contrasts in breeding systems on population dynamics. Future studies of vertebrates will need to assess the relative roles of genetic control and learning in the development of the complex social strategies and the impact of animal breeding systems on population dynamics and demography. Many important questions cannot be satisfactorily answered in captive animals and it is clear that studies that can monitor the behaviour, reproduction and survival of large samples of individuals in naturally regulated populations over several generations will continue to have a central role to play.

Looking back, I am struck by the enormous benefits I have gained from collaboration with colleagues and students. Early in my career, I learned much from David Pilbeam, Robert Hinde and John Crook and, later on, from John Maynard Smith, John Krebs, Paul Harvey, Geoff Parker, Nick Davies and Josephine Pemberton. Each project that I have worked on has involved many people at different stages of their careers who have contributed skills, focus and knowledge that I did not possess, generating insights and achieving results that I could never have managed on my own. For me, collaboration has been the key to happy, productive science and I am immensely grateful to all the many people I have worked with.

Acknowledgements

I am grateful to Lee Drickamer, Don Dewsbury, Sarah Hodge, Jo Jones, Josephine Pemberton, Dafila Scott and Paul Harvey for their comments, and to Penny Roth for her continued assistance and support.

References

Carlson, A. A., Young, A. J., Russell, A. F. *et al.* (2004). Hormonal correlates of dominance in meerkats (*Suricata suricatta*). *Hormones Behav.* **46**: 141–50.

Clutton-Brock, T. H. (1974). Primate social organisation and ecology. *Nature* **250**: 539–42.

Clutton-Brock, T. H. (1984). Reproductive effort and terminal investment in iteroparous animals. *Am. Nat.* **123**: 212–29.

Clutton-Brock, T. H. (1988). *Reproductive Success*. Chicago, IL: Chicago University Press.

Clutton-Brock, T. H. (1991). *The Evolution of Parental Care*. Princeton, NJ: Princeton University Press.

Clutton-Brock, T. H. (2004). The causes and consequences of instability. In *Soay Sheep: Dynamics and Selection in an Island Population*, ed. T. H. Clutton-Brock J. M. Pemberton, pp. 276–310. Cambridge: Cambridge University Press.

Clutton-Brock, T. (2007a). *Meerkat Manor: Flower of the Kalahari*. London: Weidenfeld and Nicolson.

Clutton-Brock, T. H. (2007b). Sexual selection in males and females. *Science* **318**: 1882–5.

Clutton-Brock, T. H. & Albon, S. D. (1979). The roaring of red deer and the evolution of honest advertisement. *Behaviour* **69**: 145–70.

Clutton-Brock, T. H. & Albon, S. D. (1989). *Red Deer in the Highlands*. Oxford: Blackwell Scientific Publications.

Clutton-Brock, T. H. & Albon, S. D. (1992). Trial and error in the Highlands. *Nature* **358**: 11–12.

Clutton-Brock, T. H. & Coulson, T. N. (2002). Comparative ungulate dynamics: the devil is in the detail. *Phil. Trans. R. Soc. Lond.* B**357**: 1285–98.

Clutton-Brock, T. H. & Harvey, P. H. (1977b). Primate ecology and social organization. *J. Zool.* **183**: 1–39.

Clutton-Brock, T. H. & Harvey, P. H. (1978c). Mammals, resources and reproductive strategies. *Nature* **273**: 191–5.

Clutton-Brock, T. & Isvaran, K. (2007). Sex differences in ageing in natural populating of vertebrates. *Proc. R. Soc. Lond.* B**274**: 3097–104.

Clutton-Brock, T. H. & Langley, P. A. (1997). Persistent courtship reduces male and female longevity in captive tsetse flies *Glossina morsitans morsitans* Westwood (Diptera: Glossinidae). *Behav. Ecol.* **8**: 392–5.

Clutton-Brock, T. H. & Lonergan, M. E. (1994). Culling regimes and sex ratio biases in Highland red deer. *J. Appl. Ecol.* **31**: 521–7.

Clutton-Brock, T. H. & McComb, K. (1993). Experimental tests of copying and mate choice in fallow deer. *Behav. Ecol.* **4**: 191–3.

Clutton-Brock, T. H. & Parker, G. A. (1992). Potential reproductive rates and the operation of sexual selection. *Q. Rev. Biol.* **67**: 437–56.

Clutton-Brock, T. H. & Parker, G. A. (1995a). Punishment in animal societies. *Nature* **373**: 209–16.

Clutton-Brock, T. H. & Parker, G. A. (1995b). Sexual coercion in animal societies. *Anim. Behav.* **49**: 1345–65.

Clutton-Brock, T. H. & Vincent, A. C. J. (1991). Sexual selection and the potential reproductive rates of males and females. *Nature* **351**: 58–60.

Clutton-Brock, T. H., Albon, S. D., Gibson, R. M. & Guinness, F. E. (1979). The logical stag: adaptive aspects of fighting in red deer (*Cervus elaphus* L.). *Anim. Behav.* **27**: 211–25.

Clutton-Brock, T. H., Albon, S. D. & Guinness, F. E. (1981). Parental investment in male and female offspring in polygynous mammals. *Nature* **289**: 487–9.

Clutton-Brock, T. H., Albon, S. D. & Guinness, F. E. (1982a). Competition between female relatives in a matrilocal mammal. *Nature* **300**: 178–80.

Clutton-Brock, T. H., Guinness, F. E. & Albon, S. D. (1982b). *Red Deer: the Behaviour and Ecology of Two Sexes*. Edinburgh: Edinburgh University Press.

Clutton-Brock, T. H., Albon, S. D. & Guinness, F. E. (1984). Maternal dominance, breeding success and birth sex ratios in red deer. *Nature* **308**: 358–60.

Clutton-Brock, T. H., Major, M., Albon, S. D. & Guinness, F. E. (1987). Early development and population dynamics in red deer. 1: density-dependent effects on juvenile survival. *J. Anim. Ecol.* **56**: 53–67.

Clutton-Brock, T. H., Green, D., Hiraiwa-Hasegawa, M. & Albon, S. D. (1988). Passing the buck: resource defence, lek breeding and mate choice in fallow deer. *Behav. Ecol. Sociobiol.* **23**: 281–96.

Clutton-Brock, T. H., Albon, S. D. & Guinness, F. E. (1989a). Fitness costs of gestation and lactation in wild mammals. *Nature* **337**: 260–2.

Clutton-Brock, T. H., Hiraiwa-Hasegawa, M. & Robertson, A. (1989b). Mate choice on fallow deer leks. *Nature* **340**: 463–5.

Clutton-Brock, T. H., Price, O. F., Albon, S. D. & Jewell, P. A. (1991). Persistent instability and population regulation in Soay sheep. *J. Anim. Ecol.* **60**: 593–608.

Clutton-Brock, T. H., Price, O. F. & MacColl, A. D. C. (1992). Mate retention, harassment, and the evolution of ungulate leks. *Behav. Ecol.* **3**: 234–42.

Clutton-Brock, T. H., Deutsch, J. C. & Nefdt, R. J. C. (1993). The evolution of ungulate leks. *Anim. Behav.* **46**: 1121–38.

Clutton-Brock, T. H., McComb, K. E. & Deutsch, J. C. (1996a). Multiple factors affect the distribution of females in lek-breeding ungulates: a rejoinder to Carbone and Taborsky. *Behav. Ecol.* **7**: 373–8.

Clutton-Brock, T. H., Stevenson, I. R., Marrow, P. *et al.* (1996b). Population fluctuations, reproductive costs and life-history tactics in female Soay sheep. *J. Anim. Ecol.* **65**: 675–89.

Clutton-Brock, T. H., Illius, A., Wilson, K. *et al.* (1997). Stability and instability in ungulate populations: an empirical analysis. *Am. Nat.* **149**: 195–219.

Clutton-Brock, T. H., O'Riain, M. J., Brotherton, P. N. M. *et al.* (1999b). Selfish sentinels in cooperative mammals. *Science* **284**: 1640–4.

Clutton-Brock, T. H., Brotherton, P. N. M., O'Riain, M. J. *et al.* (2000). Individual contributions to babysitting in a cooperative mongoose, *Suricata suricatta. Proc. R. Soc. Lond.* B**267**: 301–5.

Clutton-Brock, T. H., Brotherton, P. N. M., O'Riain, M. J. *et al.* (2001a). Contributions to cooperative rearing in meerkats. *Anim. Behav.* **61**: 705–10.

Clutton-Brock, T. H., Brotherton, P. N. M., Russell, A. F. *et al.* (2001b). Cooperation, conflict and concession in meerkat groups. *Science* **291**: 478–81.

Clutton-Brock, T. H., Coulson, T., Milner-Gulland, E. J., Armstrong, H. M. & Thomson, D. (2002a). Sex differences in emigration and mortality affect optimal management of deer populations. *Nature* **415**: 633–7.

Clutton-Brock, T. H., Coulson, T. N., Milner-Gulland, E. J., Thomson, D. & Armstrong, H. M. (2002b). Sex differences in emigration and mortality affect optimal management of deer populations. *Nature* **415**: 633–7.

Clutton-Brock, T. H., Russell, A. F., Sharpe, L. L. *et al.* (2002c). Evolution and development of sex differences in cooperative behavior in meerkats. *Science* **297**: 253–6.

Clutton-Brock, T. H., Russell, A. F. & Sharpe, L. L. (2003). Meerkat helpers do not specialize in particular activities. *Anim. Behav.* **66**: 531–40.

Clutton-Brock, T. H., Coulson, T. & Milner, J. M. (2004a). Red deer stocks in the Highlands of Scotland. *Nature* **429**: 261–2.

Clutton-Brock, T. H., Russell, A. F. & Sharpe, L. L. (2004b). Behavioural tactics of breeders in cooperative meerkats. *Anim. Behav.* **68**: 1029–40.

Clutton-Brock, T. H., Russell, A. F., Sharpe, L. L. & Jordan, N. R. (2005). 'False-feeding' and aggression in meerkat societies. *Anim. Behav.* **69**: 1273–84.

Clutton-Brock, T. H., Hodge, S. J., Spong, G. *et al.* (2006). Intrasexual competition and sexual selection in cooperative meerkats. *Nature* **444**: 1065–8.

Cockburn, A. (1998). Evolution of helping behaviour in cooperatively breeding birds. *A. Rev. Ecol. Syst.* **29**: 141–77.

Coulson, T. N., Catchpole, E. A., Albon, S. D. *et al.* (2001). Age, sex, density, winter weather and population crashes in Soay sheep. *Science* **292**: 1528–31.

Emlen, S. T. (1991). Evolution of cooperative breeding in birds and mammals. In *Behavioural Ecology: an Evolutionary Approach*, ed. J. R. Krebs & N. B. Davies, pp. 301–37. Oxford: Blackwell Scientific Publications.

Emlen, S. T. (1995). An evolutionary theory of the family. *Proc. Nat. Acad. Sci. USA* **92**: 8092–99.

Gomendio, M., Clutton-Brock, T. H., Albon, S. D., Guinness, F. E. & Simpson, M. J. (1990). Mammalian sex ratios and variation in costs of rearing sons and daughters. *Nature* **343**: 261–3.

Grenfell, B. T., Price, O. F., Albon, S. D. & Clutton-Brock, T. H. (1992). Overcompensation and population cycles in an ungulate. *Nature* **355**: 823–6.

Grenfell, B. T., Wilson, K., Finkenstädt, B. F. *et al.* (1998). Noise and determinism in synchronized sheep dynamics. *Nature* **394**: 674–7.

Hallett, T. B., Coulson, T., Pilkington, J. G. *et al.* (2004). Why large-scale climate indices seem to predict ecological processes betrer than local weather. *Nature* **430**: 71–5.

Hamilton, W. D. (1963). The evolution of altruistic behavior. *Am. Nat.* **97**: 354–6.

Hamilton, W. D. (1964). The genetical evolution of social behaviour. I. II. *J. Theor. Biol.* **7**: 1–52.

Harvey, P. H. & Pagel, M. (1991). *The Comparative Method in Evolutionary Biology.* Oxford: Oxford University Press.

Jewell, P. A., Milner, C. & Boyd, J. M. (1974). *Island Survivors: The Ecology of the Soay Sheep of St Kilda*. London: Athlone Press.

Kokko, H., Johnstone, R. A. & Clutton-Brock, T. H. (2001). The evolution of cooperative breeding through group augmentation. *Proc. R. Soc. Lond.* **B268**: 187–96.

Kruuk, L. E. B., Clutton-Brock, T. H., Albon, S. D., Pemberton, J. M. & Guinness, F. E. (1999a). Population density affects sex ratio variation in red deer. *Nature* **399**: 459–61.

Kruuk, L. E. B., Slate, J., Pemberton, J. M. & Clutton-Brock, T. H. (2003). Fluctuating asymmetry in a secondary sexual trait: no associations with individual fitness, environmental stress or inbreeding, and no heritability. *J. Evol. Biol.* **16**: 101–13.

Kutsukake, N. & Clutton-Brock, T. H. (2006). Aggression and submission reflect reproductive conflict between females in cooperatively breeding meerkats *Suricata suricatta*. *Behav. Ecol. Sociobiol.* **59**: 541–8.

Lewontin, R. C. (1983). Gene, organism and environment. In *Evolution from Molecules to Men*, ed. D. S. Bendall, pp. 273–85. Cambridge: Cambridge University Press.

Manser, M. B., Bell, M. B. & Fletcher, L. B. (2001). The information that receivers extract from alarm calls in suricates. *Proc. R. Soc. Lond.* **B268**: 2485–91.

McComb, K. & Clutton-Brock, T. H. (1994). Is mate choice copying or aggregation responsible for skewed distributions of females on leks? *Proc. R. Soc. Lond.* **B255**: 13–19.

Moorcroft, P. R., Albon, S. D., Pemberton, J. M., Stevenson, I. R. & Clutton-Brock, T. H. (1996). Density-dependent selection in a cyclic ungulate population. *Proc. R. Soc. Lond.* **B263**: 31–8.

Nussey, D., Kruuk, L. E. B., Donald, A., Fowlie, M. K. & Clutton-Brock, T. H. (2006). The rate of senescence in maternal performance increases with early-life fecundity in red deer. *Ecol. Lett.* **9**: 1342–50.

Reyer, H.-U. (1980). Flexible helper structure as an ecological adaptation in the pied kingfisher (*Ceryle rudis rudis* L.). *Behav. Ecol. Sociobiol.* **6**: 219–27.

Reyer, H.-U. (1984). Investment and relatedness: a cost/benefit analysis of breeding and helping in the pied kingfisher (Ceryle rudis). *Anim. Behav.* **32**: 1163–78.

Russell, A. F., Carlson, A. A., McIlrath, G. M., Jordan, N. R. & Clutton-Brock, T. H. (2004). Adpative size modification by dominant female meerkats. *Evolution* **58**: 1600–7.

Russell, A. F., Sharpe, L. L., Brotherton, P. N. M. & Clutton-Brock, T. H. (2003). Cost minimization by helpers in cooperative vertebrates. *Proc. Nat. Acad. Sci. USA* **100**: 3333–8.

Russell, A. F., Clutton-Brock, T. H., Brotherton, P. N. M. *et al.* (2002). Factors affecting pup growth and survival in cooperatively breeding meerkats *Suricata suricatta*. *J. Anim. Ecol.* **71**: 700–9.

Stillman, R. A., Clutton-Brock, T. H. & Sutherland, W. J. (1993). Black holes, mate retention, and the evolution of ungulate leks. *Behav. Ecol.* **4**: 1–6.

Thornton, A. & McAuliffe, K. (2006). Teaching in wild meerkats. *Science* **313**: 227–9.

Trivers, R. L. & Willard, D. E. (1973). Natural selection of parental ability to vary the sex ratio of offspring. *Science* **179**: 90–2.

Williams, G. C. (1966). *Adaptation and Natural Selection: a Critique of some Current Evolutionary Thought*. Princeton, NJ: Princeton University Press.

Young, A. J., Carlson, A. A., Monfort, S. L. *et al.* (2006). Stress and the suppression of subordinate reproduction in cooperatively breeding meerkats. *Proc. Nat. Acad. Sci. USA* **103**: 12005–10.

Zuckerman, S. (1981). *The Social Life of Monkeys and Apes*. London: Routledge and Kegan Paul.

6

Birds, butterflies and behavioural ecology

NICHOLAS B. DAVIES

Early years

In retrospect, my scientific career has been an excuse to spend as much time as possible outside and watching birds. I was born in 1952, in the village of Formby, some fifteen miles north of Liverpool on the Lancashire coast, north-west England. One of my earliest memories is of watching a chaffinch in our garden through opera glasses, from a makeshift hide of wooden chairs. I must have been about six at the time and by the age of nine I was keeping notebooks with neat lists of the species I had seen. The sea was just a mile away, a

Leaders in Animal Behavior: The Second Generation, ed. L. C. Drickamer & D. A. Dewsbury. Published by Cambridge University Press. © Cambirdge University Press 2010.

walk along Fisherman's Path through pinewoods and sand dunes, and the wildlife there thrilled me. During World War II, the beach had been covered in wooden posts to prevent enemy landings and I sat on these to watch divers and sea-duck offshore and flocks of wading birds fly past at high tide. The pinewoods used to be floodlit during night time bombing raids to trick the Luftwaffe pilots into dropping their bombs away from Liverpool docks. Natterjack toads now bred in the old crater ponds, nightjars churred in the dune slacks and there were red squirrels in the pines. In autumn, pink-footed geese arrived from their Icelandic breeding grounds to winter on the farmland behind the village and skeins flew over our house at dusk, on their way to coastal roosts. I lay in bed listening to their cries when they became disoriented on foggy nights. I have no idea where this early passion for natural history came from; my parents encouraged me but neither they nor my siblings (two younger brothers and a younger sister) developed this obsession.

From the age of nine to eighteen, I attended Merchant Taylors' School, Crosby, where there was an enthusiastic biology teacher who organised a Field Club, with weekly meetings after school and regular field trips. My favourite expedition was to Hilbre Island in the estuary of the River Dee. We walked across the mudflats at low tide, stayed on the island during the high tide period, and then walked off again the next low tide. Thousands of waders congregated on the island as the incoming tide covered their feeding grounds and by crouching behind the rocks we could get within a few metres of them. In the holidays, we visited Bardsey Island, off the tip of the North Wales peninsula, where we stayed at the bird observatory for one or two weeks at a time. This was my first taste of a wild place free from human sounds and my ambition then was to become a warden of a nature reserve. During term time, Monday to Saturday was filled with work and sport (I was a keen cricketer and rugby player), which left just Sunday for birds. Two or three friends and I used to visit the Formby pines and dunes every Sunday morning and I kept notes on the changing seasons. I contributed records to the annual school Field Club report and to the magazine of the junior section of the Royal Society for the Protection of Birds, where I had the thrill of first seeing my name in print for an early spring record of a wheatear.

Curiously, the school's science curriculum included only chemistry and physics up to the age of sixteen, so I had to wait until the last two years (the sixth form) before I was taught any biology. By then, I had been inspired by two books by David Lack, *The Life of the Robin* and *Swifts in a Tower*, and by Niko Tinbergen's *The Herring Gull's World*, and was eager to learn more about a scientific approach to natural history. But the school's sixth form biology course was largely plant and animal anatomy, with nothing on behaviour or ecology. I don't remember being frustrated by this, because I continued to find stimulation from my weekly bird watching trips and BBC television programmes, such as the "Look" series by Peter Scott. I bought an Everyman edition of Darwin's *The Origin of Species*, which fitted neatly in my school blazer pocket, and then read *The Voyage of the Beagle*. Darwin now joined Lack and Tinbergen as my trinity of heroes. With money from a school essay prize, I bought *Animal Dispersion in Relation to Social Behaviour* by V. C. Wynne Edwards, which was praised in an article in *New Scientist*. I thought that was marvellous too and longed to combine my bird watching with new discoveries about behaviour. I began to spend more time watching rather than just listing, and made notes on the systematic way in which a flock of siskins and redpolls exploited the cones in an alder wood, wondering whether depleting one section of the wood before moving on would increase the efficiency with which the flock foraged over the whole winter. I think this was my first scientific hypothesis but it never occurred to me that I could test it or, indeed, that it would be worth testing. I watched shelducks displaying in the dunes and longed to interpret their displays as Tinbergen had done for herring gulls. In my last year at school I bought David Lack's *Population Studies of Birds* (1966) and was dismayed to read his vigorous criticisms of Wynne Edwards. This was my first inkling that not everything in scientific books was necessarily true.

Cambridge

My biology school friends went to University to study either medicine or biochemistry. Despite my parents' worries that I would never get a job as a birdwatcher, I applied to Cambridge to read Natural Sciences with the aim of specialising in zoology. I was the first member of our family to go to University; when I was offered a place by Pembroke College, my grandfather gave me a pound note as congratulation. A week later, I passed my driving test and he gave me five pounds. I loved my three years at Cambridge (1970–73) but did not distinguish myself academically and found most of the courses disappointing. The emphasis then was firmly on physiology and neuroscience, and while I enjoyed lectures by Robert Hinde and Patrick Bateson on development and motivation of behaviour, I longed for more on Lack's evolutionary approach to ecology. I spent most of my time on the cricket and rugby fields, with weekend trips cycling out to Wicken Fen, twenty miles north of Cambridge, where I learnt to catch and ring birds with the local ringing group. If anyone had suggested then that I might one day return to Cambridge as a professor, I would have considered this as ridiculous and as unlikely as my prospects of playing for the England cricket team.

A fellow student and bird fanatic, Rhys Green (now also a Professor and colleague at Cambridge) was also frustrated by the lack of field work on the course and together we sought permission to do our own research project on the foraging behaviour of warblers on Wicken Fen, instead of attending laboratory practicals. Patrick (Pat) Bateson encouraged us and allowed us to use some of his aviaries at Madingley. Rhys and I had just read David Lack's latest book (1971: *Ecological Isolation in Birds*) and began by comparing the foraging niches of reed and sedge warblers to see how they might avoid competition. We then hand-raised reed warblers from the nestling stage to study how they developed foraging skills. This led to my first scientific paper (Davies & Green 1976). I remember Pat's comment on an early draft, that my writing had a "Victorian expansiveness", and the published paper is now somewhat embarrassing, but this project gave Rhys and me our first taste of original research and we learnt a lot from Pat's advice and encouragement. I still remember, with horror, long hours hammering in data with a primitive calculator, simply to work out correlation coefficients or to compare means by t-tests – tasks which these days could be completed in minutes. I have disliked doing statistics ever since.

Watching birds continued to dominate the vacations. In 1971 I hitch-hiked to Istanbul and spent a month watching the autumn migration across the Bosphorus. This was an awesome experience, as thousands of storks and birds of prey from Europe, relying on thermals for soaring flight and hence reluctant to cross the wide sea, were funnelled over the narrow straits on their way to wintering grounds in the Middle East and Africa. At the end of my undergraduate course in 1973, I again hitch-hiked to Istanbul, and then continued east by local buses through Turkey, Iran, Afghanistan and Pakistan to India, where I trekked in Kashmir, and then in Nepal. This three month trip was an excellent break before the next, and most formative, experience of my career.

Oxford

The Edward Grey Institute (E.G.I.), part of the Zoology department at Oxford University and directed by David Lack, was a Mecca for bird research and in my last year at Cambridge my ambition shifted from wardening a nature reserve to a doctorate with Lack's group. A talk on my undergraduate reed warbler research at the annual E.G.I. Student Conference in January 1973 helped me to secure a PhD place there the following October. Sadly, David Lack died in March that year, so I never had the chance to meet him. The next director, Chris Perrins, became my mentor and arranged for me to be supervised by Euan Dunn, who had just been appointed to the post of Demonstrator at the E.G.I. Euan was the perfect supervisor – full of encouragement but allowing me the freedom to develop my own ideas. I entered the Oxford zoology department for the first time in early October 1973 to find myself in the middle of a champagne reception. Thinking that this might be the normal lavish welcome for new Oxford PhD students, I made my way to the front of the crowd just as some speeches were about to begin and found myself in the middle of the celebrations for Niko Tinbergen's Nobel prize. Tinbergen's activities in the department declined soon after then, so I also never got the chance to speak to the second of my great schoolboy heroes, though I did sit in some seminars he attended and was mesmerised by the growing ash from his cigarette, held low by the side of his chair, wondering whether it would collapse to the floor before the next puff.

In those days there were no formal courses for PhD students, though we were encouraged to attend seminars, so my three year studentship (1973–76) was for full-time research. I wanted to continue the theme of my undergraduate project, studying the foraging behaviour of insectivorous birds. Most of these were summer visitors, and as it was now the autumn I was advised to spend the winter in the library, reading the literature and developing a detailed proposal. Having spent most of my undergraduate years reading, this was the last thing I wanted to do, so instead I went out to seek inspiration from the field. I can still remember the exhilarating sense of freedom as I cycled in the opposite direction to the morning rush hour traffic, out of Oxford and into the countryside. On the first day I discovered Port Meadow, an area of grassland along the River Thames, on the edge of the city. I sat on the river bank and noticed pied wagtails walking along the water's edge and picking up small invertebrates washed up by the river onto the muddy banks. Individuals were obviously defending stretches of the river as territories, because neighbours displayed at the same boundaries on successive days. I caught some of the birds at an evening roost on the edge of the meadow, colour-ringed them, and decided to watch individuals from the minute they arrived on their territories at dawn, until they departed for the roost at dusk. I collected a notebook full of data on feeding rates, and found that individuals ate, on average, one small insect every three seconds throughout the day. Calculations suggested this remarkable rate was necessary to balance their energy budget during the eight hours of daylight in mid-winter.

I took these results to my supervisor for discussion, and proudly showed him my first notebook full of data. "What's your hypothesis?", he asked. This stopped me in my tracks; I realised I had rushed into data collection without stopping to think exactly why I was measuring all these feeding rates. All I had shown was that in a short winter's day, the wagtails had to work hard to find enough food. Interesting, perhaps, but not very exciting. This was a tough first lesson; a project's interest depends on the interest of the questions, not the amount of data. I needed to spend more time thinking and less time measuring.

Oxford was a thrilling place for a young research student in the mid-1970s. I shared a flat with Tim Birkhead, a year ahead of me as a PhD student at the E.G.I., and we soon realised that attending seminars of the various research groups in the zoology department provided a better education than any formal courses. It wasn't necessary to travel the world to meet the new stars of behaviour – we simply had to wait in Oxford for them to visit. My initial disappointment at not being able to interact with Lack and Tinbergen was soon replaced by the excitement of new heroes. Richard Dawkins was writing *The Selfish Gene*, and tested drafts of his chapters as undergraduate lectures. I joined the class and this became my introduction to a gene's eye view of social behaviour. Robert Trivers came for a seminar on economics and behaviour. The previous speaker ran over time, leaving Trivers just twenty minutes before the lunch break. Trivers put his notes aside and gave a brilliant extempory explanation of W. D. Hamilton's theory of social evolution, with Trivers' own extensions to parent–offspring conflict. It was a stunning performance and taught me that deep under-standing was necessary before you could explain ideas so clearly and simply. Ric Charnov visited for a month and generously gave a talk not on his own work but on the now famous Trivers and Hare (1976) paper on haplodiploidy and sex ratio conflict in social insects. John Maynard Smith and Geoff Parker came to tell us how game theory could illuminate contest behaviour. Geoff's ideas were so novel and far ahead of the time that he was interrupted after the first sentence to be asked: what exactly do you mean by that? There followed a long and involved discussion and I suspect that Geoff never managed to even start the talk he had prepared.

I didn't follow all the arguments in these various seminars but the lively discussions that they provoked left us all with the feeling that we were on the cusp of a new wave that was about to sweep through behaviour studies. It was these seminars, more than any other experience at Oxford, which taught me the importance of clear thinking and clear exposition in both questions and answers. The questions were often tough, but there was humour too. I remember one talk on a motivational model of courtship, which included a factor "hope" to explain when individuals changed behaviour from one part of the courtship sequence to the next. Mike Cullen suggested that the model might provide an even better fit if it included "faith" and "charity" too. Mike had been one of Tinbergen's first PhD students and was soon to leave his lectureship at Oxford for a chair at Monash University in Australia. His comments would often astonish speaker and audience alike. I remember him once explaining how some puzzling data made perfect sense after all, given the magnitude of two regression slopes mentioned earlier in the talk, forty minutes apart, which Mike had remembered and combined to provide a novel explanation. The questions of Richard Dawkins and John Krebs were an inspiration too, so incisive and clearly phrased that they often provoked a contented murmur among the audience as we all suddenly grasped the main point of the seminar.

The talk that had the most immediate influence on my own work was by John Krebs, who had just returned to the Oxford zoology department, where he had previously been a research student in Tinbergen's group, after spells as a lecturer in Vancouver and Bangor. Typical seminars in the Edward Grey Institute in those days involved detailed analyses of aspects of the life history of particular species, for example clutch size in great tits or prey choice by sparrowhawks. John's title was "Optimal foraging by herons and chickadees", and I was intrigued to discover how such different species could be discussed in the same breath. John showed how foraging behaviour could be analysed as a series of decisions, each with various costs and benefits. Herons and chickadees lived in different habitats, of course, and ate very different food. But both faced the same fundamental decisions, for example when to leave a patch with diminishing foraging returns. This reductionist approach was a revelation to me and I immediately saw how my work on the wagtail territorial behaviour could be improved by thinking more clearly about hypotheses and by a cost–benefit approach to decision making.

I returned to the river bank for a fresh start, with a new notebook and new enthusiasm, and began to wonder exactly why my pied wagtails were defending territories. My all-day watches had shown that the owners exploited their territories systematically, walking along one river bank to the territory boundary, then crossing the river and returning back down the other side to complete the circuit. I suddenly realised that this was the key; prey washed up by the river formed a renewing food supply, so systematic foraging enabled the birds to allow sufficient time for prey renewal between successive visits to the same stretch. Territory defence prevented other birds from depleting patches and interfering with the owner's foraging at profitable prey renewal times. A fellow PhD student, Alasdair Houston, who is more mathematically inclined than I am, helped to formalise these ideas into an equation to predict how feeding rates varied with prey renewal and efficiency of defence. We went on to show how various features of territorial behaviour, including territory size, occasional territory sharing, and territory abandonment, all made adaptive sense given the ways in which the costs and benefits of defence varied with prey abundance (Davies & Houston 1981, 1983).

My PhD thesis was completed in 1976 and was rather a rag bag of examples of foraging analysed in terms of costs and benefits. Apart from wagtail territoriality, the only other idea

I was pleased with was that of considering the transition to independence of fledgling birds as a decision between begging for food from parents and self feeding. As the young became older, parents became more reluctant to feed them, so the profitability of begging declined, while the profitability of self feeding increased with growth and experience. I showed that the young became independent when they could gain food more profitably by self feeding than by begging, and parents could control this transition by varying their generosity (Davies 1976, 1978a).

In 1976 I was appointed Demonstrator at the E.G.I., succeeding my supervisor Euan Dunn in this three-year fixed-term appointment. I moved out of my digs in Oxford and spent the next three years living in a wooden chalet in the middle of Wytham Woods, on a hill to the west of the city, home of the long-term study of great tits started by David Lack in the 1940s and still going strong under the leadership of Chris Perrins. This was heaven: the wood was full of badgers, foxes and deer; tawny owls hooted as I cycled home at night; and in the spring there were nightingales singing and carpets of bluebells.

The wood was a forty minute cycle ride away from the zoology department, across Port Meadow. I wanted to continue the wagtail study on the meadow, especially the problem of long-term territory defence through the winter. I noticed that even when food was so scarce that the owners were forced to leave their territories for long spells, they still periodically returned to advertise their presence by calls, and to evict any intruders. I wondered whether they did this to prevent any newcomer from having the time to learn the territory characteristics sufficiently well that it could begin to exploit the renewing prey in a profitable manner. If it could do so, then perhaps the owner would be faced with a more intense contest when it returned. To test this, I wanted to remove owners experimentally for various lengths of time, and then release them to see whether they were less able to win the territory back if newcomers had been given sufficient time to settle in. However, it proved too difficult to catch the birds on territory and I had to abandon this.

The summer of 1976 was one of long spells of sunny days and Wytham Woods was full of butterflies. I saw male speckled wood butterflies defending sunlit patches on the woodland floor, engaging in brief spiral flights with intruders, and wondered whether I could test the effects of owner absence on contest outcome in this system instead, where individuals would be much easier to catch. As with the wagtail study, I began by marking individuals (in this case with dabs of coloured felt-tip pen on the wings) and watched them all day long to first get a good idea of their behaviour. I found that individuals returned to the same territory each day, for up for three weeks. Throughout the day intruders descended from the tree canopy above and were quickly chased off by the owners. Owners approached passing females and attempted to mate with them, and the sunny patches appeared to be defended as they were the best places for encountering potential mates. To my surprise, when I caught owners and replaced them back on their territories, they were quickly chased off by any newcomer who had landed in the meanwhile, even if the newcomer had been in residence for just a minute or so. I was puzzled by this, having expected a gradual reversal of dominance with increased residence time by the newcomer. The original owner's immediate retreat wasn't simply due to any adverse effects of being held in my net, because it was able to quickly reclaim its territory when released back if no newcomer was present. Finally, when I gently introduced a second male onto an already occupied territory, unnoticed by the first male, then an escalated contest ensued when the males encountered each other, both behaving as if they regarded themselves as the rightful owner.

A recent theoretical model by Maynard Smith & Parker (1976) had shown that if contests are costly, and if all contestants have a good chance of eventually gaining access to the resource, then "owner wins" could evolve as an asymmetry to settle contests quickly, without the need for prolonged fights. My observations showed that individuals observed as intruders did, indeed, often gain ownership of a sunspot territory at a later time, so I suggested that my experiments provided evidence for this asymmetry in the speckled wood butterflies (Davies 1978b). In the years since, conditions for the stability of "resident wins" for contest resolution have been shown to be restrictive and unlikely to apply often in nature (Grafen 1987). So further studies of speckled woods have investigated whether there are any physical or physiological attributes which correlate with contest success, and hence ownership, including body size, energy reserves, flight performance and body temperature. However, there is still no convincing evidence for a role for any of these, and "resident wins" remains the best proximate predictor of contest outcome in speckled woods (Bergman *et al.* 2007), although the ultimate explanation remains to be uncovered.

I continued studies of contests with Tim Halliday, who had recently completed his PhD at Oxford on newt courtship behaviour. We found a wonderful pond for common toads on the edge of the city, marked individual males with numbered waist bands and found that larger males were more likely to win fights in the scramble to grasp females and be in position (amplexus) to fertilise the eggs during spawning. Larger males had croaks of a lower pitch and we showed by experiment in laboratory trials that call pitch was one of the cues used to settle contests, with playback of deeper croaks being more likely to deter a potential attacker. We suggested that croak pitch was an example of honest assessment in displays, in the sense that a small individual was physically unable to produce a deeper croak (Davies & Halliday 1978).

Behavioural ecology

At the same time that I began as Demonstrator at the E.G.I., John Krebs was appointed to a tenured lectureship. He was approached by Robert Campbell from Blackwell Scientific Publications (based in Oxford), who was keen for a book to mark the recent research developments in behaviour. John generously invited me to join him to edit this volume, so within a year of completing my PhD I was forced to think more broadly about what we were all doing. In *The Story of Art*, E. H. Gombrich begins with the observation that there really is no such thing as "Art", there are only artists. In the same way, our field was best defined by what we and our colleagues were studying at the time. We began by listing potential authors and topics first, and only then did we decide on a title which would best link the three main themes of the research: behaviour, ecology and evolution. E. O. Wilson's *Sociobiology* (1975) had just been published and though widely admired was causing a furore in some quarters in the USA because the final chapter on humans was interpreted by some as proposing "genetic determinism" of behaviour, and providing ammunition for those with illiberal political views. We avoided the word sociobiology in the title not just to distance our book from this controversy but because we envisaged our subject as broader, including not only social behaviour but all aspects of behavioural decision making.

The first edition of our edited book *Behavioural Ecology: An Evolutionary Approach* was published in 1978 and was warmly received. One kind reviewer of a later edition said it had defined the field. What was new? We saw the subject as emerging from four schools of thought developed in the 1960s and 1970s. The first two themes involved new theory. One was the cost–benefit approach to decision making, pioneered by Robert MacArthur (MacArthur & Pianka 1966; MacArthur 1972), and now being developed for foraging behaviour (Krebs *et al.* 1977; Pyke *et al.* 1977), mating (Orians 1969; Parker 1974; Emlen & Oring 1977) and territoriality (Brown 1964, 1975). Natural selection could be viewed as an optimising agent, selecting those strategies that led to maximum net benefit. This approach led to testable, quantitative predictions about the costs, constraints and currencies involved in an animal's choice of action.

The second theme involved new theory for understanding the evolution of social behaviour. W. D. Hamilton (1964) introduced the idea of considering costs and benefits of interactions not only as they influenced one's own personal reproductive success but also that of kin. His concept of inclusive fitness illuminated the evolution of altruism and selfishness in social groups and inspired a "gene's eye" view of behaviour (Dawkins 1976). Robert Trivers (1972, 1974) and Geoff Parker (1970, 1979) highlighted the importance of conflicts of interest between breeding partners and between other family members. Sexual reproduction was no longer regarded as a harmonious venture, and we were led to expect a world of manipulation and deceit. John Maynard Smith and Geoff Parker applied game theory to the analysis of the outcome of social conflicts and introduced the idea of seeking evolutionarily stable solutions (Parker 1974; Maynard Smith & Parker 1976). The stable solution was sometimes for there to be variability in the population, for example the coexistence of alternative strategies.

The other two themes were methodological. One was the value of Tinbergen's experimental approach for testing hypotheses in the animal's natural environment. Behavioural ecology continued the Tinbergen tradition of watching and wondering, as exemplified by his inspirational book *Curious Naturalists* (1974), but now combined this with the new theory linking decision making to evolution. The other theme was the comparative method, linking the evolution of differences between species in life history to differences in their ecology. In effect, this was an analysis of the outcomes of evolutionary experiments. This had been pioneered by John Crook (Crook 1964; Crook & Gartlan 1966) and David Lack (1968) and, with improved methodology, by Tim Clutton-Brock and Paul Harvey (1977).

The field flourished. A new International Society for Behavioural Ecology was founded, together with the launch of a new journal, *Behavioural Ecology*. Our Krebs & Davies edited volume had three more editions (1984, 1991, 1997), each with a completely new set of chapters and often new authors too, to mark the changing developments. During the past thirty years, the improved quality of field studies and the greater rigour of testing alternative hypotheses, both with more detailed data and more sophisticated statistical analysis, has already made some of the earlier work seem naïve and wildly optimistic. But those of us lucky enough to be present at the birth of a new discipline will always remember the initial excitement of seeing the natural world afresh through eyes illuminated by new theory.

Back to Cambridge

In 1979 I married Jan Parr, gave my first plenary lecture, at the International Ethology Conference in Vancouver (*Territorial defence in a bird and a butterfly*), and returned to Cambridge to take up an appointment as Demonstrator in the Department of Zoology. Unlike the Oxford post, this was not of fixed term, and after five years there would be the chance of being upgraded to a tenured lectureship. I also became a Fellow of Pembroke College. The Professor of Zoology at Cambridge, Gabriel Horn, asked me what I needed by way of start up funds. I spent an evening thinking about this and the next day told him not to worry, because I already had a good pair of binoculars and should be ready to start right away. If there was any astonishment in his reaction, he hid it well. My reply would seem ridiculous today, when research is so dominated by large grants. But, with the advent of new theory, I was confident that all one had to do was to find an interesting species, mark individuals so they could be recognised, and then watch what they did. Exciting discoveries would surely emerge.

I prepared my first lecture course which, together with John Krebs's course in Oxford, formed the basis for our student text *An Introduction to Behavioural Ecology* (first edition 1981, third edition 1997). I also began to teach on the newly established field course for final year undergraduates where students undertook field projects on the north Norfolk coast, for example investigating how the foraging behaviour and daily energy budgets of wading birds were influenced by the tidal cycle. I continue to enjoy this teaching today, every bit as much as my research. One of the privileges of an academic career is the arrival each year of a new intake of students to challenge current ideas and old preconceptions.

A garden for dunnocks

Over the next three years, our two daughters, Hannah and Alice, were born and I was keen to develop a research project locally so there would be no need to spend long spells away from home. On a walk round the Cambridge University Botanic Garden I found myself in a beautiful nature reserve in the heart of the city, a short cycle ride from both home and the zoology department. The Garden was originally designed by Darwin's teacher, John Henslow, who championed the idea of expanding the collection to show not just plants of medicinal importance but a whole range of plants, including trees and shrubs, to illustrate plant diversity. In the early 1900's it was used not only for taxonomic studies but also for pioneering experiments in genetics (by William Bateson) and ecology (by A. G. Tansley).

Figure 6.1. Female dunnock soliciting to the beta male in a polyandrous trio. The alpha male is just about to fly in and interrupt. Drawing by David Quinn. From Davies 1992.

I realised that the Garden would be a marvellous place for studying bird behaviour too and noticed some dunnocks *Prunella modularis* chasing round the bushes in groups of three. This was just what I needed to get started. Bird watchers had long suspected that dunnocks had interesting social behaviour (Campbell 1952) and two recent studies had revealed that dunnock mating systems varied even within a small population, including: pairs (monogamy); a male with two females (polygyny); a female with two unrelated males (polyandry); and more complex combinations involving two unrelated males sharing two females (polygynandry) (Birkhead 1981; Snow & Snow 1982). I decided to begin a long-term study of the Botanic Garden population (some seventy breeding adults each year) to try to discover the causes of this variability. The prevailing view then was that to understand mating systems you needed to know all about a species' ecology: its food distribution, habitat structure, predation pressure and so on. I was keen to investigate how the social conflicts, predicted by Parker and Trivers, interacted with these ecological selection pressures, which could be viewed as setting the stage on which individuals played out their behavioural strategies. The dunnock, with its variable mating system, seemed an ideal model species.

The dunnock study was to last fifteen years. I didn't realise how it was dominating my life until I saw how Hannah had filled in a form at school; for "father's occupation" she simply wrote "dunnocks". Keeping the population colour-ringed was hard work and involved regular early morning mist netting, but being able to recognise individuals was the key to our behavioural observations and it opened up a whole new world to enjoy as we began to relate individual behaviour to reproductive success. Over the years I was lucky to be joined by excellent collaborators to work on this project, including: Tim Birkhead, Mike Bruford, Terry Burke, Philip Byle, Ian Hartley, Ben Hatchwell, Alasdair Houston, Naomi Langmore, Arne Lundberg, Tim Robson and Haven Wiley.

There were three "eureka" moments during the dunnock study. The first was the realisation that the variable mating system reflects the different outcomes of sexual conflict. This suddenly occurred to me as I cycled home one evening and I'm embarrassed now that it took so long to realise what, in retrospect, should have been obvious (Davies & Houston 1986; Davies 1992). A female's reproductive success varied with the amount of help she gained from males in chick feeding. A female did least well in polygyny (where she had one male's part-time help), better in monogamy (one male's full-time help) and best in polyandry, provided she shared matings between both males, in which case she gained full-time help from two males. These reproductive payoffs made good sense of female behaviour: females were aggressive to other females and encouraged copulations from second males. For a male, however, reproductive success was maximised by polygyny; despite the cost to each female from shared male care, the summed success of two females exceeded that of a monogamous female. A male did worst of all in polyandry; although more chicks were raised, he suffered costs of shared paternity and did better with sole paternity of a smaller brood in monogamy. This made good sense of male behaviour too; in direct opposition to the aims of a female, males tried to retain two females and to exclude second males. Polygynandry (two males sharing two females) could be viewed as a "stalemate" to the conflict: neither female could evict the other and so claim her best option (polyandry) and neither male could evict the other to claim his best option (polygyny).

The second eureka moment was discovering that shared matings between two males often resulted in mixed paternity in a brood. In 1988, Ben Hatchwell and I had begun to collect blood samples from adults and offspring so our collaborators Terry Burke and Mike Bruford (then at Leicester University) could analyse paternity by DNA fingerprinting. This new technique was soon to revolutionise field studies of mating systems, revealing many surprises in supposedly "monogamous" systems. The paternity analysis was done blind to the behavioural observations and it was thrilling to match the two and see how well shared paternity was predicted from shared matings (Burke *et al.* 1989). Further experiments revealed that the males themselves did not have such an exact paternity measure as a DNA fingerprint, but used their mating share as an indirect, but imperfect, cue to their success in the sperm lottery, increasing their share of parental effort in polyandry in response to an increase in their mating share (Davies *et al.* 1992).

The third eureka was the function of the strange mating display, in which the female elevates her tail to expose her cloaca for the male to peck, prior to copulation. This was first described by Edmund Selous (1933) who noted the "very wanton-looking pecks" and "rather lecherous" nature of the display, but it had remained a puzzle. I still remember my excitement when I first saw the female eject a droplet during the pecking, and suddenly realised that the male was stimulating the female to eject sperm from previous matings to make room for his own insemination (Davies 1983).

The most important finding of the dunnock study was that the life of these little brown birds was so dominated by sexual conflict, now a major theme in behavioural ecology (Arnqvist & Rowe 2005). Conflicts were sometimes easy to observe; for example, during the mating period the dominant male in a polyandrous trio attempts to guard the female to gain sole paternity of her brood, while she tries to escape his close attentions to give a share of the matings to the subordinate male too, to secure his additional help in chick feeding. This conflict provokes endless chases around the territory for ten days or so, until the female completes her clutch and begins incubation. Once the chicks hatch, provided both males gained matings, both help to feed the brood and there are no more visible squabbles. Nevertheless, we began to realise that conflict underlies these cooperative ventures too. This was revealed by a simple experiment; if one member of a polyandrous trio was temporarily removed during chick feeding, the others immediately increased their effort. If individuals have the capacity to work harder, how do they all come to an agreement on how much each should do? Alasdair Houston and I modelled this as a game in which each individual had a best response to the effort played by others. We showed that the key to stable cooperation by a pair or a trio was incomplete compensation to any reduction in effort by other individuals (Houston & Davies 1985). This model has since been developed further by McNamara *et al.* (1999) and by Johnstone and Hinde (2006).

One of the problems of working in a public place, like a Botanic Garden, is that people often come up and ask what you are doing. When I was on my hands and knees searching for sperm droplets, a truthful reply would be so time-consuming (and seem so unlikely, too) that I usually announced that I was weeding. However, explaining what was happening occasionally led to useful advice. I once was stopped by a nun. In this case I felt obliged to tell the truth and explained about the dunnock's polyandrous mating system. "Which male fathers the offspring?" she asked. It was during the early days of the study and at that stage I had to admit I didn't know. "Why not try protein polymorphisms as markers?" she suggested, helpfully. (I learnt then that she was a chemistry teacher at the nearby convent school). Another enlightening encounter was with a student gardener. I explained to him how in polygynandry, when both males shared matings with both females they usually divided their parental effort by one male helping full time at one female's nest, and the other helping full time at the other female's nest. "Perhaps that's what you'd expect from game theory," he suggested. He transpired to be a maths scholar from Oxford, and his suggestion turned out to be correct (Sozou & Houston 1994).

Alpine accentors in the Pyrenées

During the dunnock study, we tried not to forget the ecological stage for these behavioural conflicts and began to wonder whether, in harsher environments, food might become so scarce that the advantage to males from cooperating to raise offspring might increase sufficiently to offset their costs of shared paternity. In this case polyandry would be advantageous to both sexes (Gowaty 1981). This became the excuse to study a congener of the dunnock, the alpine accentor *Prunella collaris*, which inhabits high mountains in Europe through to Asia. For five summers (1990–94), I escaped the confines of the Cambridge Botanic Garden and lived in a mountain hut at 2000 m in the Reserve du Mont Valier, in the French Pyrénées, at various times together with André Desrochers, Ben Hatchwell, Ian Hartley, Naomi Langmore and Jackie Skeer.

We arrived in early June, just as the snow began to melt, and it was both exhilarating and a challenge to catch and follow the alpine accentors over their large territories, in the company of golden eagles, griffon vultures and lammergeiers soaring overhead. On some magical days the clouds, which formed at night in the valleys far below, would rise with the early morning sun and then remain just beneath us throughout the day, so we appeared to be marooned on mountain islands in a foamy white sea. The weather would often change suddenly; our mist nets once froze solid for two weeks and remained draped round the mountains like a ghostly shroud.

Our study site was dominated by an unnamed mountain which we referred to as "Accentor Peak". Here we discovered that the accentors breed in large polygynandrous groups in which up to four unrelated males and four unrelated females shared a territory. Each female had her own nest and, as in the dunnocks, females solicited matings from all the males on her territory. Several males sometimes helped to feed a brood and DNA finger-printing revealed that paternity was often split between two or even three males. However, we observed exactly the same male conflicts as in the dunnocks, with dominant males attempting to guard females to gain sole paternity while females tried to gain matings from subordinate males too. During the summer months, insect food became surprisingly abundant even on the high mountain tops, and for a male alpine accentor there were net costs of shared paternity, just as for dunnocks in the more benign conditions of the Botanic Garden (Davies *et al.* 1995, 1996; Hartley *et al.* 1995; see also Nakamura 1998).

The most striking difference from the dunnocks was that female alpine accentors sang to compete for male attention during their fertile period (Langmore *et al.* 1996). We interpreted this as a result of the increased direct competition among females for male care in the larger breeding groups compared with dunnocks. We subsequently showed that dunnocks in the Botanic Garden had the potential to adopt alpine accentor-like behaviour. When we increased competition between female dunnocks, by experimental removal of males, we induced the females to call and occasionally sing for male attention (Langmore & Davies 1997). Experimental manipulation of the dunnock's food supply to provide more widely dispersed patches of abundant food, as experienced by the alpine accentors, also led the dunnocks to adopt larger polygynandrous groups (Davies & Hartley 1996). These results show that individual behavioural strategies are flexible and modified by ecological conditions.

Three years after our alpine accentor project ended a new edition of the reserve map was published and to our delight we discovered that we had named a mountain; the unnamed peak was now officially "Pic des accenteurs". Long may these tough and wonderful birds enjoy the freedom of their mountain.

Cuckoo versus hosts

A second bird has also dominated my time in Cambridge during the past twenty years (Davies 2000). The common cuckoo *Cuculus canorus* is a harbinger of spring in northern Europe and is one of nature's most famous cheats. It never cares for its eggs or young but tricks other species into doing the work. The female cuckoo lays just one egg in each host nest. Soon after hatching, the cuckoo chick ejects the host eggs or young and so is raised alone. I saw my first cuckoo chick in a reed warbler's nest on Wicken Fen when I was an undergraduate. I was amazed that a warbler would feed the cuckoo even as it grew to ten times the warbler's own mass. Reed warblers can do some marvellous things: navigate by

Figure 6.2. A newly hatched cuckoo chick ejecting an egg from a reed warbler nest. It has already ejected a host chick. The host parent does nothing to interfere as it witnesses the destruction of its own reproductive success. Drawing by David Quinn. From Davies 2000.

the stars from African winter quarters to the Fens; select mates on subtle differences in song; pick out choice prey items based on cues such as colour and size. Why, in this context, were they apparently being so stupid?

I read Stephen Rothstein's elegant papers on cowbirds in North America (Rothstein 1975), where he had used experiments with model eggs to investigate host responses to brood parasitism, and decided to start an experimental study of host responses to cuckoos. In the 1920s, an egg collector and fine observer, Edgar Chance (1940), had shown that the common cuckoo has several host-races. Females of each race specialise on one host species and are characterised by a particular egg type that tends to match the eggs of its host in colour and spotting. We have now discovered from molecular genetic analysis that these host-races are restricted to female cuckoo lineages, with cross-mating by males maintaining the common cuckoo as one species (Gibbs *et al.* 2000). Chance found that the cuckoo timed its laying carefully, so as to coincide with the host's own laying period, and the laying was remarkably quick – within a ten second visit to the host nest. I wanted to discover whether these egg laying tactics had evolved to beat host defences. Cuckoo–host interactions seemed a promising model system for studying co-evolution by experiment.

In the first years of the study, I was joined by Mike Brooke and we began on Wicken Fen with the cuckoo host-race that specialised on reed warblers. We made model eggs and placed them in reed warbler nests to examine the significance of the cuckoo's egg laying procedure. We found that reed warblers were more likely to reject eggs that differed from their own in size or colour, and that rejection was enhanced further by the sight of a stuffed cuckoo on their nest. This explained why this cuckoo host-race has evolved a mimetic green spotted egg, and why it is so quick during laying, to avoid alerting the hosts (Davies & Brooke 1988). But why do the reed warblers need to be alerted to increase their rejection? We found that they sometimes made recognition errors and rejected one of their own eggs rather than the cuckoo egg. Calculations suggested that these costs were worth incurring only above a certain threshold level of parasitism. This explained the effect of the stuffed cuckoo; when parasitism probability increased, rejection became more worthwhile. Rejection costs also explain natural variation in reed warbler rejection rates of foreign eggs. We found that more heavily parasitized populations showed stronger rejection of our model eggs than nearby unparasitised populations. Furthermore, on Wicken Fen rejection rates have declined over recent years as cuckoos have declined. This change is too fast to reflect genetic change in the host population and is more likely a result of individual flexibility in their behaviour. Reed warblers must monitor local cuckoo abundance and vary their egg rejection accordingly (Davies *et al.* 1996; Brooke *et al.* 1998).

Mike and I extended our simple experiments with model eggs to investigate egg rejection by other species. We were pleased when one of Britain's most experienced nest recorders, Bruce Campbell, unwittingly came across one of our experimental nests and recorded our model as a real cuckoo egg. On another occasion, we were on remote moors in Derbyshire when a ghostly tall figure came out of the mist. As it approached we saw the high helmet and realised it was a policeman in full uniform. He suspected we were egg collectors and was reluctant to believe us when we explained that far from removing eggs we were putting extra eggs into birds' nests. Our permit eventually reassured him. These experiments with a range of species enabled us to reconstruct the stages of the egg arms race between cuckoo and host (Davies & Brooke 1989a, b). First, before parasitism, hosts showed no egg rejection (species unsuitable as hosts, hence with no history of parasitism, exhibit no rejection). Second, in response to parasitism hosts evolve rejection (species subject to parasitism tend to reject). Third, in response to host rejection cuckoos evolve egg mimicry (cuckoo egg mimicry is better in host-races with more strongly rejecting hosts; Brooke & Davies 1988). Finally, we found some species, rarely used as hosts, which had very strong rejection and suggested that these were old cuckoo favourites which retain rejection as a ghost of past adaptation – a legacy of the arms race their ancestors ran in previous generations.

We were left with the puzzle of why hosts so fussy at the egg stage accept a cuckoo chick so unlike their own. We argued that early defences are more beneficial, and recognition of foreign chicks might be more difficult given that hosts expect rapid changes in the appearance of their own young. Nevertheless, there are simple, foolproof rules that a reed warbler could use, for example "only feed chicks with tongue spots" (their own young have these, cuckoo chicks do not). Furthermore, some hosts of other cuckoo species have evolved chick rejection – why not reed warblers? The puzzle remains unsolved.

We did, however, make progress with the problem of how the cuckoo chick stimulates the reed warbler foster parents into bringing as much food as for a whole brood of their own young. Becky Kilner, David Noble and I showed by experiment that the cuckoo chick's remarkably rapid begging call, which sounds like many hungry host young, provides the key stimulus. When we placed a single blackbird chick in a reed warbler nest, the warblers did not provision it as much as a cuckoo of the same mass. However, when we gave the blackbird a helping hand and broadcast cuckoo begging calls every time it begged, the warblers immediately boosted their provisioning to cuckoo-like levels (Davies *et al.* 1998). Solving a practical problem during this experiment gave me as much pleasure as this final result. When we first put a blackbird into a reed warbler nest, it crouched and would not beg. We realised that the blackbird was used to a more stable nest and was distressed by a nest that swayed in the reeds (where reed warblers feel quite at home). Once we attached the supporting reeds to a stable bamboo cane, the blackbird begged beautifully and we were able to perform the experiment.

We then went on to show experimentally how the provisioning effort of the reed warbler parents depended on both visual stimuli (gapes) and vocal stimuli (begging call rate) from the young. This effort was calibrated in relation to what the warblers would normally expect in their nest, namely a brood of four young reed warblers. The cuckoo chick's problem was that its visual stimulus of just one gape was deficient compared with that of a reed warbler brood; to compensate, the cuckoo had to boost its calling to supernormal rates to persuade the foster parents to bring it as much food as for four reed warbler chicks. The cuckoo's key trick, therefore, was to tune into the way the foster parents integrated begging signals from their own brood (Kilner *et al.* 1999).

We have recently discovered that cuckoos have host-specific adaptations at the nestling stage as well as at the egg stage. These include becoming attuned to their hosts' parental alarm calls, given to silence their own young when a predator is nearby (Davies *et al.* 2006), and developing begging calls suited to exploit the provisioning of their particular host species (Madden & Davies 2006). Remarkable tricks are being found in other brood parasites (Langmore *et al.* 2003; Kilner *et al.* 2004; Takeda & Ueda 2005). I continue to be intrigued by cuckoo–host interactions; they provide a curious mixture of subtle adaptation and surprising lack of adaptation. Is the latter a case of evolutionary lag in a continuing arms race? Or are some defences so costly that apparent maladaptation is an equilibrium set by the balance of costs and benefits? Further molecular genetic analyses will enable us to calibrate when cuckoos and their host-races evolved from a common ancestor and to resolve the time course of these marvellous interactions.

Looking back

My career might seem an unadventurous one: from Cambridge to Oxford and back again; and from reed warblers via wagtails, butterflies, toads, dunnocks and cuckoos, right back full circle to reed warblers. But the scientific journey has been exciting and these familiar species, within a cycle ride of home, have provided endless fascination. When replying to an invitation to give a plenary lecture at the 1992 International Conference in Behavioural Ecology at Princeton, I offered either a talk on dunnocks or one on cuckoos. The organisers replied in apparent panic: "Oh no, we want something much broader in scope". So I relented and gave a talk on both dunnocks and cuckoos together. For me, these (and the other species) have provided models to study broader problems (territory economics, contest resolution, mating and parental conflicts, coevolution), but within systems where I could get to know the natural history first before testing theory. I feel lucky to have begun research at a time when so many new ideas were emerging to inspire my bird watching. The encouragement of established figures in the field has also been an important stimulus. I treasure, in particular, memories of visits to Staffan Ulfstrand's group at Uppsala, and to Frank McKinney's group in Minnesota. At international conferences I now look around the sea of young faces and wonder who will be the next Parker or Trivers, Dawkins or Krebs, Hamilton or Maynard Smith, to inspire a future generation of field workers.

Cambridge continues to provide wonderful colleagues and students. I am particularly proud of my scientific family of PhD students (38 so far), many of whom have gone on to distinguished careers in biological research, teaching or conservation. They have provided friendship and inspiration, and have ensured I escape the Cambridge gardens and fens for wilder and more far-flung places, from time to time.

I have also valued the sense of history from working in an ancient University. I think back to William Turner, who became a fellow of my College, Pembroke, in 1530 and was the first major student of natural history in this country. His tutor, and Master of the College, Nicholas Ridley, was burnt to death as a martyr during Catholic Queen Mary's reign. Turner, also a Protestant, fled to continental Europe until Queen Elizabeth I ascended the throne. Turner (1544) wrote the first book on birds to treat them in a modern scientific spirit. For example, he described the differences between robins and redstarts in plumage, voice and behaviour and used these to refute the idea that one species could transmutate into the other in the autumn. I wonder whether my current concerns about cuckoos will seem equally quaint in another five hundred years.

Looking back, I am struck by two recurring themes. The first is that the simplest and most satisfying ideas, which seem so obvious in retrospect, often come only after a struggle. The second is to treasure data that don't fit the theory. Over the years I have been exasperated in turn by birds and butterflies that so easily forsake hard-won territories, by dunnocks that share mates in complex arrangements and by hosts that accept cuckoo eggs unlike their own. But in each case these contradictions have been the key to new, interesting ideas.

While the scientific adventure has been fun, for me the aesthetic pleasures of watching birds will always be equally important. I am still as thrilled as ever by the first wheatear of spring and by skeins of pink footed geese against a winter sky. My watching has been inspired by artists, particularly Eric Ennion (1900–1981), his student John Busby (born 1923), and Busby's student, the Norfolk artist James McCallum (born 1967). With a few bold lines, they capture a moment of behaviour and convey the essential "jizz" of a living bird. Their art has introduced me to a new way of looking at nature, to complement the fresh look inspired by new scientific ideas. I hope the discoveries of behavioural ecology will combine with ideas from art and literature to bring increased respect for our diminishing natural world and help ensure its survival, both for its own sake and for our future generations.

References

Arnqvist, G. & Rowe, L. (2005). *Sexual Conflict*. Princeton, NJ: Princeton University Press.

Bergman, M., Gotthard, K., Berger, D. *et al.* (2007). Mating success of resident versus non-resident males in a territorial butterfly. *Proc. R. Soc. Lond.* B**274**: 1659–65.

Birkhead, M. E. (1981). The social behaviour of the dunnock, *Prunella modularis. Ibis* **123**: 75–84.

Brooke, M. de L. & Davies, N. B. (1988). Egg mimicry by cuckoos *Cuculus canorus* in relation to discrimination by hosts. *Nature* **335**: 630–2.

Brooke, M. de L., Davies, N. B. & Noble, D. G. (1998). Rapid decline of host defences in response to reduced cuckoo parasitism: behavioural flexibility of reed warblers in a changing world. *Proc. R. Soc. Lond.* B**265**: 1277–82.

Brown, J. L. (1964). The evolution of diversity in avian territorial systems. *Wilson Bull.* **76**: 160–9.

Brown, J. L. (1975). *The Evolution of Behaviour*. New York: W.W. Norton and Co.

Burke, T., Davies, N. B., Bruford, M. W. & Hatchwell, B. J. (1989). Parental care and mating behaviour of polyandrous dunnocks *Prunella modularis* related to paternity by DNA fingerprinting. *Nature* **338**: 249–51.

Campbell, B. (1952). *Bird Watching for Beginners*. Harmondsworth: Penguin.

Chance, E. P. (1940). *The Truth About the Cuckoo*. London: Country Life.

Clutton-Brock, T. H. & Harvey, P. H. (1977). Primate ecology and social organisation. *J. Zool. Lond.* **183**: 1–39.

Crook, J. H. (1964). The evolution of social organisation and visual communication in the weaver birds (Ploceinae). *Behaviour Suppl.* **10**: 1–178.

Crook, J. H. & Gartlan, J. S. (1966). Evolution of primate societies. *Nature* **210**: 1200–3.

Darwin, C. (1839). *Voyage of the Beagle*. London: Murray.

Darwin, C. (1859). *On the Origin of Species*. London: Murray.

Davies, N. B. (1976). Parental care and the transition to independent feeding in the young spotted flycatcher (*Muscicapa striata*). *Behaviour* **59**: 280–95.

Davies, N. B. (1978a). Parental meanness and offspring independence: an experiment with hand-reared great tits (*Parus major*). *Ibis* **120**: 509–14.

Davies, N. B. (1978b). Territorial defence in the speckled wood butterfly (*Pararge aegeria*): the resident always wins. *Anim. Behav.* **26**: 138–47.

Davies, N. B. (1983). Polyandry, cloaca-pecking and sperm competition in dunnocks. *Nature* **302**: 334–6.

Davies, N. B. (1992). *Dunnock Behaviour and Social Evolution*. Oxford: Oxford University Press.

Davies, N. B. (2000). *Cuckoos, Cowbirds and Other Cheats*. London: T. & A. D. Poyser.

Davies, N. B. & Brooke, M. de L. (1988). Cuckoos versus reed warblers : adaptations and counteradaptations. *Anim. Behav.* **36**: 262–84.

Davies, N. B. & Brooke, M. de L. (1989a). An experimental study of co-evolution between the cuckoo *Cuculus canorus* and its hosts. I. Host egg discrimination. *J. Anim. Ecol.* **58**: 207–24.

Davies, N. B. & Brooke, M. de L. (1989b). An experimental study of co-evolution between the cuckoo *Cuculus canorus* and its hosts. II. Host egg markings, chick discrimination and general discussion. *J. Anim. Ecol.* **58**: 225–36.

Davies, N. B. & Green, R. E. (1976). The development and ecological significance of feeding techniques in the reed warbler (*Acrocephalus scirpaceus*). *Anim. Behav.* **24**: 213–29.

Davies, N. B. & Halliday, T. R. (1978). Deep croaks and fighting assessment in toads, *Bufo bufo*. *Nature* **274**: 683–5.

Davies, N. B. & Hartley, I. R. (1996). Food patchiness, territory overlap and social systems : an experiment with dunnocks, *Prunella modularis*. *J. Anim. Ecol.* **65**: 837–46.

Davies, N. B. & Houston, A. I. (1981). Owners and satellites: the economics of territory defence in the pied wagtail, *Motacilla alba*. *J. Anim. Ecol.* **50**: 157–80.

Davies, N. B. & Houston, A. I. (1983). Time allocation between territories and flocks and owner-satellite conflict in foraging pied wagtails, *Motacilla alba*. *J. Anim. Ecol.* **52**: 621–34.

Davies, N. B. & Houston, A. I. (1986). Reproductive success of dunnocks *Prunella modularis* in a variable mating system II. Conflicts of interest among breeding adults. *J. Anim. Ecol.* **55**: 139–54.

Davies, N. B., Hatchwell B. J., Robson, T. & Burke, T. (1992). Paternity and parental effort in dunnocks *Prunella modularis*: how good are male chick-feeding rules.? *Anim. Behav.* **43**, 729–45.

Davies, N. B., Hartley, I. R., Hatchwell, B. J., *et al.* (1995). The polygnandrous mating system of the alpine accentor *Prunella collaris*. I. Ecological causes and reproductive conflicts. *Anim. Behav.* **49**: 769–88.

Davies, N. B., Hartley, I. R., Hatchwell, B. J. & Langmore, N. E. (1996). Female control of copulations to maximize male help : a comparison of polygynandrous dunnocks *Prunella modularis* and alpine accentors *P. collaris*. *Anim. Behav.* **51**: 27–47.

Davies, N. B., Brooke, M. de L. & Kacelnik, A. (1996). Recognition errors and probability of parasitism determine whether reed warblers should accept or reject mimetic cuckoo eggs. *Proc. R. Soc. Lond.* **B263**: 925–31.

Davies, N. B., Kilner, R. M. & Noble, D. G. (1998). Nestling cuckoos *Cuculus canorus* exploit hosts with begging calls that mimic a brood. *Proc. R. Soc. Lond.* **B265**: 673–8.

Davies, N. B., Madden, J. R., Butchart, S. H. M. & Rutila, J. (2006). A host-race of the cuckoo *Cuculus canorus* with nestlings attuned to the parental alarms of the host species. *Proc. R. Soc. Lond.* B **273**: 693–9.

Dawkins, R. (1976). *The Selfish Gene*. Oxford: Oxford University Press.

Emlen, S. T. & Oring, L. W. (1977). Ecology, sexual selection and the evolution of mating systems. *Science* **197**: 215–23.

Gibbs, H. L., Sorenson, M. D., Marchetti, K. *et al.* (2000). Genetic evidence for female host-specific races of the common cuckoo. *Nature* **407**: 183–6.

Gombrich, E. H. (1972). *The Story of Art*. London: Phaidon.

Gowaty, P. A. (1981). An extension of the Orians-Verner-Willson model to account for mating systems besides polygyny. *Am. Nat.* **118**: 851–9.

Grafen, A. (1987). The logic of divisively asymmetric contests: respect for ownership and the desperado effect. *Anim. Behav.* **35**: 462–7.

Hamilton, W. D. (1964). The genetical evolution of social behaviour. I., II. *J. Theor. Biol.* **7**, 1–52.

Hartley, I. R., Davies, N. B., Hatchwell, B. J. *et al.* (1995). The polygynandrous mating system of the alpine accentor *Prunella collaris*. II Multiple paternity and parental effort. *Anim. Behav.* **49**: 789–803.

Houston, A. I. & Davies, N. B. (1985). The evolution of cooperation and life history in the dunnock *Prunella modularis*. In *Behavioural Ecology: Ecological Consequences of Adaptive Behaviour*, ed. R. M. Sibly & R. H. Smith, pp. 471–87. Oxford: Blackwell Scientific Publications.

Johnstone, R. A. & Hinde, C. A. (2006). Negotiation over offspring care – how should parents respond to each other's efforts? *Behav. Ecol.* **17**: 818–27.

Kilner, R. M., Noble, D. G. & Davies, N. B. (1999). Signals of need in parent-offspring communication and their exploitation by the common cuckoo. *Nature* **397**: 667–72.

Kilner, R. M., Madden, J. R. & Hauber, M. E. (2004). Brood parasitic cowbird nestlings use host young to procure parental resources. *Science* **305**: 877–9.

Krebs, J. R. & Davies, N. B. (eds.) (1978). *Behavioural Ecology: An Evolutionary Approach*. Oxford: Blackwell Scientific Publications. (2nd edn 1984; 3rd edn 1991; 4th edn 1997.)

Krebs, J. R. & Davies, N. B. (1981). *An Introduction to Behavioural Ecology*. Oxford: Blackwell Scientific Publications. (2nd edition 1987; 3rd edition 1993.)

Krebs, J. R., Erichsen, J. T., Webber, M. I. & Charnov, E. L. (1977). Optimal prey selection in the great tit, *Parus major*. *Anim. Behav.* **25**: 30–8.

Lack, D. (1956). *Swifts in a Tower*. London: Methuen.

Lack, D. (1965). *The Life of the Robin*. London: Witherby.

Lack, D. (1966). *Population Studies of Birds*. Oxford: Clarendon Press.

Lack, D. (1968). *Ecological Adaptations for Breeding in Birds*. London: Methuen.

Lack, D. (1971). *Ecological Isolation in Birds*. Oxford: Blackwell Scientific Publications.

Langmore, N. E. & Davies, N. B. (1997). Female dunnocks use vocalizations to compete for males. *Anim. Behav.* **53**: 881–90.

Langmore, N. E., Davies, N. B., Hatchwell, B. J. & Hartley, I. R. (1996). Female song attracts males in the alpine accentor *Prunella collaris*. *Proc. R. Soc. Lond.* B**263**: 141–6.

Langmore, N. E., Hunt, S. & Kilner, R. M. (2003). Escalation of a co-evolutionary arms race through host rejection of brood parasitic young. *Nature* **422**: 157–60.

MacArthur, R. H. (1972). *Geographical Ecology*. New York: Harper & Row.

MacArthur, R. H. & Pianka, E. R. (1966). On the optimal use of a patchy environment. *Am. Nat.* **100**: 603–9.

McNamara, J., Gasson, C. & Houston, A. (1999). Incorporating rules for responding into evolutionary games. *Nature* **401**: 368–71.

Madden, J. R. & Davies, N. B. (2006). A host-race difference in begging calls of nestling cuckoos *Cuculus canorus* develops through experience and increases host provisioning. *Proc. R. Soc. Lond.* B**273**: 2343–51.

Maynard Smith, J. & Parker, G. A. (1976). The logic of asymmetric contests. *Anim. Behav.* **24**: 159–75.

Nakamura, M. (1998). Multiple mating and cooperative breeding in polygynandrous alpine accentors. I. Competition among females. *Anim. Behav.* **55**: 259–75.

Orians, G. H. (1969). On the evolution of mating systems in birds and mammals. *Am. Nat.* **103**: 589–603.

Parker, G. A. (1970). Sperm competition and its evolutionary consequences in the insects. *Biol. Rev.* **45**: 525–68.

Parker, G. A. (1974). The reproductive behaviour and the nature of sexual selection in *Scatophaga stercoraria*. IX. Spatial distribution of fertilization rates and evolution of male search strategy within the reproductive area. *Evolution* **28**: 93–108.

Parker, G. A. (1979). Sexual selection and sexual conflict. In. *Sexual Selection and Reproductive Competition in Insects*, ed. M. S. Blum & N. A. Blum, pp. 123–66. New York: Academic Press.

Pyke, G. H., Pulliam, H. R. & Charnov, E. L. (1977). Optimal foraging: a selective review of theory and tests. *Q. Rev. Biol.* **52**: 137–54.

Rothstein, S. I. (1975). Evolutionary rates and host defences against avian brood parasitism. *Am. Nat.* **109**: 161–76.

Selous, E. (1933). *Evolution of Habit in Birds*. London: Constable.

Snow, B. K. & Snow, D. W. (1982). Territory and social organisation in a population of dunnocks, *Prunella modularis*. *J. Yamashina Inst. Ornithol.* **14**: 281–92.

Sozou, P. D. & Houston, A. I. (1994). Parental effort in a mating system involving two males and two females. *J. Theor. Biol.* **171**: 251–66.

Takeda, K. D., Ueda, K. (2005). Horsfield's hawk-cuckoo nestlings simulate multiple gapes for begging. *Science* **308**, 653.

Tinbergen, N. (1953). *The Herring Gull's World*. New Naturalist Series. London: Collins.

Tinbergen, N. (1974). *Curious Naturalists*. Revised edition. Harmondsworth: Penguin Books.

Trivers, R. L. (1972). Parental investment and sexual selection. In *Sexual Selection and the Descent of Man*, ed. B. Campbell, pp. 139–79. Chicago: Aldine.

Trivers, R. L. (1974). Parent-offspring conflict. *Am. Zool.* **14**: 249–64.

Trivers, R. L. & Hare, H. (1976). Haplodiploidy and the evolution of social insects. *Science* **191**: 249–63.

Turner, W. (1544). *A Short and Succinct History of the Principal Birds Noticed by Pliny and Aristotle*. Edited by A. H. Evans (1903). Cambridge: Cambridge University Press.

Wilson, E. O. (1975). *Sociobiology: The New Synthesis*. Cambridge, MA: Belknap Press of Harvard University Press.

Wynne-Edwards, V. C. (1962). *Animal Dispersion in Relation to Social Behaviour*. Edinburgh: Oliver & Boyd.

7

King Solomon's herring gull's world

MARIAN STAMP DAWKINS

I can remember the exact day of the exact year when I knew for certain that what I wanted to do more than anything else in life was to study animal behaviour, although at the time I had no idea that it would be possible to make a career out of it. I was 11 and we were living in London in a large house that was a great place to grow up in apart from the fact that it had no proper guest room. This meant that whenever my parents had anyone to stay (which was quite often) I was required to move out of my own room into a tiny box bedroom at the top of the house to make way for the guest. On these occasions, I would be under strict instructions to take as many of my animals with me as could be moved. Fish in tanks could be left behind but everything else – stick insects, snails, silkworms, etc. – had to be removed. In particular, all hamsters had to be taken upstairs because of the noise they made during the night.

That year, 1956, my parents must have had a particularly full social calendar because it seemed to me that that I was forever being told to pack up my menagerie to make way for the next guest. So when I was told, yet again, to vacate my bedroom, I decided to take action, although, in retrospect, it was somewhat misdirected action. I obediently moved out of my room but I deliberately left my hamsters behind, hidden so that they escaped my mother's notice. That night, tucked away in the attic, I had a perverse pleasure at the thought of the guest occupying *my* room but kept awake all night by *my* hamsters.

Leaders in Animal Behavior: The Second Generation, ed. L. C. Drickamer & D. A. Dewsbury. Published by Cambridge University Press. © Cambridge University Press 2010.

Figure 7.1. Aged three. Photograph by my father, Max Stamp.

But I had reckoned without the guest, who turned out to be Leonard Waight, from the British Treasury. Far from being annoyed or even apparently disturbed by his night with the hamsters, he seemed positively delighted by the experience. Two days later I received a book with the inscription: 'To Marian, from one animal lover to another'. The book was *King Solomon's Ring* by Konrad Lorenz and that, you might say, was that.

The book was about the behaviour not only of hamsters but of water voles and jackdaws and geese. It was about what their worlds were like and how they communicated with each other. The message of the book was that if you were prepared to be patient and watch and listen, you could enter their worlds and communicate with them, not as fiction, like Dr Doolittle communicating with animals in perfect English, but as fact. Here was someone saying that it was possible to enter the lives of real animals and communicate with them in their own language.

I knew then what I wanted to do but I was repeatedly told by my parents, my school, uncles and aunts that it was not possible to study animal behaviour or animal psychology or animal minds because these were not proper subjects. Genetics was a subject. Physiology was a subject. But animal behaviour was definitely not. I therefore resigned myself to becoming a veterinary surgeon to earn a living but having a houseful of animals, just like Konrad Lorenz, as a sideline.

Figure 7.2. Niko Tinbergen. Photograph by Lary Shaffer.

Then, when I was 14, I came across another book, this time in the public library. This one was called *Herring Gull's World* and it was by a Dutchman called Niko Tinbergen. Evidently people were more enlightened in the Netherlands because over there animal behaviour was seen as a proper subject, at least to judge by the amount of time Tinbergen seemed to manage to spend watching gulls. I initially did not take in that he had spent much of this gull-watching time in the north of England and it was only as I was taking the book back to the library that I noticed a single electrifying sentence inside the front dust jacket to the effect that Tinbergen was teaching animal behaviour at Oxford University. The Dutchman was living in England! To my astonishment, I discovered that it was possible to go to Oxford, spend three years reading a whole degree in Zoology and go to lectures in animal behaviour given by Niko Tinbergen. It was my idea of heaven.

Heaven at that time seemed a very long way off. I was at an all-girls' school where I had only ever heard of one person getting into Oxford and everyone else seemed more interested in Cliff Richard than Charles Darwin. My physics was not good and my chemistry not much better. Fortunately at that time, you had to sit an entrance exam to get into Oxford and this put less emphasis on facts than on getting you to think about all sorts of interesting questions such as evolution and mimicry and even behaviour. I remember writing an essay on whether behaviour could evolve and even bringing in Lorenz's ideas about ducks and geese.

Somehow, I got into Somerville College, then one of only 5 colleges that admitted women to Oxford. What a place! Oxford was full of people who didn't think it was at all odd to be interested in animals and plants and philosophy and indeed almost anything else you cared to think about. In fact, my own knowledge of animals (gathered largely from trips to the Natural History Museum, the London Zoo and summer holidays in Ireland spent combing rock pools, ponds and peat bogs for anything that was alive) was rudimentary compared with not just that of my lecturers and tutors but that of my fellow students as well, one of whom was John Krebs. I could never take John at all seriously in those days as he spent so much of his time planning and carrying out the most extraordinarily elaborate practical jokes, mostly on the people who lectured to us, notably E. B. Ford.

We learnt geology and chemistry and botany, as well as being expected to have a detailed knowledge of every single phylum in the animal kingdom. At the end of my first year at Oxford, my college tutor, Wilma Crowther (Wilma George) solemnly told me that she hoped that by now I was thoroughly confused. I was simultaneously outraged and thrilled – outraged because I wanted things explained and tidied up and thrilled because I felt liberated from a year of being lectured to. So it was alright not to believe everything that Ernst Mayr or Theodosius Dobzhansky said! I had thought that because they wrote such big books and wrote with such authority, what they said must be true. For the first time in my life I was being encouraged to disbelieve everything. Confusion was not a failure of the intellect but actually the start of possibly finding out something new. If there is one thing I regret about modern university education it is this fear of confusing people. Lecture handouts, with neat bullet points listing the 'take-home' messages that need to be remembered, are the very opposite of the floundering confusion that Wilma wished on me. How cross I was at the time and how grateful ever afterwards.

During my second year, I interspersed Zoology with being President of the Oxford University Humanist Group, going on various marches in London and Oxford in support of homosexual and abortion law reform and acting in plays. I was never very good at acting but it was a lot of fun. Once, I played the goddess Athena in an open-air production of Aeschylus' *Oresteia* in Keble College gardens and had to deliver speeches into the gathering darkness from the top of a scaffold. More importantly during my second year, the day eventually came when we were to have lectures in animal behaviour from Niko Tinbergen.

The lectures were all I had hoped they would be, and more. Niko, a neat figure with a shock of white hair, always lectured in what looked like field gear – khaki shirt and trousers with (as I remember it) gym shoes or sneakers. It was as if he had dropped in to lecture to us but was really longing to get back to the sand-dunes, which he probably was. As Hans Kruuk documents in his biography (2003), Niko never particularly liked the college system in Oxford and was always much happier out in the field. His lectures were probably a necessary chore for him. For us, they were magic.

Niko described his own experiments in the field. He told us how you could present a cardboard cut-out of a gull head to a herring gull chick and it would peck at it as if it were the head of its own parent. He even described experiments in which he could make 'super-normal' stimuli. He could make the gull chick respond more to his cardboard models than they did to the real thing. How extraordinary, I thought. Why should that happen? I walked back from the lecture towards college with a feeling of pity for all of my fellow students who were reading law or history or indeed anything except Zoology and animal behaviour.

What impressed me most about Niko's lectures, apart from the sheer number of fascinating questions they threw up, was that he had found ways of testing hypotheses in the field. It wasn't just that he speculated about what might be the case and leave it at that. He carried out

Figure 7.3. Mike Cullen. Photograph by Marian Dawkins (taken in 1989, after Mike had moved to Monash).

experiments to see whether he was right. He distinguished between different hypotheses by making predictions from each of them and finding out which one best fitted the facts. Experiments done out on the sand-dunes were thus just as scientific as experiments done in the laboratory but they were more difficult to do. It actually took more skill to do a really good field experiment because there were so many things that could go wrong and so many factors that could affect the behaviour than in a clean tidy laboratory. The reward was understanding what real animals did out there in the real world.

Teaching us how to go about doing experiments in animal behaviour for ourselves was the aim of a 'block practical' in which we all spent two weeks carrying out research projects of our own under the watchful eye of J. M. (Mike) Cullen. Mike is hardly known outside the circle of people who knew him but everyone he worked with held him in deep and affectionate awe. He never published much himself but his name should have been on many tens of papers to which he made a major (sometimes the major) contribution. He appeared at the block practical in one of the bright orange sweaters knitted by his mother that were his trademark, hunched in his characteristic listening posture as he quizzed us all individually on what we were doing, gesticulating with his large and expressive hands. In the space of two weeks he instilled in us an understanding of experimental design and statistical methods together with the realisation that questioning your own results was not a sign of weakness but a sign of progress. I very much hoped that I would be able to have tutorials (the interactive teaching method that Oxford emphasizes for all undergraduates) with either

Mike or Niko himself. I was somewhat disappointed to find that I was going to have to make do with an inexperienced graduate student called Richard Dawkins.

Tutorials with Richard turned out to be very inspiring and they opened my eyes to yet more aspects of animal behaviour such as development, details of mechanisms and the role of behaviour in speciation. One week he gave me a whole thesis to read and criticise. In the course of reading for another set of tutorials in ecology (this time with Kitty and Mick Southern), I came across the concept of 'the search image' – the idea that birds and other predators might be searching for cryptic prey and at first fail to see them. Then, with a little practice, they would 'get their eye in' and be able to see what they had simply failed to see before. There was supposed to be an 'ah ha!' moment when a predator broke the camouflage and the prey was revealed. All this made perfect sense to me but I was disconcerted to find that there was very little evidence for it at all. Niko's brother, Luuk Tinbergen, had suggested that it might be involved in explaining why bird predators in the wild may not eat a new type of prey when it first appears (1960). The prey enjoys a honeymoon period when it appears to be immune from attack and there can be quite a time lag before it becomes a regular part of the predator's diet. I realised, however, that just because a predator does not take a prey type at first that does not mean that it failed to see it. It might have seen it but not realised that it was food. Or it may not have been looking in the right place. Or it may have been looking in the right place, realised it was food, but not known how to attack or deal with it. All of these alternative mechanisms could cause the predator to fail to eat prey at first but none of them should strictly be called 'search image', which implies a change in what the predator *sees*. I searched the literature for evidence that someone, somewhere had made this distinction experimentally and that there was direct evidence for search image formation. Surely people wouldn't continue to talk about search images unless they had evidence for them, would they? It appeared that they would and that there was no direct evidence for search image formation in any species, despite the widespread use of this term in both the behavioural and the ecological literature.

After discussing this anomaly with anyone who would spare me the time, including Niko, I decided that this would be a suitable subject for a doctoral thesis and was thrilled when Niko said he would take me on as his research student.

During the summer after taking my final exams, I visited Niko's famous field site at Ravenglass in Cumbria in the north of England. Out on the sand-dunes near the black-headed gull colony there was a collection of small tents and a trailer that served as a kitchen and shelter for the worst weather. Harvey Croze with his wife Nani and a new born baby were living in one of the tents; Harvey was already working on search images so it was his work I most wanted to see. He had devised a technique of hiding pieces of meat under different coloured mussel shells laid out on the sand-dunes and then waiting for carrion crows to fly down and find them. The shells were cryptic when seen against a background of pebbles and sand and so it was possible to study the behaviour of the crows when they first encountered a new colour of mussel and then see how this changed as they learnt what they were looking for. Perhaps this would be a way of making the distinction between search image formation and other kinds of learning that I knew needed to be made.

In the autumn of 1966, I started work at 13 Bevington Road in Oxford, the home at that time of Niko's research group (the ABRG or Animal Behaviour Research Group), although Niko himself had an office in the main Zoology building next to the University Museum. No. 13 Bevington Road was a tall, rambling Victorian house about 10 minutes' walk away from the main Zoology Department. Both its distance and its interior gave the ABRG a very distinct identity. The basement was lined with fish tanks filled with sticklebacks, minnows and pike and there was a garage out the back full of extremely aggressive gulls belonging to Juan Delius. Nothing was ever thrown away, so cupboards were full of useful bits and pieces and old electrical equipment that would be instantly condemned by today's health and safety rules. Bird cages, recording equipment and tanks of fish were to be found on every floor and at the very top of the house was an empty room bearing the sign: 'Nelson is on the Bass Rock', which referred to the fact that Bryan Nelson was away watching gannets. The heart of the building was a minute kitchen into which we all squeezed every lunchtime, with someone inevitably having a fry-up and everyone else trying to make coffee at the same time. Once a week, we all went off to the main Department for a seminar presided over by Niko and timed especially early so that he could get home in time to watch Z Cars on television.

I think it was a considerable disappointment to Niko that I decided in the end not to study search images in the field. Days spent waiting for crows and great tits to come to baits that I laid out for them convinced me, however, that it was not going to be possible to distinguish true search image formation from all the other things predators can learn about their prey with such a set-up. I needed to be able to exert much more control over my experiments and so I reluctantly decided that, at least as a first step, I would use domestic chicks as my predator and work with them in small controlled pens. This turned out to be the right

Figure 7.4. Chick learning to see cryptic food. Photograph by Lary Shaffer.

decision. By varying the degree of resemblance between prey and background, I was able to show that chicks showed changes in their prey-finding ability that were quite specific to cryptic prey and did not occur when prey were conspicuous. This strongly suggested a change in what they saw – real search image formation (Dawkins 1971a, b). I also believed at the time that these changes were occurring even when the bird was looking directly at something and that 'learning to see' could therefore not be put down to 'learning to look in the right place', although subsequent work on problems of showing exactly where a bird is looking made me wonder whether this conclusion is completely watertight (Guilford & Dawkins 1987).

There was another reason why the choice of chicks as the animal to work on turned out to be a good one. In the summer of 1967, Richard Dawkins and I decided to get married and, as Richard had been offered a job as Assistant Professor at the University of California in Berkeley, we set off a couple of months later to start a new life in California. There, the great hospitality and kindness of George Barlow allowed me to continue my doctoral work under the long-distance supervision of Niko. Californian chicks turned out to have just the same problems in seeing my camouflaged prey as the ones I had left behind in Oxford, and to be just as good at learning to break the camouflage too.

Berkeley in the late 1960s meant hippies, flower power, anti-war demonstrations, and the shock of the assassinations of Robert Kennedy and Martin Luther King. It was a time of great idealism as many of the graduate students we knew decided they would rather leave the country than fight in the Vietnam war. It was a time of excitement and going to work past lines of police with helicopters buzzing overhead and the smell of teargas lingering everywhere. But it was also the time when we joined George Barlow's research group and attended seminars by his students. Richard had to lecture to large numbers of students but

Figure 7.5. George Barlow (left) with Richard Dawkins in Berkeley, 1968.

I was able to get on with my research. It was a good time, unique in both time and space, as we discovered whenever we ventured outside Berkeley. Berkeley was not typical of the United States. We could feel at home there but not necessarily want to settle permanently in America. So when the opportunity arose to go back to the UK, we took it. We watched the moon landing on TV, packed up the car and drove across the States from Berkeley to New York, taking in various friends, relatives and tourist spots on the way. On the quayside in New York, we sold the car and boarded the SS *France* for England.

So it was that, for the second time in my life, I found myself sailing past the shining white cliffs of southern England and into Southampton Water after a period of exile in the States. The first time had been when I was 9 years old and my family were returning from 3 years in Washington, D.C., where my father had been posted on secondment from the Bank of England. At that time, I hated leaving America. I hated leaving my school; my friends and I spoke (at least at school) with a strong American accent. Now, sixteen years later, it was coming home.

Richard and I settled back into Oxford. He was now a lecturer in animal behaviour and I had a job as Niko's research assistant, helping him to prepare two volumes of his papers for publication (*The Animal in its World*, 1972, 1973). I submitted my thesis and my two examiners, Mike Cullen and Aubrey Manning, were very encouraging about the search image work. But Bevington Road days were numbered. A huge new concrete building was now to house the entire Zoology Department, including its subgroups such as the ABRG and the Edward Grey Institute for Ornithology. All of us, I think, regretted having to move out of Bevington Road, and feared that the eccentric, idiosyncratic group we all loved being part of would cease to exist. Of course it was the end of an era but to this day the ABRG retains a distinct identity and stubbornly refuses to either split or merge with other research groups.

The 1970s was, however, a rather unsettled, transitional time in the study of animal behaviour. Gone were the old certainties, such as that the Lorenz–Tinbergen psycho-hydraulic model of motivation was a good explanation of much, if any behaviour. Gone too were the research questions that had occupied many people in the previous two decades, such as whether displacement activities were best explained by the 'sparking over' hypothesis or the disinhibition hypothesis. What was less clear was what were going to be the big questions of the future and whether ethology would hold together as the single '4 question' discipline that Niko had defined it as in 1963 (Dawkins 1989a). There were, however, signals in the wind. Bill Hamilton's papers on kin selection (1964a,b) and Robert Trivers' paper on reciprocal altruism (1971) ushered in a whole new way of looking at social behaviour. Advances in neuroethology, particularly of locusts and crickets, were making it possible to link behaviour to its neurophysiological basis in ways that ethology had only ever dreamt of. Once, ethology had seen itself as including everything in animal behaviour from social organisation to the activity of neurons but now ethology itself, set beside its two thriving offspring, neuroethology and sociobiology, began to look distinctly old-fashioned.

Richard and I responded to these changes in somewhat different ways. He initially turned to the analysis of hierarchical organisation at a behavioural level (Dawkins, R. 1976a) and we did some joint work on decision-making in chicks and flies (Dawkins & Dawkins 1973,

1976), which no-one took much notice of but which I still think had some original ideas. He was then inspired by the advances in evolutionary behaviour and started working on what was eventually to become *The Selfish Gene* (1976b). When asked what he was doing at this time, he would often answer in complete seriousness, "Writing a best seller", which everyone took as a joke. The real joke was, of course, that he was.

I, on the other hand, retained more of an interest in mechanisms of behaviour, although I had to acknowledge that the behaviour I had studied so far (search images, drinking behaviour in chicks and grooming behaviour in flies) did not lend itself readily to physiological analysis. Even an attempt to study the feeding behaviour of water snails (Dawkins 1974), which I thought might lead directly to explanation at a neuronal level, left me feeling dissatisfied and restless. I had no wish to become a physiologist and open animals up to study their behaviour and I had a strong suspicion that there would be a gap of decades before there would be a physiological explanation of any of the behaviour that really interested me. Crickets singing and snails rasping their radulae might be becoming accessible to analysis at the neuronal level, but I was more interested in whole animals wielding symphonies of neurons to produce complex behaviour. What one neuron did, particularly in an invertebrate, was, if I was honest about it, only of very marginal interest. I was still a great believer in the analysis of behaviour at the whole animal level and in what Niko reputedly called (although I have never found the source) 'physiology without breaking the skin'.

There was another reason, too, for wanting to study behaviour 'without breaking the skin'. It was better for the animals. My interest in animal behaviour had always come from wanting to understand the world of animals – what they saw, what they responded to and even how they managed to achieve simple tasks such as eating food and running away from predators. Studying search images, and actually being able to provide evidence that sometimes a bird would see something and at other times the same object would be completely overlooked, had taken me some way to achieving this. But anaesthetising animals or inserting electrodes was not the way forward for me. I wanted to use their behaviour to tell me what their worlds were like. I wanted to study them but I did not want to damage them. Furthermore, there was a growing interest among the public at large in issues to do with animal welfare. Ruth Harrison's book *Animal Machines* (1964), Richard Ryder's *Victims of Science* (1975) and Peter Singer's *Animal Liberation* (1975) had all raised disturbing questions about the way in which animals were treated. People were becoming interested in the issue of animal suffering, and whether it was possible to tell from looking at an animal's behaviour whether it was suffering or not. The problem was that 'suffering' did not appear at that time to be included in any studies of behaviour that I was aware of. In fact, it was considered to be far too subjective to be included in any textbook of the time.

All my previous training, very much influenced by Niko, had been to avoid words with emotional overtones when describing behaviour. Even 'fear' or 'aggressiveness' had to be firmly enclosed in quotation marks before being applied to behaviour, and then defined in strictly behavioural terms such as 'number of bites per minute'. I understood the reasons for this and still think they are valid: if you don't define terms clearly and objectively then nobody knows what anybody else is talking about and dialogue rapidly slips into an unscientific debate where anything goes. If there are no standards for testing hypotheses you rapidly find yourself outside the realms of objective science. It followed that subjective terms such as 'suffering' had to be treated with great caution, not because animals are incapable of suffering but because, even if they do suffer, we cannot study that suffering objectively. The young discipline of ethology, desperate to be accepted as a real science concerned with what is objectively testable, therefore had nothing say about suffering. Just at the time when the public were beginning to see the importance of animal behaviour as a study in its own right, ethology appeared to be turning its back on the very aspects of animal behaviour the public most wanted to know about.

There was, however, one notable exception. W. H. Thorpe (1965) at the University of Cambridge had written a remarkable Appendix at the end of a UK Government report on the welfare of farm animals in which he quite clearly said that animal behaviour was central to issues of animal welfare. In particular, he said, understanding the natural behaviour of farm species was crucial to ensuring their welfare in the conditions in which they were now kept. He showed a complete lack of shyness in stating clearly his opinion that birds and mammals were capable of suffering as much as we do.

I didn't agree with everything Thorpe said but I was struck with the way he assumed that animal behaviour would be at the heart of decisions about animal welfare in the future. He even laid out a sort of research programme for how this would happen: studies of development would help us understand how early experience affected the welfare of adult animals, studies of communication would help us understand their social needs, studies of natural unrestrained behaviour would help us provide the right environments for the animals in our care, and so on. He was putting classical ethology at the heart of this new programme. However, all this was very much at odds with the way in which the study of animal behaviour was actually going in practice. Behavioural ecology was getting going, exploring questions about adaptation in the light of the new impetus from game theory and kin selection, and had turned away from questions about development and motivation, which were seen as the concerns of old-fashioned ethology, despite being central to the study of animal welfare. Animal welfare, in other words, was not mainstream animal behaviour. It was marginal, not very exciting and with highly dubious overtones of subjectivity and lack of scientific rigour. It was not a proper subject to be studying.

By 1974, I had secured a job as a Departmental Demonstrator (which would now be called a lecturer) in the Zoology Department and so had more freedom to pursue my own research interests independently of what anyone else thought. Fortunately, there were some people who had also concluded, like me, that animal welfare was an important area of study and that animal behaviour was crucial to its success. David Wood-Gush in Edinburgh, Paul Siegel in Virginia and Glen McBride in Australia were among a growing band of people who had shown that you could use animal behaviour to study welfare issues. Then, of course, there was the watershed 1975 IEC conference in Parma, Italy, at which Donald Griffin got up and said how unnecessarily inhibited everyone was being in refusing to ask questions about animal minds and animal awareness. Subjective they might be, he said, but they are part of biology and part of ethology and we should start investigating them. Because Griffin came from the 'hard' end of biology (the physics of bat echolocation), no-one could accuse him of being vague and sentimental and so the impact he had was enormous. It was now much more acceptable to talk about animal cognition and even animal awareness. Cognitive ethology had been born and it seemed to me that the study of animal emotions should be getting going too. After all, if animals did have the capacity to suffer (meaning, yes, that they did subjectively experience pain, fear and other unpleasant emotions) then that capacity must have evolved by natural selection and therefore should be studied as part of biology.

I was now becoming really concerned that the refusal of mainstream ethology to interest itself in animal welfare might have damaging effects if it resulted in recommendations or even legislation being enacted without any ethological input. To this day, the term etho- logical or behavioural need is still found in widespread use among non-ethologists, based on Lorenz's psychohydraulic model of behaviour. Many people were and remain convinced that animals have such needs to perform all the behaviour in their natural repertoire, otherwise frustration will build up and suffering results. One of the Five Freedoms, used throughout the world as the basis for good animal welfare, emphasizes this as a universal need in animals.

But how valid were the assumptions being made about behavioural needs? For example, what about being chased by a predator? That is a natural enough occurrence, so should it be mandatory for animals to be able to exhibit anti-predator behaviour just because it is natural? Does 'freedom to perform natural behaviour' include being chased by a predator and if not, why not? Surely, it is not the naturalness of the behaviour that determines whether an animal suffers if it cannot perform a behaviour. That will be determined by a whole range of other factors, such as early experience, whether or not there are adequate substitutes and above all, the animal's own responses. So many of the things that were being assumed to affect welfare should, I thought, be empirical matters subject to experimental test and the right sort of evidence – including ethological evidence.

There seemed no way out of the conclusion that I would have to put this all down on paper and write a book to explain how the study of animal welfare could be scientific. I started writing without realising at the time what an extraordinarily difficult task that was going to be. Not for me the easy, deathless prose that Richard has always been able to turn out. On the contrary, it was writing and rewriting, cutting, inserting and neglecting a lot of the other things I was supposed to be doing. Unanswered letters piled up. Dust accumulated on everything not directly connected with the book and at the end of this struggle was a very slim volume indeed: *Animal Suffering: the Science of Animal Welfare*. In it, I tried to lay out the case for a science of animal welfare and at the same time to point out how important it was to be critical of the evidence. There was no single litmus test of whether or not an animal was suffering but by taking different kinds of evidence – its health and its physiology as well as its behaviour – we could arrive at a reasonably objective view. The book was eventually published in 1980, the same year as I was elected a Fellow of my old College, Somerville. Administration and teaching then hit home in a big way.

Within two years of being elected and becoming the first biology fellow the College had ever appointed, I had a fleet of undergraduates to look after and teach and I was also made

Dean, which means I was in charge of the discipline of the whole College. A College ball, a police investigation and a visit from Margaret Thatcher were some of the many nightmares of this period. Mrs Thatcher was then Prime Minister and, as a former student of the College, paid us a visit to unveil a statue of herself. There were Secret Service men everywhere looking after her security but I was responsible for many of the arrangements and making sure the students could meet her but not cause too many problems. The compensation for all the disruption was a quite unforgettable moment when the Principal declared "I now call upon the Prime Minister to unveil her bust". I never understood why it was only me and one Secret Service man (who unfortunately caught my eye) who struggled not to dissolve into uncontrollable giggles.

Although my period as Dean came to an end after two years, the responsibilities of teaching did not. Designing a lecture course, running practicals and looking after graduate students took a lot of time, and as a result of the *Animal Suffering* book I was asked to be on a lot of committees. These included being on the Farm Animal Welfare Council (FAWC) that advised the UK Government on welfare issues, the Farm Livestock Advisory Committee of the RSPCA, the Animal Experiments Committee of the Royal Society and the Ethical Committee of ASAB, which helped to advise the Government on the new legislation on animal experiments (eventually brought into law as the Animals (Scientific Procedures) Act (1986)).

Like many another before me, I decided that if I had to spend time writing lectures and giving tutorials, I might as well write a book, but instead of aiming for a textbook, I wrote a series of essays explaining to a wider audience the issues I had found myself explaining over and over again in tutorials to my own undergraduates. I originally wanted to call this collection *Stumbling Blocks in Ethology* but was fortunately talked out of this by my helpful publisher, Michael Rodgers, and by Richard, who suggested the title I eventually used, *Unravelling Animal Behaviour* (1986). This book was slightly easier to write than the previous one but I resigned myself to the fact that I might never be able to write a real best seller. I was much too fond of using phrases such as "But on the other hand" and pulling apart an argument I had just made for this to happen. The whole book was about problems and difficulties and how to think straight about issues such as inclusive fitness and adaptation. It was much too careful to hit the headlines and not nearly comprehensive enough to become a required textbook. Still, it filled a niche of its own for undergraduate teaching and I found that I really enjoyed writing. Perhaps one day I could reach out to a wider audience.

In the meantime, I had been trying to develop the empirical, experimental side of animal welfare research. In particular, I felt that, in order to decide whether animals were happy or suffering, it was necessary to devise ways of getting them to tell us how they responded to different situations. They might not be able to tell us in so many words what they felt, but they could 'vote with their feet' and tell us by their actions what they wanted and what they disliked. After all, even with people, it isn't always what they say that convinces us of what they feel or believe. It is what they do. Actions, in other words, often speak louder than words, even for us, and many of those same actions, such as escaping or pressing a lever to obtain a reward, were just as available to other species as they are to us. All that was necessary was to provide animals with the right opportunities for expressing their point of view.

I chose to work on the welfare issues associated with keeping laying hens in small wire cages, at that time the way the majority of hens in the UK and much of the rest of the world were kept commercially. In the 'battery hen' system, 4–5 hens are kept in small wire cages with sloping wire floors so that their eggs roll forwards to the front of the cage, where they can be easily collected. The cages are stacked into 'batteries" or tiers, one on top of one another. The hens have nowhere to scratch or dustbathe, no perches, no way of escaping from one another, and above all, so little space that they can hardly turn round. These cages have now been banned in the European Union from 2012 but are still widely used in the US and the rest of the world. I chose laying hens partly because I could keep reasonable numbers in the Zoology Department (whereas trying to keep pigs or cows would have been quite out of the question) and partly because the conditions in which they were kept on farms lent themselves so easily to experimental test. It was possible to ask separate questions about the importance of space, flooring, absence of perches, etc. Their plight exemplified a serious animal welfare issue and could be addressed through the study of behaviour.

I started with very simple choice tests and showed that what was on the floor (such as wire or litter to scratch in) was actually more important to a hen than the size of the cage itself (Dawkins 1978, 1981). I even attempted to give hens a choice between a battery cage and an outside pen to see which they chose. The 'battery' cage was encased within a small hut on wheels that had only a passing resemblance to being in a bank of cages in a commercial hen house. The pen contained grass and was open to the elements – a sort of miniature free-range. The choice was hardly a realistic one, but nevertheless it showed that hens chose the outside run even if this was the first time they had ever seen it and had up to that moment only ever been kept in a cage (Dawkins 1978).

The next step was to make the link between preference and welfare, since choosing one thing over another could be a choice between two luxuries or, conversely, a choice of the lesser of two evils. A preference for smoked salmon over caviar does imply suffering if you have to 'make do' with the caviar. Similarly, a preference for slightly mouldy bread over very mouldy bread does not mean that either is a good diet. I decided to see whether hens' preferences (for more space, litter floors, etc.) were ' luxury' options that they would go for if available but not of much importance to them, or 'necessities', things that were really important to them and the presence of which might therefore contribute directly to their welfare. I borrowed some ideas from economic consumer demand theory to define luxuries and necessities by whether hens would continue to work or pay a price for a commodity when it became more costly for them to do so. For example, even if hens preferred a litter to a wire floor when offered the two as an easy, cost-free choice, would they still show this preference when they had to work for the litter by pecking a key many times to get what they wanted (Dawkins and Beardsley 1986)? Key pecking was just one of the ways in which we made hens work for what they wanted. Other ways included pushing spring-loaded doors, walking down a long corridor, and squeezing through a narrow gap (Bubier 1996).

The connection between preference and welfare was that if animals had preferences that were so strong that they would work very hard to get to (or away from) what they wanted or if they would forgo something that was known to be important to them, such as food, then this was evidence that whatever it was they were working for mattered to them. And if it mattered a great deal to them, then being deprived of it would cause suffering. The point of all these methods was to get the animals themselves to tell us what they really needed. This was an animal-centred approach to animal welfare (Dawkins 1990).

There was a surprising amount of opposition to this approach, including the one that animals do not always choose what is good for them, which of course they don't. Children would choose not to go to the dentist. People choose to smoke. 'Welfare' is more that just what animals choose. However, I never claimed that preferences and what animals want should be used in isolation from other more conventional ways of assessing welfare, such as using corticosteroid or 'stress' hormones, or taking note of what made animals healthy or unhealthy. What the animals themselves want is just one, albeit a very powerful, measure of welfare. I was delighted when Georgina Mason subsequently joined the ABRG and applied the consumer demand ideas to the welfare of a different species, fur farmed mink, with much better equipment and much more clear-cut results than I had ever been able to achieve (Mason *et al*. 2001).

To further reinforce the point that preference is not everything, I used as many other ways as I could think of to look at the welfare of hens. To look at their needs for space, I undertook a detailed analysis of exactly how much space they used for each of their behaviours, such as turning around, wing flapping or just standing. By filming from above and measuring the space a hen occupied at each second throughout the performance of a behaviour pattern, I came up with precise figures (means and variation) of space used in comparison with the space available. Needless to say, the space used was considerably greater than the space the birds had in a battery cage (Dawkins and Hardie 1989). Christine Nicol, who was one of my graduate students, extended the methods and looked at the extent to which hens were deprived of the opportunity to perform behaviour using the 'rebound effect' (whether hens would indulge in a behaviour they had been deprived of in a cage when finally allowed to do it).

One of the most pleasurable studies at this time was a round the clock/round the year study of the behaviour of red junglefowl at Whipsnade Zoo. Whipsnade, the country site of the London Zoo in Bedfordshire, had at that time a grove of tall trees with an understorey of rhododendron bushes, where a population of free-ranging junglefowl (the ancestors of all our domestic chickens) lived. Although they were regularly fed, they were otherwise feral and reproduced freely without interference. Their behaviour was therefore semi-natural and it was fascinating to find that for nearly 70% of the time when they were not roosting in trees, the birds would be pecking and scratching in leaf litter and loose soil. Although I was careful not to say that this implied that all farmed chickens should be allowed to scratch around for 70% of their time because this was 'natural', I did say that this was a figure we should keep in mind when designing high-welfare housing for chickens. It was a baseline, a starting point from which to work (Dawkins 1989b).

Figure 7.6. Aubrey Manning in 1991, during the writing of the 4th edition *of An Introduction to Animal Behaviour*. Photograph by Marian Dawkins.

Having (I hoped) started a ball rolling and shown, together with other like-minded colleagues such as Ian Duncan and Barry Hughes in Edinburgh, that the animal's point of view should take its place along with other measures of welfare, I did something that I suppose in retrospect could be seen as somewhat arrogant. I took a long hard look at the amount of time I was spending on various welfare committees (it turned out to be about 35 working days a year), and decided I had done my bit for animal welfare. I hoped that what I had done had been useful and that other people could now fill in the details and use the methods to ask more detailed questions. I wanted to do something different, So in 1989 I resigned from every single committee I was on, wrote what I saw as my final paper on animal welfare as a target article for *Behavioral and Brain Sciences* (1990) and started on a new tack. I wanted to finally get down to thinking about the issue that had really fascinated me all along – that of animal consciousness – and I had received an irresistible letter from Aubrey Manning asking me to collaborate with him on a new edition of his classic textbook *An Introduction to Animal Behaviour*. Besides, I had had a moment of truth on the Great Barrier Reef.

In 1988, I went to Thailand with Nigella Hillgarth to look for wild red junglefowl in the dry forests up near the Burmese border. The plan had been to use the same techniques of round-the-clock watching I had used on the feral population in England on really wild birds in Thailand, as a check that the behaviour I had observed was really natural. It was a memorable trip and the dry forest vegetation (tall trees, bamboo thickets and leaf litter

Figure 7.7. In 1988, Nigella Hillgarth and I went to the dry forest of north-west Thailand in search of wild Red Junglefowl. Photograph by Marian Dawkins.

underfoot) was structurally very similar to that at Whipsnade. But the birds were very shy. We would get up at dawn and wait patiently in bamboo bushes for the junglefowl to descend from the trees. We could always hear them but only if we were very lucky would we actually see them.. Even then it would be a brief glimpse before they melted silently into the bushes (Dawkins & Hillgarth 1992).

The following year, I took a sabbatical term with Glen and Helen McBride at the University of Queensland, Australia, and stayed at the Womens' College where Helen was Principal. By this time, Richard and I were divorced and so I was single again and beginning to really enjoy it. I took a train up to the rain forest north of Cairns where I found Esther Cullen, now separated from Mike, living in a tent. In one corner was a huge spider that Esther left in companionable peace. A planned trip to the barrier reef was clearly going to be the opportunity of a lifetime and I resolved that I would go snorkelling for the very first time even if I lost my contact lenses in the process. Being extremely short-sighted, I had never been snorkelling before because I could never see anything. This time, I left my lenses in, put on a mask and looked under the water. A feeling of fascination and deep regret at what

I had been missing all my life swept over me. A whole new world opened up. Not only could I hear the parrot fish munching the coral but I could swim up to them and they didn't fly away or try to escape. It was possible to hang there in the water, watching their every move, and they were not bothered at all. Why were so many behavioural studies done on birds, which were, comparatively, so difficult to observe? Fish watching was so much easier!

I returned to the UK determined to learn to dive and had the unusual experience of being taught by my own undergraduates. The Oxford University Dive Club was run both for and by students who gave all the tuition, some of whom I had been teaching animal behaviour. A memorable trip to Cornwall along with all the other trainees (mostly undergraduates) convinced me that British diving could be truly spectacular. We dived on wrecks and on rocks that were sheer walls of bright jewel anemones. It was tough, cold and wonderful. At the end of it, I felt prepared for a planned trip to the Caribbean to take a more detailed look at the behaviour of reef fish.

Tim Guilford, who had once been my research student, had been working on the evolution of warning signals and mimicry. What fascinated both of us was how the evolution of signals of any kind – warning signals, sexual signals, territorial signals – could be influenced by what the animals on the receiving end of those signals could best see and hear and smell. We called this the evolution of signal design by 'receiver psychology' – effective signals would be those that matched what the receiver's sense organs and brain were tuned to receive (Guilford and Dawkins 1991). Such ideas had once been common-place in the ethology of the 1950s and 1960s but had fallen out of fashion in favour of alternatives such as the 'handicap' theory. Zahavi (1975) had suggested that animal signals became large and conspicuous as a handicap to the animals giving them. By doing some-thing extravagant, such as growing a long conspicuous tail, the signaller would be paying a considerable cost, which was itself a guarantee that the signaller was strong, healthy and of high quality. It was the equivalent of a rich man demonstrating how rich he is by spending money on giving an extravagant party, something only a genuinely rich man could afford to do. The handicap theory had become widely accepted as an explanation of why some animal signals become conspicuous. We just wanted to explore the alternative explanation that conspicuousness evolved not as a handicap but simply as the best way of getting a message across to a receiver by tapping into its 'receiver psychology'. We talked to Bob Warner when he visited Oxford and were so impressed with what he told us about his work on the Blue-head wrasse (*Thalassoma fasciatium*) that we decided to visit his study site in the American Virgin Islands.

Floating around the Caribbean watching male Blue-head wrasse defending a coral head, courting females and chasing off gangs of marauding males cunningly disguised as females was pure observational ethology. The fish seemed to take no notice of us snorkelling overhead and we could watch these underwater dramas unfold. We watched the males signal to females to attract them over long distances and then change colour as a female came closer, suddenly develop conspicuous spots on the pectoral fins and flutter them in front of her, just before the two of them dashed to the surface for a brief moment of mutual spawning. The colour of the male changed with the distance of the female and the black dots served to emphasize his whizzing fins, so that they could synchronize their spawning (Dawkins & Guilford 1993). A couple of years later, I returned to the Caribbean, this time to Barbados, with an underwater video camera to test this idea further (Dawkins & Guilford 1994).

In pursuit of what animals see, I studied not just fish but chickens (Dawkins 1996; Dawkins & Woodington 1997, 2000) and homing pigeons (Dawkins *et al.* 1996), but I could never escape the ultimate fascination with animal consciousness and managed to write a book, *Through Our Eyes Only? the Search for Animal Consciousness*, which was published in 1993. This was a critical look at the supposed evidence for animal consciousness. The fourth edition of *An Introduction to Animal Behaviour*, written with Aubrey Manning, appeared in 1992, and the second edition of *Unravelling Animal Behaviour* in 1995. Then I was asked to write a review for the *Quarterly Review of Biology* (1998) on animal welfare. I accepted only reluctantly. This was a subject I thought I had left behind.

Having to look once again at the animal welfare literature after a few years away from it made me realise that nothing much had changed. The same controversies were still going strong, the same objections to this or that method of assessment were still being raised and, above all, applied behavioural work was still not being taken up and used by commercial farmers. There was something wrong with the whole way in which applied research was conceived by governments, the funding bodies and the farming industry. It was simply not true that good research would be 'taken up' by industry when that industry was itself struggling and had so little spare cash to spend on risky new ventures. There was too much of a gap between the academic research and commercial farming and perhaps it was the academic research that was partly to blame. If 'applied' animal welfare research was mostly done on a small scale in universities and research institutes, maybe it just wasn't applicable to real farming. To make a real difference to the welfare of farm animals, maybe it was going to be necessary to change the whole way in which research was carried out in the first place as well as the way scientists interacted with farmers.

Through a series of happy coincidences, I met Malcolm Pye and Paul Cook from Premier Poultry, then one of the major poultry producers in the UK, producing millions of meat (broiler) chickens each year, and was delighted to find that they were very open to trying out a range of new ideas for improving broiler welfare. Malcolm had just persuaded his company to invest millions of pounds in state-of-the-art free-range broiler farms, where I studied the behaviour of the birds outside (Dawkins *et al.* 2003). Two of my students, Abigail Hall and Lesley King, also did farm-scale studies on Premier farms, co-supervised by Paul Cook. Working with commercial farmers right from the outset seemed a much better way of improving chicken welfare than expecting commercial farmers to take much interest in what I or other academics do on a small scale.

Then, one day, having given a talk on some of our work, I found myself sitting around a table with the managers of most of the major broiler producers in the UK, representing about 70% of the industry. They were very concerned about proposed EU legislation on stocking density limits for broiler chickens and they understood that there was government funding for a study on just this subject. Would I be prepared to take this on? They would be willing to do whatever I asked of them but they wanted a proper objective study on commercial farms. I told them how difficult it would be to do a farm-scale study, how it might lose them money and how the results might not come out the way they wanted. They said they just needed to know what I wanted them to do and the would agree to my protocol. I felt my bluff had been called. I had spent a lot of time talking about animal welfare in theory and discussing methods of assessment of good and bad welfare in academic papers. Now here was the opportunity to do something in practice, to actually make a difference to the lives of millions of birds. It was an awesome task. Was I up to it? I wrote the grant proposal, obtained the funding and then had to get down to carrying out one of the biggest animal welfare experiments ever done.

Some 2.7 million chickens and 11 different commercial companies in the UK and Denmark were involved in the study, which I could never have carried out without the extraordinary organisational skills of Tracey Jones. Our measures of welfare were simple and straightforward: mortality, health, stress hormone levels and behaviour. The results showed that although stocking density did affect these basic measures of welfare, far more important were the factors affecting the air and litter quality, such as good ventilation (Dawkins *et al.* 2004; Jones *et al.* 2005). It was gratifying to find that this conclusion seemed subsequently to be recognized by the European Union, which in 2007 issued a Directive covering the welfare of broiler chickens which specifically avoided the idea of a single stocking density as the prescription of good welfare and stressed instead the importance of "ventilation systems that keep the ammonia, CO_2, temperature and humidity levels within strict parameters."

Despite the logistical difficulties of working on such a large scale, I became convinced that such practical involvement with commercial farmers was the only way forward if political decisions on animal welfare were to be evidence-based, as I believed they should be. We need evidence on the actual conditions in real farms, not small-scale models that might have very little bearing on what a commercially kept farm animal actually experiences. My current research is now directed at ways of measuring welfare, and particular animal preferences, in context – that is, on farms. For example, we then showed that the spacing patterns of broiler chickens (how close or how far away they are from each other) can be seen as an expression of social preference (Febrer *et al.* 2006), just as surely as if the birds had been given a formal choice test, but with the advantage of being measurable in the context of a commercial broiler house. The sounds they make tell us about their emotional state (Bright *et al.* 2006) and their movement patterns alert us to possible welfare problems (Dawkins *et al.* 2010).

When I first met him, Malcolm Pye confided in me that his ambition was to find a farm somewhere and run it commercially, but to make sure that the welfare of the animals was paramount. He was convinced that farming with high standards of animal welfare made good business sense and that, far from costing money, as most people supposed, could actually make money. He had two colleagues, Ruth Layton, a veterinarian, and Roland Bonney, then a sheep farmer, who shared this ambition. They just needed somewhere to try out all the ideas they had and show that they worked. Some time later I heard that the

University Farm at Oxford was in trouble financially and was looking for a tenant. Almost as an afterthought at a meeting we were both attending, I asked Ruth whether they were still looking for a farm. After a lot of legal wrangling and the personal intervention of the Vice Chancellor, FAI (Food Animal Initiative) moved into the University Farm at Wytham in 2001 as a unique experiment in a commercial–academic partnership.

The partnership is unique because FAI is run as a fully commercial farm and so, unlike most university or research farms, is not in any way subsidized by the University (in fact FAI pay a fully commercial and very hefty rent to the University). This means that all the results obtained there are directly relevant to commercial farming. There is no need for 'technology transfer' because the work is already done in the arena where it will be used. At the same time, FAI see the need for research and are very open to the needs of research and so (provided it can all be paid for from grants, etc.) will find ways in which commerce and research can coexist. Farmers, politicians, schoolchildren, representatives from animal welfare organisations and big business, journalists and many others are constant visitors. FAI have credibility with farmers because they are commercial farmers themselves and, like them, have to make a living, and they have credibility with the research community because they work closely with the universities of Bristol, Oxford and Warwick. The publicity and educational side of FAI is sponsored by some very big names indeed – Tesco and McDonald's – so that the results of the research and development at FAI can be fed out into their suppliers and into their own standards (Dawkins and Bonney 2008). It is very exciting to be part of an enterprise that seems to be genuinely changing the way farmers treat animals.

I consider myself to be an immensely lucky person. To have known from an early age what I wanted to do and to be able to do it is lucky enough. To do what I want to do and to discover that it is even more exciting and absorbing than I imagined and that its horizons expand as I ask more and more questions is fortune indeed. I live in one of the most beautiful and stimulating cities in the world, the FAI farm is set in classic English countryside, and at weekends I have a view over a shining sea and white cliffs that I share with a perfect partner. The sea and the view allow us to write books in the morning and sail, walk and windsurf in the afternoons. *Observing Animal Behaviour* (2007) was a direct result of having such a peaceful place to write in. It is a book about the importance of observational techniques in animal behaviour and it is dedicated to Niko Tinbergen on the centenary of his birth. With that out of the way, there are now new writing projects. On a train between Oxford and London, Aubrey Manning and I recently shook hands and agreed to start working on the sixth edition of *An Introduction to Animal Behaviour*. And I still hanker after the idea of one day writing that elusive best-seller, on animal welfare, perhaps, or animal minds.

In the meantime, research on animal welfare becomes both more theoretically interesting and more practical. I spend a lot of time talking to farmers, retailers, engineers and commercial companies and an equally large amount of time on farms watching animals, mostly chickens. I could never give up the connection with collecting data and actually doing the research myself, or at least as much of it as I can fit in between teaching and being Vice-Principal of Somerville. I still value the excited remark made to me by a farm worker who told me, confidentially, that the chicken house in which we were standing was part of an experiment "by an Oxford Professor". He couldn't believe that the dishevelled, boiler-suited figure in front of him could possibly fit that description, but, to me, spending time with animals in chicken houses or under the sea or in a bamboo thicket is what I started my career for and what I still feel is the most valuable and rewarding part of what I do. What I can't quite understand is how, having spent such a long time doing it, there still seems so much yet to do.

References

Bright, A., Jones, T. A. & Dawkins, M. S. (2006). A non-intrusive method of assessing plumage condition in commercial flocks of laying hens. *Anim. Welfare* **15**: 113–18.

Bubier, N. E. (1996). The behavioural priorities of laying hens: the effect of cost/nocost multi-choice tests on time budgets. *Behav. Proc.* **37**: 225–38.

Dawkins, M. (1971a). Perceptual changes in chicks: another look at the 'search image' concept. *Anim. Behav.* **19**: 566–74.

Dawkins, M. (1971b). Shifts of 'attention' in chicks during feeding. *Anim. Behav.* **19**: 575–82.

Dawkins, M. (1974). Behavioural analysis of coordinated feeding movements in the gastropod *Lymnaea stagnalis* (L.). *J. Comp. Physiol.* **93**: 255–71.

Dawkins, M. (1976). Towards an objective method of assessing welfare in domestic fowl. *Appl. Anim. Ethol.* **2**: 245–54.

Dawkins, M. (1977). Do hens suffer in battery cages? Environmental preferences and welfare. *Anim. Behav.* **25**: 1034–46.

Dawkins, M. (1978). Welfare and the structure of a battery cage: size and cage floor preferences in domestic hens. *Br. Vet. J.* **134**: 469–75.

Dawkins, M. S. (1980). *Animal Suffering: the Science of Animal Welfare*. London: Chapman and Hall.

Dawkins, M. S. (1981). Priorities in the cage size and flooring preference of domestic hens. *Br. Poultry Sci.* **22**: 255–63.

Dawkins, M. S. (1983). Battery hens name their price: consumer demand theory and the measurement of ethological 'needs'. *Anim. Behav.* **31**: 1195–205.

Dawkins, M. S. (1986). *Unravelling Animal Behaviour*. Harlow, Essex: Longmans. (2nd edn 1995).

Dawkins, M. S. (1989a). The future of ethology: how many legs are we standing on? In *Perspectives in Ethology*, Vol. 8, *Whither Ethology?* ed. P. P. G. Bateson & P. H. Klopfer, pp. 47–54. London, New York: Plenum Press.

Dawkins, M. S. (1989b). Time budgets in Red Junglefowl as a baseline for the assessment of welfare in domestic fowl. *Appl. Anim. Behav. Sci.* **24**: 77–80.

Dawkins, M. S. (1990). From an animal's point of view: motivation, fitness and animal welfare. *Behav. Brain Sci.* **13**: 1–9, plus Open peer commentary.

Dawkins, M. S. (1993). Are there general principles of signal design? *Phil. Trans. R. Soc. Lond.* **B340**: 251–5.

Dawkins, M. S. (1995). How do hens view other hens? The use of lateral and binocular visual fields in social recognition. *Behaviour* **132**: 591–606.

Dawkins, M. S. (1996). Distance and social recognition in hens: implications for the use of photographs as social stimuli. *Behaviour* **133**: 663–80.

Dawkins, M. S. (1998). Evolution and animal welfare. *Q. Rev. Biol.* **73**: 305–28.

Dawkins, M. S. (2002). What are birds looking at? Head movements and eye use in chickens. *Anim. Behav.* **63**: 991–8.

Dawkins, M. S. (2007). *Observing Animal Behaviour: Design and Analysis of Quantitative Data*. Oxford: Oxford University Press.

Dawkins, M. S. & Beardsley, T. (1986). Reinforcing properties of access to litter in hens. *Appl. Anim. Behav. Sci.* **15**(4): 351–64.

Dawkins, M. S. & Bonney, R. (eds.) (2008). *The Future of Animal Farming: Renewing an Ancient Contract*. Oxford: Wiley Blackwell.

Dawkins, M. and Dawkins, R. (1974). Some descriptive and explanatory stochastic models of decision-making. In *Motivational Control Systems Analysis*, ed. D. J. McFarland, pp. 119–68. London: Academic Press.

Dawkins, M. S. & Guilford, T. (1991). The corruption of honest signalling. *Anim. Behav.* **41**: 865–74.

Dawkins, M. S. & Guilford, T. (1993). Colour and pattern in relation to sexual and aggressive behaviour in the Bluehead wrasse *Thalassoma bifasciatum*. *Behav. Proc.* **30**: 245–52.

Dawkins, M. S. & Guilford, T. (1994). Design of an intention signal in the bluehead wrasse (*Thalassoma bifasciatum*). *Proc. R. Soc. Lond.* **B257**: 123–8.

Dawkins, M. S. & Guilford, T. (1996). Sensory bias and the adaptiveness of female choice. *Am. Nat.* **148**: 937–42.

Dawkins, M. & Hardie, S. (1989). Space needs of laying hens. *Br. Poultry Sci.* **30**: 413–16.

Dawkins, M. S. & Hillgarth, N. (1992). The two-pronged approach to studying Galliformes. *J. World Pheasant Ass.* **15–16**: 107–11.

Dawkins, M. S. & Woodington, A. (1997). Distance and the presentation of visual stimuli to birds. *Anim. Behav.* **54**: 1019–25.

Dawkins, M. S. & Woodington, A. (2000). Pattern recognition and active vision in chickens. *Nature* **403**: 652–5.

Dawkins, M. S., Guilford, T., Braithwaite, V. B. & Krebs, J. R. (1996). Recognition and discrimination of photographs of locations by homing pigeons. *Behav. Proc.* **36**: 27–38.

Dawkins, M. S., Cook, P. A., Whittingham, M. J., Mansell, K. A. & Harper, A. (2003). What makes free-range broilers range? *Anim. Behav.* **66**: 151–60.

Dawkins, M. S., Donnelly, C. A. & Jones, T. A. (2004). Chicken welfare is influenced more by housing conditions than by stocking density. *Nature* **427**: 342–4.

Dawkins, M. S. Lee, H. -J., Waitt, C. & Roberts, S. (2010). Optical flow patterns in broiler chicken flocks as automated measures of welfare and gait. *Appl. Anim. Behav. Sci.* doi: 10.1016/j.applanim.2009.04.009.

Dawkins, R. (1976a). Hierarchical organisation: a candidate principle for ethology. In *Growing Points in Ethology*, ed. P. P. G. Bateson & R. A. Hinde, pp. 7–54. Cambridge: Cambridge University Press.

Dawkins, R. (1976b). *The Selfish Gene*. Oxford: Oxford University Press.

Dawkins, R. & Dawkins, M. (1973). Decisions and the uncertainty of behaviour. *Behaviour* **45**: 83–103.

Dawkins, R. & Dawkins, M. (1976). Hierarchical organisation and postural facilitation: rules for grooming in flies. *Anim. Behav.* **24**: 739–55.

Febrer, K., Jones,T. A., Donnelly, C. A. & Dawkins, M.S. (2006). Forced to crowd or choosing to cluster? Spatial distribution measures social attraction and aversion in broiler chickens. *Anim. Behav.* **72**: 1291–1300.

Guilford, T. & Dawkins, M. S. (1987). Search images not proven: a reappraisal of recent evidence. *Anim. Behav.* **35**: 1838–45.

Guilford, T. & Dawkins, M. S. (1991). Receiver psychology and the evolution of animal signals. *Anim. Behav.* **42**: 1–14.

Guilford, T. & Dawkins, M. S. (1993). Are warning colours handicaps? *Evolution* **47**: 400–16.

Hamilton, W. D. (1964a). The genetical evolution of social behaviour I. *J. Theor. Biol.* **7**: 1–16.

Hamilton, W. D. (1964b). The genetical evolution of social behaviour II. *J. Theor. Biol.* **7**: 17–32.

Harrison, R. (1964). *Animal Machines*. London: Stuart.

Jones, T. A., Donnelly, C. A. & Dawkins, M. S. (2005). Environmental and management factors affecting the welfare of chickens on commercial farms at five densities. *Poultry Sci.* **84**(8): 1155–65.

Kruuk, H. (2003). *Niko's Nature. A Life of Niko Tinbergen and his Science of Animal Behaviours*. Oxford: Oxford University Press.

Lorenz, K. Z. (1952). *King Solomon's Ring*. London: Methuen & Co.

Manning, A. & Dawkins, M. S. (1992). *An Introduction to Animal Behaviour*, 4th edn. Cambridge: Cambridge University Press. (5th edition 1998).

Mason, G. J., Cooper, J. & Clareborough, C. (2001). Frustrations of fur-farmed mink. *Nature* **410**: 35.

Ryder, R. (1975). *Victims of Science*. London: Davis-Poynter.

Singer, P. (1975). *Animal Liberation*. London: Jonathan Cape.

Thorpe, W. H. (1965). *Brambell, F. W. R. (Chairman) (1965) Report of the Technical Committee to Enquire into the Welfare of Animals kept under Intensive Livestock Husbandry Systems*. London: Her Majesty's Stationery Office.

Tinbergen, L. (1960). The natural control of insects in pinewoods: factors influencing the intensity of predation by song birds. *Arch. Neerl. Zool.* **13**: 265–343.

Tinbergen, N. (1953). *Herring Gull's World*. New Naturalist Series. London: Collins.

Tinbergen, N. (1972, 1973). *The Animal in its World*, vols. I and II. London: George Allen & Unwin.

Trivers, R. L. (1971). The evolution of reciprocal altruism. *Q. Rev. Biol.* **46**: 35–57.

Zahavi, A. (1975). Mate selectin: a selection for a handicap. *J. Theor. Biol.* **53**: 205–14.

8

Growing up in ethology

RICHARD DAWKINS

Childhood and school

I should have been a child naturalist. I had every advantage: not only the perfect early
environment of tropical Africa but what should have been the perfect genes to slot into it.
For generations, sun-browned Dawkins legs have been striding in khaki shorts through the
jungles of Empire. My Dawkins grandfather employed elephant lumberjacks in the teak
forests of Burma. My father's maternal uncle, chief Conservator of Forests in Nepal, and his
wife, author of a fearsome 'sporting' work called *Tiger Lady*, had a son who wrote the

Leaders in Animal Behavior: The Second Generation, ed. L. C. Drickamer & D. A. Dewsbury. Published by Cambridge
University Press. © Cambridge University Press 2010.

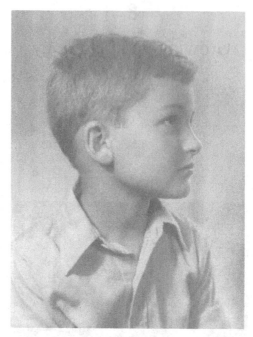

Figure 8.1. Richard Dawkins as a child.

definitive handbooks on the *Birds of Borneo* and *Birds of Burma*. Like my father and his two younger brothers, I was all but born with a pith helmet on my head.

My father himself read Botany at Oxford, then became an agricultural officer in Nyasaland (now Malawi). During the war he was called up to join the army in Kenya, where I was born in 1941 and spent the first two years of my life. In 1943 my father was posted back to Nyasaland, where we lived until I was eight, when my parents and younger sister and I returned to England to live on the Oxfordshire farm that the Dawkins family had owned since 1726.

It was through my father's middle brother that I met the young David Attenborough, already famous but not yet a household name. This uncle chose Sierra Leone for his enactment of the khaki-shorted family tradition, and David Attenborough was his guest on a filming expedition up country. When my uncle and aunt moved to England and I happened to be staying with them, David brought his young son Robert to visit, and he had us wading all day in shorts through ditches and ponds with fishing nets and jam jars on strings. I've forgotten what we were seeking – newts or tadpoles or dragonfly larvae, I expect – but the day itself was never to be forgotten. Even that experience with the world's most charismatic zoologist, however, wasn't enough to turn me into the boy naturalist that I should have been from the start.

My father's youngest brother was an innovative forest ecologist in Uganda. He later moved to Oxford, where he lectured in biological statistics – a teacher of genius with an unmatched ability to explain difficult ideas in simple language. It was for this that I later dedicated a book, *River Out of Eden*, to him. The worst he could say of a young man was 'Never been in a youth hostel in his life'; a stricture, which, I am sorry to say, describes me to this day. My young self seemed to let down the traditions of the family.

I received every encouragement from my parents, both of whom knew all the wildflowers you might encounter on a Cornish cliff or an Alpine meadow, and my father amused my sister and me by throwing in the Latin names for good measure (children love the sound of words even if they don't know their meanings). Soon after arriving in England, I was mortified when my tall, handsome grandfather, by now retired from the Burma forests, pointed to a blue tit outside the window and asked me if I knew what it was. I didn't and miserably stammered, 'Is it a chaffinch?' Grandfather was scandalized. In the Dawkins family, such ignorance was tantamount to not having heard of Shakespeare: 'Good God, John' – I have never forgotten his words, nor my father's loyal exculpation – 'Is that *possible*?' If Grandfather were alive today, I would explain that I learned late to love watching wild creatures: my original interest in biology came not from the woods and moors but from books.

For I became a secret reader. In the holidays from boarding school, I would sneak up to my bedroom with a book: a guilty truant from the fresh air and the virtuous outdoors. And when I started learning biology properly at school, it was still bookish pursuits that held me. I was drawn to questions that grown-ups would have called philosophical. What is the meaning of life? Why are we here? How did it all start? Biology comes closest to answering these deep questions, but that wasn't the reason I ended up in the biology stream at Oundle School. It was probably a bit of following-in-father's-footsteps, but also a genuinely inspirational young teacher. I. F. Thomas deliberately set out to teach in the tradition of Oundle's great headmaster, F. W. Sanderson (there was Arnold of Rugby and Roxburgh of Stowe... and there was Sanderson of Oundle). Sanderson died in 1922 so Ioan Thomas never met him, but he lived up to Sanderson's ideals, as I recounted in my inaugural Oundle Lecture in 2002, later reprinted in *A Devil's Chaplain* (2003):

Some 35 years after Sanderson's death, I recall a lesson about *Hydra*... Mr. Thomas asked one of us "What animal eats Hydra?" The boy made a guess. Non-committally, Mr. Thomas turned to the next boy, asking him the same question. He went right round the entire class, with increasing excitement asking each one of us by name, "What animal eats Hydra? What animal eats Hydra?" And one by one we guessed. By the time he had reached the last boy, we were agog for the true answer. "Sir, sir, what animal *does* eat Hydra?" Mr. Thomas waited until there was a pin-dropping silence. Then he spoke, slowly and distinctly, pausing between each word.

"I don't know..." *(Crescendo)* "I don't know..." *(Molto crescendo)* "And I don't think Mr. Coulson knows either." *(Fortissimo)* "Mr. Coulson! Mr. Coulson!"

He flung open the door to the next classroom and dramatically interrupted his senior colleague's lesson, bringing him into our room. "Mr. Coulson, do you know what animal eats Hydra?" Whether some wink passed between them I cannot say, but Mr. Coulson played his part well: he didn't know. [Again] the fatherly shade of Sanderson chuckled in the corner, and none of us will have forgotten that lesson. What matters is not the facts but how you discover and think about them: education in the true sense, very different from today's assessment-mad exam culture.

With such a teacher, it isn't difficult to see why I chose biology. Unfortunately I didn't shine at that or any other subject. I spent too much of my time in Oundle's Music School, fooling around on the clarinet or saxophone, or indeed any other instrument that I might come upon unguarded. I wasn't good at music, but I had always been drawn to musical instruments and I had (still have) the ability to play, correctly and without practice, any tune almost as easily as one might whistle or hum it. This facile gift provided a constant temptation – and I readily succumbed – to dispense with reading music. The result was that, although I spent an inordinate amount of time with musical instruments, I didn't play them so much as tootle. Not time well spent. For whatever reason, my performance in science examinations at school was no better than average.

I won't say my time at Oundle was wasted, but I cannot claim to have made the best of it. My love of poetry probably came mostly from my parents, who gave me Yeats and Housman and Rupert Brooke, although my form master in my first year, Snappy Priestman, moved me with his readings from Shakespeare and Kipling. Oundle had the finest workshops of any school in the country and a unique tradition, dating back to Sanderson, of sending every boy into the workshops for a whole week in every term. All day, every day during the Week in Workshops, normal lessons were suspended; we donned brown overalls over our grey suits and – in theory at least – worked at becoming good with our hands. But only in theory. Part of the problem was that the workshops were *too* well equipped and we were too closely supervised – not by proper teachers but by workshop technicians with no idea of pedagogy at all. We did exactly what we were told, on advanced and expensive machines, and each of us ended up making something – a 'marking gauge' one term, a 'drill stand' the next – that looked exactly like what everybody else was making. I didn't even know what a marking gauge was. Like labourers on a factory production line, we learned how to follow instructions when operating a lathe or other large piece of advanced machinery. Maybe some of us learned ingenuity, inventiveness, improvisation, resourcefulness, design, but I certainly didn't, and there was no incentive to. It never occurred to me at the time, but Sanderson must have been spinning in his grave.

Undergraduate

My father and grandfather were keen for me to follow nine earlier members of the Dawkins family into Balliol College, Oxford. My parents went to see Mr. Thomas, who did his best to be cheerful. 'Well, he might just scrape into Oxford, but Balliol is probably aiming too high.' Nevertheless I applied to Balliol, and Mr. Thomas, despite – or more probably because of – his misgivings, had me round at his house in the evenings for extra coaching – for which he would have received no extra payment and no recognition from the school. He was just a great teacher, doing what Sanderson would have done. And he got me into Balliol.

I was well into my second year at Oxford before my interest in the deep questions of existence, and biology's contribution to solving them, really found room to flourish. If I have made anything of my life, it was the Oxford tutorial system that first made me. Imagine the effect on an impressionable nineteen-year-old. Textbooks became a thing of the past,

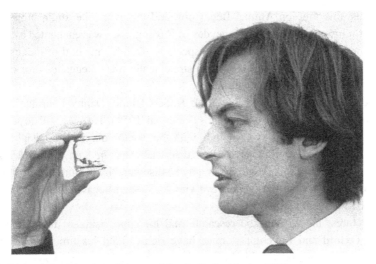

Figure 8.2. Richard Dawkins examining a cricket in Oxford in the 1970s.

together with the whole notion that there existed a received answer to every question. I had the run of one of the world's great libraries. I was sent there each week with a list of readings from the original research literature, and required to write an essay, evaluating the evidence to make my own mind up about what might often be a controversial question. What a heady experience. My later panegyric was published in a variety of places, including David Palfreyman's (2001) collection, *The Oxford Tutorial: 'Thanks, you taught me how to think'*. By way of example, I mentioned my essay on the abstrusely detailed subject of the starfish water vascular system:

I remember the bare facts about starfish hydraulics but it is not the facts that matter. What matters is the way in which we were encouraged to find them. We didn't just mug up a textbook, we went into the library and looked up books old and new; we followed trails of original research papers until we had made ourselves as near world-authorities on the topic at hand as it is possible to become in one week. The encouragement provided by the weekly tutorial meant that one didn't just read about starfish hydraulics, or whatever the topic was. For that one week I remember that I slept, ate and dreamed starfish hydraulics. Tube feet marched behind my eyelids, hydraulic pedicellariae quested and seawater pulsed through my dozing brain. Writing my essay was the catharsis, and the tutorial was the justification for the entire week. And then the next week there would be a new topic and a new feast of images to be conjured in the library. We were being educated ...

Niko Tinbergen entered my life as the lecturer on Molluscs. He announced no special affinity for that group save a fondness for oysters, but he played along with the department's tradition of handing out to each lecturer a phylum, more or less at random. From those lectures, I recall his swift blackboard drawings; his deep voice (surprisingly deep for a small man), accented but not obviously Dutch, and his kindly smile (avuncular as I thought it then, although he must have been much younger than I am now). In the following year he again lectured to us, this time on Animal Behaviour, and the avuncular smile broadened with enthusiasm for his own subject. In the heyday of Ravenglass, I was enchanted by his film on eggshell removal by blackheaded gulls. I especially liked his method of making graphs – laying out sticks on the sand for axes, with strategically placed eggshells for data points. How very Niko. How very un-Powerpoint.

Niko had by then, under the influence of Robert Hinde, Danny Lehrman and others, disowned much of *The Study of Instinct* (Tinbergen 1951). He was still loyal to *Social Behaviour in Animals* (1953) even though, with the 'sociobiological' hindsight that came later, much of that book now seems nearly as disownable as *The Study of Instinct*. I wonder how much of our present theory will eventually be disowned by the hindsights of the future. I suspect not much, where the 'gene's eye view' of social behaviour is concerned, but I would say that, wouldn't I?

Niko's lectures, his writing and research, and his supervising of numerous graduate students in Oxford and Ravenglass, must have richly filled his time and he didn't do much tutoring. My mentor at Balliol, the incomparable Peter Brunet, somehow managed to persuade him to give me four tutorials in my penultimate term. Niko carried the principle of the Oxford tutorial to a quirky extreme. Where other tutors gave out a reading list that covered a topic, Niko would hand me nothing more than an unpublished doctoral thesis by one of his graduate students. I was to write an essay around the thesis, criticize it, go into the library to sleuth down its bibliography, and plan future research to carry it further. In effect, my undergraduate task was to play at being a doctoral examiner for a week, and then again the following week with a different thesis. (Later, when I myself started tutoring as a graduate student, I once or twice experimented with the Tinbergen formula of handing out a single thesis as an essay topic – but, unlike Niko, I dared to do it only with exceptionally gifted students.) I had just four tutorials with Niko and that was all it took. I threw out my plans to do biochemical research and applied instead to Niko: could I join the Animal Behaviour Research Group? I could. And it was the turning point in my career.

Graduate student

Perhaps all scientists recall their graduate student years as an idyll. But surely some research environments are more idyllic than others, and I think there was something special about the Tinbergen group. Hans Kruuk (2003) has captured the atmosphere in his splendid biography, *Niko's Nature*. He and I arrived too late for the heroic 'hard core' period described by Desmond Morris, Aubrey Manning and others, but I think our time resembled it. We saw less of Niko himself, because his room was in the main Zoology Department while all the rest of us were housed in the annexe at 13 Bevington Road with Mike Cullen. And it was Mike who was by then the dominant influence upon the Tinbergen group. My eulogy at his Memorial Service in Wadham College said as much, and I wanted to quote it at length here, for I believe Mike Cullen deserves a place of high honour in any history of ethology. Unfortunately, there wasn't enough space, but I have placed the complete text on my website (Dawkins 2008b).

Figure 8.3. Mike Cullen.

The Friday evening seminars were the highlight of the week for the Tinbergen Group. They lasted two hours and frequently spilled over into the following week, but the time flashed by because, instead of the soporific formula of listening to a speaker's voice for an hour followed by questions at the end, our two hours were enlivened by argument throughout. Niko set the example by interrupting almost before the speaker could complete his first sentence. It wasn't as irritating as it sounds, because Niko's interventions aimed at clarification and it was usually necessary. Mike's questions were more formidable and more feared. He was the intellectual powerhouse of the seminar. Other penetrating contributors were Juan Delius and David McFarland, but the rest of us chipped in without inhibition too, almost from the first day we were there. Niko encouraged that. He insisted on absolute clarity about the question we were asking in our research. I recall how shocked I was on visiting our sister research group at Madingley in Cambridge, and one of the graduate students began to tell of his research with the words 'What I do is …' I had to restrain myself from imitating Niko's voice: '*Ja ja*, but what is your *question?*' Years later, I related this story when I gave a research seminar at Madingley. I refused to identify the culprit to a mock-scandalized Robert Hinde, and my lips are sealed to this day.

The question Niko gave me (he must have been writing his 1963 'Four Whys' paper for Lorenz's sixtieth birthday at the time) concerned the ontogeny of behaviour, and my research method was the deprivation experiment. What is meant by the 'innate' and how does it mesh with learning in the development of the young animal? The theoretical stance I adopted was Lorenzian. Maybe Lehrman the developmentalist was right that behaviour itself could in principle not be innate, because you could never deprive a developing animal of everything (1953). But Lorenz's (1965) reply (which, with my interest in evolution, I had arrived at myself before reading his book) was that you could deprive the young animal of the *specific environmental features to which the behaviour was adapted*. So you could demonstrate that the *adaptive fit* of the behaviour was innate, even if not the behaviour itself. At least in principle. How about in practice? That was what I set out to discover with my chicks (Dawkins 1968).

Newly hatched chicks peck at small objects such as spots of dirt on a wall, presumably a food-seeking response. Understandably, they prefer solid to flat objects, and this carries over to photographs. But by what cues do they recognize a photographed object as solid? Humans use surface shading cues. Because the sun shines from above not below, upper surfaces tend to be lighter than lower surfaces, with a gradient between. Telescopic photographs of moon craters can look like hills depending on the direction from which the light falls. Predators use shading cues of solidity in hunting, which is why so many camouflaged animals employ countershading: the dorsal surface is pigmented darker than the ventral, thereby cancelling the expected solid appearance. The upside-down catfish *(Synodontis nigriventris)* is the exception that proves the rule (for once, the expression is spot on). This fish habitually swims upside down and, fascinatingly, it is reverse countershaded. The ventral surface has the dark pigmentation; the dorsal surface is light coloured like the ventral surface of a normal fish.

Back to the chicks: I used grain-sized photographs of top-lit hemispheres mounted at beak height on the wall of the cage, and compared them with the same photographs inverted so that the light appeared to come from below. Chicks strongly preferred to peck at correctly oriented photographs over inverted ones. Apparently, then, chicks used the same surface shading cues of solidity as we do: they seem to 'know' that sunlight shines from above. Now for the deprivation experiments. Day-olds hatched in total darkness, who had never seen anything before, gave their first (sighted) pecks indiscriminately to inverted and correctly oriented photographs equally. Did this mean they normally *learn* the surface shading cues of solidity – learn, in effect, that the sun is overhead? Not necessarily. It could be that the naive day-olds, having never before seen so much as a chink of light, were too startled or dazzled to discriminate. So I did the definitive experiment. I reared and tested chicks in a special cage in which light came from *below*. They would be accustomed to light, and not startled or dazzled when they came to be tested. If learning is important, these chicks should if anything learn that solid objects are lighter on the underside, and hence prefer inverted photographs when tested. In fact, the chicks behaved like normal chicks. They overwhelmingly preferred the uninverted photograph, the one illuminated from above, the one that looked solid to human eyes. In Lorenzian terms, this showed that the adaptive information is innate: my chicks were telling me that they are born with the 'advance knowledge' that the sun shines from overhead. No doubt there are loop-holes in the logic, but I still think the experiment is a nice teaching aid: a neat demonstration of the *kind* of logic we employ when distinguishing the innateness of behaviour itself from the innateness of the *adaptive information* whereby behaviour fits its environment as a key fits a lock.

I think much of the research that came out of the Tinbergen group, including my own – maybe ethological research generally – could be called 'For example, as it might be' research, rather than 'This is the way it actually is' research. Watson and Crick's double helix was a discovery about the way things are. DNA is a double helix, and that's that. End of story, it will never be superseded. Sir Ronald Ross's discovery that mosquitoes carry malaria is another example of 'This is the way it actually is' research. Our research wasn't like that. If I am right, the main use of such ethological experiments is to illustrate a textbook principle. Textbooks are probably correct that animals can be fooled by some restricted part of the natural stimulus situation into behaving in a way appropriate to the whole. Sticklebacks responding to red dummies are a nice *illustration* to get the textbook point across, but the true story of sticklebacks may not be so straightforward – did Tinbergen ever try a blue dummy? Similarly, my conclusion that chicks are born with the 'advance knowledge' that the sun shines from overhead may be too simple. But if there are some animals that are born with advance knowledge of some aspects of their future environment (it would be surprising if there were not), then my work illustrates the *kind* of experiment that can, in principle, be done to test it. I at least demonstrated that developmentalists were missing something if they claimed that you could never, *in principle*, do an experiment to test the innateness of behavior.

I don't know whether the camaraderie of 13 Bevington Road was exceptional, or whether all groups of graduate students nurture a similar *esprit de corps*. I suspect, at least, that being housed in a separate annexe rather than in a large university building improves the social dynamics. When the Animal Behaviour Research Group (and other outliers such as David Lack's Edward Grey Institute of Field Ornithology and Charles Elton's Bureau of Animal Populations) eventually moved into the present concrete monster on South Parks Road, I believe something was lost. But it may be that I was by then just older and more weighed down by responsibilities. Whatever the reason, I retain a loyal affection for 13 Bevington Road and my comrades of those times who foregathered at the Friday evening seminars, or in the lunch room, or over the bar billiards table in the Rose and Crown: Robert Mash, whose epidemic sense of humour I later recalled in my Foreword to his book *How to Keep Dinosaurs* (2003); Dick Brown; Juan Delius, whose deliriously eccentric brilliance entertainingly complemented Mike Cullen's; Juan's supernormally delightful wife Uta, who gave me German lessons; Hans Kruuk, who later wrote Niko's biography; Ian Patterson; Bryan Nelson the gannet man, known to me in my first six months only from the enigmatic notice on his door, 'Nelson is on the Bass Rock'; Cliff Henty; David McFarland, Niko's eventual successor, who, although based in the Psychology Department, was a sort of honorary member of our group because his vivacious wife Jill was Juan's research assistant, and the couple had lunch in Bevington Road every day; Vivienne Benzie, who introduced the sunny New Zealand girls Lyn McKechie and Ann Jamieson as yet other honorary members of the lunch group; Lou Gurr, also from New Zealand; Robin Liley; the jovial naturalist Michael

Figure 8.4. Niko Tinbergen filming with Lary Shaffer.

Robinson; Michael Hansell, with whom I later shared a flat; Monica Impekoven; Marian Stamp, to whom I was later married for fourteen years; Heather McLannahan; Robert Martin; Ken Wilz; Michael Norton-Griffiths and Harvey Croze, who later formed a consulting partnership in Kenya; John Krebs, with whom I later wrote three papers; Michael and Barbara MacRoberts; Iain Douglas-Hamilton, unwilling exile from Africa while he wrote his thesis; Jamie Smith, with whom I wrote a paper on optimal foraging in tits; Tim Halliday; Lary Shaffer; Sean Neill; and others whom I apologize for omitting.

The Bevington Road population was tidal, emptying during the breeding seasons of the great northern seabird colonies, then filling up as the field workers returned to Oxford to write and think. Spring high tide was the fortnight of the annual 'Block Practical' when we were suddenly over-run by a swarm of undergraduates, learning how to do research under Mike Cullen's guidance with help from several of us. Once again, the emphasis was on the clear formulation of discrete questions, and more especially on quantifying the answers. For example, one pair of undergraduates working with me noticed that baby chicks utter rhythmic, piercing cheep cries. These have been labelled 'distress calls', but could the students pin down the conditions under which they are uttered? The guiding didactic was not to tell the students what to do but to encourage them to suggest it for themselves: 'How are you going to quantify cheep calls?' 'Count them.' 'OK, but how do you decide when a call is too quiet to count as a true cheep?' The students might at this point suggest some kind of decibel meter, but we didn't have one so, instead, we might steer them towards the idea of inter-observer correlation: 'If you and your research partner can't see each other counting, do you end up with the same score?'

'OK, so we have our method of quantification. Now, what hypotheses are you testing? What provokes cheeping?' 'They seem to be lonely.' 'All right, but what is "lonely"? Don't trust subjective language; look for something you can demonstrate experimentally. How might you manipulate loneliness experimentally?' 'Pick up a chick and put it by itself.' 'OK, but what are you going to compare it with?' 'The same chick while in a group of six.' 'And how many chicks will you test, alone versus in a group of six?' 'Ten chicks, for one minute in each condition.' 'And will you always test each chick alone first and then in company?' 'Oh, no, I guess we ought to control for order effects.' 'Good, why don't you get on with that experiment and I'll come back in an hour and see how you are getting on. Don't stop until you have done the full number of trials specified in advance by your experimental design.'

'OK, how did that go?' 'A Matched Pairs Test, with each chick as its own Control, showed a statistically significant effect of loneliness.' 'Well done. Now be more specific. How lonely is lonely? What if, instead of five companions, they have only one?' 'Oh yes, that's a good idea. And then, if one companion is enough to reduce the cheeping rate, why don't we try a mirror, compared with the non-reflecting back side of the mirror?' (You see how Tinbergenian these experiments were.) 'How about little balls of yellow cotton wool on stalks, instead of real companions? Do they have to be yellow? Does it help if you give them "eyes"? How about dyeing real chicks different colours instead of yellow? How about no visual stimuli at all but a loudspeaker playing the noise of an invisible clutch of chicks?' The students raced off to do these experiments, which always had to be properly controlled for order effects and other confounding variables, and – one of the things Mike Cullen added to the original Tinbergen version of the block practical – they always had to be analysed with proper statistics. By the way, although they were surely not cruel, most of these experiments could not be done nowadays without a government licence, which would probably not be granted for student experiments. In other words, this kind of education in the quantitative research mentality is impossible in Britain today.

Bevington Road, and especially its satellite research stations in the gull colonies, ran a system of 'slaves' – young unpaid volunteers who wanted a brief taste of the Tinbergen experience before going to university. Among them were Fritz Vollrath (who later returned to Oxford to head a flourishing group working on spider behaviour, and remains a close friend), and (also from Germany) Jan Adam. Jan and I found an immediate affinity, and we worked together. He had remarkable workshop skills and, fortunately, these were the days before 'Health and Safety' existed to protect us from ourselves and sap our initiative. Jan and I had the freedom of the departmental workshops: lathes, milling machines, bandsaws and all. We (that is to say Jan, with me as willing apprentice) built an apparatus to automate the counting of chick pecks, using delicately hinged little windows and sensitive microswitches. Previously, when working on the surface shading illusion, I counted pecks by hand.

Suddenly, I was in a position to collect huge quantities of data. And this opened the door to a completely different kind of research.

I early knew the name of Peter Medawar because he was an exact contemporary of my father in the biology stream at Marlborough College, and then at Oxford (Medawar in Zoology, my father in Botany). Medawar, as British biology's star intellectual, gave a visiting lecture at his old Oxford department, and I remember the excited buzz in the waiting audience (again, no bossy Health and Safety to prohibit standing-room-only). The lecture led me to read Medawar's essays, later anthologized in *Pluto's Republic* (1982) and it was from them that I learned about Karl Popper. I became intrigued by Popper's vision of science as a two-stage process: first the creative dreaming up of a hypothesis, followed by attempts to falsify predictions deduced from it. Just as Niko's stickleback experiments served to illustrate textbook principles, so I became fascinated by the idea of a textbook Popperian study: dream up a hypothesis that might or might not be true, deduce precise mathematical predictions from it, and then try to falsify those predictions in the lab. Jan's apparatus for counting massive numbers of pecks gave me the opportunity.

Medawar elsewhere made the point that scientific research doesn't develop in the same orderly sequence as the final polished 'story'. Real life is messier than that. In my own case it was so messy that I can't remember what gave me the idea for my 'Popperian' experiments. I remember only the finished story, which, as Medawar would have expected, gives an implausibly tidy impression.

The finished story is that I dreamed up a model of what might be going on inside a chick's head when it decides which of several alternative targets to peck at, did some algebra to deduce predictions from the model, then tested them in the lab. It was a 'Drive/Threshold' model, with affinities to the classic Lorenz psychohydraulic model but more precise (Dawkins 1969a), somewhat along the lines of Bastock and Manning's threshold model of *Drosophila* courtship behaviour. It is good Popperism only because the predictions I deduced from it, by simple algebra, were mathematically precise. I mean, they didn't just predict that a measured quantity should be larger, say, than some other measured quantity. The prediction was that a measured quantity should be *exactly equal to* some function of other measured quantities. Several predictions, and more elaborate deductions from extended versions of the model (Dawkins 1969b), were upheld – less accurately, I suppose, than experimental physicists expect, but with an accuracy which, for ethologists, seems spectacularly good: we would normally expect points to be vaguely clustered on a graph, not lined up like soldiers on parade as mine were. I used data not just from my own experiments but also from Monica Impekoven's on blackheaded gulls (Dawkins & Impekoven 1969) and various studies in the literature, including preferences for composers among the members of four leading American symphony orchestras. An odd consequence of this research was that, because I had derived my very own 'formula' (using nothing more advanced than school algebra), I somehow acquired an undeserved (and unsought) reputation for mathematical expertise.

I gave a talk on this work at the 1965 International Ethological Conference in Zurich. For the talk, I built a physical model of my theory, incorporating a rubber tube filled with mercury that I jiggled up and down to represent fluctuating 'drive'. The rubber tube was attached to the bottom of a vertical glass tube, into which were let three electrical contacts at different depths, representing 'thresholds'. Mercury is an electrical conductor, so when the jiggling column hit any of these contacts (the 'drive' exceeding the 'threshold') a circuit was completed. Coloured lights flashed to indicate 'pecks'. The assumption of the model (from which I had derived my formula by algebra) was that pecks are distributed randomly between all colours whose threshold is exceeded at the time (obviously if mercury was in contact with any electrode, it was automatically in contact with all lower electrodes too), and I implemented this rule by means of a noisy system of electromechanical relays switching on coloured lights to represent pecks at different colours. The whole Heath Robinson (American translation: Rube Goldberg) affair was calculated to bring the house down just as, at an earlier Ethological Conference in Oxford, a spoof hydraulic simulation devised by Desmond Morris and friends reputedly had. How I managed to transport it from Oxford to Zurich evades my memory, and indeed my comprehension. There's not a chance that today's airport security would allow anything remotely like it, bristling as it was with amateurishly soldered wires, relays, batteries and mercury.

Alas, just as I was about to go on the big stage (there were only plenary sessions at the IEC in those days) something went wrong and my contraption didn't work. I was in a sweat of panic and couldn't think straight, frantically tinkering on the floor outside the theatre, when I suddenly heard an amused Austrian accent barking out peremptory instructions at great speed behind me. The rapid-fire voice told me exactly what to do. As in a dream I obeyed, and it worked. I turned to look at my saviour, and beheld Wolfgang Schleidt, whom I hadn't previously met. Without any prior knowledge of what my infernal machine was supposed to do, this highly intelligent rising star of continental ethology had come upon my panic, instantly sized up the problem and dictated the solution to me. I have always been grateful to Dr Schleidt, whom I later was not surprised to learn had a reputation for technical ingenuity. I bore my strange device up into the theatre and at the end of my talk its spluttering coloured lights received something akin to a standing ovation. In the audience was George Barlow, rising star of American ethology, and he was sufficiently impressed by my talk to get me invited to the University of California at Berkeley for my first proper job, as a very junior Assistant Professor of Zoology.

Berkeley

I was torn between this flattering offer and another one from Schleidt himself at the University of Maryland, whither he had recently moved from Germany. Though tempted by Maryland, which, besides Schleidt, had other clever ethologists in the shape of Jack Hailman and Keith Nelson, the lure of California was strong. I was by then engaged to Marian Stamp, another Tinbergen graduate student, and we looked forward to an exciting new life by the Pacific, epitomised by the current hip anthem: 'And if you're going to San Francisco, Be sure to wear some flowers in your hair…' Niko rightly decided that Marian needed no hands-on supervision. She would simply get on with her experiments on search images at Berkeley instead of Oxford. That is exactly what she did, with great success, in the excellent scientific atmosphere provided by the Berkeley Zoology Department.

I taught the undergraduate course in Animal Behaviour jointly with George Barlow, basing my lectures largely on my experience from Oxford. During my last year there before I left for Berkeley, Niko had taken a sabbatical leave, and asked me to stand in for him and deliver the 1966 Animal Behaviour lectures to the undergraduates. He had offered me his lecture notes but, under the influence of Mike Cullen, I had become fascinated by the ideas of Bill Hamilton (whom I had not then met) on inclusive fitness, published two years before. When I came to write my lectures on social behaviour, I pushed the 'gene's-eye view' to centre stage. I was in some trepidation at my decision to depart so far from the Tinbergen canon, and also at the extravagant flourishes of my rhetoric: the body as mortal throw-away receptacle for the immortal genes, tripping like chamois from body to throwaway body down the generations. Seeking reassurance, I showed my lecture notes to Mike Cullen. He immediately took the allusion to Hamilton, and wrote 'lovely stuff' in the margin. That was enough for me. I cast hesitation aside and went to town on the gene's-eye view. I did the same thing in my Berkeley lectures, and I like to think that the undergraduates of Oxford and Berkeley in the late 1960s were among the first to hear of the new ideas that were to become fashionable, in the seventies, as 'sociobiology' and 'selfish genery'.

Other close friends at Berkeley along with George Barlow were David Bentley the neuro-ethologist, Michael Land, now the world's leading authority on eyes throughout the animal kingdom, and Michael and Barbara MacRoberts, who later came to Oxford as spirited additions to the Bevington Road circle, as, later, did David Noakes, who was George Barlow's leading graduate student during my Berkeley years. George hosted a weekly ethology seminar for interested graduate students at his house in the Berkeley hills, and those evening meetings recaptured for Marian and me the wonderful atmosphere of Niko's Friday evenings at Oxford. My research at that time was a continuation of my chick pecking work. In the departmental workshops (again, no initiative-crushing 'Health and Safety') I made a Skinner Box for chicks using a heat lamp for reward instead of food, in which I tested further predictions of my 'threshold' model, looking at actual sequences of pecks rather than just total numbers of pecks per minute. This work was later published (Dawkins & Dawkins 1974).

Some time during our second year at Berkeley, Marian and I were visited by Niko and Lies Tinbergen. Niko wanted to persuade us to return to Oxford, where he had an attractive research post to offer me, and Marian could write up her doctoral research, which, as Niko could see, was going well at Berkeley. He returned to Oxford, leaving us to think about the offer. We decided to accept it, but meanwhile Niko had written of a new opportunity. Oxford had decided to initiate a University Lectureship in Animal Behaviour accompanied by a Fellowship at New College, and Niko wanted me to apply. This teaching job would not preclude the research opportunity he had earlier promised me. I agreed to apply for the Lectureship, and Oxford flew me over for the interview.

It was a magical time, with what seemed like all before me. Once again, music stamped the memory. This time it was Mendelssohn's Violin Concerto, which I listened to on the plane, spellbound by the Rocky Mountains below and by exciting prospects ahead. Oxford put on its very best performance, which is the May time blossoming of cherry and laburnum all along the Banbury Road. New College, too, played its golden fourteenth century part and I was happy, my exuberance not dimmed by the news that Colin Beer had put in an unexpected late application for the Lectureship, and Niko had excitedly switched his allegiance from me to Colin. If Niko had decided that Colin was a better bet, that was good enough for me. I would still have the research position and, as I told the interviewing committee, if Colin were there in Oxford too, so much the better. They indeed gave the job to Colin, and I indeed took up the research grant.

Back to Oxford

Marian and I left Berkeley with mixed emotions. Our time there, from 1967 to 1969, was rather politically active. Like most of our friends, we became heavily involved in the anti-Vietnam War movement, and various other less reputable political issues such as the 'Peoples' Park, locally manufactured and later to be satirized by David Lodge (as the 'People's Garden' at 'Euphoria State College') in *Changing Places*. Berkeley remained an episode of magic in memory, a happy/sad dreamtime of lost youth, of clever and friendly colleagues, of clear, bright sunshine over the Golden Gate, of gentle people putting flowers in the rifle barrels of the California National Guard. We crated and dispatched our few belongings from our Berkeley apartment and drove right across the continent in our old Ford Falcon, thickly encrusted with anti-war slogans and Eugene McCarthy election stickers, to New York. We sold the Ford on the quayside, boarded the liner *France* for Southampton, and prepared to resume our life at Oxford with many of our old friends still there and Colin Beer newly arrived. In the event, Colin preferred to spend his time in New College and was scarcely seen in the Department, much to everyone's disappointment. He stayed only a year. Danny Lehrman had far-sightedly kept his position at Rutgers warm for him and, when it became clear that Oxford could not find a position in Medieval French to match the professorship his wife held in America, Colin decided to return. Once again, the Lectureship in Animal Behaviour was advertised, once again the long-suffering New College agreed to associate a Fellowship with it, and once again Niko urged me to apply. This time, however, I really wanted it, and this time I got it.

The life of a Tutorial Fellow of an Oxford College is in many ways a charmed one. I got a teaching room in a mediaeval building surrounded by famous gardens, a book allowance, housing allowance, research allowance, and free meals (not free wine, contrary to envious rumours) in the stimulating and entertaining company of leading scholars of every subject except my own. The stimulating scholars of my own subject were to be found in the Zoology Department – where I spent the majority of my time. The glory days of 13 Bevington Road came to an end and the Animal Behaviour group moved to the battleship-like horror on South Parks Road, then informally known as HMS *Pringle* after the ambitious Linacre Professor who persuaded the university to build it (I have mixed feelings about my part in getting it recently named the Tinbergen Building, for it is widely reputed as the ugliest building in Oxford). The ironically named 'Laughing John' Pringle was further immortalised in the mock-German past participle *'abgepringelt'* which became the Department's in-word for any kind of ruthless modernization.

In the same spirit of modernity, my research grant paid for a PDP-8 computer, my pride and joy and a valued resource for everybody in 13 Bevington Road. I was already acquainted with large mainframe computers, both at Oxford during my doctorate and at Berkeley, a fact that may have contributed to my undeserved reputation for mathematical sophistication. Jobs were submitted to the mainframes on miles of punched paper tape (British computers) or barrowloads of punched cards (American computers) and came back the following day, so there was a premium on not making trivial mistakes in programming. Unlike today, we did our own programming, and results came back not on screens but clumsily printed on paper. Before I left for Berkeley, a major feat of *Abgepringelheit* had been the luring to Oxford of David Phillips's molecular biophysicists from London. They brought with them an Elliott 803 computer, and its keeper Dr Tony North kindly allowed me to use it at night. This was when I became fully aware of the addictive lure of computer programming. I really did literally – and frequently – spend all night in the warm, glowing computer room, entangled in a spaghetti of punched paper tape, which must have resembled my insomnia-tousled hair. The Elliott had the charming habit of beeping an acoustic rendering of its inner processing. You could listen to the progress of your computation through a small loudspeaker which hummed and hooted a rhythmic serenade, doubtless meaningful to Dr North's expert ear but merely companionable to my nocturnal solitude.

When I returned to Oxford from Berkeley, the singing Elliott had gone the way of all silicon, but my 'own' PDP-8 was more than a substitute, and programming became even more of an addiction than before. Previously I had used only high-level compiler languages but, in order to use the PDP-8 as a research tool, I had to master its 12-bit machine code, and I threw myself into this with zest. My first machine-code project was the 'Dawkins Organ', a system for recording behaviour using a keyboard. Engineers had developed rival systems, using sophisticated and expensive hardware. I wanted to replace almost all the hardware with software. Following a brilliant suggestion by my Oxford colleague Roger Abbott, I succeeded, and over the next few years Dawkins Organs were used by numerous members of the Oxford Animal Behaviour Group, and even – for I published a paper on the design (Dawkins 1971), and supplied the software free of charge – by some ethologists elsewhere in the world, for example Canada.

Marian soon obtained her doctorate, and we started to collaborate on research. We had always been good colleagues, thinking aloud to each other at mealtimes and on walks in the country. Our doctoral theses had both, in their different ways, approached the question of what it means to talk of decision-making in animal behaviour. We now planned a study that would – yet again – serve to illustrate a textbook point. The point was that decisions occur between fixed action patterns (FAP) not within them. Once a FAP had begun it would continue to completion, after which there would be a new decision on what to do next. That was the textbook ideal, but could we demonstrate it statistically? We decided to film behaviour (it happened to be drinking in chickens, an elegant *glissando* of a movement) and analyse it frame-by-frame. Any frame that was highly predictable from preceding frames would be deemed part of a fixed action pattern. Moments of high unpredictability would be our 'decision points' (Dawkins & Dawkins 1973, 1974). We were, in effect, testing the hypothesis that frames were bimodally distributed: either very predictable or very unpredictable but not intermediate. Using an information theory metric, we plotted a graph of predictability, measured in bits of information, against time. The research could almost be described as philosophical: helping scientists to clarify what they mean (in this case by 'decision') rather than actually finding something out about animals. Marian and I followed this with a similar, but more elaborate, piece of work on grooming behaviour in flies (Dawkins & Dawkins 1976). We used a Dawkins Organ to record the behaviour, and it then occurred to us to *listen* to the tape to see whether any pattern emerged. It did. We could *hear* the rhythms and melodies of the flies' Fixed Action Patterns as well as analyse them statistically! The tunes reminded me of the Elliott computer, companion of my youthfully mis-spent nights.

In 1974, I became the British Editor of *Animal Behaviour*. I took over the job from David McFarland who, in turn, had succeeded Pat Bateson. I can't say I enjoyed it but it was made enormously more bearable by the cheerfully efficient Jill McFarland, my assistant. Jill had been doing the job with David, so the office simply went on running as a well-oiled machine. I probably didn't make much difference as Editor, although I occasionally tried to improve the clarity of the writing. My particular bugbear was the formulaic scientific paper with its standard headings: Introduction, Methods, Results, Discussion. The rubric's limitations were especially glaring when – as was common – the author had done a series of experiments, each one prompting the next. I tried to persuade authors that the proper sequence for such a paper was: Question 1; Methods 1; Results 1; Discussion 1 leading to Question 2; Methods 2; Results 2; Discussion 2 leading to Question 3 ... and so on. You'd be amazed how many people arranged their paper in the following way: Introduction; Methods 1, Methods 2, Methods 3, Methods 4 ... ; Results 1, Results 2, Results 3, Results 4 ... ; Discussion. Could anything be more obviously calculated to confuse and bore? But I don't think I had any lasting impact on the standard format of the scientific paper, and I was glad to pass the Editorship into Peter Slater's more capable hands in 1978.

In 1974, the year after he shared a Nobel Prize with Konrad Lorenz and Karl von Frisch, Niko retired and the coveted position of Reader in Animal Behaviour was advertised. There was a stiff competition in which I played no part (recounted by Hans Kruuk in his biography of the Maestro) and eventually David McFarland emerged as the new Reader. David had an idiosyncratic approach to the study of behaviour, highly original but difficult for non-mathematical ethologists to understand. He surrounded himself with bright people, including Richard Sibly and Robin McCleery (who tragically died in 2008) and the evening seminars continued stimulating and interesting, as in Niko's era. The work that Marian and I were doing on the statistics of decision-making fitted into the mathematical atmosphere rather well. And I became increasingly addicted to programming computers.

The Dawkins Organ was only one of several programming projects to which I devoted more time than was good for me or my career. While still at Bevington Road, I had devised my own language, BEVPAL, to speed up machine code programming for the PDP-8 (indeed the Organ program was itself partly written in BEVPAL). I now wrote a program to translate any program from one language to another. Not a very useful exercise, but it taught me a lot about the theory of syntax and enabled me to read Chomsky with far more comprehension than I otherwise could have mustered. Again using syntactic principles, I wrote a program to simulate the song of any cricket. Punningly called Stridul-8, this PDP-8 program was planned for an ambitious series of experiments on a laboratory population of crickets of the Pacific Island species *Teleogryllus oceanicus*, which my graduate student Ted Burk maintained from stocks supplied by my old Berkeley friend David Bentley. David had persuaded me that crickets were the animals that everybody should be working on. Although Ted went on to complete a good doctorate on other aspects of cricket behaviour, my project on behavioural responsiveness of crickets to computer-generated song was never completed, and my song-simulation software was scarcely used in earnest, though it worked well and was versatile enough for anybody to synthesize the song of any cricket in the world.

One of my most ambitious programming projects was inspired by daily exposure to the ethos of the McFarland research group. The talk in the coffee room and the evening seminars was of control theory models of animal behaviour: boxes and arrows and feedback loops. The natural way to simulate a control model is with an analog computer. You set up a model by patching together a circuit of modules with names like integrator, subtractor, adder, multiplier, comparator. Then you switch on the analog computer and see what happens. We didn't possess an analog computer, but I was taken with the idea that a digital computer can be programmed to do anything that any other machine can do, so I set out to program the PDP-8 to behave like an analog computer. Writing my GenSim program probably wasn't a good use of my time, and, as with the language translation program and the cricket song simulator, the software was scarcely ever used, but the exercise taught me a lot about integral calculus, and it equipped me to keep abreast of Robin McCleery, Richard Sibly, Alasdair Houston, Ivor Lloyd and the other control theorists of the McFarland group who dominated my world at the time.

For me, the decision-making work, which I had begun with my thesis and continued in the joint papers with Marian, came to a climax – and closure – in 1975. On their 25th anniversary, our sister group at Madingley decided to hold a birthday conference, to which both David McFarland and I were invited as the Oxford contingent. I threw myself into producing something completely new, and spent a year, with Marian's constant advice and encouragement, working on a paper on Hierarchical Organization (Dawkins 1976a). I won't try to summarise it here, but it is of all my papers perhaps the one to which I devoted the most hard thought and concentrated effort. A red-letter year for me and the end of an era (that's what I meant by 'closure'), 1976 was also the year in which my first book, *The Selfish Gene*, was published (Dawkins 1976b).

Selfish genery

Once again, the real story of *The Selfish Gene* was less tidy. It wasn't really a new departure after 'closure'. I had actually written the first chapter two years earlier. The winter of 1973 was a time of industrial unrest in Britain under Edward 'Grocer' Heath, and there was a period known as the three-day week when electricity was available only intermittently. This made my normal research impossible and I turned to something that needed only sunlight (or, at most, candles). I would write a book, expounding the 'gene's-eye view' that had dominated my lectures at Oxford and Berkeley in the 1960s. I was moved to do so by a spate of popular books of the time, which promoted a kind of group selection. The authors of these books, for example Konrad Lorenz and Robert Ardrey, never properly realised they were promoting group selection. They seem to have so misunderstood Darwinian natural selection as to think that it actually *was* a theory of group selection! My main motive in writing *The Selfish Gene* was to counteract this by expounding neo-Darwinism from first principles, and that meant – so I maintained – the 'gene's-eye view'.

The three-day week came to an end. I went back to the laboratory and put my unfinished chapter away in a drawer until 1975, when I took a sabbatical leave, dusted off the chapter again, and finished the whole book rather swiftly. By then new theoretical ideas by Robert Trivers and John Maynard Smith were available to supplement those of Bill Hamilton and George Williams, which had inspired the first chapter. On Desmond Morris's advice, I showed some chapters to Tom Maschler, doyen of London publishers, in his spacious, book-lined room at Jonathan Cape. He liked the book but not the title: 'selfish' is a 'down word'; I should call the book *The Immortal Gene*. Perhaps he was right. He might have published the book but meanwhile Roger Elliott, the professor of Theoretical Physics and a colleague at New College, had introduced me to Michael Rodgers of Oxford University Press and from then on there was never any question about who would publish it. Michael simply had to have the book, and he would not rest until he got it: 'I haven't been able to sleep since I read it. *I must have that book.*'

There is something exhilarating in having your first book published, especially when you have a publisher like Michael Rodgers. In joking reference to ecological theory, he has described himself as a '*K*-selected publisher', and that is exactly right, with the addition of the adjective 'obstinate'. If your publisher is *K*-selected and obstinate, and if he really likes your book, there are no lengths to which he will not go on its behalf. One anecdote. The International Ethological Conference in 1977 was at Bielefeld. I was invited to give a plenary talk, and I used it to introduce the idea of the extended phenotype as an outgrowth from *The Selfish Gene*. The conference bookshop had a few copies of that book, and they instantly sold out. The bookseller frantically telephoned Oxford University Press to try to get emergency reinforcements and she received a brush-off from one of the suits who infest such organizations: 'Whom do we have the honour of addressing? Well, you must understand, we have procedures to go through, you might get the books in three weeks if you are lucky.' Three weeks would have been much too late: the conference would be long over. The German bookseller appealed to me for help, and I telephoned Michael Rodgers. My memory still hears the slam of his fist on the desk: 'Good! You've come to the right man!' I don't know how he did it but, the very next day, a large box of books arrived in Bielefeld. I had indeed gone to the right man. Not just on that occasion, but in the first place. If you have a book to sell, go to a *K*-selected publisher. An obstinate one.

The Selfish Gene was well received by the critics, almost the only major exception being Richard Lewontin, a man of fabled intelligence who completely missed the point and thought it was an advocacy of panglossianism and genetic determinism! It was an especial joy for me to receive a rave review from my intellectual hero Bill Hamilton, lately migrated from London, where he felt unappreciated, to Michigan, where he was deservedly appreciated by the flourishing group of Darwinian social theorists under Richard Alexander. I met Bill in a London underground train, shortly before his departure for Michigan. He told me he was reading *The Selfish Gene* and far preferred it to *Sociobiology*, which, of course, delighted me. His review in *Science*, when it eventually appeared, was all that I could have hoped for. Not just complimentary but deeply and idiosyncratically *thoughtful*. Bill characteristically ended by quoting two poems, one by Wordsworth (the famous lines on the statue of Newton in Trinity College, Cambridge) and one by Housman (I don't know whether Bill consciously identified with the melancholy protagonist of *A Shropshire Lad*, but it would have been in character):

> From far, from eve and morning
> And yon twelve-winded sky,
> The stuff of life to knit me
> Blew hither: here am I
>
> …
>
> Speak now, and I will answer;
> How shall I help you, say;
> Ere to the wind's twelve quarters
> I take my endless way.

What 'stuff' and what 'I' was Housman referring to, Bill wondered: memes or genes? Bill later became my Oxford colleague and dear friend. His premature loss remains hard to bear. I organized his secular funeral in New College chapel, and spoke a eulogy, which is reproduced as Chapter 4.1 of *A Devil's Chaplain*.

After *The Selfish Gene* and the Bielefeld conference, my research interests took a new turn with the arrival in Oxford of Jane Brockmann as a post-doc. Her PhD under Jack Hailman at Wisconsin was a field study of digger wasps, *Sphex ichneumoneus*, American solitary wasps similar to those made famous by Tinbergen and Baerends in Holland. Jane's study was a classic of field observation. She colour-marked all the individual females digging and provisioning nests in two study areas. She possessed precisely timed records of exactly when each marked individual started digging a burrow, left it, returned with prey, left again, entered another wasp's burrow, fought with another individual, sealed up a burrow, started another burrow, etc. She had used her data for a completely different purpose in her thesis. But it was Alan Grafen at Oxford who clearly saw the econometric possibilities opened up by such carefully *timed* individual data as Jane had gathered.

In mentioning Alan, I need to digress for a moment. I have taught, and learned from, some good students in my time, both undergraduate and postgraduate. But I am sure they would not mind my singling out Alan Grafen, Mark Ridley and Yan Wong as ex-students from whom I have learned *hugely* more than I ever gave them as a teacher. Alan was, at the time of

Jane Brockmann's year in Oxford, doing a master's degree in mathematical economics after his BA in experimental psychology (during which I had tutored him). He was greatly valued as a kind of honorary member of the Animal Behaviour Research Group, in advance of joining it formally to do his D.Phil (Oxford-speak for PhD) with me. Alan's formidable intellect devoured Jane's data like a swarm of locusts. Alan and Jane showed me how a really good collaboration can be one of the great pleasures science has to offer. What Alan taught Jane and me during that collaboration (Brockmann *et al.* 1979) was the importance of *time* as an economic variable in animal behaviour. I gladly pay tribute to the earlier mentoring I received from Niko Tinbergen and Mike Cullen. But what I later learned from the young Alan Grafen during our wasp work was, I think, at least as influential on me. Alan wasn't so directly involved in the two further papers that Jane and I wrote on digger wasps (Brockmann & Dawkins 1979; Dawkins & Brockmann 1980), but his extraordinary biological and economic intuition continued to guide us. The second of those two papers, by the way, was an intriguing demonstration that digger wasps behaved as if committing the 'Concorde Fallacy', a term that I had introduced in *The Selfish Gene* (and in Dawkins & Carlisle 1976).

The learning process carried on after Jane took up her faculty position at the University of Florida at Gainesville. I went there for a sabbatical term, taking Alan, who was by now my graduate student, with me. Jane and I ran a graduate seminar and it was almost embarrassing how, in the nicest possible way, Alan seemed naturally to take command. He played the same role as he had during our Oxford collaboration, but now he was playing it for the benefit of a whole class of students, most of them older than he was. I would summarise his gift as a quite extraordinary economic and biological intuition. Mathematics comes into it, but Alan is like R. A. Fisher in that, although he can do the mathematics explicitly (and does so in his published work) it is his biological intuition that lifts him over mathematicians who think they are qualified to take over biology. As Marian put it in the Preface to one of her excellent books, Alan has 'the annoying habit of always being right.' He is now my colleague at Oxford.

My sabbatical term at Gainesville was my opportunity – enriched by numerous discussions with Alan and Jane – to break the back of my second book, *The Extended Phenotype*. It is the only one of my ten books aimed primarily at a scholarly audience of scientific colleagues, but I like to think that all of them can be read with profit by professionals as well as amateurs. I finished it when I returned to England, and it was published in 1982. Of all my books, I think it is my most original, although the novel part doesn't really start until Chapter 11, 'The genetical evolution of animal artefacts' where I introduce the idea of the extended phenotype itself. There isn't space to expound it here, of course. I reprised the main argument in the second edition of *The Selfish Gene*. Earlier chapters consist of replies to critics of *The Selfish Gene* and various other exercises in clarification and cleaning up. The book benefited from a month I spent in Panama. The Smithsonian Tropical Research Institute had the good habit of inviting biologists to 'interact with' the resident field researchers on Barro Colorado Island and my stint overlapped, happily, with John Maynard Smith's. To spend time with John anywhere would be a privilege. To do so in a tropical jungle, in the company of locally expert naturalists, was a bonus. How I miss him, but of course biologists all over the world are saying that.

I never worked in the same department as John, but I revered him as a mentor in the same class as Niko, Mike Cullen and Bill Hamilton at Oxford. I first got to know him at the BBC in London, where Peter Jones was producing a documentary version of *The Selfish Gene* for *Horizon* (*Horizon* documentaries were frequently rebranded with an American accent as 'Nova'). I was too nervous to accept Peter's invitation to present the show, which was a good thing as John was better than I could ever have been. The documentary undoubtedly helped sales of the book. Some eight years later, I was approached by another *Horizon* producer/director, Jeremy Taylor, who was planning a documentary on the 'prisoner's dilemma', with special reference to Robert Axelrod's computer tournament (1984, reprinted 1990). Jeremy wanted me to present the show, and this time I had the courage to say yes. It was released under the title 'Nice Guys Finish First'. As a result of this title, I became briefly regarded as a champion of niceness instead of selfishness, which was a welcome, if temporary, change. I was wooed by three great corporations, each eager to demonstrate their niceness. The Chairman of Marks and Spencer invited me to lunch in the boardroom so he could explain to me how nice his company was to its employees (a fact that I have no reason to doubt). A woman from the public relations department of Mars Bars gave me lunch in order to persuade me that her company's motivation was not to make money but to distribute sweetness to consumers. And a man from IBM flew me to Brussels to preside over a day of prisoner's dilemma-style gaming for middle executives undergoing a refresher course. The idea was to instil in them a spirit of amicable cooperation, but unfortunately it backfired. The suited executives were divided into three teams: the Reds, the Blues and the Greens. They played for notional money, not real, but shortly before the game ended (and, of course precisely because it was about to end), the Reds suddenly betrayed a whole day of cooperative trust by reneging on the Blues. The bitterness was so palpable and personal, they all had to have counselling for fear they would be unable to work together in the future. As for the idea of *The Selfish Gene* being an advocacy of either selfishness or niceness, both were absurd, and good examples of the inflated importance of titles. The 'selfishness' we are talking about is of *genes*. From selfish genes, either altruism or selfishness at the individual organism level might flow, depending on the economic conditions that obtained. That was the whole point!

As with *The Selfish Gene* a decade earlier, British sales of *The Blind Watchmaker* received a boost from a BBC Horizon documentary that shared its name. Like 'Nice Guys Finish First' it was directed by Jeremy Taylor and presented by me. A new bout of programming addiction fed into the book, my 'computer biomorphs' simulation of evolution by artificial selection, the word 'biomorphs' being borrowed from Desmond Morris's paintings. I described in my book the surreal exhilaration that the emergent biomorphs induced in me:

When I wrote the program, I never thought that it would evolve anything more than a variety of tree-like shapes. I had hoped for weeping willows, cedars of Lebanon, Lombardy poplars, seaweeds, perhaps deer antlers. Nothing in my biologist's intuition, nothing in my 20 years' experience of programming computers, and nothing in my wildest dreams prepared me for what actually emerged on the screen. I can't remember exactly when in the sequence it first began to dawn on me that an evolved resemblance to something like an insect was possible. With a wild surmise, I began to breed, generation after generation, from whichever child looked most like an insect. My incredulity grew in parallel with the evolving resemblance… I still cannot conceal from you my feeling of exultation as I first watched these exquisite creatures emerging before my eyes. I distinctly heard the triumphal opening chords of *Also sprach Zarathustra* (the '2001 theme') in my mind. I couldn't eat, and that night 'my' insects swarmed behind my eyelids as I tried to sleep.

My next writing project was the second edition of *The Selfish Gene*, commissioned by Oxford University Press and published in 1989. Right from the start, the publishers agreed that the original text should remain unchanged, warts and all. Revisions would take the form of extensive endnotes, some of them quite long. And there would be two completely new chapters, 'Nice guys finish first' (the title came from my BBC television documentary, and the subject matter mostly from Robert Axelrod) and 'The long reach of the gene' (a potted version of the last four chapters of *The Extended Phenotype)*. In writing all three new portions, I was hugely helped by Helena Cronin; in return, I offered modest help to her with her own beautiful book, *The Ant and the Peacock*.

I have always enjoyed collaborating, and wish I had done more of it. My three joint papers with John Krebs showed me how immensely valuable joint thinking can be – like a kind of mutual tutorial. In 1978, we wrote a chapter on 'Animal signals: information or manipulation' (Dawkins & Krebs 1978). This paper, and its sequel (Krebs & Dawkins 1984) were influential in redirecting the attention of students of animal communication, in two ways. Instead of treating communication as a mainly cooperative enterprise, we stressed deception (mainly under John's influence) and manipulation (this was my contribution). It is often not realised that deception and manipulation are not the same thing, although both are important. In 1979, John Maynard Smith and Robin Holliday convened a meeting of the Royal Society on the evolution of adaptation by natural selection, and they invited John and me to present a joint paper. The subject we chose was evolutionary arms races (Dawkins & Krebs 1979), and I think it was at least as valuable a contribution as our joint papers on animal signals. The contribution to that symposium that I mostly remember was the paper on comparative studies of adaptation by Tim Clutton-Brock and Paul Harvey. It pulled the rug from under Gould and Lewontin's 'critique of the adaptationist programme' (1979). Characteristically, however, Gould blithely went ahead and ignored Clutton-Brock and Harvey when he delivered his 'Spandrels' paper later in the same day. Even the ludicrously over-rated written version of that paper didn't deign to mention Clutton-Brock and Harvey (1979).

My life became increasingly driven by commissions rather than by my own initiatives. I was invited to give the 1991 Royal Institution Christmas Lectures for Children, five one-hour lectures for a 'juvenile auditory' (the words of Michael Faraday, founder of the lectures) and since 1966 televised in their entirety. After much justified hesitation, I agreed. Not the least humbling aspect was the list of my predecessors, which began with Faraday himself and ran through just about every household name in British science ever since. The Christmas Lectures are apt to take over a scientist's life in a way no other commitment easily does. I won't say the lecturer's every footstep and gesture is choreographed for the benefit of the television audience, but there is some truth in the exaggeration.

My series of five lectures was headed 'Growing Up in the Universe'. 'Growing up' had three meanings, on different timescales: first, on a timescale of decades, the growing up of an individual's understanding of the world – especially appropriate for a series of lectures for children; second, on the historical timescale of centuries, the growth of humanity's understanding of the universe; and third, on the geological timescale of millions of years, growing up in the sense of evolution. At one time I thought of writing a book with the same title, and it even mysteriously found its way onto an Amazon list. However, as it turned out, the theme was too big for one book and it spread into two: *Climbing Mount Improbable*, published in 1996, and *Unweaving the Rainbow* in 1998, whose publication was delayed by a shorter book, *River Out of Eden* (1995), commissioned by my literary agent John Brockman and the publisher Anthony Cheetham. *Unweaving the Rainbow* owed more than just its subtitle (*Science, Delusion and the Appetite for Wonder*) to my Richard Dimbleby Lecture, commissioned by the BBC and televised in 1996. *Climbing Mount Improbable* incorporated a colour version of the biomorphs program (mostly written during a gloriously productive fortnight in Los Angeles as the guest of Alan Kay and his pioneering group within the Apple company). It also included 'arthromorphs', perhaps the most biologically interesting program I have written, in which I collaborated with Ted Kaehler, one of Alan Kay's most brilliant colleagues. I think *Climbing Mount Improbable* is a candidate for my favourite of all my books, and it is surely the most under-rated. The arthromorphs are the centrepiece of a chapter called 'Kaleidoscopic Embryos' which, together with 'The museum of all shells', I think captures something of the flavour of what later became known as 'evo-devo'.

  River Out of Eden and *Climbing Mount Improbable* were both illustrated by Lalla Ward, who had recently become my wife. Though an accomplished artist, Lalla's main profession was acting. She is best known as Dr Who's companion, but was distinguished by her role as Ophelia to Derek Jacobi in the BBC television production of *Hamlet*. Her beautiful speaking voice came into its own again in 1996 when I was touring the USA, promoting *Climbing Mount Improbable*. I got laryngitis and I lost my voice… in San Francisco. The show must go on, and Lalla stepped into the breach, doing elegant readings from the book, after which I croaked out answers to questions from the audience. We continued like that for the rest of the tour and her readings were such a success that, even after my voice came back, we maintained the formula as a double act when promoting all my later books, and also when recording audio books. Audiences agree that the two-voice pattern makes for easier listening. I think, too, that I have learned from Lalla how to read aloud, and it stood me in good stead recently in my difficult task of recording an audio version of *The Origin of Species* (2007). In that undertaking, I made no attempt to act the part of Darwin but instead worked hard on the phrasing and stressing, to make his prose as understandable as I possibly could.

Charles Simonyi Professor

By the early 1990s I had become fretful at my long service as University Lecturer in Animal Behaviour and Tutorial Fellow of New College. Much as I had loved the tutorial system as an undergraduate, I couldn't help feeling that as a tutor, with the best will in the world, I was becoming a little jaded. Promotion to the distinctively Oxford rank of Reader was a gratifying honour, but the workload remained the same. Approaching my last decade before retiring, I felt that I needed a complete change in order to give of my best. At the same time, the professional fundraisers at Oxford University's Development Office had the idea of using my reputation as an author to raise the money for a new Professorship in the Public Understanding of Science. They briefed Oxford's New York office, which set about exploring various possibilities until they found the perfect benefactor.

Charles Simonyi, originally from Hungary, was one of Microsoft's most brilliant software designers (he has now left to found his own company, Intentional Software). He has a deeply informed interest in science, and a dedicated commitment to its promotion. A tentative approach from Oxford's New York office revealed that he admired my books, and prompted an invitation for Lalla and me to visit Seattle. He invited about 30 people to dine with us from the high-tech electronics, software and biotech industries, including Bill Gates, Nathan Myhrvold and other luminaries. Charles presented every dinner guest with a copy of *River Out of Eden*, which had just been published. When, at the end of the dinner, he called on me to speak, I became alarmingly aware that this was my audition (as Lalla, with her acting background, put it) for a part that I very much wanted to play. Even more alarming was the question and answer session afterwards. A lifetime among university academics had hardened me to intelligent and searching questions, but the high-tech whiz kids of Charles's West coast circle conferred new meaning on both 'intelligent' and 'searching'. I felt lucky to come through in one piece. The next day Charles piloted us in his helicopter for a breathtaking ride towards the Canadian border, jostling the skyscrapers of Seattle on the homeward journey. Back on *terra firma*, a ten-minute meeting between him and Michael Cunningham of Oxford's New York office apparently clinched the deal. There were details to be worked out, but from our homebound plane window the dreamlike vision of Mount Rainier rearing up through the clouds seemed to promise an exciting new career: Charles Simonyi Professor of the Public Understanding of Science.

Oxford has a good rule that no individual should be *promoted* as a result of a benefaction. Therefore, although Charles had endowed a full Professorship, I gratefully accepted the job initially at the lower level of Reader, which was the same level as my existing job, and I even took a cut in salary. A year later, after undergoing Oxford's customarily rigorous process of peer-reviewed vetting for promotion, I was raised to the rank of Professor. Charles was generous enough to endow the post in perpetuity, so I am only the first of what I hope will be many Simonyi Professors at the University of Oxford. Charles's manifesto, which he wrote for the guidance of future Oxfords, seems to me a model of generous far-sightedness, and I have reproduced it on my website (Dawkins 2008c). One of the first things I did, on assuming the Chair, was to make a modest endowment to New College to finance an annual lecture in Charles's honour. Lecturers have included two Nobel Prizewinners, the President of the Royal Society and many other distinguished scientists.

After *Unweaving the Rainbow*, my next writing project was the largest and most demanding of my career: *The Ancestor's Tale* (2004). Once again it was a commission, this time by Anthony Cheetham of Orion Books. Anthony and his wife Georgina had been friends since he published *River Out of Eden* in 1995. Lalla and I were staying with them in their country house at Paxford in the Cotswold Hills, on the very day that little book first hit Number One in the British bestseller list, so it was a house of good omen. Some time in 1998, in the same house, Anthony urged me to write a big book, a *magnum opus* on the evolutionary history of the whole of life. To say that I had cold feet would be to overstate their temperature. I understood well what a lot of work it would be. At the same time, something told me that I needed a major challenge. Having given up the daily round, the common task of a tutorial fellow for the comparative luxury of the Simonyi Professorship, I felt that I owed it to Oxford, and to Dr Simonyi, not to relax but to push myself to a new limit. I think I quoted Yeats, 'The fascination of what's difficult ...' On a subsequent weekend, in an effort to help me resolve my shilly-shallying, Lalla drove over to Paxford alone, to discuss the matter with Anthony further. When she returned full of enthusiasm, I took the plunge and signed the contract. It would take five years to complete and, as it seemed, nearly killed me. Looking back on those five years, I now do not understand how the project was ever finished. Several times I approached the brink of giving up and handing back the (frighteningly generous) advance. I think the only things that kept me going were Lalla's unswerving moral support, and the talented and hard work of my research assistant Yan Wong.

Yan had been my student at New College, one of the best undergraduates I ever had. I suppose he was my grandstudent too, since Alan Grafen supervised his doctoral thesis. Before he had finished his thesis, I hired Yan to work with me on *The Ancestor's Tale*. I can't imagine anybody more suitable. He had the breadth of biological knowledge, the diligence and dedication, the computer expertise, and, not least, the literate wit and good humour to see me through the desponds. Somehow we finished the book, and it has to be the achievement in my life that was the hardest to complete. It is a comprehensive history of life written backwards, in the form of a Chaucerian pilgrimage from the present to the evolutionary past.

Despite Yan's help, I became so embarrassed at the slow progress I was making that in 2002 I offered Anthony Cheetham, and the American publisher Houghton Mifflin, another book, as a sop. This was a collection of essays, mostly previously published but gathered together into some kind of coherent order. It was published in 2003 as *A Devil's Chaplain*. Apart from the title essay, which I wrote specially for it, and some connecting notes mustering the essays into themed sections, no new writing was required, so it could all be assembled quite fast, with the knowledgeable assistance of my Editor, Latha Menon. It distracted me for a while from the big book, and I think it really did enable me to return to that book, refreshed for the task of completing it by 2005.

I suppose that in an autobiographical chapter I should say something about honours. In 2001 I was overjoyed to be elected a Fellow of the Royal Society, the highest accolade my country can bestow on a scientist. As a scientist interested in communicating the love of science, I am also gratified that my first two Honorary Degrees were Doctorates of Literature (rather than Science as all the others have been). The first of these was from St Andrews, which is Britain's third oldest university after Oxford and Cambridge, and one of its most distinguished. It was a great pleasure to me when my daughter Juliet enrolled there to begin her training as a doctor. I have also been fortunate during my career to win some valued prizes, including the 1994 Nakayama Prize for Achievement in Human Science, the 1997 International Cosmos Prize, the 2001 Kistler Prize, and the 2005 Shakespeare Prize for Contributions to British Culture. On one of the expeditions to Japan to accept a prize, Lalla and I were accompanied to the ceremony by the British Ambassador and his wife, Sir John and Lady Boyd, who had become dear friends since we first met them on a previous visit when I reprised the Christmas Lectures in Japan. After the prize giving there was a group photograph, for which we all had to sit in a very neat row. A neat-suited young woman employed by the photographer bustled along the chairs, adjusting our knees to perfect symmetry and our shoes to flawless alignment. When she reached Julia Boyd and Lalla, well-bred politeness struggled but failed to restrain their giggles. The photographer's assistant literally reached inside their skirts to straighten their tights. Gospel truth.

My next book after *The Ancestor's Tale* was *The God Delusion* (2006). It is not so detached from my scientific career as some might think, but I shall say little about it here, nevertheless. It sold more than a million copies in English even before the American paperback was published, and it is being published in more than 30 languages. Its success has prompted me to start my own charitable foundation for the promotion of Reason and Science – actually two sister foundations, one in Britain and one in the USA. Most recently, my latest book (2008a) is *The Oxford Book of Modern Science Writing*, an anthology again assembled with the help of Latha Menon.

That brings me to the present. I have entered my retirement year, but I can't detect in myself any desire to wind down. Quite the contrary. My charitable foundations, and the consciousness-raising exercise associated with my website (www.RichardDawkins.net) run from California by the incomparable Josh Timonen, seem to have rejuvenated me and filled me with enthusiasm for the future. In March 2007, Josh and the website regulars secretly conspired to put together a sixty-sixth birthday tribute for me (two thirds of a century) to which more than 3000 people from around the world contributed (http://richarddawkins.net/happybirthdayRD). The previous year, Helena Cronin organized a celebration at the London School of Economics for the thirtieth anniversary of *The Selfish Gene*. The overflow audience heard speeches by the philosopher Daniel Dennett, the novelist Ian McEwan, and the biologists John Krebs and Matt Ridley. And, again at about the same time, Oxford University Press published a *Festschrift* volume, edited by Alan Grafen and Mark Ridley, under the title *Richard Dawkins: How a Scientist Changed the Way we Think*, with all new essays specially written for the book by 26 authors: the biologists Andrew Read, John Krebs, Michael Hansell, Marian Stamp Dawkins, David Haig, Alan Grafen, Patrick Bateson, David Barash and Matt Ridley, the historians Helena Cronin, Ullica Segerstråle and Marek Kohn, the physicist David Deutsch, the philosophers Daniel Dennett, A. C. Grayling, Michael Ruse and Kim Sterelny, the psychologists Martin Daly and Margot Wilson, the psychiatrist Randolph Nesse, the publisher and editor Michael Shermer, the memeticist Robert Aunger, the linguist Steven Pinker, the computer scientist Seth Bullock, the novelist Philip Pullman and the theologian Bishop Richard Harries. To be presented with a dedicated essay by any one of these distinguished individuals would have been an honour in itself. To receive 26 essays all bound together, and edited by my two most outstanding pupils... well, what can I say? My cup runneth over.

But now, move on, there's work to be done!

References

Axelrod, R. (1990). *The Evolution of Cooperation*. London: Penguin. (Foreword by Richard Dawkins. British edition of Basic Books (1984) New York edition.)

Bastock, M. & Manning, A. (1955). The courtship of *Drosophila melanogaster*. *Behaviour* **8**: 86–111.

Brockmann, H. J. & Dawkins, R. (1979). Joint nesting in a digger wasp as an evolutionarily stable preadaptation to social life. *Behaviour* **71**: 203–45.

Brockmann, H. J., Grafen, A. & Dawkins, R. (1979). Evolutionarily stable nesting strategy in a digger wasp. *J. Theor. Biol.* **77**: 473–96.

Clutton-Brock, T. H. & Harvey, P. H. (1979). Comparison and adaptation. *Proc. R. Soc. Lond.* B**205**: 547–65.

Darwin, C. *On the Origin of Species*. Abridged and Read by Richard Dawkins. 2007. Audiobook Download. CSA Word Classic.

Dawkins, M. & Dawkins, R. (1974). Some descriptive and explanatory stochastic models of decision-making. In *Motivational Control Systems Analysis*, ed. D. J. McFarland, pp. 119–68. London: Academic Press.

Dawkins, R. (1968). The ontogeny of a pecking preference in domestic chicks. *Z. Tierpsychol.* **25**: 170–86.

Dawkins, R. (1969a). A threshold model of choice behaviour. *Anim. Behav.* **17**: 120–33.

Dawkins, R. (1969b). The attention threshold model. *Anim. Behav.* **17**: 134–41.

Dawkins, R. (1971). A cheap method of recording behavioural events for direct computer access. *Behaviour* **40**: 162–73.

Dawkins, R. (1976a). Hierarchical organization: a candidate principle for ethology. In *Growing Points in Ethology*, ed. P. P. G. Bateson & R. A. Hinde, pp. 7–54. Cambridge: Cambridge University Press.

Dawkins, R. (1976b). *The Selfish Gene*. Oxford and New York: Oxford University Press.

Dawkins, R. (1982). *The Extended Phenotype*. Oxford and San Francisco: W. H. Freeman.

Dawkins, R. (1989). *The Selfish Gene*, 2nd edn. Oxford: Oxford University Press.

Dawkins, R. (1995). *River Out of Eden: A Darwinian View of Life*. New York: Basic Books; London: Weidenfeld.

Dawkins, R. (1996). *Climbing Mount Improbable*. London: Viking Penguin; New York: W. W. Norton.

Dawkins, R. (1998). *Unweaving the Rainbow*. London: Allen Lane, Penguin Press; New York: Houghton Mifflin.

Dawkins, R. (2003). *A Devil's Chaplain*. London: Weidenfeld & Nicolson; New York: Houghton Mifflin.

Dawkins, R. (2004). *The Ancestor's Tale*. London: Weidenfeld & Nicolson; New York: Houghton Mifflin.

Dawkins, R. (2006). *The God Delusion*. London: Transworld; New York: Houghton Mifflin.

Dawkins, R. ed. 2008a. *The Oxford Book of Modern Science Writing*. Oxford: Oxford University Press.

Dawkins, R. (2008b). www.richarddawkins.net/article,2623, Tribute-to-a-Beloved-Mentor, Richard-Dawkins.

Dawkins, R. (2008c). Charles Simonyi Professorship in the Public Understanding of Science. www.richarddawkins.net/article,2246, Charles-Simonyi-Professorship-in-the-Public-Understanding-of-Science, Richard-Dawkins.

Dawkins, R. and Brockmann, H. J. (1980). Do digger wasps commit the Concorde fallacy? *Anim. Behav.* **28**: 892–6.

Dawkins, R. & Carlisle, T. (1976). Parental investment, mate desertion and a fallacy. *Nature* **262**: 131–2.

Dawkins, R. & Dawkins, M. (1973). Decisions and the uncertainty of behaviour. *Behaviour* **45**: 83–103.

Dawkins, R. & Dawkins, M. (1976). Hierarchical organization and postural facilitation: rules for grooming in flies. *Anim. Behav.* **24**: 739–55.

Dawkins, R. & Impekoven, M. (1969). The 'peck/no-peck' decision-maker in the black-headed gull chick. *Anim. Behav.* **17**: 134–41.

Dawkins, R. & Krebs, J. R. (1978). Animal signals: information or manipulation. In *Behavioural Ecology*, ed. J. R. Krebs & N. B. Davies, pp. 282–309. Oxford: Blackwell Scientific Publications.

Dawkins, R. & Krebs, J. R. (1979). Arms races between and within species. *Proc. R. Soc. Lond.* B**205**: 489–511.

Gould, S. J. & Lewontin, R. C. (1979). The spandrels of San Marco and the Panglossian paradigm: a critique of the adaptationist programme. *Proc. R. Soc. Lond.* B**205**: 581–98.

Grafen, A. and Ridley, M. (eds.) (2006). *Richard Dawkins: How a Scientist Changed the Way We Think*. Oxford: Oxford University Press.

Krebs, J. R. & Dawkins, R. (1984). Animal signals: mind-reading and manipulation. In *Behavioural Ecology*, ed. J. R. Krebs & N. B. Davies, pp. 380–402. Oxford: Blackwell Scientific Publications.

Kruuk, H. (2003). *Niko's Nature: The Life of Niko Tinbergen and his Science of Animal Behaviour*. Oxford: Oxford University Press.

Lehrman, D. S. (1953). A critique of Konrad Lorenz's theory of instinctive behaviour. *Q. Rev. Biol.* **28**: 337–63.

Lorenz, K. (1965). *Evolution and Modification of Behavior*. Chicago, IL: Chicago University Press.

Medawar, P. B. (1982). *Pluto's Republic*. Oxford: Oxford University Press.

Mash, R. (2003). *How to Keep Dinosaurs*. London: Weidenfeld and Nicolson.

Palfreyman, D. (2001). *The Oxford Tutorial: Thanks, you taught me how to think*. Oxford: OxCheps.

Tinbergen, N. (1951). *The Study of Instinct*. Oxford: Clarendon Press.

Tinbergen, N. (1953). *Social Behaviour in Animals*. London: Chapman and Hall.

Tinbergen N. (1963). On aims and methods in ethology. *Z. Tierpsychol.* **20**: 410–33.

9

A passion for primates

FRANS B. M. DE WAAL

Although I am prepared to give human artifacts my undivided attention for an hour, or so, show me any animal and time stands still. My career is a direct outgrowth of a life-long fascination with animals and their behavior. No external incentives, such as the prospect of discovery, the attraction of an academic career, or even money, have been necessary. Although all of those surely have helped, they have been secondary to the basic attraction to animals since I was young.

Leaders in Animal Behavior: The Second Generation, ed. L. C. Drickamer & D. A. Dewsbury. Published by Cambridge University Press. © Cambridge University Press 2010.

As soon as I was allowed to go off on my own on weekends, I'd take my bike and go fishing. Not with a rod, but a net. I would bring home live sticklebacks, salamanders, glass eels, mud loaches, bitterlings, or whatever aquatic life I'd caught. I turned the sunroom of our backyard into a little zoo with buckets and tanks full of animals. I caught daphnia or dug up worms to feed them. I bred mice, adopted cats, and raised jackdaws, which are still my favorite birds: very smart and social, just like the primates that I study now. The high-point of this period was that I bred my own sticklebacks and released them back into the ditch from which their parents had come. All of this made for a natural attraction to ethology (the biological approach to animal behavior as opposed to the psychological approach popular in America), as the founders of ethology studied sticklebacks (NikoTinbergen) and jackdaws (Konrad Lorenz).

Born in 1948 in s'Hertogenbosch (also: den Bosch), in The Netherlands, I am the fourth of six sons (no daughters) of a bank director's family. We grew up in a large, lively home. But I was the only one keenly interested in animals, a "gene" that must have skipped a generation since one of my grandfathers loved songbirds and owned a pet store. I spent most of my youth in Waalwijk, a town in the southern part of The Netherlands. This Catholic region is culturally and historically closer to Flanders, in Belgium, than to the Calvinist north, including Holland. I received an excellent education at public schools, and joined the boy scouts. This had the great advantage that I spent almost every Saturday in the extensive forests south of Waalwijk, where I learned about plants and animals as well as male bonding and rivalry.

Apart from keeping animals and learning from direct observation and interaction, I had a few books with wonderful illustrations of wildlife and descriptions of their habits. I was extremely fond of these books, and read them over and over, the way other boys did with books on cars, airplanes, or the national football team. I feel I have been a naturalist since birth.

When I went to high-school, Dutch children still learned three languages in addition to their native Dutch. I hated languages, though, and kept saying that I would never need them, which in retrospect is ironic given that most of my textbooks as a university student were in German, I ended up emigrating to an English-speaking country, and married a Frenchwoman.

I liked the natural sciences a whole lot better than languages. But unfortunately, I had a boring, unpleasant biology teacher, who could only talk with any enthusiasm about the citric acid cycle, which he did in class after class. The result was that I spent most of my time systematically unscrewing my desk until one day it fell apart, upon which the teacher turned purple, and said many unflattering things about me. Our poor relationship affected me such that upon graduation, when everyone recommended that I go to university, I wanted to do mathematics or physics rather than biology.

My grades for the final high-school exam were exceptionally high, so high in fact that everyone – both at school and at home – was surprised. Suddenly, I could pick and choose any career I wanted. I had studied diligently for the exam, and had in fact enjoyed the process. The outcome showed me what I could do if I applied myself.

In view of my love of animals, my mother didn't think that mathematics or physics could be the right direction. In The Netherlands, one chooses the major before entering the university. So, despite my teacher and thanks to my mother, I chose biology, and moved to Nijmegen, a much larger city. I was seventeen.

Undergraduate studies

Starting as a student at Nijmegen University, in 1966, I developed a circle of buddies and grew my hair until I was the sort of bohemian that scared the establishment. Apart from growing up, I took classes in animal anatomy, cell biology, biochemistry, and so on. I rarely heard anything about animal behavior, though. Most of the animals I saw smelled of formaldehyde and had no life left in them.

My eyes were opened during a stint in the psychology department of the same university, where I accepted a paid job to work with two pre-adolescent male chimpanzees, named Koos and Nozem. Those two apes were fun! Every day I took them to a room where we spent hours together doing cognitive tasks. Even though they were far stronger than me, I was in charge most of the time. Another student conducted the same cognitive task with macaques, and in comparing notes we reached the counter-intuitive conclusion that apes do worse than monkeys because of their superior intelligence. Monkeys will perform thousands of trials in a row without ever slowing down, whereas Koos and Nozem found the task (a simple tactile discrimination) boring. After a good start that demonstrated their knowledge, their performance would rapidly drop off. They used any excuse to lure me into playing games with them, and one of their tricks was making lots of errors in a row. Still today, whenever I see widely publicized negative results from ape studies, I think back to this time, knowing how hard it is to motivate these wily characters. Often, such results mean less than people think (de Waal 2009).

My fellow student and I had noticed that Koos and Nozem were keenly interested in women, whether students or secretaries. They would drop anything they were doing and stare at the visitor with obvious sexual arousal. We wondered how they could tell a human male from a female, which led to my first primate experiment. Both of us dressed up in drag. We walked in with skirts, talking in high-pitched voices, and... nothing happened. The chimps showed no interest in us at all, least of all sexual interest. It was not the best-controlled experiment, obviously, and the chimps may simply have recognized us, but it suggested that there was more to sex discrimination than we had thought.

My job with the chimps taught me that I wanted to study *live* animals. I followed a course in so-called "comparative psychology," in which I learned about Skinnerian behaviorism with its emphasis on closely controlled learning conditions and its Cartesian perspective (i.e. animals lack emotions and mental states, or if they do have them, we'd better not speculate about them). This outlook baffled me. My interactions with the chimpanzees had re-confirmed my deeply held conviction of continuity between humans and animals. In this, I had an ally in the department: Paul Timmermans. Paul was one of the course's teachers, who sometimes took me aside to sow doubt about the behaviorist doctrine. With a sly smile, he would ask if I really could believe that everything Koos and Nozem did was based on trial-and-error learning, that they were devoid of intentions and emotions? Even for rats, Paul wasn't sure. He thus helped me think critically about the mechanistic view that was popular at the time, and that lingers in certain corners still.

Graduate studies

In 1970, I moved to the University of Groningen in the north of The Netherlands, where I finally received the education I wanted. I observed my beloved jackdaws as a graduate project (a large free-ranging colony had settled in nest boxes on the outside wall of the zoological laboratory), conducted neuroscience on rats with electrodes the size of telephone poles (they must have been a hundred times thicker than modern electrodes), and learned from such masters as Gerard Baerends, Jaap Kruyt, and Piet Wiepkema. It was the heyday of Dutch ethology and I took full advantage.

In the meantime I had fallen in love with a French girl, Catherine. She would occasionally hitch-hike from Nantes to Groningen to share my tiny apartment, and later moved in with me when I lived in Utrecht. This was my third Dutch university, where I conducted dissertation research under Jan van Hooff, a specialist of primate facial expressions. Jan was a student of Nico Frijda and Desmond Morris. The latter was a student of Niko Tinbergen, so that my academic blood line was firmly within Dutch ethology.

My first project was the aggressive behavior of long-tailed macaques in Jan's laboratory in Utrecht. In those days, aggression was all that ethologists could talk about. Books by Robert Ardrey (*The Territorial Imperative*, 1967) and Konrad Lorenz (*Das sogenannte Böse*, 1963, translated as *On Aggression*, 1966) had set the stage. Human evolution was always phrased in terms of our aggressive background. This focus was perhaps natural after World War II. In the monkeys, I saw plenty of aggression, of course, but I was also struck by how they nevertheless managed to maintain good relations. One minute they would be chasing each other around with loud, angry grunts, and next they would be settling down. If one waited long enough, they'd even groom and play as if nothing had happened. This intrigued me so much that these observations may have planted the seeds of my later discovery of reconciliation, a behavior that I never noticed in the macaques (even though we now know that they engage in it). Discoveries are often guided by unconscious intuitions.

One of the issues I began to address more explicitly was the way in which the dominance hierarchy is structured in a monkey group. The literature emphasized the winning of fights and the obtaining of advantages, such as food and mates, but I noticed that my monkeys were constantly communicating status in more subtle ways. Subordinate individuals would bare their teeth in a wide submissive grin even while approaching a dominant individual. If the grin reflected fear, as the literature claimed, why did it go together with approach behavior? Perhaps signaling about status removes the need for open conflict? I called it a signal of *formalized dominance*, comparing its function to that of stripes on military uniforms indicating rank. In this idea I was supported by Walter Angst, a primatologist working with the same species at the Basel Zoo, in Switzerland. I visited Walter, he visited us, and we were pretty sure we had it all worked out. At the time, however, few people bought into the idea of formalized dominance. It was too much at odds with the view of dominance as an expression of raw competition. Whenever I mentioned formalized signals in my writings, reviewers would call it an "idiosyncratic" view (translation: it doesn't appeal to anyone but the author).

Part of my formation in Jan's laboratory was my time spent in the small wood-paneled library, where I browsed the journals on animal behavior. There were very few in those days (the main ones were *Behaviour, Animal Behaviour,* and *Zeitschrift für Tierpsychology,* which later became *Ethology*). It was easy to keep up. Instead of copying the papers, which was expensive, I kept large notebooks in which I summarized every paper by hand. My favorite author at the time was Hans Kummer, another Swiss primatologist, whose work provided food for thought, and helped me operationalize notions of triadic (three-way) relations and complex social knowledge.

My English improved with time. Even though in those days many papers and textbooks were also in German, English was gradually getting the upper hand. Learning to read and write in English was an essential part of my education. I still remember sitting at my large table at home, pipe in hand, working with pencil and paper, dictionary open, looking up almost every word that I wasn't sure about. This took a lot of concentration. Native English speakers don't have the slightest idea how easy they have it writing in their own language, and how much of a struggle it is writing in another.

Jan also organized journal club meetings, one of which greatly influenced my thinking. One evening in 1972, we went over Robert Trivers' seminal reciprocal altruism paper, which had appeared the year before. The discussion was lively, but what stuck in my mind most is that the author had found a theoretical problem worth addressing. Why do animals help each other? Do they return favors that they have received? Instead of documenting a behavior, which is something I was used to, he focused on its evolutionary explanation. What appealed to me, no doubt, was also the move away from aggression. Here we were dealing with two new elements that I hadn't considered much before: (a) cooperation, and (b) the evolutionary reasons for behavior.

Most of my readings until then had contained a lot of arm-waving when it came to evolution. Authors would throw in the word "evolution" every now and then for good measure – they'd say that "humans evolved to become territorial," or "our instincts derive from animal instincts," and so on – but rarely would they specify the selection pressures involved. Trivers' paper opened my eyes to this issue. While the founding ethologists in Austria, Germany, and the Netherlands were experts at describing and analyzing natural animal behavior, and had exported this specialty to the United Kingdom and America (especially with the move of Tinbergen from Leiden to Oxford, in 1947), these other countries were now giving us something in return. They had developed a more theory-driven, evolutionary approach to animal social behavior, soon to become known as "sociobiology."

Some of my fellow students worked with orangutans in the field, so that it was logical that one day Catherine and I took our first airplane flight to Indonesia. Upon arrival in Medan, a driver picked us up to take us to Gunung Leuser, the reserve on Sumatra where the orangs lived. The driver's home was about half-way on the trip, and we stopped there for dinner and a good night's sleep. His wife had prepared days for our visit, spreading out dozens of little plates in the traditional "rijst-tafel" (Dutch for "rice table"). And although I had had rijst-tafels before, this one was different. Every little dish we tried tasted hotter than the one before, so hot in fact that normally we might have declined. But under the circumstances, we politely ate from all of the dishes, closely watched by the driver and his wife. We have been a fan of hot food ever since.

In the forest, I developed great admiration for people like Herman Rijksen, Chris Schürmann, and Carel van Schaik, all students of Jan, who followed wild orangs around from dawn to dusk every day. They would note every little behavioral detail, while being eaten alive by insects and the ubiquitous leeches. Even though I enjoyed my stay – and have made many brief visits to other field sites since – this experience confirmed that what really interests me are the details of social interaction, and that these details are almost impossible to collect in the field. Orangs are probably the worst example, because most of the time one watches a single individual or a mother with her dependent offspring, eat, travel, eat, and travel. The greatest excitement occurs when several orangs meet in a fruit tree, or when two adult males meet. But I never got to see any such charged moments.

Back home in Utrecht, I followed a political drama among the Java monkeys (*Macaca fascicularis*) that set the stage for my later work on chimpanzees. Akim, the old alpha male, dominated his son, Conrad, even though the latter was twice his size. It was curious how this was possible, but I think I know how. Males who grow up under their father's rule never truly challenge them. In the wild, this situation doesn't typically occur since male macaques leave their natal group at puberty. Young males from the outside try to enter established groups: they have no inhibition whatsoever to kick out an older male. But Conrad had grown up with Akim as supreme leader and it somehow never occurred to him to take advantage of his superior size.

We don't know exactly how it had happened, but one day we found Akim with a deep gash, most likely caused by one of Conrad's huge canine teeth. Probably, Akim had miscalculated, had attacked Conrad and gotten a quick defensive action back that was more than he could handle. So, now we had an alpha male who was extremely nervous each time his son came close. He even submitted a few times to his son, showing him the bared teeth grin that normally signals subordination. But Conrad was not ready to claim Akim's position: he never reasserted himself. After a while, Akim regained his confidence and resumed alpha status. I wrote up these observations in my very first scientific paper, in 1975, under the title *The Wounded Leader*.

In Utrecht, I learned to think about animal behavior, ask pertinent questions, and connect to many different fields, ranging from social insects (the laboratory maintained a large honeybee colony), to electric fish, and primates both in captivity and the field. And the incident with Akim and Conrad taught me that systematic data are one thing, but that they need to be supplemented with qualitative observations. One can never understand social dramas like this without intimately knowing all of the actors and keeping a diary about their daily social affairs. I became interested in a narrative approach (i.e. a full description, with personal impressions and all, of social events), something I put to good use when I started at the Arnhem Zoo.

Chimpanzee politics

From 1975 through 1981, I worked at Burgers' Dierenpark, in English usually referred to as the Arnhem Zoo. My project was facilitated by the zoo director, Antoon van Hooff, being my advisor's brother. In 1971, they had established a colony of around 25 chimpanzees on a two-acre island – still the world's largest such colony today.

Every morning, rain or shine, I would cycle to the zoo to spend hours watching chimps. I did little else. Working at my desk, writing up my thesis on macaques, I would keep an ear to the chimps. At the slightest noise, I had my binoculars in hand to follow the spectacle. Over this six-year period, I supervised more than twenty students, who helped with observations.

Most of my ideas about primates, such as those relating to power, sex, conflict resolution, empathy, cooperation, and reciprocity, were formed during this period. I was directly inspired by the chimps in front of me. Perhaps because of my closeness to the apes, I developed an aversion to the simplifications of theoretical biologists, who in those days depicted both humans and other animals as thoroughly nasty and selfish. I saw a much more complex picture, including a variety of genuinely positive tendencies.

Prepared by my work on macaques (see above), I was fascinated by how chimpanzees handle conflict. One day, when the colony was locked indoors in one of its winter halls, the highest-ranking male fiercely attacked a female. This caused great commotion as other apes came to her defense. After the group had calmed down, an unusual silence followed, as if the apes were waiting for something to happen. This took a couple of minutes. Suddenly the entire colony burst out hooting, and one male produced rhythmic noise on metal drums stacked up in the corner of the hall. In the midst of this pandemonium, two chimpanzees kissed and embraced.

I was puzzled by what I had seen, but it took hours before the term *reconciliation* came to my mind. This occurred when I realized that the two embracing individuals had been the same male and female of the original fight. My hand-written diaries documented reconciliation, expressed in the Dutch term "verzoening," on November 19th, 1975. In the same month, another entry mentions the Dutch term "troost," translated as *consolation*, for reassuring contact provided by bystanders to victims of aggression. Once recognized, both interaction patterns were seen on a daily basis. It became indeed hard to imagine that they had gone unnoticed before.

When several new graduate students came to work at Arnhem, in the spring of 1976, I proposed reconciliation as a research topic. The student put onto this project was Angeline van Roosmalen (younger sister of well-known Amazonian conservationist Marc), who came to us from the Psychology Department of the University of Amsterdam. This is an interesting detail as my supervisor and I were both in the Biology Department of the University of Utrecht. In other words, the student's thesis committee was beyond our control.

The committee members were not altogether enthusiastic about the proposed topic of study, even though they had never watched chimpanzees, and refused to come to Arnhem to make up for this deficit. They were *a priori* convinced that there could be no such thing as reconciliation in nonhuman animals. Their chief alternative hypothesis, developed extemporaneously at one of our meetings, was that social animals go through systematic alternations between positive and negative interactions so as to find a balance between the two: hence every positive encounter must be followed by a negative one, and vice versa. This could account for the aggression–affiliation sequences that we (erroneously) interpreted as reconciliation behavior.

The discussion with these psychologists, although frustrating at times, did help us design a systematic approach to the phenomenon. The question we posed was whether post-conflict encounters are influenced by previous conflicts, or not. For this we needed to compare behavior after fights with baseline behavior at other times. The study was carried out in the spring and summer of 1976, and published as *Reconciliation and consolation among chimpanzees* (de Waal & van Roosmalen, 1979).

At the same time, I further developed ideas about formalized dominance, especially since in chimpanzees the distinction between the outcome of aggressive confrontations and the communication of status is far more pronounced than in macaques. Chimpanzees have elaborate greeting rituals in which the subordinate bows and pant-grunts, literally groveling in the dust in front of the dominant, who stands with his hair on end. These rituals serve to clarify the hierarchy, and as long as they take place the relationship between dominant and subordinate is relaxed, often even positive. Thus, formalized dominance facilitates bonding and cooperation (de Waal, 1986a).

The third topic that drew my attention, and remained a lifelong interest, is altruism, especially reciprocal altruism. This goes of course back to the paper by Trivers mentioned before. The study of spontaneous food-sharing, coalitions, and grooming in chimpanzees offered great prospects to test this idea. The approach that I followed at the time was entirely observational and correlational: i.e. how often does A do a favor for B compared with how often B does a favor for A? My studies hinted at reciprocal exchange.

In 1977, at the age of 29, I received my Ph.D. in biology in a ceremony in the old "academiegebouw" of Utrecht University, right next to the Cathedral Tower in the heart of Utrecht. Soon thereafter, I began writing *Chimpanzee Politics*, a popular account of the power struggles among the Arnhem males. The chimps had conveniently waited to challenge the existing order until about one year into my study, when I had a good grasp of their personalities and behavior. I had carefully kept track of developments in my diaries, and on top of this had a wealth of systematic data to draw from. In order to make sense of their behavior I had turned to Niccolò Machiavelli. During quiet moments of observation, I read from a book published four centuries earlier. *The Prince* put me in the right frame of mind to

Figure 9.1. Cover of the author's dissertation, which he defended at the University of Utrecht, in 1977, comparing the strategic maneuvering of long-tailed macaques with a chess game (drawing by the author).

interpret what I was seeing on the island, though I'm pretty sure that the philosopher himself never envisioned this particular application of his work.

I was risking my career ascribing intelligent social maneuvering to animals. I had been trained to avoid any such talk. But since no one follows this rule with regards to human behavior, I questioned the wisdom of a different standard for a species with which we share so much evolutionary history. If two closely related animals show similar behavior, I felt that the most parsimonious approach was to come up with a single explanation for both.

Desmond Morris facilitated publication of *Chimpanzee Politics*, translated into English from my Dutch original. It first appeared in London, in 1982. The Dutch edition was presented in the same year at the Arnhem Zoo, where I stood in front of the ape island flanked by the van Hooff brothers. A copy of the book was thrown across the moat to its leading characters, who acted as if they were eager to read it, but then tore it apart.

With its Machiavellian flavor, the book drew wide attention. The US Speaker of the House, Newt Gingrich, even put the Arnhem saga on the recommended reading list for freshmen Congressmen. But the German edition came out under the condescending title *Unsere haarigen Vettern* ("*Our Hairy Nephews*"), while the French publisher decided to make fun of politicians François Mitterrand and Jacques Chirac by putting them on the cover holding a grimacing chimp between them. Both publishers held out for the opposite of what I had tried to achieve: to instil respect for the apes. Nowadays, book contracts offer me a veto over cover images and title translations.

In 2007, the twenty-fifth anniversary edition of *Chimpanzee Politics* was published by the Johns Hopkins University Press. The book has never been out of print, and the animal behavior literature is still full of references to planning, manipulation, social strategies, reciprocity, and triadic awareness, so the book's themes remain up to date. But even though *Chimpanzee Politics* introduced the Florentine philosopher to primatology, I remain skeptical about the label of "Machiavellian intelligence" for anything related to social intelligence (Byrne & Whiten 1988). Rightly or wrongly, this term implies a cynical, the-ends-justify-the-means exploitation of others, thus ignoring a vast amount of social knowledge and understanding that has little to do with one-upmanship.

Why did I decide to write a popular book, and why have I been writing such books ever since? In fact, it amounts to a second career, one that runs parallel to my traditional scientific work and publications. The idea may have started with presentations to groups of zoo visitors, such as nurses, lawyers, housewives, psychiatrists, and so on. They would come to the Arnhem Zoo to get out of their work environment, relax, but also be educated. I noticed how little the general public knew about primates and their behavior, and how incredibly eager they were to learn. They had absolutely no interest in the specific theoretical issues we academics get excited about, and even less in graphs and data. They just wanted to hear about the social life of the primates and how it compared with that of their own species.

The writing of popular books requires stepping back, taking the perspective of those with little background, and highlighting the human implications of our work. It is a combination of communicating science through simplification and engaging the reader by giving lots of lively examples. It is a very satisfactory activity even though some of my colleagues grumble at the "vulgarization" of research, and some readers get the false impression that primatology is just story-telling.

Before leaving the Netherlands, two events took place that greatly affected the rest of my career. One was extremely sad. In 1980, two adult males in the colony attacked and gruesomely injured a third, named Luit, who later died of his injuries. To make matters worse, Luit was my favorite personality in the colony. I have described this incident in several places (e.g. de Waal, 1986b). At the time, it was interpreted by some as a product of captivity – hence as an artifact that could be safely ignored – but now we know that similar incidents, including the castration of the victim that it entailed, also occur in nature. I am not referring here to lethal intergroup encounters – amply documented by Goodall (1986) and others – but to the fact that wild chimpanzees, too, occasionally kill within their own community (e.g. Watts, 2004).

Figure 9.2. The essence of "chimpanzee politics" is the formation of coalitions among males in the pursuit of power. The two grooming males on the left dominated the colony together while excluding the male on the right, Luit, who was individually stronger than either one of them. Years after this picture, the coalition permanently eliminated Luit (photograph by the author).

Figure 9.3. Two bonobos at the San Diego Zoo, standing upright. Given their relatively long legs, the body proportions of this species have been compared with those of our direct ancestors (photograph by the author).

Until then, I had looked at conflict resolution as merely an interesting phenomenon. The shock of standing next to the veterinarian in a bloody operating room, handing him tools for the hundreds of stitches he placed, opened my eyes to how absolutely essential it is for chimpanzees to keep their relationships in decent shape. Without these mechanisms, things can get ugly. So, the event impressed upon me the critical importance of peacemaking, and played a major role in my decision to continue work in his area.

The second event was a visit to Wassenaar Zoo. This now-defunct Dutch zoo owned several pygmy chimpanzees, now known as bonobos. After having watched chimpanzees for so many years, the differences were obvious. During my visit I was struck at how bonobos differed in behavior, demeanor, physical appearance, and so on. They seemed much more sensual and sensitive. At the time, almost nothing was known about this species, and I decided on the spot that this had to change. Having been taught that careful comparisons between related species often yield fresh insights into the evolution of behavior, I felt there was a great need to see how bonobos differed behaviorally from chimpanzees.

Across the Atlantic

After one of my first lectures at an international conference in England, Robert Goy, then director of the Wisconsin Regional Primate Research Center, said he wanted me to join him.

Initially, I went on my own to Madison, Wisconsin, since the position was supposedly for one year only. This was in 1981. Within weeks, however, Bob talked me into staying. Later Catherine, by then my wife, joined me.

Before moving to the US, I had thought long and hard about which research direction to take, and decided on the twin topics of reconciliation and social reciprocity. The first topic I had developed myself in Arnhem. At the time no one worked on animal conflict resolution – nowadays a major research area. It led to my second book, and the first one written directly in English: *Peacemaking among Primates* (1989).

With my assistant, Lesleigh Luttrell, I studied macaques at the small Vilas Park Zoo, in Madison, Wisconsin. We did so for ten years, day in and out, and I greatly enjoyed it. I was in the enviable situation that I could devote all of my time to research and had a director who fully supported behavioral work. Most primate center directors watch the bottom-line of how much money research brings in and have a bias against behavioral work as "soft" (as opposed to health-related science, which is considered "hard"). Bob was not at all like this, and simply looked for "quality science," as he liked to call it. I remain grateful for the freedom that he offered me.

I delved deeper into formalized dominance with papers on status signals in monkeys (de Waal & Luttrell 1985), and by showing that the two species that I studied – rhesus and stumptail macaques – might have similar formal dominance orders, but nevertheless had quite different "dominance styles" (de Waal & Luttrell 1989). Dominant rhesus monkeys severely punish any transgression against the rules of priority, thus creating rather terrified subordinates. Dominant stumptails, in contrast, rarely punish anyone and quickly reconcile after fights. The result is a more relaxed, more cohesive society based on bonding rather than intimidation. This work allowed me to integrate ideas about dominance hierarchies, social tolerance, and conflict resolution into a single framework. It also led to my first social experiment, in which we housed members of the two species together to see how this might affect their behavior (de Waal & Johanowicz 1993).

In those days, scientists at the Wisconsin Primate Center enjoyed camaraderie through the lively parties organized by our paterfamilias, Bob, who was a great cook and generous supplier of fresh homemade bread, all kinds of delicious foods as well as local beers. For me it was also a period of discovery at a cultural level: becoming familiar with American life, and understanding the country, its cultural biases and taboos. Compared to the Dutch – who are brutally honest and direct – I felt there were many topics Americans were excessively shy about. I was learning to be more circumspect in my opinions about politics, sex, or private affairs. On the other hand, I appreciated very much the American attitude towards work: no one quibbled with the connection between effort and success. The land of opportunity indeed!

One important additional influence should be mentioned. A German-born student of law, Margaret Gruter, had founded the Gruter Institute for Law and Behavioral Research, in California, which brought together people from all sorts of disciplines interested in the evolution of behavior. We had regular meetings in Squaw Valley, in California, where I would meet biologists, economists, anthropologists, law professors, psychologists, and so

Figure 9.4. Increasingly, capuchin monkeys are becoming the favored model for studies on behavioral economics and prosocial behavior. They readily cooperate with each other, share food, and are easy to work with. Here a juvenile begs for food from an adult male (photograph by the author).

on. This was extremely stimulating. Through the institute I got to know Robert Trivers, William Hamilton, Robert Frank, Lionel Tiger, Elizabeth Loftus, Richard Alexander, and many others, all of whom were brilliant and inspiring. My contacts outside my discipline have grown over the years, but Margaret showed the way with her message of intellectual integration.

Something was missing in my work, however: I missed apes. Working with monkeys did not fully satisfy me. Fortunately, several opportunities presented themselves. I wrote and received my very first grant: a small "expedition" funded by the National Geographic Society to the San Diego Zoo to study its bonobos. I also visited the Yerkes Regional Primate Research Center's field station, northeast of Atlanta, for one summer to document food-sharing in chimpanzees. And a third remedy against my monkey malaise was another monkey, sometimes known as the South American chimpanzee: the brown capuchin monkey. These primates are tool-users with large brains, and I thought that working with them might give me a better handle on the issue of cooperation, which isn't as well developed in macaques.

The chief veterinarian of the Center knew of my interest in capuchins, and one day he called me to tell me of 13 capuchins available on campus. They belonged to the medical school, which had planned a terminal study that was cancelled due to a researcher's sudden resignation. This was lucky for me and even luckier for the capuchins.

I housed the monkeys at the Center. They lived socially, but I trained them on voluntary separation. My inspiration was Hans Kummer's laboratory, in Zürich. In the final decade of his career, Kummer had set up a situation of group-housed macaques, who could be separated for individual and pairwise testing. The great advantage over traditional single-housed conditions in which monkeys are often kept is that they have an actual social life. This allows one to test all sorts of parameters based on what one knows about their social relationships, such as dominance, bonding, and kinship.

My most satisfying and instructive project of this entire period, however, was the one conducted in the winter of 1984 on bonobos at the San Diego Zoo. Every morning, I would stand at the enclosure with a photo camera, a video camera, and a tape recorder, ready to record absolutely everything that could happen. This wasn't too difficult given that the groups were small. Sometimes new individuals were introduced to each other, producing lots of excitement. I had an excellent rapport with the two main animal caretakers, who made sure I knew of every new development and scheduled feeding. I did nothing else than record data, and in the evening kept an extensive diary, so that by the end I had a wealth of material that back in Wisconsin took years to be analyzed (e.g. de Waal, 1987).

More than a decade after my San Diego stay, I met celebrated photographer Frans Lanting, a fellow Dutchman living in Santa Cruz, California. We decided to produce a book on these "make love, not war" apes using Frans' unique pictures from the wild: *Bonobo: The Forgotten Ape* (1997). In this and other writings, including a full-blown ethogram of bonobo behavior (de Waal 1988), I contrasted bonobos with chimpanzees, and used them to make the point that comparisons between humans and apes are incomplete without including this gentler relative of ours. Inevitably, bonobos became the darlings of feminists and the gay community, but the main value I saw in them was as a tool to challenge the prevailing notion of nature "red in tooth and claw." My claim was never that aggression is absent in bonobos – they have in fact quite a few conflicts – but that they have evolved particularly effective means of coping with it. Increasingly, I was taking an explicit stance against the bleak, cynical view of nature, including human nature, advocated since the 1970s by a great many evolutionary biologists.

Developed out of my fascination with conflict resolution and reciprocal altruism, this message needed additional research support, which I planned to provide at my next position, at Emory University. The themes I wished to address were unchanged, but I felt a strong need for a more experimental approach. I do not consider experiments more important than observations, and feel the study of animal behavior needs both in equal measure (de Waal 1991), but experiments offer superior controls and as such are uniquely capable of "nailing" the behavioral causes suggested by observational work. My next step was to become an experimentalist, therefore.

Emory University

In 1992, I joined the faculty of the Psychology Department as well as Yerkes National Primate Research Center both at Emory University, in Atlanta, Georgia. Immediately, I was surrounded by more primate experts than I had ever been used to. The area has more ongoing primate research than any other part of the world: primatologists can be found at a range of local institutions. This was obviously very stimulating. During my time in Wisconsin, I had never had a faculty position, and my team had been incredibly small, with one technician and an occasional student. I moved to Emory to change this situation, to teach courses, and be closer to the apes, of which the Yerkes Center had plenty. My wife and I settled in a suburban home, near Stone Mountain Park, surrounded by magnificent tall trees and lots of wildlife, providing an inspiring environment for my home office where most of my writing takes place.

At Yerkes my team grew to around 15 people, and has stayed roughly this size throughout the years. I love having an active team of graduate students, postdocs, technicians, and undergraduate students. We are spread out between the capuchin laboratory at the Yerkes Main Center, near the university campus, and outdoor-housed chimpanzees and macaques at the Yerkes Field Station, which is a one hour drive northeast of Atlanta. I do my teaching on campus, including a class entitled "Primate Social Psychology," which is quite popular and helps us recruit interested undergraduate students into the lab.

I had brought my capuchin monkeys from Wisconsin, and split them into two groups, named the "Nuts" and the "Bolts," both of which began reproducing right away. Brown capuchins are friendly, easy to handle, and incredibly smart. I was delighted to discover that they share food, making them ideal for tests on cooperation and reciprocity. They are in fact ideal primates for the study of behavioral economics, as this field is now called.

Having so many eager workers, our research began to diversify. Inevitably, I now define the areas of work that we do by the students or postdoctoral associates who first moved into them. One of my first graduate students, Lisa Parr, trained chimpanzees to use a joystick to move a cursor over a computer screen. This permitted all sorts of new tests, such as on face recognition and facial expressions, including the demonstration that chimpanzees recognize facial similarity between kin (Parr & de Waal 1999). Sarah Brosnan explored economic issues by exploiting the fact that capuchin monkeys understand barter: we give them a token and reward them for returning it to our open hand. This led to work on inequity aversion, such as that two monkeys will barter with us without problems if both get identical rewards, but that one will stop exchanging if the other gets a better deal. Like humans, they seem to compare pay scales (Brosnan & de Waal 2003).

Together with Peter Judge and later Filippo Aureli, we worked on the connection (or lack thereof) between crowding and aggression. Interest in this topic goes back to my Dutch background. Being from one of the most crowded nations in the world with a relatively low rate of violent crime, I have always been skeptical about the claimed effects of crowding. One dramatic picture in a popular book that comes to mind shows a mass of rhesus monkeys gathered on a high rock with some of them jumping off. The caption says something like "monkeys want to escape crowding." This was the popular opinion: crowding can only lead to misery and death. Our studies, however, indicated that primates have ways of dealing with social frictions, and actually handle crowding quite well (de Waal *et al.* 2000). The monkey picture in question was probably taken on a hot day when monkeys like to jump into the water.

This was also a time of summarizing over two decades of research into reconciliation behavior. Since my initial observations at the Arnhem Zoo, so many other studies had been conducted by others, and so many interesting new ideas had been developed and tested. The chief idea was and remains that of the "valuable relationships hypothesis" according to which peacemaking will occur especially between parties who stand much to lose if conflict continues. Tested in many species, both observationally and experimentally, both in captivity and in the field, both on nonhuman primates, humans, and non-primates, the field was ripe for a literature review and a full-sized volume edited together with Filippo, who had become a champion of this field (Aureli & de Waal, 2000).

My own studies focused mostly on cooperation, reciprocity, empathy, and related phenomena. I set up experiments in which monkeys or apes had to work together to obtain rewards, or introduced food to see whether food sharing was affected by previous grooming. After hundreds of such tests, the data demonstrated partner-specific returning of favors to those individuals from whom previous favors had been received (de Waal 1997; de Waal & Berger 2000). I delved deeper and deeper into this field, which in psychology goes by the label of "prosocial" behavior. It resulted in my book *Good Natured* (1996) on the origin of morality as well as a theoretical paper with Stephanie Preston.

Stephanie had joined us as an undergraduate, and later left to get her graduate education at Berkeley. While there, she delved into the literature on empathy going back to the beginning of the twentieth Century. She added original thinking on the possible neural basis of empathy, resulting in an almost book-length paper that drew wide attention (Preston & de Waal 2002). Later I tried to formulate what these ideas about empathy might mean for altruistic behavior (de Waal 2008). I wrote partially in response to the, in my opinion, lamentable tendency in the literature to confuse the ultimate reasons for the evolution of altruistic behavior (such as reciprocity and kinship) with the proximate reasons why animals behave a certain way. That animals help each other for "selfish reasons" may be true at the evolutionary level, but it is unlikely that they know much about the long-term benefits of their behavior, such as inclusive fitness. Their actual motivation must come from another source, and empathy seems the perfect candidate. While I was preparing this theoretical paper, other research teams demonstrated pain contagion in rodents (Langford *et al.* 2006) and spontaneous helping in chimpanzees (Warneken *et al.* 2007), thus supporting my point. Similarly, an experiment in our own laboratory showed prosocial tendencies in capuchin monkeys (de Waal *et al.* 2008a).

We also returned to previous themes, such as formalized dominance, when Jessica Flack studied its manifestation in pigtail macaques. She showed that "policing" (i.e. breaking up fights) by high-ranking macaques has a pronounced effect on the level of social cohesion of the entire group (Flack *et al.*, 2006).

Finally, it was a special pleasure for me to visit another student, Amy Pollick, at the San Diego Zoo, and meet old friends, both human and bonobo, and see that twenty years after my original study there still was plenty to do. Amy set out to document hand gestures in bonobos, and compare them with those of chimpanzees against the background of language evolution – a natural connection for someone raised with American Sign Language (Pollick & de Waal 2007).

Culture studies

Based on the remarkable behavioral diversity of wild chimpanzees, field workers speculated that variation from group to group is attributable to social learning. It is notoriously hard in the field, however, to exclude alternative explanations, such as ecological or genetic differences between groups. The claim that "cultural" variation depends on learning was heavily debated at the time, including even some who denied altogether that apes were capable of aping (Tomasello *et al.*, 1993).

In order to write a popular book on this topic, I went to the country where it had all begun. Japanese primatology has given so much to our field, including Kinji Imanishi's first formulation, in 1952, of the concept of animal culture (Itani & Nishimura, 1973). My secondary aim was to compare Western and Eastern primatology, with the unsurprising conclusion that the latter places more emphasis on integration and harmony than the former (de Waal 2003). I stayed in Kyoto for three months, and from there traveled the entire country, from Hokkaido in the north to Kuyushu in the south. Visiting Koshima Island (where Japanese scientists documented the spread of potato-washing) and discussing the topic with many field workers put me in the right frame of mind to write *The Ape and the Sushi Master* (2001). The book made strong claims about the capacity of primates to learn from each other, so that afterwards I felt a need to put my money where my mouth was by putting these ideas to the test.

Imitation had grown into the holy grail of primate cognition studies. The perfect occasion to study this phenomenon came along in the form of a collaboration with Andrew Whiten of St Andrews University. Andy had perfected the "two-action method" to test for social learning in apes and children. The technique uses a device, a sort of puzzle box containing food, that can be accessed in two different ways, which are equally effective. You teach one way to a model, then see what observers do after having seen the model. If their solution is based on individual learning, they should use both available methods to access the device. If they learn socially, on the other hand, they should favor the model's method. Together with Victoria Horner, who joined us at Yerkes from St. Andrews, and a student of mine, Kristin Bonnie, we set out to test cultural learning ideas.

Our culture studies worked remarkably well, no doubt largely because of Victoria's skill in dealing with chimpanzees (Whiten *et al.* 2005; Bonnie *et al.* 2006; Horner *et al.* 2006). The results were greatly appreciated by field workers, who were pleased to see their suspicion confirmed that chimpanzees are excellent imitators. The main *forte* of our research, I believe, was that all of our experiments used a "chimp-to-chimp" design, and the fact that we had two separate groups to work with. Previous studies had presented chimpanzees with human models, hence suffered from a species barrier. Having criticized this approach (de Waal 1998, 2001), it remains an issue today, especially with regard to comparisons between apes and human children (Boesch 2007; de Waal *et al.* 2008b). Our culture studies leave no doubt that when apes deal with their own kind they pay better attention, and are perfectly willing to copy the model's behavior. This is, of course, also what one would expect if social learning evolved to acquire habits from relevant others.

In 2003, I went to see chimpanzees in the field. Toshisada Nishida, a colleague and friend, was about to retire and warned me of a now-or-never opportunity to join him in the Mahale Mountains, in Tanzania. It was a wonderful experience. Even though I have worked all of my life with captive apes, and like the detail and control such work provides, I do enjoy being in the field, and seeing what apes do there that I don't get to see in captivity. For example, I found the chimpanzees at Mahale much more vocally active than captive apes – in the forest, vocalizations help them stay in touch. I also saw several monkey hunts. I had read about this behavior many times, but seeing and experiencing the excitement is clearly something else. But I also saw scenes that I was fully familiar with, such as the political manipulations of an aging male, and occasional reconciliations and helping actions.

Conclusion

Throughout the years, my topic of study has remained more or less the same (i.e. the social behavior and social cognition of primates), but the specific problems have kept changing: from power politics to conflict resolution, and from cooperation to empathy and cultural learning. Overall, my inclination has been to go for the positive side of primate sociality, not because I don't recognize the negative side, but because I grew up in a time when competition and aggression were the only topics anyone could talk about. It is more exciting and challenging to deal with an unpopular topic than join everyone else talking about the same ideas. In fact, it was the brutal killing of Luit that put me onto this path, precisely because I wanted to see how animals could live together despite the destructive tendencies they so clearly possess.

Over the years, I have seen the resistance against comparisons between humans and other animals subside, seen Darwinism become more widely accepted – at least among academics – and also seen a growing interest in the positive side of animal sociality. Some of these changes have to do with general changes in culture, from the "gospel of greed" period of Ronald Reagan and Margaret Thatcher to the present time in which social responsibility and business ethics receive more press. Scientists like to think that their work occurs independently of general contextual influences, but it never does.

This cultural influence may also apply to my own work. I come from a nation that traditionally places less emphasis on confrontation and more on consensus building and tolerance. Perhaps this is due to the high population density of the Netherlands combined with its common enemy: the North Sea. In fact, historian Simon Schama (1987) has ascribed the inequity aversion of the Dutch, reflected in his book's title *The Embarrassment of Riches*, to their need to stand shoulder-to-shoulder at the dykes. In past centuries, the Dutch had to do so on a regular basis. Whatever the ecological explanation, this national background may have made me more sensitive to what binds individuals together rather than what drives them apart. It also may explain my skepticism about a sharp dichotomy between "selfish" and "altruistic" behavior.

Before one concludes that this makes me culturally biased, let me quickly add that my larger claim obviously is that all scientists are culturally biased, including those who believe they are not. This is why I have enjoyed, and still enjoy, exploring a side of primate sociality that used to be seriously neglected, and that gradually has become more mainstream. This has been true for so many developments: animals could not think, now they have minds; human behavior was entirely cultural, now it is at least partly genetic; animals are instinct machines, now they have some degree of culture. For those of us who were witness to and participated in these debates it has been an exciting ride, and for those who are new to them, brace yourself, because the debate about humanity's place in nature is far from over.

Author's Books (chronologically ordered)

Chimpanzee Politics (1982). London: Jonathan Cape.
Peacemaking among Primates (1989). Cambridge, MA: Harvard University Press.
Good Natured (1996). Cambridge, MA: Harvard University Press.
Bonobo: The Forgotten Ape (1997). Berkeley, CA: University of California Press.
The Ape and the Sushi Master (2001). New York: Basic Books.
My Family Album (2003). Berkeley, CA: University of California Press.
Our Inner Ape (2005). New York: Riverhead.
Primates and Philosophers (2006). Princeton, NJ: Princeton University Press.
Age of Empathy (2009). New York: Harmony Books.

References

Ardrey, R. A. (1967). *The Territorial Imperative*. London: Collins.
Aureli, F. & de Waal, F. B. M. (2000). *Natural Conflict Resolution*. Berkeley, CA: University of California Press.
Boesch, C. (2007). What makes us human (Homo sapiens)? The challenge of cognitive cross-species comparison. *J. Comp. Psychol.* **121**: 227–40.
Bonnie, K. E., Horner, V., Whiten, A. & de Waal, F. B. M. (2006). Spread of arbitrary conventions among chimpanzees: a controlled experiment. *Proc. R. Soc. Lond.* B**274**: 367–72.
Brosnan, S. F. & de Waal, F. B. M. (2003). Monkeys reject unequal pay. *Nature* **425**: 297–9.
Byrne, R. W. & Whiten, A. (1988). *Machiavellian Intelligence*. Oxford: Oxford University Press.
de Waal, F. B. M. (1975). The wounded leader: a spontaneous temporary change in the structure of agonistic relations among captive Java-monkeys. *Neth. J. Zool.* **25**: 529–49.
de Waal, F. B. M. (1986a). Integration of dominance and social bonding in primates. *Q. Rev. Biol.* **61**: 459–79.
de Waal, F. B. M. (1986b). The brutal elimination of a rival among captive male chimpanzees. *Ethol. Sociobiol.* **7**: 237–51.
de Waal, F. B. M. (1987). Tension regulation and nonreproductive functions of sex in captive bonobos (*Pan paniscus*). *Natn. Geogr. Res.* **3**: 318–35.
de Waal, F. B. M. (1988). The communicative repertoire of captive bonobos (*Pan paniscus*), compared to that of chimpanzees. *Behaviour* **106**: 183–251.
de Waal, F. B. M. (1991). Complementary methods and convergent evidence in the study of primate social cognition. *Behaviour* **118**: 297–320.

de Waal, F. B. M. (1997). The chimpanzee's service economy: food for grooming. *Evol. Human Behav.* **18**: 375–86.

de Waal, F. B. M. (1998). No imitation without identification. *Behav. Brain Sci.* **21**: 689.

de Waal, F. B. M. (2000). Primates: a natural heritage of conflict resolution. *Science* **289**: 586–90.

de Waal, F. B. M. (2003). Silent invasion: Imanishi's primatology and cultural bias in science. *Anim. Cogn.* **6**: 293–9.

de Waal, F. B. M. (2008). Putting the altruism back into altruism: the evolution of empathy. *A. Rev. Psychol.* **59**: 279–300.

de Waal, F. B. M. (2009). The need for a bottom-up account of chimpanzee cognition. In *The Mind of the Chimpanzee: Ecological and Experimental Perspectives*, ed. E. V. Lonsdorf, S. R. Ross & T. Matsuzawa. Chicago, IL: University of Chicago Press. (In Press.)

de Waal, F. B. M. & Berger, M. L. (2000). Payment for labour in monkeys. *Nature* **404**: 563.

de Waal, F. B. M. & Johanowicz, D. L. (1993). Modification of reconciliation behavior through social experience: an experiment with two macaque species. *Child Devel.* **64**: 897–908.

de Waal, F. B. M. & Luttrell, L. M. (1985). The formal hierarchy of rhesus monkeys: an investigation of the bared-teeth display. *Am. J. Primatol.* **9**: 73–85.

de Waal, F. B. M. & Luttrell, L. M. (1989). Toward a comparative socioecology of the genus *Macaca*: different dominance styles in rhesus and stumptail monkeys. *Am. J. Primatol.* **19**: 83–109.

de Waal, F. B. M. & van Roosmalen, A. (1979). Reconciliation and consolation among chimpanzees. *Behav. Ecol. Sociobiol.* **5**: 55–66.

de Waal, F. B. M., Aureli, F. & Judge, P. G. (2000). Coping with crowding. *Scient. Am.* **282**(5): 76–81.

de Waal, F. B. M., Leimgruber, K. & Greenberg, A. R. (2008a). Giving is self-rewarding for monkeys. *Proc. Nat. Acad. Sci. USA* **105**: 13685–9.

de Waal, F. B. M., Boesch, C., Horner, V. & Whiten, A. (2008b). Comparing children and apes: not so simple. *Science* **319**: 569.

Flack, J. C., Girvan, M., de Waal, F. B. M. & Krakauer, D. C. (2006). Policing stabilizes construction of social niches in primates. *Nature* **439**: 426–9.

Goodall, J. (1986). *The Chimpanzees of Gombe: Patterns of Behavior*. Cambridge, MA: Harvard University Press.

Horner, V., Whiten, A., Flynn, E. & de Waal, F. B. M. (2006). Faithful replication of foraging techniques along cultural transmission chains by chimpanzees and children. *Proc. Natn. Acad. Sci. USA* **103**: 13878–83.

Itani, J. & Nishimura, A. (1973). The study of infrahuman culture in Japan: a review. In *Precultural Primate Behavior*, ed. E. W. Menzel, pp. 26–50. Basel: Karger.

Langford, D. J. *et al.* (2006). Social modulation of pain as evidence for empathy in mice. *Science* **312**: 1967–70.

Lorenz, K. Z. (1966). *On Aggression*. London: Methuen.

Parr, L. A. & de Waal, F. B. M. (1999). Visual kin recognition in chimpanzees. *Nature* **399**: 647–8.

Pollick, A. S. & de Waal, F. B. M. (2007). Ape gestures and language evolution. *Proc. Natn. Acad. Sci. USA* **104**: 8184–9.

Preston, S. D. & de Waal, F. B. M. (2002). Empathy: its ultimate and proximate bases. *Behav. Brain Sci.* **25**: 1–72.

Schama, S. (1987). *The Embarrassment of Riches: An Interpretation of Dutch Culture in the Golden Age*. New York: Knopf.

Tomasello, M., Kruger, A. C. & Rander, H. H. (1993). Cultural learning. *Behav. Brain Sci.* **16**: 495–552.

Trivers, R. L. (1971). The evolution of reciprocal altruism. *Q. Rev. Biol.* **46**: 35–57.

Warneken, F., Hare, B., Melis, A. P., Hanus, D. & Tomasello, M. (2007). Spontaneous altruism by chimpanzees and young children. *PLoS Biol.* **5**: e184.

Watts, D. P. (2004). Intracommunity coalitionary killing of an adult male chimpanzee at Ngogo, Kibale National Park, Uganda. *Int. J. Primatol.* **25**: 507–21.

Whiten, A., Horner, V. & de Waal, F. B. M. (2005). Conformity to cultural norms of tool use in chimpanzees. *Nature* **437**: 737–40.

10

Taking my cues from nature's clues

STEPHEN T. EMLEN

Every organism is the product of the interplay between its genes and its environment. In my case, nature and nurture conspired to predispose me to a career in animal behavior.

I was born in August 1940 in California, the second of three sons of John Thompson Emlen, Jr., and Virginia Merritt Emlen. As a child, my father, who came from a long line of Philadelphia Quakers, developed a passion for natural history, particularly bird-watching. After graduating from Haverford College, he obtained his Doctorate in Ornithology at Cornell University. By the time I was born, he was a zoology professor at the University

Leaders in Animal Behavior: The Second Generation, ed. L. C. Drickamer & D. A. Dewsbury. Published by Cambridge University Press. © Cambridge University Press 2010.

Figure 10.1. The author as a young boy, aged three, with his parents, John T. Emlen, Jr. and Virginia Merritt Emlen, Baltimore, MD, 1944.

of California, Davis, but not for long: during World War II, UC-Davis closed so that the agriculture campus could house the US Army Signal Corps. We moved to Baltimore, where my father, as a conscientious objector, performed his alternative service, helping bring the city's rat populations under control. Subsequently, Dad accepted a position as professor of zoology at the University of Wisconsin, Madison, where he remained from 1946 until his retirement.

It was there, in the farmlands, marshes and woodlots of southern Wisconsin, that my older brother, John, my younger brother, Woody (for James Woodruff) and I spent much of our childhood. Our parents valued nature and prized curiosity. We were given a long leash, and encouraged to explore the outdoors and to collect and catalog all sorts of natural objects. In my case these ranged from owl pellets to butterflies to abandoned birds' nests.

In our big old house, we cared for a continuous series of unusual pets – among them, at various times, a loggerhead shrike, a screech owl, a raccoon named Scrappy and a great-horned owl named Melvin. Once we filled an empty room with sand a half-meter deep and moved in a dozen prairie dogs so a film crew could shoot close-ups for a Walt Disney documentary. The most memorable animal in our menagerie, though, was a talking mynah, Harvey, whose cage sat right next to the dining room table. Harvey's repertoire of vocalizations was large: he spoke, wolf-whistled and did a credible imitation of a flushing toilet. Over the years he entertained numerous distinguished visitors.

During summers, our family often traveled. A pivotal place was my paternal grandparents' summer home in Pennsylvania's Pocono Lake Preserve. Most summers, we visited for two weeks, which gave us a chance to interact with our Philadelphia-based relatives.

Figure 10.2. Four-fifths of the Emlen family relaxing on the dock at the Pocono Lake cabin, summer 1945. The author, second from front, lies behind his three-month old brother, Woody.

Pocono was idyllic, a safe place for children to go wandering. My cousins, brothers and I stayed outdoors for hours everyday. Pocono's homes were set back from the lake a bit, and when I fished or swam, I felt I was in Nature's playground. There were two tiny islands we could reach by canoe, and whenever we camped overnight on one, I felt immersed in wilderness, although truthfully we were only 200 meters from the dock. Often we kids camped alone, enjoying a great sense of independence, while our parents slept in the house less than a quarter-mile away, knowing they could walk through the water to reach us if necessary.

I spent many hours picking blueberries on the lakeshore, often with bald eagles or ospreys flying overhead. There were wild bogs within easy hiking distance. And I learned just where to paddle in the evening to watch beavers build their dams and lodges – and slap their tails in alarm when I came too close. At Pocono, I learned what it means to know intimately a wild place. I imprinted to the maple, birch and hemlock forests, and over time I grew familiar with the habitat's salamanders and frogs, mosses and ferns, and spots where I could find such rarities as cardinal flowers, lady slipper orchids and wild wintergreen.

When Dad began teaching at various biological field stations, we traveled more extensively. We spent three summers at the University of Minnesota station on Lake Itasca, and another at the University of Michigan station on Douglas Lake. But the most memorable of these postings were the two glorious summers – 1951 and 1952 – that we spent at the Wyoming Field Station in spectacular Grand Teton National Park.

Our home was a rustic three-room cabin midway between the hamlet of Moran and the field station, about a mile away. Since it was the only house outside the station, and Jackson Lake Lodge didn't yet exist, we had a vast area to ourselves. The habitat was wet, with willow and alder bushes extending from the lake to the Snake River backwater.

I don't recall seeing much of my father during those summers, but my mother and siblings and I had a great time. My brothers and I imagined that we were cowboys as we rode horses,

Figure 10.3. Author (center), age ten, riding horseback with his mother and older brother at the University of Wyoming Biological Field Station in the Teton Mountains near Jackson Hole, WY, 1951.

hiked, and backpacked in the mountains. I fished for cutthroat trout. Most Saturday nights, we went to the rodeo in Jackson. Inspired, I practiced for hours lassoing the fence posts outside our cabin. Toward the end of those summers, real cowboys came through our valley, driving herds from their summer pastures. In 1951, sitting astride our horses, my mother, brothers and I watched as many scenes of the classic Western film "Shane" were filmed practically in our backyard.

My brothers and I became interested in biology without realizing it. Walking around, I might find myself standing in 10 inches of water or on the high ground of a tussock. The wetland supported an incredibly dense population of toads, and we caught and brought home literally hundreds, which we would later release.

Charles Carpenter, a graduate student at the station (later a herpetologist at the University of Oklahoma) kindly incorporated our efforts into his research. I believed we were being useful when we brought in dozens of toads and Chuck obligingly recorded their temperatures and measurements. I also learned to hypnotize a toad so it would lie motionless belly-up in my palm. Thus I spent my tenth and eleventh summers immersed in the glamour of the cowboy lifestyle and reveling in wilderness.

Adventure abroad

During the 1953–54 school year, my father took a sabbatical to pursue research in East Africa. We flew to Europe, and then, on DC3s, hopped our way to Cairo, Khartoum, Entebbe and Nairobi, our intended destination. Unexpectedly, the outbreak of civil war and the Mau

Mau uprising against British colonial rule made staying in Kenya too dangerous. My father must have had to scramble to find a research situation elsewhere; after just two days we left for Southern Rhodesia.

We spent most of that year in Bulawayo, a city at a political crossroads. Southern Rhodesia had recently joined with Northern Rhodesia and Nyasaland to form the Central African Federation, or CAF, which was working to initiate inter-racial democracy as a step toward dismantling apartheid. (Unfortunately, it was a short-lived experiment: The CAF would dissolve in 1963.) Our relocation to Bulawayo made our year in Africa especially meaningful to my mother, who had majored in political science and was intensely concerned about civil rights. Even as my father was establishing research links with the National Museum of Southern Rhodesia, she was befriending Masotsha Mike Hove, a black politician who would soon be elected to CAF's first legislative assembly.

We rented a house in a Bulawayo suburb, and my brothers and I attended nearby Milton School. The school followed the British system, which meant not only that we wore uniforms that included short pants and ties, but also that we studied a vast number of subjects. Back in the States, students might learn algebra one year, geometry the next and trigonometry the year after, but Milton School pupils studied all three year after year. I took upper-level classes in subjects I had never studied. My grades were no surprise: I did well in subjects I had studied before and failed several that I hadn't, like third-year Latin. No matter:

Figure 10.4. The Emlen family during their sabbatical year in Bulawayo, Southern Rhodesia (now Zimbabwe), in 1953. The author (front left, age 12) and his two brothers wear their Milton Boys School uniforms.

my learning came from living in another country, becoming close friends with my Rhodesian schoolmates, witnessing apartheid and the nascent attempts to resist it, and being exposed, through our diverse acquaintances, to the culture, history and politics of Africa.

We, in turn, were of great interest to others. As the first American family in our neighborhood, we were invited to all sorts of events, and my parents frequently reciprocated, inviting neighbors, scientists from the museum, and Mr. Hove and other black politicians to dine at our home. We children were encouraged to join the conversations. At one gathering, Mr. Hove presented me with a copy of Alan Paton's *Cry, the Beloved Country*, in which he inscribed a message of inter-racial hope. That gift is a remembrance of my awakening to issues of race and equality – issues I still care deeply about today.

Our family traveled to numerous local attractions. On collecting trips for the museum, I was given the opportunity to "collect" birds using a shotgun, a task that proved remarkably easy: because birdshot spreads out, it was difficult to miss my target. I learned how to skin and make taxidermy mounts of the few birds I collected, among them a scarlet-chested sunbird, a lilac-breasted roller, a mousebird and several masked weaverbirds. Back in Bulawayo I studied the weaverbirds' nest-building behavior for a science project. I still have the notebook containing my drawings and crude field notes describing how males "weave" nests out of grasses.

During school vacations our family roamed further, to Victoria Falls, and to the ruins of the ancient civilization of Zimbabwe. We traveled to the Western Cape, Basutoland, and the Zulu areas of South Africa. During visits to Wankie and Kruger National Parks and the Hluhluwe Wildlife Reserve, I saw the continent's great megafauna, observing elephants, giraffe, sable and kudu antelope, rhinoceros, hippopotami, and such large mammalian predators as lion, cheetah, and African wild dogs.

By the end of our year in Africa, I was in many ways transformed – excited about politics, other cultures, different peoples, new habitats and magnificent wildlife. On our way home, we spent a month in Switzerland – the International Ornithological Congress was meeting in Basel – and in England. In Switzerland, I hiked, rode gondolas to spectacular mountaintops and met such giants of animal behavior as Gustav Kramer, David Lack, Konrad Lorenz and Niko Tinbergen. The highpoint of my European experience, though, was the week we spent on tiny Lundy Island, known as the "island of puffins," off England's west coast. Only a handful of people lived there, clustered in a few buildings and the active lighthouse, where we slept. Biologically, Lundy was astonishing. Tens of thousands of seabirds – puffins, guillemots, auks and kittiwakes – nested on its many cliffs. I loved walking around, nearly deafened by the noise and almost overwhelmed by the density of birds nesting virtually on top of one another. I would sit atop a cliff watching the birds on the facing cliff come and go, threatening, feeding and vocalizing to one another. Quite possibly, the excitement of that experience influenced my adult decisions to study the costs and benefits of colonial living, first in bank swallows and then in white-fronted bee-eaters and grey-capped social weavers.

Teenaged naturalist

Back in Madison after our year of travel and adventure, I began high school. Previously, I hadn't worked hard, but now, fueled by new experiences and enthusiasms, I was off to a fresh start. Moreover, poor grades notwithstanding, my raft of Milton School courses had given me a leg-up in many subjects new to typical Wisconsin ninth-graders. I began receiving A's on many assignments, a stunning rise in my academic fortunes. Thus rewarded, I started applying myself and became an excellent student.

My interest in nature kept growing. Throughout primary school, my closest friend had been Johnny McLeod, a fellow outdoor enthusiast with whom I had biked to local parks and birding spots, armed with binoculars and field guides. We had joined the Junior Audubon Club, receiving encouragement – and cookies – from an elderly woman birder who lived nearby. Now a new friend, Steve Martin, joined our bike trips to wild areas. A favorite destination was Dunn's Marsh, several miles from home. Ditching our bikes and carrying our lunches, we would walk the railroad tracks along one edge of the wetland, adding to our bird lists and gaining a deep feeling for the habitat and creatures that dwelled within.

On one such visit, we discovered a sunken boat. With my father's help, we took it home, dried it out, patched and painted it and christened it "Mud Hen" – probably the heaviest boat for its size ever to float. Back at the marsh, we hid it among the cattails, and for several years we would paddle or pole among the reeds, discovering a world not visible from the wetland's edges. Dunn's Marsh was large enough to support diverse bird species, including redwing blackbirds, sedge and marsh wrens, sora rails, Virginia rails, coots, common moorhens, pied-billed grebes, blue-winged teal and both American and least bitterns. We discovered nests of grebes, rails, and coots hidden in the vegetation. Once I watched a grebe, leaving its nest, pull a small blanket of vegetation across its eggs, presumably for camouflage. Another time, an American bittern stood frozen, neck erect, while I approached to within a few meters. How many behaviors and events I witnessed that I would never have seen had we stuck to walking the railroad tracks, peering in from the outside!

During the summers of 1955 and 1956, I attended Camp Manito-wish, an institution dedicated to canoe-camping in the wilderness of Northern Wisconsin. There I learned to steer and portage a canoe and navigate white water, and I studied survival skills – finding and preparing food, avoiding dangerous animals, search-and-rescue and first aid. The next two summers, I began earning money for college as a counselor at Red Arrow Camp, which was near and similar to Manito-wish, but geared to less-experienced campers.

There, I became Mr. Nature Study, taking campers on hikes where we might see an active beaver lodge, spot birds on their nests or find such "predatory" flora as pitcher plants and sundews. I loved to lead my charges across a bog and watch them react as the ground "moved" beneath their feet. I also accompanied campers on multi-day canoe trips. Working with two age groups, 8- and 12-year-olds, I strove to find ways to inspire them – excellent training, I see now, for my future as a parent and teacher. At the time, of course, that future was only a hazy notion. But much that would influence my path had become clear. I knew that I loved being intimately involved with and learning to understand the natural world and that I wanted to spend my time doing so. I realized that my senses, indeed, my spirit, comes alive when I am at one with nature.

Some of the leading figures of ethology and ornithology that I had met in Europe now visited my father in Wisconsin. I had developed an appreciation of who these people were and could discuss with them the study of wildlife. During one visit, Konrad Lorenz personalized my copy of his book, *King Solomon's Ring*, by sketching a likeness of Harvey, our mynah, as a frontispiece. He also sketched a portrait of Skippy, our dog, inside the cover of *Man Meets Dog* for Woody.

Whenever possible, I tagged along on the field expeditions my father led. One trip was to Horicon Marsh, a staging area for tens of thousands of migrating geese. Another was the annual trek to Plainfield, Wisconsin, to observe prairie chickens on their courting grounds, called leks. Wildlife biologists Fred and Fran Hamerstrom conducted their long-term study with the help of visiting volunteers every breeding season. The ritual was always the same. Staying in the Hamerstrom's farmhouse, we woke at 3 AM, downed breakfast and headed to our blinds. At dawn, the male prairie chickens arrived and began "booming" – puffing out specialized air-sacs and stomping their feet – to attract females. Every prairie chicken within miles had been color-banded, and we kept notes on all the behaviors we could see: which birds got into fights, and especially which females visited or copulated with which males. For hours, we stayed immobile in the freezing cold, fascinated by the behavioral spectacle before us. Hiking out after our subjects flew off, we volunteers would re-assemble at the Hamerstroms' to eat, report our data, and compare notes. The Hamerstroms treated every volunteer as an equal, and even as a teenager, I felt like a full contributor.

Rite of passage

During my teen years, my father established a wonderful tradition. For the summer before our senior year of high school, Dad let each of us boys propose a one-on-one experience in which he would do whatever we wished (within reason) for two weeks. Each son was responsible for planning all aspects of his adventure. Brother John took Dad hiking and camping in Puerto Rico. When my turn came, I proposed a canoe trip in the Quetico-Superior wilderness area, sometimes called the International Boundary Waters Park. The Quetico spans a swath of Northeastern Minnesota and Southwestern Ontario. The area, with its labyrinth of clear-water lakes and streams, was set aside for wilderness canoeing and hiking. No public-access roads crossed the area, and motorboats were prohibited.

I planned menus, purchased food and gathered equipment and maps. Dad and I agreed to three rules: no wristwatches; short of an emergency, no contact with the outside world; and no shaving. In August 1957, we drove the family station wagon, our canoe tied to the roof, to Ely, Minnesota, the launching point for most Quetico trips.

To avoid commonly used areas, I planned a route that included many portages at the start, and we encountered almost no one for the 11 days of our journey. In our aluminum canoe, we rode over many small waterfalls and easily portaged around others. Beaver were every-where, and we saw porcupines and muskrats, too. We shared many lakes with loons, and when they sang, I yodeled back. Once we spotted a bull moose swimming and paddled toward him, getting close enough to see the velvet still covering its antlers. Several times we heard wolves howling at night, although we never saw even their tracks. There were bears in the area, so each night we suspended our food in a bag slung over a high branch.

It was a near-perfect trip. The only flaw in my plan was that we didn't catch enough fish. I had packed lots of dried fruit, Rye Crisp and chocolate, but we were shy on protein. When we got back to Ely, Dad, now bearded, phoned home and reported that we were fine but had lost considerable weight. To me he feigned confusion: Had I greatly underestimated the amount of food we required, or had I greatly overestimated my ability to catch fish?

Finding my path

In 1958, I enrolled at Swarthmore College in Philadelphia. I had taken my college applications seriously, focusing on small liberal arts schools, where I hoped to find faculty committed to the progress of every student. I chose Swarthmore because it was far enough from home to allow me some independence, yet I had comforting ties to the Philadelphia area. Dad's mother still lived in Germantown, and my brother John was enrolled at nearby Haverford.

Swarthmore's freshman class, slightly smaller than my high school graduating class, seemed full of driven students. I found it difficult to participate in activities like wrestling and cross-country running and still succeed academically. Two of my first three semesters, I was on the Dean's list – which at Swarthmore identifies students in danger of flunking out! To survive, I gave up a number of favorite social and athletic activities.

Initially, I majored in psychology, thanks largely to a stimulating introductory class taught by Professor Henry Gleitman. I took several psychology courses during my first two years, and then abruptly became disillusioned by the way a certain class project was run. I never took another psychology course, although I have remained excited about the study of human behavior. Searching for a new major, I gravitated back to biology.

As I had hoped, Swarthmore's professors fostered the development of students in and out of class. Several were crucial to my growth as a biologist. Most notable was Professor Robert K. Enders, who encouraged me to apply for a small grant that enabled me to conduct independent research at the Rocky Mountain Biological Laboratory (RMBL) near Crested Butte, Colorado, during the summer following my sophomore year. In retrospect, I'm not impressed with what I accomplished scientifically, but the experience showed me that I could combine my love of wild places with my developing passion for conducting research and interacting with like-minded scientists.

My project was a simple study of the population density cycles of small rodents; I collected standardized data using live traps. At summer's end, Dr Enders suggested I travel to Teton National Park and run trap-lines there, using grids he had used many years earlier. I wonder now if he was more interested in the data or in enabling me to revisit the park he knew meant so much to me, this time with a challenging scientific mission of my own. I hitchhiked with my traps from Colorado to Wyoming and back to Madison. For two weeks I

Figure 10.5. Author at the Rocky Mountain Biological Laboratory (RMBL), in Gothic, CO, summer 1960. In addition to conducting his first (unimpressive) independent field research project, the author and two fellow students refurbished, and then lived in the ghost town's only two-story building which we dubbed "Lee's Tavern".

lived in a tent in bear country. It was a character-forming experience, a test of my ingenuity and dependability.

Dr Kenneth S. Rawson, another professor who took a keen interest in my development, was interested in the sensory basis of habitat selection and homing in small mammals. I worked as his assistant on a project at RMBL the summer after my senior year. From Dr Rawson I learned how science is done when it is question-driven. Working with the same species of voles and deer mice that I had studied previously, I learned how a quantitative scientist designs experiments and data-collection regimes to test specific hypotheses and thus contribute meaningfully to scientific knowledge.

Under Dr Rawson's supervision, a Swarthmore undergrad, William Kem, and I performed a series of lab experiments testing the responsiveness of deer mice to loud noises throughout their circadian cycle. The results had significance for timing the administration of medical therapies. We wrote up our results, and thus my first scientific paper was published in *Science* (Emlen & Kem 1963).

By the time I received my B.A. with distinction in 1962, Swarthmore had delivered on my hope for rich experiences and caring professors. I knew I wanted to pursue graduate studies. I didn't know what I wanted to study, although I knew it would involve fieldwork: I wanted to be outdoors, studying animals in their natural environments. My eventual specialty – sociobiology and behavioral ecology – didn't really exist yet. Several doctoral programs accepted me. The University of Michigan took me on the basis of my research potential, and I accepted the offer.

Senior year, when I wasn't focused on grad school applications, I had been focused on my relationship with Katharine Peckham, a classmate majoring in art history. We had dated throughout college, and like many of our contemporaries, we were eager to get married right after graduation. Just two weeks after receiving our Swarthmore degrees, Katy and I were wed in New York City's magnificent Riverside Church.

Present at the Creation

Two months later we were in Ann Arbor. Graduate school was a new world. Whereas Swarthmore had emphasized breadth, the focus at Michigan was narrow, and my colleagues spent virtually all their time reading about and debating issues arising from then-current evolutionary thinking about organismal biology.

My time at Michigan shaped me in three profound ways. First, I learned not to be afraid to ask "big" research questions – questions of broad general importance. I learned this from the theoretical population ecology group, led by Professors Lawrence B. Slobodkin, Nelson G. Hairston and Frederick Smith.

Slobodkin taught a course based on his recent textbook, *Growth and Regulation of Animal Populations* (Slobodkin 1961). His approach was a revelation. He saw ecology as more than describing habitats and naming innumerable facets of natural history. He sketched the subject with sweeping strokes, asking major theoretical questions and seeking general, even universal, answers.

Slobodkin, Hairston, and Smith jointly taught informal seminars fostering discussion of current ecological questions. They had recently co-authored a controversial paper about community structure and population control (Hairston *et al.* 1960). Making radical simplifying assumptions, they derived global predictions about the role resources and predators play in regulating population numbers of organisms at different levels of the food chain. I found it stunning that questions of this magnitude were even being asked, and exhilarating to think there might be general conceptual answers that could explain large-scale patterns of diversity in nature.

Second, I learned that colleagues can respectfully disagree, and that frank discussion and constructive criticism are essential to clarify complex issues and design optimal research protocols. Discussion forums were egalitarian; my perspective was valued even when I disagreed with a professor. My years in Ann Arbor entailed many, many hours of theoretical debate. These exchanges were especially exciting because the topics included the first unifying principles of what would become sociobiology.

Indeed, the third and most important influence on my intellectual development was the birth of sociobiology. In this regard, it was my good fortune to be in the right place at the right time.

Sociobiology originated when several nearly simultaneous advances heralded a synthesis of animal behavior with population ecology and genetics. On the behavioral–ecological side, the seminal contributors were Jerram L. Brown (1964), John H. Crook (1964, 1965) and David Lack (1968), who argued that ecological factors play a major role in shaping an animal's society. The degree of gregariousness, the presence of territorial defense, even the basic form of the mating system – all are profoundly influenced by the abundance and distribution of key resources, predators and competitors. The structure of an animal society can be viewed as an adaptive solution to the ecological problems its individual members face, and animals faced with similar problems should exhibit a predictable convergence in the solutions shown in their social organizations. This premise is the cornerstone of the ecological side of sociobiology, which seeks to understand the distribution pattern and functioning of different types of societies.

On the genetic side, population biologists were increasing our understanding of natural selection. George C. Williams (1966) sounded the death knell to the idea that selection could promote traits "for the good of the species". William D. Hamilton (1964) elegantly demon-strated that in calculating the genetic fitness of an individual, one must incorporate the effect of the individual's behavior on the fitness of genetic relatives, because they have a meas-urable probability of sharing the same genes by virtue of common descent. In essence, biologists had been omitting a component of fitness in their models of animal behavior. The staggering implications of this new thinking led to the second surge of interest in sociobiology.

Each of these seminal papers was thoroughly debated at Michigan. Professor Richard D. Alexander, a brilliant animal behaviorist, had a significant impact on me. Instantly recognizing the importance of Hamilton's ideas, he led several groups that began decipher-ing and discussing the role of kinship and inclusive fitness in the evolution of altruistic, cooperative, selfish and spiteful behaviors.

Follow the drinking gourd

In the midst of this intellectual fervor, I was selecting my thesis topic. I wanted to study an important behavioral issue, in hopes that my findings would be generalizable to other systems or species. I wanted the topic to be interesting to a wide spectrum of people. And I wanted my project to be tractable – I wanted to be reasonably certain I could answer the question I was asking. The topic I chose: How do nocturnally migrating songbirds find their way between their breeding and wintering destinations?

The seasonal migrations of birds are among the most impressive phenomena in nature's annual calendar. The sheer numbers that migrate (billions), the distances they travel (often many thousands of kilometers) and the risks and hardships they encounter are reasons enough to study migration. Furthermore, most bird migration occurs at night. And, of the many species that travel at night, most migrate *only* nocturnally.

Why do most birds migrate at night? How do they know in what direction to fly? What orientation cues do they use and what information do they obtain from them? How does a young bird find its way along a route it has never traversed? To me, there was something compelling about these unsolved questions.

As a young bird-watcher, I had been captivated by the arrivals of spring migrants in Wisconsin. I made a game of searching for male warblers in full breeding plumage as they traveled north after over-wintering in the neo-tropics. I had also witnessed many spectacular waterfowl migrations. To this day, I thrill to see long skeins of geese flying in V-formations high overhead.

The study of the directional orientation of migratory birds had its breakthrough in the 1950s, with the work of Gustav Kramer. It had been known that when migrant songbirds were kept in captivity, they displayed intense motor activity at night, but only during spring and fall, their migration seasons. German researchers called this activity *Zugunruhe*, which translates as "travel unrest." Kramer was the first to realize that caged migrants, under appropriate conditions, will spontaneously orient *Zugunruhe* activity in the correct migratory direction (Kramer 1949, 1950).

Another German ornithologist, Franz Sauer, studied the spontaneous *Zugunruhe* orientation of European warblers. He placed birds in cages under the night sky and later under planetarium skies. It was Sauer who first hypothesized that European warblers determine their migratory direction from the stars (Sauer 1957, 1961).

Whereas Kramer worked mostly with diurnal migrants, I decided to study only nocturnal migration, so *Zugunruhe* would not be confounded by behaviors that occur only in daytime, such as fighting, feeding and courtship. I planned to bring my study "into the laboratory", and experimentally alter information available to caged birds while noting any changes in their *Zugunruhe* orientation.

I started by repeating some of the observations of Kramer and Sauer. Like them, I lay in darkness under a clear night sky, looking up through the Plexiglas bottom of a cylindrical cage with a wire mesh top. I had put a "nocturnally restless" indigo bunting inside. What I saw was astonishing.

The bunting moved constantly, quivering its wings, jumping up the side of the cage, returning to the perch and quivering its wings again. Throughout these antics it seemed to peer up at the sky. It gave occasional single-syllable call notes, identical to those buntings give during migratory flights. All night, hour after hour, the bunting jumped.

Convinced that this nocturnal jumping was a clear manifestation of migratory motivation, I embarked on my studies. After settling on indigo buntings as the appropriate species, I needed to design and build cages for recording *Zugunruhe*. My father and I collaborated on a design that would record *Zugunruhe* footprints. Reliable, inexpensive, and easy to replicate in large numbers, the cage, wider at the top than at the bottom, was fashioned from an inkpad floor, a cone-shaped piece of blotting paper for the sides, and a black wire-mesh top. Others have since dubbed this simple invention the Emlen funnel, which honors my father as well as me. Because of its low cost, portability and absence of moving or electrical parts, it has become the standard, used by researchers throughout the world.

By autumn, 1964, my research was well under way. I had moved to the Edwin George Reserve, a University field station located northwest of Ann Arbor. There I housed my buntings in outdoor aviaries, exposing them to the sky and to changing weather conditions.

Initial experiments consisted of placing individual buntings that showed high levels of nocturnal restlessness in funnel cages and variably exposing them to clear night skies, to overcast night skies, and to artificial overcast, created by covering the cages with frosted glass. When the stars were clearly visible, buntings showed spontaneous preferences to orient in their known migratory direction. Under overcast skies and simulated overcast, they exhibited less activity and the direction of their jumps generally did not differ significantly from random.

My project depended on my being able to manipulate the celestial cues available to the birds. Fortunately, there was a superb facility, the Robert Longway Planetarium, in nearby Flint, Michigan. I am immensely grateful to its director, Dr Maurice Moore, and his staff. They taught me how to run the Spitz B star projector and gave me a key and permission to use the planetarium all night, after closing hour.

Like my birds, I became nocturnal. Every evening, I waited in the parking lot until the patrons had gone, then dragged in stepladders and planks, set up my funnels and put in the birds. I would alter the star projector, simulating a variety of natural and manipulated skies, and record any changes in the orientation of individual birds. In one series of experiments, I blocked out portions of the planetarium sky by blocking off portions of the projector.

Ultimately I demonstrated that, whether flying north in spring or south in fall, indigo buntings derive directional information from a single portion of the night sky – an area within 35 degrees of the North Star. In contrast to Sauer, who thought European warblers use a true navigation system to know their precise global position, I concluded that buntings use a relatively simple system to determine the direction of geographic north. The critical information is configurational: buntings recognize star patterns – constellations, if you will – and from them extrapolate the direction of the axis of celestial rotation (Emlen 1967).

We humans learn the same thing as children. If we can find the Big Dipper, we can extrapolate the position of the North Star. Prior to the Civil War, slaves seeking to escape to Canada often traveled under cover of darkness. They advised one another to "follow the drinking gourd" toward north, the direction of freedom.

My thesis advisor, Professor Harrison "Bud" Tordoff, was a highly respected ornithologist who gave me just the right amount of independence. He was excited by my project. He had been a British-based fighter pilot in WWII, and he told me of using landmarks, more than stars, to find his way at night into and out of Germany. By coincidence, Franz Sauer had also been a fighter pilot in WWII, flying for the German Luftwaffe. He wrote that his interest in celestial navigation stemmed from his using the stars to orient his aircraft. Years after graduate school, I introduced the two to each other and they traded stories about their wartime flying experiences.

Cornell calling

Right place, right time: by 1966, when I was finishing my doctorate, many universities were hiring new PhDs to introduce the perspectives of behavioral ecology into their curricula.

Skipping over post-doctoral fellowships, I applied for several faculty positions. I was far along in negotiations with Michigan State University when I received a call from William T. Keeton of Cornell University. Explaining that Cornell was reorganizing its biological sciences, he invited me to interview for a new position as assistant professor in the Neurobiology and Behavior (NB&B) group.

I agreed without hesitation. Cornell had a long history of excellence in organismal biology, and I had extensive family links to the school and to Ithaca, New York. My grandfather, Ernest G. Merritt, had been a physics professor and the first dean of Cornell's Graduate School; my mother grew up in Ithaca, and my grandparents lived there when I was young. My mother earned her BA at Cornell, and my father received his PhD there, under Dr Arthur A. Allen, arguably the first professor of ornithology in North America and later founder of Cornell's Laboratory of Ornithology.

My interview there was unlike any other. Arriving at night, I was met at the airport by a grad student who took me to campus to meet William Keeton. He was working intently on a lab exercise for an introductory biology class, and one of his first questions was how I thought the exercise might be improved.

Over the next three days I met innumerable people, attended lectures and gave a research seminar on nocturnal orientation by indigo buntings. Having described the importance of the circumpolar region of the sky to birds in both spring and fall, I concluded by joking that autumn migrants had to look over their shoulders as they flew to know where they were going. I felt energized by the enthusiasm and intellectual intensity I experienced. Bill Keeton drove me to the airport and said a warm goodbye. He told me there was already great interest in having me join the NB&B faculty.

Nonetheless, Cornell's offer, when it came, was a disappointment. I wasn't being hired to teach animal behavior, but to be in charge of advising and tutoring incoming students in a new program in which students would receive both their undergraduate and doctoral degrees in six years. Cornell was where I wanted to be, but I didn't even approve of an accelerated PhD program, which I thought would push young scientists to narrow their focus prematurely. Reluctantly, I telephoned the chairman of NB&B and declined the job. I resumed negotiating with Michigan State.

I soon got another call from NB&B's chair. Surprisingly, he offered to create a position tailored to my requirements, including the courses I most wanted to teach: animal behavior at both introductory and cutting-edge advanced levels, as well as related graduate seminars. It was the ideal position at an exceptional institution, and I happily accepted. In July 1966, one month before my twenty-sixth birthday, I drove a crammed U-haul to Ithaca to assume my new position. It was and still is my first job as a professor.

My new nest

Cornell's reorganization of biological sciences, completed just as I arrived, streamlined and standardized a previously fractured program, with all faculty merged into a new Division of Biological Sciences. The plan also entailed adding positions to ensure expertise in emerging disciplines – hence the creation of my new professional home, the Section of Neurobiology and Behavior. The glue binding the members of NB&B was an interest in animal behavior, whether in the neural basis of behavior or the evolution of behavior. Combining the disciplines of neurobiology and behavior was a bold initiative in 1966 and remains so today.

I instantly formed close bonds with Bill Keeton and another colleague, Thomas Eisner. Widely considered the founding father of chemical ecology, Tom is passionately curious about the role of chemicals in the communication, courtship, and defense of insects. He had volunteered to develop our section's signature undergraduate course, and one of my primary responsibilities was to help him formulate and teach it. Our Introduction to Neurobiology and Behavior premiered in September 1966 to an unexpectedly large enrollment of over 400 students!

Tom and I have remained involved with this team-taught course, and its successor, Introduction to Behavior, for four decades. One of the most popular and influential classes at Cornell, it has served as the model for courses at many other universities. I have taught more than 12,000 students in animal behavior. I seldom lecture away from campus without some former students re-introducing themselves and reminiscing about the course's impact on their careers.

Bill Keeton, my other instant friend and collaborator, originally had been hired as a specialist in millipede taxonomy. But Bill had raced homing pigeons as a boy, and his secret scientific passion was to understand their homing abilities. The restructuring of Biological Sciences let him join NB&B and change his focus.

Joining forces to critique one another's research designs, we discovered that we both viewed solving the mysteries of bird orientation as a giant detective game. We asked our "questions" by manipulating the cues the birds had available to find their way; the birds revealed their "answers" through changes in their directional behavior. A series of National Science Foundation (NSF) grants would enable Bill and me to continue our investigations together for almost 15 years.

Bill was at the height of his research creativity when he died of sudden heart failure in 1980. He was known and respected by thousands for his innovative Introductory Biology textbooks and course. In 2008, Cornell named one of its program-based residence halls the William T. Keeton House, an apt tribute to my friend, who had such an impact on under-graduate life at Cornell and beyond.

Good for 13,000 years

With set-up money from Cornell, I equipped numerous rooms – built to house free-flying birds – with temperature and day-length controls. I could capture wild birds and hold them in these rooms, yet bring them into the appropriate physiological state of migratory readiness by having the indoor light cycles mimic those outdoors. Moreover, by manipulating the indoor day length, I could accelerate or delay the birds' progression into and out of

Figure 10.6. The author, inside the Cornell research planetarium, monitoring indigo buntings in the Emlen funnel cages, which are arranged around the star projector. The projector beams a night sky onto the planetarium ceiling. See text for experimental details.

migratory condition. I also designed my own research planetarium, outfitted with a basic Spitz projector. I wanted to delve further into the way the northern circumpolar area of the sky sufficed for orientation for both northern and southern migrations.

At the end of one breeding season, I captured adult male indigo buntings and divided them randomly into two groups, each housed in its own flight room. I carefully manipulated the day length differently in the two rooms, and by May, one group was coming into spring migratory restlessness in full breeding plumage, while the other had already passed through this phase and was exhibiting *Zugunruhe* in the dull brown feathers of fall.

Under a stationary planetarium sky set for the latitude intermediate between the winter and summer home ranges, I simultaneously tested both groups in individual funnel cages. The results were clear. Birds in spring physiological condition spontaneously oriented their footprints to the northeast, whereas those in autumnal physiological condition hopped consistently to the south (Emlen 1969). This demonstrated that differences in hormonal state caused buntings to interpret identical star information in very different – but seasonally appropriate – ways.

Curious as to how the birds acquired or interpreted the information present in the northern area of the sky, I also initiated a series of experiments studying the ontogenetic development of indigo buntings' orientation capabilities. Removing young birds from the nest before their eyes opened, I hand-reared them in three randomly assigned groups. These birds were housed in flight rooms that duplicated local daylight conditions. As they matured, I exposed each group, at night, to different sky conditions. One group was exposed three nights a week to a planetarium sky that mimicked a typical clear night in Ithaca, with full rotation of the stars. A second was exposed to a sky that has never occurred naturally, one that rotated around the bright star Betelgeuse in the constellation Orion. The third group was not exposed to any natural or planetarium sky until the birds came into migratory condition in the autumn.

During fall migration, when all the birds showed *Zugunruhe*, I tested them under stationary planetarium skies. The birds that had been exposed to a sky that rotated with Polaris at the apex spontaneously took up appropriate orientation toward the south. The group that had been exposed to Betelgeuse as its pole star oriented in the direction away from Betelgeuse – opposite from the northern apex of the axis of rotation to which they had been exposed. The third group showed lower levels of activity overall and no consistent orientation (Emlen 1970).

I consider these among the most important results I obtained. They indicated that stars by themselves had no inherent directional meaning to the buntings; there was no star map embedded in the genome. Rather, young birds were primed to pay attention to the angular rotation of the night skies they saw prior to their first fall migration, particularly the northern circumpolar area of the sky. Star patterns took on meaning only after they were coupled to the directional north–south axis of rotation. Then, even under a stationary sky, the birds could use constellations near the northern apex of this axis to determine geographic north and take up appropriate migratory directions.

This programmed learning, analogous to imprinting, is important, given the astronomical phenomenon known as "the precession of the earth's axis." Over a period of 26,000 years, the earth wobbles in its axis; 13,000 years from now, Vega, not Polaris, will be the pole star. Learning two-dimensional constellations close to the circumpolar area is an adaptive mechanism that enables birds of each generation to determine the north–south axis of rotation, now or 13,000 years from now. Such flexible adjustments are important on a lengthy evolutionary timescale (Emlen 1975a).

Immersed though I was in orientation research, I was also encountering new occasions of wonder on the home front. Soon after our move to Cornell, Katy and I had started a family. We had two children: Douglas John Emlen, born April 1967, and Katharine Merritt Emlen, born June 1968. Ithaca was a great place to raise children, and I loved being a father. Rifts developed in our marriage, however, and in 1970, Katy and I separated and subsequently divorced. A key provision of our settlement was joint custody, because we wanted our children to have deep, full relationships with each of us. This agreement, rare at the time, would prove especially significant when Katy relocated to Tennessee, and Douglas and Katharine continued to live for prolonged periods, including most school years, with me.

Migrants as meteorologists

By the early 1970s, it was clear that neither migratory birds nor homing pigeons relied on a single orientation system. While I focused primarily on birds' star-compass abilities, other researchers suggested the importance of the sun, magnetic cues, landmarks on the ground and winds aloft as sources of directional information. I realized that the secrets of migratory orientation would be fully revealed only by experiments on free-flying birds.

I started by putting miniature radio transmitters on white-throated sparrows and Swainson's thrushes, then releasing them in a woodlot. Evenings, I would listen from a tower near Ithaca's airport, hoping to pick up the signal as one of my subjects began a migratory flight. These experiments were largely failures because the birds would wait days for optimal departure conditions. I needed a way to "persuade" a migrant to depart under conditions of my choosing so I could manipulate the information available to the bird when it left.

I and researchers elsewhere began devising ways to release birds in midair, in hopes that a bird finding itself suddenly aloft would initiate a migratory departure even under conditions that might not have prompted it to do so voluntarily. One German team tried tossing birds from helicopters. Numerous strategies failed. Then my research assistant, Natalia Demong, hit upon a new one: send a bird aloft inside a cardboard box suspended beneath a helium-filled weather balloon; have a simple fuse melt a loop of monofilament holding the floor of the box closed. The bird initiates flight as the floor literally drops out from underneath it. It was a crazy idea – but it worked!

Around this time, I learned that the National Aeronautics and Space Administration (NASA) had tracking radars that could follow an object as small as a songbird. No device need be attached; the tracking radar would detect radar energy reflected by the water content of the bird's body. Some radars could potentially follow a songbird dozens of kilometers.

I received permission from NASA to conduct experiments at Wallops Island, Virginia, with the help of a radar crew that would use large-dish SPANDAR and FPS-16 radars to track individual birds we released 500 to 1,500 m aloft. Such radars usually tracked rocket and satellite launches, but budget problems were crimping the launch schedule. NASA willingly turned over an idle crew and radar to my project.

The radar was impressive. By tracking a reflector tied to a helium balloon, it could obtain information on wind direction and speed at all altitudes relevant to migration. After it "locked on" one of our released birds, it accurately recorded the bird's position every second in three-dimensional space. An analog plot-board mapped the bird's altitude and "track" – its direction and speed over the ground. By knowing the wind direction and speed at each altitude, we could subtract the wind vector from the track vector and obtain the bird's actual "heading" and air speed. In this way, we learned how migrating birds compensate for adverse winds.

For several weeks in each of three migration seasons, an affable radar crew shifted its schedule to work evenings for the benefit of my team. We would arrive at the NASA base and capture locally migrating white-throated sparrows, which we housed in activity cages in an abandoned Coast Guard station. We used *Zugunruhe* records, coupled with fat-index scores, to choose the birds most likely to initiate migration when released aloft. We tracked several sparrows each evening, and most continued to fly. The radar tracked their departures for 10–25 km – once to 40 km – as the birds selected the direction, speed, and altitude of their departure (Demong & Emlen 1978).

White-throated sparrows proved to be astute meteorologists, predictably changing their heading and airspeed when flying in differing wind conditions, and electing to fly at the altitude where the winds were most favorable for their known direction of migration (Emlen & Demong 1978).

Sparrows released in the spring with a view of both stars and landscape took up meaningful northeasterly tracks over the ground. Sparrows released with only a view of stars overhead – when fog or low clouds obscured landmarks below – adopted northerly *headings* but could not compensate for how the winds caused their *tracks* to drift (Emlen & Demong 1978). An accomplished sailor knows how to adjust her boat's heading to compensate for wind drift, thus achieving her desired direction of travel. Similarly, a night-migrating sparrow – providing it can see stars and note its progress over the ground – can do the same thing.

When I started what became 15 years of orientation research, most researchers focused on one cue system – the sun, stars, magnetic fields, landmarks or olfaction – and assumed it would eventually explain, completely, avian navigation. From the outset, I disagreed. Bird migrants traverse thousands of kilometers, and a navigational error can be fatal; migratory behavior has been shaped by intense selection pressure. The environment contains numerous cues that could give directional information to a migrant, and natural selection should foster abilities to use such information. A system involving multiple cues could provide redundant information, enabling a bird to maintain its orienting ability when one cue is lacking or provides equivocal information (Emlen 1975a,b).

Although my primary focus was star-compass orientation, I also performed experiments demonstrating the importance of the position of sunset, the earth's magnetic field and wind patterns aloft as directional cues. By the time Bill Keeton and I wrote our respective comprehensive reviews of avian navigation (Keeton 1974; Emlen 1975b), the idea that orientation is based on multiple cue systems was becoming widely accepted.

The realization that different cues can function as components of a complex navigational repertoire opened a new line of questioning. What are the advantages of one cue over another? Does the simultaneous use of multiple cues result in increased accuracy? How are different cue systems integrated with one another?

By the time my review was published, my own orientation was shifting: behavioral ecology was claiming an increasing amount of my attention. I saw the potential for a small number of ecological factors to profoundly influence the forms of animal societies. I was eager to begin collecting empirical data and testing ideas related to ecological aspects of animal behavior.

Insights from anurans

My first field study of social behavior dated back to 1965–66, at Michigan, when the seasonality of my orientation research left my summer evenings relatively free. Crane

Pond on the George Reserve harbored a large population of bullfrogs, which intrigued me, perhaps because of my childhood memories of bullfrogs at Pocono.

I devised a method for marking individual frogs using color-coded elastic waistbands. Every night, from a rowboat, I observed the marked males and females, monitoring the presence and movements of virtually the entire breeding population. I recorded the males' wrestling matches and territorial fights, plotting their changing locations on a map of the pond. When females arrived, I noted which males they visited and, whenever possible, which they chose for amplexus, the protracted mating act. This was possibly the first study to track the spatial and temporal movement patterns of all members of an individually marked population of amphibians (Emlen 1976).

Males arrived at the pond on different dates, but tended to remain active on most nights afterward. Larger, older males arrived earlier and stayed longer than smaller, younger ones. Females, by contrast, came only when they were ready to lay their eggs. Females showed great asynchrony in the timing of ovulation. Sexual activity for any female generally lasted only one night.

The bullfrog mating system, then, was characterized by an extended breeding season of continuous male sexual activity and extremely brief, asynchronous periods of individual female receptivity. This produced a strong male bias in what I termed the Operational Sex Ratio, or OSR: the ratio of sexually active males to sexually receptive females at any particular time (Emlen 1976). Nightly OSR values for the study averaged 8:1 and ranged as high as 18:1. This resulted in intense male–male competition for the few receptive females on any given night.

To gain a quick appreciation of OSR, consider a contrasting example: a hypothetical species of desert frog that estivates in the soil at the bottom of dried-out ponds. This frog emerges with the first heavy rains and breeds explosively and synchronously over a season lasting a few days. All adults of both sexes are active and receptive on the same days, the pond an orgy of amplexing amphibians. The OSR will not differ from the adult population sex ratio. Assuming an equal number of adult males and females, OSR = population sex ratio = 1:1. There is little potential for any males to monopolize matings. Male–male competition and sexual selection are predicted to be weak.

The Operational Sex Ratio was the key concept to emerge from my bullfrog study. Easily measured in the field, it gives a quick indication of the expected level of male–male competition and, by implication, the intensity of sexual selection, for a wide variety of animals.

In the early 1970s, the focus of animal mating-systems research was avian polygyny. Gordon H. Orians and two of his doctoral students, Jared Verner and Mary F. Willson, had realized the importance of environmental resources in their classic polygyny threshold model (Verner & Willson 1966; Orians 1969). This model emphasized the female's role in choosing whether to mate with an already mated male on a high-quality territory (polygynous bonding) or pair with an unmated male occupying a lower-quality territory (monogamous bonding).

The bullfrogs had made me appreciate how mating systems are shaped by several environmental factors, including the length of the breeding season, the temporal pattern of receptivity of individual females and the defensibility of resources important for reproduction. I believed that these three factors would be important predictors of the full spectrum of anuran (frog and toad) mating systems. But I wanted to conceptualize these thoughts in a broader framework with relevance for birds as well. This meant incorporating long-term pair bonds, and parental, if not bi-parental, care of dependent young.

Some of the stranger avian mating systems are forms of polyandry, in which females lay eggs for more than one male. Lewis W. Oring, then at the University of North Dakota, published the first paper on polyandry in the spotted sandpiper in 1972. His goal was to tease apart environmental factors as determinants of polyandry. I contacted Lew, and we agreed to work together to find a unifying set of principles to describe the full spectrum of avian mating systems.

We joined forces in 1975–76, when Lew came to Cornell for a sabbatical. Together we developed a preliminary framework for the role various environmental factors might play in shaping mating systems. We proposed that a male's ability to control access to females would be strongly influenced by the OSR and by seasonal variation in the spatial and temporal distribution of receptive females. Polygamy occurs only when some individuals can monopolize an unequal share of mates. The greater the possibility for monopolization, the greater what we called "the environmental potential for polygamy." The precise form of polygamy depends on which sex is limiting and the manner in which the limited sex controls access to mates.

Our ideas were a logical extension of Jerram Brown's concept of economic defensibility. Brown (1964) had argued that the distribution pattern of a resource determined the benefit-to-cost ratio of defending that resource, and used this approach to model the evolution of territorial behavior. We extended the concept to encompass monopolizability, recognized that receptive mates – usually females – are the critical resource for understanding mating systems, and used the approach to develop an ecological classification of mating systems.

In autumn, 1975, Lew, Ruth Buskirk, and I organized a seminar around this topic and used it to test our ideas. Joined by three other faculty and 12 doctoral students, we reviewed the literature from as many taxa of organisms as possible. As the semester progressed, Lew and I grew increasingly confident that our ideas had broad generality. We wrote our mating-systems synthesis the following semester.

Science published "Ecology, sexual selection, and the evolution of mating systems" in 1977. "Emlen & Oring", as it quickly became known, pioneered the integration of sexual-selection theory with economic defensibility to develop one of the first predictive models of animal mating systems. It received wide acceptance and has served as a foundation for many refinements and extensions in the theory of mating-system evolution. In 1988 it was named a *Current Contents* "Citation Classic."

Few people know that the synthesis originated with insights I gained studying Michigan bullfrogs or that its framework was first used to explain the diversity of mating systems among anurans. The success of this approach in explaining the diverse mating systems of

Figure 10.7. The author with Lewis Oring (center) and Sidney Gauthreaux (left) birding in South Carolina, spring 1989. Lew and the author collaborated to develop a general theory integrating ecology, sexual selection and the evolution of mating systems (Emlen & Oring 1977). Sidney Gauthreaux, a close friend and colleague, studies regional patterns of bird migration using networks of surveillance and weather radars.

organisms as disparate as frogs and birds, insects and coral-reef fishes, tropical bats and humans falls into the domain of answering big questions using a small number of unifying principles.

Cliff-face cooperation

Since my days as a doctoral student, I had been intrigued by the paradox posed by the evolution of altruism: why do animals engage in behaviors that benefit others but are costly to themselves? I had been introduced to this conundrum in Hamilton's (1964) paper on inclusive fitness. In 1965, when I was the teaching assistant in Richard Alexander's first animal-behavior class at Michigan, we devoted considerable time to Hamilton's ideas. Both Alexander and Mary Jane West-Eberhard, a fellow doctoral student, later published major reviews exploring the implications of kinship in the evolution of social behavior (Alexander 1974; West-Eberhard 1975). Their syntheses strongly influenced my future research trajectory.

Early on at Cornell, the evolution of social behavior, especially of altruistic and cooperative behaviors, informed a significant portion of what I taught in my advanced classes. Naturally, I focused on an ornithological manifestation: cooperative breeding in birds. Cooperative breeding refers to situations in which three or more individuals provide care to dependent young. In addition to the breeding pair, one or more helpers regularly incubate the clutch or, more commonly, provide nestlings with food.

Such helping-at-the-nest had been observed in a number of species, but in the early 1970s only a few researchers were actively studying it. The emerging pattern was that helpers were usually unpaired offspring fledged the previous year. Rather than disperse, they remained on their natal territories, and when their putative parents reproduced again, they helped defend the parental nest and feed the nestlings, which were, in all likelihood, their younger siblings. Seemingly, an individual became a helper-at-the-nest when its opportunity for successfully dispersing and breeding was poor. Researchers pointed to ecological constraints imposed by habitat saturation and a shortage of breeding vacancies.

The evolutionary "dilemma" posed by helpers-at-the-nest seemed to be resolved: mature offspring that temporarily delayed dispersal were "making the best of a bad job" and, while waiting for better opportunities, could gain inclusive fitness by assisting their parents' subsequent reproductive attempts.

Cooperative species, however, exhibit an array of social systems spanning many ecological situations. The classical helper-at-the-nest pattern could not explain all avian cooperative breeding. As my first sabbatical approached, I knew I wanted to initiate a field study of a cooperative system where ecological constraints seemed not to apply – where helping persisted in the absence of restrictions on dispersal or independent reproduction. Searching the literature, I learned that Hilary Fry, then at Aberdeen, Scotland, had found helping to be commonplace in colonial-nesting red-throated bee-eaters in equatorial Nigeria (Fry 1972).

As Fry described them, the bee-eaters excavated colonies into earthen cliffs along riverbanks and gullies. Using their long, slender bills and short legs and feet to dig, the bee-eaters clustered their burrows together. The resulting cliff-face, riddled with burrow entrances, resembled the cut surface of a block of Swiss cheese. The bee-eaters roosted in colonies of 50–100 individuals throughout much of the year and nested in the burrows when it came time to breed.

What intrigued me was that these birds lived in colonies! Sitting at my desk, never having seen a bee-eater, I decided that a study of helping behavior in a colonial species of bee-eater could be the ideal sabbatical project. Ecological constraints and habitat saturation should not apply, I thought; a young bee-eater had only to find a mate and excavate a burrow to create its own nesting opportunity. I was optimistic such a study would uncover an alternate evolutionary route to cooperative breeding.

Even as I was developing these plans, I was courting a wonderful young woman. Natalia Jean Demong, a recent Cornell graduate, shared my passions for the outdoors, nature and field research. I first met Natalia when she joined the migration research team; it was she who invented the bird release box we used so successfully in the NASA radar tracking studies. She and I were married in 1973, in an outdoor ceremony at Pocono Lake Preserve, the scene of so much childhood happiness. Douglas and Katharine participated, signaling to them and all present Natalia's and my commitment to being a devoted family of four. A Coopers' hawk circled low and called as Natalia and I shared our vows. In the years ahead, Natalia would play pivotal roles as good-will ambassador, equipment designer, computer programmer and documentary photographer on countless field projects; editor and author of

Figure 10.8. The author and Natalia Jean Demong are married at Pocono Lake Preserve, PA, summer 1973, with Katharine and Douglas, his children, participating.

numerous publications; and generous, fun-loving partner and stepparent. Our marriage is now 35 years strong, and we still enjoy working, playing and traveling together.

As I planned my study of bee-eaters in Kenya, I was especially excited – about the project, about returning to Africa, and about sharing travel and field adventures with Natalia. With support from the John Simon Guggenheim Foundation and the National Geographic Society, she and I spent six months initiating a study of white-fronted bee-eaters in the Rift Valley. On our way, we visited Hilary Fry in Aberdeen. He generously shared ideas and field techniques that helped get our study off to an excellent start.

We arrived in Nairobi, rented a Volkswagen "kombi" minivan and headed to the bottom of the Rift Valley. Making the tiny town of Gilgil our base, we identified several colony locations. After marking birds with individually identifiable wing tags, we observed their interactions and monitored their reproductive success. For a clear view of eggs and nestlings deep inside each burrow, we used a long, lighted, optical scope that we had designed and built in Ithaca with help from some talented physicists.

We kept track of the many roosting and nesting burrows by labeling enlarged photographs of the colony faces. We assigned each active burrow a number and noted all the birds that visited, defended or participated in nesting behaviors at each. Activity was so hectic that we needed tape recorders to take accurate notes.

Sitting in blinds, outfitted with binoculars and telescopes, we spent most days making observations, then spent evenings transcribing them and creating "flash cards" summarizing which individual birds belonged to each burrow. I likened our techniques to those of Jimmy

Stewart's voyeuristic character in Alfred Hitchcock's film *Rear Window*. We watched the unfolding drama of the lives of the bee-eaters, making inferences about their relationships based on what we saw at the entrances to their burrows, just as Stewart's character drew conclusions about his neighbors based on what he saw through their windows.

Each day in the final hour before dusk, the bee-eaters huddled on branches in the slanting light, in small, tight groups. Then as many as 200 descended to the cliff face. There was a cacophony of call notes and frenzied activity as multiple birds, clinging at the tunnels' entrances, excitedly vocalized and greeted one another with wing and tail quivers. With loud, aggressive call notes, some made threatening gestures with their bills to prevent certain birds from entering particular burrows. Many individuals visited numerous entrances. Sometimes the occupants greeted them with bill quivers; other times they aggressively repelled them. This period of intense socialization was unlike anything I had ever seen and unlike anything reported in the literature for other colonial-nesting species.

At the end of our pilot season, Natalia and I were enthusiastic about the project's prospects. We loved life in the bush. Our research findings were a success: helping proved common among the bee-eaters, with helpers providing significant amounts of food to nestlings. Nests with helpers produced more fledglings than those without, primarily by suffering less nestling starvation. Surprisingly, we witnessed numerous cases in which a breeder, even a breeder with its own helpers, changed status to become a helper at a different nest. This happened when its nesting attempt failed but also could occur soon after it successfully fledged young. The explanation for this puzzling "redirected helping," as we called it, would elude me for years. Ultimately, it was the key to understanding the bee-eaters' complex form of cooperative society.

Back in the USA, I secured NSF funding to continue the project, arguing that our preliminary results suggested that colonial bee-eaters indeed represented a completely different route to cooperative breeding. In 1975, I returned to Kenya to set up a larger, more intensive study at colonies primarily inside Lake Nakuru National Park. Given my university and family responsibilities, I sought competent people to help run the project when I couldn't be there. Still, for years, often for two and sometimes for as many as seven months, I, and often Natalia, joined various assistants to conduct the fieldwork.

No-one contributed more than Dr. Peter H. Wrege, who worked with me for two decades on data collection and analyses on the bee-eater study and a subsequent study of jacanas in Panama. Our collaboration enabled me to maintain intensive long-term studies abroad while fulfilling my other duties at home. I am grateful to Peter and the many others who helped conduct this research through the years.

Kenya, in particular Lake Nakuru National Park, became a second home to all of us. There is a thrill that comes with doing research on foot while sharing the savanna with a magnificent array of birds and other wildlife. For more than ten years, the Bee-eater Research Project (BERP) rented a small house within the Park. BERP House had telephone service and city water but no electricity. We used camping lanterns for light and propane cylinders to heat the stove and power the fridge. Hot water from the outdoor, wood-fired "bush boiler" was a once-a-day treat.

Figure 10.9. The author, together with Natalia, conducting fieldwork in Lake Nakuru National Park, Kenya. This was the site of our long-term study of cooperation and conflict among White-fronted bee-eaters.

The bee-eater project would run for 14 years. Slowly, we built our knowledge of the birds' pedigrees, and eventually we also used DNA analysis to strengthen our genealogy database. In 1987, as the study was concluding, the BBC filmed an award-winning half-hour documentary, *The Bee Team*.

Nakuru proved a wonderful place for our family. When they were 10 and 11, Katharine and Douglas spent seven months at BERP House. They donned school uniforms (just as I had, at a similar age, in Rhodesia) and attended nearby Greensteds school as weekly boarders; weekends and long holidays they spent with us. Game drives through Lake Nakuru National Park were a favorite pastime. Because our Renault 4 was tiny, Doug and Katharine would stand on the back bumper while I slowly drove along the dirt tracks, giraffe and waterbuck curiously gazing our way.

Mana Katula, BERP House's elderly Kenyan watchman, taught Douglas archery with a bow and intricate hand-carved arrows. The night before we left Kenya, he gave Katharine a live chicken, pointing to the sky and pantomiming that the little *ndege* – the bird – should travel in the big *ndege* – the airplane. The little *ndege* did spend the night tied to Katharine's bedpost, but we found her a new home on our way to the airport.

Douglas and Katharine each returned to Kenya years later. In 1982, Douglas assisted for several weeks on the bee-eater project. Our Maasai field assistant, Moses Kipelian, taught him warrior skills, and together they collected honey from the nests of angry wild bees. Katharine, now a freelance writer, poet, and photographer, didn't return until 2002, when

Figure 10.10. The author's children, Katharine and Douglas, ages ten and eleven, standing on the front steps of BERP House, our family home inside Lake Nakuru National Park, Kenya, spring 1979. Note their Greensteds school uniforms (see Figure 10.4 for comparison).

Natalia and I were working on a new project elsewhere in Kenya. The three of us revisited Lake Nakuru National Park and BERP House, now a hostel for the Wildlife Clubs of Kenya. At Greensteds School, Katharine surprised us by singing Kenya's national anthem in Swahili. I believe that both our children have a love of Kenya and the Kenyan people imprinted into their psyches.

Kinship counts

In 1980–1981, I spent my second sabbatical as a Fellow at the Center for Advanced Studies in the Behavioral Sciences (CASBS) in Palo Alto, California. I used the solitude of the Center to synthesize and reflect on our project's findings.

I had studied bee-eaters because I wanted to examine an exception to the classical helper-at-the-nest situation. After observing many instances of redirected helping, I assumed that most helping was not kinship-based. After observing many young pairs excavate burrows within the colony, I assumed there were no ecological constraints preventing independent breeding. On both accounts, I was mistaken.

I originally misinterpreted the data because I failed to consider that a cooperative unit might include more than one breeding pair. Slowly it dawned on me that bee-eaters lived in *extended* families. Within a family, several pairs were potential reproductives in any given year, but different pairs and different numbers of pairs actually bred, depending on ecological conditions. The extended family, shaped by female dispersal and male philopatry – the tendency to remain in the natal family – potentially contained grandparents, parents, offspring, uncles, aunts, nephews, nieces, cousins and unrelated female in-laws. At the time, this level of social complexity was unknown for any colonial or cooperatively breeding bird. With the insight that numerous *extended* families comprise a bee-eater colony, many conclusions fell into place.

The presence of several pairs of potential breeders within each extended family provided the opportunity to test the limits of kin discrimination among bee-eaters. Given a choice between breeding pairs of differing relatedness, helpers overwhelmingly preferred to assist those breeders to whom they were most closely related. In-laws – the unrelated female mates of younger males – were the least likely to "lift a feather" to assist others. Similarly, a replacement mate that joined a family following the death or divorce of a previous breeder rarely helped until it had successfully raised young with its new partner (Emlen & Wrege 1988; Emlen *et al.* 1995).

Rather than refute the importance of kinship among bee-eaters, we demonstrated that bee-eaters possess excellent kin-recognition abilities. Conforming to Hamilton's Rule, they practice sophisticated kin-favoritism in choosing which relatives to help (Emlen & Wrege 1988; Emlen *et al.* 1995).

Hedging their bets

White-fronted bee-eaters live in arid savannas where rainfall varies notoriously, in timing and amount, from year to year. They need an abundance of insects, produced by good rains, to successfully fledge their young. Their breeding prospects are thus closely linked to the rains, which they cannot predict.

At CASBS I placed this uncertainty of upcoming breeding conditions into the developing framework of ecological constraints theory. The constraint imposed was not a shortage of breeding vacancies or mates, but a constraint on the ability of pairs, in harsh times, to successfully reproduce without helpers. I considered the erratic changes in rainfall as creating the functional equivalents of breeding openings and closures (Emlen 1982a). By maintaining social ties within the extended family, pairs could coalesce to rear young during harsh times (fusion), but split apart to breed independently when conditions were benign (fission).

Bee-eaters, I proposed, had evolved a flexible "bet-hedging" strategy to cope with this temporal unpredictability (Emlen, 1982a,b, 1984). By forming a stable pair bond and excavating a potential nesting burrow, each member of a mated pair represented a "ready reproductive". The pair could rapidly initiate breeding under favorable conditions, but if conditions were harsh or deteriorated a "ready reproductive" was socially integrated into its extended family and could easily forego its breeding to become a helper at the nest of a close relative. By the same reciprocal links, an actively breeding pair was positioned to receive help, including redirected help, from other extended family members. When insects were plentiful, most bee-eater pairs could reproduce, but when the rains failed, bee-eaters became obligate cooperative breeders.

This bet-hedging flexibility – to rapidly adjust breeding group size to changing environmental conditions – is facilitated when genetic relatives are socially available and when they benefit by helping at a relative's nest. Bee-eaters fulfill both conditions: family members remain in close social and spatial proximity to one another, and helpers gain inclusive fitness when they help fledge a relative's young. Fledging success increases with increasing breeding group size in bee-eaters, primarily because nestling starvation declines. This effect is greatest when conditions are harshest (Emlen & Wrege 1991).

Promise of a unified science

With the discovery of multiple "ready reproductives" in the same extended family came a new line of inquiry. How is reproduction partitioned among group members: who gets to breed?

Within any multigenerational social group, asymmetries exist in relatedness, age, and dominance. Conflict over who will reproduce is expected. To understand the allocation of reproduction within such groups, I modeled the conditions favoring retention of young in natal groups from the different perspectives of a dominant parent and a subordinate offspring. I calculated the change in direct fitness that would accrue to each depending on whether the offspring left or stayed. I assumed that while at home, offspring, like classical helpers-at-the-nest, helped but did not breed.

Plotting these changes on fitness-space diagrams helped me visualize when continued association was and was not mutually beneficial. I focused on a conflict zone where the offspring benefited by leaving but the parent benefited if the offspring stayed.

One way a parent might induce an offspring to stay would be to "allow" it to reproduce within the group. I modeled the more general conditions when a dominant would benefit by "forfeiting" or "conceding" fitness to a same-sex subordinate as a means of achieving multi-year group stability. The model predicted that a dominant is likely to concede more fitness when relatedness is low, ecological conditions are benign, and the difference in social dominance is small (Emlen 1982b, 1984).

One of my early doctoral students, Sandra Vehrencamp, studied groove-billed anis in Costa Rica (Vehrencamp 1977). Anis are another atypical cooperative breeding species, in which several unrelated paired females lay eggs in one nest. Vehrencamp documented reproductive conflict: females frequently tossed each other's eggs from the nest. After completing her degree, Vehrencamp began modeling the partitioning of reproduction in anis. Concurrently, each of us was trying to predict when and how much reproduction would be shared among members of plural breeding groups.

It was Vehrencamp who coined the term "reproductive skew" to describe the degree of asymmetry in the partitioning of reproduction within groups (Vehrencamp 1979). In high-skew societies, reproduction is concentrated in a small set of individuals; in low-skew societies, reproduction is shared more equitably. She and I pooled our ideas in a co-authored review chapter (Emlen & Vehrencamp 1983). Our early models provided the original impetus for concession models that are today part of the transactional framework of reproductive skew theory.

Reproductive skew theory is exciting because it offers the promise of a unifying frame-work for viewing much of social evolution. Skew models incorporate ecological factors (the difficulty of breeding independently and the benefits of group association), genetic factors (the degree of relatedness) and social factors (relative dominance) to predict not only the partitioning of reproduction, but also when groups will form and what the stable group size will be. More than thirty years ago, E. O. Wilson wrote in *Sociobiology*, "…when the same parameters and quantitative theory are used to analyze both termite colonies and troops of rhesus macaques, we will have a unified science of sociobiology" (Wilson 1975, p. 4). I believe such a goal is within reach.

Predicting female infanticide

Mating-system research is a subject I have revisited at intervals throughout my career. In 1987 and again beginning in 1990, Natalia and I examined polyandry in wattled jacanas. As I often do, I chose to study an outlier species. Among wattled jacanas there is a reversal of both sexual selection and traditional sex roles. Male jacanas are sole caretakers of the eggs and young, while females compete to obtain harems of males. Females are bigger, heavier and totally dominant over males. By conducting an intensive behavioral, ecological and genetic (paternity) study of a marked population, I hoped to contrast the costs and benefits of polyandrous associations for each sex in a species in which selection operates more intensively on females than on males.

Animal infanticide had leapt into the research community's consciousness when Sarah Blaffer Hrdy published her dissertation on infanticide among the langurs of India (Hrdy 1974). She hypothesized that killing dependent young was part of an adaptive strategy of reproductive males.

The mating system of langurs is harem polygyny: females form social kin groups, and males compete to monopolize sexual access to the groups. In such societies, competition often leads to one male usurping a female group from a previously breeding male. It was at these junctures that Hrdy witnessed infanticide. Males that supplanted a breeding male frequently killed the group's dependent young – presumably sired by the former male. The infants' deaths caused mothers to become reproductively available sooner than if they had cared for their young until weaning.

Hrdy's hypothesis seemed logical to most evolutionary behaviorists. Before long, cases of conditionally expressed infanticidal behavior were reported for other harem-polygynous primates, as well as lions and rodents. The common thread was that killing occurred when a new male took over a group of females, some of which were caring for dependent young.

In 1979, I began predicting in class that female-perpetrated infanticide would someday be observed in sex-role-reversed jacanas. Females would be the infanticidal sex, showing such behavior when they took over a territory containing one or more males that were incubating eggs or tending chicks. As a result, the males would more rapidly accept eggs laid by the incoming female.

In 1984, while visiting the Smithsonian Tropical Research Institute (STRI) in Panama, I noticed a high density of wattled jacanas breeding on the floating vegetation of the Chagres River. I began contemplating my own study of polyandry.

My opportunity came in 1987, when I received a short-term STRI fellowship for a trip to determine the feasibility of a long-term investigation of jacanas on the Chagres. Douglas, on spring break from Cornell, joined me in the field, followed by Natalia. In addition to testing field methods, a key task was to obtain tissue samples from a small number of jacanas to determine, back at Cornell, whether there was sufficient DNA variability to conduct paternity studies.

Toward the end of our visit, armed with permits and a borrowed shotgun, I collected birds. Purposefully, I first collected a female I knew was mated to two males, one with small, dependent chicks and the other with larger, older young. With stunning speed, a new female moved onto the territory, searching out the small chicks and pecking and shaking each one to death. The smaller male occupant squawked in protest, but could not stop her. Soon after killing his chicks, the female crouched to sexually solicit the male, and within a short time he responded, jumping onto her back in the typical pre-copulatory position. Later she chased off the older chicks. I felt shock at the violence of the killings, but also intellectual excitement: the incoming female had behaved exactly as Hrdy's hypothesis predicted.

I subsequently shot and collected tissue from a second female, paired to three males, two with dependent chicks. This time, two females soon trespassed onto the territory and proceeded to fight one another, as well as to kill three young chicks. Over three days, having collected just two females, I observed three female intruders kill five of seven dependent chicks and drive off two older chicks. My predictions of jacana infanticide had been confirmed, though I had not anticipated the speed of the females' responses nor the degree of brutality.

These were extraordinarily important observations, extending the applicability of Hrdy's sexually selected infanticide hypothesis to a new taxon (birds) and to the opposite sex (females). I ceased collecting females and, for the remainder of the samples needed, collected males that were not caring for eggs or young.

When the results of our preliminary study were published (Emlen *et al.* 1989), they provoked a critical response from Marc Bekoff, a behavioral colleague, challenging the ethical legitimacy of such research. In response, I explained why I think testing scientific hypotheses sometimes can justify suffering. I believe I made the right decision in conducting those experiments, and the results were so swift and absolute that I felt no need to expand the sample. Our exchange of views (Bekoff 1993; Emlen 1993) gives readers a chance to ponder the difficult ethical issues that institutional animal-welfare committees face regularly.

Empowering lessons for humans

Inevitably, it dawned on me that the multi-generational, extended family structure of bee-eater society was remarkably similar to that of human societies. For millennia, humans, too, have lived in extended family networks, interacting frequently and preferentially with relatives. Cooperative care in child rearing has long been the human norm, with most care provided by kin. Successful child rearing often required more care than two parents alone could provide, perhaps making us obligate cooperative breeders (Hrdy 1999). Recognizing these parallels gave added significance to my search for patterns in bee-eater society.

The study of human family systems has historically been the domain of social scientists. Knowing little about the theoretical constructs of sociology, I sought help from Cornell colleagues, particularly Elaine Wethington and the late Urie Bronfenbrenner. The dominant view of social scientists is that cultural, rather than genetic, influences are of overriding importance in mediating human behavioral interactions. As an evolutionary behaviorist, I believe that genetic, as well as cultural, factors influence social behavior. Many social behaviors are governed, at least in part, by heritable assessment procedures and decision rules that have been shaped by natural selection.

Where should one look for model animal systems to provide insights about the types of decision rules we humans might be predisposed to employ? Organisms living in similarly structured societies are those most likely to have evolved similar sets of predispositions. In my opinion, birds and mammals that live in multigenerational, pair-bonded, kin-structured societies provide the best model systems for understanding human family social interactions. Birds have the additional advantage of being largely culture-free.

In 1995 I published a paper titled "An evolutionary theory of the family" (Emlen 1995). Building on general principles of ecological constraints, life-history theory and inclusive-fitness theory, I offered 15 testable predictions concerning when families will form, the determinants of their stability and the patterns of interactions expected among family members. My initial goal was to provide a unifying framework for animal behaviorists studying non-human family systems.

What types of predictions emerged? Because families are composed of close genetic relatives, a high degree of cooperation is predicted. Avian data strongly support this prediction. More than 95% of bird species that live in multi-generational family groups exhibit cooperative care of young. Such cooperative care is extremely rare in any other type of avian social groups (Emlen 1995, 1997a).

Conflict is also expected, because the reproductive interests of parents, offspring and other family members are rarely identical. The death or divorce of a breeding parent and its subsequent replacement alters the basic genetic and dominance structure of the family. I predicted that conflict would intensify in replacement families (the equivalent of step-families) because replacement mates are unrelated to offspring of the previous pairing, and extant offspring are less related to future young of the new pairing.

Several predictions specifically addressed behavioral differences expected among members of intact family groups – those with both biological parents present – and replacement or stepfamily groups. These include: (i) increased competition and aggression between parent and same-sex offspring over sexual access to the replacement mate; (ii) reduced investment by the replacement mate in offspring remaining from the previous pairing; (iii) reduced cooperation by offspring of the previous pairing in rearing offspring of the replacement pair; and (iv) increased likelihood of breakup of the replacement family – because offspring of the previous pairing are more likely to disperse. Each of these predictions finds robust support in data from a number of family-dwelling bird and mammal species (Emlen 1995, 1997a; Emlen *et al.* 1995).

How are these findings relevant to human family systems? Numerous sociological studies have reported that: (i) human stepparents invest less time and effort in offspring from their partners' previous marriages than in children produced in the new marriage; (ii) stepchildren are statistically at increased risk of abuse – including sexual abuse – from a stepparent; and (iii) children report more conflicts with half-siblings and stepsiblings than with full brothers and sisters. Further, (iv) stepfamilies are less stable than intact families – children are likely to leave home at an earlier age (Daly & Wilson 1985, 1988; Emlen 1997b).

The behavioral changes reported in human stepfamilies thus closely mirror those found in avian replacement families. The value of the avian model is that it provides a window to observe the effects of heritable social predispositions in the absence of confounding cultural effects. The remarkable similarity of changes of behavior in both avian and human families following divorce and re-pairing suggests that similar heritable predispositions do indeed influence human family interactions.

If one accepts this proposition, the evolutionary perspective can offer insights into behavioral changes accompanying recent rapid changes in human family composition. Three major changes concern social scientists. The first is the transition from extended families to smaller nuclear families. The evolutionary insight is the concomitant reduction in the number of available relatives predisposed to assist with child-care. Second is the rise in rates of divorce and remarriage. For stepfamilies, the evolutionary perspective predicts a statistically greater risk of family conflict. Third is the increase in single parenting. If cooperative child-care has been the norm in human history, then single parenting is a recent invention. Supportive social networks must be found to replace the assistance previously provided by the extended family.

The study of genetic predispositions in avian families provides empowering lessons. As intelligent humans, we can learn to anticipate problems and deal more effectively with altered family composition in our own lives. Consider the prediction of increased strife in stepfamilies. The evolutionary theory of the family robustly predicts both the contexts and participants of likely conflicts. Equipped with such knowledge, family members can diligently work to promote harmony and stability. It's like taking informed, preemptive action after learning one has a genetic predisposition for a nutritional disorder. Unlike bee-eaters, we humans can use our intellectual resources to modify behaviors we judge undesirable (Emlen 1997b).

In the 1990s, I became increasingly interested in creating bridges between evolutionary biology and the social sciences. I accepted more speaking engagements and gave interviews to publications that target broad social-science audiences. In 2001, I taught my first course devoted to evolutionary perspectives on human social behavior.

In 2003, Cornell's provost requested interdisciplinary proposals addressing important societal problems. Ten faculty from eight disciplines and four colleges, including me, jointly submitted "The evolving family: family processes, contexts, and the life course of children." It was chosen as the first three-year Theme Project for Cornell's new Institute for the Social Sciences. Among ourselves, and at the workshops and conferences we hosted, we discussed such topics as the role of biology in families, intergenerational exchange and support, and fathers and fatherhood. The project strengthened bridges among scientists who shared a common interest but approached the subject from very different initial perspectives.

Full circle

In 2000, Natalia and I initiated a project on another multigenerational, colonial, cooperatively breeding bird species in Kenya, the grey-capped social weaver. My goal, unabashedly, is to study another model system for understanding the evolution of human social behavior. I injured my back in 1991 while working in Panama, and since then have faced rather serious mobility limitations. I am indebted to Bernard Amakobe Ocholi and many other Kenyan field assistants who helped me accomplish the grey-cap project. We concluded active data collection in June 2007, but analyses continue.

My scientific path has come almost full circle since my days as a psychology major at Swarthmore, when I first became interested in human behavior. Although I shifted to animal behavior with an emphasis on field biology, my study of questions ranging from migratory orientation to mating-systems evolution to cooperative breeding in birds has brought me back to humans and the evolution of their family dynamics.

Having finished graduate school just as sociobiology and behavioral ecology was emerging as a discipline, I feel fortunate to have ridden the crest of the wave of conceptual advances as both the field and I have matured. I am thankful that my work has enabled me to immerse myself in the natural world, to work in stunning environments, and to experience life in other cultures. I am grateful to have been able to combine my love of wild places with my passion for understanding natural history and animal behavior. Wilderness has always provided me with both spiritual and intellectual inspiration. I have always taken my cues from nature's clues.

References

Alexander, R. D. (1974). The evolution of social behavior. *A. Rev. Ecol. Syst.* **5**: 325–83.

Bekoff, M. (1993). Experimentally induced infanticide: the removal of birds and its ramifications. *Auk* **110**: 404–6.

Brown, J. L. (1964). The evolution of diversity in avian territorial systems. *Wilson Bull.* **76**: 160–9.

Crook, J. H. (1964). The evolution of social organisation and visual communication in the weaver birds (Ploceinae). *Behaviour Suppl.* **10**: 1–178.

Crook, J. H. (1965). The adaptive significance of avian social organizations. *Symp. Zool. Soc. Lond.* **14**: 181–218.

Daly, M. & Wilson, M. (1985). Child abuse and other risks of not living with both parents. *Ethol. Sociobiol.* **6**: 197–210.

Daly, M. & Wilson, M. (1988). Evolutionary psychology and family homicide. *Science* **242**: 519–24.

Demong, N. J. & Emlen, S. T. (1978). Radar tracking of artificially released migrants: a new technique allowing orientational manipulation of free-flying birds. *Bird Banding* **49**: 342–59.

Emlen, S. T. (1967a). Migratory orientation of the indigo bunting, *Passerina cyanea*, I. Evidence for use of celestial cues. *Auk* **84**: 309–42.

Emlen, S. T. (1967b). Migratory orientation of the indigo bunting, *Passerina cyanea*, II. Mechanism of celestial orientation. *Auk* **84**: 463–82.

Emlen, S. T. (1969). Bird migration: influence of physiological state upon celestial orientation. *Science* **165**: 716–18.

Emlen, S. T. (1970). Celestial rotation: its importance in the development of migratory orientation. *Science* **170**: 1198–201.

Emlen, S. T. (1975a). The stellar-orientation system of a migratory bird. *Scient. Am.* **223**: 102–11.

Emlen, S. T. (1975b). Migration: orientation and navigation. In: *Avian Biology*, Vol. 5, ed. D. S. Farner & J. R. King, pp. 129–219. New York: Academic Press.

Emlen, S. T. (1976). Lek organization and mating strategies in the bullfrog. *Behav. Ecol. Sociobiol.* **1**: 283–313.

Emlen, S. T. (1982a). The evolution of helping. I. An ecological constraints model. *Am. Nat.* **119**: 29–39.

Emlen, S. T. (1982b). The evolution of helping. II. The role of behavioral conflict. *Am. Nat.* **119**: 40–53.

Emlen, S. T. (1984). Cooperative breeding in birds and mammals. In *Behavioural Ecology: An Evolutionary Approach*, 2nd edn, ed. J. Krebs & N. Davies, pp. 305–35. Oxford: Blackwell Scientific Publishers.

Emlen, S. T. (1993). Ethics and experimentation: hard choices for the field ornithologist. *Auk* **110**: 406–9.

Emlen, S. T. (1995). An evolutionary theory of the family. *Proc. Nat. Acad. Sci. USA* **92**(18): 8092–9.

Emlen, S. T. (1997a). Predicting family dynamics in social vertebrates. In *Behavioural Ecology: An Evolutionary Approach*, 4th edn, ed, J. R. Krebs & N. B. Davies, pp. 228–53. Oxford: Blackwell Scientific.

Emlen, S. T. (1997b). The evolutionary study of human family systems. *Soc. Sci. Info.* **36**: 563–89.

Emlen, S. T. & Demong, N. J. (1978). Orientation strategies used by free-flying migrants: a radar tracking study. In *Animal Migration, Navigation and Homing*, ed. K. Schmidt-Koenig & W. T. Keeton, pp. 283–94. Berlin: Springer-Verlag.

Emlen, S. T. & Kem, W. (1963). Activity rhythm in Peromyscus: its influence on rates of recovery from nembutol. *Science* **145**: 1682–3.

Emlen, S. T. & Oring, L. W. (1977). Ecology, sexual selection, and the evolution of mating systems. *Science* **197**: 215–23.

Emlen, S. T. & Vehrencamp, S. L. (1983). Cooperative breeding strategies among birds. In *Perspectives in Ornithology*. ed. A. H. Brush & J. G. A. Clark, pp. 93–133. Cambridge: Cambridge University Press.

Emlen, S. T. & Wrege, P. H. (1988). The role of kinship in helping decisions among White-fronted Bee-eaters. *Behav. Ecol. Sociobiol.* **23**: 305–15.

Emlen, S. T. & Wrege, P. H. (1991). Breeding biology of White-fronted bee-eaters at Nakuru: the influence of helpers on breeder fitness. *J. Anim. Ecol.* **60**: 309–26.

Emlen, S. T., Demong, N. J. & Emlen, D. (1989). Experimental induction of infanticide in female wattled jacanas. *Auk* **105**: 1–7.

Emlen, S. T., Wrege, P. H. & Demong, N. J. (1995). Making decisions in the family: an evolutionary perspective. *Am. Scient.* **83**: 148–57.

Fry, C. H. (1972). The social organisation of bee-eaters (Meropidae) and co-operative breeding in hot-climate birds. *Ibis* **114**: 1–14.

Hairston, N. G., Smith, F. E. & Slobodkin, L. B. (1960). Community structure, population control and competition. *Am. Nat.* **94**: 421–5.

Hamilton, W. D. (1964). The genetical evolution of social behaviour. I, II. *J. Theor. Biol.* **7**: 1–52.

Hrdy, S. B. (1974). Male-male competition and infanticide among Langurs (*Presbitis entellus*) of Abu, Rajahsthan. *Folia Primatol.* **22**: 19–58.

Hrdy, S. B. (1999). *Mother Nature: A History of Mothers, Infants, and Natural Selection.* New York: Pantheon Books.

Keeton, W. T. (1974). The orientational and navigational basis of homing in birds. *Adv. Study Behav.* **5**: 47–132.

Kramer, G. (1949). Uber Richtungstendenzen bei der nachtlichen Zugunruhe gekäfigter Vogel. In *Ornithologie als Biologische Wissenschaft*, ed. E. Mayr & E. Schuz, pp. 269–83. Heidelberg: Winter-Unniversitätsverlag.

Kramer, G. (1950). Weitere Analyse der Faktoren, welche die Zugaktivität des gekäfigten Vogels orientieren. *Naturwissenschaften* **37**: 377–8.

Lack, D. (1968). *Ecological Adaptations for Breeding in Birds*. London: Methuen.

Orians, G. (1969). On the evolution of mating systems in birds and mammals. *Am. Nat.* **103**: 589–603.

Sauer, E. G. F. (1957). Die Sternorientierung nachtlich ziehender Grasmucken (*Sylvia atricapilla, borin*, und *curruca*). *Z. Tierpsychol.* **14**: 29–70.

Sauer, E. G. F. (1961). Further studies on the stellar orientation of nocturnally migrating birds. *Psychol. Forsch.* **26**: 224–44.

Slobodkin, L. B. (1961). *Growth and Regulation of Animal Populations*. New York: Holt, Rinehart and Winston.

Vehrencamp, S. L. (1977). Relative fecundity and parental effort in communally nesting anis *Crotophaga sulcirostris*. *Science* **197**: 403–5.

Vehrencamp, S. L. (1979). The roles of individual, kin, and group selection in the evolution of sociality. In *Handbook of Behavioral Neurobiology*, ed. P. Marler & J. G. Vandenbergh, Vol. 3. pp. 351–94. New York: Plenum.

Verner, J. & Willson, M. (1966). The influence of habitats on mating systems of North American passerine birds. *Ecology* **47**: 143–7.

West-Eberhard, M. J. (1975). The evolution of social behaviour by kin selection. *Q. Rev. Biol.* **50**: 1–33.

Williams, G. C. (1966). *Adaptation and Natural Selection*. Princeton, NJ: Princeton University Press.

Wilson, E. O. (1975). *Sociobiology: the New Synthesis*. Cambridge, MA: Belknap Press of Harvard University Press.

11

A most unlikely animal behaviorist

BENNETT G. GALEF

Prologue

My invitation to participate in this volume, gratifying though it is, reflects not only decades of hard, thoroughly enjoyable work, but also more than a little good fortune. I became interested in animal social learning long before appreciation of the importance of social acquired information in the development of animals' behavioral repertoires became widespread. Social learning could as easily have remained of interest to few students of animal behavior. If so, someone other than I would have been invited to write this autobiographical sketch.

Early years

I was born in midtown Manhattan and, until I was 17 years old, lived in a twelfth-floor apartment just off Park Avenue, in the heart of New York City and about as far from the

Leaders in Animal Behavior: The Second Generation, ed. L. C. Drickamer & D. A. Dewsbury. Published by Cambridge University Press. © Cambridge University Press 2010.

biological world as one can get and remain on this planet. My privileged early years owed much to my paternal grandfather, who as a child in the late 1880s came to the United States from Russia. At age 15, he left school and took a job as a janitor in a sporting-goods firm that he was to own by the time he was 23. At its height, the family business had offices in New York, Berlin, and Paris, and earned my grandfather a truly extraordinary amount of money in the late 1920s and early 1930s.

My father was raised in luxury: a 10-room Park Avenue apartment, a 90-acre, 20-room, beach- and river-front summer estate on the New Jersey coast with more than a dozen in help, a racing yacht, string of polo ponies, and other accessories of the very wealthy. The summer property even had a private train stop where my father and grandfather began and ended their 45-minute, weekday commute to and from work in New York each summer.

My father was a man of many interests and accomplishments: a prize-winning sculptor and photographer, one of the first recreational downhill skiers in North America, and an explorer in South America. A patron of the arts and author of an unpublished novel with a Master's degree in psychology earned at night, he worked all of his too-short adult life (he died at 48) as an unhappy second fiddle in the business my grandfather controlled. My mother was a very attractive, possibly beautiful socialite, descended on her mother's side from the less successful branches of two very prominent Jewish families whose names are familiar to everyone.

By the time I was old enough to understand such things, my parents had divorced and the family fortune was in decline. The Second World War had seriously undermined a business that depended largely on imports, and although the summer home was to remain until I was 7 or 8 years old, the polo ponies were gone, the help was reduced by half, and the racing yacht leaked badly.

Despite my family's somewhat reduced circumstances, my mother, my younger brother Jack and I were always more than comfortable. Still, I understood from an early age that, sooner or later, I would have to make my own way in the world.

Like my father before me, I was under considerable pressure to join the family business. However, to jump ahead a decade, following my graduation from college and a few months working at buying and selling things in my grandfather's firm, I became convinced that such was not the life for me, though I had no idea what was.

The greatest benefit that I received as a result of my paternal grandfather's financial success was a first-rate education, first for 9 years at the Riverdale School in the Bronx, and then for 4 years, at Princeton. At Riverdale, I was one of a class of about 50 that contained some of the brightest, most highly motivated people that I have encountered in a life in academia. Eight of us were National Merit Scholarship Finalists, an honor that, if I remember correctly, was awarded to fewer than 1 in 1000 of those who took the qualifying examination.

The Riverdale faculty was excellent and the material presented to us in the last couple of years of high school was, as I later learned, generally at university level. Several of my classmates went on to become successful academics at Yale, MIT, Stanford, University

College London, Georgetown, and Tufts. Others were to be doctors, lawyers, architects, and journalists.

Perhaps the most lasting effect of my years at Riverdale came from writing a senior paper on the battle of Gettysburg for Mr Gartner's American history class. I chose a civil war battle as my subject because Mr Gartner despised military history, and because the more military history I wrote, the more military history Mr Gartner would have to read, I wrote a very long paper indeed. Such was rebellion in the 1950s.

The project, completed over spring vacation, was based on primary-source material held in the great collections of the New York public libraries. Although the resulting manuscript was surely no great contribution to Civil War scholarship, it did introduce me to the pleasures of historical research. To this day, I spend what time I can reading about the history of the life sciences, particularly biographies of Victorian naturalists, though I have published work on only one historical figure, Edward Thorndike, the first Ph.D. in animal psychology in North America (Galef 1998).

Throughout my high-school years, I was underweight and undersized for my age, and because I had entered first grade when 5 rather than 6 years of age, I was also 12–18 months younger than most of my classmates. Always one of the last to be chosen for any sports team, assigned to spend my time on the baseball diamond languishing in right field, massacred in dodge ball, etc., I found many of my larger, more mature classmates totally intimidating. However, in my sophomore year of high school, Riverdale hired a philosophy teacher and wrestling coach, Bill Williams. Under his tutelage, I went on to enjoy great success on the high-school wrestling circuit: captain of Riverdale team as a senior, winner of each of the four or five city-wide tournaments that I entered that year, nationally ranked and named "athlete of the year" at my high school. After experiencing success at wrestling, I rarely felt intimidated by anybody or anything.

Part of the Riverdale campus was a heavily wooded hillside that lay between the main classroom building and an athletic field a hundred or more feet below. I spent many happy hours wandering the paths through that hillside looking for birds' nests, insect cocoons and anything else of interest. Although I greatly enjoyed the time I spent clambering about the undeveloped parts of the Riverdale campus, I had no intellectual interest in natural history. Indeed, in tenth grade, when I had to choose between courses in Latin and biology, my greater interest in literature and art than in science led me to choose to study Latin. The unanticipated, and as it turned out, unfortunate consequence of this decision was that I was never to take a formal biology course in high school, college or graduate school.

Although Riverdale offered a fine education, the greater influence on my intellectual development was the amazing city in which I was fortunate enough to live. New York in the 1950s was a wonderful place. The city was safe to roam day and night. Weekends could be filled doing endless things for next to nothing: museums and art galleries beyond counting, numerous bookstores where I could sit and read the afternoons away. There were second acts of Broadway shows and Philharmonic concerts to be snuck into in the evenings, all kinds of food to eat and all kinds of people to enjoy. I spent half my weekends uptown with my schoolmates from Riverdale, most the children of successful professionals, and their friends, and half in Greenwich Village amongst the less conventional of my age mates. My social life was the center of my existence, and partying until all hours of the night was my favorite occupation.

Summers of my school years were spent mostly just lying on the beach in southern Connecticut or in camps in Maine chasing snakes and frogs and fishing. For two summers I labored on a working horse ranch at the edge of the Bob Marshall Wilderness Area in western Montana. Although I enjoyed the outdoors, 6 months spent working horses and without much access to either running water or electricity convinced me that I was a city boy at heart.

Summers of my sophomore and senior years, I travelled with my family to Europe. These trips, first to France and Italy, then to England and Scandinavia, introduced me to the larger world and were the origin of a life-long passion for travel.

College years

After a childhood spent in Manhattan, Princeton University was a shock. In 1958, Princeton was a school for men; a place where, on weekends, women were imported by the busload from Vassar, Bryn Mawr and like places in what my fellow Princetonians liked to call "cattle drives." Attendance at Sunday chapel services was compulsory and monitored. The University was often caricatured, with some justice, as the northern-most southern school in the United States. In my freshman year, the fellow living across the dormitory hall from me claimed, with considerable pride, to be a member of the Knights of the White Camellia, apparently an upper-class version of the Klu Klux Klan.

After a life spent in the Big Apple, the social side of the University seemed more than a little childish. A not-so-subtle anti-Semitism and political and religious conservatism made full participation in the life of the University difficult for an agnostic, liberal, Jewish-born New Yorker. Still, there were many who shared my reluctance to participate fully in the accepted undergraduate culture. Lectures were generally excellent and the opportunity to interact with outstanding professors in small preceptorials was truly exciting.

My original plan was to major in chemistry and minor in art history with a view to becoming a forensic art historian. Freshman year I took physics, chemistry and calculus. However, the courses I loved were cultural anthropology and philosophy of science. Melvin Tumin taught one of the most engaging courses I have ever taken, and Carl Hempel's exposition of logical positivism was a revelation that has shaped my entire career in research. Sadly, in the midst of my first year in University, my father died in the crash of a commercial airliner.

The chemistry courses that I took sophomore year convinced me that I was not the least interested in chemistry. Also, that year I was diagnosed with a possibly terminal kidney disease that has left me short on stamina throughout my life. The near simultaneous loss of father, career goals, and sense of immortality left me without much of a life plan.

For reasons I can no longer recall, I decided to major in psychology. Surely, my choice did not reflect some deep-seated interest in the study of the human mind. During my first 2 years in University, I had taken only a single, half-year psychology course, and had neither much liked nor done particularly well in it. I expect I was influenced by the perception, then as now, that psychology is an easy major. I worked hard in college. However, I was basically lazy, and would do whatever was necessary to maintain my social life in New York, just over an hour by train from Princeton, while completing my undergraduate degree.

Except for a course in comparative psychology based on Maier and Schneirla's (1935) *Principles of Animal Psychology* taught by Glen Weaver and a brilliantly organized course on animal learning taught by Mymon Goldstein, I remember little of my two years as an undergraduate psychology major other than the senior thesis on human concept formation that I completed in Mymon's laboratory.

Elijah Lovejoy, a fellow undergraduate, and I devised a method that allowed us to provide immediate feedback to groups of subjects taking multiple-choice tests. We modified an old printing press to overprint the right answers to multiple-choice questions with an invisible phenolphthalein solution. We then gave our test takers pens filled with yellow ink that had a high pH. The ink turned the otherwise invisible phenolphthalein marks bright pink when applied to a correct answer. Lige and I then used the method in our thesis work. Solving the technical problems was exhilarating. However, I did not take the research on concept formation that followed very seriously.

I graduated Princeton magna cum laude, in the top 7% or 8% of my class, and with a handful of prizes (I mention these things because little of what follows would make much sense if I did not), but without the slightest idea of what I wanted to do with my life. I had fallen hopelessly in love with a Swedish girl that I had met at the University of Bordeaux and Toulouse, where I had spent a month of the summer of my junior year improving my French. We agreed to attend Trinity College Dublin together starting in the fall, after we had both finished University. She was to work on a Ph.D. in languages, and I was to study psychology. We both applied to TCD, and were duly admitted. However, in matters of the heart a few months apart can be forever. Four months later, she was engaged to a previous boy friend, and I was heartbroken. Neither of us ever got to Ireland.

A wasted year well spent

As my college years drew to a close, I was adrift. I wrote the law boards, and was admitted to Harvard Law School. However, when Harvard refused to allow me to delay matriculating for a year, I decided not to go. I took the Graduate Record Examinations and did extremely well, but did not apply to graduate schools. Instead, I spent four months working as a laboratory technician at Princeton, four more working for my grandfather in the family firm, and four on the road in France, Italy, Switzerland, and Spain.

I had rather enjoyed the months spent as a research assistant, so while working in the family firm doing rather little I found of any interest, I applied to a handful of graduate programs in psychology. I was admitted to most, and chose to go to the University of Pennsylvania, where I had been accepted to study concept formation with Frank Irwin. The financial offer was good, my friend Lije from Princeton was at Penn and making encouraging noises, and Penn had sent notification of my acceptance by telegram rather than by post. I had no idea that, at the time, Pennsylvania had one of the world's foremost programs in experimental psychology, although Frank Irwin was one of its lesser lights. He was to prove an extremely challenging, if equally austere, mentor.

Graduate school

I was a terrible, awful, vile first-, second-, and most-of-third-year graduate student. I spent much of my first year improving my skills at poker, chess, and hearts, and most of my second either playing bridge under the tutelage of Marty Seligman or engaged in a kind of blind chess called kriegspiel. I did well enough in my courses, especially the first half of the first-year course when Dick Solomon taught animal learning and Phil Teitelbaum physiological psychology. However, I can remember little from the first four semesters of lectures other than that I shall never really understand the moment-generating function. I certainly had no interest in experimental psychology.

I completed an uninspired Master's thesis investigating why humans bet more rationally when offered a single opportunity to wager than when offered a series of identical wagers. I had to write draft after draft before Dr Irwin finally found the fourteenth acceptable, and this in the days before computers and word processing.

At the time I failed to realize how generous Dr. Irwin was in reading and correcting 14 drafts of a useless manuscript. Ten years later, when I returned to Philadelphia on business, I made a point of thanking him for teaching me to write academic prose and making my career possible. My Master's degree oral examination was an embarrassing debacle best forgotten. I was awarded the degree anyway.

During my third year as a graduate student, I had to study for my Ph.D. comprehensive examination in human learning and discovered that I had absolutely no interest in the field that I had to master. Along the way, I happened to read Wittgenstein's work on concepts and came to the realization that, if Wittgenstein were right, not only was everything that I and everyone else studying concept formation doing meaningless, but also that I had no insight as to how to do things any better. I began, admittedly somewhat belatedly, to think seriously about leaving graduate school and doing something more compatible with my talents, whatever they might be.

I had audited an undergraduate course that Paul Rozin taught in comparative psychology. Like the course that I had taken with Glen Weaver at Princeton, Rozin's course seemed pretty interesting, so I asked Paul if he might accept me as a graduate student.

Paul had some wild rats that an exterminator had captured for him on the wharves of Philadelphia and was looking for someone to work with them, but had had no takers. I suspect that my more knowledgeable peers were too clever to take on such unpromising animals as subjects for their thesis research. However, I decided to give the wild rats a try, and Paul agreed to serve as my supervisor.

Paul turned out to be the near-perfect research supervisor for me. He was supportive, but not particularly directive, with a strong background in both biology and psychology, very, very bright, and with great enthusiasm for most everything from good food to the fluctuations of the stock market. He was also a superb experimentalist and remarkably integrative thinker.

In his undergraduate course, Paul had mentioned in passing some early work suggesting that wild rats raised appropriately might become as tame and easy to handle as domesticated rats (Rasmussen 1938). Aggression and domestication were hot topics in the 1960s, as was the role of the environment in shaping development. So I had a thesis topic as well as a thesis supervisor.

My early attempts to work with wild rats were laughable. I had essentially no experience handling even a nice, gentle domesticated rat, and here I was trying to study behavioral development in an animal generally considered to be both completely unmanageable and dangerous. At first, the only way I could move a rat from one place to another was to turn it loose on the floor, corner it, throw a lab coat over it, and then put it where I wanted it to go. Still, in time, I learned to breed and work with wild rats.

I found animal behavior generally interesting, and volunteered to work for W. John Smith, then the sole ethologist in the Biology Department at Pennsylvania. I became an informal teaching assistant in John's undergraduate ethology course and collected some data for him on the behavior of members of a colony of prairie dogs residing in an enclosure at the Philadelphia zoo.

At the time, I did not realize how generous John was in accepting me as a volunteer participant in both his teaching and research. However, the opportunity to work closely with a committed biologist and naturalist was to have lasting consequences, as was my enrolling in a graduate-level laboratory course in comparative psychology that John, Paul Rozin, and the late Allen Epstein organized. Students in the course formed small teams, designed an experiment, built any necessary apparatus, and collected some data. Although the project that Chuck Snowdon and I conducted on aggression in Siamese fighting fish was a flop, I admired the approach of the course to animal behavior, treating it as a field to which anyone who wanted to contribute could.

I also started to read in ethology and animal behavior, albeit unsystematically. I particularly remember Tinbergen's inspirational *Curious Naturalists* and *Herring Gull's World*, textbooks by both Hinde (1970), and Marler & Hamilton (1966), and a wonderful book, *Readings in Animal Behavior*, edited by Tom McGill (1965). However, my disorganized and sporadic reading was not nearly sufficient to prepare me for my doctoral comprehensive examinations, now in comparative and physiological psychology. I failed them cold. It wasn't even close.

By early 1968, research for my thesis on the role of stimulus novelty in the expression of aggressive behavior in wild rats was complete. I had no publications, and I had failed my comprehensives. I had also, to the great approval of my family, married Elizabeth Sarnoff, grandniece of the founder of RCA. Getting married was not one of my better decisions. At 25, I was still far too immature to marry anyone.

Getting a job

I applied to work as a post-doctoral student with Niko Tinbergen at Oxford. However, Tinbergen was accepting only graduate students, not post-docs, and my attempt to get further training went nowhere.

If I had understood a bit more about academic life, I would have realized that I had little reason to hope for a university appointment. However, in my ignorance, I applied for jobs at Tennessee, Princeton, Toronto, Purdue, and Stony Brook. All had advertised positions that seemed suitable for a budding comparative psychologist, although by 1968, I had already started to label myself an animal behaviorist in response to comparative psychologists' hesitancy to embrace ethology.

Ed Stricker, a post-doctoral fellow in Biology at Pennsylvania, whose work in regulatory physiology I greatly admired, had the previous year accepted a job at McMaster University. McMaster was then a small school of which I had never heard, home to a Psychology Department with only 15 faculty members. However, McMaster had an opening for an assistant professor in comparative psychology, so I applied there too. Perhaps, more realistically, I started looking into alternative careers with various think tanks and intelligence organizations.

I interviewed at Purdue, Stony Brook, and McMaster. After a visit to Purdue, where I was offered a sort of post-doctoral fellowship to work with Victor Denenberg, I was pretty sure that I would rather live as a shoe salesman in New York City than as a professor in West Lafayette, Indiana. I never received a firm job offer from Stony Brook. However, an hour after I finished my job talk at McMaster, I was offered a truly unbelievable opportunity. There was to be no teaching my first term and a light teaching load thereafter, good start-up money, guaranteed grant support and travel funds, a new building within 2 or 3 years, no income tax for 5 years, and McMaster was only 45 minutes from Toronto and a major international airport. As icing on the cake, the faculty was exceptionally strong in experimental animal psychology. I knew that I was not ready for an academic position. However, I wasn't crazy either. I accepted the sole real job offer that I had. In July of 1968, I shipped a couple of crates of wild rats from Philadelphia to Hamilton, Ontario, and my wife and I moved into a downtown apartment and started looking for a house.

Although I had failed my comprehensive examinations, I had a job and was married. I suspect, as much out of pity for my predicament as anything else, I was awarded a Ph.D. in psychology from the University of Pennsylvania. There were some mutterings about requiring me to return at some unspecified future date to make up for the comprehensive examinations that I had failed, but there was never any real follow-up.

A new assistant professor

Unexpectedly, shortly after arriving in Hamilton, I became fascinated with animal behavior. Every evening and weekend for the next several years I spent hours reading books and papers in the area. I was particularly fortunate that my introduction to evolutionary theory was G. C. Williams's (1966) *Adaptation and Natural Selection*. I was completely sold on the single-allele substitution model, though, of course, I knew no other. At least with regard to evolutionary theory, I was ahead of the curve.

I am not sure why I finally became serious about academics. I do remember that shortly after arriving in Hamilton, I came to the realization that this was it. This was my career and my life. Unless I did something pretty drastic, I was going to spend it as a third-rate professor at a second-rate University living in a town generally considered, albeit unfairly, to be one of the least livable in Canada.

Although I soon came to realize that the McMaster job was, in many ways, one of the best first jobs in North American academics, I had no intention of spending the rest of my life in Hamilton, Ontario. I decided that if I did not get moving, my life would be a disaster.

I had also become enamored of experiments. I filled all dozen sinks in the University Faculty Club washroom and let them drain to see whether the Coriolis force really worked in the way that people said it did. It did not. When I went up to Princeton to give a colloquium, Donna Howell and I got a bat to hover in a bottle placed on a scale to see whether the apparent weight of the bottle and its contents changed when the bat was in flight. It did, if the bottle was open.

More relevant to animal behavior, I bought a stuffed owl at an antique store and put it on a pole at the edge of the campus to observe mobbing at first hand. I tried experiments on prey selection in frogs, schooling in minnows, and taxis in sow bugs. I caught a few sticklebacks and tried to repeat some of Tinbergen's classic experiments. None of this was work. I was just having fun of a sort that would be nearly impossible today. In the late 1960s, there were no institutional ethics committees. Consequently, work with animals could have a spontaneity that the present generation of researchers will never experience.

Shortly after arriving at McMaster, I submitted my thesis for publication to Danny Lehrman, then editor of the *Journal of Comparative and Physiological Psychology*. I had no idea how a typescript became a journal article, so I just sent Danny a copy of my thesis, assuming that he would somehow turn it into one of those professional-looking publications I had been reading in the journals. Danny patiently led me through the process of preparing a manuscript for publication (Galef 1970), for which I shall be eternally grateful.

After the debacle of my first manuscript submission, I resolved to be completely professional in all future interactions with the world outside McMaster. I would do whatever I was asked to do, whether I wanted to or not, as well as I could and on time, a set of resolutions to which I managed to remain faithful for 30 years or more. I hoped that with a little luck, one thing would lead to another, and I could move on.

I was soon blessed with two great graduate students: Jeff Alberts stayed on after his B.Sc. to work on his Master's thesis, and Mertice Clark, another McMaster undergraduate, to pursue her Ph.D. The three of us were desperately short of space. My laboratory was two rooms in a tumbledown, 'temporary' building erected during the Second World War. The laboratory was so tiny that after classes were over for the day, Alberts would sneak into an adjacent classroom, turn some of the tables there upside down, tape cardboard around their legs to make arenas for testing rats, run an experiment, and then put everything back before classes started the next morning (Alberts & Galef 1971, 1973). I have never had more fun or been more productive.

The chair of the department at the time was a little too fond of drink, and fulfilled many of his administrative obligations by assigning me, as the youngest faculty member, to chair most departmental committees. With Ed Stricker, I redesigned the Ph.D. comprehensive examination system in the department. With Michael Leon, I changed forever the way that introductory psychology course was taught to our first-year undergraduates.

One of my responsibilities, along with space, the shop, the graduate program, the under- graduate program, animal care, etc., was to organize the departmental colloquia. Over the next few years, I abused the position to invite everyone in whose work I was interested to speak at McMaster. The invitees addressed the department one day and spent the next talking with me and my students about animal behavior.

I remember particularly visits of Niko Tinbergen, Ken Roeder, Vince Dethier, Phil Teitelbaum, and Danny Lehrman. They could not have been more helpful: Tinbergen invited me to spend time with his group at Oxford, and both he and Lehrman wrote letters on my behalf requesting that I be invited to the International Ethology Congress, which, at that time, was a closed meeting. Ken Roeder provided an invitation to attend a demonstration he had organized at the planetarium of the America Museum of Natural History illustrating the celestial cues available to migrating birds. I had the rather naïve notion that I should completely solve a problem before attempting publication. Phil Teitelbaum convinced me that it might be wise, if I wished to remain in academics, to publish parts of a solution as I discovered them.

And the research was going well. At first, I had extended my thesis work on the role of novelty in eliciting aggression in wild rats. However, problems with dependent variables soon caused me to question the value of the work. Compared to Ed Sticker's experiments investigating the physiological control of sodium and water intake, aggression seemed a very difficult topic on which to get a handle. With water or salt, you knew exactly what to measure: water or sodium intake and excretion. With aggression, there seemed no equally valid dependent measures. Gaining proper control of independent variables was hard enough without having to worry about the validity of dependent variables. It felt like time for a change.

In the late 1960s, John Garcia's research program on taste-aversion learning in rats was the most exciting and controversial in experimental animal psychology. The McMaster Psychology Department, then exceptionally strong in animal learning, hosted a workshop concerned in part with Garcia's work and attended by many of the day's most respected experimental psychologists. I went to many of the presentations, and because I have never been able to focus long on oral presentations of the minutiae of experiments in animal learning, I began thinking about the possibility of studying taste-aversion learning in my wild rats.

I faced a major problem. The then standard, hanging, rat cage was useless for housing wild rats. The moment that I opened such a cage, the rat it had contained would pop out of the cage and onto the floor. I had designed cages to house wild rats individually. However, it was taking time to have them built. In the interim, my animals were living in mixed-sex groups of four or five in 1 m × 2 m floor cages that filled one of the two rooms that were my laboratory.

Coincidentally, I had arranged to have a paper by Fritz Steiniger, an applied ecologist who had worked on rodent control, translated from its original German, a language that I could not begin to read despite having passed a German-language requirement in graduate school.

Steiniger (1955) had made some field observations consistent with the view that adult rats could teach the young of their colony to avoid eating poison baits. Although that seemed unlikely, it also seemed worth looking into.

Social learning

At the time, a female in one of my cages of wild rats had given birth to a litter of six pups. It was easy to put the colony of which she was a member on a feeding schedule so that all their food intake could be observed on closed-circuit television, just then becoming available for laboratory use. We gave adult members of the colony access to two food bowls each day; one bowl contained a safe but uninteresting food and the other a very tasty food that I had adulterated with a mild toxin. The adults rapidly learned to avoid the good-tasting, poisoned food and to eat only the safe alternative, and, as John Garcia would have predicted, continued to avoid the good-tasting food even when subsequently offered uncontaminated samples of it.

Christmas was approaching. My wife and I had scheduled a week vacationing at a Club Med in Martinique, and the litter of wild rats was rapidly approaching weaning age. I asked my technician to watch on television during the rat colony's daily feeding periods and keep track of which food the young ate on each of the first few days they fed on solid food, and took off for a week in the sun. When I returned, I learned, much to my surprise, that although the pups had eaten the relatively unpalatable food that the adults of their colony were eating on scores of occasions, they had never even touched the highly palatable food that the adults of their colony had learned to avoid (Galef & Clark 1971). Thirty-seven years later, I am still investigating the "hows" and "whys" of such social transmission of food choice.

For the first 10 of those 37 years, I worked on effects of relatively simple social stimuli on food choices of young rats: adult presence at a feeding site (Galef & Clark 1972), residual cues that adults leave on or near foods they are eating (Galef & Heiber 1976), flavors transferred from mother to young in mother's milk (Galef & Henderson 1972; Galef & Sherry 1973), etc.

The work went well, and I started to receive invitations to write reviews, first in 1974 from Mike Domjan to discuss my empirical work (Galef 1977), and later the same year from Jay Rosenblatt to place my studies of wild rats' social learning of food preferences in a broader context (Galef 1976). Jay's invitation led to nine months of hard work in the library searching out and organizing the then widely scattered literature on animal traditions. Those months of library research formed the intellectual foundation of my career in social learning, forcing me to think about why I was interested in the phenomenon, and why others might share that interest.

Litter cannibalism

In those early days, almost all of the research in which I was involved concerned social learning or feeding behavior in Norway rats. However, when I was growing up in New York,

I had briefly a pair of pet golden hamsters (*Mesocricetus auratus*). Shortly after the female delivered her first litter, she ate several of them, and my mother banished the hamsters from our apartment. I subsequently read a bit about litter cannibalism in hamsters and learned that it was generally attributed to a mother hamster being 'disturbed' in some way. However, I did not believe that disturbance was involved in the litter cannibalism I had observed. I had not bothered my pets during the week before the unfortunate demise of the pups.

Corinne Day joined the lab as a Master's student and agreed to look into the causes of litter cannibalism in hamsters. She discovered that female hamsters almost invariably gave birth to more young than they are prepared to rear and cull their litters to some predetermined size. If you added pups to a litter, the dam would cull pups, and not selectively the ones you added, to maintain her litter at the size that she preferred. If you removed pups from a litter at birth, cannibalism was proportionally reduced (Day & Galef 1977). I thought the work was terrific. However, the paper was published before there was much interest in litter cannibalism, and it never attracted the attention that I thought that it deserved.

Life outside the laboratory

Important events occurring outside the laboratory during my first decade at McMaster included the end of my first marriage in 1971, a sabbatical year spent in Panama in 1974–5, and the founding of the Winter Animal Behavior Conferences in 1978. I have nothing much to report about my personal experience of divorce other than that it was emotionally wrenching and resulted in a year-long hiatus in research. Both the time spent in Panama and the founding of the Winter Animal Behavior Conferences were to have positive and lasting effects on my professional life.

It seemed to me pretty clear, even in 1968, that much of the future of animal behavior lay in fieldwork, of which I had absolutely no experience. I spoke with W. John Smith, at Pennsylvania, about the possibility of my learning something about behavioral research outside the laboratory, and he made arrangements for me to spend my first sabbatical year at the Smithsonian Tropical Research Station on Barro Colorado Island, a few square kilometers of tropical forest sitting in the middle of Lake Gatun, which forms a major part of the Panama Canal.

My then graduate student, now wife, Mertice Clark and I moved onto Barro Colorado Island in May of 1974 for seven of the most intellectually valuable months of my life. I worked every morning and many evenings as a pair of free hands for any researcher on the island who wanted some help, and in the afternoons on Mertice's and my research on maternal behavior in agouti (*Dasyprocta punctata*), a large, diurnal rodent (Galef & Clark 1976). Mertice spent her mornings following agouti through the forest and her afternoons writing her Ph.D. thesis. When not watching agouti, I caught poison-arrow frogs for Kathy Toft and baby alligators with Harry Greene, and followed howler monkeys for Katie Milton and Julia Chase and army ants for Charlie Hogg. As a result, I spent 8 hours each day for 9 months roaming one of the most exciting ecosystems on earth. By the time I left Barro Colorado, I had a pretty good idea what fieldwork was like, though I was never again to do any. More importantly, I had begun to see my own work on social learning as lying at an interface between biology and psychology.

The WABC

In 1976, Jeff Alberts invited me to participate in a symposium he had organized for the week-long Winter Conference in Brain Research (WCBR). I have never been much interested in the functioning of the brain and attended the 1976, 1977, and 1978 meetings of the WCBR mostly to enjoy the excellent skiing at Keystone, Colorado. Indeed, all I remember of the dozens of presentations I attended at the WCBR was an evening of papers presented by several students from Seymour Benzer's lab. I was absolutely blown away.

To be invited to WCBR for a fourth year, it was necessary to engage in some sort of presentation, so Alberts and I proposed a symposium. However, the executive committee of the WCBR turned us down. We decided that any organization that would not accept the very exciting symposium we had proposed (if I remember correctly, we had planned to invite both Vincent Detheir and Ken Roeder to participate) was the wrong organization for us.

One evening late in the 1978 WCBR, Alberts, Mertice Clark, and I, along with three, like-minded folk, met for dinner to talk about organizing an alternative. We six compiled a list of 28 people with whom we would like to spend a week; I made a phone call to reserve 28 rooms in Jackson, Wyoming, for late in January of the following year, and the Winter Animal Behavior Conference (WABC) was born.

There were a slew of intellectually acceptable rationales for creating the WABC that Alberts and I used to bludgeon all 28 invitees into attending. However, in truth, I was interested in the WABC for largely selfish reasons. McMaster University lay well outside the usual social network in animal behavior and related disciplines. Consequently, I had found it difficult to get to know colleagues working in ethology, ecology and animal behavior.

The original intention was to hold a single meeting of the WABC. However, 31 years later, the organization without an organization still flourishes. Co-hosting for 10 years a meeting that eventually became very well known to those interested in animal behavior let me get to know and be known by many of the movers and shakers in my intellectual world. I had got to know many of my early heroes: Paul Sherman, Jack Bradbury, Jim Simmons, Randy Thornhill, Steve Emlen, Steve Arnold, John Wingfield, and Mike Ryan, as well as innumerable other first-rate scientists.

There were real benefits resulting from getting to know these people, rather than just reading their papers. When I needed to know something in an area about which I was fundamentally ignorant, and that included most fields, I could just pick up the phone and call the relevant expert. In the decades before Google, this mattered a great deal. More important, the WABC allowed me to feel part of an intellectual community. Science is, obviously, very much a social undertaking, and although working in relative isolation has real advantages when pursuing original lines of research, social isolation is also limiting. The WABC let me join the main stream. And of course, it was great fun. Take 30 bright people and put them in a beautiful place for a week and interesting things invariably happen.

A new beginning

Over the next several years, my career continued to develop nicely, I thought. I received many invitations to give talks, attend international conferences, write chapters, review books, serve on editorial boards, etc., and accepted every one. However, my research, though solid, contained little that was really new or exciting. That changed in 1982.

That winter, Barbara Strupp, then a graduate student at Cornell, came to visit. Barbara was considering leaving Cornell and coming to McMaster to complete her Ph.D. She described for me a procedure that she and David Levitsky, her supervisor, had developed in their work on nutrition that seemed to me to have considerable potential for studies of social learning.

When Barbara ultimately decided not to leave Cornell, I wrote to her and to David describing the research that I had in mind to ask whether they intended anything similar. They did not, so I asked Steven Wigmore, a new Master's student in my laboratory, to work on the experiments I had proposed to Barbara and David. The work went spectacularly well, and in 1983 Steve and I published the first of what was to become a long series of articles on the role of olfactory cues in the social transmission of food preferences in Norway rats. In brief, Steve and I found that after a Norway rat (an observer) interacts with a conspecific that has recently eaten a distinctively flavored food (a demonstrator), the observer will show a substantially enhanced preference for whatever food its demonstrator ate (Galef & Wigmore 1983). More than 25 years later my laboratory is still working with the same paradigm.

As the laboratory's studies on social transmission of food preference became better known, I started to receive invitations to participate in symposia on topics ranging from dentistry to psychiatry. However, it was not until 1985 that anyone organized a symposium dedicated to animal social learning. The talks that Tom Zentall arranged for a meeting of the Midwestern Psychological Association led to a book, *Social Learning: Psychological and Biological Perspectives* (Zentall & Galef 1988), the first of five collections of readings in social learning that I was to co-edit.

Editing books, organizing conferences

In the mid-1980s, it was still possible to bring together in a single volume chapters by all of the major contributors to the field of animal social learning. Zentall, who had taken the initiative in proposing the volume, edited the work of those taking a psychological approach to social learning, while I looked after those whose work was more in the biological tradition. The inclusion of work from diverse perspectives was the book's most unusual feature, and a similar interdisciplinary approach has been characteristic of all five of the collections in which I have participated as co-editor. My own major contribution to the 1988 volume was an introductory chapter in which I discussed problems in the definition of behavioral processes supporting social learning (Galef 1988) and is probably still the most frequently cited of my publications.

Zentall's 1985 social-learning symposium was not the last. In 1992, Paula Valsecchi, Marisa Mainardi, and I organized a meeting with social learning in animals as one of its major themes at the marvelous conference center in Erice, Italy. The conference led to a book entitled *Behavioral Aspects of Feeding* (Galef *et al.* 1994). In 1994, Celia Heyes and I convened a social-learning conference at Madingley Hall outside Cambridge, UK, attended by 46 investigators, that resulted in publication of *Social Learning in Animals: the Roots of Culture* (Heyes & Galef 1996). Further social-learning conferences that I played no role in organizing followed in London in 1996, Naples in 1998, and St Andrews in 2005. Clearly, social learning has become a coherent field of enquiry.

At the invitation of Shepard Siegel, Celia Heyes and I also edited a special volume of the journal *Learning and Behavior* that introduced the work of a number of newcomers to the field, along with several papers by veterans (Galef & Heyes 2004). Last, but surely not least, Kevin Laland and I edited a book entitled the *Question of Animal Culture* (2008). It is the most focused of the books I have worked on, and I believe, the most successful.

I am proud of the way in which the field of animal social learning has developed over the years, and of whatever role I played in that development. My goal was always to be as inclusive as possible, welcoming researchers regardless of their theoretical perspective or academic discipline, but exclusive in restricting such welcome to work of quality. Although it annoyed me at the time, in retrospect, I feel considerable satisfaction when reading the proceedings of a conference on mammalian social learning that I was expressly invited not to attend because I was "too critical." The organizers were quite right to have been concerned about my presence. I would have given many of the contributors a hard time, while vigorously applauding others. The result, I hope, is an area with scientific respectability and one that I know has attracted the interest of some of the brightest and most creative young people in academia.

Laboratory personnel

In 1986, I had the very good fortune to hire Elaine Whiskin, a recent McMaster honors graduate in Biology, to work as my research assistant. In the 18 years that I had by then been at McMaster, I had employed many laboratory technicians, some part time and some full time. However, after a year or two all had returned to school or had moved on to more lucrative jobs. Elaine was to be with me until the end of my career, always careful, cheerful, and accomplished. An admittedly inadequate indication of the magnitude of her contribution to the work of the laboratory is her co-authorship of 38 published papers describing some of the work in which she participated.

In 1986, Mertice Clark also started publishing her extraordinary series of papers on the effects of intrauterine position on the reproductive life histories of Mongolian gerbils (Clark *et al.* 1986). That work, and related studies of effects on reproductive behaviors of naturally occurring variation in prenatal exposure to hormones (Clark & Galef 1988), was to continue for 20 years (Clark *et al.* 1992, 1997). Throughout that time I was of use to Mertice in helping her to prepare the results of her labors for publication. Her perseverance in the face of innumerable obstacles and considerable disbelief in the possibility of her doing what she set out to do was extraordinary.

The patent

Results of our studies of social learning in rats had indicated that the olfactory information passing from a recently fed demonstrator rat to its observer had two components: first, an olfactory cue reflecting the flavor of the food that a demonstrator rat had eaten (a food-identifying cue), and second, an olfactory cue carried on the demonstrator's breath (a contextual cue) that when experienced by naïve rats together with the diet-identifying cue caused them to develop an instant liking for a food.

I entered into collaboration with Russ Mason and George Preti at the Monell Chemical Senses Center in Philadelphia to identify the chemicals that constituted the contextual component in rat breath. Russ and George carried out some mass spectrometry on rats' breath that suggested that carbon disulfide (CS_2) might be an important part of the contextual cue. I ran the obvious experiment, adding a little CS_2 to a distinctively scented food and presenting the combination to rats that I then tested for their preference for the food experienced together with CS_2. Amazingly, exposure to a food together with CS_2, like exposure to a food together with a demonstrator rat, markedly increased a rat's preference for the food (Galef *et al.* 1988). For the only time in 45 years in research, I experienced an emotional rush in response to data. For a few minutes, I felt that I had accomplished all that could be hoped for in analysis of a social-learning phenomenon. And on the horizon was the possibility of becoming really rich.

The interest of the people at Monell in my laboratory's work on food preferences in rats was in part practical. An observer rat's enhanced liking for its demonstrator's diet could be interpreted as indicating that naïve rats treat any food that they learn that other rats have eaten as safe to ingest. If so, adding CS_2 to poisoned baits might just induce rats to eat the baits without exhibiting their usual extreme caution when first encountering a potential food for the first time. In fact, future work in collaboration with Russ Mason was to show that out in the real world, on pig farms and pheasant-breeding facilities, wild rats' intake of a poison bait could be increased as much as four-fold by the presence of CS_2 (Bean *et al.* 1988).

In 1988, the Monell Chemical Senses Center and I received a patent for use of CS_2 as an additive to poisoned baits, and Monell's lawyers opened negotiations with four chemical corporations for the rights to use the new method in their respective portions of the globe. I shall never forget receiving a phone call from one of the lawyers involved in those negotiations who, after asking me what I was doing (I was writing a paper, of course), suggested that I should, instead, be in the south of France pricing villas. It was very heady stuff. In the end, problems with the shelf life of CS_2 and the Environmental Protection Agency brought the project to an unhappy conclusion, leaving me considerably wiser about the ways of the world and more than US$20,000 poorer.

Three theoretical papers

Together with my students, I continued work on social learning about food by rats, extending our understanding of the robustness of the phenomenon and, with Mathew

Beck, a Ph.D. student, exploring some of its potential functions (Beck & Galef 1989). However, my major contributions in the early 1990s were two theoretical papers, the first concerned with a central issue in regulatory physiology (Galef 1991a) and the second invited by Jeanne Altmann for *Human Nature* (Galef 1992) that eventually led to my engagement in "the culture wars".

Most of the reviews that I have written over the years have been concerned with various aspects of social learning in animals. However, in 1979 I was invited by David Gubernick to contribute to a volume concerned with parental behavior in mammals that he and Peter Klopfer were editing. I submitted a chapter in which I discussed the changing behavior of developing rat pups as reflecting changes in the ecological niche that they occupy as they mature, drawing analogies between various types of parasitism and the infant rat at different stages in its ontogeny (Galef 1981). Although the paper was a great deal of work, I learned that I really enjoyed exploring literatures outside my major field, and have been happy ever since to look into interesting or challenging subjects for reviews.

In the early 1970s, I had attempted to replicate some of Curt Richter's classic studies on cafeteria feeding in rats. Although the data I collected were in some ways similar to those Richter had produced decades earlier, my data were also different in important respects from Richter's, and they led me to an interpretation of the outcome of Richter's experiments quite different from that which he had proposed. I consider, Curt Richter to be a major figure in the history of behavioral research. At the time, Richter was in his late 80s, and had been very helpful to me in precisely replicating his procedures, even forwarding invoices for dietary ingredients that he had ordered in 1933. Consequently, I felt that it inappropriate during Richter's waning years to call into question one of his major contributions. So I waited until Richter had passed away before submitting for publication "A contrarian view of the wisdom of the body as it relates to food selection," a paper I consider my most important contribution to the literature (Galef 1991a). I have always been a bit surprised that it did not attract more attention than it did. I suppose it concerned an issue whose time had passed.

"The question of animal culture", published a year later (Galef 1992), was, I thought, a relatively minor piece of work. However, it received considerable attention, and seems to have earned me the only enmities in the academic world of which I am aware. It might be argued, and it was, that it was inappropriate for me to write a chapter calling into question some of the field work on animal traditions. I had never engaged in field research and never worked with primates, the animals with which the paper was largely concerned. I knew work on traditions in animals only through the literature. Indeed, when I first met Toshisada Nishida, he remarked that I wrote about sweet-potato washing in the Koshima-Island macaques as though I had been there. I have never been quite sure whether he intended the remark as compliment or criticism. However, there is no question about how some others felt about the paper.

Although most seemed to accept the manuscript as it was intended, as a challenge to provide more convincing field evidence of traditions in animals, there were a few who seemed bent on transforming what I felt was a legitimate intellectual question into a personal matter. The only really good thing to have come out of the paper (Galef 1992) is the book with the same title that Kevin Laland and I edited (Laland and Galef 2008).

Work with Burmese jungle fowl

At the same time that I was writing the two reviews, I initiated a series of studies of social learning about foods in Burmese jungle fowl (*Gallus gallus*), the wild progenitors of the domestic chicken. The birds are very different from rats in that their feeding behavior is largely controlled by visual rather than by olfactory cues, and I expected there might be some interesting consequences of that difference for the type of social learning about foods that they would show.

Jerry Hogan, at the University of Toronto, had a colony of Burmese fowl and was good enough to provide me with a few birds to start my own flock. I purchased some commercial incubators, lots of chicken wire and some chicken batteries from a local farm-supply house, and 6 months later, had a full-fledged chicken farm set up on the second floor of the Psychology Building at McMaster. A new graduate student, Laurel McQuoid, and I began to explore visually mediated social influences on foraging in fowl. We had considerable success with the work, using video images to direct chickens' feeding behavior to one type of food or another. After "observer" chickens had seen conspecifics on television feeding from a food bowl marked red and ignoring a food bowl marked blue or vice versa, they preferred to feed from the food bowl they had seen conspecifics exploiting (McQuoid & Galef 1992, 1993).

Another theoretical paper

The years passed rapidly, and unremarkably. My only publication of particular note was the result of an invitation from Meredith West to contribute a review to *Animal Behaviour* (Galef 1995). I made a major mistake when I wrote that paper that resulted in a public disagreement with Kevin Laland, whose work in the field of social learning I greatly admire. Indeed, the kind of sophisticated intellectual tools that Kevin and others of his generation bring to the study of animal social learning is the best indication of the tremendous progress of the field during the last 30 years.

My mistake was that before submitting the manuscript for publication, I failed to send it to either Robert Boyd or Peter Richerson, whose work I both criticized and, unfortunately, misinterpreted. I am moderately mathematically challenged and did not realize that by changing the time scale in Boyd and Richerson's models, my criticism of their work would be invalidated.

As I often tell my students, the only way to be sure you never make a mistake is to not do anything. Still, there is no excuse for not having checked with Boyd or Richerson before putting my foot in my mouth. I knew better. In the end, Laland and I engaged in a published exchange in *Animal Behaviour* in which he made explicit the error in my interpretation of Boyd and Richerson's work, and I pretended that it did not really matter. I imagine we both feel that we got the better of the subsequent argument in print. I surely got a well-deserved comeuppance.

Working for the Animal Behavior Society

Executive Editor of Animal Behaviour

Through the mid-1990s I continued publishing reviews and empirical papers, editing books and contributing to meetings, but nothing much new happened until 1997, when I became American Executive Editor of *Animal Behaviour*, the joint publication of the Association for the Study of Animal Behaviour (ASAB) in the UK and the Animal Behavior Society (ABS) in the US. The appointment was challenging in unexpected ways.

The financial arrangements between ABS and ASAB had resulted in ASAB, the owners of *Animal Behaviour*, and Academic Press, its publisher, each keeping half of the surprisingly large profits resulting from library sales of the journal. ABS bore half the costs of production, but received no financial return from sales. The result of this peculiar arrangement was wealth for ASAB and financial disaster for ABS. Indeed, the financial situation was so bleak at ABS that both the membership and executive committee had voted to start an independent journal, if ASAB refused to allow ABS to share in the profits from *Animal Behaviour*.

Lee Drickamer negotiated a contract with Indiana University Press for a new journal, of which I was to be executive editor. The contract lacked only the signature of Sue Reichert, then President of ABS, to take effect. A tense transatlantic teleconference between the ABS and ASAB executive committees ensued. In the event, ASAB agreed to share future profits from *Animal Behaviour* with ABS, the new ABS journal never saw the light of day, and I became the American Editor of *Animal Behaviour*. Still, there were hard feelings on both sides of the Atlantic. The late Chris Barnard, then European Editor of *Animal Behaviour*, graciously agreed to come to Toronto to try to smooth things out.

I made considerable efforts to facilitate our discussions, hiring a chauffeured car to carry Chris and me from the airport to Chris's suite in the finest hotel in Toronto and arranging, though the good offices of Mart Gross, for an office at the University of Toronto where Chris and I could work in peace. The first few hours of our discussions were not promising. Accusations and recriminations flew. However, the atmosphere improved through the day, and Chris and I eventually agreed to a list of 17 changes that we wanted to make in the journal. These involved everything from the cover, page size, and typeface to the addition of reviews and commentaries. Over the next 3 years, while we were, respectively, European and American executive editors of the journal, we managed to implement 16 of the 17 changes, failing only to add paid advertisements to the journal. Working with Chris was a great pleasure.

Financial Advisor to the ABS

My stepfather and maternal grandfather had both been stockbrokers, and I had been raised reading the financial pages of the *New York Times*. Further, when my father died in 1959, I received a modest insurance payment that I had been investing ever since. Consequently, I felt comfortable shepherding the financial resources of the ABS, certainly more so than a previous financial advisor to the Society, who confided to me an unwillingness to talk with stockbrokers because they were out to steal the Society's money. Although I could not disagree with his premise, it did not seem to me to provide a sufficient rationale for keeping substantial amounts of the Society's assets in savings accounts.

In 1997, I assumed the role of financial advisor to the ABS, and had the pleasure of investing the considerable surpluses that had started to flow from the journal. The society's investments have done well, and ABS now has both sufficient funds to survive almost any financial emergency as well as guidelines for investing that should enable anyone interested to look after the Society's money in a responsible way without spending more than 60 minutes a year doing so.

President of the ABS

The year following the end of my 3-year editorship of *Animal Behaviour*, I ran for election as President of the ABS, expecting to win, and much to my chagrin, lost, ran again the following year, and won. The four years spent in the ABS presidential succession were relatively uneventful. Money was plentiful and ideas were easy to implement. While President, the only major change I tried to make in the Society, other than keeping the annual Executive Committee meeting to a bearable length, was to move the annual meetings of the Society from college campuses to destination resorts. My view is that if you hold meetings in interesting places, senior researchers will want to come. They will both bring their students and attract other senior researchers to do likewise, enabling ABS to compete successfully with other societies for attendees.

Hosting the annual meeting

As one of the stronger proponents of the move to destination resorts as venues for the ABS annual meeting, I felt an obligation to host at least one of the Society's annual meetings, and volunteered to host the 2006 meeting in Snowbird, Utah. Jim Ha had hosted the ABS meeting at Snowbird in 2005, and he had done such a fine job that there was little left for me to do other than line up some plenary speakers and to reduce, in so far as possible, the costs to students of attending the meeting. I was fortunate to recruit two very eminent plenary speakers, Tim Clutton-Brock and Robert Trivers. Tim gave his keynote address on his sixtieth birthday, and after giving his talk, served birthday cake to students waiting in line for their morning coffee. He was the perfect guest.

The last 15 years of research

Mate-choice copying

In the late 1990s administrative chores of one sort or another were becoming increasingly important in my life, though I still had at least one first-rate research program left in me. In 1992, Lee Dugatkin had provided evidence that female guppies copy one another's mate choices. Although the phenomenon Lee described had been repeatedly demonstrated in his laboratory, others were struggling to replicate his findings, suggesting that mate-choice copying in guppies might not be a particularly robust phenomenon. I had heard Elizabeth Regan from Cornell talk at the WABC about her research on sexual behavior in Japanese quail. The birds sounded like ideal animals for studies of mate-choice copying, and quite by luck there was a commercial quail farm just 30 minutes from my lab, where Japanese quail were reared in the thousands for the gourmet market.

A new graduate student, David White, was interested in the project. In short order, he produced convincing evidence of mate-choice copying in quail, and went on to explore the range of conditions under which it occurs (White & Galef 1999, 2000). After David left McMaster to accept a post-doctoral fellowship, Kamini Persaud and Alex Ophir took up the quail project. Kamini worked brilliantly from a feminist perspective to ask questions about the causes of female preferences among males and extended the analysis of social effects on mate choice from affiliation to actual reproductive success (Persaud & Galef 2003, 2005). Alex developed some very clever video techniques that he used to look at the relationship between male aggression and female preference among males (Ophir & Galef 2003a, b).

Taste aversion learning in vampire bats

Another research project from my later years of which I am particularly proud is a study conducted in collaboration with John Ratcliffe and Brock Fenton on poison-avoidance learning in vampire bats. The idea is simple. However, completing the experiment was a 26-year-long nightmare.

It has often been argued that taste-aversion learning is an adaptively specialized cognitive process that allows animals to learn to avoid eating any toxic substances that they happen to encounter while foraging. There is, however, little evidence that directly supports that hypothesis. When I was on Barro Colorado in 1974, it struck me that vampire bats, obligate feeders on mammalian blood, were unlikely to encounter toxic foods, and should therefore have no need for a specialized cognitive mechanism supporting taste-aversion learning. I had tried twice, once in 1974 with Jeff Waage on Barro Colorado Island and again in the 1980s with Julia Chase at Columbia University, to carry out the relevant experiments. However, at the last minute, Waage and I were denied the blood we had been promised to feed the vampire bats we had captured, and we had to release them. Julia went into labor on the way to the airport to pick up a shipment of vampire bats for our work, and they perished before she could get to them. I received a shipment of vampire bats that had frozen to death on their way to the lab.

John Ratcliffe, who had completed his undergraduate honors thesis research in my laboratory and joined Brock Fenton's laboratory at York University, was looking for a research project for his Master's thesis. I mentioned the vampire-bat problem to John in 2000, and he completed the experiment brilliantly, travelling around the Americas to examine taste-aversion learning in four species of bat and finding that only one, the common vampire bat, did not learn flavor aversions (Ratcliffe *et al.* 2003).

Learning socially to love alcohol

I had always operated on a relatively small budget, partly because it was all I needed, and partly because I feel sorry for the average taxpayer some of whose money is being confiscated so that I, and others like me, can do whatever amuses us without much in the way of direct return or accountability to those footing the bill. Still, I am a competitive soul, and as the size of one's grants increasingly became the measure of research accomplishment at McMaster, I succumbed to temptation and sought some big bucks.

Skip Spear at SUNY Binghamton and I applied to the National Institute for Drug Abuse for a grant to use Norway rats as a model system in which to explore the development of appetence for alcohol, and I signed on a new graduate student, Lynne Honey, with an interest in both alcoholism and animal research. The result was a series of papers extending the laboratory's earlier work on development of preference for flavors like cinnamon and cocoa to the development of a preference for alcohol (Honey & Galef 2003).

Testing formal models of social learning. For the past few years my undergraduate students, Elaine Whiskin, and I have been engaged in a series of experiments using social learning of food preferences in rats as an empirical system in which to test predictions from formal models as to the circumstances under which animals should increase their reliance on social information and the characteristics of individuals whose behavior should be copied (see, for example, Galef *et al.* 2008; Galef & Whiskin 2008). Our data suggest that, at least in rats using socially acquired information to select foods, predictions from formal theory as to when copying should occur hold up pretty well, whereas predictions as to who should be copied are next to useless. The last paper I intend to publish will be a review of this work.

Summary of research

Both editors who reviewed an earlier version of this manuscript suggested that the draft they read did not provide sufficient information about the research that I had conducted to permit a naïve reader to understand why I might have been invited to participate in this volume. I had decided not to discuss my research in detail because, over the past 40 years, I have already written a couple of hundred papers in which I did so. All are available as pdfs on my website at www.sociallearning.info. Furthermore, telling you that my work is important simply isn't my style. In my view, you are the one best suited to decide whether my publications are important or run of the mill. Still, the editors know better than I what *they* want, so here is my view of why my autobiography has been solicited for this volume.

First, as indicated in the first paragraph, I was fortunate to be the first both to see social learning as a distinct field of enquiry and to engage in sustained study of social learning phenomena. Why? For two reasons: first, I am a convinced reductionist, and second, I like to watch animals behave. I realized early on that the behavior of individuals allows reductionist analysis only of social interactions. Social learning fit my predispositions perfectly.

Second, I had the good luck to discover two extremely robust instances of social learning: social learning of food preferences of Norway rats and social influences on mate choices of Japanese quail. Work in both of these experimental systems provided convincing evidence that social interactions can play a major role in the development of behavioral repertoires.

Third, my interest in both biological and psychological approaches to the study of behavior allowed me to formulate questions that were of interest to a broad range of investigators. Consequently, I may well have published more negative findings than anyone else in the history of experimental psychology. And although we are not supposed to count publications, we all do.

Fourth, I decided quite early that my job as a scientist was first to convince myself that something was true and only then to try to convince the rest of the scientific community.

Fifth, I was fortunate to receive invitations at critical points in my career to write reviews of the literature that forced me to think critically about what I was doing. I have already mentioned papers that I wrote at the invitation of Jay Rosenblatt, Mike Domjan, David Gubernick, and Merideth West. There were numerous others that, similarly, required thinking conceptually about my areas of interest and my way of doing research (see, for example, Galef 1989, 1991b, 2006; Galef & Beck 1990).

Last, my editorial and administrative activities have made substantial contributions to my visibility. Co-editing the first three major volumes of contributed articles in social learning, and serving on the editorial boards of a dozen journals in areas from feeding behavior to communication, like involvement in the WABC, helped to bring my work to the attention of the academic community. In fact, twice during conversations that took place while I was interviewing for part-time jobs prior to my retirement in 2004 (see below), someone suggested that my major contribution to animal behavior was creating and sustaining the WABC.

The end of the trail

When Lynne Honey left my laboratory in 2003, I was 62 years old and facing mandatory retirement at age 65. Although the government of Ontario was about to put an end to compulsory retirement of its professorate, I was to be one of the last to be required to retire under the old rules. I had looked into the possibility of moving to schools in the United States where there would be no mandatory retirement, and although I had a couple of interviews, there were to be no offers of the kind of part-time position I was seeking.

Being a contrary sort of person, and quite unhappy about being forced to retire, or forced to do anything else, for that matter, I negotiated a contract with McMaster that allowed me to keep all the important things – my laboratory, office, and parking privileges – until my scheduled age of mandatory retirement, and retired. Essentially, I had arranged to spend my last 3 years at McMaster on sabbatical leave.

The Psychology Department and Dean of Science at McMaster were good enough to fund a "Galefest" to celebrate my retirement, Mertice Clark and David Sherry put in an inordinate amount of work to make the occasion the great success that it was, and on my 63rd birthday, colleagues from Paul Rozin, Morris Moscovitch and Jeff Alberts to David White and John Ratcliffe travelled to Hamilton for a day of talks, fellowship, and fine dining at my favorite sushi restaurant. I was officially retired.

Since then, I have spent almost as much time at work as ever, except that now I take Saturday and Sunday afternoons off to pursue hobbies. McMaster has permitted me to keep my laboratory and an office even after my negotiated deal concluded in 2006. However, my position is tenuous, and 40 years in the business feels like about enough. My days as an experimentalist will end by September of 2008.

I have agreed to co-edit a special issue of a journal with Rachel Kendal to introduce the work of the next generation of students of social learning, and I am currently editing a section on social learning for an encyclopedia of animal behavior for Elsevier, but the focus of my life will soon change.

Life outside the laboratory

While at McMaster, I always worked 7-day weeks to compensate for the short work days my chronic kidney condition imposes on me. However, I have had a rich and rewarding life outside the laboratory, enjoying a wonderful second marriage and engaging in activities that will surely occupy an increasing amount of my time in future years.

My early interest in art has never waned, and for many years, I was a habitué of the McMaster Museum of Art, eventually serving on its Board of Directors. I also had the great pleasure of selecting works of art to decorate the five-story building that houses the McMaster Psychology Department. I like to think that we have the only animal quarters in the world with reproductions of great art on its walls.

I have also managed to visit most of the great museums of Europe and Asia during my travels, and I have been fortunate to have been able to travel a great deal. Invitations to visit laboratories abroad have been regular occurrences: Australia twice (first for a month in 1975 as a guest of Justin Lynch at the University of New England, then for a month in 2005 with Chris Evans at Macquarie University), a month with Joseph Terkel at the University of Tel Aviv, five sabbaticals at the University of Colorado at Boulder totaling 3 years, and short working trips to England, Scotland, France, Italy, Germany, Holland, Switzerland, Austria, Portugal, China, Japan, New Zealand, Spain, and Mexico.

Mertice and I have also been travelling for pleasure for more than 20 years now: seven trips to Africa, five to Asia, six to central and South America, a dozen to Mexico and the

Caribbean, at least a dozen more to Utah, Wyoming, Nevada, etc. For most of my life, I have enjoyed fishing whether with dry fly or lure, and have snorkeled everywhere from the San Blas and Belize to the Great Barrier Reef and north shore of Bali. Skiing a week or two a year in the Rockies remains a regular part of my schedule.

Ever since my first sabbatical leave in Panama in 1975, I have been interested in photography, and you are cordially invited to visit the gallery at my web site (www.social-learning.info) to see a few of my photos. When I retired, I took up Tai Chi, which I greatly enjoy. I have also started to make time in my schedule to play chess on the web for an hour or two a week, watch a couple of movies each weekend, and read a book or two a month.

However, the hobbies that I pursue most vigorously involve food. Mertice and I cook seriously most nights, and we have had the pleasure of eating in many of the great and near-great restaurants of the world. For both of us, eating is more a reason for continued existence than a way of staying alive.

Some last thoughts

So that is the upbeat version of my life. No mention of the many sources of frustration and anxiety: hundreds of experiments that produced garbage as data, students who, after accepting offers to work in the lab, for one reason or another, never showed up; having to respond positively to reviewers' comments on manuscripts even when the reviewers didn't have a clue; worrying about placing students finishing their degrees; the constantly increasing cost in money, time, and energy of complying with often meaningless regulatory requirements; ten years of hassles with a University veterinarian one of whose major goals in life seemed to have been closing my lab. I could go on for a couple of paragraphs. Still, on balance, life as an animal behaviorist has been terrific. For at least the first 25 years, I would happily have done the research, teaching, and administration without monetary reward. Both the freedom to spend my time as I saw fit and the opportunity to interact on an almost daily basis with interesting and interested people were beyond valuation.

I am particularly lucky both to have worked with a seemingly endless stream of first-rate students and to have been active in research when the kind of "mom and pop" laboratory that I enjoy overseeing was, at least in Canada, both an accepted *modus operandi* and relatively easy to fund with a single grant at an almost satisfactory level.

There is currently an emphasis, at least in my University, on participation in collaborative mega-grants, large laboratories, publication in a handful of journals with the highest citation indexes, and pursuit of coverage by the popular press. None of these things have been of particular interest to me. Consequently, I wonder how I would have fared in today's academic environment. I see the scientific method as a process of slow accretion of solid information, like the growth of a stalagmite. Not exactly the sort of approach likely to reap accolades today.

My preference has always been for a laboratory with one technician (preferably Elaine Whiskin), one or two graduate students and one or two undergraduates where I could review the data each day as they were collected and could be intimately involved in the design and fine-tuning of experiments, analysis of data and structuring of publications. Even a minimal crew of three could turn out more interesting data then I could hope to write up for publication, and I write pretty quickly.

As I have told my students, probably far too often, I see a life in science as a marathon, not a sprint. Ask simple questions arising from clearly stated hypotheses. Use simple experimental designs, and transparent statistical analyses. One step at a time, experiment after experiment, frequently replicating your main effect, until you understand what you set out to understand and can be quite sure that when others attempt to repeat your procedures they will get the same results that you did. And if not, you will know why not.

So that's it. By the time you read these lines, I shall have closed the door to my laboratory for the last time, and my career in academia will have ended. Any research papers still in the publication pipeline are but a first look at issues that I would have pursued in greater depth, if as all things must, this phase of my life had not come to an end. I shall miss it.

References

Alberts, J. R. & Galef, B. G. Jr. (1971). Acute anosmia in the rat: a behavioral test of a peripherally-induced olfactory deficit. *Physiol. Behav.* **6**: 619–21.

Alberts, J. R. & Galef, B. G. Jr. (1973). Olfactory cues and movement: stimuli mediating intraspecific aggression in the wild Norway rat. *J. Comp. Physiol. Psychol.* **85**: 233–42.

Bean, N. J., Galef, B. G. Jr. & Mason, J. R. (1988). At biologically significant concentrations, carbon disulfide both attracts mice and increases their consumption of bait. *J. Wild. Mgmt* **52**: 502–7.

Beck, M. & Galef, B. G. Jr. (1989). Social influences on the selection of a protein-sufficient diet by Norway rats. *J. Comp. Psychol.* **103**: 132–9.

Clark, M. M., & Galef, B. G. Jr. (1988). Effects of uterine position on rate of sexual development in female Mongolian gerbils. *Physiol. Behav.* **42**: 15–18.

Clark, M. M., Spencer, C. A. & Galef, B. G. Jr. (1986). Reproductive life history correlates of early and late sexual maturation in Mongolian gerbils. *Anim. Behav.* **34**: 551–60.

Clark, M. M., Tucker, L. & Galef, B. G. Jr. (1992). Stud males and dud males: intrauterine position effects on the reproductive success of male gerbils. *Anim. Behav.* **43**: 215–21.

Clark, M. M., De Sousa., D., Vonk, J. & Galef, B. G. Jr. (1997). Parenting and potency: alternate routes to reproductive success in male Mongolian gerbils. *Anim. Behav.* **54**: 635–42.

Day, C. S. D. & Galef, B. G. Jr. (1977). Pup cannibalism: one aspect of maternal behavior in golden hamsters. *J. Comp. Physiol. Psychol.* **91**: 1179–89.

Galef, B. G. Jr. (1970). Aggression and timidity: responses to novelty in feral Norway rats. *J. Comp. Physiol. Psychol.* **71**: 370–81.

Galef, B. G. Jr. (1976). Social transmission of acquired behavior: a discussion of tradition and social learning in vertebrates. *Adv. Stud. Behav.* **6**: 77–100.

Galef, B. G. Jr. (1977). Mechanisms for the social transmission of food preferences from adult to weanling rats. In *Learning Mechanisms in Food Selection*, ed. L. M. Barker, M. Best, & M. Domjan, pp. 123–48. Waco: Baylor University Press.

Galef, B. G. Jr. (1981). The ecology of weaning: parasitism and the achievement of independence by altricial mammals. In *Parental Care*, ed. D. J. Gubernick & P. H. Klopfer, pp. 211–41. New York: Plenum Press.

Galef, B. G. Jr. (1988). Imitation in animals: history, definition and interpretation of data from the psychological laboratory. In *Social Learning: Psychological and Biological Perspectives*, ed. T. R. Zentall & B. G. Galef, Jr., pp. 3–28. Hillsdale, NJ: Lawrence Erlbaum.

Galef, B. G. Jr. (1989). An adaptationist perspective on social learning, social feeding and social foraging in Norway rats. In *Contemporary Issues in Comparative Psychology*, ed. D. A. Dewsbury, pp. 55–79. Sunderland, MA: Sinauer.

Galef, B. G. Jr. (1991a). A contrarian view of the wisdom of the body as it relates to food selection. *Psychol. Rev.* **98**: 218–24.

Galef, B. G. Jr. (1991b). Innovations in the study of social learning in animals: a developmental perspective. In *Methodological and Conceptual Issues in Developmental Psychobiology*, ed. H. N. Shair, G. A. Barr & M. A. Hofer, pp. 114–25. Oxford: Oxford University Press.

Galef, B. G. Jr. (1992). The question of animal culture. *Hum. Nature.* **3**: 157–78.

Galef, B. G. Jr. (1995). Why behaviour patterns that animals learn socially are locally adaptive. *Anim. Behav.* **49**: 1325–34.

Galef, B. G. Jr. (1998). Edward L. Thorndike: revolutionary psychologist, paradigm-bound biologist. *Am. Psychol.* **53**: 1128–34.

Galef, B. G. Jr. (2006). Theoretical and empirical approaches to understanding when animals use socially acquired information and from whom they acquire it. In *Essays in Animal Behaviour: Celebrating 50 years of Animal Behaviour*, ed. J. R. Lucas & L. Simmons, pp. 161–82. San Diego, CA: Academic Press.

Galef, B. G. Jr. & Beck, M. (1990). Diet selection and poison avoidance by mammals individually and in social groups. In *Handbook of Neurobiology*, Vol. 11, ed. E. M. Stricker, pp. 329–49. New York: Plenum Press.

Galef, B. G. Jr. & Clark, M. M. (1971). Social factors in the poison avoidance and feeding behavior of wild and domesticated rat pups. *J. Comp. Physiol. Psychol.* **75**: 341–57.

Galef, B. G. Jr. & Clark, M. M. (1972). Mother's milk and adult presence: two factors determining initial dietary selection by weaning rats. *J. Comp. Physiol. Psychol.* **78**: 220–5.

Galef, B. G. Jr. & Clark, M. M. (1976). Non-nurturent functions of mother-young interaction in the agouti (*Dasyprocta punctata*). *Behav. Biol.* **17**: 255–62.

Galef, B. G. Jr. & Heiber, L. (1976). The role of residual olfactory cues in the determination of feeding site selection and exploration patterns of domestic rats. *J. Comp. Physiol. Psychol.* **90**: 727–39.

Galef, B. G., Jr. & Henderson, P. W. (1972). Mother's milk: a determinant of the feeding preferences of weaning rat pups. *J. Comp. Physiol. Psychol.* **78**: 213–19.

Galef, B. G. Jr. & Heyes, C. M. 2004. Social learning and imitation. *Learning & Behav.* **32** (1): 1–140.

Galef, B. G. Jr. & Sherry, D. F. (1973). Mother's milk: a medium for the transmission of cues reflecting the flavor of mother's diet. *J. Comp. Physiol. Psychol.* **83**: 374–378.

Galef, B. G. Jr. & Whiskin, E. E. (2006). Increased reliance on socially acquired information while foraging in risky situations? *Anim. Behav.* **72**: 1169–76.

Galef, B. G. Jr. & Whiskin, E. E. (2008). Use of social information by sodium- and protein-deficient rats: a test of a prediction (Boyd and Richerson 1988). *Anim. Behav.* **75**: 627–30.

Galef, B. G. Jr. & White, D. J. (1998). Mate-choice copying in Japanese quail, *Coturnix coturnix japonica. Anim. Behav.* **55**: 545–52.

Galef, B. G. Jr. & Wigmore, S. W. (1983). Transfer of information concerning distant foods: a laboratory investigation of the "information-centre" hypothesis. *Anim. Behav.* **31**: 748–58.

Galef, B. G. Jr., Mason, J. R. Preti, G. & Bean, N. J. (1988). Carbon disulfide: a semiochemical mediating socially-induced diet choice in rats. *Physiol. Behav.* **42**: 119–24.

Galef, B. G. Jr., Mainardi, M. & Valsecchi, P. (1994). *Behavioral Aspects of Feeding: Basic and Applied Research in Mammals*. Chur, Switzerland: Harwood Academic.

Galef, B. G. Jr., Dudley, K. E. & Whiskin, E. E. (2008). Social learning of food preferences in 'dissatisfied' and 'uncertain' rats. *Anim. Behav.* **75**: 631–7.

Heyes, C. M. & Galef, B. G. Jr. (1996). *Social Learning and Imitation: the Roots of Culture*. New York: Academic Press.

Hinde, R. (1970). *Animal Behavior: a Synthesis of Ethology and Comparative Psychology*. New York: McGraw-Hill.

Honey, P. L. & Galef, B. G. Jr. (2003). Ethanol consumption by rat dams during gestation, lactation and weaning enhances voluntary ethanol consumption by their adolescent young. *Devel. Psychobiol.* **42**: 252–60.

Laland, K. N. & Galef, B. G. (2008). *The Question of Animal Culture*. Cambridge, MA: Harvard University Press.

Lorenz, K. (1966). *On Aggression*. London: Methuen.

Maier, N. R. F. & Schneirla, T. C. (1935). *Principles of Animal Psychology*. New York: McGraw-Hill.

Marler, P. & Hamilton, W. J. III. (1966). *Mechanisms of Animal Behavior*. New York: Wiley.

McGill, T. E. (1965). *Readings in Animal Behavior*. New York: Holt Rinehart and Winston.

McQuoid, L. M. & Galef, B. G. Jr. (1992). Social influences on feeding site selection by Burmese fowl (*Gallus gallus*). *J. Comp. Psychol.* **106**: 137–41.

McQuoid, L. M. & Galef, B. G., Jr. (1993). Social stimuli influencing feeding behaviors of Burmese fowl: a video analysis. *Anim. Behav.* **46**: 13–22.

Ophir, A. G. & Galef, B. G. Jr. (2003a). Female Japanese quail that eavesdrop on fighting males prefer losers to winners. *Anim. Behav.* **66**: 399–407.

Ophir, A. G. & Galef, B. G. Jr. (2003b). Female Japanese quail affiliate with live males they have seen mate on video. *Anim. Behav.* **66**: 369–75.

Persaud, K. N. & Galef, B. G. Jr. (2003). Female Japanese quail aggregate to avoid sexual harassment by conspecific males: a case of conspecific cueing. *Anim. Behav.* **65**: 89–94.

Persaud, K. N. & Galef, B. G. Jr. (2005). Female Japanese quail mated with males that harassed them are unlikely to lay fertilized eggs. *J. Comp. Psychol.* **119**: 440–6.

Rasmussen, E. W. (1938). Wildness in rats: heredity or environment? *Acta Psychol.* **4**: 295–304.

Ratcliffe, J. M., Fenton, M. B. & Galef, B. G. Jr. (2003). An exception to the rule: vampire bats do not learn flavour-aversions. *Anim. Behav.* **65**: 385–9.

Steiniger, F. von (1955). Beitrage zur Sociologie und sonstigen Biologie der Wanderratte. *Z. Tierpsychol.* **7**: 356–79.

Tinbergen, N. (1958). *Curious Naturalists*. London: Country Life.

Tinbergen, N. (1963). *The Herring Gull's World: a Study of Social Behaviour of Birds*. London: Collins.

White, D. J. & Galef, B. G., Jr. (1999). Affilitative preferences are stable and predict mate choices in both sexes of Japanese quail, *Coturnix japonica*. *Anim. Behav.* **58**: 865–71.

White, D. J. & Galef, B. G. Jr. (2000). Differences between the sexes in direction and
    duration of response to seeing a potential sex partner mate with another. *Anim. Behav.*
    59: 1235–40.
Williams, G. C. (1966). *Adaptation and Natural Selection*. Princeton, NJ: Princeton
    University Press.
Zentall, T. R. & Galef, B. G. Jr. (1988). *Social Learning: Psychological and Biological
    Perspectives*. Hillsdale, NJ: Erlbaum.

12

Watcher: the development of an evolutionary biologist

PATRICIA ADAIR GOWATY

I took it that a *scientific* autobiography is meant to be about "what I was thinking about then". I wrote about what is neither too aversive for me to think about nor too personal, and I am aware that much is not here.

First observations

A hazy print captured from an old 16 mm home movie shows me as a happy, strong toddler on my grandmother Julia Tomkiewicz Gowaty's farm in McKeesport, Pennsylvania. I am carrying a wriggling Muscovy duck. My first word was "Duck".

Leaders in Animal Behavior: The Second Generation, ed. L. C. Drickamer & D. A. Dewsbury. Published by Cambridge University Press. © Cambridge University Press 2010.

Watching tadpoles. I was five. They were in an ephemeral pond in the woods near our house in East Gadsden, Alabama. Squatting, I watched and watched. When I told my mother about them, she asked if I wanted to bring some home and she helped me make tiny aquaria of dozens of pimento jars. Tadpoles in jars sat on my bookshelves, but they also entered my dreams. As I watched, the tadpoles changed; one night I dreamt that frogs were everywhere, and indeed, in the morning, to my delight frogs were everywhere, on the floor, the bookshelves, under my bed. Mother never mentioned that everyone already knew that tadpoles were changelings. Thus, when I was small, my mother's wisdom and generosity fixed something in me, for then, as now, I am never happier than when I am watching animals.

When I was six, my job was to make sure the rabbits the hospital used for pregnancy tests always had food and water, and I also had to keep their hutches clean. I watched the mothers make nests from their fur. I watched their babies nurse. I noticed that rabbits were silent. When one of the white rabbits I tried to dress in doll clothes escaped my arms into the garage, a neighbor's cat chased her into a corner. A shovel fell over, startling the already terrified animal, who let out a high pitched yelp. Family stories claim that I later announced, "Today I discovered that rabbits do talk sometimes".

My father trained bloodhounds to track people. He would send me to the woods to hide. I would wait sometimes almost all day, oblivious to the time because I was in the woods, until the dogs lucked out (the woods were small) and came my way. My favorite dog, Brownie, was a mutt, a "swamp dog", a volunteer recruit to our pack. Watching him copulate was an extraordinary thing. It stimulated many questions, all answered by my urologist father and matter-of-fact former-nurse mother.

When a calico cat followed me home one winter day from piano lessons, I begged unsuccessfully that she be let in. Nevertheless, that cat did give birth in our basement; and I was able to watch her move her kittens to a snug corner and nurse them. Later I watched her in heat solicit toms. And, since my parents enjoyed my questions, I learned to continue to ask them.

A sense of place, if nothing else

Before school, I spent most of my free time outdoors; sometimes with my siblings, Cissie (two years younger) and Frankie (seven years younger), more often alone. I built forts in wisteria tangle-vines, played with dolls and toy guns in my Dale Evans holster, but most often I was watching other animals. An active child (Figure 12.1) I spent whole summers in my father's large vegetable gardens and his peach orchards watching black snakes and racers moving through the squash vines; once I saw an indigo snake, more often corn snakes climbing pine trees. I was always hoping to see a copperhead or a rattlesnake. As a teenager, I spent countless hours in a boat on the Coosa River, at a bend known locally as "Buzzard's Roost". The Coosa turns because it hits a sandstone ridge. My father owned land on the west side of the deep bend, where Native Americans had placed fishing weirs and where, 200 years after they left, I collected scores of their arrow heads. On the Coosa, I watched pileated woodpeckers, kingfishers, snipe, great blue herons, mocking birds, and mud turtles. I often

Figure 12.1. Aged six, outdoors playing. Photo by H. J. Gowaty, deceased.

thought about the fish that we sometimes caught, seldom saw, but knew were there through the stories told by local naturalists. I knew they were true, because my father once brought home a still living giant (as long as our bathtub in which he put it), a spoonbill sturgeon, caught from the Coosa or maybe from the more-distant Alabama River. That fish in my bathtub is what I think of whenever I am struck dumb by some bit of wild nature. No one could imagine such a creature, yet it lived in our local, still un-dammed rivers.

In contrast to the relative freedom of my pre-school years, school was prison. I didn't feel quite alive except when school was out. My father was Catholic, as were 3% of the rest of the population of Alabama. So my school from first through eighth grade was St. James Catholic School. There were 11 other students in my class, but some years our teachers, nuns of the Missionary Servants of the Most Blessed Trinity (an order founded in Alabama in 1918 whose women never covered their hair and often wore regular street clothes) taught three grades in a single classroom. I knew how to read before St. James, but I learned to write and some arithmetic. I yearned to be outside. Indoors, my "very best friend" Kathy Hoffman and I read voraciously, mostly teaching each other things that interested us, again abetted by my mother, who never denied me any book I asked for, and Kathy's mother Mary Hoffman, who was always ready with advice for us about what to read next. At St. James Catholic School what I was taught was catechism. Even as a child the inconsistencies bothered me. I asked often why there were two Catholic Churches in Gadsden, one for black people and one for whites. *Two* Catholic Churches with a combined membership under 900 people in a town of 30,000 awakened my civil-rights activism – while I was in first grade – which turned into a major preoccupation during high school.

The first week of college I abandoned Catholicism, and sometime in graduate school started to identify myself when asked, as "an agnostic with strong atheist tendencies". Now, I am an atheist, unmodified. In grade school the catechism provided an answer to questions about why life is as it is. Trouble was, it didn't make sense and satisfied me only under the vigilance of the Sisters and the threat of hell. My transition to atheist was completed during my first week of college by an event that marked me like a sacrament. During Confession at Loyola University's The Holy Name of Jesus Church on St. Charles Avenue, New Orleans, I asked the priest, "How does the church explain the origin of evil?" He did not answer me, but gave me a penance of "Our Fathers and Hail Marys". That did it! I was done. During my earliest grade school years religious training took up so much time that it restricted the time I had to pursue *my* questions; and, worse, some of the nuns, like that Jesuit priest at Loyola, stifled my earliest and smartest questions when I did ask them. The focus on religious training retarded the natural expression of my earliest passions to observe life around me. Thus, I have come to resent the poverty of my early formal education. I have worried if that forced rigidity killed something in me, just as I wonder now, if rote memorization of anything inhibits creative, adaptively flexible minds.

Though Gadsden High School was not much of an intellectual improvement, the friendships I forged there have remained steadfast. My most memorable lesson came from Miss Heath, the biology teacher, when she asked us to study "nature/nuture". Debating that question gave me my first inkling that there was an earthly answer to my question, "Why is life as it is?"

And, there were Saturdays, when I was by myself in the woods or on the river climbing over rocks and under Nocollula Falls, a time when I thought granite outcrops were as common as dirt. I've never been much of a ticker, but I started bird watching when a friend, Susan Wilson, decided to be a bird painter. With her, I started hunting birds – not killing them, but stalking them–to get serious looks at the ones that were harder to see. To bird watch on Susan's farm, we would drive NE out of Gadsden north to Turkeytown Gap Road. From there, we climbed the east and south escarpment of a small plateau that is part of the Ridge and Valley province of the Appalachians, where we once snuck up on a whip-poor-will resting along the branch of a pine tree deep in woods. We watched it for a long time: it sat still, so still that it tricked our eyes momentarily into thinking it was not there. To this day, that was my best and longest look at a whip-poor-will.

With other friends, I became a spelunker, who shimmied into caves that littered the hills we wandered. The ridges give northeast Alabama a northeast–southwest bent; and the shallower ridges shaped some of the Coosa's windings. Gadsden sits in the valley of the Coosa River alongside the most southern extension of Lookout Mountain, part of the Cumberland Plateau. Our landscape was stunningly beautiful, diverse, and interesting, something I understood only as I saw more of the world. Whether beautiful or not, where I grew up marked me forever – I am imprinted, which in the hierarchy of "soul marks" is greater than sacrament: nowhere else on Earth is as satisfying. Nowhere else have I had as strong a sense of place.

Over a hundred years after Darwin and Wallace

My alienation with H. Sophie Newcomb College of Tulane University, less than a mile from the Mississippi River, began the first week I was there. The day I moved into the Josephine Louise dorm, I signed a petition for the integration of the downtown New Orleans movie theatres. The next weekend someone burned a cross on my dorm room door and filled my room – floor to ceiling – with toilet paper. Socially, things started badly and did not get better for four years.

I thought I would be a Biology major; my friend from Gadsden, John Bass, who also went to Tulane and also had a strong interest in Biology, wrote to me when he started classes, "College is harder than high school. You might need to study". In my first college biology course I memorized everything and forgot most right after the test. Despite Professor E. Peter Volpe's charisma, which was considerable, there was sometimes a tedium to those classes in which DNA molecules, organelles, and cells were disembodied from the organisms they were in. My first biology course was mostly about *how* life works, and occasionally it was fascinating, but what excited me then and now is the question, *why is life as it is?*

And so, one class on a single day was life-altering. That was the day that an Assistant Professor, William Franklin Brandom, PhD, using colored chalk, explained Darwin's idea about evolution through natural selection. I can remember most of the detail of that lecture; I even remember the gray herringbone dress I was wearing, and how the light filtered in the open window on that hot, fall day on the third floor of the then un-air-conditioned Newcomb Hall. I had no inkling that day how central to my future thoughts Darwin's idea would be. Darwin's idea is like arithmetic (Dennett 1996). As two and two are four, if individuals vary in traits, and some individuals survive or reproduce better in the environments they inhabit because of the trait variants they carry, that is natural selection. Natural selection is differential survival and/or reproduction of individuals because of trait variants favoring some individuals over others in given environments. Couldn't be simpler. I was enthralled.

And, another question dawned: Why did my education not expose me to natural selection until I was eighteen years old? Worse, my later Biology courses didn't teach me more about natural selection. Rather, my job was to memorize the accumulated bounty of facts, details that others had discovered. No one expected me to know the organisms that lived all around me. No one even expected me to know who were the organisms from which those cells came. No one taught me the value of systematic observations. No one asked me to learn to think like a Darwinian. No one expected me to do an experiment. No one expected me to become a scientist.

Irritatingly, the powers at Newcomb College, including many of the students, focused on what I would wear to the dining hall (a dress was required), and whether I arrived as a freshman with the seven pair of white gloves Newcomb coeds were said to need (I did, and as a monument to the ache and pain of the trivialities of social conditioning, I still have them). School rules did require me to memorize *exactly* when I was to be in my room, especially during Mardi Gras.

I did learn how to eat an artichoke, to second line, and to appreciate the Blues. So, my Sophie Newcomb education was not quite "no education", but not exactly a set-up for academic success either.

Being alone and liking it

The day I graduated from Sophie Newcomb, Mother cashed in an insurance policy and gave me US$1700. Two weeks later I left for Europe, the first significant travel of my life. For eight months I visited museums, a few zoos, and a lot of old buildings, while living in student hostels. I was independent, and so, by myself, I visited Israel and Turkey, Yugoslavia, Greece, Italy, Portugal, Belgium, Sweden, Norway, Denmark, and Germany, occasionally seeing friends from home. The day my friend Kirk Follo and I accidentally met in Athens, Greece, he introduced me to yogurt. Still incredibly naïve, I delighted at new food and wondered at big cities. I learned about fear the first time on a bus from Tel Aviv to Bar Sheba in the winter after the Six Day War, when the army preceded us, mine-sweeping our road. I observed with pride that "girls" (as I would have said then) my age were in the Israeli army and carried guns just like the boys did. After a while, cities no longer fascinated; I wanted to be in the woods. In Italy I happened upon Jay Rosenblat, a former US consular officer in China. For me Jay at eighty was an ancient teacher. With him, for the first time, I read Marx and Engels. When I asked why political life was so contentious, Jay responded, "No one understands human nature". *That* was interesting. Since it was so novel for any adult but my mother to expect me to do anything, his suggestion that I study human nature was interesting too.

When I returned to the US, armed with a few months' travel that taught me more than my BA in Biology, I realized I was prepared to do… nothing. But, without much effort and with absolutely no appropriate training, I was hired to teach fifth grade in Marrero Junior High School in Jefferson Parish, Louisiana, a ferry-ride across the Mississippi River from New Orleans. I was flying blind, trying to teach the thirty-nine (!) students in my classroom. Even then I was interested in how our symbols worked to manipulate, deceive, and sometimes guide, so my classroom had peace signs. I also tried to help them learn to read. After six months, I left New Orleans for New York, where, because I asked, I was given a "provisional teaching license", a piece of paper that made me eligible to teach in the New York City Public Schools, but which did nothing to increase what I knew about teaching. I taught fourth grade in the Two Bridges District of lower Manhattan. There were fewer students, but more ethnic diversity and conflict in the classroom: I was challenged. I was in a district funded by the Ford Foundation to increase participation of parents in the public school system. One of my assignments was to visit each of my students' homes once a semester. I loved that part: everyone was bi-lingual (except me), and most were recent immigrants with stories of remarkable achievement. I desperately wanted to be an adequate teacher for these children, but with one emotionally difficult child, the classroom dynamics were beyond my training. I lasted until summer. For reasons I never understood, the administration wanted me to continue my contract the following year. But, I gave away all the accouterments of primary school teaching, and resigned.

Because I was living in the very interesting lower east side of Manhattan, I embraced a life of financial struggle, "drugs and rock and roll", when two-sided fortune struck.

Feminist consciousness induced

That summer of 1969 I was invited into a "consciousness-raising group". All were women, all southerners, all reasonably well-educated (at least we were all college graduates), who got together weekly to talk about what it meant to be women in late twentieth century "patriarchy". We discussed the disappointments of the civil rights movement, in which human rights often had meant men's rights, especially in the politics of those movements where women were seldom the designated leaders. We talked about birth control methods, how to avoid STDs, sexual assault, and rape; what we might do if lovers or employers sexually threatened or hurt us. Our summer-long conversation helped me know my mother, not in the radically particular way I had known her up until then, but also as a person who lived her life, thriving despite the near-universal social limitations of her day. It dawned on me that much of what I had resisted as a teenager and young adult were the narrow expectations of others, which I had interpreted as being about *me*. That summer I realized that those narrow expectations might have had less to do with *my* capabilities, *my* talents, *my* creativity and *my* energy, and more to do with my *gender*.

Escape

Late that same summer, no one was more surprised than I was when I got a job with the New York Zoological Society (NYZS) in the Education Department of the Bronx Zoo! In the first three months, I learned more than in three years in college. Herb Knobloch, the Curator of Education at NYZS, had been during WWII a para-medic working with my father in Puerto Rico teaching "jungle survival" to US soldiers. My father had introduced me to Herb when I was in high school. When Herb heard I was in NYC, he called and asked if I would apply to work in his department. This was the first bit of "friendship bias" I consciously experienced and which later fascinated me (Gowaty 2007). So began my years on subways from the lower east side of Manhattan to the burned-out rubble of the south Bronx. I suppose that my walk from the subway station at the south end of the Zoo to my office on the north end taught me as much as working at the Zoo, but the non-human animals, all in cages, interested me more than the humans in socially constructed cages. I didn't think much about poverty then, even through I lived in a poor neighborhood and worked in a sea of privilege surrounded by the poorest in NYC.

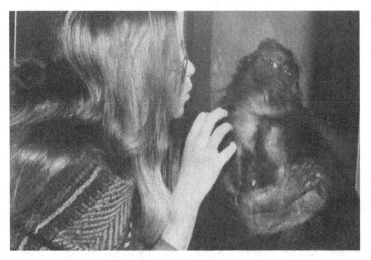

Figure 12.2. At the Bronx Zoo with Sacki, a Moor Macaque.

My lifestyle was different from that of my parents, but I never thought of myself as anything but well off. Despite not having much money, and living alone in what we called a "slum" on E. 10th Street on the Lower East Side of Manhattan, in an apartment that often was without heat, I was doing what I wanted to do. By not buying lunches for a year, I managed to buy a Nikon F and a lens, which I still use. By not buying new clothes, I was able to take three international trips. My parents would have helped me had I asked, but I wanted to be independent.

I was paid US$6,000 and my job was fabulous. Preparing lectures and animal demonstrations was easy. Of course, there was no chance of failure because of the expertise all around me, not to mention "the real thing". Having a snake in one's hands rivets audience attention. None of the students ever misbehaved. I felt really alive, and I was having fun; and so were the students. I asked students to design a better zoo exhibit, the first active learning project I thought of. The students loved it, and I was learning (Figure 12.2) more rapidly than since my mother suggested I bring tadpoles home. As I look back on teaching at the Zoo, each lesson was about "natural selection thinking", each was about adaptation, in which I talked about environmental selective pressures that favored the traits that made these creatures special – opossums, kestrels (sparrow hawks, then), hedgehogs, black snakes, corn snakes, indigoes.

Once I asked Joe Davis, one of the mammal curators, a question about mandrills that he couldn't answer. He suggested I watch them to find out for myself! So, I began to watch the mandrills, something I continued for two years. I did not yet know how to make systematic observations, and I did not know enough to worry about that then. What was important to me was that I seemed to be learning more about these individuals than I knew before. I saw things others had not yet reported and assumed it was because the behavior was induced by captivity and not necessarily a part of their wild repertoire. I do not believe their marking behavior was yet in the literature, as the papers I did find were about their visual signals. Joe told me he thought that the development of facial color signals and variation in fur color of young male mandrills was inhibited by interactions with alpha males. This was 1969, long before there were compelling, widespread discussions of flexible phenotypes. I still don't know whether that interesting idea, social inhibition of development, has been investigated in mandrills. I noticed that when I was watching mandrills, I was happy in a way I was when I was very small. And, looking back, at those happy years I spent at NYZS, I was happiest when watching mandrills.

I was part of the team that trained the first docent class at the NYZS. Unlike me, these volunteers were accomplished, well-educated, rich, most a bit older than me, but like me, they were interested in the lives of animals. Martin Flauteau, who was in the first class of NYZS docents, was older than all of us and a man, a 70+ year old Holocaust survivor, a designer of camera lenses in Germany for Leitz and in New York for Kodak. He knew more than any of us about bears and almost everything else. He was a naturalist, whose passion was photography. I spent Saturdays with Martin. He believed I could do anything, and, as he criticized and praised my photographs, he encouraged me to watch ever more closely as he taught me photography, touching my psyche as deeply as any earlier or later teacher.

The mission of the NYZS was education, research and conservation. The greatest living students of wildlife were associated with the zoo. George Schaller remains for me the quintessential student of animal behavior. Tom Strusaker asked me to learn about the exploitation of tropical forests, and his studies of east African primates were beyond exotic and interesting. The Institute for Research in Animal Behavior, a collaborative effort between NYZS and The Rockefeller University, was at the Zoo, on the east side of the Bronx River, so Peter Marler, Don Griffin, and their students were often around. Steve Green once gave me a half-day tour of the primate collection, telling me that the things I was reading in the old texts might not be "just as they say". He speculated that the alpha males might not really father all of the offspring, and I remember thinking, "*That* is interesting". Haven Wiley was collecting data on common grackles *Quiscalus quiscula* living in the forests of the Bronx Zoo; and the few times I met Carl Hopkins at seminars or dinner parties, his quiet, electric intensity about his studies of animal communication was riveting. Don Griffin and Jocelyn Crane both made enormous impressions on me, particularly Jocelyn. Konrad Lorenz, whom I met once briefly at the Zoo, wrote about watchers, and about Jocelyn as "the watcher of fiddlers". Lorenz' words resonated with me, since as a child I knew *I was watcher*, though not yet in Lorenz' category "the watcher of...". Jocelyn's kind personality, her attentive interest in the lives of women, and the connection I felt for another watcher inspired me. I thought, it might be possible for *me* to study animal behavior. I decided I would go to graduate school.

To get to graduate school, I had to take chemistry, physics, and calculus, courses "girls didn't need" and from which I had been tracked out since high school.

So, carrying my organic chemistry text, all five pounds of it, I went to Brazil to visit my good friend from college, Roy Fleming, who worked for the US Consular Service in Brasilia. On the way, I went to the Amazon, where I saw very few vertebrates (black vultures, turkey vultures, river porpoises, herons, piranha), but many nameless insects. What really impressed – in Manaus, Belem, Brazilia, and Rio – was that there were very poor people, some of whom lived in cardboard boxes on the sides of muddy hills. That trip's most conscious lesson was about profound poverty: "poverty" that was not in Etowah County, Alabama, where everyone had a plot of land on which to grow okra and tomatoes; "poverty" that was not in New Orleans, where the poorest neighborhoods spawned the most exciting food and music; "poverty" that was not in the Bronx, where everyone, or almost everyone, lived indoors. Since Brazil, I've never quit thinking about poverty. Poverty is gripping wherever it is, but relative nonetheless. I made a "B" in organic chemistry.

One day at lunch Bill Conway, the Director of NYZS, asked me what my favorite book about Africa was, and I said *The Long African Day* by Norman Myers. Bill then and there introduced me to Myers, who was his guest for lunch, which was very nice, because Kathy Ralls, my new boss and Curator of Education, and I were then planning our first trip to East Africa. Bill and others introduced us to their friends and associates in East Africa, making parts of that spectacular trip truly easy. Since I had traveled to Brazil the year before, and gone by boat from Manaus east on the river to Itacuaciara, and visited the great natural history collections in Rio, I knew the tiniest bit about the New World tropics. Getting ready to go to East Africa to the Old World tropics the first time, I finally thought my real life had begun; I was doing what I wanted to do and seeing some of the world that mattered. I had just turned 26 when Kathy and I visited a tourist Lodge in Masai Mara to have a cold Tusker, when she asked me the scientific name for the vervets peopling the porch. Not having a clue, I joked, "black-faced grey vervetus". A deep voice resounded, "*Cercopithecus aethiops*". Irv Devore introduced himself and then Bob Trivers. I told Bob that as part of my job at NYZS I taught NYC teachers to teach *Man: A Course of Study*, a fifth grade anthropology curriculum that used life history variation of salmon, herring gulls, common chimpanzees, and Netsilik Eskimoes to illustrate how environmental selection pressures shaped traits, particularly social behavior traits. He told me he had written it.

"It's the questions!"

I entered graduate school in 1974 in the Department of Zoology at the University of Georgia, which I picked as an appropriate place because many UGA students did ecological field studies. Crucially, UGA was in the south, close to Alabama, to my family, and that Alabama landscape on which I am imprinted. The other students in my cohort thought I wouldn't last. I had re-entered a prison, where others controlled, at least for a time, what I was to think about. I complained bitterly about some courses and some teachers. Yet, as I look back, there were many exceptions, people, and experiences that made graduate school for me one of the most exciting and satisfying times of my life.

John Avise's graduate evolution course introduced me to strong inference (1964). One can get to answers with fewer missteps using strong inference. Avise emphasized crucial predictions that if tested well could simultaneously reject one hypothesis and support an alternative. I learned from him, without at first knowing the words, to beware confirmatory science. I wrote a term paper on the evolution of mating systems, which became my first published scientific paper (Gowaty 1981a). In it I used the logic of Orians (1969) to predict fitness variation for the "least advantaged individuals" in other types of mating systems besides polygyny. The novel part of my argument was to extend Orians's resource axis to the left. When resources were difficult to come by, my extension predicted social polyandry and helping, which I expected to be flexibly expressed when environments changed. My interests in mating systems and the adaptive significance of individual flexibility were sealed.

Wyatt Anderson asked me to read Fisher on sex ratios. He also ran a seminar course on population genetics, which I took every term while at UGA. Often at Wyatt and Margaret Anderson's house, we met with seminar speakers. I asked an unbearable number of "dumb" questions. One evening Wyatt assured me that my questions were not "dumb", or at least not as dumb as I thought they were. He said my questions needed to be aired because other students needed to know the answers, but he said they were probably too shy to allow others to know they didn't know the answers. Nevertheless, I remember being dazzled by fellow graduate students, including John Keeley, Sterling Keeley, Mike Douglas, Rich Michod, and Justin Schmit. Thus, Wyatt, ever gentle, reinforced my questioning so that the seminar course facilitated my continuing growth. It was there that I met John Maynard Smith. I described an alternative hypothesis to his for the evolution of polyandry in Tasmanian Native Hens, and his response was to ask me to join him as a postdoc. "But, I just got to graduate school." Later, Rich gave me a note John had written to him saying, "Gowaty's creativity distinguished her among the graduate students at UGA". I was puzzled, "Me, creative?" but somehow that little note made my struggles with graduate school easier. It remained for many years stuck to a succession of my refrigerators in Georgia and South Carolina.

Irwin Bernstein took me to the Yerkes Field Station in Lawrenceville, Georgia. Given that he wanted me to watch macaques, he probably should not have shown me the geladas, *Theropithecus gelada*. I was struck dumb by their vocalizations. One can swoon from pleasure in the sigh of a gelada. I watched for five months straight, almost not leaving my high perch over their outdoor cage. Irwin also, around the same time, gave me Altmann (1974). I read it and, then, reread it. In a profound way it represented for me something that I had probably wished for since I was a child. Like Platt (1964), it held real, practicable information about how to know the nature of things. Years later, I learned from Steve Hubbell, whose entomologist father encouraged the young butterfly collector to "collect series, Son" and whose mother was a professional statistician, that he first knew about systematic sampling as an eight year old. When I first knew, I was 27. Had I, like Steve, known earlier the simplicity of systematic sampling as a path to knowledge, would I have known more sooner? Would my college education have been more valuable? Would I have pursued the study of animal social behavior sooner and in more efficient ways? Would I have been a happier teenager and young adult? I am convinced that we should teach very small children how to make systematic observations, perhaps starting as my mother did, when she gave me all those pimento jars for tadpoles, but then continuing and reinforcing the habit by example and encouragement.

Geladas were exotic enough, as elephants are big enough (*sensu* Romain Gary in *The Roots of Heaven*). I anticipated great adventures studying geladas in the wild. So, I thought that my dissertation project would be a study of social behavior of geladas in Ethiopia. As I considered wars and famine there, as I thought about the difficulties of doing observations, much less experiments, on big, complex primates and the financial hurdles to studying geladas in the wild, most often I was sitting on a high observation platform overlooking the outdoor gelada cage at Yerkes. From there, I also often saw and heard eastern bluebirds, *Sialia sialis*. It occurred to me that I could ask my questions using organisms in my back yard. I started to think about birds of North America. Geladas showed me that my passion was not in exotic landscapes or in charismatic monkeys, but in my *questions*.

Irwin also took me to my first Animal Behavior Society (ABS) meeting. I couldn't believe how much fun I had or that there were so many people who shared my interest in under-standing the lives of animals, or that so many were as interested in my question of why life is as it is. Intellectually, I had found my niche. Since my first time in 1975, I've missed only a handful of ABS meetings, and when I think back on those many meetings during which so much was so new, I realize that ABS kept me going in the long years after my PhD and before my first tenure track job.

Carl Helms was wonderfully hands-off. Giving me advice when asked, he was the perfect major advisor for me. He participated in my quest to find the "right species" to use as a model for studying "ecological correlates of mating system variability" by taking me to Sapelo Island to look at clapper rails, and he revised my early education about birds, and introduced me to ecological physiology. Carl was present, always available, and supportive, making sure that I had RA or TA support during every semester I was in graduate school. When he decided to leave UGA to go to Clemson University to join his wife, Dori Helms, now Clemson's Provost, I went to South Carolina too, one of the best decisions I ever made.

Clemson's Zoology Department did not have the depth of faculty expertise that UGA did, but it did have Sid Gauthreaux, a passionate naturalist of all things avian, and then and now the world's expert on descriptive study of neotropical migration using radar ornithology. Sid was as excited then about sociobiology as he had been about birds as a teenager. He created class after class of fertile and challenging discussion of the new evolutionary biology. He stimulated a successful cohort of students, including Frank Moore, who was a major part of the critical mass that made my remaining graduate education so heady.

Clemson also had farms and forests, the perfect places to study bluebirds. Having learned from Verner and Willson (1969) that, unlike any other North American passerine, eastern bluebirds were sometimes monogamous, sometimes polygynous, sometimes polyandrous, and sometimes had helpers, I realized they were as I have said so often, "the species of choice for a study of ecological correlates of mating system variation". At the time, I knew they were beautiful, but the ease of studying cavity nesters had not yet consciously occurred to me, and besides, the buzz was that eastern bluebirds were on the brink of extinction. Thinking bluebirds were rare, I set up study sites on Clemson farms and forest gaps as well as on my father's farm in Alabama. Because sampling in two places was impossible without help, my mother volunteered to do nest censuses in the Alabama pastures on which I had laid grids before randomly assigning nest boxes to them.

The bluebirds of happiness

As I remember it, the first year in the field with eastern bluebirds was the happiest of my life. The intimacy of getting to know wild-living individuals is like no other, uniquely pleasure-producing and almost indescribable (Figures 12.3 and 12.4). During my first field season, I caught and color-banded every adult bluebird that recruited to the hundreds of nesting boxes I had put up and every single nestling they produced. Carroll Belser, then a graduate student in Forestry at Clemson who became a friend, was watching bluebirds to describe time-budgets, and with her as inspiration, I described an ethogram, too, though mine was far less extensive than hers. I censused nests, watched individuals, and recorded behavior to describe local breeding phenology. Under the influence of Altmann (1974), I did focal nest watches, focal female watches, and focal male watches during every stage of all three nesting periods. Sometimes I just watched for hours and hours and hours. I developed sampling protocols that have served my students, post-docs, and me to this day. Half way through the season, which in Clemson lasts from early March until the end of August, I knew bluebirds weren't threatened: there were just too many of them recruiting to my newly established study sites. I abandoned formal observations at my Alabama study sites, though my mother continued to census nests there for more than ten years. And, despite my pleasure in them, I considered abandoning eastern bluebirds altogether after my thesis, because I discovered that, far from the variability I had expected from the promises in the literature,

Figure 12.3. A female eastern bluebird on her nest. I took this photograph in Clemson, SC.

Patricia Adair Gowaty

Figure 12.4. A clutch of eastern bluebird eggs hatching. I took this photograph in Clemson, SC.

they were milk-toast monogamous, offering me very little opportunity to study why some were monogamous, some polygynous, some polyandrous, and some had helpers-at-the nest.

Disappointed in their apparent lack of mating system variation, I was nevertheless exhilarated by all I was seeing. Almost no one else was studying socially monogamous species from the perspective of social behavior evolution. A common logic then was that because social monogamy was the norm in birds, we should study social polygyny to learn efficiently from the exceptions, so I considered changing to another species. But, as far as I knew then, no other passerine in North America had more variable breeding behavior than eastern bluebirds; relatively few socially monogamous species are as sexually dichromatic as bluebirds; and for very few bird species can one tell the sexes of nestlings by plumage variation as one can in bluebirds. Bluebirds, even though they were mostly monogamous, provided me the opportunity to ask other questions.

My questions about monogamy and eventually the answers were different enough to earn special attention and sometimes ridicule. I thought monogamy offered many disadvantages for females, at least compared with the solitary lives of females of most species. So, my questions, mostly about females, were: Did female eastern bluebirds *really* need help from males to raise their offspring? Was male parental care *really* necessary for female reproductive success? Were social fathers the genetic fathers? Did female eastern bluebirds multiply mate? Did fathers guard their paternity as Trivers (1972) suggested? Was that why males were aggressive early in nesting attempts? If so, how did that work? Were these males keeping "their" females home and away from other males, or were males keeping rivals away from the females? I had already seen deadly fights between female bluebirds. But, why did females fight? Fascinated, I decided to stay with bluebirds, to study a monogamous species. I expected, as did members of my advisory committee, that my study would be a minor one about a mating system few seemed interested in, unlikely to draw much attention, but one that would answer a few new questions–making my plan a good one for a PhD. I designed my thesis (Gowaty 1980) around experiments on aggression in males and females (Gowaty 1981b), the value of male parental care to female reproductive success (Gowaty 1983), and whether or not females mated multiply (Gowaty 1985). The hard part was genetic parentage testing. At the time, the method of choice was protein electrophoresis of blood samples. I asked John Avise to teach me. He told me, "It's like a soufflé recipe. Just do it". I wasn't much of a cook then, so I collected blood and looked for a collaborator.

For the aggression experiments (Gowaty 1981b; Gowaty & Wagner 1987), I first used skins of bluebirds mounted on a wire frame that looked fairly life-like. These were often destroyed when aggressive resident bluebirds attacked them. Later, I used the bodies of bluebirds I found dead in the field. I removed their viscera and injected their muscles with formalin, then stuck wires through them to use to anchor them on perches near bluebird cavities. They worked (Figures 12.5 and 12.6). These models were considerably more durable than the first version of models, but they always got pretty ratty after ten or so tests. I focused on the fact that aggression is potentially costly to aggressors, sometimes even lethal, so I expected that aggression was not always inducible, but probably had a pattern associated with the risks of exposure to conspecifics during the nesting cycle. The experimental protocol included putting the models within a meter of a cavity during females' fertile periods just before and during egg-laying, during incubation, and when adults were feeding nestlings. The patterns were unambiguous: males were aggressive to male models; females were aggressive to female models. Males were most aggressive to male models when "their" females were fertile, and their aggression declined precipitously as incubation began. Females were most aggressive to other females during egg-laying, which was

Figure 12.5. A stuffed skin of a female bluebird induces aggression from a resident female. I took this photograph in Gadsden, Alabama.

Figure 12.6. A stuffed skin of a male bluebird induces aggression from a resident male. I took this photograph in Clemson, South Carolina.

difficult to explain based on what was known at the time, but I speculated that females, like males, were protecting their parentage, decreasing the likelihood of conspecific nest parasitism ("laying eggs in neighbors' nests"). These experiments were easy to do, and the photographs I made of Alabama bluebirds (to avoid disturbing my experimental populations) were easy to make, because aggressive individuals appear blinkered, so that I was able to sneak up and photograph the action from 15 feet away. I thought photographing aggressive bluebirds was fun; later it was important that I had, because, as Don Jenni told me at an American Ornithologists' Union meeting, "I wouldn't have believed it, if I hadn't seen the pictures". I didn't know whether he was saying he wouldn't have believed my work, or wouldn't have believed that females were aggressive, a general disbelief reflected even in the comic books. What ever he meant, it was a cutting remark, the cutting edge of many I've heard since. It shocked me then. But not everyone was so damning: my presentation at the Animal Behavior Society meeting in New Orleans, "The experimenter as Iago? Aggression of eastern bluebirds to conspecific intruders during the nesting cycle", received second place Allee honors.

To understand whether male parental care had an effect on female reproductive success, I removed males, held them in outdoor aviaries, and released them after the end of the nesting cycle, so that "their" females had to feed the nestlings without the help of a male. In this experiment (Gowaty 1983) I compared the number of young fledged for experimental and control females, whom I matched for age, nesting cycle, and habitat. Even though I measured **many** reproductive success variables, experimental females who raised their offspring without male help fledged as many offspring as control females. *That* was interesting, because I expected that there would be some deficits; others expected that lone females would fail entirely. I interpreted the results conservatively, and presented them as a challenge to the common assumption that monogamy predominated when male help was *essential* for female reproductive success. This experiment, the first of many that repeated it, opened that assumption up to further challenge. The unavoidable inference was that perhaps there were other selective pressures favoring social monogamy besides the incompetence of females.

Because of different results in Ontario, Raleigh Robertson, Greg Ball, Al Dufty, and I decided to do a cross-latitude male-removal study. The results of that study (unpublished) showed that in South Carolina experimental females did as well as control females, as they did in Ontario, but not in New York. The most compelling interpretation of the many male removal studies that followed mine was that female dependence on males was often not a species-specific characteristic, but an individually flexible trait, depending on variation in individual females and the social and ecological environments that they lived in. So, I concluded that females vary (Gowaty 1996a, b). This seemed simple enough to me, but it was apparently challenging for others. For years, I have wondered why. No doubt part of the reason is that sociobiological sex difference theory emphasized that reproductive success variation among females was much smaller than in males. The implication for some might have been that females did not vary. Despite this expectation, the fact remains that in bluebirds and many other species there is between-female variation in their abilities to raise their offspring without male help. Selection will act on any appropriate variation, and so, I started to wonder whether between-female variation was the source of fitness variation in all mating systems. This too turned out to be a question from the outfield, and for a while it felt like a minefield.

Of course, in 1980 most discourse in evolution of social behavior was about males, so Happ Wheeler asked me during my defense seminar, "Why do males stay?" It took me more than a decade to come up with what I considered a plausible answer (Gowaty 1996c).

Costs and benefits of daughters versus sons

Beginning with a bluebird study (unpublished), I have been interested in progeny sex ratios (Gowaty and Lennartz 1985; Gowaty 1990, 1993). In 1978, Mike Lennartz, who was studying helping in red-cockaded woodpeckers, *Picoides borealis* (RCWs), told me that unlike the adults whose plumage appears identical until angry males raise their usually invisible red cockades, nestling and juvenal males are obviously sexually dichromatic. Young males have a distinctive cap of always-apparent red-feathers, while the heads of their sisters are black. We talked a lot about why strong sexual dichromatism occurred in juvenal RCWs, but seemingly disappeared after the post-juvenal molt. I wondered out loud about the sex ratio of the chicks, and Mike opened his deep desk drawer and pulled out a file with data on the sex ratios in RCW nests. I analyzed these data and called Mike, "The sex ratio is 59% sons." From that followed one of the most enjoyable and productive collaborations of my life. We systematically rejected then-known Fisherian explanations for the sex ratio skew, until only one explanation was left standing. I said, "Mike, do you also have weights on these chicks?" Mike pulled open his deep desk drawer and handed me the records of nestling weights. Nothing arresting there. We thought from these data that it was unlikely that differential parental investments in sons versus daughters could explain the sex ratio. "Mike, do you have data on dispersal?" Mike pulled open his deep desk drawer and produced a file with sparse formal records of dispersal but he emphasized anecdotal observations of juvenal males being aggressive to their sisters, followed by daughters' disappearance from natal territories before sons'. I thought this meant that daughters had to be less expensive to parents than sons, so we could provisionally reject the idea that interactions with daughters were more costly to parents than interactions with sons. In contrast, we knew from Mike's long-term thesis data that sons were far more likely to be helpers than daughters, and that led us to ask other questions, and so, evolved the first paper showing skewed progeny sex ratios in an avian helper's system (Gowaty & Lennartz 1985).

The worm turned

Questions about females have had moments of ascendancy. One was just after Sarah Blaffer Hrdy's *The Woman That Never Evolved* (1981). The book was not just inspiring, it buoyed me. Even though my first studies of females were mostly completed before I knew about Sarah's work, her ideas resonated deeply with me and her ideas in that book and beyond have affected my work ever since. As recently as the 2007 European Society for Evolutionary Biology meeting in Uppsala, I was stunned when once again I realized that some scholars continue to think that female–female competition is unremarkable and evolutionarily uninteresting. Despite the progress of the last thirty years, Hrdy's metaphor has as much punch today as it did then.

Whether paternity was certain in socially monogamous animals was equally far from our standard perseverations about mating systems. From contemporary perspectives it is hard to imagine that it was ever doubted that socially monogamous females might "mess around", mating with multiple males. However, controversy dogged me when we reported (Gowaty & Karlin 1984) for the first time in a socially monogamous bird that females had offspring in a single nest from more than one sire. Famous male ornithologists openly sneered when we predicted more such observations to come (Gowaty 1985). One now famous woman ornithologist declared, "My birds wouldn't do that". I had no response to her then, but now can say, "Of course, they do", because "her birds" are prominent in the list of 90% of tested socially monogamous bird species that do have females that mate multiply. Almost an eternity later, in 1990, when Lisle Gibbs presented his data on multiply mating females in socially polygynous red-winged blackbirds, a colleague remarked, "You must be so smug". Smug? Well, no. Vindicated? Yes. My report that female eastern bluebirds mated with more than one male was the tip of an iceberg of observations that followed showing that many females are genetically polyandrous in socially monogamous birds. The worm had turned in a relatively short period.

Twelve and a half years without a job

However, to me it didn't feel short; the time from my PhD in August 1980 until my first tenure track job in January 1993 seemed infinite. I stayed in Clemson, asking questions of bluebirds, analyzing data, and publishing papers. Doug Mock offered me a postdoc the same year I was awarded an NSF grant from the "Visiting Professorships for Women Program". I spent 1983–1984 with Doug and 1984–1985 using my VPW funds at Cornell visiting Steve Emlen and watching robins, *Turdus migratorius*, discovering that robins had extra-pair paternity and conspecific nest parasitism (Gowaty & Davies 1986; Gowaty & Plissner 1987). With Jon Plissner, who later became my graduate student, I tested a novel hypothesis for so-called "mate guarding" (Gowaty & Plissner 1987). Despite my empirical successes at Cornell, I failed to get a job, so my husband, Gabe Acebo, and I moved back to South Carolina, planning to support my work with "Acebo Grants".

We lived in Cateechee, an old mill town along Twelve Mile Creek that had carved a gorge that rivaled Ithaca gorges for beauty and fast-moving water, and it was only a few miles from Clemson, surrounded by my bluebird study sites. My work thrived. Even without a job, I was productive. Two more NSF grants aided me in my exploration of the correlates of extra-pair paternity, conspecific nest parasitism, and social behavior. Working out of a cold room (yes, really) in the Department of Biology at Clemson, postdoc Dale Droge and I reported that father bluebirds fed their daughters more than they fed their sons, while mothers were egalitarian and fed sons and daughters equally (Gowaty & Droge 1991). I hypothesized that parent–parent conflict over the sex ratio explained these patterns. With Wes Weathers we discovered that, despite having more visits from their feeding fathers, daughters' metabolic rates, growth rates, and growth curves were similar to those of their brothers.

Despite successes and the fun I was having in the field (Figure 12.7), I worried incessantly about from where the next US$15,000 would come. Having discovered my grail early, it is true I enjoyed *my* work. I did realize that I was contributing significantly to major debates. I did know that I had changed my field, more than once; nevertheless, I didn't have a job and wasn't contributing to a retirement fund, and I was flummoxed.

Females seek and accept extra-pair fertilizations

At the time, many investigators were convinced that males alone were responsible for extra-pair mating behavior, because they assumed that females would seldom or never seek extra-pair mates: after all, most still thought females were "coy" and "discriminatingly passive" (Darwin's words). The first bird data to challenge this view suggested that females, perhaps as much or more than males, seek extra-pair fertilizations: fertile females that were off-territory and away from their territorial partners, "stay-away" females, were more likely to have extra-pair offspring than "stay-at-home" females (Gowaty *et al.* 1989; Gowaty & Bridges, 1991a,b). At the same time that I was reporting these observations, younger ornithologists, presumably with less of the received wisdom of earlier years, began to

Figure 12.7. At Clemson in the field weighing bluebird eggs. Photo by permission from Jon Plissner.

document remarkable variation among females and a suite of behavior associated with females seeking extra-pair fertilizations. I hope my earlier bird work helped to produce the climate that fostered their success. Today, it is remarkable to me that it was ever controversial to imagine that females were active participants in extra-pair matings. But it was. So, my observations on active females, the first in socially monogamous birds, were once again greeted with skepticism; for some they were counter-intuitive. These observations not only showed that frequencies of extra-pair offspring varied with female behavior, but also that males mate-guarded those "stay-away" females more than "stay-at-home" females (Gowaty & Bridges 1991a). Trivers (1972) said that males guarded females from the sexual advances of rivals, so the prediction was that extra-pair offspring would be less frequent with strong mate guarding. We observed just the opposite. Strong male mate-guarders had more extra-pair offspring in their nests. Weak guarders had no extra-pair offspring in their nests. My interpretation was that the behavior of females *induced* male guarding, and that males did not guard their females unless females behaved in ways that suggested extra-pair behavior. I hypothesized that male guarding behavior was a *conditional*, individually flexible response, perhaps induced by female behavior. This observation deepened my interest in within-population, between-female variation.

From 1986 to 1991 I threw myself into work on social correlates of genetic parentage in eastern bluebirds, and perhaps made the mistake of publishing most of the work in just two papers (Gowaty & Bridges, 1991a,b) instead of the least-publishable-unit approach that many others were taking. My field experiments showed that when the density of breeding neighbors was highest, rates of extra-pair paternity were highest; when densities were lowest, extra-pair fertilizations were lowest. I concluded that, given the opportunity, many males *and* females seek or accept extra-pair copulations. We also did manipulative field experiments that induced conspecific nest parasitism (Gowaty & Bridges 1991b). I was interested in the effect of catastrophes on what fertile and gravid females would do, so we experimentally created a catastrophe when we removed nesting sites half way through the nesting season, reducing the availability of nesting sites to birds who previously bred there. If females respond flexibly and adaptively to such disaster, they should lay their eggs in the nests of conspecifics, which is exactly what they did. Residents who retained nesting boxes on their territories suffered significantly higher rates of conspecific nest parasitism in subsequent nests, perhaps no surprise in a secondary cavity-nesting species, but I enjoyed having experimental data. These two experiments were the first of their kind.

In the companion paper, (Gowaty & Bridges 1991a) we reported other correlates of genetic parentage. Three of the twelve ecological and behavioral variables we measured – number of nesting sites per territory, female age, and nestling sex – showed no effect on genetic parentage. But, second year males cared for more offspring not theirs (53.2%) than older males (53.2% vs 19.8% respectively). Pairs that had been together longer had fewer genetic mismatches with nestlings they cared for than newly formed pairs. Opposite to standard expectations, males that guarded females the most were most likely to care for nestlings not theirs than males that guarded less strongly. Fertile females who spent the most time off their territories were significantly more likely to have extra-pair offspring than fertile females who spent the least time off their territories. Despite the fact that males that guarded females the most strongly had the most extra-pair offspring in their nests, these males later seemed to adjust their paternal investments downward. Males who stayed close to incubating females and their nests had fewer extra-pair offspring in those nests than males who stayed further away. Males with the highest number of extra-pair offspring in their nests spent the most time off territory during incubation. In addition, female breeding synchrony was unrelated to extra-pair parentage. Even today these two papers provide one of the most complete examinations of the social correlates of genetic parentage for any species.

In the meantime, even though I had already had three NSF grants and been influential in my field, I still didn't have a job. Desperation galvanized me against the sting of criticism, so I applied and applied and applied, finally getting a Research Career Development Award, a K-award, from NIH. Funding for five years began in 1990, and that is when I finally seemed to be truly considered for jobs, but in a series of four interviews other candidates (Nancy Burley, Lynn Carpenter, Ellen Ketterson, and Patty Parker) got the jobs. Yes, it was demoralizing, but after thinking carefully about my situation, I realized that I was finally paying into a retirement fund, and I was being paid to do my research, which is what I really wanted to do anyway.

My first tenure-track job

In 1992 after the publication of Gowaty (1992) Irwin Bernstein called and said, "Your job has just been advertised at UGA". The advertisement was for a joint appointment in a science, to be determined by the discipline of the successful candidate, and the Women's Studies Program. I applied, got an interview, and was offered the job as an Associate Professor of Zoology and Women's Studies with two years' credit towards tenure. I joined the University of Georgia (UGA) faculty after a second competitive review of the K-award, which extended the award length to a total of seven years. Around the same time, I was awarded an NIH R01 and another NSF grant.

Adjusting to my new job wasn't difficult, and my home was an interesting old house built in the 1880s. The landscape had neither the drama of Alabama nor the gorges of Cateechee, but it was familiar, and there were bluebirds, not to mention colleagues, fondly remembered teachers and friends. The sad part was that Gabe stayed in Cateechee, and we divorced in 1994.

For the first time in a dozen years, I was no longer worried about from where my next salary would come. I realized that my earlier grants spoke not just to my livelihood, but also to the interest of my colleagues in my work, whether I got those earlier jobs or not. I cannot say that getting my first tenure track job 12.5 years after the completion of my PhD was transformative, but it was different. Two years later, I was a full professor with tenure; and shortly after, I was traded out of Women's Studies for a full time position in the Institute of Ecology. In 1996, I had a successful review of an application for what became my third K-award. For all but five of the fifteen years I was at UGA, I brought my own salary to UGA. Even though most of my administrators never remembered it, *that* was pretty satisfying. It might have had something to do with why Ron Pulliam nominated me as a Distinguished Research Professor, a rank to which I was appointed in 2003.

Weird Athens bluebirds

The intellectual conundrum was the decision to study bluebirds in Athens rather than in Clemson. I thought: Athens is close to Clemson. The birds are plentiful in Athens, so I set up study sites in and around Athens. Jon Plissner, who had completed his PhD on natal dispersal in bluebirds in Clemson, joined me, and the fun started. We amazed each other with our stories of how different from our typical expectations (Gowaty & Plissner 1998) Athens bluebirds were. We kept a running list on the white-board of "weird" things bluebirds were doing, the weirdest being that Athens males were aggressive to females whereas we had seldom seen such male behavior in Clemson. Athens bluebirds also were significantly less productive than Clemson bluebirds. Males guarded females more and male parental care was highly variable in Athens, where males helped females much more than Clemson males did.

Walking past me one day in the hall, Bob Mathews grinned and asked if I meant to put a bluebird box on every fence pole that had an introduced fire ant mound under it. Of course,

I hadn't meant to do that. And, of course, we had noticed how rapacious fire ants are. They even ate bluebird nestlings alive. Observing this horror more than once, no matter how ecologically interesting, was enough for us. From then on, we excluded fire ants from nesting boxes. It wasn't a stretch to imagine that the differences between Athens and Clemson bluebirds had to do with food competition with fire ants that ate the same ground arthropods that bluebirds do. Critically, fire ants were long-ago invaders of Athens but remained absent from Clemson. So began the "difficult" field years, as we tried to understand the ecological correlates of differences in Athens and Clemson bluebirds. Our best failure was a field experiment in which we randomly assigned territories to a treatment – removal of fire ants – and a sham control in which we visited control territories with our "removal equipment", a mrymicide applicator which we did not turn on in the control sham treatments. I had hoped that experimental treatment would allow local ground arthropods to bounce back, which I predicted would translate into behavioral and reproductive changes in bluebirds on experimental territories. I had hoped that experimental removal of fire ants would transform Athens bluebirds into the familiar creatures I had studied in Clemson. The results were distressing: we had fire ants on our experimental as well as our control territories. We suspected leakage of fire ants from control territories that were contiguous with randomly assigned experimental territories. We decided, "It's ecology, we have to pseudo-replicate". So, we assigned whole study sites instead of just randomly chosen territories to fire ant removal and sham control treatments. I was worried about the obvious sample size issues, but we forged ahead and to my great dismay, we observed significantly more fire ants on fire-ant-removed study sites. Inadvertently, our removal of intra-specifically aggressive monogyne fire ant colonies had opened habitat for settlement of intra-specifically pacific polygyne fire ant colonies. Well, *that* was very interesting. Nevertheless, I have never had the heart to write this result up, because until recently I was not entirely convinced that fire ants were the key to the differences in Athens and Clemson bluebirds. Instead, I began a five-year comparative project in which we monitored changes in behavior, demography, and productivity of Clemson bluebirds as fire ants invaded and swept through the upper piedmont, while simultaneously monitoring Athens bluebirds. This too-big project, underfunded by an NSF LTREB grant, turned into Jason Lang's dissertation (still in progress).

Evolutionary biology and feminism

In between field seasons I was teaching "The Biology and Politics of Reproduction", and, in the wake of my article in *Human Nature* (Gowaty 1992), I thought it would be interesting to organize a symposium on feminism and evolutionary biology. I received a State of the Art grant from UGA, which provided the bulk of funding for the symposium. The Society for the Study of Evolution (SSE) meeting was to be held in June, 1994 at UGA, and Jim Hamrick, one of the co-hosts, suggested that the SSE might be interested in co-sponsoring it. To my surprise, SSE did provide funding. After I cajoled, John Maynard Smith came. I still have his hand-written acceptance note, "Gowaty, you are a hard woman to say "no" to". I felt

Figure 12.8. John Maynard Smith (sitting on the end of the sofa left) and Toshihide Hamazaki (sitting in chair at right) with me (on the floor) at Karen and Jim Porter's home in Athens, Georgia during a reception for participants and attendees at the SSE Evolutionary Biology and Feminism Symposium 1994. Photo by Karen Porter.

I needed him, not for his feminism, which was, as he admitted, "flexibly expressed in convenient moments" (Figure 12.8), but because he had simultaneously been scientist and Marxist. I invited everyone I thought had interesting things to say about evolutionary biology and feminism. I meant the symposium to be a place to discuss the similarities in *philosophy and ideas* of evolutionary biology and the *philosophy and ideas* in feminism. I had not meant it to be about hiring and salary discrepancies, so was surprised when that was what some expected. I aimed to talk about those scientific hypotheses that had a similarity to feminist ruminations. I aimed to talk about bedrock similarities between evolutionary biology and feminism, which to me meant: (i) the centrality of variation, (ii) the emergence of language in science that sounded like feminist rallying cries, e.g., "the manipulation and control by others of female reproductive capacities and sexuality"; (iii) the communicative functions of "femininity"; and (iv) the mechanisms of manipulation and control through deceit and self-deception. At least, that was *my* short list. I suspected others had other theoretical agendas. Several of the papers met my goals and I should have been grateful and pleased about that. However, some participants were antagonistic to other participants; they were obstructive and got in the way of open discussion, and I was greatly disappointed, so that I felt at the time that the conference was a failure. Natalie Angier bylined a report of the conference in the Tuesday Science Section of *The New York Times*. Thus, we got some attention. But, I was not happy about the level of discourse, nor was I happy in subsequent years as I edited contributions from contentious contributors and worked hard on the publication of the book (Gowaty 1997b) that fell from attention almost as soon as it was published. The failure of the book taught me not to put "feminism" in the title of anything.

Variation in females

My paper at the conference and in the book was about how variation in female life histories affected between-population and between-species female vulnerability to social control (Gowaty 1997c). Here I hypothesized that it was not just the distribution of females in space and time (Emlen & Oring 1977) but variation in females' abilities to remain in control of their own reproductive capacities that was central to the dynamics of social systems. I first made this point at the 1992 International Society Behavioral Ecology (ISBE) conference at Princeton and I elaborated on it in a series of articles that followed (Gowaty 1996a, b, 1997a; Gowaty & Buschhaus 1998; Gowaty 1999, 2003a,b).

Gowaty (1996a) and Gowaty & Buschhaus (1998) are the key citations in this line of thought, whether I usually get credit for it or not. First, I reviewed the background of variation in avian females' dependence on males in raising offspring to make the point that there was much unanticipated between and within species variation in females' abilities to raise their offspring alone. I emphasized that the variation was likely due to intrinsic variation among females and extrinsic environmental factors. Second, I discussed the hypothesis that within population, between-individual variation in females' vulnerabilities to control determined why they sought extra-pair matings. At the time I was focused on social constraints such as variation in female neediness, so that I thought of some male help as "helpful coercion" (Gowaty 1996a, b), which put females in the bind of trading-off components of fitness in order to have full access to the resources they needed for reproduction. The female constraint hypothesis predicts variation in rates of extra-pair paternity, female productivity, variance in male fitness, and offspring viability (Gowaty 2003a). The ratio of input variables to output predictions is favorable, and I thought – naively – that others would find the range of these predictions interesting. I expected more tests of these ideas than have so far appeared. I hoped this idea would stimulate more work.

Gowaty and Buschhaus (1998) used a similar approach to theoretically explain the co-evolution of male aggression against females, including forced copulation, and the evolution of social monogamy in ducks. The key predictive variable in this version of the female constraint hypothesis was variation in females' vulnerabilities to male aggression.

I extended these ideas to human social systems in a discussion of variation among women in their vulnerability to male aggression, forced copulation, and trades of sexual access for resources essential to reproduction (Gowaty 2003a). And, I stressed that this key predictive variable – variation among women in their vulnerability to coercion of their reproductive decisions – predicted not only women's productivity (number of offspring surviving to reproductive age), but also variance in reproductive success of men, and offspring viability. The thematic thread came from my attempts to express ideas current in modern feminism as *testable* scientific hypotheses of fitness dynamics of individuals, in particular mating and social systems. I think I succeeded, because the scientific hypothesis contains within itself the means of its own rejection, which is a value I've held for my scientific work ever since my first course with John Avise.

What kept me awake at night during the winter 1993–1994

Shortly after settling into UGA, one question kept plaguing me: Why are the conflicts of interest between the sexes never resolved? Or are they? Are not the interests of the sexes sometimes congruent? I didn't know then, nor do I now, but I suspect sometimes that the interests of the sexes are the same, though not necessarily even in strict monogamy. What ever is the truth, I am impressed with how common and dynamic sexual conflicts are. I see the dynamism as the back and forth of fitness variation sometimes favoring females, some-times males, something I called "the dialectics of sex" or "sexually antagonistic selection pressures". So, the question that kept me awake at night was, why is sexual conflict so ubiquitous? What keeps the selective dynamic alive? Put another way: why has male dominance, which seems to be so common in mammal societies at least, not favored universal "passivity" in females?

Perhaps the answer was in the Red Queen. I was impressed with the possibility that mate preferences predicted offspring health, but by the early 1990s it was clear that Hamilton and Zuk's (1982) prediction that fancy male traits predicted the health of their offspring was not holding up. Yet, it was also clear even then that the showiest males were healthiest. Could it be that many mate preference studies were incomplete? Did females who were asked to discriminate between fancy and less fancy males make reproductive decisions under the constraint of being dazzled or tricked? Could those females most vulnerable to sensory exploitation be making reproductive decisions that did not favor the long-term interests of their offspring? Was the best measure of fitness the first fitness components we could measure (number of eggs laid or offspring born)? Was a better fitness measure the number of offspring that survived to reproductive age? I wondered what would happen if one did preference studies *at random with respect to male phenotypes* and then enforced matings to compare fitness outcomes when choosers were with males that females did and did not like. I wondered whether females constrained to mate with partners they did not prefer had offspring of lower viability than females not so constrained. I was anxious for the answer, because the association of constraints with offspring viability was a pivotal assumption of the compensation hypothesis, which I had been thinking about during the previous winter in relation to human fertility variation (Gowaty 2003a). What was needed was an experiment.

Doing an experiment on captive birds didn't seem like much fun given the cage cleaning involved. So, in December I asked Wyatt Anderson to teach me to culture flies. When I told him of my interest in offspring viability, he told me about his efforts to repeat Linda Partridge's (1980) germinal study of offspring viability and mate choice. If I had ever known of Wyatt's earlier study, I had forgotten it, but it was shortly clear to both of us that we would collaborate on my questions that bore important similarity to his own.

In Vienna at the 1994 International Ornithological Congress, Cindy Bluhm and I discussed constrained mate choice, and we decided to do mate preference experiment on mallards, *Anas platyrhynchos*, a species that Cindy is a world's expert on. In September 1995 in Lexington, Kentucky, at the "Sexual Conflict Workshop", my talk was "Sexual dialectics, sexual selection, and the evolution of mating systems", and Allen Moore and I discussed his joining our group to study fitness effects of mate preferences expressed under constraints in *Nauphoeta cineria*, Tanzanian cockroaches. Then I asked Lee Drickamer, the most energetic, competent, all-round experimentalist I know, to also join our effort to ask the same questions of feral house mice, *Mus musculus*. Gunilla Rosenqvist joined our effort to work on guppies, and I started writing grant proposals to attract funding. These generous colleagues were tolerant of but also skeptical of my ideas, and I am convinced that Wyatt, Allen, and Lee didn't understand me for years, but it seemed to make little difference to their willingness to collaborate or to the main question we wanted to ask: did constraints on expressed mate preferences matter to choosers' fitness? Of course, NSF turned us down. Frustrated, Allen and I visited NSF to discuss the panel summary and the reviews. Randy Nelson, the Panel Officer, did reconsider; we were funded finally in September 1996, and so began an intense three years of work by my amazing collaborators in which some interacted almost daily to do an experiment that was the same in all of our study species. I cannot thank these generous collaborators enough, for they were ever-vigilant skeptics, always forcing our experimental design in ways that made it ever more conservative relative to my expectations.

Wyatt convinced me we had to do mate preference trials, not just with female choosers, but male choosers in *Drosophila pseudoobscura*. At first I considered this a silly suggestion because *D. pseudoobscura* is a species with the largest known asymmetry in gamete size, so reasoning from Parker *et al.* (1972) and Trivers (1972) I expected that males would not be choosy, wouldn't trade off components of fitness, and wouldn't compensate. Wyatt insisted. Yong-Kyu Kim joined us as a postdoc; without him, Wyatt and I would probably have been unable to complete the large scale experiment that we did.

Our experiments told us a lot (Drickamer *et al.* 2000; Moore *et al.* 2001; Gowaty *et al.* 2002; Drickamer *et al.* 2003; Gowaty *et al.* 2003a, b; Moore *et al.* 2003; Bluhm & Gowaty 2004a, b; Kim *et al.* 2005; Anderson *et al.*, 2007; Gowaty *et al.* 2007). In all but one species (Allen's cockroaches) offspring viability was significantly lower when focal females or males were constrained to reproduction with partners they did not prefer, compared with when focals were with a partner they did prefer. Our collaboration brought us to generality in three years. Because we studied flies, cockroaches, fish, birds, and mice, I am confident of two conclusions: (i) mate preferences predict offspring viability, and (ii) constraints on the free expression of mate preferences matter.

For me, though, the most important results were about the compensation hypothesis (Gowaty 2008). The radical prediction of the compensation hypothesis is that constrained individuals increase the numbers of eggs laid or offspring born, at some survival cost to themselves, in an effort to increase the number of offspring that survive to reproductive age. Wyatt, Yong-Kyu, and I met weekly to look at our fly data. Week after week, the number of eggs laid was higher for constrained than unconstrained females. Wyatt thought this was odd, so he remarked on it frequently, and I, in turn, drew a graph with the main prediction of the compensation hypothesis. We repeated this conversation so many times – almost weekly for two years – that I thought Wyatt did not believe me, or, perhaps found the simplicity of my explanation unreasonable.

One of the most satisfying moments in my career was during a meeting of our group in Norway in 1999 during a discussion of components of fitness in all our study species. In all, offspring viability was lower for constrained individuals, but for none was the number of eggs laid or offspring born higher for unconstrained. It was just the opposite. Constrained individuals gave birth to more young or laid more eggs than unconstrained, consistent with the most challenging prediction of the compensation hypothesis. I still think my collaborators didn't understand me then. None the less, it was at this meeting that we agreed to a group paper (Gowaty *et al.* 2007).

Enter Steve Hubbell

In the midst of the excitement of these experiments, Steve Hubbell "whirlwinded" into my life and I married him in 1997, and we both began traveling between Princeton and Athens. I worried when we married that there would be less time for my scholarship and I feared that if we ever collaborated people would see my contributions as secondary to Steve's. I still worry about that, especially now that we do collaborate.

During the early years of our marriage, it was difficult for me to find the balance between being supportive of Steve's work while doing my own. Even though my new life with Steve was exciting and fun, the years following our marriage were filled with family trauma, and there was never enough time. My brother was caring for his HIV-infected partner. My father died in 1998 after eight septicemias associated with an antibiotic-resistant staphylococcus infection. My grandmother and mother were both ill by 1999 when the "free female choice consortium" had our answers. That year Steve was offered a job in the Botany Department at UGA and moved to Athens. It was the year we began a renovation on my old farmhouse; the year I began to be curator not just of all my stuff and my parents' stuff, but of Steve's stuff. It was the year I began constant travel between the US and Panama, Japan, and, it seemed, everywhere else. In 2000 my 104 year-old maternal grandmother died; the week before she died I was in Norway trying to talk to her on the phone. I thought, "I don't need to be in Norway now". *The Unified Neutral Theory of Biodiversity and Biogeography*, Steve's important book, was published in April 2001. In June, my mother died in my arms. Steve was writing a grant proposal and almost did not go to Gadsden to be with me, while I was with her. The week after Mother died, I was in Perth for a small conference; the week after Perth, I was the presiding President of the ABS annual meeting; and the week after that, I was at the American Ornithologists' Union meetng serving as Vice President. In November Steve took me away from the TV, where I had been stuck since the world changed on 9/11, to spend a month on Barro Colorado Island, in the middle of the Panama Canal, where his long-term 50 ha forest dynamics plot is. There I prepared for a keynote address for a conference in Germany in December (Gowaty 2003b). That month on BCI gave me my intellectual life back. It was quiet: I had time to think; but my health was in serious decline. I had acute arthritis and I was in terrible pain. While losing the spongy discs in my lower spine, I also lost 1.5 inches in height. When I returned from Germany, my favorite cat Catharine was near death, and she died two days after I returned. By January 2002 I was sicker than I have ever been in my life. I was having trouble walking, and by 2005 I thought I'd never be able in my life to again walk a half mile.

During the time of not being able to walk

But not being able to walk did slow me down, did give me a semblance of the quieter life I prefer. I have always been a contemplative scholar, often a lone field biologist, who likes nothing more than the privacy of a Sunday morning on my field sites watching bluebirds. Not being able to walk did force me to say "no" to more "professional opportunities". It allowed me the time to identify and begin to treat a B12 deficiency. It forced me to find the help I needed to get my strength back, so that I can now lift more weight on the back machine than anyone I know and I again walk fluently. And, it gave me time to think hard about fascinating questions.

At the beginning of the time of not being able to walk in the fall of 2001, I was quibbling about my concerns over the calculation of potential reproductive rate (Clutton-Brock & Parker 1992), when Steve gently suggested I read Hubbell and Johnson (1987). So, pulling myself away from fear-filled rhetoric and images on TV, I closed the door to my study to read it. Sick as I was, I emerged seven hours later completely confused, but intrigued. Their model provides an analytical solution for the calculation of chance effects on fitness variances due to time available for mating, a goal originally inspired by Sutherland's (1985) seminal work also on chance effects on fitness variances. Fitness variances can be due entirely to chance. *That* was very interesting, but the more important point, I thought, was that if one were able to partition fitness variances into components – those due to chance, and those due to adaptation – one would have a far better stab at understanding sexual selection. I emphasized this in my talk in Germany in 2001 and in the paper (Gowaty 2003b).

My second thought was more creative. Hubbell and Johnson's model was a static model of how selection in one generation could lead to frequencies of choosy and indiscriminate individuals in the offspring generation – at least that is how I would test their model and how I interpreted it. A novel idea, to be sure, but not yet the model I had been waiting for. It wasn't dynamic enough and its assumptions about mate quality were too restrictive, I complained. I wanted a model of adaptively flexible, induced, sex-neutral behavior of individuals. If we could produce the mathematics of such a theory, I thought we could add substantially to debates about the evolution of behavioral sex differences, and possibly even encourage crucial testing that would allow some real resolution to questions about whether sex differences are ecologically and developmentally induced or hard-wired by fixed sex differences. Our first real collaboration resulting in work that neither of us could do alone, began that day in November 2001. In May 2004, I wrote our first published paper (Gowaty & Hubbell 2005) about chance, time, and sexual selection. For the first time, I described DYNAMATE, our individual based simulation model that uses the old and now the new Switch Point Theorem to predict individually flexible reproductive decisions and the resulting fitness dynamics from sexually antagonistic selection pressures. Since 2004, our models have gotten leaner, predicting more with less input; we are working on the manuscript of a book now.

And, because these new models make predictions often alternative to anisogamy and parental investment theories about the reproductive decisions of individuals and the fitness variances in their populations, we have gone back to the basics. Wyatt, Steve, Yong-Kyu, my new postdoc, Brant Faircloth, and I, with funding from NSF, are now testing these predictions in three fly species (all wildtype) that vary in gamete size asymmetries. In concert with this new work, we have repeated Bateman's (1948) study using the same mutants he did (Kim *et al.* 2005). In preparation for that difficult study with mutants, we reread Bateman's paper very carefully. In the process we reevaluated its conclusions, and showed that the fitness variances he reported were no different from those expected from chance effects on time available for mating (Snyder and Gowaty, 2007). I am extremely grateful to my collaborators, for there is no way one person could ever do these ambitious studies, which compel me as much as any of my previous work, going as they do to the heart of the fitness debates that have focused my scientific thought from the beginning.

As Steve calls it, "the screen-play: Patty in LA"

In 2005, Steve and I started to interview for new jobs. Two senior scientist offers brought us to the Department of Ecology and Evolutionary Biology, UCLA, in July 2007.

Our move incited the difficult decision to stop my bluebird field research. Even so, I am not done with bluebirds or genetic parentage studies. Set-up funds have guaranteed that I can now finish analyses of thousands of bluebird DNA samples. I have started what I hope will be a long-term field project studying avian life-histories in Madagascar. And, I hope to soon have my own fly lab.

Steve was surprised when I said, "I'll move to LA". However, here I am. I'm still mindful of my history: I don't have an adequate retirement fund, so I'm happy to have a job I love. Sitting in my study in our home in Topanga, I am looking at the dramatic north-facing slope of Dix Canyon in the Santa Monica Mountains. I am unlikely to imprint on these mountains or on chaparral, but I am enjoying this gorgeous landscape, the new intellectual opportunities in front of me, and the fact that I am now on a faculty with five other behavioral ecologists and several wonderful tons of evolutionary ecologists. Lately, sometimes I feel almost as strong and confident about life as I did as a child (Figure 12.9), still happiest when watching animals.

Figure 12.9. Me on a rock in 1990 on the Blue Ridge Parkway, Virginia. Steve Wagner took this photograph.

Acknowledgements

I wish my mother were alive to hear me thank her. I also thank the two people most important to me today, Steve Hubbell and Frank Adair, both of whom commented on a draft. I thank my extraordinarily generous collaborators, including Wyatt Anderson, Lee Drickamer, Yong-Kyu Kim; and my students Jon Plissner, Judy Guinan, and Jason Lang. I thank Sarah Hrdy, Mary Jane West-Eberhard, Jeanne Altman, and Teri Markow for all I've learned from them. I thank Steve Hubbell, Lee Drickamer, Don Dewsbury, Jane Peak, Justine Tally, J. P. Drury, B. J. Ard and Brant Faircloth for helpful comments. I thank JoAnn Burdette, whose work for me helps create an environment in which I can work. I acknowledge twelve years of funding from NIMH K-awards, three years of funding from an NIH R01, and continuous NSF funding from 1982 to 2009.

References

Altmann, J. (1974). Observational study of behavior: sampling methods. *Behaviour* **48**: 227–65.

Anderson, W. W., Kim, Y.-K. & Gowaty, P. A. (2007). Experimental constraints on female and male mate preferences in *Drosophila pseudoobscura* decrease offspring viability and reproductive success of breeding pairs. *Proc. Natn. Acad. Sci. USA* **104**: 4484–8.

Bateman, A. J. (1948). Intrasexual selection in Drosophila. *Heredity* **2**: 349–68.

Bluhm, C. K. & Gowaty, P. A. (2004a). Reproductive compensation for offspring viability deficits by female mallards, *Anas platyrhynchos*. *Anim. Behav.* **68**: 985–92.

340 *Patricia Adair Gowaty*

Bluhm, C. K. & Gowaty, P. A. (2004b). Social constraints on female mate preferences in mallards *Anas platyrhynchos* decrease offspring viability and mother's productivity. *Anim. Behav.* **68**: 977–83.

Clutton-Brock, T. H. & Parker, G. A. (1992). Potential reproductive rates and the operation of sexual selection. *Q. Rev. Biol.* **67**, 437–56.

Dennett, D. (1996). *Darwin's Dangerous Idea: Evolution and the Meaning of Life*. New York: Simon & Schuster.

Drickamer, L. C., Gowaty, P. A. & Holmes, C. M. (2000). Free female mate choice in house mice affects reproductive success and offspring viability and performance. *Anim. Behav.* **59**: 371–8.

Drickamer, L. C., Gowaty, P. A. & Wagner, D. M. (2003). Free mutual mate preferences in house mice affect reproductive success and offspring performance. *Anim. Behav.* **65**: 105–14.

Emlen, S. T. & Oring, L. W. (1977). Ecology, sexual selection, and the evolution of mating systems. *Science* **197**: 215–23.

Gowaty, P. A. (1980). The origin of mating system variability and behavioral and demographic correlates of the mating system of Eastern Bluebirds (*Sialia sialis*). *PhD thesis*, Clemson University, Clemson, South Carolina.

Gowaty, P. A. (1981a). An extension of the Orians-Verner-Willson model to account for mating systems besides polygyny. *Am. Nat.* **118**(6): 581–9.

Gowaty, P. A. (1981b). The aggression of breeding Eastern Bluebirds *Sialia sialis* toward each other and intra- and inter-specific intruders. *Anim. Behav.* **29**: 1013–27.

Gowaty, P. A. (1983). Male parental care and apparent monogamy among Eastern Bluebirds (*Sialia sialis*). *Am. Nat.* **121**: 149–57

Gowaty, P. A. (1985). Multiple parentage and apparent monogamy in birds. In *Avian Monogamy*, ed. P. A. Gowaty & D. W. Mock pp. 11–17.

Gowaty, P. A. (1990). Facultative manipulation of sex ratios in birds: rare or rarely observed? *Curr. Ornithol.* **8**: 141–71.

Gowaty, P. A. (1992). Evolutionary biology and feminism. *Human Nature* 3(3): 217–49.

Gowaty, P. A. (1993). Differential dispersal, local resource competition, and sex ratio variation in birds. *Am. Nat.* **141**: 263–80.

Gowaty, P. A. (1996a). Battles of the sexes and origins of monogamy. In *Partnerships in Birds*, ed. J. L. Black, pp. 21–52. Oxford: Oxford University Press.

Gowaty, P. A. (1996b). Field studies of parental care in birds: new data focus questions on variation in females. In *Advances in the Study of Behaviour*, ed. C. T. Snowdon & J. S. Rosenblatt, pp. 476–531. New York: Academic Press.

Gowaty, P. A. (1996c). Multiple mating by females selects for males that stay: a novel hypothesis for monogamy in birds. *Anim. Behav.* **51**: 482–4.

Gowaty, P. A. (1997a). Females' perspectives in avian behavioral ecology. *J. Avian Biol.* **28**: 1–8.

Gowaty, P. A. (ed.) (1997b). *Feminism and Evolutionary Biology: Boundaries, Intersections, and Frontiers*. New York: Chapman & Hall.

Gowaty, P. A. (1997c). Sexual dialectics, sexual selection, and variation in mating behavior. In *Feminism and Evolutionary Biology: Boundaries, Intersections, and Frontiers*, ed. P. A. Gowaty, pp. 351–84. New York, Chapman Hall.

Gowaty, P. A. (1999). Extra-pair paternity and parental care: differential fitness among males via male exploitation of variation among females. In *Proceedings of 22nd International Ornithological Congress*, ed. N. Adams & R. Slotow, pp. 2639–56. Durban: University of Natal.

Gowaty, P. A. (2003a). Power asymmetries between the sexes, mate preferences, and components of fitness. In *Women, Evolution, and Rape*, ed. C. Travis, pp. 61–86. Cambridge, MA: MIT Press.

Gowaty, P. A. (2003b). Sex roles, contests for the control of reproduction, and sexual selection. In *Sexual Selection in Primates: New and Comparative Perspectives*, ed. P. M. Kappeler & C. P. Van Schaik, pp. 163–221. Cambridge: Cambridge University Press.

Gowaty, P. A. (2007). Perception bias, social inclusion and sexual selection: power dynamics in science and nature. In *Modern Controversies in Evolutionary Biology in Science and Technology in Society: from Space Exploration to the Biology of Gender*, ed. D. L. Klienman & J. Handelsman, pp. 401–20. Madison, WI: University of Wisconsin Press.

Gowaty, P. A. (2008). Reproductive compensation. *J. Evol. Biol.* **21**: 1189–200.

Gowaty, P. A. & Bridges, W. C. (1991a). Behavioral, demographic, and environmental correlates of uncertain parentage in eastern bluebirds. *Behav. Ecol.* **2**: 339–50.

Gowaty, P. A. & Bridges, W. C. (1991b). Nest box availability affects extra-pair fertilizations and conspecific nest parasitism in eastern bluebirds, *Sialia sialis*. *Anim. Behav.* **41**: 661–75.

Gowaty, P. A. & Buschhaus, N. (1998). Ultimate causation of aggressive and forced copulation in birds: female resistance, the CODE hypothesis, and social monogamy. *Am. Zool.* **38**: 207–25.

Gowaty, P. A. & Davies, J. C. (1986). Uncertain maternity in American Robins. In *Behavior and the Dynamics of Populations*, ed. L. C. Drickamer, pp. 65–70. Toulouse, France: Privat.

Gowaty, P. A. & Droge, D. D. (1991). Sex ratio conflict and the evolution of sex-biased provisioning. *Acta XX Congressus Internationalis Ornithologici*, Vol. **2**: 932–45.

Gowaty, P. A. & Hubbell, S. P. (2005). Chance, time allocation, and the evolution of adaptively flexible sex roles. *J. Integr. Comp. Biol.* **45**: 247–60.

Gowaty, P. A. & Karlin, A. A. (1984). Multiple parentage in single broods of apparently monogamous Eastern Bluebirds (*Sialia sialis*). *Behav. Ecol. Sociobiol.* **15**: 91–5.

Gowaty, P. A. & Lennartz, M. L. (1985). Nestling and fledgling sex ratios of Red-cockaded Woodpeckers favor males. *Am. Nat.* **126**: 347–53.

Gowaty, P. A. & Plissner, J. H. (1987). Association of male and female American robins (*Turdus migratorius*) during the breeding season: paternity assurance by sexual access or mate-guarding. *Wilson Bull.* **99**: 56–62.

Gowaty, P. A. & Plissner, J. H. (1998). Eastern Bluebirds. In *The Birds of North America*, ed. A. Poole & F. Gill. pp. 1–32. Philadelphia, PA: The Birds of North America, Inc.

Gowaty, P. A. & Wagner, S. J. (1987). Breeding season aggression of female and male Eastern bluebirds (*Sialia sialis*) to models of potential conspecific and interspecific eggdumpers. *Ethology* **78**: 238–50.

Gowaty, P. A., Plissner, J. H. & Williams, T. (1989). Behavioral correlates of uncertain parentage: Mate guarding and nest guarding by eastern bluebirds, *Sialia sialis*. *Anim. Behav.* **38**: 272–84.

Gowaty, P. A., Steinichen, R. & Anderson, W. W. (2002). Mutual interest between the sexes and reproductive success in *Drosophila pseudoobscura*. *Evolution* **56**: 2537–40.

Gowaty, P. A., Drickamer, L. C. & Holmes, S. -S. (2003a). Male house mice produce fewer offspring with lower viability and poorer performance when mated with females they do not prefer. *Anim. Behav.* **65**: 95–103.

Gowaty, P. A., Steinechen, R. & Anderson, W. W. (2003b). Indiscriminate females and choosy males: within- and between-species variation in *Drosophila*. *Evolution* **57**: 2037–45.

Gowaty, P. A., Anderson, W. W., Bluhm *et al.* (2007). The hypothesis of reproductive compensation and its assumptions about mate preferences and offspring viability. *Proc. Nat. Acad. Sci. USA* **104**: 15023–27.

Hamilton, W. D. & Zuk, M. (1982). Heritable true fitness and bright birds: a role for parasites? *Science* **218**: 384–7.

Hubbell, S. P. & Johnson, L. K. (1987). Environmental variance in lifetime mating success, mate choice, and sexual selection. *Am. Nat.* **130**: 91–112.

Kim, Y.-K., Gowaty, P. A. & Anderson, W. W. (2005). Testing for mate preference and mate choice in *Drosophila pseudoobscura*. In *Proceedings of the 5th International Conference on Methods and Techniques in Behavioral Research*, ed. L. J. J. Noldus, F. Grieco, L. W. S. Loijens & P. H. Zimmermann, pp. 533–5. Wageningen, The Netherlands: Noldus Information Technology.

Moore, A. J., Gowaty, P. A., Wallin, W. J. & Moore, P. J. (2001). Sexual conflict and the evolution of female mate choice and male social dominance. *Proc. R. Soc. Lond.* **B**: 517–23.

Moore, A. J., Gowaty, P. A. & Moore, P. J. (2003). Females avoid manipulative males and live longer. *J. Evol. Biol.* **16**: 523–30.

Orians, G. H. (1969). On the evolution of mating systems in birds and mammals. *Am. Nat.* **103**: 589–603.

Parker, G. A., Baker, R. R. & Smith, V. G. (1972). The origin and evolution of gamete dimorphism and the Male–female phenomenon. *J. Theor. Biol.* **36**: 529–53.

Partridge, L. (1980). Mate choice increases a component of offspring fitness in fruit flies. *Nature* **283**: 290–1.

Platt, J. R. (1964). Strong inference: certain systematic methods of scientific thinking may produce much more rapid progress than others. *Science* **146**: 347–53.

Snyder, B. F. & Gowaty, P. A. (2007). A reappraisal of Bateman's classic study of intra-sexual selection. *Evolution* **61**: 2457–68.

Sutherland, W. J. (1985). Chance can produce a sex difference in variance in mating success and account for Bateman's data. *Anim. Behav.* **33**: 1349–52.

Trivers, R. L. (1972). Parental investment and sexual Selection. In *Sexual Selection and the Descent of Man*, ed. B. Campbell, pp. Chicago, IL: Aldine.

Verner, J. & Willson, M. F. (1969). Mating systems, sexual dimorphism, and the role of male North American passerine birds in the nesting cycle. *Ornithol. Monogr.* **9**: 1–76.

13

Myths, monkeys, and motherhood: a compromising life

SARAH BLAFFER HRDY

Definition of an Anthropologist: "(Someone) who studies human nature
in all its diversity."   *Carmelo Lisón-Tolosana (1966)*

Maternal effects (1946–64)

From a young age, I was interested in *why* humans do what they do. With little exposure to
science, certainly no inkling that there might be people in the world who studied other
animals in order to better understand our species, I decided to become a novelist. Born in

Leaders in Animal Behavior: The Second Generation, ed. L. C. Drickamer & D. A. Dewsbury. Published by Cambridge
University Press. © Cambridge University Press 2010.

Texas in 1946, right at the start of the postwar baby boom, I was the third of five children –
four daughters and finally the long-awaited son. My father's father, R. L. Blaffer, had come
to Texas from Hamburg via New Orleans in 1901 at the time oil was discovered at
Spindletop. He recognized that fortunes would be made in the oil business. He married
Sarah Campbell from Lampasas, whose father was in that business. I was named for her,
Sarah Campbell Blaffer II. My mother's father's ancestors, the Hardins, French Huguenots
from Tennessee, arrived earlier, in 1825, before Texas was even a state. They settled in East
Texas in what later became Hardin and Liberty Counties.

I know more about the Hardins than I otherwise might because after my father died, my
mother wrote a book, *Seven Pines*, based on old family letters. Like her mother before her,
my mother was a compulsive scholar and stickler for accuracy. My grandmother Davis was a
woman of tremendous determination, one of the first women from Texas to attend Wellesley
College. She had a passionate love of literature, and after college ran a bookstore in Dallas
while waiting for my grandfather Davis, a Yale graduate and a bank president, to marry her.
After marriage, with the same determination, she assumed the role of strait-laced *grande
dame*, but never lost her love of books. When widowed, she went to graduate school in
English, and later to Paris to learn bookbinding.

When my own mother died, I inherited part of grandmother Davis' extraordinary library,
including the bound copy of her 1944 Master's thesis about the poet and actress Adah Isaacs
Menken, a self-made woman if ever there was one. Menken made up just about everything,
not just her poetry, but also her parentage, date of birth, and the authorship of her poetry,
since some verses were plagiarized. My great-grandfather's mother had met Menken when
she passed through East Texas with Victor Franconi's Hippodrome in 1850. This "liberated-
to-the-point-of-scandalous" woman made an impression so deep that my grandmother,
and later my mother, were still talking about her when I was growing up. Years later when
I decided to study anthropology, when I first went to India, it was my mother who funded
the work, and later, when against family opposition I married a fellow anthropologist, my
mother and maternal grandmother stood up for me.

In *Mother Nature: A history of mothers, infants and natural selection* I acknowledged my
debt to these remarkable women. "There is an old saying that *sons branch out, but one
woman leads to another*," I wrote.

Perhaps its author was aware of sex-specific parental effects. In my case, this book owes its existence to
my mother, Camilla Davis Blaffer Trammell, and to her mother, Kate Wilson Davis, for passing on to
me their dogged temperaments (probably genetic) combined with a love of learning (more likely a
maternal effect). Both women were closet bluestockings ... [although like] other women of their class
and time they were determined to 'marry well.' How else to achieve an acceptable social status?
Alternative options in those days were not obvious. Yet these women imparted to me their love of
books and ideas, and stood up to support an iconoclastic kinswoman in her defection from tribal
customs. ...

Born in Dallas, I grew up in Houston when it was still a fairly sleepy city with graceful
oaks, long lazy gar and alligators swimming in the bayous, and cattle grazing along Buffalo

Speedway. Prevailing values were distinctly "Southern", generating genteel manners, extreme segregation, and patriarchal institutions. It was no accident that I would later become interested in the evolutionary and historical origins of patrilocal marriage, male-biased inheritance, female sexuality and peoples' obsessive concerns with controlling it. Elsewhere the women's movement may have been getting under way, but no one I knew talked about it.

Reared by a succession of nannies, I was a case study in "insecure attachment" and, except with friends, quite shy. I was simultaneously bookish, rambunctious, and imaginative. I dreaded school and was inattentive, doodling and daydreaming through classes. I loved horses though, devouring Walter Farley novels, and in ninth grade worked my way through Moyra Williams' *Horse Psychology*. Together with Dusty, a hunter of undistinguished conformation but with a tidy way of folding his front legs when he jumped, we traveled to horseshows around Texas and as far away as Tennessee. When I was fifteen my mother arranged for me to go to a school in Maryland known for its riding program.

Fortunately, St. Timothy's was something more. Back home the term "bluestocking" was a pejorative, but this small, girls-only school took women's education seriously. Miss Watkins, the headmistress, was sensible and humane. Given my shaky academic record, I was promptly assigned to the remedial quarter of my class. We called ourselves "the dupes." Foreseeing trouble, Miss Watkins also assigned herself as my advisor. During monthly meetings I did my best to reveal as little as possible, but Miss Watkins somehow knew it all. She expressed unwavering confidence in my abilities.

For all its good points, like most girls' schools, St Timothy's did not offer much science. However there was Mrs. Cross' biology course in which I was promptly nicknamed by my classmates "queen of the Bio dupes." Too lacking in either worldliness or self-awareness to view doing well in school as a route to anything, I was motivated purely by a lust to learn. This included a passion for *Scientific American* (at that time a magazine even a neophyte could enjoy), an avocation with important consequences later on. Academic prizes and medals came as sur-prizes, as if by some odd happenstance. It was like being nearly six foot (which by then I was) and not realizing that other people considered me tall. I honestly had no idea I was becoming a scholar. Years later, groping to explain how feminist ideas began percolating into my writing, I compared myself to "some savage on the fringe of civilization" dimly and awkwardly rediscovering the wheel (Hrdy 1986: 151). It was like that.

"The Savage Mind" (1964–69)

In 1964, I enrolled in Wellesley College where my mother and grandmother had gone. My favorite courses were creative writing, Mary Lefkowitz's Greek mythology, and a course in geology from the novelist Erskine Caldwell's son because the language used to describe the deep history of the earth struck me as beautiful. At the end of my sophomore year, I transferred to Radcliffe, the women's part of Harvard. To this day I don't understand how I managed it. As far as my father was concerned, Cambridge was a den of immorality and radicalism. Fortunately I was the third daughter, the heiress to spare, and by then there was a fourth, followed finally by the long-desired son. No one, I suspect, paid much attention.

Tuition bills were simply sent to my father's secretary. The handful of transfer students were supposed to arrive early. Unable to find the off-campus house to which I had been assigned, and not knowing whom to ask, I spent my first night in Cambridge at the Central Square YWCA, where someone swiped the copy of Dostoyevsky's *The Idiot* which had been given me by a friend for luck.

My reason for transferring was simple. I had begun a novel about modern Mexicans of Mayan descent who were torn between their contemporary worlds and ancient heritages. It occurred to me that it would be helpful to actually learn something about Mayans and their mythology. I decided to study under Evon Vogt, world's expert on Mayan cosmology. This meant entering Harvard as a junior, changing my major from English to anthropology, and learning about "structuralism."

Beginning with *Le Cru et le Cuit*, in 1964 (the year I entered college), one by one, volumes in Claude Lévi-Strauss' ambitious and highly speculative Introduction to the Science of Mythology were published. After *The Raw and the Cooked* came *The Origin of Table Manners, From Honey to Ashes*, and finally, *Man Naked*. I devoured the work of the structuralists, especially Mary Douglas' elegant 1966 classic, *Purity and Danger*. Prior to that, I had thought of myths as Jungian archetypes or perhaps grist for Freudian mills. Then Professor Vogt (everyone called him "Vogtie"), exposed me to this grand explanatory framework. According to Lévi-Strauss, folktales were the products of human minds attempting to make sense of the animals, plants, seasons, and social relations in the worlds they lived in. Using techniques inspired by linguistics, Lévi-Strauss compiled vast networks of myths from North and South America and then broke these complex narratives into their component parts, seeking recurring patterns and the logic that linked them. He used his version of the comparative method to reveal the "mental adaptations" of those devising the stories. The emerging categories ("living" versus "dead," Natural versus Cultural, etc.) often involved binary oppositions, which Lévi-Strauss considered fundamental to the architecture of human cognition, the structuralists' version of "core knowledge."

When critics pointed out that Lévi-Strauss' interpretation of how "savage minds" worked was tainted by his own Sorbonne-educated French mind, he famously retorted that since he was dealing with universals, it scarcely mattered. It made no difference whether "the thought processes of the South American Indians take shape though the medium of my thought, or whether mine takes place through the medium of theirs." This was 1966. I was hooked.

For an undergraduate interested in Maya-speaking peoples, the Harvard Chiapas Project offered unparalleled opportunities. Every undergraduate or Ph.D. student working on the Maya (and there were dozens of us) deposited copies of our fieldnotes and publications in the Harvard–Chiapas files, a common library open to all, with an avuncular but firm Vogtie riding herd on rampaging egos. It was a model of scholarly collaboration.

My undergraduate honors thesis, published in 1972 as *The Black-Man of Zinacantan: A Central American Legend*, was a structural analysis of folktales about "anomalous" animals, creatures that failed to fit established categories of living or dead, natural or cultural. I was interested in learning how and why human imaginations invented demons. I had at my disposal several hundred folktales collected from Tzotzil-speaking Maya in

Chiapas by Robert Laughlin, now at the Smithsonian, as well as stories from across Central America collected by Vogtie's students, including myself. By that time I was spending summers as a volunteer medical technician on projects in Honduras and Guatemala sponsored by Radcliffe's Education for Action. I knew only a smattering of the various Mayan dialects, but my Spanish was good. Women would come to the clinic early in the morning before it opened and share stories. My day job was teaching hygiene, setting up vaccination clinics, and arranging matters for weekly visits by a visiting dentist or doctor. But my predawn and evening hours (just the times when they were most prone to prowl) were reserved for asking questions about *espantos* (literally "spooks"), learning about the natural history of imaginary creatures, about *characotel*, the man who turns into a dog at night to go off and make mischief, *auitzotl*, the spectral water animal that lurks near river crossings to consume the nails and hair of drunkards, or *x'pakinte*, really tree bark dressed as a woman to deceive men. Most fascinating of all was the creature called *h'ik'al*, a tiny, black, supersexed winged demon who punishes women careless of their sexuality – women who deviate from normative sex roles. Transgressors would be seized, carried off by *h'ik'al* to his cave, raped with his super-long penis. Impregnated victims would swell up and then give birth, night after night, until they died. No wonder women cowered at the prospect of night-time assignations. People were so terrified of *h'ik'al* that they avoided mentioning his name, for fear he might come.

The Black-man of Zinacantan was a conventional exercise in structural analysis except that work by Maya archeologists and cryptographers allowed for comparisons far back in time. Combined with what ethnographers, ethnobotanists and ethnozoologists were learning, knowledge about ancient Maya belief systems added a new dimension. Transcriptions of codices compiled by early Spanish explorers were also available at Harvard's Tozzer Library and at the University of Texas in Austin, so I was able to trace the origins of the *h'ik'al*, a contemporary chastiser of sexual sins, back to the ancient Maya bat demon *Camazotz*. To test the validity of my interpretations, I generated predictions about how (if my analysis was correct) contemporary Maya subjects should respond to different propositions derived from them. It was the beginning of an abiding interest in "deep history" – though sociobiology's explanatory framework would push the time depth back millions of years, and broaden the comparisons to include other species.

Formative detours

In truth, my metamorphosis from structuralist to sociobiologist was actually more convoluted. After the first semester of senior year, I took time off. My return coincided with Woodstock and Kent State. I graduated as a member of the infamous "Class of 1969." It was my classmates, including friends from Education for Action, who took over University Hall in protest against the Vietnam War. I will never know whether I would have joined them because that fateful day was also my first up after spending two weeks bed-ridden with mononucleosis. Eager to understand what was happening, I went to Harvard Yard. My classmates were already inside the building, but I saw Professor Irven DeVore, a fellow

Texan whom I knew slightly. I had taken his undergraduate course on primate behavior and was dating one of his advisees. It was a beautiful spring day, and when I asked Irv what was happening, he replied, "I'm not sure. But in my day we would have called it a panty raid."

That June I graduated Phi Beta Kappa, receiving my degree *summa cum laude*, but skipped Commencement. None of my family planned to come – which was in fact a relief, since I was not sure how my volatile and conservative father would react to the political turmoil. The main reason, though, was my inability to make up my mind about red arm bands. Some classmates planned to wear them to signal protest against the war. I opposed the war but felt politics and scholarship should be kept separate – a tenet of anthropology in those days, albeit not today. My indecision was symptomatic of inner conflicts I experienced throughout 1968–69.

I had spent the "time-off" traveling first to Chiapas to see for myself a real-life *karnaval* ceremony in which a man would paint himself black and act like a mischievous *h'ik'al*, then to Yucatan and Central America to collect more tales, then on to Kenya, Tanzania, Zambia, South Africa, and Ethiopia. On my return, before returning to college, I worked at a meat processing plant in Watertown. "Jack-Pack", as it was called, specialized in "proportioned meats" for restaurants, ensuring that every diner received the same size portion. I wanted to learn what it was like to make a living. Clearly, I was beginning to deviate from the anticipated trajectory of a former debutante. My growing (still limited) political awareness brought with it unease on various fronts, especially over what was happening in Central America.

Recollecting my naiveté while working among villagers in Honduras and Guatemala, now shamed me. I recognized that the Guatemalan military had been using our services as part of its own public relations campaign. How could I have failed to register why the machine guns on the backs of the jeeps transporting the medical volunteers were there? A moderately accomplished artist, I used to make teaching posters for my hygiene students, big colored demonstrations about parasites and why it was important to wear shoes, boil water, eat a balanced diet. In retrospect, my admonitions to people who could not possibly afford to follow my advice were painful. I loved anthropology, but my travels in Chiapas and Guatemala had convinced me that if I lived among people there I would have to become some sort of revolutionary, which I had neither the desire nor the temperament to do. Better change course while I still could.

It was in this frame of mind that I decided to give away my anthropology books and apply to a graduate program in communications at Stanford University to learn to make educational films for people in developing countries. To prepare, I enrolled in a seminar on television. One partner on the required film project was Al Gore, already way ahead of the rest of us in his understanding of television and its potential.

So far as I was concerned, Stanford's communications program had little to offer. I began auditing Paul Ehrlich's population ecology course. He had just published *The Population Bomb* (1968), and as I listened, I was reminded of John Calhoun's *Scientific American* article about "Population density and social pathology." It had made a vivid impression when I first read it early one Sunday morning before church in an empty dining hall at St Timothy's

School. When kept at high densities, Norway rats experience a "behavioral sink," exhibiting such pathological-seeming behaviors as maternal neglect, infanticide, and cannibalism (Calhoun 1962). I was also reminded of something I had read in Professor DeVore's undergraduate course. Primate behavior then was still too young a science to have a textbook, so we used edited collections of field studies, including one by Yukimaru Sugiyama and his colleagues. They had reported infanticide among the high density population of langurs at Dharwar Forest in South India. It occurred to me that these monkeys might provide a model for the behavioral effects of crowding. By switching to nonhuman primates, I could also avoid (or so I thought) ethical issues raised by studying people.

Naive again – on so many counts, I made up my mind to go to India to study infanticide in crowded monkeys. Before the end of the term, I dropped all my courses, even those for which I had completed the work, to make certain that I would have no stake in staying. I applied to graduate school at the University of California-Berkeley, where langurs were already being studied, and to Harvard. Harvard (but not Berkeley) admitted me in the middle of the year. Thus in January 1970 I joined Harvard's fledgling program in behavioral biology within the Anthropology Department. (At that time, primate behavior, if taught at all, was taught by anthropologists.)

I had no idea how I was going to learn to study monkeys. Serendipitously, that summer the former husband of one of my aunts asked me to join him and two of my favorite cousins on a hunting safari in the Masai Mara and Kenya's northern frontier district. Although not a hunter, tagging along seemed the opportunity of a lifetime. On the way to Nairobi, I bought my first pair of binoculars at the Frankfurt airport, the battered Leica 10×40s I still use today. When the safari was over, I stayed on. Louis Leakey sent me to the Tigoni Primate Center outside of Nairobi, where a young primatologist named Neil Chalmers was in charge. In this way my uncle's invitation provided my first chances to see monkeys in the wild and to briefly study them in captivity.

In addition to running the colony and playing housemother to a succession of young women Leakey deposited, Neil was doing research on comparative infant development in African cercopithecine monkeys. I did my best to be useful and in return received as good an introduction to Old World monkeys as would have been possible anywhere in the world at that time. Neil loaned me Pru and John Napier's *Handbook of Living Primates* and generously oversaw my pilot study of allomaternal behavior (then still called "aunting") among caged patas monkeys.

Choosing a study site – and a mate

By the time I set out for India in June 1971, I still had no training in field methods beyond the weeks observing monkeys at Tigoni and reading a copy of Jeanne and Stuart Altmann's just-published *Baboon Ecology*. However, as I fulfilled course requirements, I took every opportunity to work langurs in. I convinced the always amiable Professor William Howells to let me write about colobine taxonomy (in chronic and bewildering flux) in his course on fossil man. But when I turned in a paper on "Infant biting and deserting among langurs" to the cocky graduate student, Robert Trivers, who was co-teaching the *Evolution of Sex Differences* with Irven DeVore, he acidly pointed out that "this paper has nothing to do with sex." Trivers and I got off to a less-than-promising start. Even so, I had a dawning awareness he might be someone worth learning from. In fact, of course, Trivers would be the most inspirational teacher I ever had. Like a shaman, he dove deep inside himself, resurfacing with extraordinary insights – often at great personal cost, occasionally requiring hospitalization. By the end of my first field season at Mount Abu, I would be ready to set aside the social pathology hypothesis I started with. I was just beginning to understand how important Trivers' stunningly original ideas about the connection between parental investment and Darwinian sexual selection were for understanding infanticide. Let me explain how I got to Abu.

At the end of spring term, financed by my mother, I headed for the forest of Dharwar in south India where Yukimaru Sugiyama had first reported infanticide among langurs. By then a new report of infanticide had been published by S. M. Mohnot (1971), who was studying langur behavior near Jodhpur, in Rajasthan, north India, so Jodhpur became my first stop. Another Harvard anthropologist, Dan Hrdy, met me there. Something else happened that year in Professor Howell's course on fossil man – I fell in love.

Dan had a travel fellowship and planned to spend that summer in Peru. When he decided he could just as well use the fellowship to work in India, I was glad. A year later, we were married in Kathmandu, Nepal, meeting up there as Dan was on his way to the Pacific to join the Harvard Solomon Islands Project and I was on my way back to Abu. The ceremony was held in the garden of the American consul to Nepal, Carleton Coon, son of the anthropologist of that same name. I wore white cotton slacks and the Coon children put plastic monkeys on top of the wedding cake, painting estrous swellings on the female of the pair with red nail polish. Dan and I only narrowly avoided missing the ceremony. The day before we had gone by motorcycle toward the border between Nepal and China to look for langurs. On our return we encountered torrential rains and took shelter in a cave beside the road. A shepherd was already there, urgently trying to tell us something about the location of our motorcycle. Minutes after Dan moved the only possible transportation back to Kathmandu, a flashflood swept across the spot where it had been parked. Later, as we slid and swerved down the steep, muddy road back to the capital, me clinging to Dan's back as he muttered something about "If I had known what a backseat driver you were. ...". When the news reached Irv DeVore back in Cambridge, it evoked one of his more memorable one-liners: "I expect he married her for her vowels."

I fantasized that Dan would become a professor of anthropology somewhere. Together we would lead a life of shared research. But prompted by medical anthropologist Al Damon ("Why don't you go to medical school, young man, and really learn something?") Dan enrolled in the Harvard-MIT joint M.D.–Ph.D. program. Instead of an anthropologist, I had married an infectious disease doctor. We published only one scientific paper together. But over time, as our mutual devotion deepened, maintaining our partnership became one of my life's main goals, requiring compromises I did not then anticipate. Thirty-six years into the enterprise, I can only second what I wrote in the dedication to *Mother Nature*: selecting Dan was "the wisest choice this female ever made."

So, headed for Dharwar, how did we end up at Abu? Return with me to Jodhpur, a bustling North Indian city set in the middle of the Great Indian desert, unlikely habitat for leaf-eating colobine monkeys. A population of around 1,000 langurs survives there, thanks to 13 committees composed of devout Hindus who supplement their diets with fresh produce on a daily basis. The designation "Hanuman" derives from the monkey who helped Lord Rama in *The Ramayana*, a Hindu sacred text, while "langur" comes from the Sanskrit *langulin*, "having a long tail". At Jodhpur, S. M. Mohnot and his mentor, Professor M. L. Roonwal, a towering figure in Indian zoology, greeted us warmly.

Early in the morning (by noon it would be 120 degrees), S. M. took me to the outskirts of Jodhpur. After scouring the rocky crags, I encountered my first langur, a female about the size of a springer spaniel with the slender-waisted elegance of a greyhound, an extraordinarily elegant silver-grey creature with a black face and dainty black gloves, inexplicably separated from her troop, making her way back to them as I scrambled behind. It was S. M. who advised me to go to Mount Abu. He promised that langurs at Abu would be no less crowded than at Dharwar, and at 4,600 feet, Abu would be healthier. We headed for Mount Abu to check it out.

The langurs of Abu (1971–1980)

Even beyond the sheer beauty of this town atop the Aravalli hills, Abu had much to recommend it. Langurs there were spread along a gradient from town-dwelling groups already habituated to humans, to wild groups on the forested hillsides. While Dan surveyed langur populations at Dharwar and elsewhere, I remained at Abu to map home ranges and learn to identify individuals. Because time was short, I focused on groups near town, including an unusually small group with a single adult male accompanying six females. There were four females with infants, a very distinctive female missing part of her forearm accompanied by twins, and a very old-looking, childless and periperhalized female that I named "Sol" (Hrdy 1974, 1977). Although I did not know it at the time, its small size and location made this group particularly prone to male takeovers. It would be the following year before I actually witnessed an adult male repeatedly stalk, attack, and wound infants in this troop, but observations that first summer were already leading me to reassess my starting hypothesis.

The langurs at Abu lived at relatively high density, in close proximity to humans, yet intra-troop relations were calm, just as Phyllis Jay had reported in her pioneering studies of the "peaceful" langurs she watched at northern Indian sites at Orcha and Kaukori. She had been aware of observations of fighting among langur males recorded by nineteenth century and early twentieth century naturalists, but dismissed them as "anecdotal, often bizarre, certainly not typical behavior" (Jay 1963: 8). At Abu, males were tolerant if aloof, and extremely protective of infants in their troop. Inter-troop encounters were tense affairs with mostly ritualized aggression. Even though males became agitated by the approach of all-male bands, nothing suggested "pathological" aggression. Yet in August of that year, a month after the monsoon began, the resident male in the Bazaar troop was replaced by a new and very distinctive looking male with a chunk missing from his left ear. All six infants were suddenly gone. Two people living in Hillside Troop's range independently told me that they had each seen something inexplicable, a monkey killing an infant. It dawned on me that infanticide might be more widespread and normal than I had assumed. Infanticide was not just occurring at Dharwar but also at Jodhpur and probably Mt. Abu. Fall term was about to begin, but I knew I had to return.

Becoming a sociobiologist

Spring of 1972, Ed Wilson and Irv DeVore co-taught a seminar posing the question: could there be a science of sociobiology? Wilson had just completed *The Insect Societies*. A great visionary of boundless optimism, Wilson was preparing to lay out his ambitious blueprint for integrating ecology, demography, genetics, development, behavior, and evolutionary theory in one grand explanatory framework, *Sociobiology: The new synthesis* (1975). His sense of mission was infectious. These were heady times to be anywhere near the Life Sciences at Harvard. In that seminar, I also made several lifelong friends. Peter Rodman, just back from fieldwork in Borneo, was using the seminar to write up his data on male and female foraging strategies among wild orang utans, while Martha McClintock, still primarily interested in the effects of pheromones on reproduction, was taking the opportunity to survey the (mostly rodent) literature on how latitude affects reproduction. My paper explored how "Hamilton's Rule" could help explain the evolution of allomaternal care in primates.

Still a prophet unrecognized in his own country, British evolutionary theorist W. D. Hamilton's ideas were being reverently explicated by just-minted Ph.D. Robert Trivers. Between 1971 and 1974, Hamilton's Harvard "bull-dog" was in the throes of producing his own classic trilogy on reciprocal altruism, parental investment, and parent-offspring conflict, articles that would transform the way I (and many others) thought about the evolution of social relationships.

My seminar paper was titled "The care and exploitation of Infants by conspecifics other than the mother." It was about costs and benefits of shared care from the perspectives of the various parties concerned: mothers, infants, and allomothers. By semester's end, I had not finished, but Wilson urged me on. When completed he submitted the manuscript on my behalf to Robert Hinde at *Advances in the Study of Behavior*. Written in 1972, this was my first scientific paper, although a delay in publication meant it did not appear until 1976. The acknowledgement read: "Without the advice and encouragement of Professor E. O. Wilson, I never could have completed this paper. Without the input of Dr. R. L. Trivers, it would not have been worth writing; in his writings and private discussions he has exposed me to a theoretical construct that I believe begins to make sense of the problems with which anthropologists must deal." By 1972 then, I already felt a profound debt to Trivers and Wilson and considered myself a sociobiologist.

The infanticide controversy (1974 to the present)

I returned to Mount Abu as soon as term ended. I would do so nine times between 1971 and 1980. Findings from 1,500 hours of observations from the first five field seasons were summarized in *The Langurs of Abu: Female and male strategies of reproduction* (1977). In chapter 8, "Infanticidal males and female counter-strategists," I explained how – far from being pathological – infant-killings were the outcome of goal-directed male behavior. After invading a troop from outside, a langur male would target unweaned infants and relentlessly stalk them in a process that sometimes continued over days. Male attacks on infants were not seen at other times, and a strange infant kidnapped from another group would not be attacked, so long as it was carried by a familiar female – that is, one with whom the male had mated. Although the tenure of resident males was highly variable, the average was about 27.5 months. I hypothesized that by eliminating his competitor's offspring, an usurping male enhanced his own opportunities to breed since females who lost infants resumed cycling sooner than if they had continued to lactate.

In 1974 I had proposed that infanticide at Abu could be explained as a variant of classic Darwinian sexual selection. By canceling the female's last mate choice, the new male reduced the reproductive success of a rival while improving his own chances to mate. The hypothesis generated very specific predictions. Following Trivers (1972), attackers should belong to the sex investing least in offspring (in this case, male). Victims should be unrelated, and also unweaned, with the effect of compressing a female's fertility into the limited period when the killer had access to her. Not only were my observations of attacks on infants consistent with these predictions, but new observations from other species were conforming as well. By early 1977, even before the book appeared, I was sufficiently confident that my hypothesis applied more broadly to publish an article in *American Scientist* entitled "Infanticide as a primate reproductive strategy." I proposed that infanticide by males was a highly conserved behavioral trait widespread in the subfamily Colobinae, but also cropping up (perhaps through convergent evolution?) *throughout* the Primate Order in prosimians, Old and New World monkeys, and great apes. The ensuing controversy caught me by surprise.

At the time I embarked on my research on langurs, primatologists (who, remember, initially came mostly from the social rather than the biological sciences) were profoundly influenced by social theorists like Durkheim and Radcliffe-Brown. Primate social organization was assumed to be a "functionally integrated structure" in which each individual had a role to play in the life of the group and all group members functioned together to ensure the group's survival. Thus early reports of infanticide by male langurs had been dismissed as "dysgenic." My proposal in the January–February 1977 issue of the *American Scientist* provoked a series of rebuttals, beginning in the May–June issue. They began when Phyllis (Jay) Dolhinow, the first Western primatologist to study langurs in the wild, wrote that "It comes as a great surprise that infanticide might be considered a normal adaptive evolutionary strategy…" because "Normal" langurs "do not kill infants." Because it "shows destruction not adaptation" the behavior I described had to be abnormal. Furthermore,

Dolhinow asserted, "Incredible powers of memory and reason are attributed to the langur monkey (how else could a male recognize paternity and recall events that occurred six months or more in the past?)" (1977: 266). These were early salvos in a long-running debate, which persisted long after the sexual selection hypothesis and other adaptive explanations for infanticide were accepted by biologists. Within anthropology, they continue to this day.

Looking back, I divide the saga into two phases. The first phase began with exchanges involving Dolhinow, her mentor Sherwood Washburn, and their students from Berkeley, followed by critical articles by other social scientists like political scientist Glendon Schubert (1982) and the eminent physical anthropologist Christian Vogel of Göttingen University, who in 1982 published "Der Hanuman-Langur (*Presbytis entellus*),[1] ein Parade-Exempel für die theoretischen Konzepte der 'Soziobiologie'?" These critiques occurred at an early stage in the study of this phenomenon, and partly grew out of the ongoing paradigm shift from selection at the level of groups to selection on individuals. In my opinion these were useful, ultimately constructive, debates.

On both sides, everyone agreed that we needed more and better data. Over the course of the controversy I learned to be more self-critical about assumptions. I still remember sitting down to correct by hand each reprint from my 1974 paper before mailing them out. In Table VI, where I summarized available data on "Political changes and infanticide at Dharwar, Johdpur and Mt. Abu" I added to the caption of the last column "Infants Killed *or Missing*." Although I had counted any infant attacked by a male who subsequently disappeared as killed by *that* male, this was only a probability, not a fact.

The criticisms also made me think harder about my main underlying premise. Though only expressed under specific circumstances, I assumed that infanticidal responses such as infant-biting were heritable traits. To this day, however, there is no definitive evidence of a genetic basis for this behavior in primates, although by the time I wrote the Preface to the new paperback edition of the *Langurs of Abu* (1980), evidence for heritability was emerging for rodents and in time grew stronger (Parmigiani and vom Saal 1994).

The controversy also pressured me to think more about "human disturbance." In collaboration with Jim Moore and two Berkeley-trained primatologists, Naomi Bishop (one of Dolhinow's students who had studied langurs at very low densities high in the Himalayas) and Jane Teas, we devised "Measures of human disturbance in the habitats of South Asian monkeys" (Bishop *et al.* 1981), published the same year that Oxford's Paul Newton reported infanticide among langurs in a disturbance-free North Indian tiger sanctuary. Finally, the controversy made me think much more critically about whether an infanticidal heritage among monkeys had *anything* to do with infanticide in our own species.

Whereas in nonhuman primates infanticide typically involves unrelated males (or occasionally, as in chimps or marmosets, rival females), human infanticide most often involves the closest of relatives, an infant's own mother. Yet with very few exceptions, maternal infanticide does not occur in wild monkeys and apes. In 1979 I devised a classification of

---

[1] *Presbytis entellus* was the nomenclature in use at the time, though taxonomists have now returned to the older name, *Semnopithecus entellus*.

infanticide according to different explanatory hypotheses, each generating its own set of (testable) predictions regarding the age of the victim; the age, sex and degree of relatedness of the infant to the killer; nature of gain (if any) to the killer, and so forth. The five main classes were: *Sexual Selection; Exploitation of the Infant as a Resource*; *Competition for Resources*; *Parental Manipulation*; and *Social Pathology*. Cases of infanticide in humans are far more likely to fit predictions from "Parental Manipulation" or "Resource Competition" than "Sexual Selection" even though, as Margo Wilson and Martin Daly would soon conclusively demonstrate, infants with unrelated males in the household were at greater risk. It was not until 1999, in *Mother Nature*, where I reviewed anthropological and historical evidence on maternal retrenchment, abandonment, and infanticide in humans, that I felt able to discuss how infant-killing by unrelated men fit in. Even then I tread cautiously and was tentative (in Chapter 10). My point here is that by and large these early exchanges were part-and-parcel of healthy scientific debate. They made me more cautious.

Without my realizing it, the second, far less constructive, phase of the debate began with the publication of *Infanticide: Comparative and Evolutionary Perspectives* (Hausfater & Hrdy 1984). The volume grew out of the First International Conference on Infanticide in Animals and Man, funded by the Wenner-Gren Foundation and held at Cornell University, August 16–22, 1982. The primate section consisted of chapters by primatologists from zoology as well as anthropology, including Hausfater himself (an Altmann student), Carolyn Crockett, Lysa Leland, Tom Struhsaker, Anthony Collins, Jane Goodall, Dian Fossey, and Yukimaru Sugiyama. Dolhinow declined, but her student Jane Boggess provided a detailed critique of the sexual selection hypothesis. Zoologists Craig Packer and Anne Pusey summarized evidence from lions and other carnivores. Ornithologist Doug Mock discussed siblicide in birds, and there were also chapters on exploitation of infants as a resource (mostly through cannibalism) in fish and invertebrates. Questions of proximate causation were discussed in chapters on controlled experiments with rodents by Fred Vom Saal, Bob Elwood, Bruce Svare and Craig Kinsley, Jay Labov, and others. The final section contained overviews by demographers and historians, as well a sociobiological overview on human infanticide by Daly and Wilson. I held up publication of the volume to include Bugos and McCarthy's extraordinary case study of maternal infanticide among the Ayoreo of Paraguay. It provided the first empirical demonstration that probability of neonaticide declines with maternal age and reproductive value. In retrospect, it is fortunate we included it, as shortly afterwards anthropologists began refusing to sanction publication of data on infanticide in traditional societies (discussed in Hrdy 1999: 293–7).

My 1984 Preface reveals how convinced I was that the volume would end the controversy. "Over the past decade," I wrote,

the intellectual pendulum... has swung from an earlier view that infanticide could not possibly represent anything other than abnormal and maladaptive behavior to the current view that in many populations infanticide is a normal and individually adaptive activity. ... Quite possibly, readers ten years from now may take for granted the occurrence of infanticide in various animal species and may even be unaware of the controversies. ..." (p. xi).

So far as biologists were concerned, that is what happened. The book was well reviewed, selected by *Choice* as one of the best academic books of that year, and my research on infanticide played a role in my election to the California Academy of Sciences in 1985 and to the National Academy of Sciences in 1990, as well as to the American Academy of Arts and Sciences in 1992. So far as anthropology was concerned, my optimism proved ill-founded.

In 1993 a long critique of my hypothesis that infanticide could be an adaptive reproductive strategy appeared in the *American Anthropologist* (Bartlett et al. 1993). An abridged version, "Infant killing as an evolutionary strategy: reality or myth?", by the same authors (anthropologist Robert Sussman, his student Thad Bartlett, and geneticist James Cheverud) followed. They claimed that "Most witnessed cases of infant killing appear to be simply genetically inconsequential epiphenomena of aggressive episodes" (1995:150). Publication was accompanied by a press release summarizing interviews with Dolhinow and other critics, generating articles in the popular press with titles like "Monkey 'murderers' may be falsely accused" (e.g. Mestel 1995).

The critics deemed comparative evidence from rodents irrelevant. This meant there was *no* evidence for a genetic basis for infanticidal behaviors, since everyone agreed that we had not shown this for primates. By then papers from the second international conference on infanticide held in 1990 at the Ettore Majorana Center in Erice, Sicily, had begun to circulate (Parmigiani and vom Saal 1994), including Volker Sommer's summary of 18 years of data from Jodhpur based on a long-term collaboration between Christian Vogel's team from Göttingen with Mohnot and others at the University of Jodhpur. That population of roughly 1,000 langurs had been continuously monitored by more than ten full-time grad students and post-docs, resulting in tens of thousands of observation hours. They had observed numerous takeovers accompanied by 13 "witnessed", 7 "likely", and 21 "presumed" cases of infanticide. In 95% of cases where paternity could be assigned, the killer had not been in a position to be the father. (DNA analyses demonstrating that males were not attacking their own infants would not be available until later; Borries et al. 1999). After reading his student Volker Sommer's Ph.D. thesis, even Vogel had reversed his earlier position, becoming Europe's strongest advocate of my hypothesis to explain infanticide by males.

Nevertheless, the article in the *American Anthropologist* included a half-page pie chart showing that 43.75% of all observed cases of infanticide derived from *Presbytis entellus*, while other species such as red colobus and blue monkeys living in the Kibale Forest of Uganda accounted for only a fraction of observed infanticide (Bartlett et al. 1993, Fig. 1). No mention was made of how many more hours many more individuals at Jodhpur had been monitored for far longer with excellent visibility. Instead, the authors suggested that the disproportionate number of cases meant there was something abnormal about langurs, and Jodhpur in particular. Furthermore, they claimed, killers were often fathers of their victims (Sussman et al. 1995: 149). This of course is not what the Jodhpur data they cited showed.

By then I was working on human inheritance patterns. Reluctantly, I joined Carel van Schaik and Charles Janson, then actively working on infanticide, to publish a brief reply (Hrdy et al. 1995). Against Carel's advice (and I regret not taking it) I inserted what I intended as a conciliatory passage about two different approaches to science. In the first more cautious and deductive approach, researchers proceed from known facts without imaginative leaps. In the other, relying on "strong inference", a hypothesis is devised even before all facts are in and researchers then test the predictions it generates (Platt 1964). "The continuing debate over infanticide among primates" I wrote,

reflects two different world views, both of them defendable. ... While some are interested in emphasizing the uniqueness of each case – a valid position – others are driven to seek general patterns and to use theory to explain them. ... The latter derive their greatest pleasure from noting that so many findings could have been correctly predicted on the basis of pitifully incomplete data sets merely by relying on logic, comparison, and extrapolations guided by evolutionary theory.

I had not meant to imply that I thought evidence was irrelevant. Nevertheless, for years afterwards, I would encounter in the pages of the *American Anthropologist* and elsewhere statements to the effect that Hrdy believes in "powerful models regardless of data" (Fuentes 2002: 696). Reading these still evokes a visceral, sick feeling.

Historically the American Anthropological Association's flagship journal has not devoted much space to nonhuman primates, and virtually none to nonprimates. But once Sussman became editor, for the first time, the journal published an article by Canadian zoologist Anne Innis Dagg titled "Infanticide by male lions hypothesis: a fallacy influencing research into human behavior" (1999). The manuscript had previously been turned down by biology journals. When Sussman learned of it he phoned the author and told Dagg that if she added some references to primates, it would be publishable in the *American Anthropologist*. As Dagg subsequently told a reporter for *Lingua Franca*, she was "astonished" but pleased (Shea 1999: 25). Zoologist Craig Packer immediately replied that "Infanticide is no fallacy" (2000), as did a consortium of (mostly) primatologists led by Joan Silk and Craig Stanford, although I do not believe it was ever published. So the controversy rolled on – but without me.

Once before, in 1976 when the American Anthropological Association entertained a motion to "ban" sociobiology, I had resigned my membership in disgust, but later rejoined. This time, I felt as if I occupied some separate reality with nothing more that I could usefully or appropriately say. I returned to the debate over the sexual selection hypothesis again only once, just long enough to write the preface for the third volume on infanticide, this one edited by Carel Van Schaik and Charles Janson (2000). The title, *Infanticide by Males and Its Implications*, signaled their intention to ignore the controversy in anthropology and finally move on. It was a source of pleasure to me that two fine, innovative chapters were written by young anthropologists who had taken a 1984 seminar on infanticide I gave at UC-Davis in my first year teaching there. One, by Leslie Digby, reviewed infanticide by female mammals and its implications for the evolution of social systems; the other, by Ryne Palombit, explored "Infanticide and the evolution of male–female bonds in animals."

Fieldwork, politics and lost opportunities

I discussed the infanticide controversy at some length because it may be of more interest to animal behaviorists than other aspects of my anthropological career. It was a minor skirmish in the broader critiques of science by post-modern deconstructionists known as the "Science Wars" and in the larger controversy surrounding sociobiology. During much of this time, I was still going back and forth to India. Expenses rarely amounted to more than my air, rail, and bus fares, and the cost of renting a little bungalow within walking distance of the langurs. Only in the final years, as the research became better known, did we receive significant outside funding from the Smithsonian and National Science Foundation. This funding, along with an ambitious expansion of the project to include a number of researchers, turned out to be the kiss of death. By the end of the first funded year we found outselves embroiled in a different sort of controversy, one which would bring fieldwork in India to a close.

The field of primatology was developing fast. My research on langurs was not keeping pace. In June of 1979, Irv DeVore and Dan submitted a joint Anthropology Department/ Harvard Medical School proposal to do an "Integrated Field Study of the Behavior and Biology of the Hanuman Langur." (As an unsalaried post-doctoral researcher, I lacked standing to be a principal investigator.) It was an ambitious and, for its time, innovative proposal. We planned to integrate behavioral observations with epidemiology and genetics. While Sylvia Howe focused on maternal and allomaternal behavior, I would study female sexual behavior and Irv's grad student Jim Moore would tackle the roving "all-male bands" – langur male bands containing anywhere from 2 to 60 or more males, sometimes temporarily joined by females (Figure 13.1). Although fascinating, the fast-moving male bands were difficult to study. A pilot study on the steep hillsides around Abu indicated that Jim had the physical stamina to keep up with them. Meanwhile, Dan (by then working on double-stranded RNA viruses) and Rob Negrin from Harvard Medical School would study the epidemiology of rotavirus in langurs and the other animals (including humans) in their ecosystem. For this research, we would all collect stools and also briefly trap animals for measurements, tooth casts, and blood samples. Preliminary research (D. Hrdy *et al.* 1975) indicated that there were sufficient blood protein polymorphisms in langurs so that we could use blood samples combined with behavioral observations to work out relatedness and do some paternity exclusions. (This was in the days before we had less invasive methods for DNA analyses.) Had we succeeded, this would have been the first primate field study to integrate behavioral and genetic data.

From the outset there were problems. Our collaborators at Jodhpur University were eager to have us work among the provisioned langurs at Jodhpur. However, Jim and I worried about criticisms over human disturbance. Furthermore, Vogel and his team were understandably not happy to have us trap Jodhpur langurs that they were studying, nor did we want to. To Jim and me, Abu was the obvious choice. We had long-term records yet there was still much to learn, especially about the troops and male bands out on the relatively undisturbed hillsides. Sylvia Howe, on the other hand, wanted to work at Ranthambhore, a tiger sanctuary with a charismatic and highly effective field director (Fateh Singh Rathore) eager to help with the research.

Other problems had to do with public relations. In the ongoing political drama in South Asia, the U.S. had just announced that it would not continue supplying fuel to the Indian nuclear power plant at Tharapur, thus appearing to "tilt towards Pakistan." There was also

Figure 13.1. Langurs are found in breeding troops and in roving "all-male" bands containing anywhere from 2 to sixty or more males, sometimes, as in this rare photograph, temporarily joined by mothers who join with ousted males to avoid usurping males back in their "home troop". (S. B. Hrdy/Anthro-Photo.)

tension over U.S. efforts to remove a ban on exporting monkeys for medical research. Photographs of macaques being used in U.S. radiation experiments had just been prominently published throughout India. All this meant that Americans seeking to trap monkeys that Hindus considered sacred, even if briefly and with no harm to them, had some explaining to do. Our problems made us vulnerable, but with tact and the support of our Indian colleagues, we should be able to satisfactorily explain our activities (Figure 13.2). However, there was another obstacle no one anticipated.

Our research permissions were granted at the federal level, through the Ministry of Education and Culture and the Ministry of Agriculture. Within Rajasthan, our contacts were with local forest officials at Abu and Ranthambhore. However, the Chief Wildlife Warden at the state level was the former director of the New Delhi Zoological Park, a politically well-connected expert on tigers, known for his tiger photographs. Unknown to us, this was the same individual who had undermined efforts to study Indian wildlife by other American researchers, including George Schaller (which is why Schaller did his classic study of snow leopards in Nepal instead of India). Only later, during the months Dan was fruitlessly commuting back and forth between Delhi and Jaipur in an effort to have our research permissions reinstated, with time on my hands to read old files in the American Embassy, did I learn about these previous – virtually all aborted – American projects, and notice the recurring patterns.

Figure 13.2. Studying animals regarded as sacred meant that many langurs were already habituated to people. But it also meant we had a lot of explaining to do if we wanted to mark or trap them. Here a langur visits one of the saddhus living in the hillsides around Mount Abu. (S. B. Hrdy/ Anthro Photo.)

Why should we have wandered so blindly into this morass? I wanted to write about it. However, Smithsonian officials were concerned that I would further complicate the situation and requested I not do so. I complied, but in retrospect think it was a mistake. Many individuals in India and its government care deeply about Indian wildlife. Had what was going on become more generally known, the situation might have improved.

I first met the Chief Wildlife Warden of Rajasthan during his visit to Mount Abu in February of 1980. He was all charm. Later, after dinner at a local friend's home, we spoke late into the night and he told me that he had many admirers and many enemies but I was his "only friend." Still, I could not help but notice that whatever paperwork we provided, he always requested something more. He never confronted me directly. The full range of his mercurial personality was reserved for the students. When interacting with Sylvia Howe, he would alternate between great warmth, scathing attacks, and threats to have us all thrown in jail. It became increasingly clear that, for whatever reasons, this man did not want foreigners studying wildlife he regarded as *his*. When federal and local officials continued to support our work, sensational stories, originating from Jaipur, authored by a journalist friend of the Chief Warden, began to appear in the Indian press. Dan was accused of working for the

American "Defense Pathology" organization, with hints of involvement in germ warfare (which fortunately, after the matter came up in Parliament, the Indian Ministry of Defense opted to ignore, pointing out that no such organization existed). I was accused of wanting to export monkeys for profit. In a quote that I am sure was distorted and taken out of context, Professor Vogel was quoted as saying that our work would render langurs dangerous to humans.

About that time, a troop of langurs was trapped and then bludgeoned to death by villagers, probably in response to crop-raiding. On September 21, 1980, a story datelined Jaipur and titled "100 Langurs Killed" appeared in *The Statesman*. At the bottom appeared a gratuitous passage about how "American researchers on the Hanuman langur… had created a row… The issue was raised in Parliament and also in the State Assembly." The hint dropped there was picked up in the *National Herald* on September 23, which ran a story headlined **"FOREIGN HAND IN LANGUR KILLINGS?"** From then on, our fate was sealed. Though never actually revoked, our research permissions were suspended. Given the cultural sensitivity of the buttons pushed – U.S. conspiracies to get around bans on exporting monkeys, foreign agents, germ warfare, the murder of sacred monkeys – no official dared help us. About this time, Christian Vogel's team ran into similar problems, but the German government made foreign aid contingent on one German researcher a year being able to work at Jodhpur. The Indo-German langur project at least limped on, ultimately yielding important knowledge.

For Dan, me, and the students stranded in India but unable to watch monkeys, it was a nightmare of many months' duration. Hardest to bear was the sheer waste of it. No one gained. On June 10 of that year, before the storm but as clouds were massing, Irv DeVore wrote a masterful letter to Professor Roonwal, "grand-old-man" to "grand-old-man". Irv acknowledged current and past difficulties – international, local, and personal – then got to his point. If this American effort – by far the most carefully orchestrated of several previous efforts – failed, it was unlikely there would be another in the foreseeable future. After three decades, I still stay in touch with Indian colleagues, and every so often am reminded of Irv's letter when I read a proposal from a young Indian primatologist for whom, without much contact with developments outside, time seems to have stood still. Even though India has produced some of the world's finest scientists in highly competitive fields, and in spite of the existence in India of a rich array of lorises (two species), macaques (7), langurs (5), or the Hoolock gibbon (Roonwal & Mohnot 1977), few of these remarkable creatures have been well studied in the wild.

Occasionally a journalist writes about me and mentions my early research on "lemurs" (sic!).

Given how widespread langurs are, ranging from sea level in the south to high altitudes in the Himalayas, and given how relatively terrestrial and easy to observe these elegant and fascinating monkeys are, it is staggering how little we know about even this best-studied of all Indian primates. African baboons, also widespread and terrestrial, have become the best-known primates in the world. Yet many people don't even know what langur monkeys are. Several Indian species are liable to disappear before they are ever studied.

Raising Darwinian consciousness

It was the phenomenon of infanticide that drew me to study langurs. Increasingly, however, I found myself drawn into the "great Colobine soap opera" unfolding before me in female-centered social groups. Nor could I help empathizing with the plight of fellow females. Every 27 months or so, a strange male would burst into a mother's world and stalk her infant. If he succeeded in killing it, within days, the mother would sexually solicit the killer. So why didn't female langurs, Lysistrata-like, simply sexually boycott infanticidal males, eliminating this noxious trait from the gene pool? That mothers did not forced me to rethink selection pressures on females (Figure 13.3).

Most Darwinians still took for granted that natural selection weighed more heavily on males than on females. Even top textbooks still presumed that "most adult females... are likely to be breeding at or close to the theoretical limit" while "among males by contrast there is the probability of doing better" (Daly & Wilson 1978: 59). The *Langurs of Abu: Female and male strategies of reproduction* was the first book on wild primates to devote equal attention to the reproductive strategies of *both* sexes. Yet, as I ruefully noted in the Preface to the 1980 paperback edition, responses focused almost entirely on males. Male behavior, I complained, has this power to rivet attention.

The chapter on female–female competition described a novel form of female dominance hierarchy, one in which young females gradually rose in rank, occupying the top positions in the hierarchy when they were at peak reproductive value, and then declining with age. It was because I was unsure about this interpretation that after submitting my thesis in 1975, I had rushed back to Abu. When I left the season before, three subadult females in Toad Rock Troop – if my model based on reproductive value was correct – were poised to rise to the top of the female hierarchy. But would they? On my return, I was amazed by how unanimously

Figure 13.3. A "sisterhood" of langurs resting in a *Grevillea* tree at Abu. (S. Hrdy/Anthro-Photo.)

the monkeys confirmed my predictions. All three young adult females routinely displaced older females who weighed more, monopolizing the top three rungs of the female hierarchy. As usual, the oldest females remained at the bottom (Hrdy 1977, Fig. 6.9). Only later would members of the Indo-German team confirm the existence of this same pattern in langurs elsewhere, first at Jodhpur, and then at the new langur study site Paul Winkler and Volker Sommer founded in Nepal (see Borries *et al.* 1991). It was a never-before-documented type of female dominance hierarchy, consistent with George Williams' ideas about the importance of reproductive value. I waited in vain for some feedback.

Similarly, the chapter on infant-sharing provided the most detailed examination to that point of exactly which females engaged in allomaternal care and why. Yet almost all the commentary focused on males (Figure 13.4). In the preface to the paperback edition, I actually pleaded with readers, expressing the hope that "this paperback edition will ... call attention to a facet of the behavior of female primates too lightly brushed aside the first time around: The extent to which females are competitive creatures of strategy whose preoccupations extend far beyond 'mothering' and the traditional boundaries of 'maternal behavior'" (1980:vi). I knew that we could not understand reproductive strategies of either sex without taking into account the other, but in reaction to the androcentric response to *The Langurs of Abu*, I decided to make females the focus of my next book. There was also another reason. The intellectual ferment that was transforming the intellectual landscape within the social sciences, and which was also just beginning to transform the genderscape at American universities, was trickling into my consciousness.

Without my being very aware of it, the Women's Movement had been picking up steam. In fields such as history and anthropology, reactions against "top-down" history, along with

Figure 13.4. Two mothers spar at the interface between their home ranges. When the *Langurs of Abu* was first published, I received a letter from sociologist Jesse Bernard telling me that this was the first image she had ever seen of female–female aggression in nonhuman primates. Overall though, most responses focused on male behavior. (D. Hrdy/Anthro-photo.)

a new interest in marginalized peoples, were already under way. In the emerging field of women's studies, critiques of science were actually pitting "feminist" scholars *against* science, sociobiology in particular. This was partly in response to what was happening at roughly the same time in the life sciences. Genetics was moving to center stage, including among those seeking to understand the evolution of sex differences. The "gendering" of the social sciences was on a collision course with "the gening of America" and especially with what social scientists regarded as the twin evils of "genetic determinism" and "reductionism" inherent in sociobiology. It was a time of tremendous intellectual tension. I was caught smack in the middle.

To me the outstanding problem was Darwin's brilliantly original theory of sexual selection, so powerfully explanatory in some spheres (as when applied to explain the evolution of infanticide by males), yet so short of the mark in explaining how females also strive for reproductive success. Important assumptions underlying the theory, especially those being used to extrapolate to humans, were based on misleading Victorian stereotypes. After Ernst Mayr alerted Trivers to the "key" reference for his 1972 paper, the rediscovery of Bateman's *Drosophila* experiments highlighted this supposed dichotomy between an "undiscriminating eagerness" to mate in males, and "discriminating passivity" in females. Bateman had extrapolated from observations of a single species of fruit-fly to humans. From there, often invalid stereotypes about "The Reluctant Female" and "The Ardent Male" passed unchallenged into sociobiology (see, for example, Daly & Wilson 1978: 55), and later (long after animal behaviorists knew better), into evolutionary psychology.

In *The Woman that Never Evolved* (Hrdy 1981) I surveyed both power relations between males and females and the role played by female–female competition. I identified polyandrous tendencies universally present among primates and explored reasons why females in so many species of monkeys and apes expend so much energy and take such risks to mate with more males than are needed to ensure conceptions –one of the reasons male primates have to try so hard to control females (my first foray into the origins of patriarchy). I reviewed emerging evidence for cyclical libido as well as lapses from it (as in situation-dependent sexual receptivity) and explored selection pressures shaping sexual swellings at estrus, functional clitorises and erratic orgasmic reward systems, explaining why primates (humans included) eschewed conspicuous advertisements at ovulation. My goal was to demonstrate that the sexually passive female in Darwinian stereotypes could not have evolved within the Order Primates. But this was still some years before our current open discussion of female sexuality. Even the great Masters and Johnson only broached topics like female orgasm wearing white lab coats and grave faces. I was more than a little anxious. When a friend cautioned me that "it sounds like a woman looking between her legs with a mirror" I deleted the subtitle *"a primatologist examines her sex"*.

The book was well received, selected by the *New York Times Book Review* as one of the Notable Books of the Year. Furthermore, the backlash I anticipated from feminists never came. Through a stroke of good fortune, an early review appeared in the radical Washington, D.C., newsletter *Off Our Backs*. The reviewer had background in biology, grasped my intentions, and declared that "every aspect" of the book reflects a feminist perspective. This gave some of sociobiology's fiercest feminist critics cause for pause. Thereafter, the most common responses from those in women's studies and related fields were polite nods, but initially not a whole lot more. By now, however, gender studies programs were more open towards biologically based explanations of behavior.

This transformation was beginning to be apparent during the period between 1995, when I first spoke in the Gender and Science program at Princeton, and 1999 when I spoke there again. On the first occasion I gave a talk called "Raising Darwin's consciousness: female sexuality and the prehominid origins of patriarchy." Questions following my lecture were tinged with hostility. I was pointedly asked why evolutionists were so fixated on reproduction? Didn't our bias "privilege" heterosexuality? When in 1999 I lectured there again, this time on maternal love and ambivalence (for feminists an even more politically charged topic than sexuality is) I sensed genuine interest in the biology of women as well as recognition of a need to engage with evolutionary concepts.

I was hardly the only Darwinian interested in revising longstanding biases. Later, I learned more about women writers – marginalized from science and largely ignored – who a century earlier had politely proposed that evolutionary theories would benefit from greater consideration of selection pressures on females (Hrdy 1999: 12–23). Women's contributions have this way of fading from view. All the more reason not to forget this humbling history. Even though some contemporary biologists remain resistant to the notion that evolutionary theory ever had a problem with androcentric bias, that concepts needed to be revised, or that inclusion of women researchers had anything to do with fixing them ("Females were just harder to study," a prominent British biologist told me with a perfectly straight face), what stands out most in my memory is that from the early 1980s on there was a lot of support from such legendary evolutionists as George Williams and Bill Hamilton. George went beyond moral support, lending his eminence, offering to co-author a critique of "Darwin and the Double Standard" with a younger, far less distinguished colleague (Hrdy & Williams 1983). Later, John Maynard Smith and Bill Eberhard would also urge more open discussion about how "inadvertent machismo" affected the way sexual selection theory was applied. The 1994 Symposium on Evolutionary Biology and Feminism, conceived by Patricia Adair Gowaty and sponsored by the Society for the Study of Evolution, marked the culmination of this trend (Gowaty 1997).

Far from resistance, from the 1980s onward there was a small stampede among animal behaviorists to study female reproductive strategies, especially phenomena related to female mate choice. One of the ironies of the charge "sexist" so thoughtlessly leveled against sociobiology in the 1970s was that it was sociobiology's relentless focus on selection at the level of individuals that, after more than a century of neglect, ushered in an expansion of evolutionary theory to include both sexes.

Correctives proposed by women had nothing to do with females "doing science differently." In an essay entitled "Empathy, polyandry and the myth of the coy female" I traced the impact that questions raised and pursued by women researchers have had in behavioral biology, especially regarding sexual selection. Far from rejecting Darwinian theory, we sought to refine and expand it so as to better encompass selection pressures on both sexes.

The essay appeared in a volume on *Feminist Approaches to Science* (1986). Although I was explicitly *not* using feminist perspectives to arrive at any ideologically preordained conclusion, the presence of the "F-word" ensured that few biologists ever read it. I was grateful therefore when, two decades later, Elliott Sober reprinted it in the third edition of his *Conceptual Issues in Evolutionary Biology* (2006).

Women, science and compromises

Up to this point, the mentors mentioned have all been men. There is a reason. The year I graduated from Radcliffe, I only recall one woman professor, Cora DuBois. She retired that year. By 1970, there was not a single female full professor in Harvard's Faculty of Arts and Sciences, leading in April of that year (according to a June 11, 1970, story in the *Crimson*) to Harvard becoming the first school in the nation to come under (preliminary) federal investigation for sex discrimination. Yet there *were* talented women about. One day while lost in the labyrinthine basement of the Peabody Museum, as if in a fairytale, I encountered an elf-like, very intense, chain-smoking woman in a tiny office down there. She was Tatiana Proskouriakoff, the museum's "honorary curator" of Mayan art, at that time probably the greatest living Mayanist. (Proskouriakoff discovered that hieroglyphs carved on monuments actually recorded historical events.)

So far as primatology was concerned, I had to look far beyond Harvard for female role models, to Jane Lancaster, and especially Alison Jolly, then in an adjunct position at Sussex. I vividly remember holding Alison's path-breaking 1966 monograph on *Lemur Behavior* and scrutinizing the photograph on the back, staring into the woman's eyes. Later, when Jolly visited Harvard, I surprised myself by impulsively blurting out from within a throng of graduate students clustered about her: "But what is your life like?" I admired her as a researcher, writer, and as a person, and also knew that she had a husband and children. How did she manage? There was no real opportunity for her to answer then. Later I would learn that (like me) she never held a full-time tenured professorship. Alison was probably my primary role model.

In 1975, the year I received my Ph.D., molecular biologist Nancy Hopkins published an essay entitled *The high price of success in science*. Nancy went on to become a professor at MIT, as famous for her activism on behalf of women scientists as for her research on genes involved in cancer. Except for a brief union in graduate school, she never married, becoming a self-described "nun of science" until suddenly at age 60 she fell in love and married a financier (Sipher 2007). Hopkins questioned whether it was possible for a woman to be a successful wife and mother as well as earn a living as a full-time professor at a top research university. "Serious science with its long hours and energy absorbing quality is barely compatible with motherhood and being a professor … it is barely even compatible with a sustained husband/wife relationship." In her view, "the intellectual processes involved in 'real' science are as natural (or unnatural) to women as they are to men. But 'professional' science was constructed by and for men (a certain type of man), and a woman who chooses to conquer this world at its higher echelons usually requires a major overhaul of self and world views" (1975: 16).

Such views impacted me, though not always in ways I was aware of at the time. Years after the fact, the anthropologist William Irons ruefully told me how, shortly after I received my Ph.D., he and Napoleon Chagnon had sought to recruit me to their university. Because the anti-sociobiology campaign was in full swing, Penn State's was virtually the only Anthropology Department in the United States actually *looking* to hire a sociobiologist. In line with current practice, they first approached my advisor, not me. According to Irons, he simply said "Oh, she's married" and proposed another recent Ph.D. with similar training in primate sociobiology, male, and also married, whom they hired.

Nancy Hopkins had in mind scientists who ran big labs, who every few years would have to confront stiff competition in order to get grants to keep them running. Still, my advisor had a point – married women in academics confront special challenges. Since institutions were not going to change in time, I had to make my own adjustments. As a primatologist familiar with spending long periods of time at Abu by myself, I already had an idea about how incompatible fieldwork might be with family life.

After our research permits in India were suspended, instead of trying to start up from scratch in another country I shifted to archival research. I loved fieldwork, and regretted the shift. Yet, I reminded myself, none of my compromises were so great as those made by intellectually ambitious women before me, women like St. Hildegard of Bingen, the abbess who first described estrous cycles in monkeys. She really was a "nun of science." Plus, as I saw it, after the debacle in India, it was Dan's turn. He had made massive sacrifices to share my work in India. Thus in 1981 I followed Dan to Houston, where he could complete his medical residency at Baylor University. I took a part-time position across the street as a visiting associate professor of Anthropology at Rice University.

Spring of 1982 we returned to Cambridge so Dan could complete his Infectious Diseases fellowship at Peter Bent Brigham hospital and his doctoral research in Bernie Fields' lab at Harvard Medical School. By then I had two children. Assessing the three strands to my life then – *family*; *research and writing*; and *teaching* – I figured I could manage two of the three. But given institutional expectations and support systems then available, committing to all three involved cutting more corners than I wanted. I remained an (unsalaried) research associate at the Peabody Museum while working as a volunteer at Sasha and Katrinka's daycare center. In 1984, Dan accepted a position at the University of California-Davis medical school. Benefiting from a special program to target "outstanding women", simply by flying across the country, I went from being a volunteer in a daycare center to being a full professor permitted the option of working part-time.

Work, life, balance, and compromise

Conflicts between women's aspirations and the needs of their children are complex and very nearly irreconcilable. Still, some compromises work out better than others. Whenever some administrator or journalist holds me up as an example of a woman who combined a career in science with rearing three children, I feel obliged to point out how misleading that term "career" is. I never worked "full-time", or more precisely, was never paid for working full-time. I expended far more on allomaternal assistance than I ever earned from teaching. I could only afford to provide what my children needed, forgoing medical benefits and pension plans, because I benefited from inherited wealth. It's worth asking then how many women far more talented than I am had to forgo careers in science not just for the usual reasons (subtle and not-so-subtle discrimination) but because they were offered situations on terms that few mothers would be willing to accept? Nor do I exemplify the "opt-out revolution," the term used by *New York Times Magazine* writer Lisa Belkin in 2003 to describe highly educated professional women who step off the "fast track" to "stay home" with children. It's true that in 1996 I abruptly retired, but as I tried to explain to the NYT magazine's fact-checker when she called, I was *not* opting out.

By the 1990s, I had begun archival research on human inheritance patterns and "parental investment after death" in collaboration with Debra Judge, another "lapsed primatologist". We combined classic ethnographic research with empirical tests of hypotheses generated by sociobiological theories (Judge & Hrdy 1992; Hrdy & Judge 1993). Ours was the most detailed, longitudinal analysis of American probate records ever undertaken. It was tedious work undertaken in dusty courthouses, but interesting enough once you got into it. More importantly, it enabled two mothers to flexibly schedule research close to home.

Like many women, I entered this period of my life sandwiched between work (research, teaching, and administrative duties) and being a wife and mother as well as having a natal family that needed me back in Texas. My irreconcilable dilemmas involved work versus family as well as questions about *which* family? Commuting back and forth between the university and home, California and Texas, I developed such classic stress symptoms as migraines and crippling neck and back pain.

After a prolonged illness, my mother was dying. Her son, my brother, was killed the year before. Both she and I were convinced, but never able to prove, that he was murdered. In the wake of their deaths, problems in my natal family ballooned out of control. My life was taken hostage by, among other things, a series of court cases and lawsuits. My university allowed me time off. But when the dust settled, I questioned whether after such a hiatus I would be able to re-engage in creative work in a fast-developing and competitive field?

In 1986 I had received a Guggenheim fellowship to write a book on "the natural history of mothering", but my third child, Niko (named for Niko Tinbergen), was born that year. Several papers, but no book, got written. I either needed to write the book, or abandon the idea. In 1996 I submitted a proposal to publishers, and was surprised by the interest, which resulted in an auction. Instead of going with the highest bidder, I selected Pantheon because of their fine science editor, Dan Frank. Even so, I emerged with an advance that over the three years it took to write the book paid more generously than any university salary I had ever received (which at quarter-time was US$15,000 annually when I retired). For the next three years my time was divided between *Mother Nature* and being a mother to talented, strong-willed, complicated human youngsters. After the book was done, it was awarded the Howells Prize for outstanding contribution to Biological Anthropology and chosen by both

Publisher's Weekly and *Library Journal* as one of the best books of 1999. UC, which had accorded me the title of *emerita* when I retired, awarded me their "Panunzio Award" for post-retirement productivity. Leaving the university was scarcely "opting out."

By this time, Dan was engaged in commercial walnut-growing, and we were living on a farm near Davis, intensely involved in our offspring's lives as well as habitat restoration, consciously seeking to lead "balanced" lives. It was 1997, while working on *Mother Nature*, that I reread Nancy Hopkins' essay and picked up the phone to ask her what she thought decades later. By then, of course, female footholds in science were considerably more secure. Yet as Nancy put it: "Each generation of young women thinks it is an issue of the past and then has to discover for themselves … (how hard it is) to create an environment where their own way of being is allowed" (pers. comm. July 16, 1997). The elephant in the lab is the tremendous responsibility rearing children entails, even more daunting perhaps for parents familiar with humankind's evolutionary heritage. Left alone, even for minutes, infant primates are well within their rights to exhibit distress. In the hominid line infants seek even more than tactile reassurance, perpetually monitoring for signs of emotional commitment from mothers and allomothers alike. For reasons explained in *Mother Nature* and elsewhere (Hrdy 2005), human infants are "connoisseurs" of such commitment, requiring more than other primates. Nor (as some administrators seem to imagine) does parental investment end with infancy or when children are "school-age" – neither in the Pleistocene, nor today.

Science and motherhood

My first child had been born in 1977. Twice we took Katrinka to remote parts of Rajasthan. I doubt I would have done so had Dan not been an infectious disease specialist (Figure 13.5 and 13.6). Even so, it was harrowing for all concerned. The first year I brought along a young artist/*au pair* eager to experience Rajasthan, but alas, not interested in childcare. Throughout that field season, Katrinka suffered chronic diarrhea and virulent diaper rash. I would come back from watching langurs dawn-to-dusk to find an unhappy toddler and problems needing to be solved, ranging from laundering diapers to preparing healthy meals. Exhausted, I became ill. On the plane home I lay prostrate on a row of seats, suffering from pneumonia, with a temperature of 104 degrees, sucking on one of Katrinka's baby bottles to stay hydrated. Next day, the *au pair* quit.

Another time, I left Katrinka in Cambridge with her father and a housekeeper – also hard on her. My second daughter, Camilla ("Sasha"), was only a week old when she flew with me to Ithaca for the First International Conference on Infanticide – no place for a baby, my co-organizer declared (Figure 13.7). I was told not to bring her into the building where the conference was held. This is why in the days before flying to Ithaca I enlisted another mother with a year-old baby and lots of milk to nurse Sasha daytimes while I suckled her only at night so as not to build up my milk supply so much I could not handle hours at the conference. I was taking a risk. Back at the hotel, Sasha was given expressed milk from a

Figure 13.5. Katrinka was 14 months old when she first traveled with me to Rajasthan. Taking young children to the field can be logistically challenging, and Katrinka herself often found her strange new environment daunting. (D. Hrdy/Anthro-Photo.)

Figure 13.6. Dan joined us when he could take time off from his own research on rotaviruses in Bernie Fields' lab at Harvard Medical School. (S. B. Hrdy/Anthro-Photo.)

Figure 13.7. Sasha was born well past her due date, arriving a week before the 1982 conference on Infanticide in Animals and Man. This panel from a comic strip by my friend Susan Meddaugh, author of *Martha Speaks*, presents a child's-eye view of her mother's obsessions, which may or may not explain Sasha's late arrival. (Courtesy of Susan Meddaugh.)

bottle by a nurse. She might have learned to prefer the more rapid flow of the bottle and thereafter refuse my breasts. Fortunately, Sasha was an easy-going baby.

My third child, Niko, also experienced premature exposure to academia. Joan Silk, Dorothy Cheney, and I decided to organize a session at the American Anthropological Association meetings to demonstrate why anthropology departments needed to hire young primatologists. Such events are planned far in advance. As it happened, we all three turned up nursing new babies. From the outside it may have looked like we easily combined motherhood with professional responsibilities. From the inside – speaking now only for myself – I was too preoccupied to learn or offer much of interest. And what of infants treated this way?

By this time I was busy learning everything I could about infant development and attachment theory, as well as the biological, evolutionary, and historical bases of maternal love, and to combine this with what I knew about maternal ambitions and the ambivalence that I knew from personal experience tensions between maternal love and ambition could produce. Increasingly convinced that, unlike other apes, human mothers had not evolved to care for their children alone, and that our species had evolved as "cooperative breeders" (Hrdy 1999, 2005), I looked for ways to build up a stable network of as-if extended family, composed of daycare providers and resident allomothers. One of these allomothers, Lupe de la Concha, has lived with us now for more than twenty years. She has her own career but continues to be as involved in the lives of far-flung Hrdy progeny as I am. Also, for the first time I was beginning to have collegial support from women evolutionary biologists, a commodity hard to come by early on.

In graduate school, Martha McClintock had been my only "age mate." I grieved when she left Harvard to complete her Ph.D. at Penn. I felt increasingly alienated by "post-modern" intellectual trends within anthropology that struck me as anti-scientific. But the closer I got to Harvard's evolutionists, the scarcer women became. At Davis, fellow professors in biological anthropology were also all male, albeit preternaturally supportive ones. As at Harvard, my women friends tended to be younger, often my own or Peter Rodman's students, Joan Silk, Meredith Small, Amy Parish, and Debra Judge. It was the 1990s before I connected with a "critical mass" of female colleagues my own age with whom I could explore ideas about how to combine productive careers in science with healthy human lives – a vital, ongoing discussion.

Separated by vast distances, we found innovative ways to meet (e.g. house parties). These friendships made a difference to me personally as well as to my work. Patty Gowaty's bold theoretical footprints are all over my writings from the late 1990s, Jeanne Altmann's on *Mother Nature*, and after 1998, nothing I wrote was ever *not* influenced by Mary Jane West-Eberhard's ideas about development, Kristen Hawkes' and Polly Wiessner's views on early human food sharing, or Sue Carter's coaching on proximate causes. A 2003 Dahlem conference in Berlin that Sue Carter spearheaded was the first scientific meeting I ever attended where discussions about infant emotional needs and optimal daycare were incor-porated right into week-long sessions on the neurobiology of parental behavior (Carter *et al.* 2005). The resulting synthesis provides the groundwork for my current book, *Mothers and Others: The Origin of Emotionally Modern Humans*.

If previously I identified with "a lonely castaway throwing message bottles from a desert island," increasingly I felt like a swimmer "gently uplifted by a pod of supportive dolphins – fellow Darwinians and fellow feminists" (as I mused in 1999: 599). Evolu-tionary theory, which had for so long overlooked or miscast selection pressures on females and left out altogether consideration of the special cognitive as well as emotional needs of human infants, was being revised and expanded in ways that brought me closer to the old-fashioned anthropological mission I signed on for: "studying human nature in all its diversity."

Acknowledgements

Thanks to Jeanne Altmann, Dan Hrdy, Camilla Alexandra Hrdy, Bernard von Bothmer, Jim Moore, June-el Piper, Volker Sommer, and M. J. West-Eberhard for advice.

References

Altmann, J. (1980). *Baboon Mothers and Infants*. Cambridge, MA: Harvard University Press.

Altmann, S. A. & Altmann, J. (1970). *Baboon Ecology: African Field Research*. Chicago, IL: University of Chicago Press.

Bartlett, T. Q., Sussman, R. W., & Cheverud, J. M. (1993). Infant killing in Primates: a review of observed cases with specific references to the sexual selection hypothesis. *Am. Anthropol.* **95**: 958–9.

Bishop, N., Hrdy, S. B., Teas, J. & Moore, J. (1981). Measures of human influence in habitats of south Asian monkeys *Int. J. Primatol.* **2**: 153–67.

Blaffer, S. C. (1972). *The Black-man of Zinacantan: A Central American Legend*. Texas Pan-American Series. Austin, TX: University of Texas Press.

Borries, C., Sommer, V. & Srivastava, A. (1991). Dominance, age, and reproductive success in free-ranging Hanuman langurs (*Presbytis entellus*). *Int. J. Primatol.* **12**: 231–57.

Borries, C., Launhardt, K., Epplen, C., Epplen, J. T. & Winkler, P. (1999). DNA analyses support the hypothesis that infanticide is adaptive in langur monkeys. *Proc. R. Soc. Lond.* B**266**: 901–4.

Calhoun, J. (1962). Population density and social pathology. *Scient. Am.* **206**: 139–48.

Carter, C. S., Ahnert, L., Grossmann, K. E., *et al.* (eds.). (2005). *Attachment and Bonding: A New Synthesis*. Cambridge, MA: MIT Press.

Dagg, A. I. (1999). Infanticide by male lions hypothesis: a fallacy influencing research on human behavior. *Am. Anthropol.* **100**: 940–50.

Daly, M. & Wilson, M. (1978). *Sex, Evolution and Behavior*. North Scituate, MA: Duxbury.

Dolhinow, P. (1977). Normal monkeys? *Am. Scient.* **65**: 266.

Fuentes, A. (2002). Review of infanticide by males and its implications. *Am. Anthropol.* **104**: 693–6.

Gowaty, P. A. (ed.). (1997). *Feminism and Evolutionary Biology*. New York: Chapman and Hall.

Hausfater, G. & Hrdy, S. B. (eds.). (1984). *Infanticide: Comparative and Evolutionary Perspectives*. Hawthorne, NY: Aldine.

Hopkins, N. (1975). The high price of success in science. *Radcliffe Quarterly*, July issue.

Hrdy, D. B., Barnicott, N. A. & Alper, C. A. (1975). Protein polymorphism in the Hanuman langur (*Presbytis entellus*). *Folia Primatol.* **24**: 173–87.

Hrdy, S. B. (1974). Male-male competition and infanticide among the langurs (*Presbytis entellus*) of Abu, Rajasthan. *Folia Primatol.* **22**: 19–58.

Hrdy, S. B. (1976). The care and exploitation of nonhuman primate infants by conspecifics other than the mother. *Adv. Study Behav.* **6**: 101–58.

Hrdy, S. B. (1977). 1980 (revised paperback edition). *The Langurs of Abu: Female and Male Strategies of Reproduction*. Cambridge, MA: Harvard University Press.

Hrdy, S. B. (1979). Infanticide among animals: a review, classification, and examination of the implications for the reproductive strategies of females. *Ethol. Sociobiol.* **1**: 13–40.

Hrdy, S. B. (1981). 1999 (second edition with new preface). *The Woman that Never Evolved*. Cambridge, MA: Harvard University Press.

Hrdy, S. B. (1986). reprinted 2006. Empathy, polyandry and the myth of the coy female. In *Conceptual Issues in Evolutionary Biology*, 3rd edn., ed. Elliott Sober, pp. 131–59. Cambridge, MA: MIT Press. (Page numbers refer to 2006 reprint.)

Hrdy, S. B. (1997). Raising Darwin's consciousness: female sexuality and the prehominid origins of patriarchy. *Human Nature* **8**: 1–49.

Hrdy, S. B. (1999). *Mother Nature: A History of Mothers, Infants and Natural Selection*. New York: Pantheon.

Hrdy, S. B. (2002). The past, present and the future of the human family. *The Tanner Lectures on Human Values* **23**: 57–110. Salt Lake City, UT: University of Utah Press.

Hrdy, S. B. (2005). Comes the child before man: cooperative breeding and the evolution of prolonged post-weaning dependence. In *Hunter Gatherer Childhoods*, ed. B. Hewlett & M. Lamb, pp. 65–91. New Brunswick: Aldine/Transaction.

Hrdy, S. B. (2009). *Mothers and Others: The Origin of Emotionally Modern Humans*. Cambridge, MA: Harvard University Press.

Hrdy, S. B. & Hrdy, D. B. (1976). Hierarchical relations among female Hanuman langurs. *Science* **193**: 913–15.

Hrdy, S. B. & Judge, D. S. (1993). Darwin and the puzzle of primogeniture: an essay on biases in parental investment after death. *Human Nature* **4**: 1–45.

Hrdy, S. B. & Williams, G. C. (1983). Behavioral biology and the double standard. In *Social Behavior of Female Vertebrates*, ed. S. Wasser, pp. 3–17. New York: Academic Press.

Hrdy, S. B., van Schaik, C. & Janson, C. (1995). Infanticide: let's not throw out the baby with the bath water. *Evol. Anthropol.* **3**: 151–4.

Jay, P. (1963). The social behavior of the langur monkey. Ph.D. dissertation, University of Chicago.

Judge, D. S. & Hrdy, S. B. (1992). Allocation of accumulated resources among close kin: inheritance in Sacramento, California, 1890–1984. *Ethol. Sociobiol.* **13**: 495–522.

Levi-Strauss, C. (1966). *Du Miel aux Cendres*. Paris: Plon.

Mestel, R. (1995). Monkey "murderers" may be falsely accused. *New Scientist* (July 15): 17.

Mohnot, S. M. (1971). Some aspects of social changes and infant killing in the Hanuman langur, *Presbytis entellus* (Primates Cercopithecidae) in Western India. *Mammalia* **35**: 175–98.

Packer, C. Infanticide is no fallacy. Commentaries. *Am. Anthropol.* **102**: 829–57.

Parmigiani, St. & vom Saal, F. (eds.). (1994). *Infanticide and Parental Care*. Ettore Majorana International Life Sciences Series vol. 13. Chur, Switzerland: Harwood Academic.

Platt, J. R. (1964). Strong inference. *Science* **146**: 347–53.

Roonwal, M. L. & Mohnot, S. M. (1977). *Primates of South Asia: Ecology, Sociobiology and Behavior*. Cambridge, MA: Harvard University Press.

Schubert, G. (1982). Infanticide by usurper hanuman langur males: a sociobiological myth. *Soc. Sci. Inform.* **21**(2): 199–244.

Shea, C. (1999). Motive for murder. *Lingua Franca* **9**(6): 23–5.

Sipher, D. (2007). Nancy Hopkins and Dinny Adams. *New York Times*, July 29, 2007.

Sommers, V. (1994). Infanticide among the langurs of Jodhpur: Testing the sexual selection hypothesis with a long-term record. In *Infanticide and Parental Care*, ed. S. Parmigiani & F. vom Saal, pp. 155–98. Chur, Switzerland: Harwood Academic.

Sugiyama, Y. (1965). On the social change of hanuman langurs (*Presbytis entellus*) in their natural conditions. *Primates* **6**: 381–417.

Sussman, R. W., Cheverud, J. M. & Bartlett, T. Q. (1995). Infant killing as an evolutionary strategy: reality or myth? *Evol. Anthropol.* **3**: 149–51.

Symons, D. (1979). *The Evolution of Human Sexuality*. Oxford: Oxford University Press.

Trivers, R. L. (1972). Parental investment and sexual selection. In *Sexual Selection and the Descent of Man 1871–1971*, ed. B. Campbell, pp. 136–79. Chicago, IL: Aldine.

Van Schaik, C. & Jansen, C. (eds.). (2000). *Infanticide by Males and its Implications*. Cambridge: Cambridge University Press.

West-Eberhard, M. J. (2003). *Developmental Plasticity and Evolution*. Oxford: Oxford University Press.

Wilson, E. O. (1975). *Sociobiology: The New Synthesis*. Cambridge, MA: Harvard University Press.

14

Luck, chance and choice

JOHN R. KREBS

Speak what we feel, not what we ought to say   *King Lear*

Setting out to write an autobiography, I am immediately confronted with two challenges: the unreliability of memory and the universal human compulsion to create a narrative.

Milan Kundera exposes the first of these dilemmas like this:

> without much error I could assume that the memory retains no more than a millionth, a hundred millionth, in short an utterly infinitesimal bit of the lived life…We will never cease our critique of those persons who distort the past, rewrite it, falsify it, who exaggerate the importance of one event and fail to mention some other; such critique is proper (it cannot fail to be), but it doesn't count for much unless a more basic critique precedes it: a critique of human memory as such
>
> *(Kundera 2002).*

Story-telling is probably one of the oldest and most universal facets of human culture, as a means of entertaining, educating, and preserving or instilling values and traditions.

Narrative is central to telling a good story: there has to be a plot with a beginning, middle and end, and most if not all autobiographies follow the narrative tradition. Mine is no exception and I acknowledge that in creating a narrative the 'facts' have to be arranged to make the story work.

So let me at least begin on solid ground with some recorded historical facts, beyond the reach of Kundera's critique of human memory.

Family background and early life (1945–56)

Parents

I was born on April 11th, 1945, in Sheffield, the youngest of three children of Hans Adolf and Margaret Cicely Krebs. My father was Professor of Biochemistry at the University of Sheffield, and my mother, who had originally trained as a Domestic Science (Cookery) teacher, did not follow a career once she had children.

My father, born in 1900, had come to Britain in 1933 as a Jewish refugee from Nazi Germany. His story has been told in his autobiography (H. Krebs 1981), and in a two-volume work by F. L. Holmes (Holmes 1991, 1993) that covers the first 37 years of his life.

In brief, he had trained as a medical doctor (following in his father's footsteps) and embarked on a research career at a Kaiser Wilhelm Institut in Berlin under Otto Warburg. He was working as a doctor in the University of Freiburg, when he was dismissed, shortly after Hitler came to power and enacted legislation that prohibited Jews from holding public posts. My father's reputation as a research scientist, including his discovery of the urea cycle in 1932, enabled him to move to Cambridge to work under Sir Frederick Gowland Hopkins, the discoverer of vitamins. In 1937, by which time he was at Sheffield University, he published the crucial discovery (the synthesis of citrate from oxaloacetate and pyruvate; the final formulation of this step awaited an insight from stereochemical principles by A. Ogston in 1948) that completed the Krebs Cycle, for which he was awarded a Nobel Prize in 1953. Although I was too young to appreciate the full significance, I do remember the splendid ceremonies and formal banquet, and my conversation with King Gustav Adolf VI of Sweden about my favourite football club, Sheffield United.

My mother was from south Yorkshire, and her family owned a chain of shoe shops in and near Rotherham.

Sheffield

In Sheffield we lived near an extensive area of woodland (Eccleshall Woods), where I spent many hours exploring, perhaps an early indication that I would take to field-work. At school I was, as far as I know, pretty unexceptional. In fact at the age of about 6 I developed a habit of skipping afternoon school to play in the woods nearby, and had not spotted in my innocence that the teachers would report my absence to my parents! My most vivid memory of primary school in Sheffield was flicking ink on a girl's dress and being told by the teacher that I would have to go to purgatory and spend many years walking on red-hot bricks.

Decades later when I went to an exhibition of Botticelli's illustrations of Dante's *Inferno*, I realised where she might have got her ideas, but nevertheless, it was hardly a suitable admonition for an 8-year-old child.

Oxford

In 1954 we moved to Oxford, where my father became Whitley Professor of Biochemistry, a post he held until his retirement in 1967. We lived in the village of Iffley in a house with a very large garden. This must have been a factor in my emerging interest in birds and bird watching.

I also started hand-rearing birds. Two particularly noted individuals in the family were a blackbird called Wilfred and a nameless robin. My mother must have been tolerant because both these birds spent at least part of their time flying free in the breakfast room. On one notable occasion the robin hid in a large pot plant during Sunday lunch in the middle of our large circular dining table, and emerged briefly to deposit a large, shiny, gelatinous dropping close to my father's dinner plate, before disappearing back into its hiding place.

One of my early recollections of bird-watching on holiday was when we went as a family to Dale Fort Field Studies Council centre in West Wales. The Warden, John Barrett, took a kindly interest in my nascent bird-watching ambitions and I particularly remember him talking me through a large glass case of stuffed birds, pointing out the key field identification traits. Family summer holidays in the 1950s, usually in the Bavarian, Austrian or Swiss Alps, also had a strong botanical element – my father was a keen botaniser, as was my sister Helen, who went on to read Botany at Somerville College, Oxford.

High school 1956–63

I passed the 11-plus exam to enter the City of Oxford High School for Boys (COHS), one of the two local grammar schools in Oxford at the time. Oxford was then, as it is now, well supplied with Independent Schools, and very few sons of academics went to COHS. The school had academic pretensions but modest achievements.

I was usually amongst the top half dozen boys in the top stream up to O-level (public exams taken at age 15 or 16): very good but not outstanding. My best subjects were modern languages and history, my weakest subjects maths, physics and chemistry.

In spite of this, and as a result of some considerable paternal pressure, I chose science A-levels (in the English system, then as now, pupils have to specialise in a small number of subjects for their last two years), and my marks were pretty abysmal.

Perhaps part of the reason I did so poorly was the poor quality of the teaching. Our chemistry teacher ("Slosh" Coleman) simply sat at this desk at the front and read out notes, which we subsequently had to copy into a fair copy notebook. The physics teacher ("Pongo" Bodey) made his disdain for those doing A-level biology all too apparent ("The ignorant biologists will find this topic difficult to understand...."); and the biology teacher ("Spud" Taylor), whilst very supportive and encouraging, was still teaching the topics he had learned at Cambridge in the 1930s. Genetics and evolution, not to mention cell and molecular biology, had not come onto his radar screen.

When Spud Taylor took a term's sick leave, the school employed a graduate student from Oxford, Robert Mash, to teach us some aspects of biology.

Bob Mash was then a student of Niko Tinbergen and studied the displays of black-headed gulls. Luckily he largely disregarded the syllabus and in his laconic and witty way encouraged the half a dozen or so of us doing A-level biology to read and write about animal behaviour, evolution, and related areas.

Alas, the topics we covered were of less relevance to the exam than being able to draw spongy mesophyll in a TS of a leaf or name the cranial nerves of the dogfish, so although they were both stimulating and enjoyable they did not help me to do well in A-levels.

Bavaria

Nevertheless, these topics did become the platform for my later career. Seizing on my nascent interest in ethology, my father arranged for me to spend the last two summer vacations of my school years working as a technician at the Max Planck Institutes in Bavaria directed by Konrad Lorenz and Jurgen Aschoff.

This was a truly formative time for me, simultaneously learning German, learning about the science of ethology, and falling in love with Bavaria as well as with Aschoff's youngest daughter, Annette.

My first job at the Max Planck was to collect material to put in the aviaries where Jurgen Nicolai kept his wydahs for research on brood parasitism. Soon my lack of German became apparent when I was instructed to fetch branches for the aviaries. I understood the bit about fetching branches, but missed the fact that they had to be forked, so my first batch were all rejected. I did not forget the word "Gabel" (fork) in a hurry. Another early task was to collect pine needles for the aviary floor. Setting off into the forest with a large sack, I was delighted after a few minutes to come across a neatly piled up mountain of needles. Without further ado I knelt down and plunged my arms into the pile to scoop it into my sack. Wood ants don't take kindly to having their nest ripped apart, and I ran off through the forest yelping!

I stayed with the Aschoff family in a rather grand nineteenth century house in Erling-Andechs, where Aschoff had his Institute next door. In my second year I worked as a technician for Aschoff, one of my key tasks being to paste the daily activity rhythms (recorded on chart recorders) of the experimental subjects on boards, one day under another, so that the pattern of circadian activity over some weeks could be grasped at a single glance. I used to hold up my boards during lunchtime discussions, during which Aschoff would ruthlessly interrogate his subordinates. Each work day included a break in which we all piled into one of the institute's VW minibuses to head off for a swim in the Pilsen See, Ammer See, or a small peat lake known as the Moorweihe. Weekend excursions to the Alps were military exercises with strictly no drinks allowed on the day-long walk, to encourage consumption of large quantities of beer or Radler (shandy) at the end of the day. Evening entertainment often consisted of Aschoff on the accordion leading the family and many other hangers-on in singing traditional German songs. In spite of this enthralling time, the fact that I ended up as a research scientist was more accident than design.

At school, two of my principal hobbies were music and coin-collecting. I played the violin, not with great talent but on reflection with considerable persistence. Although I did not keep up my violin-playing very assiduously, it has, over the years, given me much pleasure in playing chamber music with colleagues and students. Perhaps one indication of the fact that the pleasure was more for the players than the listeners was the time when I was playing trio sonatas with Marian Dawkins and Richard Sibly. Richard Dawkins, then married to Marian, sat down to listen to us. After about five minutes he coughed indistinctly and muttered about having some work to do as he slipped out of the room!

I don't recall when I started collecting coins, but one of my favourite teenage excursions was to B. A. Seaby, the coin dealer in London, to purchase Roman, or English hammered silver coins. I also used to search at Romano-Celtic sites around Oxford for pieces of pottery and other artefacts: my favourite location was a field at Woodeaton near Oxford, where there had been a temple in Roman times.

University: Oxford 1963–70

Undergraduate degree in zoology

Given my interests, I suggested to my parents that I might read archaeology at university. This was not received with great warmth, and my father pressed me hard to apply to Oxford to read medicine. At the interview at Trinity College, Oxford (my father's College), when Charles Phillips, the medical tutor, asked why I wanted to study medicine, I gave the honest answer "I don't". This was probably my biggest moment of rebellion against parental domination! The letter from the President of Trinity to my father explaining their decision not to take me was deeply apologetic and understated.

After what would be called in diplomatic circles a "full and frank exchange of views" (although I suspect I was more recipient than discussant), plan B was hatched. After a delay of a year (and following my second summer in Bavaria) I would try for zoology. Samuel Johnson defined patriotism as the "last resort of a scoundrel". Zoology was probably nudging into second place in my father's eyes. Even then things did not go smoothly: I applied for a scholarship at St John's, failed to get in, and was finally accepted by Pembroke College. Without good old-fashioned nepotism I would surely not have made it to Oxford. My first-year tutor, a remarkable character called J. R. P. (Percy) O'Brien, who plied his students with sherry while teaching them about entropy and free energy, once said to me "Young Krebs, I have more neurones in my glans penis than you have in your head".

But Oxford was a revelation and a transformation. I was lucky enough to arrive in a Zoology Department populated by some of the greatest scientists of the middle twentieth century in ecology and behaviour: David Lack, Niko Tinbergen, Charles Elton, Arthur Cain, George Varley. I was awestruck to be actually taught by the people who had written the literature they were teaching! Just as important were the other students. Tony Sinclair, now at UBC, and I would spend hours in furious debate about natural selection, density dependence, maintenance of genetic variation and so on. From the class of about 35 in my year, many have achieved distinction in research or other careers, including five Fellows of the Royal Society (Brad Amos, Mike Bate, Ron Laskey, Tony Sinclair and me).

Tony Sinclair was the one who introduced me to the girl whom I later married, Katharine Fullerton. Kathy, a Welsh girl, was two years below us reading zoology, and when Tony celebrated his engagement with a dinner party at the Fox at Boars Hill, I was between girlfriends, so Tony asked Kathy on my behalf. We were married three years later. Kathy went on to do a D. Phil. in developmental psychology, before switching to training as a biology teacher, a career she followed (with a short interlude working with Richard Dawkins as assistant editor of *Animal Behaviour* when our first child was very young) until she retired a few years ago. Kathy has helped me enormously at all stages of my career, by hand-raising birds, correcting English (including in this essay), and advising me on choices of jobs. She has also produced two lovely daughters, Emma, who has followed in her mother's footsteps, and George, a clinical psychologist.

Post-graduate research for a D. Phil.

By the third and final year at Oxford I had focussed on the idea of staying on to do a D. Phil. One of the raging controversies in the mid-1960s in population ecology was between David Lack, who argued in his 1954 book (Lack 1954) that bird (and implicitly other) populations are regulated by density-dependent mortality inflicted by winter food shortage, and V. C. Wynne Edwards, who in his 1962 book (Wynne-Edwards 1962), proposed that populations are self-regulating, using a range of behavioural mechanisms from communal roosts to territorial conflict, to maintain numbers within the limits of the environment. Such was the ferocity of the debate that in his 1966 book (Lack 1966), which came out just before I started my thesis research, Lack dedicated a 100-page appendix to a more or less line-by-line surgical dismembering of Wynne-Edwards' argument. One of the many ways in which Wynne-Edwards shot himself in the foot (at least as far as Oxford orthodoxy was concerned) was by espousing the notion of group selection as the mechanism behind self-regulation of populations.

I wanted to build my thesis topic around this debate. I was much influenced by J. M. ("Mike") Cullen, who was Tinbergen's right-hand man and an utterly brilliant mind (Krebs & Dawkins 2001). He and I discussed the idea of territorial limitation of numbers, without resorting to group selection. Armed with these ideas, I trotted off to the Edward Grey Institute of Field Ornithology, then housed in a beautiful neoclassical building overlooking the Botanical Gardens, along with Elton's Bureau of Animal Population. Lack listened closely to my hesitant and ill-constructed thoughts on territoriality and population limitation. He asked "and what species do you think of working on?", to which I replied that I had not yet figured this out, but wondered vaguely if Mute Swans might be good, because you could at least have a chance of spotting any "floating birds" that had been excluded as a result of territoriality.

Lack took me downstairs to the coffee room to introduce me to his group – people such as Chris Perrins, Ian Newton, and Mike Harris. "This is John Krebs. He wants to do a D. Phil. but doesn't know what he wants to work on", was his opener. Lack's approach to research had always been to choose a species and then let the problems emerge, so a typical thesis of his students at the time would be "Some observations on the winter feeding ecology of ..." or "The breeding biology of the ...". My approach of defining a problem and not a species did not fit comfortably in the Edward Grey Institute of the mid-1960s.

So I returned to Mike Cullen and he agreed to supervise me in the Animal Behaviour Group, if I obtained a good enough degree to get a research studentship. He suggested that I should study territorial limitation of numbers in David Lack's own species, the Great Tit, in his study area, Wytham Woods.

The inspiration for the work came from so-called "removal experiments" in which territorial pairs are taken out of their breeding territories to test whether there is a surplus population of non-breeders waiting to fill up the gaps. This idea had been published by Stewart and Aldrich (1951) and developed by Adam Watson (1967) in his work on Red Grouse.

My thesis was about 50 pages long and it had just two chapters. One described a series of removal experiments that demonstrated the limitation of breeding density in oak woodland by spring territorial behaviour, whilst the other involved a "key factor" analysis and very simple population model. G. C. Varley, who taught us insect ecology, was a master of key factor analysis, and Mike Hassell, then a student of Varley, helped me to construct a model of the Wytham Great Tit population, which we ran on a Facit electro-mechanical calculator, taking perhaps a minute to do each run. It was by no means rocket science, but it was very new in bird ecology.

Because of my choice of a contentious topic and a well-studied system (arguably the most famous vertebrate population study at the time), my thesis work (Krebs 1971) gained rapid recognition (although when I submitted it to *Ecology*, it was appropriately lambasted by R. B. Paine for careless presentation). Eventually it became a citation classic (Krebs 1971).

The Animal Behaviour Group was a very stimulating environment in which to start a research career. Housed in a tall thin Victorian House in central North Oxford, the senior citizens (postdoctoral scientists) when I started out were Mike Cullen, Juan Delius and Cliff Henty, with Richard Dawkins, Mike Norton Griffiths and Harvey Croze as the prominent D. Phil. students. Lunchtimes were always lively and the intellectual atmosphere open, challenging and enthusiastic. Niko himself occupied an office in the main Zoology building half a mile away. The weekly seminars were a jaw-dropping opportunity to see how he and the other senior members of the group could dissect a visiting speaker. I remember John Crook starting a seminar and never getting beyond his first point, because of Niko's insistent "what exactly do you mean by...?" and the follow-up questions from others.

Niko's and Mike Cullen's absolutely rigorous approach to asking and answering questions was one of the key things I learned as a D. Phil. student, along with an appreciation, from Mike and others, of quantitative methods in research, and the value of forming connections between disparate pieces of literature to weave them into a tapestry, which was one of David Lack's great talents.

My research in population ecology was on the periphery of the Animal Behaviour Group's interests. The big issues at that time (mostly now having rather a dated feel to them were): displacement activities (Ken Wilz and Juan Delius), search images and area-restricted search (Harvey Croze, Marian Dawkins, Jamie Smith), analysing sequences of behaviour using techniques such as autocorrelation and Markov Chain analysis (Juan Delius, who once said that a graduate student should spend 24 hours collecting data and the remaining 2 years 364 days analysing it), and various studies of behaviour ontogeny (Richard Dawkins, Mike Norton-Griffiths, Cliff Henty). Desmond Morris also put in a brief appearance to study preening behaviour of finches.

My D. Phil. grant was due to run out in the autumn of 1969 but I had not finished, or even started, writing. As a result of some discussions between Mike Cullen and David Lack, I found myself in the running for a junior, non-tenured, faculty position (Departmental Demonstrator) in the Edward Grey Institute. The duties were light – teaching ornithology practical classes, editing the *Ibis Abstracts*, and as much or little tutorial teaching as I wanted, and I moved from the ABRG to the EGI for one year. It could have been longer, but I had already decided to move, in the summer of 1970, to the University of British Columbia. This came about because a year or so earlier Dennis Chitty had given a talk in Oxford extolling the virtues of an experimental approach to population ecology, as opposed to the Oxford tradition of collecting very long runs of data. I thought this sounded interesting enough, and close enough to my own approach, for me to write to Dennis and ask whether there were any postdocs going in his group. He said there weren't, but that UBC was hiring several new faculty members to join an Institute of Resource Ecology that evolved out of the old Institute of Fisheries, and was led by CS "Buzz" Holling. I ended up applying for an assistant professorship, flying out to give a job seminar (which involved me getting locked out on the roof of the Zoology building 5 minutes before it was due to start and having to shout until rescued – don't even ask) and being offered the job.

In addition to finishing my thesis, I used that year (1970) at the EGI to embark on a new research topic, inspired by David Lack's *Ecological Adaptations for Breeding in Birds* (Lack 1968). Lack, building on the work of Ian Patterson and John Crook, had argued that group living in birds could have two selective advantages: enhanced foraging efficiency (through some form of information transfer; loosely one might call it copying) or better defence against predators. Ian Patterson had shown convincingly in Black-headed Gull colonies that there was an anti-predator advantage, but as far as I knew no one had looked at the foraging benefits of group living. I teamed up with a postdoc in Niko's group, Michael MacRoberts, and my supervisor Mike Cullen to devise experiments to test out the foraging advantage in flocks of Great Tits. We showed that birds in flocks learn from the success of other birds not only the locations of food sources but also the kinds of places in which food is found (Krebs *et al.* 1972). Here my childhood experience of hand-raising and keeping birds stood me in very good stead.

In the same year a garrulous and argumentative Israeli scientist called Amotz Zahavi came to stay with me, and he was thinking on the same lines, with his idea of communal roosts as information centres. As usual Amotz was rich in anecdote but low on rigour, and probably right!

Vancouver, Bangor and back to Oxford, 1970–75

University of British Columbia

UBC was an intellectual culture shock. The cosy assumptions of Oxford orthodoxy – the value of long-term data sets, the importance of density dependence, and the central dominance of natural selection thinking – were by no means accepted by Dennis Chitty, Charley Krebs, Buzz Holling, Carl Walters and the others in the Institute. It was as though I had to start again and build a new set of foundations. Nothing could have been more challenging and stimulating: I would recommend this kind of cold bath for anyone at the equivalent career stage. As far as my research was concerned, I was keen to carry on the successful line of flocking and foraging that I had started in Oxford. I even exported a set of hand-raised Great Tits to Vancouver (when one escaped and flew to the top of the Engineering building, and I lured it down with a mealworm to land on my hand, I became an instant legend on the campus). I also extended the original experiments to mixed flocks of Chickadees.

Herons

Rudi Drent had arrived at UBC just before me, and he became a great colleague, helping me to set up a field study of Great Blue Herons that encompassed both a test of the idea that a nesting colony acts as a Zahavian information centre and the role of flocking in enabling birds to locate good feeding patches (Krebs 1974). Part of the fun for Rudi was in designing and constructing a platform some 30 m off the ground in the crown of a huge Douglas Fir, to enable us to look down on the surrounding heron nests in a stand of alders. The platform was accessed by 1" by 3" wooden slats nailed into the trunk, at ever-increasing intervals as you went up the tree.

At Oxford, Richard Dawkins was in the forefront of computer technology (he had a PDP 8 that occupied a whole room and had a fraction of the computing power of a mobile phone) and amongst other things he had developed an event recorder – the Dawkins Organ – that encoded behavioural events as tones on a portable tape recorder. My research assistant Brian Partridge and I used this new device in the field to collect data on foraging success and intake rates of herons in flocks of different sizes feeding in the inter-tidal mudflats near Vancouver airport. Decoding the tapes was an arduous task, mainly undertaken by Brian.

Optimal foraging theory

The British political satirical magazine *Private Eye* (number 994, 28 January 2000) once did a profile of me which included the following "….under the influence of brilliant theoretical biologist Eric Charnov, he found a new subject for research, optimal foraging, the question of whether birds tend to feed in the areas where – er – feeding is best". Private Eye's choice of words may be debatable, but the thrust of what they reported was remarkably accurate: Ric was certainly a big influence on my science at UBC.

Ric was a graduate student at the University of Washington, but spent much time at UBC hanging around Holling's group. He was always looking for someone with whom to share ideas (i.e. tell them about his latest thought), and because of his interest in natural selection, not widely appreciated in the Institute, I became the focus of Ric's attention. I was in the middle of writing a review article about foraging behaviour and describing John Gibbs' idea of "hunting by expectation", referring to the exploitation of eucosmid moth larvae in pine cones, by Coal Tits. Ric had been talking about MacArthur's and Pianka's work on optimal diets, and I asked him to figure out what the optimal (rate-maximising) strategy for Gibbs' Coal Tits would be. Within a few days he was back at my house on West 16th with the answer – the marginal value theorem (independently derived by Geoff Parker in relation to copula duration in yellow dungflies). We recruited a final year undergraduate, John Ryan, and my hand-raised Chickadees, to devise an experimental test of the marginal value theorem (Krebs *et al.* 1974)). Although in retrospect the test was weak and the experimental design left much to be desired, the results were consistent with the model and we were jubilant.

Ric and I also wrote two short papers that came out of coffee time chats. One was on life history theory, predicting that the optimal clutch size in birds should be smaller than the "Lack optimum" (the one leading to the greatest number of surviving young) because of the trade-off between reproduction and survival (Charnov & Krebs 1974). The other, on alarm calls (Charnov & Krebs 1975), did two things. First it posited a model of group selection that hinted towards the models later developed by D. S. Wilson, and second it proposed that alarm calls might be a form of manipulation, an idea that Richard Dawkins and I generalised a few years later to encompass all animal communication (Dawkins & Krebs 1978).

Other research at UBC

In my second year at UBC, Bruce Falls came as a sabbatical visitor. I had actually met Bruce before I started my D. Phil., at the International Ornithological Congress in Oxford, 1966, and I was a great admirer of his papers on Ovenbird territoriality. Bruce suggested that we work together on the songs of Western Meadowlarks, and we spent many happy mornings collecting recordings, followed by various kinds of sequence analysis (Falls & Krebs 1975). This was the start of a long-term collaboration: Bruce came twice to Oxford on sabbatical and I went to Toronto, and on each visit we did field-work and wrote papers together.

At that time, in the early 1970s, UBC felt as though it was *the* exciting place to be for ecological research. The Institute had expanded under Holling and there was both a palpable sense of interdisciplinarity (Buzz was one of the earliest people to realise that environmental problems needed a mix of natural and social science, politics and planning) and a wonderful social atmosphere.

Rudi Drent, in addition to helping me to get started on herons, introduced me to the Canadian Arctic, when together with Wayne Campbell, we conducted bird surveys in the North West Territories in anticipation of developments associated with oil exploration. The trip include many hilarious and exciting moments, including not one, but two, seaplane crash landings, the second of which led to us wading several hundred metres to the shore in waist-deep icy water with heavy packs.

Bangor

Given that UBC was such a wonderful place to be, academically and socially not to mention the scenery and surroundings, why did I leave after three years, in the summer of 1973? Well, there were downsides. Although I had been promised a "light teaching load" to start my career, I ended up teaching Bio 101 more or less from the moment I arrived. Further, in my impatience to get on with research I found the delays in getting my aviaries built very frustrating. However my move from UBC to Bangor in North Wales was not planned in advance. I suppose I had always assumed that at some point I would return to the UK. My mother sent me the job advert for a lectureship in Zoology at Bangor. Kathy is from Wales, so we thought it might be nice to live there, and without much thought I applied for the job. Shortly after, I had a phone call from Jimmie Dodd, Head of Zoology at Bangor, inviting me for an interview, which after some debate I accepted. To cut a long story short I ended up moving to Bangor, although Pete Larkin, Head of Zoology at UBC, cunningly offered to keep my job there open for me – a gambit designed to unsettle me. Bangor, although it had some famous ecologists on the staff, including T. B. Reynoldson, J. L. Harper, D. J. Crisp, and P. Greig-Smith, was not well equipped for the kind of work I wanted to do, nor was there money to develop facilities such as aviaries.

Furthermore, living in a small North Wales community, in a rented house in which the salt deliquesced in the damp atmosphere, after the cosmopolitan vibrancy of Vancouver, was a definite culture shock.

A fortuitous return to Oxford

After about six months, I was on a visit to Oxford to give a seminar, when I chanced to meet David McFarland in the University Parks. David had, a year earlier, been the surprise choice to succeed Niko Tinbergen as Reader in Animal Behaviour. Many of us assumed that Mike Cullen, who had in effect run Niko's group for many years, would get the job. But even in those days, Mike's lack of publications (a consequence of his totally unselfish dedication to his students) told against him. David, with his background in experimental psychology and his work on control of motivational state, seemed to be destined to take the Behaviour Group in a new direction. He asked me whether I would be interested in coming back to Oxford as a named postdoctoral research assistant on a grant that he was about to submit to the Science and Engineering Research Council. David explained that he was planning to develop a new approach that would link the kind of optimality modelling that I had become involved with to his interest in feedback mechanisms and control theory.

In retrospect it seems that I took a big gamble, but without much hesitation, Kathy and I agreed that we would leave my permanent job at Bangor to return to Oxford on a three-year contract research post. David was awarded the grant and we duly moved to Oxford in January 1975, after just 18 months in North Wales. My three research students, John Pickstock, Richard Cowie and Pete Garson, moved with me.

Oxford: 1975–88

The Animal Behaviour Group under David MacFarland

During my five years away from Oxford, much had changed in the Animal Behaviour Group. For one thing, it had now moved into a spanking new building, along with the main Zoology Department and the other outliers – the Edward Grey Institute and Animal Ecology Research Group, and, in the other half of the building the Department of Experimental Psychology. David McFarland had brought in new ideas and a different kind of student to follow them up. Richard and Marian Dawkins were back in Oxford after a short spell at Berkeley, and Mike Cullen was still involved, although as a lecturer in the Department of Experimental Psychology. I found that the new ideas and approaches that I had picked up at UBC meshed well with the new, quantitative and theoretical approach of McFarland's group, and I was set for a very exciting phase of my research career. Above all, one of the joys of Oxford was, and still remains, the quality of the graduate students. Soon, I had become particularly involved with several students who were already there: Alex Kacelnik was visiting from Argentina on a British Council Scholarship to work with David, Jon Erichsen, a Marshall Scholar from Harvard, who was doing a D. Phil. on bird vision with Mike Cullen, a bearded, cricket-loving student in the Edward Grey Institute called Nick Davies, and Mac Hunter, a Rhodes Scholar from Maine looking for a project and a supervisor.

Research in my nascent group expanded on many fronts simultaneously: Richard Cowie and Alex quickly established laboratory experiments on optimal foraging, Richard using a setup similar to the one I had developed in Vancouver, and Alex building our first primitive version of the experimental psychologists' operant laboratory using logic modules cadged from Nick Rawlins and Jeffrey Gray. I focussed initially on bird song (as did Mac Hunter), and other students in my rapidly expanding group worked on a range of problems including sexual selection, flocking, and territoriality. Also at that time, Richard Dawkins was working hard on a new book, the title of which was still to be determined, and for which he sought suggestions. Mine (*Immortal Coils*) made it as a chapter heading, but *The Selfish Gene* was, I believe, Richard's own suggestion! I had come into close contact with Bob Trivers' work on parent – offspring conflict while I was at UBC, and I remember Ric Charnov sending me an advance copy of the 1976 *Science* paper by Trivers and Hare, which I showed with great excitement to Richard, and which, of course, became part of *The Selfish Gene*.

Moving to the Edward Grey Institute

Within my first year back at Oxford a permanent job was advertised: a university lectureship in Ornithology, in the Edward Grey Institute, one floor below the Animal Behaviour Group in the same building. I duly applied and was short-listed and interviewed. After a delay of a day, I was offered the job by the then Head of Department, John Pringle. Later I learned how David McFarland had skilfully intervened to get me the job. John Pringle favoured another candidate and the committee was unsure whether or not to follow Pringle's lead. Oxford faculty search committees always include two representatives of the College with which the post is affiliated. David guessed that these two, being non-specialists, would tend to follow the Head of Department and that this would be sufficient to swing the vote against me, so he suggested that the committee slept on the matter before deciding first thing the next morning, at 8.00am. This was a carefully judged ploy, based on the assessment that the College representatives would not make it to the meeting. David was right, and in their absence, the vote went my way. I sometimes wonder how my career would have worked out had David not been so shrewd in manipulating the process.

With longer-term prospects in Oxford, my research group built up two principal lines of research: bird song and optimal foraging, later linked through studies of daily routines and the dawn chorus.

Foraging theory

By this stage, in the mid-1970s, optimal foraging was becoming very fashionable as well as controversial, with a number of robust critiques, for instance from Russell Gray and John Ollason. Looking back, I would highlight the following as the most significant contributions from my group at this time.

First, experimental tests of what became known as "classical optimal foraging models" – rate-maximising models of prey selection and patch use: Richard Cowie's neat experimental test of the relationship between travel time and patch residence time (Cowie 1977), and the test of prey selection model using an apparatus developed by Jon Erichsen for studies of avian vision (Krebs *et al.* 1977). As with the Vancouver experiments, by modern standards both the design and the interpretation are dated, but at the time it was stunning to be able to make quantitative predictions about animal behaviour and test them in experiments. Ten years earlier, as a graduate student, I had complained to anyone who would listen that animal behaviour had no coherent theoretical framework, and here it appeared that we had found one!

Second, the theoretical work and associated experiments that Alex Kacelnik and I did, together with Peter Taylor who was visiting the Mathematics Institute, on optimal sampling. In this we developed a Bayesian model of sampling and a test that showed a good fit to predictions (Krebs *et al.* 1978a).

Over the next dozen years, between the late 1970s and the early 1990s, our work on foraging expanded in many different directions. Importantly, Alasdair Houston developed a dynamic, state-dependent approach to foraging and decision-making, using stochastic dynamic programming. This not only shed new light on many aspects of foraging decisions (for instance the state-dependence of risk-sensitivity) but also enabled us to build foraging decisions into the broader context of daily routines and foraging versus vigilance trade-offs, followed up by students such as Ron Ydenberg and Ruth Mace. Alex Kacelnik, following a period in Holland in Rudi Drent's group, came back to work with me in the early 1980s and together with Innes Cuthill and others began to uncover some the mechanisms underlying foraging decisions in a mixture of laboratory and field experiments, as did Paul Schmid-Hempel in his ingenious experiments with foraging bees. Dave Stephens originally came to Oxford to do a second undergraduate degree, but I persuaded him to join my group as a D. Phil. student, co-supervised with Alasdair Houston. Some parts of Dave's thesis eventually found there way into the *Princeton Monograph* we co-authored (Stephens & Krebs 1986). Dawn Bazely, and subsequently Jonathan Newman, extended our foraging work to encompass grazing animals, and together with Mark Avery I developed a field project on foraging in European Bee-Eaters in the Camargue in Southern France.

This last project also developed into a study of helpers at the nest, jointly with Kate Lessells. We were among the first groups to use the then new technique of DNA fingerprinting to establish the genetic relations of helpers and those they were helping. The Camargue was also a wonderful time for musical events: Luc Hoffmann, who owned the Station Biologique de la Tour du Valat where we worked, invited me to numerous concerts and operas that were performed in Provence as part of the Festival.

Bird song and signalling

Meanwhile the bird song work was also developing. I invested several years of intensive winter and spring field work at Wytham Woods, testing the role of song in Great Tits as a "keep out signal" by removing territorial pairs and "occupying" their territories with loud-speakers (Krebs *et al.* 1978b). Mac Hunter, whom I mentioned earlier, and I spent several exciting and enjoyable field trips to different parts of Europe and the middle-east studying the match between the acoustic properties of the habitat and geographical variation in the song of Great Tits. This built on the pioneering work of Gene Morton and Haven Wiley, suggesting that song structure is adapted to the acoustics of the environment. Mac and I found that regardless of geographical location, open country Great Tits converged on one set of parameters while forest populations were characterised by another (Hunter & Krebs 1979). In the course of collecting the data we were chased by bandits in the Zagros mountains of Iran, stuck in a sand dune in Morocco, stalked by border guards on the Polish–Soviet border in the Bialowiesza Forest and given Easter eggs by the Greek police. Each of these is a story of its own.

In the late 1970s, together with Richard Dawkins, I wrote an article that looked more generally at the evolution of animal signals (Dawkins & Krebs 1978). The prevailing view at that time was that communication is an essentially cooperative venture in which signaller and receiver share a joint interest in information transmission. We turned this on its head, proposing that communication is essentially manipulative, with signallers gaining advantage by altering the behaviour of reactors. This, along with Amotz Zahavi's related idea of honest signalling, soon became the mainstream orthodoxy for thinking about animal communication.

Food storing

In 1979 I was joined by a postdoc from Toronto, David Sherry. David had done his PhD with Jerry Hogan on incubation behaviour in domestic fowl. When he arrived I had been pondering over the question of food storing in tits. While baiting feeders for mist netting birds in Wytham I had noticed how Marsh Tits would come repeatedly to the feeder to take away sunflower seeds. I went back to the library to read up on food storing. The key reference was Sven Haftorn's 1956 (Haftorn 1956) paper, in which he concluded that tits do not hoard for their own exclusive use, but that rather the hoarded food was communal property for times of hardship. He reached this view because he assumed that the birds could not possibly remember the locations of the huge number of seeds they stored, and therefore must recover them by chance encounter, and this could not be restricted to an individual's own seeds.

Steeped in *Selfish Gene* thinking, this did not make sense to me. I did a mini-experiment that consisted of watching a Marsh Tit hide seeds, locating the hidden items and placing my own seed in an apparently identical spot nearby. Hey presto, when I came back a few days later the bird's seeds had gone and mine were still there. I later carried out the same experiment more rigorously together with David Sherry (Cowie *et al.* 1981).

Meanwhile, on a visit to Sweden to teach a Nordic Council for Terrestrial Ecology course, I had discussed my ideas with Malte Andersson, and we produced a simple model that indicated, as one would have suspected, that unless the hoarding individual gains more than others from its efforts, hoarding is not an evolutionarily stable strategy (Andersson & Krebs 1978).

David Sherry's great contribution at that time was to bring the system into the laboratory and carry out a most ingenious experiment based on lack of inter-ocular transfer of memory, to demonstrate that Marsh Tits rely on memory to locate their stores (Sherry *et al.* 1981). Soon after this, another visitor from Toronto, Sara Shettleworth, came on sabbatical and together we took up the food storing story – a collaboration that lasted about 15 years and that often involved carrying out parallel experiments in Toronto (on Chickadees) and Oxford. One of our main aims was to understand whether or not food-storing species have features of memory that are "special" and distinguish them from non-storing species. Nature has provided a number of "natural experiments", because within the Paridae, some species store food whereas others do not. Many years of experiments with different designs suggest that food-storing species may be distinguished by their spatial memory.

David Sherry took the story along another route when he turned up in Oxford in 1988 in great excitement to show me sections of the brains of non-storing (Great Tit) and storing (Chickadee) parids that he and Anthony Vaccarino had prepared in Toronto. These seemed to indicate that the hippocampus (the part of the brain that Vaccarino had shown by lesion experiments to play a key role in memory for cache sites) of storers was larger than that of non-storers. We agreed that this was worth proper investigation, both in terms of the measurements and in terms of sample size and analysis. So we agreed to obtain as much data as we could to see if the relationship was real and general. Sue Healy, who was working with me as a research assistant, and I sought advice from a neuroanatomist, Hugh Perry. With Hugh's help we rapidly built up a data-base to test the hypothesis that food-storers, once you have factored out all the possible confounding variables such as body size and taxonomic position, have a larger hippocampus, and within about a year we were ready to write up and submit a paper. David was pursuing a parallel study in Canada and eventually we published the two studies separately (Krebs *et al.* 1989; Sherry *et al.* 1989) .

Relatively little was known about the structure of the avian hippocampus, or dorsomedial forebrain, so substantial effort went into establishing some basic anatomical descriptions. These included an extensive study of the immunocytochemistry, carried out in conjunction with my old colleague and friend Jon Erichsen, a study using Golgi staining by Catherine Montagnese, and pathway tracing work by Andrea Szekely.

The three-way link between food-storing, memory and brain structure was taken much further by Nicky Clayton, when she joined my group in the early 1990s. I had taught Nicky as an undergraduate at Oxford, and she came back after a PhD at St Andrew's and a postdoc at Bielefeld. Nicky is one of those scientists with green fingers, as well as incredible experimental skills. In the space of a few years she had completed a wonderful set of experiments on the development of the avian hippocampus, showing that experience of food-storing, or spatial memory, triggers growth of the hippocampus in food-storing species. This is a beautiful example of how an inborn predisposition is revealed through specific environmental experience (Clayton & Krebs 1994).

The food-storing system soon became well-known as an example of the evolutionary approach to brain and cognition, and is still an active area. In August 2008 Tom Smulders and others organised a workshop at Cornell to celebrate the 30th anniversary of Andersson and Krebs. The speakers and attendees came from over 20 different institutions.

Behavioural ecology

Every so often, within a field of research, a new agenda is created and defined. For a variety of reasons this is what happened with the field of behavioural ecology in the 1970s.

I think that several influences came together to create the environment in which this could happen.

First, both the fields of animal behaviour (ethology) and ecology were themselves in a very rapid and exciting period of development.

In ethology, Lorenz, Tinbergen and von Frisch had shared a Nobel Prize in 1973, which was quite remarkable in the context of the kind of work on cellular and molecular mechanisms that normally attracts this recognition. Robert Hinde, in his masterly synthesis published in 1966, had swept away much of the old theorising of Lorenz and Tinbergen and with it the conflicts between experimental psychologists and zoologists, and pointed to new linkages between ethology, psychology endocrinology and neuroscience. The whole field was pregnant with possibility and opportunity. Although not seen as mainstream, W. D. Hamilton had developed the theory of kin-selection.

At the same time, something similar was happening in ecology, especially in North America. Robert MacArthur and his followers, as well as others such as Gordon Orians, Robert Paine and Jerram Brown, were creating a coherent, broad brush theoretical framework for ecology, ranging from individuals to communities, and offering accounts of population dynamics, species diversity, habitat distribution, niche width and so on. In Britain, a group of young ecologists, including John Lawton, Mike Hassell, John Beddington, Roy Anderson, Mick Crawley, and me used to meet every six months or so, for a curry and results discussion. In this way I was fortunate to be plugged into both the ethological and ecological communities.

Secondly, and not unrelated to the first point, ecology and animal behaviour were the newly fashionable subjects for graduate students. This was partly because of the tangible sense of exciting and rapid progress, but also because of the emerging awareness, triggered by Rachel Carson's *Silent Spring*, among other books, of environmental issues.

Thirdly, new ideas from evolutionary biology were infiltrating both ecology (e.g. optimality theory) and animal behaviour (kin selection, reciprocity and game theory, introduced by Hamilton, Trivers and Maynard Smith, respectively).

Finally, and of huge significance in my view, the fields of ecology and ethology were both small enough for an industrious reader to be on top of all the literature and know all the key people personally. At the International Ethological Conferences (size restricted and by invitation only), it was a reasonable expectation to know more or less every individual there. Lee Smolin makes a similar point with regard to physics (Smolin 2006): "It is said that there are more scientists working now than in the whole history of science. This is certainly true of physics; there may be more professors of physics in a large university department today than there were a hundred years ago in the whole of Europe, where almost all the advances were being made."

One outcome of this hubris was that Nick Davies and I hatched a plan to edit a book on what we saw as the newly emerging field of "Behavioural Ecology" (Krebs & Davies 1978). We were encouraged in this thought by Bob Campbell, the commissioning editor at Blackwell Science. Nick and I shared research interests, including foraging and territoriality, both liked doing simple field experiments (Nick was rather better at it than I was) and we appreciated a similar style of writing. The edited book was built around the exciting ideas and people that we knew – Geoff Parker, Paul Harvey, Tim Clutton-Brock, Henry Horn, Richard Dawkins, Steve Emlen, John Maynard Smith and others. It was also to be radically different from the main textbooks on Animal Behaviour, by John Alcock, Robert Hinde, and Peter Marler & William J Hamilton. We aimed to weave together strands from ethology, ecology, population biology and evolution. Already two books, in very different ways, had attempted a similar task: E. O. Wilson's monumental *Sociobiology* and Richard Dawkins' *The Selfish Gene*, a brilliantly original polemic and synthesis for a wider audience. We did not see these as in any sense competitors, because of their different intended audiences and compass.

Where did the title *Behavioural Ecology* come from? I cannot be absolutely sure, but I do recall that we tried out several titles including *Evolutionary economics* and *Behaviour, evolution and ecology*. The title *Behavioural Ecology* had nearly been used by Peter Klopfer in his short book *Behavioural Aspects of Ecology* and a new journal edited by Jim Markl, *Behavioural Ecology and Sociobiology* had appeared a year or two earlier. One view is that we chose the title to deliberately distance ourselves from the sociobiology furore in the USA (conspicuous by its absence in Europe), but I do not recall this as a motivation. We did not focus on social behaviour, so that the title *Sociobiology* would in any case, have been misleading. Although it has been reported that I "spearheaded a hostile takeover of E. O. Wilson's new synthesis almost before the paint could dry, successfully usurping and greatly improving the emerging field we now perceive" (Parker 2006) I do not recall this as a motivation that Nick and I shared at the time.

The immediate success of the edited book stimulated the two of us (again encouraged by Bob Campbell) to write a shorter, more coherent, introductory textbook. Luckily Nick and I saw eye-to-eye on writing style and content, and we divided the book in half. We each wrote seven chapters, at one chapter per week, in less than two months. I had prepared lecture notes and examples, but then wrote without reference to source material. The fact that the book was written so quickly, and in a sense effortlessly, gave it fluency and coherence (Krebs & Davies 1981).

Job offers

Job offers came in the 1970s and 1980s, but the attractions of Oxford were too great to relinquish. The two most tempting offers were from Harvard, where Ed Wilson and others had decided that I was the right person to bring as a Professor to the Museum of Comparative Zoology, and the Max Planck Society, where I was offered the Directorship of one of the Institutes where I had worked as a schoolboy. When I mentioned the Harvard job to Dick Southwood, my Head of Department at Oxford, he acted swiftly and successfully to seek money from Sir Edward Abraham, the discoverer of cephalosporin, through one of his funds set up to help academics at Oxford. The fund endowed a Fellowship for me at my former undergraduate College, Pembroke. At the time, Dick said to me: "If I do this for you, you must agree not to go to Harvard, or I will end up with egg on my face".

The Max Planck offer was tempting indeed, with the generous resources and the emotional bond from my teenage years. It would have enabled me to transfer my group from Oxford. Weighed against this was the relative isolation of the Institute, from both a professional and a family perspective, and in the end I suggested a split post between Oxford and Bavaria, but the President of the Max Planck Society, Heinz Staab, and I realised, after a short but cordial conversation in Munich, that we would not be able to agree on a split that would work for both of us. Instead I was offered an External Scientific Membership of the Institute, and through this developed a most fruitful and enjoyable collaboration with Ebo Gwinner and Herbert Biebach. I spent several summers there with my family, and made frequent visits at other times.

Herbert Biebach and I returned to an old problem from the bee literature, "Zeitgedachtnis", referring to the idea that bees are able to associate particular feeding locations with different times of day. We showed that Garden Warblers are capable of time-place learning and that this is dependent on the circadian clock, persisting in "free running" conditions (Biebach *et al.* 1994). The work with Ebo Gwinner built on the results of our food storing – hippocampus story, by examining whether or not there was a similar effect in migratory birds. We compared migratory and non-migratory warblers, and different populations of rose finches with different migratory behaviour. Sadly, Ebo died before we had written up a large chunk of work. The published work, comparing a migratory and non-migratory species of *Sylvia* warbler, showed that in experienced adults but not juveniles the relative hippocampal volume is larger in the migratory species, but that this was a result of differential shrinkage of the telencephalon (excluding the hippocampus) (Healy *et al.* 1996). The unpublished work included two extensive experiments on homing pigeons to test whether or not different kinds of homing experience affected the hippocampus, and a large-scale study of migratory and non-migratory populations of rose finches. The pigeon results did not show any effects of experience on the hippocampus, whereas the rose-finch data showed that changes in the hippocampus occurred primarily in the caudal region: it appears that this region grows before birds migrate, but only remains large if they have experience of migration.

Royal Society Research Professor: 1988–94

Arguably the most desirable job in scientific research outside the biomedical field in the UK is a Royal Society Research Professorship, both because of its prestige value and because it brings with it automatic research support. In 1988 I applied for, and was lucky enough to be awarded, one of these jobs: the application was entitled "A study of animal memory". There is a fixed number of these posts (in 1988 there were 14 general Professorships and a further 4 in designated subjects). In 1988, the Department of Zoology at Oxford held four of these Chairs, the others being Richard Gardner, Bill Hamilton and Bob May.

From pharmacology to farm ecology

Not long after I had taken up this job, a new and very different funding opportunity arose. The UK Research Councils were planning to set up a small number of large research centres with long-term funding (so-called Interdisciplinary Research Centres). One of these was to be supported jointly by the Agricultural and Food Research Council and the Natural Environment Research Council, on an ecological theme. Dick Southwood proposed that Oxford should bid for a Centre and that I should lead the bid, which would focus on Agriculture–Environment interactions. When Dick, Bob and I went to the interview and I gave my presentation, one of the questions, from the plant ecologist John Harper, mindful of my recent interest in the workings of the avian brain, was "John, are you more interested in Pharmacology or Farm Ecology?"

Our bid was unsuccessful (the Centre for Population Biology at Imperial College came out top), but in compensation, both Research Councils offered to fund a Unit with a five-year renewable grant. In spite of my efforts to merge the two, the Councils insisted on separate funding, so I became simultaneously Director of the NERC Unit for Behavioural Ecology and of the AFRC Unit for Ecology and Behaviour! The NERC money enabled me to employ Alasdair Houston for a further five years and the AFRC money enabled me to bring together a group of postdocs including Richard Griffiths, Sean Nee, Pat Doncaster and Jonathan Wright, to work on vertebrate populations in farmland settings.

By this stage, in the early 1990s, I found myself running a group of about 30, including 10 postdocs, a dozen D. Phil. students and various visitors and research assistants. At the same time, I was gradually becoming more involved in Research Council affairs, as a member of the Agriculture and Food Research Council, as well as through writing a report on the Future of Taxonomy in the UK, for NERC, that led to major new funding in this area (Krebs 1992). Although I did not recognise it at the time, I was pulled in too many different directions, which may in part explain why I made what to many was an incomprehensible career change.

Natural Environment Research Council: 1994–99

One evening in the early autumn of 1993 I was working late in the lab, when my secretary's phone rang in the adjoining office. When I answered it, the voice at the other end explained that he was from a firm of Head Hunters, and working for the Government's Office of Science and Technology. My name had been mentioned as a possible Chief Executive of the Natural Environment Research Council, the body responsible for funding environmental research in universities as well as running major Institutes throughout the country. I heard a voice saying "tell me more" and realised it was my own: until that very moment I had not thought of such a job.

I did indeed let my name go forward and eventually reached the final stage, with a daunting interview in the Cabinet Office at 70 Whitehall. The Panel included Sir William Stewart, the Chief Scientific Advisor, Sir David Phillips, the Chairman of the Advisory Board for the Research Councils, industrialist Sir Robert Malpas, the Chairman of NERC, Sir Richard Mottram, Permanent Secretary in the Cabinet Office, and Bob May. One of the questions I was asked was "Given that you have never run anything larger than a research group of 30, do you have any relevant experience for running an organisation with 3000 staff and a budget of £220 million?" My answer, which I genuinely believe, is that the same skills are required for leading small or large groups: an ability to inspire, energise and reward those who work for you, a clear sense of direction, and the right balance between delegation and authority. I also added that in an organisation of 3000, I would interact on a regular basis with about the same number of people as in my research group.

Nevertheless there were dramatic differences. On my first day in the office I was confronted with a pile of unanswered mail about 30 cm high. I quickly figured out that the way to deal with this was to initial each one, to show that I had seen it (and therefore could be held accountable if anything went wrong) and to indicate who should deal with it – if Finance, then the Finance Director, if Strategy, then the Director of Corporate Affairs, and so on. The trick in this situation is to develop an eye for the potential dangers and to skim over the routine. In my subsequent post as Chairman of the Food Standards Agency my mail arrived in a folder, neatly indexed and divided into sections such as Action and Note. Whenever a letter from Greenpeace or Friends of the Earth about GM food or pesticide residues appeared, even if only in the Note section, I gave it extra attention because of the scent of possible trouble ahead! I also had to get used to having my life planned to the nearest few minutes every day, and the rhythm of many short meetings, often with disparate agendas. The attention space and rhythms of running a large organisation are very different from those of research.

I loved the new job, even though it had meant giving up most of my research time (I had negotiated a half day a week in Oxford, but usually this was Friday afternoon, by which time I was usually more or less brain dead). It was exciting to be able to learn about such a broad range of science (from oceanography, to earth science, to meteorology, to hydrology, to ecology), as well as to feel involved in policy decisions at the national and international level.

TB and Badgers

England (and parts of South Wales) has a problem with bovine tuberculosis. As a result of pasteurisation of milk and testing of cattle, the risk to human health is negligible, but the consequences for affected farms is severe: meat and dairy production are halted, and cattle slaughtered.

In the autumn of 1996 I was asked to review the science behind the Government's policy on bovine tuberculosis, in light of the fact that TB was on the increase. Without quite realising what I was letting myself in for, I agreed and duly went to discuss the plan with Douglas Hogg, the Agriculture Minister. The big issue was whether or not the increase in bovine TB in cattle was linked to increases in the wild badger population. Two previous reviews, by Lord Zuckerman in the 1970s and George Dunnet in the 1980s, had concluded that badgers are part of the problem and that one way to treat the problem was to kill badgers in affected areas. However, killing badgers in Britain is a hot potato. Perhaps thanks to children's stories such as Wind in the Willows and Rupert Bear, badgers are seen by wildlife lovers as cute and cuddly.

The temptation when asked by Ministers for scientific advice is to say "I'm an expert, I've looked into the problem, and here is your answer". Both Dunnet and Zuckerman fell into this seductive trap. When we scrutinised all the evidence, we concluded that it was highly likely that badgers did transmit TB to cattle, but it was by no means clear that killing badgers was an effective or cost-effective way of controlling the disease. So I recommended to Jack Cunningham, by then Agriculture Minister following the 1997 election, that the government should undertake a massive field experiment to test the efficacy of badger killing as a control strategy (Krebs *et al.* 1998).

The experiment, which came to be known as the "Krebs Trials", involved three treatments, set up as triplets in a region: control (no culling), reactive culling (kill badgers when there is an outbreak of bovine TB in the area) and proactive culling (scorched earth policy within the area). Each treatment was to occupy a 10 km × 10 km square. The trials were completed in 2007. The conclusion was that reactive culling actually makes the problem worse, possibly by breaking down the territorial system and enabling non-territorial, TB-affected badgers to come into an area (the so-called perturbation effect). Proactive culling works, but only if the area is large enough for the positive effects of removing badgers in the middle of the area to outweigh the negative perturbation effects around the edges. The culling also has to persist over many years. One estimate is that the minimum area to achieve a net benefit would be between 200 and 300 square kilometres. Adding up all the badly affected areas ("hotspots"), a proactive, or scorched earth, policy would lead to the culling of roughly half the badgers in Britain. Eventually, in July 2008, the Secretary of State for Environment, Food and Rural Affairs, announced that the English Government would not proceed with culling as a policy. But this was not before the Government's Chief Scientific Advisor, Sir David King, had muddied the scientific waters by producing a hurried, and ill-thought-out, report in which he recommended that culling should go ahead.

The issue of bovine TB was my first introduction to the sharp end of the science–policy interface, and it neatly encapsulates many of the problems: the science is often not clear-cut, the pressure groups interpret any ambiguities in the science to support their case, and the policy decisions may be influenced by political acceptability as much as by science. This already complex situation was exacerbated by conflicting advice from scientific experts, making life especially tough for Ministers looking for expert help. These were challenges that I faced on a day-to-day basis in my next job.

The Food Standards Agency 2000–05

I was at NERC for five and half years. The Royal Society, with exceptional generosity, had not only kept my job open for me to go back to, but also continued to pay my research expenses for my secretary and a postdoc. However the large group that I had built up at the time of my departure to NERC had dispersed. The senior people were in tenured posts elsewhere: Alasdair Houston in Bristol, Nicky Clayton at UC Davis, Sean Nee in Edinburgh, Mark Pagel at Reading and so on. The remainder of the group had focussed on the impacts of agricultural intensification on farmland bird populations (Krebs *et al.* 1999), and the work was increasingly collaborative with the Royal Society for Protection of Birds (several people from my group at that time now work for the RSPB).

Towards the end of my time at NERC, the head-hunters persuaded me to consider taking the job as a four-day-a-week Chairman of the UK Food Standards Agency (FSA). The FSA was to be created, as a result of an Act of Parliament, as a new Non-Ministerial Government Department. This meant that the Chairman would have quasi-ministerial position and take over responsibility for food safety and nutrition from Health and Agriculture Ministers. The Agency would in the first place be staffed by transferring about 600 civil servants from the Ministry of Agriculture and the Department of Health. The selling point to me was that here was an opportunity to create a new organisation more or less from scratch.

The history of the FSA lay in the bovine spongiform encephalopathy (BSE) crisis of the 1990s. Ministers were seen to have failed the public both by their secrecy and by their false reassurances on the safety of beef (everyone who was in the UK at the time recalls the iconic photo of the Agriculture Minister feeding his young daughter a hamburger with the slogan "British beef is perfectly safe". It was not), and by putting their own survival, as well farmers' interests, ahead of safety. Scientists as well as politicians came out of the BSE crisis badly: the expert advisers had underplayed the uncertainties in the science and given Ministers enough room to claim that beef was safe. Hence the new Agency, a promise made by Labour in its 1997 election manifesto, was to be charged with putting things right. It was to be impartial and evidence-based, honest, and transparent as well as free from political or industry pressure, to put the interests of the consumer first. It was responsible for regulation and enforcement, advice to the public, and policy development. As the Secretary of State for Health Frank Dobson put it to me with disarming honesty, the job was: "A poison chalice without the chalice" and "A defensive shield for Ministers". The general mood of the media was one of distrust in Government, and of course food is a topic in which everyone considers themselves to be an expert.

GM and organic food

If you look in Wikipedia, you will read that my time as Chair of the Food Standards Agency was marked by my criticism of organic food. Early on in the life the Agency I was painted, by some of the pressure groups, as "pro-science, pro GM (genetically modified food), anti-organic". This was code for "goes with the objective evidence rather than with worries and assertions of certain pressure groups". The "anti-organic" statement was one that I made on a BBC TV show called Country File, in which I said that the Agency had looked at the scientific evidence and could find no support for the claim that organic food is healthier or safer than conventional food. Of course the FSA was not alone in reaching this conclusion: every independent organisation that has looked at the evidence has drawn the same conclusion. But my comments, unsurprisingly, unleashed a furious response from the organic food industry, which has, in its brilliant marketing, persuaded consumers that organic food is worth the 50% or so premium because of the health benefits.

People who like organic food tend to be against GM, and the fact that I took the scientific stance, that the GM products approved for human consumption by the FSA were as safe as their conventional counterparts, meant that I was clearly "pro-GM". There was no room for impartiality in this territory.

BSE (bovine spongiform encephalopathy)

However, these two issues, although they filled a great deal of media space, were not the most important matters for the Agency. The biggest challenge in the early days was to deal with BSE, or mad cow disease.

No-one knows how BSE started, although the Horn review (2001) concluded that it might have been a mutation of the related disease, scrapie, in sheep, that has been endemic in the UK since the early eighteenth century and is apparently harmless to humans. Following a long standing agricultural practice, farmers in Britain and elsewhere gave their cattle high-protein food supplements (meat and bone meal, MBM) derived from rendered carcases of dead livestock, including sheep. In the 1970s the rendering process was changed to a lower temperature, and this may have allowed a mutated prion protein (the infective agent) to survive more readily. Once the disease had infected cattle, the custom of using MBM would have accelerated its further spread (this practice was banned in the late 1990s as a part of the BSE controls).

In the late 1990s, epidemiological modellers had forecast that over the next 20 or so years, several hundred thousand people could be affected by the human form of mad cow disease, known as new-variant Creutzfeld–Jakob Disease (vCJD). This estimate was based on calculations of the number of infected cattle (perhaps over a million) that had gone into the food chain in the 1980s and early 1990s, and assumptions about the amount of infectivity in each animal and the susceptibility of humans to the infective agent, thought to be the prion protein. Luckily, these projections have turned out to be far too pessimistic: the total number of deaths from human BSE is only 164 (August 2008) and the number per year has fallen from a peak of 28 in 2001 to around one or two. Although there are still uncertainties because the disease has such a long incubation period, and some scientists argue that there could be second and third waves, as different genotypes with different incubation periods succumb, it seems that humans do not catch BSE from meat very easily. At the time, of course, we did not know this and the role of the Food Standards Agency was to ensure that scientifically based control measures to keep infectivity out of the food chain were effectively enforced.

When I started at the FSA, no-one knew whether BSE had jumped into sheep in the UK. Sheep were known to be susceptible to BSE because they could, under experimental conditions, catch it from the same infected meat and bone meal that generated the epidemic in cattle. Furthermore, if BSE had infected the sheep population it would have been possible to distinguish it from scrapie from the external symptoms.

It now seems unlikely that BSE did get into sheep to any significant extent, because a biochemical test has been developed for the BSE prion and many thousands of sheep have been tested with negative results. However, in the early days of the FSA the story looked very different. At the Institute of Animal Health at Compton a sample of pooled sheep homogenate collected in the early 1990s sat in a deep freeze ("the brain pool"). We suggested to the Institute Director that it would be worth testing this pooled sample for any sign of BSE. At this stage the only available test was a very tedious one of injecting the suspect material into different genetic strains of mice, whose pattern of succumbing to the disease would reveal whether or not the signature of BSE was in the brain pool. The experiment lasted over 12 months and the news got worse as the months went by, until eventually it was almost certain that the brain pool contained BSE. At this dreadful moment the FSA had to make a decision about the policy: should we in effect close down the

UK sheep industry by banning sheep meat? There seemed to be no alternative, unless a certification scheme for disease-free flocks could be developed.

The Agency had committed right from the start to make all its major policy decisions in public, through its Board. So we planned a special board meeting, with the TV, radio and other media assembled along with over a hundred members of the public, to debate and decide. Then, in the best traditions of a good thriller, there was a totally unexpected twist. A few days before decision day, the Institute of Animal Health let us know they were doing a final DNA test just to check that the brain pool sample had not been contaminated with other animal material. In fact it had not been contaminated at all: it had been mislabelled and it was a sample of cow brains. Chaos ensued: I went on the morning radio current affairs show, the Today Programme, to announce that there had been a "cock-up" in the laboratory. When I was reprimanded by my media team for using such vulgar language, I replied "but Bob May does it all the time!" (Bob was Chief Scientific Advisor to the Government and often in the media), to which the reply came "He's Australian".

Foot and mouth disease

In February 2001, a new, and devastating crisis hit the UK livestock industry: an outbreak of foot and mouth disease (FMD). FMD is not a human health risk, but it is potentially disastrous for farmers. When the disease was spotted, in some pigs that had been transported from the North of England to the South for slaughter, the Ministry of Agriculture, Fisheries and Food (MAFF) line was that the disease would be rapidly contained and there was no need to worry. On one of our Sunday morning runs, Bob May and I chatted about the disease and wondered how MAFF could know that it had not infected sheep (where the symptoms are hard to spot) and been transported around the country. As the days went by, more farms were affected, and I decided to call a meeting of my academic colleagues expert in disease modelling, from Imperial College (Roy Anderson and Neil Ferguson), Cambridge (Bryan Grenfell), Edinburgh (Mark Woolhouse) and Warwick (Graham Medley). MAFF said they were too busy to come along. By now it was early March, and at the meeting we concluded that if MAFF would give the modellers the locations and the times of all cases so far, it would be possible to come up with some estimate of the rate of spread of the disease. I rang the permanent secretary of MAFF to ask for the data: after initial reluctance, he agreed. The modellers were given two weeks. By this time (the third week of March) the disease was clearly not under control, and MAFF was in some disarray. The meeting at which the modellers reported their results was electric. This time, MAFF turned up in force, with the Chief Vet and their Chief Scientist. Dave King, the Government's Chief Scientific Adviser, was also there. Neil Ferguson spoke first. With a characteristically polished Powerpoint presentation, he explained his modelling approach, and the bottom line conclusion that the doubling time for the number of farms affected was about 9 days, so that by mid-May half the livestock farms in the country would be affected. He concluded that the only way (apart from vaccination) to prevent this would be to create a *cordon sanitaire* around each infected area by massive culling of livestock to prevent further transmission. I then turned to Mark

Woolhouse for his view: he said he had taken a different modelling approach but reached very similar conclusions. Dave King immediately wrote to the Prime Minister and the policy recommended by the modellers was adopted.

Risk and uncertainty

More often than not, when a new contaminant was found in food, it was not possible to quantify the risk with certainty, so the decision on what action to take was a judgement call (sometimes dignified with the title The Precautionary Principle). When Swedish scientists discovered that many baked, roasted and fried foods contained a carcinogen that results from cooking, acrylamide, we opted for advising the public that this was not a new risk (it had been there since prehistoric humans first roasted woolly mammoth steaks, or whatever), and that it was within the range of acceptable risks, given that nothing is risk-free. In a different food safety scare, in which an illegal carcinogenic dye, Sudan 1, had been added to chilli pepper, we took the view that although the risk was probably very low (not, as one newspaper reported, exactly the same as smoking one cigarette, because no one knows its magnitude), we could not be sure how big it was, and furthermore the substance was illegal and the products containing it should be recalled. In the end it turned out that more than 500 different kinds of manufactured food contained the contaminant and the recall possibly cost several hundred million pounds. When the food industry argued that we had over-reacted, I played them the alternative story: "We have found an illegal carcinogen in your food, but in order to save money for the food industry we advise you to carry on eating the products as normal".

Nutrition and health

Food safety is often a matter of crisis management, but the other part of the FSA's remit, nutrition, was a much longer-term, and in many ways more difficult area. It's not that nutrition is unimportant in terms of health. Far from it: if you calculate risks in terms of deaths per year, then the contribution of poor diet to cardiovascular disease and cancer means that diet-related risks are much bigger than risks such as BSE or food poisoning. But the role of the state in determining people's diet is not obvious: for most of us it is deemed to be a rather personal matter of individual choice.

Nevertheless, the Agency did have some successes in relation to nutrition, most notably in persuading the food manufacturers and retailers to reduce the amount of salt (often added in large quantities because it is the cheapest flavouring) in processed food. When I first starting talking about health risks of high salt intake, the industry both rejected the evidence and said their consumers were not interested in the matter. Within the space of a year, and after a campaign by the Agency to raise awareness among the public and to name and shame products, retailers and manufacturers alike were making "salt reduction" a marketing advantage issue.

At the time I left the Agency, the debate about nutrition and health became focussed on the contribution of diet to the rapidly rising prevalence of obesity in many countries, including the UK. This is a huge challenge for public health policy, as preventing obesity implies very large changes in many people's personal behaviour, and in turn these behaviours are heavily shaped by government decisions on planning and on the regulation of marketing, as well as on powerful industry interests.

Jesus College, Oxford, and the House of Lords

Jesus College

After nearly five and half years I decided to move on from the FSA to take up the job of Principal of Jesus College, Oxford, a role roughly equivalent to being the non-executive Chair of a Charitable Foundation. The Oxbridge Colleges are self-governing educational charities, responsible for admitting students to the University and undertaking significant parts of the teaching. As Head of House I have no executive authority, unless delegated by the Governing Body, but I chair every major committee as well as the Governing Body itself, which is made up of the Academic Fellows (University Lecturers and Professors) associated with the College. The other roles of the Head of House include social (entertaining students, Fellows and Old Members, often in connection with fund-raising) and representational (representing the College's interests externally, within the University and beyond).

The House of Lords

In 2007 I was appointed to the House of Lords as an "independent cross-bencher": that is, a non-party-political appointee. For those not familiar with the UK system, I should explain that the House of Lords is the second chamber of Parliament, and its primary role is to scrutinise, and suggest revisions to, legislation. It has about 750 members, of whom very roughly 200 are Labour, 200 Conservative, 200 independent cross-benchers, 75 Liberal Democrats, 25 Bishops and about a dozen others. Until the middle of the twentieth century peers were hereditary, and only in 1999 were all but 92 of the hereditary peers excluded from the House. The great majority of peers are "life peers". Those appointed to the cross-benches are chosen by an independent appointments commission and the aim is to bring a range of different kinds of expertise into the House, without alignment to any political party, so that one can speak and vote for or against the Government depending on one's own view. I am one of about half a dozen scientists. For me, as with many others, sitting in the Lords is part-time and has to be fitted around my day-job as Principal of Jesus.

Only a couple of months after I had started work in the House of Lords, Tony Blair stood down as Prime Minister. On the day before Gordon Brown took over, I had a phone call asking whether I would be willing to move from the cross-benches to the Labour benches to become Science Minister in the Brown administration. I had roughly 12 hours to decide.

After a rather sleepless night I said no, on the grounds that I did not wish to leave my current job, nor was I convinced that another spell at the sharp end of policy was what I wanted. However, I will always have some pang of regret!

Concluding remarks

Writing an autobiography is a rather narcissistic occupation, a bit like spending several hours in therapy. Nevertheless, seizing the moment to reflect on what I hope is a still-to-be-completed career, I will make a number of closing remarks.

First, about planning a career. I have changed directions many times, both in my research and in my moves from academia to government and back. None of this was a result of planning, but rather as a consequence of what former British Prime Minister Harold McMillan referred to as "events, my dear boy, events". As George Eliot wrote in the finale of *Middlemarch*, "..the growing good of the world is partly dependent on unhistoric acts". My life and career has resulted from a succession of lucky, but unhistoric acts.

Second, about supervising research students in particular and leadership more generally. Here my two role models (I may not always have lived up to them) are my own supervisor, Mike Cullen, and Dick Southwood as Head of the Zoology Department. From these two I learned that to be effective in supervising or leading, your pleasure has to come from the success of others.

Third, about being the son of a Nobel Prize winner. I have often been asked whether it was an advantage or disadvantage to go into science as the son of such a famous scientific father. I think the answer is that it was, on the whole, a benefit (for instance, I would have been far less likely to have worked as a schoolboy in Lorenz's group), but not without its downsides. Even now, I am often introduced as "the son of Hans Krebs". My friend Gustav Born, the distinguished pharmacologist and son of Max Born, the Nobel Prize-winning physicist, recounts that he was often asked "Are you the son of Max Born?". When he finally had a son he named him Max, so he could retort "No, I am the father!". I am not sure that either of my daughters would have appreciated being called Hans, to enable me to deploy the same response. When I was a first-year graduate student, and my research was not going well, nor was I committed to it, Niko Tinbergen called me into his office for a chat. His main theme was that I would always carry the burden of feeling inferior to my father as a scientist, and would therefore never take my science seriously. I am not sure why Niko felt obliged to draw this idea to my attention (it was around the time that he began to develop his interest in human psychology, including his speculations on autism, so perhaps he was indulging in general folk psychology), but it certainly was not much help to me as a struggling graduate student, even if there is an element of truth in it.

Finally, priorities. When she was about 12, my younger daughter, George, and I were walking home when she turned to me and said "What really counts in your life?". While I hesitated, she offered the answer: "I think for you it is your work. I have decided what really counts for me is my family and friends". Having been immensely fortunate in so many ways in my career, perhaps this is the best guide for the future.

References

Andersson, M. & Krebs, J. R. (1978). On the evolution of hoarding behaviour. *Anim. Behav.* **26**: 707–11.

Biebach, H., Krebs, J. R. & Falk, H. (1994). Time-place learning, food availability and the exploitation of patches in garden warblers, *Sylvia borin. Anim. Behav.* **48**: 273–84.

Charnov, E. L. & Krebs, J. R. (1974). On clutch-size and fitness. *Ibis* **116**: 217–19.

Charnov, E. L. & Krebs, J. R. (1975). The evolution of alarm calls: altruism or manipulation? *Am. Nat.* **109**: 107–12.

Clayton, N. S. & Krebs, J. R. (1994). Hippocampal growth and attrition in birds affected by experience. *Proc. Natn. Acad. Sci. USA* **91**: 7410–14.

Cowie, R. J. (1977). Optimal foraging in great tits (*Parus major*). *Nature* **268**: 137–9.

Cowie, R. J., Krebs, J. R. & Sherry, D. F. (1981). Food storing by marsh tits. *Anim. Behav.* **29**: 1252–9.

Dawkins, R. & Krebs, J. R. (1978). Animal signals: information or manipulation? In *Behavioural Ecology* (ed. J. R. Krebs & N. B. Davies), pp. 282–309. Oxford: Blackwell Scientific Publications.

Falls, J. B. & Krebs, J. R. (1975). Sequences of songs in repertoires of western meadowlarks *Sturnella neglecta. Can. J. Zool.* **53**: 1165–78.

Haftorn, S. (1956). Contribution to the food biology of tits especially about storing of surplus food: Part IV. A comparative analysis of *Parus atricapillus L, P cristatus L*, and *P ater L. Kgl. Norske Vidensk. seskl. skrift.* **4**: 1–54.

Healy, S. D., Gwinner, E. & Krebs, J. R. (1996). Hippocampal volume in migratory and non-migratory warblers: effects of age and experience. *Behav. Brain Res.* **81**: 61–8.

Holmes, F. L. (1991). *Hans Krebs: The Formation of a Scientific Life, 1900–1933.* Oxford: Oxford University Press.

Holmes, F. L. (1993). *Hans Krebs: Architect of Intermediary Metabolism, 1933–1937.* Oxford: Oxford University Press.

Horn, G. (2001). Review of the origin of BSE. www.defra.gov.uk/animalh/Bse/publications/bseorigin.pdf

Hunter, M. L. & Krebs, J. R. (1979). Geographical variation in the song of the great tit *Parus major* in relation to ecological factors. *J. Anim. Ecol.* **48**: 759–85.

Krebs, H. A. (1981). *Reminiscences and Reflections.* Oxford: Oxford University Press.

Krebs, J. R. (1971). Territory and breeding density in the great tit *Parus major. Ecology* **52**: 2–22.

Krebs, J. R. (1974). Colonial nesting and social feeding as strategies for exploiting food resources in the great blue heron *Ardea herodias. Behaviour* **51**: 99–134.

Krebs, J. R. (1992). *Evolution and Biodiversity: The New Taxonomy.* Natural Environment Research Council Report. Swindon: NERC.

Krebs, J. R. & Davies, N. B. (eds.) (1978). *Behavioural Ecology: An Evolutionary Approach*, 1st edn. Oxford: Blackwell Scientific Publications.

Krebs, J. R. & Davies, N. B. (1981). *An Introduction to Behavioural Ecology*, 1st edn. Oxford: Blackwell Scientific Publications.

Krebs, J. R. & Dawkins, R. (2001). Mike Cullen [Obituary]. *The Guardian,* 10 April 2001: 20.

Krebs, J. R., MacRoberts, M. H. & Cullen, J. M. (1972). Flocking and feeding in the great tit *Parus major* – an experimental study. *Ibis* **114**: 507–30.

Krebs, J. R., Ryan, J. C. & Charnov, E. L. (1974). Hunting by expectation or optimal foraging? A study of patch use by chickadees. *Anim. Behav.* **22**: 953–64.

Krebs, J. R., Erichsen, J. T., Webber, M. I. & Charnov, E. L. (1977). Optimal prey selection in the great tit *Parus major. Anim. Behav.* **25**: 30–8.

Krebs, J. R., Kacelnik, A. & Taylor, P. (1978a). Test of optimal sampling by foraging great tits. *Nature* **275**: 27–31.

Krebs, J. R., Ashcroft, R. & Webber, M. I. (1978b). Song repertoires and territory defence in the great tit. *Nature* **271**: 539–42.

Krebs, J. R., Sherry, D. F., Healy, S. D., Perry, V. H. & Vaccarino, A. L. (1989). Hippocampal specialization of food-storing birds. *Proc. Natn. Acad. Sci. USA* **86**: 1388–92.

Krebs, J. R., Anderson, R. M., Clutton-Brock, T. H. *et al.* (1998). Badgers and bovine TB: conflicts between conservation and health. *Science* **279**: 816–18.

Krebs, J. R. *et al.* (1999). Author to provide full reference, please.

Kundera, M. (2002). *Ignorance.* London: Faber & Faber.

Lack, D. (1954). *The Natural Regulation of Animal Numbers.* Oxford: Oxford University Press.

Lack, D. (1966). *Population Studies of Birds.* Oxford: Oxford University Press.

Lack, D. (1968). *Ecological Adaptations for Breeding in Birds.* London: Methuen.

Parker, G. A. (2006). Behavioural ecology: natural history as science. In *Essays in Animal Behaviour* ed. J. R. Lucas & L. W. Simmons, pp. 23–56. San Diego, CA: Elsevier.

Sherry, D. F., Krebs, J. R. & Cowie, R. J. (1981). Memory for the location of stored food in marsh tits. *Anim. Behav.* **29**: 1260–6.

Sherry, D. F. Vaccarino, A. L. Buckenham, K. & Herz, R. S. (1989). The hippocampal complex of food storing birds. *Brain Behav. Evol.* **34**: 308–17.

Smolin, L. (2006). *The Trouble With Physics.* London: Allen & Lane.

Stephens, D. W. & Krebs, J. R. (1986). Foraging theory. *Princeton Monographs in Behavior and Ecology 4.* Princeton: Princeton University Press.

Stewart, R. E. & Aldrich, J. W. (1951). Removal and repopulation of breeding birds in a spruce fir forest. *Auk* **68**: 471–82.

Watson, A. (1967). Territory and population regulation in the Red Grouse. *Nature, Lond.* **215**: 1274–5.

Wynne-Edwards, V. C. (1962). *Animal Dispersion in Relation to Social Behaviour.* London: Oliver & Boyd.

15

My life with birds

GORDON H. ORIANS

My life with birds began in Milwaukee, Wisconsin, when I was about seven years old. The initial stimulus was provided by a fathers-and-sons banquet at my father's church (he was the pastor). The after-dinner speaker, from the Milwaukee Public Museum, showed slides of birds he had taken from blinds erected near their nests. The slide show resurrected my father's boyhood fascination with birds. He quickly read several books on how to take pictures of birds, built a blind, and together we went into the field searching for nests at which to photograph. Thus began both a close relationship with my father, which I deeply

Leaders in Animal Behavior: The Second Generation, ed. L. C. Drickamer & D. A. Dewsbury. Published by Cambridge University Press. © Cambridge University Press 2010.

cherish, and a vital role for birds in my life. I still have little notebooks in my childish handwriting listing the birds I saw on particular days.

Thereafter, birding was my only serious hobby. I and several bird-watching buddies were "adopted" and often taken into the field by two older women. One of them, "Dixie" Larkin, was a superb cook. I still remember fondly her southern fried chicken and pecan pie. During migration seasons, I often rose from my bed early enough to go birding along the shores of Lake Michigan near our home before eating a hasty breakfast and dashing to school. My mother found herself in the remarkable position of frequently urging her teenage son to sleep longer on school mornings!

When I was about thirteen, I made a momentous discovery: some people were paid to watch birds! I thereupon determined to pursue a Zoology Major at a university and to become a professional bird-watcher. Needless to say, my comprehension of what that would entail was, to say the least, naive, but, except for a period when I imagined becoming the director of the world's greatest zoo (finances to be dealt with at some later time!), I never faltered from that trajectory.

Thus, I enrolled as a Zoology Major at the University of Wisconsin, Madison. My youthful enthusiasm for birds was sufficient for me to complete the major despite the dullness of most courses I was required to take. Among the few courses that did excite me were John Curtis's course in plant ecology and John Neese's general ecology course. Curtis inspired me because, until that time, I had found plants uninteresting. They had no behavior, and their positions were predictable. Thus, they provided no surprises. Although the plant ecology course was a rather traditional one, it did show me that plant community dynamics was an exciting topic.

John Neese, on the other hand, approached ecology in an unconventional way. I distinctly remember a lecture in which he presented a calculation of the amounts of phosphorus and potassium that had been taken out of ecological circulation by the habit of interring human bodies in decay-resistant caskets. He introduced to me some new ways to look at the world.

A favorite activity during my Madison years was trapping and banding migrating hawks on the shores of Lake Michigan in the autumn. During the 1930s depression, scientists at the Milwaukee Public Museum established a small banding station north of Milwaukee near the town of Cedar Grove. It had lain fallow for many years until Daniel Berger and Hellmut Mueller repaired it and set up mist nets and bow nets to capture hawks attracted by tethered pigeons and house sparrows that could be induced to flap at appropriate moments. On many fall weekends we left Madison for Cedar Grove on Friday afternoon, stopping at barns to capture house sparrows. Exhausted but stimulated, we returned to Madison late Sunday evening.

During my fourth year at the UW-Madison, I met Elizabeth (Betty) Newton at the Presbyterian Student Center. Our first date was a February excursion on the ice of Lake Mendota while I photographed ice fishing for a nature film my father was producing. During spring she accompanied me while I studied red-tailed hawks and great horned owls for my undergraduate research project. She often dozed at the base of a tree while I struggled to reach the nest, band the nestlings, and record remains of prey. Clearly she knew what to expect when she responded positively to my invitation to a life together.

Throughout my undergraduate years I had the fortune to have John T. Emlen as my advisor. As all who knew him appreciate, he was both a good scientist and a wonderful person. He encouraged my enthusiasm, but helped me channel it in productive ways. He also helped make possible my next, extremely important opportunity. I decided to apply for a Fulbright Fellowship to study in Europe and sought his advice concerning venues. He suggested that I apply to work with David Lack at the Edward Grey Institute of Field Ornithology (EGI) at Oxford University. He and David Lack were good friends and had collaborated on several research projects. David Lack had a rather low – and, as I found out, at least partly justified – opinion of American primary and secondary education. He had not previously accepted an American student but was willing to take a chance with me given a strong recommendation from John Emlen. Because I was good at memorizing facts to regurgitate on examinations, I had accumulated an excellent grade point; I was awarded the Fulbright. Thus, in October 1954. I found myself occupying a cold, clammy room in Lincoln College, and a more comfortable, but still very cool, office in centrally heated EGI. The prevailing view in Britain at the time was that, if a room was warm enough to be truly comfortable, it was probably unhealthy.

The "crisis moment" in my professional life came during my first month at Oxford. At my first tutorial, David Lack asked me whether I had a project I could complete while I assessed the local situation and decided on the focus of my research efforts during the year. I did have such a project – my senior thesis on red-tailed hawks and great horned owls. The study involved nothing imaginative. I simply found as many nests as I could and climbed to them at least once while they contained nestlings. The manuscript I submitted was well received by my advisors. I felt rather proud of it. Lack suggested that I give him a copy so that he could read it prior to my next tutorial.

I confidently entered Lack's office a week later, quite unprepared for the reception I would receive. He returned my manuscript, which was immersed in a sea of red ink. He informed me that he had not bothered to comment on minor matters of grammar and sentence structure, but had confined his remarks to major matters of logic, data presentation, and the like! After an hour of "*Gestalt* therapy" I staggered back to my office and slumped down into my chair. Fortunately, and I do not know why, rather than becoming angry and defensive, I looked at myself and said "Orians, you did not learn how to think!" That hour permanently affected how I approached science and how I taught courses throughout my career at the University of Washington.

That hour also launched me into a period of intensive reading, during which time I came to appreciate the power of the hypothetico-deductive approach to science. David Lack convinced me that ecology and evolution were inseparable fields. Indeed, more than once a student in my advanced ecology class at the University of Washington came to me after the first two weeks of the course to ask me when I was going to start talking about ecology! My answer, that I had been discussing ecology all the time, did not always satisfy them. Later, when I wrote a general biology textbook, I dedicated it to David Lack "who first taught me the difference between proximate and ultimate factors." He was pleased. This was the first time someone had dedicated a book to him.

During my year at Oxford, I substantially rewrote and submitted for publication my hawk and owl manuscript. With the expert guidance of P. H. Leslie, who was at Charles Elton's Bureau of Animal Populations in the same building with the EGI, I also completed and submitted – on June 24, 1955 – a capture–recapture analysis of a Manx Shearwater population. The manuscript was eventually published in the *Journal of Animal Ecology*. The next day Betty and I were married at the Oxford City Church of St Martins and All Saints. None of our parents were there, so David Lack accompanied Betty down the aisle and gave her to me. The gift has lasted!

While at Oxford I was invited to join Niko Tinbergen and his graduate students Friday evenings at the Tinbergen home. This was a time of great excitement in Tinbergen's group. Ideas about releasers, specific action patterns, and "the four questions" for which Niko became famous were discussed with great enthusiasm. In spring, I spent a week with Niko and some of his students on the Farne Islands, where pioneering studies of gulls and terns were being carried out. I observed directly how Esther Cullen determined why nesting on cliffs had molded so many features of kittiwake breeding behavior. I watched courtship behavior of common terns and herring gulls. The world of animal behavior came alive for me. The large male grey seal, somewhat unceremoniously named "Haldane," that regularly cruised past the cliffs, also stimulated my thinking about the marvelous workings of natural selection.

The year I spent at Oxford was an exciting time in the development of behavioral ecology. V. C. Wynn-Edwards had recently published the initial manuscripts leading up to his influential 1962 book, *Animal Dispersion in Relation to Social Behaviour*. While I was at Oxford, David Lack's *Natural Regulation of Animal Numbers*, and H. G. Andrewartha and L. C. Birch's *The Distribution and Abundance of Animals* were published. The dominant ecological issues of the day were whether populations were regulated by density-dependent or density-independent factors – indeed whether density-dependence was a logical necessity – and the importance of group selection. David Lack held strong opinions on both of them. He was convinced that only density-dependent factors could regulate animal populations. He also believed that natural selection acted only at the level of individual organisms.

As an impressionable young biologist who was, for the first time, really thinking deeply about science, I was naturally much influenced by him. I concluded that although group selection processes were theoretically possible, the conditions for their operation were so stringent that they could generally be ignored. That conclusion guided all my subsequent investigations of behavioral ecology, although I eventually adopted a more nuanced view of the operation of selection at different levels. By contrast, although I became an ardent proponent of density-dependence, not too many years elapsed before I came to regard the dispute as a tempest in a teapot. Too much of the argument centered on the definition of the term "regulation" that one adopted. Also, I realized that density-dependence and density-independence were not mutually exclusive concepts. Fortunately, that debate has subsided, not because one side won, but because the debate came to seem less and less interesting to most ecologists. Nevertheless, thinking about that debate focused my attention on the importance of asking the right questions. Much of the Advanced Ecology course I taught for many years at the University of Washington focused on asking questions, how to pose questions at the appropriate scale of resolution, how to distinguish poor from good questions, and how to transform a poor question into a good one.

California interlude

Prior to being awarded the Fulbright Fellowship, I had applied and been accepted to study as a graduate student with Frank Pitelka at the University of California, Berkeley. The original intent was that I would develop a thesis project on comparative behavior of breeding jaegers at Barrow, Alaska, where Pitelka had a major ongoing project on lemmings and their predators. However, as a result of my year at Oxford, followed by a six-month stint with the U. S. Army Transportation Corps to fulfill the active duty requirements of my ROTC training, I arrived at Berkeley two years later than originally planned. I was still welcomed by Frank and the department, but my attempts to talk with him about my jaeger research received strangely muted responses. I soon discovered that, during the two-year delay, another graduate student had preempted my jaeger project. I was back to the drawing board.

John Emlen and David Lack emerged again to influence my choices. Together they had conducted a study of tricolored blackbirds in California that highlighted how different their breeding system was from that of the nearly morphologically identical red-winged blackbird. This striking contrast induced me to focus my attention on relationships between avian social systems and the environments they exploited. The American blackbirds (Icteridae) were ideal subjects because they encompassed most of the social systems found in the avian world. Redwings and tricolors, sister taxa with strikingly different social systems, but breeding in the same environment, were ideal species with which to begin. When I moved to Seattle, outside the range of the tricolored blackbird, I added yellow-headed and Brewer's blackbirds to the species I studied. Although I no longer conduct research on blackbirds, tricolors eventually followed me north to Washington, becoming the most recent addition to the State's breeding birds.

My thesis project was an observational, non-experimental, comparative study of the two blackbird species. I gathered typical data on breeding biology and undertook detailed analyses and descriptions of their behavior (Figure 15.1). I also documented competitive interactions between them, as tricolors moved in and, by sheer force of numbers, displaced some male redwings from their territories (Orians 1961). My first and only "ethograms" were products of this research, published together with behavioral data I subsequently gathered on yellowheads (Figure 15.2). My thesis contained, in addition to descriptions of how tricolors in large, dense colonies exploited the surrounding environment, speculations as to why they evolved to be the most densely colonial passerine bird in North America. My hypothesis is still plausible, but unfortunately it will probably remain untested. The environment of the Central Valley of California has been so dramatically

Figure 15.1. With one display, a male red-winged blackbird claims ownership of his territory and invites females to settle there.

Figure 15.2. A male yellow-headed blackbird uses an asymmetrical song spread to advertise his possession of a territory.

altered, and tricolored blackbirds so reduced in numbers, that appropriate experiments are now impossible.

Although I was deeply disappointed with the abrupt demise of my projected jaeger project, the forced shift to blackbirds turned out to be a blessing. Blackbirds have been excellent subjects for investigations of comparative social systems and a variety of other topics as well, such as habitat selection, foods and foraging, and territoriality. Indeed, the red-winged blackbird has become the *"Drosophila"* of avian behavioral and population ecology in North America.

Our sojourn in Berkeley also witnessed our launch into parenthood. Our two daughters, Carlyn and Kristin, were born while I was a graduate student. Our son, Colin, arrived within a year after we moved to Seattle. Frank Pitelka discouraged his students from becoming parents while engaged in dissertation research. Nevertheless, I completed all requirements within four years. I would not have finished sooner had I been spared parental investment responsibilities. Moreover, I soon discovered that my graduate student years were less stressful than my years as an untenured professor.

Seattle forever

Luck was with me again as I faced the problem of finding employment as a fresh PhD. The year of my job search coincided with retirement of the mammalogist in the Department of Zoology at the University of Washington in Seattle. The department advertised the position for a vertebrate zoologist, for which I was an appropriate applicant. I interviewed, emerged victorious, and accepted the job immediately over the phone, no questions asked. I am now amazed that I did not bargain for salary or space. It is clear, however, that my neglect of these important issues has made no difference. The limits to my insightfulness and productivity have not been set by laboratory facilities or salary.

Still excited by blackbirds, I continued studying them in the Columbia Basin desert of Eastern Washington. Scoured by repeated floods that roared across the basin when ice dams that blocked the flow of the Columbia River broke, the "Channeled Scablands" today are populated with many lakes and ponds, ringed with cattails and bulrushes. The area supports dense breeding populations of blackbirds, and the open terrain, with many cliffs adjacent to the marshes, facilitates observing them.

For many years my blackbird research focused on issues of habitat selection and mate selection. Because I had chosen blackbirds primarily because of their varied social systems, I wished to develop and test hypotheses that could explain why some species were monogamous whereas others were polygynous. Retrospectively, I judge that my early paper on the "polygyny threshold model," influenced in part by work of two graduate students, Jared Verner and Mary Willson, is probably the most important one I have published (Orians 1969b). It has a conciseness and clarity that reflect the great amount of time I devoted to it. As my career unfolded and life became increasingly complex, I found it difficult to give comparable attention to individual papers. I am troubled somewhat by concluding that my most important paper may have been published so early in my career. In retirement I am attempting to restore that deficiency.

To assess the clues that birds use to select breeding habitats and choose mates, I gathered data on physical structure of nesting habitats and on blackbirds' food, particularly items they delivered to nestlings. Luckily, one of my early field assistants was a pipe smoker. After trying several ineffective methods of neck-collaring nestling blackbirds, he suggested that we try pipe cleaners. A third of a standard pipe cleaner proved to be ideal, being soft on the outside and flexible enough to allow a fit around a bird's neck that permitted normal breathing but that blocked passage of food. Pipe cleaners became a standard research tool for many years, a technology that was well matched to my overall mechanical competence.

Marshes as producers of blackbird food

The application of pipe cleaner technology revealed that blackbirds were feeding their nestlings mostly on emerging aquatic insects. To measure how many of them marshes produced, we built dozens of traps that we placed over the water to capture larval insects as they emerged from the water to metamorphose on stalks of emergent plants (Orians 1980).

Becoming a rather compulsive applier of pipe cleaners to avian necks, I gathered enough data to compute the amount of overlap in food delivered to nestlings of the four species of blackbird (the fourth was the western meadowlark) that nested in my study area. At the time, there was considerable interest in amount of "niche overlap" that was compatible with coexistence of species in the same habitat. My data fit theoretical predictions reasonably well (Orians & Horn 1969), but efforts to find general rules governing niche overlaps have failed to yield what its proponents had hoped for. Trying to make sense of the massive amount of food data I had collected induced me to think seriously about foraging theory. Together with two graduate students, Eric Charnov and Nolan Pearson, I developed some useful hypotheses about foraging in patchy habitats and about central place foraging, the specific case appropriate for data I had gathered on blackbirds (Orians & Pearson 1979).

Eric Charnov continued to make seminal contributions to foraging theory (Charnov 1976). Many foraging theories have been subjected to clever testing in the field and laboratory. We were not the only people working on foraging theory, but our work did function as a significant stimulus for development of a rich body of theories. Indeed, development and testing of foraging theories expanded explosively, becoming what is arguably the richest body of data and concepts in behavioral ecology (Stephens & Krebs 1986).

At the time I was much occupied with foraging theory, I gave a number of lectures on the topic, which I typically called "optimal foraging theory." These lectures coincided in time with a frontal attack by some biologists on optimality modeling in general, an attack that was a component of the more general assault on "Sociobiology." Spandrels and Dr. Pangloss were frequent topics of conversation; in some quarters, all optimality modeling was suspect (Gould & Lewontin 1979). Interestingly, I found that my lectures received a more favorable reception if I simply discussed "foraging theory" rather than "optimal foraging theory," even though the content of my talks was otherwise identical. I came to appreciate both the power of single words to channel thinking and the ease with which ideological issues can impinge upon and deflect scientific discussions. The same issues, alas, still populate the current debate over the emerging field of evolutionary psychology.

Emergent vegetation

Although foraging theory occupied much of our attention, we did not neglect the role of structure of the physical environment in habitat selection. Emergent herbaceous plants, to which most redwing and yellowhead nests are attached, undergo rapid within-season and between-season changes. Cattail and bulrush stems that survive winter are the best substrata for blackbird nests because they are sturdy and available early in spring when birds start nesting; and they do not grow. Fresh spring growth, on the other hand, is flexible, and because stems grow at different rates, a nest anchored to more than one of them is likely to be tipped by differential growth of its supports. In addition, cattail beds undergo a longer structural cycle. Several years of vegetative growth yields dense stands of emergent stalks that greatly reduce the intensity of light reaching the water surface. Consequently new growth is suppressed and the aging, dead emergent stalks eventually succumb to winter storms, ice scouring, and snow. The water surface then basks in sunlight and vegetative re-growth is vigorous.

By monitoring such natural changes and experimentally clipping vegetation to create reduced structural complexity, we learned that redwings are very site-tenacious. If a marsh was otherwise suitable, males faithfully returned to their seriously denuded territories, displaying vigorously on short cattail stubs. Females also returned and anchored their nests to the short stems available to them. Unfortunately, we did not monitor insect emergence from clipped areas, so we do not know whether our manipulations affected food supplies available to nesting birds. However, given that our clipping little disturbed the underwater environment, I suspect that food availability was not adversely affected.

Territorial dominance

Since Elliott Howard directed the attention of biologists to the prevalence of territoriality in birds in 1920, ornithologists have devoted considerable effort to attempts to answer several key questions about territoriality (Figure 15.3). Why do individuals defend space at all? What determines sizes of areas that are defended? Why does the territorial owner nearly always win contests? Why do individuals unable to gain territories, and thus failing to breed, concede defeat without putting up a real fight? Why do individuals of some species defend territories against individuals of other species as well (Figure 15.4)?

Hypotheses to explain why it is advantageous for individuals to defend space have focused on costs and benefits of gaining exclusive access to resources provided by the space. Gaining a nest-site easily explained defense of small spaces. Food appeared to be the resource that required defense of larger areas, so most efforts were directed toward determining which types of food were or were not economically defendable and, using a comparative approach, to determine whether patterns of territorial defense conformed to predictions generated by those models (Brown 1964). Those efforts were generally successful, but my blackbird research focused on the other questions, even though we found that redwing territory size was inversely proportional to emergence rates of aquatic insects.

Gordon H. Orians

Figure 15.3. Two male red-winged blackbirds in a territorial boundary dispute, the only situation in which this display is performed.

Figure 15.4. Male red-winged and yellow-headed blackbirds use their species-specific displays during boundary disputes, but both understand their meaning. Colored leg bands identify individuals.

At this point in time I was fortunate to develop a long-term professional association with Les Beletsky, who had previously studied redwing vocalizations. Les was able to devote full time to our blackbird research project, enabling him to be in the field nearly constantly during the extended breeding season in eastern Washington. I, on the other hand, constrained by teaching and other commitments, could spend only fragmented time periods at the project site. Thereafter, Les and I co-authored most publications resulting from our field investigations (Beletsky & Orians 1996).

Les and I decided to test two non-exclusive hypotheses to explain both dominance of the territorial owner and inhibition of challengers (Parker 1974). The Resource Holding Potential (RHP) hypothesis explained owner dominance by assuming that owners initially gained their territories because they were better fighters than other contestants. If so, it follows that individuals seeking territories typically have lower RHP than territory holders, which is why they are challengers rather than owners. Challenger inhibition could be explained by assuming that challengers "know" that owners have higher RHP and that, therefore, they are highly likely to lose escalated contests. Provoking escalated contests that one is likely to lose would not be favored by natural selection, in part because of risk of injury.

The Value Asymmetry (VA) hypothesis explains the same phenomena by postulating that a territory has greater value to its owner because, as a result of having spent time there, he has better knowledge than challengers about its resources and escape routes. Later in the breeding season he also has genetic investments in offspring, whose survival might be adversely affected should he lose the territory. In a dispute over a resource, the individual for whom the resource has greatest value should be willing to escalate a contest beyond the level that any challenger would attempt to follow. Thus, according to VA, challengers avoid initiating contests that they would inevitably abandon.

Our primary method of testing these hypotheses was removal experiments carried out early in spring before females had built any nests, so that only knowledge of the area could generate value asymmetry. We trapped territorial owners, housed them elsewhere, and then released them some time later to face new owners who had occupied their territories while they were held captive. As predicted by the VA hypothesis, most males regained their territories if they had been held captive no longer than 48 hours, but most of them failed to do so if they were held captive for about a week. None the less, since males did lose some weight while in captivity, it could be argued that they had lost RHP. Therefore, to clearly distinguish between VA and RHP, we performed some double-removal experiments in which first replacement males were also removed and held captive. Thus, when the original males were released after about a week in confinement, they faced new owners that had held their territories for only about two days. Most of these males did regain their territories, providing strong evidence in favor of the VA hypothesis.

Although these experiments strongly favored the VA hypothesis, the two hypotheses also generate other predictions that differ. For example, the RHP hypothesis predicts that a challenger should enter a territory while the owner is there to ascertain whether his RHP might have deteriorated enough to warrant a serious challenge. The VA hypothesis, on the other hand, predicts that challengers should preferentially enter territories while the owner is absent so that they can learn more about them. Our field data clearly showed that floater males entered territories almost exclusively at times when owners were present, as predicted by the RHP hypothesis.

The RHP hypothesis also predicts that floater males should peruse a large number of territories so as to increase the number of owners whose current status they could assess. The

VA hypothesis, on the other hand, predicts that floaters should confine their search to a small area about which they could gain valuable information. Our extensive observations of floater males clearly showed that, as predicted by VA, they established relatively small searching areas.

Not surprisingly, as a result of these and other observations, we concluded that floater males use a variety of information in their efforts to gain territories. Although this was not the definitive answer that we would have preferred, rarely are the phenomena that behavioral ecologists study influenced by other than a combination of factors.

Decisions and lifetime reproductive success

In the process of gathering information on habitat and mate selection, foods and foraging, and reproductive success of blackbirds, Les and I assembled a substantial database that allowed us to compute lifetime reproductive success (LRS) of many individuals (Orians & Beletsky 1989). We used a combination of extensive observational data and selected experiments to assess the LRS consequences of decisions individuals make concerning mates, places, and various activities. We treated males and females separately because the major breeding-season decisions made by male and female redwings differ strikingly and because, in that species, most of those decisions are made independently of the behavior of individuals of the other sex. For females we looked closely at when and where they decided to nest and re-nest and how much energy they allocated to reproduction in any one season. For males, we concentrated on how they established and maintained territories.

Our data, not surprisingly, suggested that a variety of factors influenced when females initiated nesting in spring. Females achieved a higher annual reproductive success if they began nesting early in spring but seasonal reproductive success was fairly constant after the first few weeks of the breeding season. Therefore, we concluded that variation in starting dates was probably influenced by a combination of the females' age, where they spent the winter, when they began their spring migration, and their condition when they arrived on the breeding grounds.

Our data also indicated that female redwings used a variety of information when making their initial settling decisions and when selecting sites for re-nesting within breeding seasons. Older females possess information they have stored from previous breeding seasons. Later arriving yearling females lack such information but they can, and do, use clues provided by already settled females. Perhaps our most surprising finding was that the presence of familiar male neighbors affected settling decisions by females more than the identity of the territory owner. Only later did we learn the value to females of familiar neighboring males.

Female redwings within harems frequently engage in agonistic behaviors that appear to function to achieve a temporal rather than a spatial distribution of nests. By using aggressive behavior a female may delay initiation of her nest by another female within the harem, thereby reducing the temporal overlap when both females are feeding nestlings.

We were much interested in determining why females terminated their breeding efforts at a time when insects continued to emerge in great numbers from the marshes for at least two more months. One possible hypothesis was that continuing to breed would delay molting and fattening for fall migration, resulting in lower over-winter survivorship, later arrival on the breeding grounds the following spring, and poorer physiological condition on arrival. Our data showed, however, that no matter how much effort females made in reproduction within a breeding season, they were either able to recover their physiological condition prior to winter or adjusted their within-season efforts so that they did not pay any between-season costs. Another possible hypothesis was that late-fledging offspring have a low probability of becoming breeders, either because they have less time to prepare for migration and winter, or because they are less competitive as adults. If so, there would be little benefit in producing them. We rejected this hypothesis because females that fledged during the final third of the breeding season actually had the highest probability of returning to breed.

Our results show the weakness of attempting to answer many questions about female breeding season decisions by using only non-experimental data. We cannot reject the hypothesis that females monitor their physiological condition and terminate breeding in time to allow them to recover, survive at high rates over winter, and become competitive breeders the following spring. But, because we have no measures of their physiological condition, we do not know whether they would have suffered had they invested more effort than they did.

It's a wise man that knows his own son

During most of the twentieth century, research in behavioral ecology was stimulated primarily by new concepts and theories, such as inclusive fitness, parent–offspring conflict, and ideal free distribution. The development of DNA technology, on the other hand, enabled researchers to measure fitness for the first time. The discovery that, in many species, social and genetic partners were often different generated a revolution in behavioral ecology. Investigators rushed to assess the extent of extra-pair copulations and fertilizations in many species of birds and to determine why such behavior was so prevalent.

Because our assessments of lifetime reproductive success in redwings had been based on an assumption that males sired all offspring on their territories, we clearly needed to determine how seriously that assumption might have misled us. One of my students, Elizabeth Gray, carried out a three-year study in which she was able to identify both participants in 375 copulations (Gray 1996). Of those, only 66 (17%) were extra-pair copulations (EPCs), whereas 33.7% of 403 nestlings she tested were the result of extra-pair fertilizations (EPFs). All but one EPF nestlings were fathered by other territory owners on the same marsh, most of them by adjacent neighbors. All copulations were between consenting adults, but clearly EPCs were more difficult to observe, indicating that females chose relatively inconspicuous places when soliciting them.

We had no way to correct for the error introduced in our calculations of lifetime reproductive success by our faulty assumption but, fortunately, the error was inconsequential. Male redwings gained, on average, as much as they lost via EPCs, and there were no "super males" in the population. Both within and across breeding seasons, equivalent numbers of males gained and lost paternity from EPCs; individual males did not consistently gain or lose from EPFs from year to year.

Another moment of concern was generated by Nancy Burley's discovery that male zebra finches gained fitness if they had red bands on their legs. We had regularly used red bands as part of our color-band combinations, so we were worried that our data might have been biased if males that had been given red bands actually had greater reproductive success than other males. Fortunately, our subsequent analysis did not reveal any benefit to male redwings from having red bands.

Ecosystem detour

Although I kept generally abreast of developments in ecosystem ecology, I never anticipated becoming an author in that field. Surprisingly, a behavioral observation generated exactly that. As part of my "duties" as a visiting scholar at Monash University in Clayton, Australia, I accompanied graduate students on their research projects. On the critical occasion I was with a graduate student who was studying sugar and squirrel gliders, marsupials very similar to flying squirrels, in a dry *Eucalyptus* forest in northern Victoria. He slept in a small rented room in town while Betty and I camped in the forest. At that time eucalypts were in flower and the canopy was literally boiling with nectar-feeding birds, some of them as large as robins and grackles. I recognized that allocation of energy to nectar production in that forest was orders of magnitude greater than anything I had seen elsewhere in the world. My curiosity eventually led to collaboration with another ecologist, Antoni Milewski, to produce a major review paper on the ecology of Australia, a continent dominated by nutrient-poor soils.

Together we developed a "Nutrient-Poverty/ Intense Fire Theory," which postulated that most anomalous features of organisms and ecosystems of Australia are evolutionary consequences of adaptations to nutrient-poverty, compounded by intense fires that occur as a result of nutrient-poverty (Orians & Milewski 2007). The theory postulates that plants growing in environments with plentiful light and periodic adequate moisture, but on soils poor in phosphorus, zinc, and other nutrients, can synthesize carbohydrates in excess of the amount that they can combine with these nutrients for metabolism and production of nutrient-rich foliage and reproductive tissues. We suggested that plants use this "expendable energy" to produce foliage heavily defended by non-nitrogenous chemicals, large quantities of lignified tissues, and readily digestible exudates. Rapid accumulation of nutrient-poor biomass, a result of low rates of herbivory, provides fuel for intense fires, which exacerbate nutrient-poverty by volatilizing certain critical micronutrients. Among the many anomalous features of organisms of Australia that we suggested could be explained by this theory are that plants allocate unusually large amounts of expendable energy to production of carbon-based exudates, such as nectar and gums, an allocation I observed that evening. Also, vertebrates pollinate an unusually high proportion of plant species of Australia (Figure 15.5a), and, as I also saw that evening, they are on average larger than pollinators

(a)   (b)

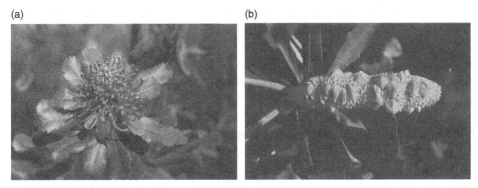

Figure 15.5. Reproductive traits of Australian plants reflect nutrient poverty and intense fires. (a) The conspicuous red flowers of the waratah (*Telopia oreades*) are pollinated by large honeyeaters. (b) All species of *Banksia* defend their seeds in large woody capsules.

on other continents. An unusual proportion of species defend the few nutrient-rich seeds they are able to produce in large, woody carbohydrate-based capsules (Figure 15.5b).

As this is a recent publication, I do not yet know where the insights I generated while preparing it will eventually lead me.

Human behavioral ecology

The central goal of behavioral ecology is to explain the ways in which individual organisms make decisions about habitat, shelter, food, and mates. Humans make choices among these features of the environment, but human perceptions of and responses to environments are imbedded in a complex nexus of symbolisms and cultural memories that have no counterpart in the responses of other organisms. These complexities make it difficult, but not impossible, to untangle the respective influences of culture and evolution on human responses to environments. None the less, my investigations of habitat selection in birds stimulated me to try to apply some behavioral ecology insights to human behavioral ecology.

It was obvious that human responses to nature affected fitness. Our ancestors lived in environments devoid of modern comforts and conveniences. Their survival, health, and reproductive success depended on their ability to seek and use environmental information wisely. They had to know how to gather information and interpret signals from the animate and inanimate environment. They needed to understand relationships between habitats and resources and how to evaluate habitats (Wilson 1984). Such responses are obviously influenced by culture and learning, but I judged that it would be surprising if adaptations to our environment were not part of the contemporary human psyche. At least it was worth a serious look (Orians 1980).

The belief that contact with nature is good or beneficial to people is ancient. The gardens of ancient Egypt, the walled gardens of Mesopotamia, and the gardens of merchants in medieval Chinese cities indicate that for centuries people have gone to considerable lengths to maintain and enhance their contacts with nature. More recently, evidence that exposure to nature fosters psychological and physical health has formed part of the justification for providing parks and other nature in cities and for preserving wilderness.

The study of emotions is a central focus of human behavioral ecology, because biologists expect emotional responses to evolve in response to conditions that influence fitness. Simply put, those of our ancestors who did not enjoy food and sex, for example, were more poorly represented genetically in future generations than those who did enjoy and, hence, seek out, food and sexual partners. Similarly, individuals who selected inferior environments in which to live should have produced fewer surviving offspring than individuals who made better habitat choices.

The approach I use to study human behavioral ecology assumes that evolution predisposed humans, as well as other animal species, to learn easily and quickly, and to preferentially retain, those associations or responses that foster survival when certain objects or situations are encountered. Moreover, even though modern societies have largely eliminated the danger posed by objects of today's fears and phobias, fear and avoidance responses may none the less persist because selection against those responses may be very weak.

An evolutionary perspective also suggests that features of environments and objects in them should be viewed in functional rather than morphological terms. Humans evaluate environments, not necessarily consciously, by the opportunities they provide for pursuing activities that contribute positively to survival and reproductive success. In this context, value means – what do they offer? What does an object or environment offer to an individual viewer at a particular moment? The perceiver assesses what could be done with that object or in that environment and evaluates the consequences of doing those things. The observer asks of an object not only – What is it? But also – What's in it for me? For example, a human observer may ask of an unfamiliar environment – How easy would it be to enter it, explore it, and find my way back if necessary? How valuable would the acquired knowledge be?

In other words, when we look at trees in a landscape, we think of them not just as objects with recognizable characteristics by which they can be named and classified (although scientists often do so), but also as objects suggesting opportunities for doing things, such as seeing further, hiding better, or climbing to safety. We see rivers not simply as structural components of the landscape, but as sources of water. Or we may see them as channels along which we can move if we have the means to do so, or as obstacles to impede our passage if we have not. Clearly, what an environment affords is not something static. Instead, opportunities vary with season and weather, and the age and current needs of an individual.

Together with a psychologist, Judith Heerwagen, I have used these concepts to explore patterns of human emotional responses to various features of the physical, biological, and social environments in which we live (Heerwagen & Orians 1993). I chose as the title of one of my papers *"Human behavioral ecology: 140 years without Darwin is too long"* because, although what the environment does to us is as important as what we do to the environment, our attention has been directed almost exclusively to the latter.

The guiding principle in our investigations is that the human mind (and the minds of all other species that have them) has been designed by evolutionary processes to assist in making decisions that enhanced our ancestors' survival and reproductive success. The psychological mechanisms that support decision-making are adaptations that have been molded by natural selection over evolutionary time.

Much relevant environmental information concerns the location of things in space. Where were prey animals yesterday? Where did I cache food I could not carry back to camp? Where are trees that bear nutritious fruit? Where are safe hiding places that I may need to use in an emergency? And so on.

Resources are clearly important components of the environment, but the relevant environment includes not only its physical and biological components that provide those resources and the conditions under which we seek them. Effectiveness in exploiting environmental resources also enhances reproductive success by increasing access to reproductive partners and by increasing survival of dependent offspring. Thus, sexual selection can help explain why certain aesthetic responses to environmental features evolved in the first place, and also why human preferences often come to express themselves as seemingly useless and random exaggerations.

Because all behavior is adapted to past environments, some components need not be adaptive under rapidly changing current conditions. Thus, Judi and I expected to find some "ghosts of environments past" in the human psyche. Considerable uncertainty surrounds the length of time over which "ghosts" are likely to persist, particularly under conditions of relaxed selection. One reason to expect that contemporary human behavior should reflect adaptations of our ancestors to conditions on African savannas is that a relatively small number of generations separate us from our ancestral "mother-Eve," who lived some 5 million years ago in East Africa (Figure 15.6). Assuming an average generation time of 20 years, no more than 350,000 individuals separate us from her, a number no larger than the

Figure 15.6. Japanese red pines in imperial gardens in Kyoto are modified by pruning to generate shapes that resemble those of acacias of African savannas.

population of a small American city. In the absence of strong selection against a trait, its persistence under today's radically altered conditions is at least plausible (Coss 1991, 1999).

Modern societies have greatly reduced the danger posed by the typical objects of fears and phobias, but fear and avoidance responses persist, probably because selection against those responses is weak (Öhman & Mineka 2001). Conversely, fears and phobias have been slow to develop to dangerous objects that are recent arrivals in our environments, such as electric wires and nuclear weapons. Some human traits, however, have evolved relatively rapidly during recent millennia because of strong selection. Among them are evolved responses to the chemical characteristics of the array of new foods.

Culture and learning clearly exert strong influences on the ways human perceive and respond to environmental information and they have important impacts on the symbolisms we attach to natural objects. But attempts to understand and interpret human aesthetic responses to environmental features without asking why they evolved have failed. For example, people who studied the concept of beauty from a non-biological perspective assumed that beauty was an intrinsic property of objects. They therefore looked for, and expected to find, correlations between the characteristics of objects and human aesthetic responses to them. This attempt largely failed because, as an evolutionary perspective immediately suggests, beauty is not an intrinsic property of the objects that we call beautiful. Rather, it is the product of interactions between traits of objects and the human nervous system that evolved so that we regard as beautiful objects having properties that, if positively attended to, result in improved performance in some aspect of living. Conversely, we regard as ugly, objects that should be avoided or destroyed. In essence, an evolutionary perspective suggests that concepts such as beauty and ugliness should be viewed from a functional perspective. Emotional responses are best studied by asking: "How did these responses help us solve problems?"

A better understanding of evolved human responses to environments may assist us in our attempts to educate the public about the importance of biodiversity and ecosystems. Ecologists base their arguments primarily on the goods and services ecosystems provide. Human societies are, of course, fundamentally dependent upon ecosystem goods and services, but these issues rarely stimulate the level of concern that leads to action. Clearly, additional motivators are needed if people are to develop the sense of commitment to the biological resources of Earth that will be required to generate the policy decisions that must be made if we are to preserve Earth's biological heritage. No magic bullet can accomplish this task, but we may improve our success if we make use of the deep psychological and physiological roots of human responses to nature as part of our approach.

Science and environmental policies

For several decades a major part of my professional activities has focused on the interface between science and public policy. My involvement was not triggered by basic research but by my activities in opposition to the Viet Nam war. In January 1969, I received a call from

Bert Pfeiffer at the University of Montana. He was calling to invite me to accompany him to Viet Nam to look at the environmental effects of the war, particularly the consequences of massive spraying of defoliants over the landscape. He knew I was an ecologist, but he asked me because of my activities as chair of the Peace Action Committee of the American Friends Service Committee office in Seattle. I carefully considered the invitation before answering him. The war was in full swing at the time. As father of three children I was especially uninterested in becoming a statistic.

But I did agree, so in March 1969 Pfeiffer and I spent two weeks in Viet Nam. After several days of delay, owing to the reluctance of the U.S. military to certify us as official correspondents, we were able to travel freely. We also were given the official version of the history of the war, which, amazingly, never mentioned the Viet Cong. When I asked why not, the reply was "No comment." I learned how information was provided "off the record," which meant that you were not allowed to attribute it to an official source, but were encouraged to report it as established fact. The official version of the war also told me that at times it is as difficult to predict the past as to predict the future!

To gather information on the environmental effects of what was basically a war against the environment, we visited rubber plantations and talked to scientists at the University of Saigon. We cruised in a patrol boat around the Saigon River delta, where vast areas of mangrove forests had been killed by herbicides. We also flew on a spraying mission, sitting between the pilot and co-pilot to photograph herbicidal spray settling down on the forest canopy. A courageous colonel also took us by helicopter to see the many large craters formed by the thousands of 500 and 750 pound bombs that were dropped on the land in an effort to kill foot soldiers hiding in foxholes (Orians & Pfeiffer 1970).

When I first arrived in Saigon, I was quite naïve about politics and how science interfaced with political dialogues. I returned to Seattle two weeks later, still relatively naïve but rudely awakened to the complexity of major events and the ways in which information about them is manipulated to serve political purposes. Soon after I returned from Viet Nam, I was appointed to the Ecology Advisory Committee of the newly formed Washington State Department of Ecology. A few years later I joined the Science Advisory Board to the U.S. Environmental Protection Agency, eventually chairing the Ecology Subcommittee of the Board.

These activities, and those I subsequently became involved with at the National Research Council (NRC), immersed me in the science–policy interface. I became aware that throughout its history, the science of ecology was enriched and stimulated by interactions with applied issues (Kingsland 1985). Nevertheless, it took me some time to learn how to develop appropriate "search processes" to extract from my mental files information that might be useful for dealing with the varied issues that the advisory committees were asked to address. I also became acutely aware that the standard scientific paper is an art form, designed to communicate with a particular audience for a specific purpose. It is not a format suitable for communicating information to managers and decision-makers. These insights greatly influenced how I taught graduate courses in science and environmental policy at the University of Washington, and how I later functioned as a member or as chair of NRC committees.

Reflections

Other than my initial, crucial decision to pursue a career as a professional avian biologist, I have not followed any grand plan. My choices of research subjects and the topics I investigated were powerfully influenced by events beyond my control. To a certain degree, my upbringing in a Midwest parsonage, which had given me a comprehensive worldview, stimulated me to develop an alternative, evolution-based perspective that provided a basis for attributing meaning to human life. As did Darwin, I found grandeur in that view of life.

Nonetheless, several themes dominated my research career. One, which I owe directly to my year in Oxford, was my appreciation of the importance of asking the right kinds of questions. Both David Lack and Niko Tinbergen clarified for me the importance of the different types of questions biologists ask, the domains of answers to each of them, and why a full understanding of any biological problem requires answers to all four of Niko's questions. During that year I also came to understand that the scientists who had made the most important advances did so not because they were brighter than others but because they had determined the best questions to ask. As a result, my Advanced Ecology class focused on the art of asking questions. My major goals were to help students tell the difference between good and bad questions, how to massage mediocre questions into good ones, and to determine the domain of applicability of answers to those questions.

Another theme that has characterized my research stems from my Oxford education on the importance of questions. I concluded that integrating behavior, ecology, and evolution would facilitate progress in all three fields. Virtually all of my research concerned how the traits of organisms came to be what they were. I did almost no research that took the characteristics of organisms as fixed inputs and used those input to compute dynamical outputs. This is not to imply that I judge dynamical models to have no value; they clearly do. Rather, I judged that progress in ecological science was hindered by its relative isolation from "selection-thinking." So I became and remain an ardent "adaptationist." I recognized that assuming a trait to be an incidental byproduct of selection for something else was clearly a "stop work order," even though the assumption might actually be true.

Finally, I became increasingly interested in the importance of understanding phenomena at multiple levels of organization. As implied by G. Evelyn Hutchinson's metaphor "The ecological theater and the evolutionary play," actors in the evolutionary play perform a variety of behaviors. Many of the properties of ecological systems are the result of their behavior. Indeed, a major goal of ecology is to determine both consequences of behavior for the structure and functioning of ecological systems and, in turn, the influence of those complex systems on the evolution of the behavioral traits of organisms.

Although what I have just stated is now generally accepted, during much of the past century many students of animal behavior paid little attention to the ecological theater, even though an ecological perspective was central to Niko Tinbergen's approach to animal behavior. Ecologists, in turn, have had an uncertain relationship with the field of animal behavior. Behavioral ecology has been an important component of ecology, but population ecologists have little incorporated the rich results of behavioral ecological research in their models. Papers dealing with behavioral ecology have, unfortunately, occupied a decreasing proportion of the programs of the annual meetings of the Ecological Society of America during recent decades.

Part of the reason for the general neglect of behavioral ecology in ecological studies is due to the difficulty of incorporating behavioral data into ecological models. Models of complex systems sacrifice details in the service of tractability. Behavior of individuals has typically been sacrificed, in part because of a lack of the necessary species-specific information, but also because behavioral responses typically occur much more rapidly than rates of change of population parameters. In addition, behavioral ecologists have, until recently, given little thought to the broader ecological consequences of the decision rules they have developed to predict and describe behavior of individuals.

Thus, a major challenge confronting investigators in the fields of behavior and ecology is to find more powerful ways to integrate studies of the ecological theater and the actors that perform the evolutionary play. In this context two components of behavioral ecology – habitat selection and foraging behavior – have been extensively explored. The implications of habitat selection and foraging behavior for the structure and functioning of ecological communities suggest great, untapped potential for behavioral research. Under which conditions do different types of behavior have the greatest influence on population dynamics and on the structure and functioning of ecological communities? Although habitat selection and foraging behavior are likely to be the most important behaviors with respect to influencing ecological communities, the ecological roles of social organization and mate selection, whose role in community structure and functioning have been little explored, may be substantial.

It is even possible that major evolutionary divergences may regularly originate as bifurcations in behavioral patterns. For example, as suggested by Mary Jane West-Eberhard (1989), behavioral novelties can explain the origin of facultative worker behavior in wasps and bees, and new behavioral phenotypes can evolve semi-independently within a single species. For several reasons, plastic behavioral traits are likely to be important initiators of new directions in evolution. First, behavior is more labile than morphology. Second, because of the great abundance of potential cues for regulating the expression of a behavioral adaptive response, adaptive behavioral plasticity is likely to evolve more readily than adaptive morphological plasticity. Third, behavior during development can greatly influence morphology. Therefore, behavior may accelerate the rate of speciation, especially among organisms with complex behaviors. This suggestion is supported by phylogenetic analyses suggesting that speciation rates are higher in clades influenced by strong sexual selection.

Although all of life's great experiments lack controls, I find it difficult to imagine a more satisfying professional life than the one I have experienced. Not that I have enjoyed all departmental faculty meetings, every moment in the field, or every class I have taught, but I know that the greater greenness of grass on the other side of the fence is certainly an illusion.

For the most part I have spent my professional life engaged in activities I would willingly have pursued without financial remuneration had I not needed cash to fund my activities and those of my family. I have been blessed with associating with stimulating colleagues and graduate students.

Curiosity, we are told, sometimes kills cats, but without curiosity little new is learned. I never lost my youthful curiosity about how nature works. The thrill of discovering something new, the rare "Aha Experience," as Konrad Lorenz called it, still motivates me. All in all, it has been, and continues to be, a great journey!

References

Beletsky, L. & Orians, G. H. (1996). *Red-winged Blackbirds: Decision-making and Reproductive Success*. Chicago, IL: University of Chicago Press.

Brown, J. L. (1964). The evolution of diversity in avian territorial systems. *Wilson Bull.* **76**: 160–9.

Charnov, E. (1976). Optimal foraging, the marginal value theorem. *Theor. Popul. Biol.* **9**: 129–36.

Coss, R. G. (1991). Evolutionary persistence of memory-like processes. *Concepts Neurosci.* **2**: 29–168.

Coss, R. G. (1999). Effects of relaxed natural selection on the evolution of behavior. In *Geographic Variation in Behavior: Perspectives on Evolutionary Mechanisms*, ed. S. A. Foster & J. A. Endler, pp. 180–208. Oxford: Oxford University Press.

Gould, S. J. & Lewontin, R. C. (1979). The spandrels of San Marco and the Panglossian paradigm. *Proc. R. Soc. Lond.* B**205**: 581–98.

Gray, E. M. (1996). Female control of offspring paternity in a western population of Red-winged Blackbirds (*Agelaius phoeniceus*). *Behav. Ecol. Sociobiol.* **38**: 267–78.

Heerwagen, J. H. & Orians, G. H. (1993). Humans, habitats, and aesthetics. In *The Biophilia Hypothesis*, ed. S. R. Kellert & E. O. Wilson, pp. 138–72. Washington, D.C.: Island Press.

Kingsland, S. (1985). *Modelling Nature*. Chicago, IL: University of Chicago Press.

Öhman, A. & Mineka, S. (2001). Fear, phobias and preparedness: toward an evolved module of fear and fear learning. *Psychol. Rev.* **108**: 483–522.

Orians, G. H. (1961). The ecology of blackbird (*Agelaius*) social systems. *Ecol. Monogr.* **31**: 285–312.

Orians, G. H. (1969a). Natural selection and ecological theory. *Am. Nat.* **96**: 257–63.

Orians, G. H. (1969b). On the evolution of mating systems in birds and mammals. *Am. Nat.* **103**: 589–603. (Has been reprinted in two collections of readings in behavioral ecology.)

Orians, G. H. (1980). Habitat selection: general theory and applications to human behavior. In *Evolution of Human Social Behavior*, ed. J. S. Lockard, pp. 49–77. Amsterdam: Elsevier.

Orians, G. H. (1980). *Some Adaptations of Marsh-nesting Blackbirds*. Princeton, NJ: Princeton University Press.

Orians, G. H. (1999). Human behavioral ecology: 140 years without Darwin is too long. *Bull. Ecol. Soc. Am.* **79**: 15–28.

Orians, G. H. & Beletsky, L. D. (1989). Red-winged Blackbird. In *Lifetime Reproduction in Birds*, ed. I. Newton, pp. 183–97. New York: Academic Press.

Orians, G. H. & Horn, H. S. (1969). Overlap in foods and foraging among four species of blackbirds in the Potholes of central Washington. *Ecology* **50**: 930–8.

Orians, G. H. & Milewski, A. V. (2007). Ecology of Australia: the effects of nutrient-poor soils and intense fires. *Biol. Rev. Cambr. Phil. Soc.* **82**: 393–423.

Orians, G. H. & Pearson, N. E. (1979). On the theory of central place foraging. In *Analysis of Ecological Systems*, ed. D. J. Horn, R. D. Mitchell & G. R. Stairs, pp. 155–77. Columbus, OH: Ohio State University Press.

Orians, G. H. & Pfeiffer, E. W. (1970). Ecological effects of the war in Vietnam. *Science* **168**: 544–54.

Parker, G. A. (1974). Assessment strategy and the evolution of fighting behavior. *J. Theor. Biol.* **47**: 223–43.

Stephens, D. W. & Krebs, J. R. (1986). *Foraging Theory*. Princeton, NJ: Princeton University Press.

West-Eberhard, M. J. (1989). Phenotypic plasticity and the origins of diversity. *A. Rev. Ecol. Syst.* **20**: 249–78.

Wilson, E. O. (1984). *Biophilia*. Cambridge, MA: Harvard University Press.

16

Reflections before dusk

GEOFF A. PARKER

Origins and early years

I was born in 1944 towards the end of World War II; an era when dogs were called 'Bonzo', grandfathers were 'Harold' (both mine were), and 'Charlotte' was the name of somebody's elderly great aunt. Though it had finished before I became aware of it, World War II dominated Britain for most of my early life; it was a world of food rationing, stoicism, and devotion to the common good. My parents, my brother and I lived in a rented house in an unpretentious terrace close to the Manchester Ship Canal in Walton, south of Warrington in North West England.

From earliest days, I was consumed with a passion for animals and natural history. This generated an obsession for pets, and for keeping such things as caterpillars and frogspawn, though attempts to rear them to adulthood invariably failed. From age four, I idolised a cat, a mouse, which the cat eventually despatched and presented proudly to the breakfasting family, and guinea pigs, which bred relentlessly. My father, who had a Ph.D. and worked in

Leaders in Animal Behavior: The Second Generation, ed. L. C. Drickamer & D. A. Dewsbury. Published by Cambridge University Press. © Cambridge University Press 2010.

industry as a research chemist/microbiologist, had an excellent knowledge of natural history in terms of British flora and (vertebrate) fauna, which diffused into my childhood almost as if by osmosis.

Quite often my paternal grandmother took me to Chester Zoo, a 20-mile bus journey, and I was captivated by its magic. In the 1950s, it housed fewer specimens in much less favourable conditions than the vastly bigger and better transformation that now captivates my grandchildren, but I was completely enchanted. My grandmother had been a farmer's daughter; her ancestors had for centuries been farmers in north Cheshire. I pestered her for stories of her father's farm animals, and my grandfather for stories of his working horses. He had worked throughout in the family business – soft drinks, often in large stone jars inscribed with 'T. W. Parker – Botanical Brewer'. T. W. (Thomas Williamson), my great-grandfather, had sailed steerage in the *Calabria* from Liverpool to Boston in August 1870 for a life in Denver, Colorado, where he met Buffalo Bill and others. On my grandfather's knee, I delighted in accounts of T. W.'s escapades. He was a trader, and I recall the tale of a torrential flood in a river, when T. W. lost his wagon, wares, and horses, and – at least as the story was told to me – only just escaped with his life. His wife and their four children had joined T. W. in1871, but she died three years and two children later. T. W. returned to Lancashire with his six children. It must have been almost impossible to raise his family unaided in what was then frontier land. Back in England T. W. remarried; my grandfather was the fifth of the second batch of six children.

Although my father had a Ph.D. and what was known as a 'good job', we lived in modest accommodation. We were very happy. My brother Antony, my senior by five years, was an exemplary and extremely clever child; in contrast I was something of a pain and caused trouble. But my mother claimed she could achieve results – I apparently responded to rational explanations and gentle reasoning. My grandparents lived only about ¼ mile away, each couple in adjacent sides of a semi-detached house. I loved them all; I now realise that I had an idyllic childhood. My mother's parents originated from Crewe; her father served in France in the First World War. He worked as a travelling salesman for a shoe company, and was a respected member of the local Council, one year serving as its Chairman (equivalent of Mayor). He had a sharp sense of humour; his wife, my mother's mother, was genial and immediate.

I well remember the huge merchant ships sailing down the Manchester Ship Canal, and the slow opening of the swing bridges, allowing their passage through. American sailors threw us packets of chewing gum, shouting things we didn't fully understand.

Schooldays

Stockton Heath County Primary School was some ¾ mile away; we walked there and back twice a day, eating dinner at home. One teacher, Mrs Jones, the headmaster's wife, influenced me greatly – she encouraged interests in natural history, and her storytelling and reading engrossed all. At age ten, during my 'Mrs Jones year', we moved to a detached house with a big garden, but only ¼ mile away from the terraced row as the crow flies. The event was carefully planned, and construction of the 'new house' was witnessed with growing excitement until the great day of the move in early 1955. This sudden change in lifestyle had a simple explanation. From school until marriage in 1938, my mother had worked in an office, where I suspect she had shown considerable ability; by the time she married my father she was capable of running everything single-handed. One of the commercial activities was money-lending. She had seen such harrowing events involving debt that she would not contemplate borrowing money – even a mortgage. House purchase was delayed until the bill could be paid. Despite pleas to return to the office, she never went back after marriage, but instead devoted her life to the family, and became the greatest mother imaginable.

In my first days at school I met Eric Richardson, the son of a Cheshire dairy farmer; our friendship still persists after six decades. My obsession with animals was quite general – Eric's farm was just as fascinating as the zoo visits, or fishing trips with my father. At weekends we would alternate: I visited him; he visited me. The farm was a source of fun and inspiration; in addition to the domestic animals there were chicks, cats, dogs and guinea pigs. We would climb the bales of hay in the barn and watch excitedly the farmyard ritual of a cow's service by the Friesian bull. Those weekends undoubtedly helped me later to choose my Ph.D. research project on yellow dung flies (Parker 2001).

In September 1955, I started at the grammar school where my father and brother had preceded me. My fascination for animals persisted (Fig. 16.1). Lymm Grammar School was

Figure 16.1. As a schoolboy in the garden of the 'new house', a year or two after moving there.

6 miles away and involved a novelty – the bus ride. If I missed the bus, the options were to walk to 'town' (Warrington) and take the regular bus, or to cycle. Like primary school, Lymm was co-educational, at that time unusual in Britain, and several children from Stockton Heath moved there. Its academic achievements were possibly not as notable as some of its competitors, but I owe much to Lymm; it had many commendable features, one of them being its liberal, good-natured, and rather unpretentious approach to life. I shall always be indebted to my spirited and unconventional biology teacher, 'Jake' Newman, and my draconian physics teacher Eric Liddle. I admired and liked the former, who I remember once remarking, I believe with some disdain, something about theorists like Haldane thinking they could explain biology by mathematics. I cannot claim to have been fond of Eric Liddle, a rigid disciplinarian who met the slightest infringement with instant, usually physical, punishment. However, I owe him a great debt. He taught us how to think logically, by meticulously demanding that each step in a process be defined in its correct sequential order, as a clear and logical consequence of the step before it. The importance of this for an incipient theoretical biologist hardly demands stressing.

I was selected, with about ten others, to take the first important examinations, 'Ordinary Levels', a year early at age 15. In many ways this was crazy: it meant giving up history, geography, and a host of other subjects that I lamented rather less. It also contributed to my failing mathematics, which I had to repeat and pass in the sixth form. So great was my obsession with biology that other subjects seemed trivial, especially mathematics. My underlying mathematical ability was probably about average for the top class of my school year (I had done well in the second year); however, I gradually slipped back because I had no interest – and worse – I began to resent and dislike the subject, something that ten years later I was to regret very deeply, and still do.

In the sixth form, I studied zoology, botany and chemistry for 'Advanced Levels', the next big exam. I boiled up rodent carcases, and bleached and reconstructed their skeletons, which I then mounted in Perspex cases. Somehow, I became beguiled into applying to read medicine at University. More natural choices would have been zoology or veterinary science. I could not apply to Oxford or Cambridge as my brother had done, because I lacked qualification in Latin through taking Ordinary Levels early. Instead, I applied to Bristol, where my biology teacher had studied zoology. Because I had missed a year and had started the sixth form at age 15, I was too young to begin the medical course. I needed Ordinary Level physics to qualify for Medicine. Undaunted, I asked Eric Liddle for a textbook, telling him that I intended to enter for the physics exam. He asked who would be teaching me – when I replied that it would be my brother, then studying physics at Oxford, he handed me a text with a very wry smile. Nevertheless, I passed the exam with my highest mark, something that caused a small crisis at the school because another pupil, also called Parker and the son of the maths teacher, had taken the exam at the same time and failed. There was an inquisition as to whether there had been a mix up between the two of us. Mercifully, I was vindicated. The result reflected more my brother's exceptional teaching skills than my true abilities in physics, but I did enjoy the logic of physics and somehow did not have an aversion to it as I did for maths.

I obtained a provisional acceptance for medicine at Bristol if I improved my A-level mark in chemistry, another subject I had neglected, by 5%. I languished a further year in the sixth form and achieved the desired objective, and probably not a great deal more, other than maybe improving my clarinet skills. My brother had gained a prestigious State Scholarship at the end of his school career. I was entered for the same award and might have achieved it had I not thrown away my chances in the Scholarship zoology exam. I had been fascinated with Baldwin's 'Dynamic Aspects of Biochemistry' and how ionic regulation was achieved across fish gills. There was a question on epithelial membranes, which required an account of their histological structure. Rather than give an easy, safe account of epithelial cells, which I realised was required, I could not resist launching into a discourse from Baldwin. It went well beyond the syllabus, but did not exactly answer the question. The risk failed; I did not realise that the examiners were bound by strict marking schemes. I received an ignominious 'unclassified' grading for the scholarship paper, a sharp contrast with my mark for the Advanced Level paper.

Bristol (1962–68)

As an undergraduate

My first days in Bristol represented a personal cataclysm. It took just a few days to appreciate that I had made a dreadful mistake. It was not the gruesome horror of the dissecting room so much as the realisation that my life was now diverting along a track that somehow did not seem its natural course. I wandered past the Zoology Department, longingly staring into its museum, wishing that I had not been so stupid as to be seduced by the power and social status of a career in medicine – if that had been its attraction; I am still not sure. I suspect that for a short time I became what would now be called 'clinically depressed'. I summoned up the courage to change from medicine to zoology. The Pre-Clinical Tutor was kind but incredulous, asking me to consider seriously what I would do with a degree in zoology – the obvious expectation would be to teach – only a very few would be likely to do research, the wish I had expressed. It was as if my request to change from medicine to zoology was like asking to change from Captain to Deck Hand. The Dean of Medicine, who had interviewed me the year before, was severe and indignant – I had denied a more committed student a place on the course. In fact, a replacement was actually found within a fortnight.

Once this was all over, Bristol became exciting and fun. Ten of us were affiliated to Wills Hall, which resembled an Oxbridge college, but annexed in a large house close by, where we had bed and breakfast. We drank in the local pub after the evening meal in Wills. It was the era of the New Orleans ('Trad') jazz revival in Britain, and I loved it. I had bought a clarinet two-three years before and had played with a few school friends. Jazz was every-where, until eventually the Beatles took over the popular music scene. I sat enthralled by its magic at one of the many Bristol jazz venues – far too often than was good for my studies – and felt withdrawal symptoms if I could not play my clarinet regularly, usually alone, but occasionally with a student band. This interest never stopped, but it was some twenty years later before I started playing gigs frequently in the Merseyside jazz scene. A further twenty years later, I am still doing so, and though I see myself as a mediocre player, to go out and play a night's jazz – something that is so immediate and spontaneous – is the perfect antidote to a day spent anguishing over the algebraic details of theoretical models, or writing difficult manuscripts. I wish I had more talent at it. At both!

Zoology was my main subject, and because I had to do a subsidiary subject, I also studied botany. My old problem arose; I loved zoology but neglected botany, preferring to listen to jazz, or to play snooker in the Students' Union... or almost anything. At the end of my first year, I came second in zoology, and failed botany by two marks. The genial professor of botany indicated that I would probably have to repeat the year and settle for an ordinary, rather than an honours, degree. I would not now be writing this autobiography had not the zoologists pleaded the case that I should be allowed to continue. I learned a valuable lesson, and have perhaps been rather more sympathetic than most academics to students who fail their exams. During my second year, I managed to trade-off effort on zoology for effort on botany, with a corresponding but more secure change in success in both subjects.

My third and final year of undergraduate study at Bristol became everything I had dreamed of. The course resembled the Cambridge system and was based heavily on research projects; each student completed four separate projects, one in the preceding summer vacation and then one per term. There were about ten specialist courses, and it was essential, if a little risky, to study only three. Suddenly I was transfixed; I was deeply fascinated by the courses, and became obsessed with the projects. The first involved the diurnal activity patterns of the predatory beetle *Nebria brevicollis*. I rigged up a primitive activity meter using a relay counter (ideas and equipment borrowed from my brother) and found dawn and dusk peaks of activity. However, what fascinated me most was my first encounter with a theoretical problem in animal behaviour: what is the optimal rate of movement of a predator, given that this depends on the movement rate of its prey? It was my first clash with a problem where the best strategy depends on what another individual does. I sought academics who might know how collisions between two bodies change with the speeds of those two objects. A physics professor suggested that the Pentagon would have the mathematics solved for meetings between submarines, and a chemist claimed that the equations should be documented for collisions of gas molecules. But the problem lay fallow until collaboration with Peter Hammerstein two decades later (Hammerstein & Parker 1987), which focussed on encounters between males and females rather than prey and predators. Even now I still have notes and plans to make a general analysis of the problem of evolutionarily stable strategies for movement rates when there are different costs and benefits of meeting. I had learned that mathematics could be necessary to answer questions in biology.

The second project concerned the sexual behaviour of the blowfly *Protophormia terrae-novae*. I reared adult flies from maggots bought at angling shops, and conducted endless experiments to probe into the mechanisms preceding mating (Parker 1968). I worked all hours, and even took the flies home with me over Christmas. They were everywhere – in bottles on the piano, the dining room table, and so on; my parents were very patient. The work made me speculate continuously about *Protophormia*'s mating system in nature, of which I knew almost nothing – I began to see the value of field studies. My next project was supervised by the now eminent biophysicist, Colin Pennycuick, and concerned experiments on pigeon bones and tendons to investigate the safety factors found in nature given the maximum power output achievable in flight (Pennycuick & Parker 1966). Colin suggested the experiments and did all the maths. After my failure to answer the predator mobility question, Colin's analytical approach showed me the power and elegance of using mathematics in biology. My animosity towards mathematics was declining rapidly, and I realised that I was mathematically illiterate.

The last project in that final year as an undergraduate was a field investigation on the reproductive behaviour of dung flies. It began in the Easter vacation on farmland close to home, and was completed in Bristol before the final examinations. The *Nebria* and *Protophormia* projects had been, at least on paper, supervised by the late Howard E. Hinton, a distinguished entomologist and a remarkable man and scientist (see Salt 1978). A Zoology Department rule limited the number of times one member of staff could supervise a given student, so Howard could not supervise the final project, but I nevertheless asked his advice for a study of insect behaviour in the field. He effortlessly reeled off many possibilities – the behaviour of dung flies was one. I immediately chose it because dung flies were familiar from my childhood farm associations. It was a choice that was to define my life as a scientist (Parker 2001).

Maybe the most significant career event during that stimulating year related to insights about the mechanism of natural selection. In the Bristol zoology course of that time, natural selection and the mechanism of evolution hardly gained a mention. A member of staff gave four or five lectures, delivered very slowly (one could write every word), which started with the evolution of the horse and Darwin's finches, and ended up with a discourse on the rate of penetration of fixative into liver. The latter was a subject on which he had published, and which represented an analogy with the rate of spread on a gene into a population. The world of ecology and ethology at that time was submerged under a sea of implicit group selection and 'survival value for the species' thinking. A fellow student, Robin Baker, and I became preoccupied with how selection worked. We discussed it endlessly, because our views were Darwinian and opposed to the all-pervading group selectionism of that time; they in fact paralleled those expounded so plausibly by Williams (1966), which I did not see until several years later (see Parker 2007).

Robin was later to make his mark first in the field of migration (Baker 1978) and human sperm competition (Baker & Bellis 1995). His approach, from earliest research days, was confrontational, even gladiatorial; he was out to slay the demons of past misconceptions. This did not always endear him to the scientific establishment, and he has been accused of advocacy beyond the facts (Birkhead 2000). He nevertheless showed great originality and flair, and I shall always recall the stimulus of our early discussions with great nostalgia and indebtedness.

Susan Wallis and I first began our relationship early in 1964. Sue was a student on the same course who later became my wife of 27 years and the mother of my children, before her tragic early death from cancer in 1994 just a few months after we celebrated our fiftieth birthdays.

Figure 16.2. Sue and myself at graduation, Bristol University, July 1965.

But at age 20, I probably spent rather too much time with my Wills Hall friends and on jazz than Sue would have preferred; our relationship was initially more turbulent than many, but nevertheless had a depth that few could perceive. Sue had been just six years old when her mother died, understandably leaving scars that persisted throughout her life. She had always anticipated an early death like her mother, and sadly her intuition proved to be correct.

My last project was completed about a week before the final examinations, and I entered a frantic spate of last-minute revision. Despite my lack of time, and to Sue's amazed horror, four days before the first exam I made the journey to Bath to hear Henry 'Red' Allen play with Alex Welsh's jazz band. It was worth it. To my total disbelief, I gained the top first and won the Rose Bracher prize for the best degree in the biological sciences; graduation was a blissful occasion (Fig. 16.2). For many years I anticipated a letter informing me that the result was erroneous.

I did not realise that I was at the dawn of the most exciting era of my life.

As a postgraduate

I owe Howard Hinton a huge debt for many things, especially for his stimulation and enthusiasm, and for suggesting dung flies. 'Supervising', in those days, and certainly to Howard, meant a sort of congenial mentoring, without undue recourse to detailed discussions of the student's research. It was perfectly acceptable to talk – passionately if required – about the ways of the world, i.e. insects, but more than one or two requests for advice or guidance per year were unbecoming. Independence ruled. One felt that one belonged, and indeed one did, but supervision in its present form did not exist, at least with Howard. We nevertheless all liked, respected and admired him.

There was much to be said for that system, though it did breed casualties, and I was nearly one of them – an autobiographical account of the dung fly as a model system and its conceptual role is given elsewhere (Parker 2001). I have been indeed fortunate to have it featured in many textbooks, and for a variety of reasons also in philosophical monographs (Kitcher 1985; Myers 1990; Sober 1993; Ruse 1996, 1999). In summary, I would make two claims for that early work, which has to be seen against the backdrop of the events of the time (Parker 2006; Birkhead & Monaghan 2009; Fox & Westneat 2009). I believe it showed, for the first time quantitatively, using detailed optimality approaches from data obtained mostly in the field, that a suite of male reproductive behaviour patterns fitted closely and in a very 'fine-grained' fashion with the predictions of Darwinian sexual selection (Parker 1978, 2006). Second, it played a role in the development of several general concepts in behavioural ecology (sperm competition, animal distributions, animal contests and sexual conflict) in the 1970s (see also Parker 2006). I was extremely fortunate to have stumbled across a model system that – although maybe not top ranking aesthetically – had great advantages scientifically.

It is probably hard for a young behavioural ecologist of today to envisage how insecure I felt as a research student pursuing an individual selection philosophy in 1965–68. Although population geneticists were clear about the mechanism of selection, most other biologists, and certainly most ecologists and ethologists, were implicit group selectionists. Though the group selection debate had barely surfaced, most adaptive interpretations followed 'advantage to the species' lines, and very few researchers in those disciplines worked on the philosophy that adaptations are shaped by benefits to individuals. Just one decade later, the discipline now known as behavioural ecology or sociobiology had exploded, and individual advantage (or its parallel, selfish gene philosophy) had become the established order (Parker 2006). The aim of behavioural ecology is to understand behavioural adaptations in terms of the selective forces that have shaped them, as I strove to do in my postgraduate work on dung flies (Figures 16.3 and 16.4).

Let me try to explain my insecurity. In Bristol, we had no David Lack or John Maynard Smith; 'survival value to the species' ruled. My rare discussions with my supervisor, Howard Hinton, only exacerbated my anxiety. I recall only two significant discussions that related directly to supervision. The first concerned the evolution of the postcopulatory guarding in dung flies. Instead of leaving the female, after copulation the male guards her from other searching males while she lays her eggs (Figure 16.3). I was convinced that this behaviour had arisen through sexual selection to protect the guarding male's paternity (Parker 1970a), and hence wanted to know how many eggs the last male to mate fertilizes. To do this, I made arrangements at the local hospital to irradiate males in order to 'label' their sperm – with a sufficiently high dose, eggs from an irradiated male do not hatch, following the 'sterile male' technique. Howard was puzzled as to why I was irradiating flies. On hearing my explanation, his comment was that I should seek an advantage *to the female* for the male guarding behaviour – *males were expendable*; the important thing was that the female should manage to lay her eggs. And to my horror, I found that a rather similar proposition had just been published in *Nature* (Foster, 1967). Howard had thought about

Figure 16.3. Male dung fly guarding the female after copulation, while she lays her eggs. One of the original photographs from my Ph.D. thesis, September 1968.

Figure 16.4. The field site at Ladymeade Farm, Langford, near Bristol, summer 1966. I had modified the Bedford van to live in during the field work (see Parker 2001).

such things and had come to the conclusion that if one thought in terms of three generations, natural selection could equate with advantage to the species, even if this initially conflicted with the interests of the individual. I was worried, but disinclined to change my mind about my sexual selection interpretation.

My other 'supervision' episode was unrelated to dung flies. In ignorance of Fisher's principle (Fisher 1930), I had become interested in why the sex ratio was typically unity, and following a suggestion from Robin Baker, devised a computer program that simulated selection for sex ratio in a population with a series of alleles determining the ratios of male to female offspring (9:1, 8:2, etc., to 1:9). In 1967–68, the programme to run this simulation for 5,000 generations took a couple of hours on the mainframe computer; these days it may take a few seconds on a PC. I wanted the allele for 5:5 to win. It didn't: though an equilibrium occurred when the population sex ratio became unity, the alleles for sex ratio then stopped changing, and their frequencies depended on the frequencies at the start of the simulation. This is perfectly comprehensible to all who understand Fisher's principle – under simplest assumptions, in a population at unity sex ratio, all sex ratio genes become neutral. Unfortunately, I hadn't heard of Fisher or his principle. I decided to abandon this project until I had completed my thesis, but Howard wanted to know why I was using 'the computer', an unusual practice for a biology research student of that time. He was immediately fascinated – he had worried previously why the sex ratio was unity, because "the population would do better if there were many more females than males". When I explained that the simulation gave a unity sex ratio, he asked me to explain why. I was less than clear, because I didn't fully understand my neutral allele result. He stopped me after a few halting sentences, with the comment that "a biological theory is a waste of time if it cannot be explained in a single sentence." Maybe he was right.

Another episode of insecurity occurred when a fellow postgraduate berated Robin Baker and myself for our individual selectionist approach: characteristics arising for 'the common good' posed no problem – a process called *group selection* explained everything; there was a huge tome full of evidence (Wynne-Edwards 1962). Robin read more of Wynne-Edwards than I did, and his readiness to fight intellectual battles, and to proclaim it misguided, probably helped me to have the courage of my (and his) convictions. We were undeterred, but I was worried that there was such opposition.

In July 1967, towards the end of my second year as a Ph.D. student, Sue and I married in her hometown of Taunton, and we lived in a flat in Redland, Bristol. I would drop her and pick her up at Temple Meads Station, from where she travelled 40–50 miles each day to Taunton to teach biology at her old school. Only now does it dawn on me how uncomplaining and accepting she was of all that commuting.

After an abortive attempt to gain a postdoctoral fellowship (Howard Hinton told me that he could fully support only one of his group for the same post, and he had decided to support Robin Baker), I applied for, but failed to secure, both a Demonstratorship in ecology at Oxford (the successful candidate was the now famous ecologist John Lawton), and an academic post at Bedford College, University of London (gained by the now eminent ornithologist Clive Catchpole).

I was interviewed for the post of Assistant Lecturer in Zoology at Liverpool University in early 1968. Professor Arthur J. Cain, the eminent evolutionary biologist, then recently appointed as Derby Professor of Zoology, chaired the panel. I was desperate to get the post, not only because I began to see the looming spectre of unemployment, but also because Liverpool would be close to friends and family. Against the odds, and to my great delight, this time I was lucky. Since Howard was not well acquainted with what I had done (he declined to review our theses on the grounds that the external examiner would be assessing him rather than his students) writing me a reference must have been difficult. My relief that he had not read my thesis, and discovered that I had adhered to my individual selection approach, had a downside: I suspect that his reference may have been decidedly off track. For some years at Liverpool, Arthur Cain would introduce me to visitors as "a new member of staff who is working on species isolation mechanisms in sympatric dung flies". My work had been a single-species study that had nothing to do with species isolation; I constantly worried whether I might be exposed as some sort of charlatan...

In summer 1968, Sue and I left Bristol to live with my parents, where I finished writing the thesis. I went down to Bristol to get it photocopied and bound, sleeping in the lab for a couple of nights while it was being completed. I handed it in on 30 September 1968, the final day of my three years' postgraduate study, and returned, drained but fulfilled, to begin work in Liverpool on 1 October, the first day of my official employment. My second meeting with Arthur Cain, my new Head of Department at Liverpool, can only be described as unfortunate. He rounded on me, red in the face, and gave me a most severe telling off for failing to attend the staff meeting the week before. Since I had not been informed about the meeting, and had been working day and night to submit my thesis before starting at Liverpool on the first day of my contract, this seemed not a little unjust. He was so irate that any form of protest did not seem an option...

Robin Baker and I had our Ph.D. vivas on the same morning in December 1968, in fairly quick succession, probably about an hour each. Our examiner was the renowned ecologist, the late T. R. E. Southwood. During the viva, Southwood made some perceptive comments, and then asked why I had been using 'the computer'. I explained my sex ratio project, and why it did not appear in the thesis. He remarked: 'There's a chap at Imperial... Bill Hamilton... do you know him? He's interested in sex ratios, and has a paper in *Science*.' When I read Hamilton's (1967) paper, and then Fisher's book, I realised I had a great deal to learn and tried to forget my sex ratio simulations; they still lie in a box file somewhere in my office.

Liverpool and the 1970s

So my start in the Zoology Department at Liverpool in October 1968 was ignominious; I had upset my Head of Department by missing a staff meeting even before I arrived. I later came to admire and respect Arthur Cain for his erudition and his scientific achievements, yet I think it was well into the 1980s before I was able to call him 'Arthur' rather than 'Professor Cain'. A sort of archetypal Edwardian English Professor, he was equally capable of showing great generosity as he was of delivering a savage verbal attack. But he was committed to scientific truth and integrity, and a devoted advocate of the power of natural selection (see Clarke 2008).

After some months of living with my parents, Sue and I moved to my grandmother's empty house, some 20 miles from Liverpool. My grandfather had died, and my grandmother was confined to hospital. Commuting was tedious, and we were saving to buy a house nearer to Liverpool. House prices were rising much faster than we could save, and so in summer 1970 we searched for a house on the Wirral, moving there in early January 1971. The vendor was a car salesman, and I suspect we paid well over the odds for our bungalow in its rural setting and 1/3 acre. Despite promises to the contrary, he drained the heating oil and on moving in we discovered that the lounge was under water: central heating pipes in the attic had burst with the frost. We put all our money into the venture, and had what then seemed a huge mortgage; initially we needed two salaries to survive. However, inflation quickly took care of that problem, and I still occupy the same house 38 years later. It feels like a comfortable old coat.

This new chapter in our lives started badly. Soon after we moved, we learned that my brother's elder child, aged only three, had leukaemia; as a close family, we were devastated. Despite pessimistic prognoses, he survived, being one of the first groups of children to be treated with the new set of drugs. But tragedy struck thirty years later; a car ran into him as he crossed the road returning from work. He died four days later, and our family lost a delightfully kind and dear member.

Although the explosion of behavioural ecology in the 1970s was one of the key events in biology in the last century, even more vital for me was the birth of my daughter Claire in 1973, and of my son Alan in 1977. I was unprepared for the deep feelings I would have. Maybe this was due to having worked on sexual selection, and realising that males of most species do not care for their offspring. Whatever the reason, I was shocked by my emotional involvement. To this day I would deny any suggestion that I was any less obsessed with our children than Sue was. They – and my grandchildren – are still the most important aspect of my life.

In revising my dung fly work for publication; I had only a few useful comments from my examiner, T. R. E. Southwood, to go on. I worked very hard the first year at Liverpool writing papers from the thesis, but found I had been given a very large amount of teaching – I suspect this was in part a consequence of missing that crucial staff meeting. Some months later, when I timidly enquired of Arthur Cain whether I might hope for a rather lighter load, he retorted peremptorily that "we have tried to keep your load as light as possible in your first year – I can't promise that it won't be increased next year". I was desperate to develop my irradiation experiments, to do further field work, to complete the papers from my thesis, and to write a review that had been developing as ideas and notes over the last three years, i.e. my *Biological Reviews* paper (Parker 1970b) on sperm competition. With so much teaching, progress was frustratingly slow.

The *Biological Reviews* manuscript demanded some library work, but was written quite swiftly; it had been long in the gestation. Howard Hinton had kindly sent me – from an abstracting service he had a connection with – details of papers using the sterile male technique, some of which looked at irradiated males in competition with normal males. While finalising details of the manuscript, I well recall phoning the ecological geneticist, Philip Sheppard, then Head of Genetics at Liverpool and one of the sharpest intellects I ever met, about some detail of the butterfly *Papilio*. He asked about the project, made me feel at ease, and as I was explaining how sexual selection might be expected to shape male behaviour, he incisively summarised my main argument. Taken aback, I remarked that, in effect, he had just paraphrased the main thrust of the paper… After replacing the receiver, I looked at the draft manuscript with considerably less confidence than before. Was it all so obvious, and was it really worth writing about? Fortunately, perhaps, I decided it probably was – it became my most cited and best known paper, often credited as pioneering the field of postcopulatory sexual selection (see, for example, Birkhead 2000).

I had completed the irradiation work, and made my first attempts at modelling in a manuscript (Parker 1970c) I submitted to *Journal of Insect Physiology* – then edited by Howard Hinton. He had decidedly eccentric views of genetics – he believed in heritability, but not that 'genes' were located in chromosomes. He required that I replaced the term 'mutant', which he disliked, with 'variant' – which I then standardised across all the early papers. The journal's policy was not to permit serial publications, so I could not include it with my others in the series "The reproductive behaviour and the nature of sexual selection in *Scatophaga stercoraria* L. (Diptera: Scatophagidae)". I probably should have been grateful – how dated that title seems now!

Several other papers in the dung fly series also came out in 1970 (see Parker 2001). Some of these, though empirical, used what could be described as using an 'incipient' evolutionarily stable strategy (ESS, Maynard Smith 1982) approach, as had Fisher (1930), Shaw & Mohler (1953) and Hamilton (1967) for sex ratio, Orians (1969) and Fretwell (1972) for animal distributions, and Trivers (1971) for reciprocal altruism. For example, my 'equilibrium position' for the distribution of male dung flies searching for females around a dropping (Parker 1970d, 1974a) is an ESS distribution ('ideal free' *sensu* Fretwell 1972). In the *Journal of Insect Physiology* paper (Parker 1970c), I explicitly calculated the fitness of mutants ('variants') varying in copulation duration from the rest of the population. In populations fixed away from the present copula duration (i.e. that observed in nature), selection would favour mutants towards the present copula duration. In contrast, in a population fixed at the present copula duration, the fitness of deviant mutants was less than that of those playing the population strategy. This is about as close to ESS logic as one can get (see Parker 1970c, pp. 1323–5). However, I later abandoned this ESS approach for a marginal value approach (Parker & Stuart 1976), which worked well because the frequency dependence around the zone of interest became very weak. Another paper asked how guarding behaviour was maintained in the present population, and compared the fertilization rate of a hypothetical mutant ('variant') male that separates from females immediately after copulation with that of guarding males, all in the present guarding population (Parker 1970a). This showed that the guarding behaviour is stable – such a mutant could not spread in the present population; again, an early ESS approach.

The dung fly work had shown me very emphatically that I needed to model processes algebraically in order to make testable predictions that could be compared with the observations. The optimality approach is a way of testing whether an adaptation could have been shaped by the selective processes envisaged in the model (Parker & Maynard Smith 1990). I became more and more fascinated with general questions – my dabblings with sex ratio and optimal mobility had been the forerunners. One was the evolution of anisogamy (Parker *et al.* 1972), which was a sort of spin-off from my failed sex ratio project and influenced by my interest in sperm competition. This began in phone discussions with Robin Baker, who had moved to Newcastle University after completing his research fellowship at Bristol. The initial model was obvious, but a problem had arisen – in my initial hand calculations, I did not have a large enough range of gamete sizes to allow anisogamy to be generated. That anisogamy could arise with a big enough range had been spotted by Vic Smith, a friend from Bristol days, then studying a higher degree in Liverpool, whom I had chatted with. Robin and I discussed it at length; I recall that Robin wrote the initial computer simulation, which I modified and ran at Liverpool. The simulations confirmed that anisogamy could develop from isogamy under a wide range of conditions, and although the paper that arose from this project (Parker *et al.* 1972) has only about one tenth the citations of the sperm competition review, I regard it as being perhaps more fundamental. I asked Arthur Cain and the now eminent geneticist Brian Charlesworth (then at Liverpool) to read the manuscript – I suspect they both thought it a disaster. Arthur gave up after writing a series of emphatic comments on the first few pages, and don't believe I ever received comments from Brian. However, I was delighted when it was accepted, I believe without revision, early in 1972. On publication, George Williams sent me a very gracious letter about the work, something I still greatly treasure.

In 1970–71, I had started two other projects. The first concerned male time investment strategies with females – the problem of how a male should optimally allocate his time courting or guarding a female, more generally the problem of when to give up on a resource that has a decreasing probability of yielding benefits (now well known as Charnov's marginal value theorem, 1976). This stretched my limited mathematical skills: I could not differentiate at that time, and the problem needed calculus. I happened to talk about it with my father. He knew from his work on industrial processes (he was working on antibiotic production) that there was a simple graphical solution to this problem, which was used to decide when to stop industrial production in a closed system that yields diminishing returns and has a fixed setting-up cost. This was exactly analogous to the problem I wanted to solve in relation to courtship duration. The paper was rejected from *American Naturalist* in mid-1972, and I used to wonder whether it stimulated the classic Smith & Fretwell (1974) paper on optimal investment in each offspring – it certainly had an American reviewer. I sent my manuscript to *Behaviour* in June 1972, and it was eventually published after a long and uncertain period in review (Parker 1974a).

Eric Charnov had derived his marginal value theorem (Charnov 1976) in relation to foraging in a patchy habitat in his Ph.D. thesis (Charnov 1973). Because I couldn't do calculus in the early 1970s, I could not obtain the solution for different patch types correctly as he had done; I achieved something rather similar in terms of matching a gradient from the particular resource (a female) to the average gradient for the habitat. In the mid-1970s, I worked with R. A. Stuart in Engineering at Liverpool, to produce an independent version (Parker & Stuart 1976) of what was to become the marginal value theorem. Our paper was actually reviewed by Ric Charnov, who recommended its acceptance. This was very gracious of him because his classic marginal value theorem paper (Charnov 1976) was still in press at the time; he could have acted other than generously at this early stage in our careers. We have been good friends ever since our meeting in 1976. I see him as one of the leading evolutionary biologists of the past half-century; his paper is rightly a citation classic.

From watching male dung flies fighting for females I knew just how intense male–male battles could be; in the early 1970s I had begun a theoretical project on animal fighting. I realised that the problem was complex: for an escalated fight, each opponent had to have a positive expectation before each 'bout' in order to continue. I envisaged that the contestants should assess the asymmetries between them in order to come to a conventional settlement without a lengthy battle. Each opponent's decision should be determined after each had assessed (i) the relative 'resource holding potentials' (RHPs – roughly, 'fighting abilities') and (ii) the relative values of the contested resource. Should a hypothetical long fight occur, the combatant that would be the first to use up his 'fitness budget for fighting' (i.e. reach a negative expected payoff even if he were to win) should withdraw immediately after assessment, leaving his opponent as victor.

This paper (Parker 1974b) took a long time to develop, but I had produced a first draft by early 1973. Brian Charlesworth remarked to me that John Maynard Smith had just written an essay on animal contests (Maynard Smith 1972). Shortly after, John's *Nature* article on animal conflict with George Price appeared (Maynard Smith & Price 1973), the paper generally seen founding the evolutionarily stable strategy (ESS) concept, and I felt deflated. However, I realised that my approach was all about asymmetries between opponents; theirs assumed opponents to be equal. I submitted to *Journal of Theoretical Biology*, and the reviewers, John Maynard Smith and (I think) Robert Hinde, both recommended acceptance; after all my thoughts and analyses, I was immensely relieved that the project would not be a total loss. It became my second most cited paper (Parker 1974b), and although not a formal ESS analysis, many years later we (Hammerstein & Parker 1982) confirmed that my assessment rule was an ESS.

Soon after he had reviewed this paper, John invited me to Sussex. He was also interested in my dung fly and anisogamy results. I quickly felt a profound respect for him, both at a personal level, and as a scientist and thinker. He has probably been the person I have most admired (Parker 2007), and, as everyone did, I found him extremely likeable, immensely stimulating, and great fun. He had begun work on asymmetric contests, and showed that a conventional settlement did not necessarily need to relate to asymmetries in RHP or resource value; a purely 'arbitrary' asymmetry could be used (Maynard Smith 1974). But could an arbitrary asymmetry be respected against a strong contradictory asymmetry in RHP or resource value? This question stimulated a collaboration concerning the relative role of arbitrary versus payoff-related asymmetries, and the paper that resulted (Maynard Smith & Parker 1976) became an ISI citation classic. I was awestruck by John's intellect and found it difficult to contribute much directly other than proposing the 'information acquired during a contest' model, which John simulated on his computer at Sussex.

In 1975, I was asked by Murray Blum, who had worked briefly in Hinton's lab during my Ph.D. era, to chair a symposium on sexual selection in insects at the XV International Congress of Entomology in the summer of 1976. I accepted; it would be my first trip to the

States. The plan was to produce an edited volume from the symposium, and I suggested the participants. However, trouble was brewing: Arthur Cain called me into his office and told me that I would be needed for the final-year field course, scheduled around the same time. The course lasted two weeks, and I had planned to spend about the same time in the USA. Arthur pointed out that I had just about enough time to go directly from the field course to the conference. The thought of leaving my wife to cope single-handed with our three-year-old daughter for a month seemed unreasonable, and a month's separation seemed unbearable. I protested, but Arthur was intransigent; he did not see any problem. I was fortunate to be promoted to Senior Lecturer in 1976 but it was little compensation. I reluctantly withdrew from the conference, but agreed to write an article for the symposium volume.

I had been working on models of sexual conflict, and became excited about the work; it eventually contained 4–5 separate analyses (including a first model of mate choice), each of which could have been written up as separate papers. I rolled them quickly into one poorly written chapter, and was a week or so late in meeting the deadline, for which I apologised profusely to Dan Otte, who had chaired the symposium in my place and who was assembling the manuscripts. I did not realise that some of the other authors would delay for two years! The paper (Parker 1979) languished in press for what seemed an interminable length of time – the book eventually appeared in late 1979.

This delay cost me dearly; competing papers began to appear. I had sent the manuscript to John Maynard Smith; Michael Rose, then a Sussex Ph.D. student (now a leading authority in the evolution of senescence), probably saw it or John discussed it with him, and produced a paper on one aspect (an arms race model) before the book appeared in print. Chatting to me at a conference in Bielefeld in late 1978, John Maynard Smith buoyantly proclaimed that he had produced a solution to "when should I [mate with] my sister?" I was able to respond: "when the fitness of an offspring from this incestuous mating is not less than a third that of an outcross, and she should agree to it if its fitness is not less than two thirds". It was all in my manuscript, and he graciously decided to suppress his analysis. Reinhardt Selten, the game theorist who later won the Nobel Prize with Nash and Harsanyi, was at the same conference and we discussed the result that my arms race model appeared to have no stable solution (as Rose 1978 had also found). He pointed out that a Nash equilibrium to the problem was likely if one applied his 'trembling hand' technique of allowing the strategy to vary randomly around its mean level, which I subsequently investigated (Parker 1983a). Richard Dawkins and John Krebs were also there, and were just finishing their classic arms race paper (Dawkins & Krebs 1979); I showed them my manuscript and fervently hoped that the book would appear soon.

I continued to see Howard Hinton until his death in 1977: after leaving Bristol we met annually at the Verrall Supper (annual exotic dinners for entomologists, amateur and professional), then at Lyon's Corner House in Piccadilly, and later at Imperial College. It was here I first met Bill Hamilton in the early 1970s. He had read my *Biological Reviews* paper and had interpreted sperm displacement by males as an example of biological spite (rather than selfishness), which we discussed. The Verrall Suppers were intense occasions, and I remember seeing (but not speaking to) many legendary entomologists, such as Miriam

Rothschild, V. B. Wigglesworth, and C. B. Williams. I finally summoned up the courage to speak to O. W. Richards, who had much influenced me by his review (Richards 1927) on sexual selection in insects; I timorously told him how much I had admired it. He muttered something and promptly wheeled away, leaving me completely crushed. I hope I have never treated a young scientist this way.

As postgraduates, Robin Baker and I often travelled to the Verrall with Howard Hinton, who gave us a lift to the Temple Meads Station, where he left his car. We would return on the last train back, excited and inebriated. One year it had snowed heavily, and in the early hours of the morning, Howard gave us a high-speed demonstration around the centre of Bristol – in those days it was a huge roundabout – of how to control a car (a 1960s Ford Zodiac) while skidding in the snow. It was a terrifying but exhilarating experience; we were far too deferential to object.

During one Verrall in the mid-1970s, Howard suddenly proclaimed that sexual selection was an inadequate explanation of bird coloration; it really all had to do with warning coloration. He announced that we would write a paper – Robin would look into the data and I would look into theory. To dissent was not an option; some time later, we stayed at Howard's house in Bristol to discuss the project. Howard died before we produced a manuscript, but we put Howard as the first author and submitted to *Philosophical Transactions of the Royal Society*. George Salt acted as correspondent; he insisted that Howard's name should be withdrawn in favour of an explanatory note in a Preface (see Baker & Parker 1979), on the grounds that authorship carried responsibility as well as credit.

I remember being impressed by Bob Trivers' paper on parent–offspring conflict (Trivers 1974); it was something I simply hadn't realised, and the proposition came as a shock. Mark Macnair, then one of Philip Sheppard's postgraduate students, came to discuss it and we decided to attempt a genetic model to gain insight into how it might work. We first approached the analysis using Taylor expansions; a generous and diligent reviewer (Richard Sibly) suggested a more obvious differentiation technique. We also ran computer simulations, and investigated how the extent of conflict depended on the mating system. I wanted to deduce how the conflict between parents and offspring would be resolved. It seems obvious enough now, but the way to solve this actually came to me in the middle of the night. The ESS for the offspring can be derived in terms of the parent's strategy, and the ESS for the parent in terms of the offspring's strategy; plugging the two together generates the ESS pair of strategies. I simply had to get up to write this down and solve it; by morning I might have forgotten everything. Sue asked me what was so important as to justify getting up for a couple of hours in the night, and I eagerly explained my finding – a parent should give more to an offspring than is ideal, yet the offspring should still protest for more. Her disdain was unconcealed: as a father, didn't I realise this without having to do mathematics during the night?

I have elaborated on some of my works from the 1970s because they have probably been my major contributions during the development of behavioural ecology. This time of the 'behavioural ecology revolution' was remarkably exciting, especially for someone motivated by natural history (for accounts of the history of behavioural ecology, see Birkhead & Monaghan 2009; Owens 2006; Parker 2006), and occurred against a backdrop of bitter controversy (Segerstråle 2000). My heavy teaching load was in persistent conflict with my research and I had to abandon many ideas through lack of time; many scientists must have encountered such constraints during their most creative period in life. Working at home was not an option once my children were born, and working in the department was equally difficult. The only way I could make progress was to go to the library of the Liverpool Veterinary Station at Leahurst, two miles from home. In the mid-1970s I bought a 1929 Morris Cowley, and used it as a second car to drive to this sanctuary. I should have walked or cycled; the Morris remains in my garage, covered in dust.

The late 1970s generated an invitation to spend a year in King's College, Cambridge. Their Research Centre was, and still is, an amazing venture – it draws together researchers in a given area to enable discussion and catalysis. So in September 1978, Sue, the children and I left the Wirral for the excitement of a year in Cambridge. King's paid for my replacement, who collected my heavy teaching load, but my salary continued from Liverpool. The arrangement seemed to suit everyone, and even my talented research student of the time, Hrefna Sigurjónsdóttir, working on dung flies, appeared to cope well with the situation.

We lived in Grantchester in a bungalow owned by King's College, an easy cycle ride away from King's. The year was a quite magical one. I had been to Oxford and Cambridge to give talks, but had no experience of living there, and was initially sceptical about the college system, which seemed unreal and privileged compared with life in a red brick university. In a sense it was, but I soon began to appreciate the immense benefits of college life; at lunch one could chat to mathematicians, philosophers, musicians, or whosoever; the diversity was fascinating. I loved it, and the stimulus of the Research Centre's behavioural ecologists was immense: I coincided with Tim Clutton-Brock, Dan Rubinstein, Brian Bertram and Robin Dunbar, and Pat Bateson had a role in running the Centre with Donald Parry. Despite this, and although the year was wonderful in terms of generating ideas and thought, I was disappointed with what I managed to achieve in terms of scientific papers.

I completed some ongoing projects, and collaborated on a couple of projects, but spent most of my time on a manuscript that never came to fruition. I had signed a contract with Wiley to write a book entitled *The Evolution of Sexual Strategy*; the idea was to start with the evolution of sex (recombination) and show that, given sex, anisogamy would usually evolve and generate a unity sex ratio. This in turn would lead to sexual selection and all its consequences: selection for recombination generates a series of evolutionary steps that follow from each other, generating the diversity of sexual strategies we now see. I wrote about two thirds of the book and eventually abandoned it – a rather stupid thing to do, but at least I cannot be accused of committing the Concorde fallacy (Dawkins & Carlisle 1976). One reason was that John Maynard Smith produced his books on evolution of sex (1978) and evolutionary game theory (1982); my efforts would have looked pretty insignificant alongside his.

Life in Cambridge had an almost dream-like quality, at least in retrospect. There were elegant college feasts, after which the cycle ride home to Grantchester seemed akin to floating effortlessly through the air, no doubt a tribute to the quality of the wine. Sue enjoyed Cambridge; my son Alan was too young to really care, and my daughter settled well into the little school in Grantchester. Towards the end of the time in King's, there seemed to be the possibility that I could stay on a further year. This appealed to me considerably, but I received a less than enthusiastic response from my Head of Department in Liverpool (by then the late C. J. Duncan), and so I thought it time to return to reality. Nevertheless, I look back on that year with immense affection for King's and the Research Centre; it was a wonderful experience.

I attended rather few conferences in the 1970s. My first research talks were on dung flies, in 1973 (Edinburgh) and 1974 (Oxford). It is hard to believe that anyone could get to around thirty without giving a research seminar – unthinkable these days. Oxford was difficult: at that time they asked questions during the lecture, which meant jumping to parts of the talk in unprepared order. But it went well, and I have greatly valued my relations with Oxford Zoology ever since. A year or so later, I received letters of encouragement to apply for a post there. I was astonished and felt unworthy; Oxford was a sort of intellectual Mecca for me. I confess that I said nothing to Arthur Cain, but asked Phillip Sheppard to act as a referee; he was entirely positive and encouraged me to apply. I prepared the required ten copies of application, and the night before the last date for posting sat with them on my knee in front of the fire at home, in emotional turmoil. In the end, I realised I could not go ahead for a number of reasons; one was that I simply didn't really think I was in the Oxford league. Over the years, similar situations have arisen, and I never followed any of them up; somehow I have never been able to leave Liverpool. I may have done the wrong thing, but have few regrets.

My first conference outside Britain was in winter 1978, while I was at Cambridge, on Evolution and the Theory of Games, at the University of Bielefeld, West Germany, where the game theorist Reinhart Selten was present. With this, and Cambridge, I felt I had walked into a different world. We returned from Cambridge in August 1979, and I soon resumed normal academic duties. A year later, I was promoted to Reader, but the dream had ended and reality had returned.

The 1980s: life after Cambridge

Return to Liverpool was a new chapter. Though family and science continued to be my main preoccupations, life began to include jazz and exhibition poultry. Strange bedfellows – indeed, non-overlapping sets – both have nevertheless been great fun. I had played the clarinet (badly) since school days, and occasionally played with a few friends in Mechanical Engineering in Liverpool. Around 1980, as I entered the local barber's, he apologetically turned off a tape of his own band playing Dixieland jazz. I protested and asked to hear more, and the resulting conversation meant that within some weeks I was playing with a local band. By the mid to late 1980s, I was playing regularly and increasingly. I have had more than just musical fun out of this hobby; it has been a wonderful antidote to a day's algebra, and has generated much camaraderie, and many hilarious episodes that cannot be recounted here. While I listen to classical music and jazz about equally, and have huge admiration for classical musicians, I would not wish to play classical music – for me the fun is being unconstrained by a written score, free to improvise and weave around the melody as the mood dictates. At various times I have dabbled with the piano, guitar, trombone, and trumpet for shorter or longer periods of time, but the only other instrument I play in bands is the tenor sax. How I'd love to play the clarinet as well as Benny Goodman and the tenor as well as Scott Hamilton...

The purebred poultry began because I imagined my children would benefit from having hens – a gesture to my own childhood. In 1980 I bought chickens from a work colleague; within three weeks the children were bored and I was obsessed. After dalliance with large fowl, I moved to exhibition bantams and found myself exhibiting Plymouth Rocks, and later Wyandottes. There followed judging engagements, two periods of service on the Council of the Poultry Club of Great Britain, and its Presidency, 2003–2006. I probably should have given up the hobby when I won Supreme Champion at the Poultry Club National Show (the poultry equivalent of Crufts, the major UK dog show) in 1997, but after decades it still persists, and I don't quite know why.

In the early 1980s, both distractions were relatively small operations; life was busy with academic duties, science and children. My rather guilty escapes, at weekends or on days when I was free of teaching, to work at the Veterinary Station library some 2–3 miles away were vital in order to work on theoretical problems, and to write up papers. Money was scarce, Sue having given up teaching to care for Claire and Alan, and the house ate up all our spare resources. We nevertheless began to take annual family holidays in the northern Lake District, something that persisted through the 1980s, and which I can still find extremely enjoyable.

Invitations to conferences abroad started to occur with a dramatic increase in frequency. Early in 1980, John Krebs, Nick Davies and I featured at a Behavioural Ecology Workshop for Stockholm University, held at their Tovetorp field station. I had better not recount the hilarious episodes – John Krebs is now a peer of the realm. After a decade of only modest interest, sperm competition as a subject area suddenly took off; a leading proponent, Bob Smith (University of Arizona) invited me to give a keynote address at a joint meeting of American Society of Naturalists/Society for Study of Evolution, in Tucson, Arizona, in summer 1980. So I eventually made my first trip to the States five years later than it should have been, landing in Phoenix to face a temperature of 120° F. This and the landscape were something totally outside my experience; it was so different from the verdant English countryside that I might have been on another planet. I loved every minute of my stay; my hosts Bob and his delightful, beautiful wife Jill were unbelievably kind and showed me some of the Arizona landscape and its natural history (Bob is an excellent field biologist).

Bob and Jill hosted a party for several conference delegates (Figure 16.5); one was the great Sewall Wright, then over 90, who had given a special evening lecture, introduced by G. G. Simpson (approaching 80). Chatting with Wright in that warm Arizona evening will always remain one of my most cherished memories. Later, they drove me to Salt Lake City,

Figure 16.5. Chatting to Don Dewsbury (left) at the party at Bob and Jill Smith's house in Tucson after the 1980 meeting.

where I would spend a few days with Ric Charnov before I flew home. On the journey, we slept under the stars in the wilderness, and stopping en route at Jill's parents in the delightful little Southern Utah town of Escalante. It was all an enchanting experience, and I have been a devout Americophile ever since.

In 1981, I attended a Dahlem conference on 'Animal Mind – Human Mind'; I mention it because I gave my worst ever presentation there. The conference contained a mix of psychologists, philosophers, evolutionary biologists, etc., and was mostly outside my area of interest. Sociobiological speculation about human behaviour had been highly contentious in the 1970s and I had serious worries that good science was in danger of being ostracised as well as bad; I could also appreciate some of the objections to intemperate speculations about our nature. At that time, I was preoccupied with the problem of why sperm should remain small under internal fertilisation (Parker 1982a) – I was repeatedly distracted from the discussions by ideas and calculations related to that work. I found myself giving a spec-ulative presentation on the evolution of motivation mechanisms. Although I had thought previously about the problem, I was unfocussed and inarticulate; somehow I had unthink-ingly constructed a three-dimensional graph with all three axes labelled x. The audience was bewildered. I suddenly felt appalled, apologised at the confusion I had created, declared that I should spare everyone further hassle, and stopped immediately – I walked back to my place amid stunned silence. Later, John Crook, the pioneering behavioural ecologist, remarked jovially that everyone had been in my position at some stage, but that I was the only person he had known do the honourable thing and quit – "most of us just carry on, despite realising we have completely lost everyone..."

At that time I was preoccupied with contest models, and did much work that was never published. I worked on an asymmetric war of attrition model in which contestants could make mistakes about their 'role' – roles were related to an index of self's resource-holding power and resource value, relative to the same index for the opponent. I ran simulations, and found a rule: a truncated distribution of contest duration choices, one for each role, and without a gap between the distributions. I then calculated the ESS analytically, based on this rule. Around that time I was fortunate to see quite a lot of the game theorist Peter Hammerstein, who read my manuscript. As a mathematician, he was horrified; one simply could not do simulations to *deduce* an ESS, and then derive the ESS distribution algebraically. He began a proper analysis, which proved to be highly complex. I was gratified, however, that the basic ESS equation remained unchanged (Hammerstein & Parker 1982), and gave a formal validation of the assessment rule I had proposed in my early contest paper (Parker 1974b). I nevertheless suppressed a second manuscript on a related problem, using the same principle.

In 1983, two young American behavioural ecologists, Doug Mock and Trish Schwagmeyer, called in to see me on their way to get married in Edinburgh. They were interested in optimality theory, and later returned to work for several months in Liverpool. Thus began an enduring and highly stimulating collaboration – initially, with Doug on intra-familial conflict (Mock & Parker 1986), and with Trish on the mating system of thirteen-lined ground squirrels (Schwagmeyer & Parker 1987). Much more was to follow; Doug and Trish remain very treasured friends and collaborators.

I became preoccupied with strategies related to phenotype (Parker 1982b, 1983b), and in particular, ESS rules for alternative mating strategies, and the formally equivalent problem of how individuals of differing competitive ability should become distributed among habitat patches (Sutherland & Parker 1985; Parker & Sutherland 1986). The latter project started following the suggestion of Bill Sutherland, who was replacing me during my tenure of a one-year Nuffield Research Fellowship, held in Liverpool, starting October 1982. I much enjoyed collaborating with Bill, now an eminent conservation biologist, and felt that the work was fundamentally useful. This, and my earlier phenotype-limited strategy work, with that of Charnov (1979), and West-Eberhard (1979), forms the basis of what is now termed status-dependent switch (SDS) theory (Gross 1996).

While flying to University of Utah in 1983, I had investigated pilot models of complex life cycles in helminths (something that came to fruition some 20 years later; Parker *et al.* 2003), but apart from this, modelling non-ESS problems was something of a novelty. I began work on clutch and egg size problems. Around 1986, Leigh Simmons, then completing his Ph.D. on crickets in Nottingham, came to see me with a view to working in Liverpool. We wrote a grant application (my first) on copula duration in dung flies, naming Leigh as postdoctoral research associate. It was successful, but by then he had gained an independent fellowship to work in Liverpool, and his place was taken by the late Paul Ward, who had just completed his Ph.D. with me. After two years, Paul moved to a prestigious post in Zürich University, and Leigh completed the dung fly project after his fellowship, and fortunately remained in Liverpool for many years. Leigh and Paul marked the beginnings of the research group that was to form so important a part of my life in the 1990s; both became internationally prominent researchers.

Conference invitations became frequent: I will mention just two. The 1986 Dahlem Conference on sexual selection (the 'Porno-Dahlem'), was notable both for its science and for a party in John Maynard Smith's room on the final night. Paul Harvey, who has an advanced sense of humour, had set all this up; the only person not to know about it was John himself. So after a lavish party funded by the conference, we drank John's mini-bar dry, not realising that he must have been feeling rather less than the beneficent host (he didn't show it). The same year, I gave a featured lecture at the first International Conference for Behavioural Ecology, in Albany, the important meeting that marked the start of the now flourishing society (ISBE). I met Doug and Trish in New York, and we stayed a night or so at Doug's brother's house in Long Island before taking the train to Albany from Grand Central Station. Unfortunately, the quality of my presentation at the conference was inversely proportional to my delight at being there; I felt so embarrassed that I left early. Giving conference presentations is like playing jazz solos – they vary. That one was probably two standard deviations below the mean.

The mid 1980s brought a sudden, awful blow: Sue was diagnosed with breast cancer. With our children aged 11 and 7, we tried to face the future as best we could. After surgery and radiotherapy, and angst and uncertainty, life gradually settled back to normality, but with the ever-persistent fear that the cancer might return. Life continued, and was interspersed with pleasant episodes (Figure 16.6) – but I realised that the cocoon of security, so much a part of my earlier life, had gone forever.

Figure 16.6. With my brother, Dr Antony John Parker (left) and my father, Dr Alan Parker (centre), on our way to a Reunion Dinner of the Warrington & District Old Friends' Association (March 1985). My father was Chairman that year; we attended as his guests.

Figure 16.7. Signing the Charter Book on Admission Day at the Royal Society, June 1989. Dr. P. T. Warren (left) and Professor B. K. Follett, Biological Secretary of the Royal Society (background right). Photo by kind permission of Terry Chambers, Photographer, London.

In 1988, biology at Liverpool underwent a major upheaval. Zoology, marine biology and half of botany merged, at least in name, into a new Department of Environmental and Evolutionary Biology; plans were made for people to move into different locations at a later date. Arthur Cain retired, and after around a century, the old Zoology Department was no more.

Later that year I was proposed as a candidate for election to the Fellowship of the Royal Society. I had no expectations of such elevation, and certainly did not anticipate success. Remarkably, I was elected in my first year, and was formally admitted in 1989 (Figure 16.7) at the same time as Arthur Cain, whose election was judged by many, myself included, to be long overdue. Liverpool had been extremely fortunate, and my stunned amazement at election to FRS was probably matched only by that of the University, which immediately promoted me to Professor. Some four years earlier my Head of Department had proposed me for a Personal Chair; the proposal had been unsuccessful and I was reconciled to remaining as Reader for the rest of my career.

FRS election generated an immense high, reminiscent of my degree result in 1965, but much more dramatic. Once again I dreamed that it would all be revealed to be a mistake. During the period of official secrecy, I gave the news in confidence to my family. My father was especially pleased; his more errant offspring had, after all, distinguished himself. His gratification was all too tragically short; only a few weeks later, on a sunny Sunday morning in May 1989, he suffered a heart attack, and he was already dead by the time my mother found him some minutes later. My brother phoned – he was already with my mother – just as I was about to leave for a jazz gig. I made a call to extricate myself from the gig, and sped home to enter the worst tragedy I had faced up to that time.

As time passed, I became thankful that he lived long enough to have the news about the Royal Society.

The 1990s

Election to the Royal Society heralded a decade that was to become my most bittersweet chapter. It began with a wonderful release from academic duties in the form of an SERC (now BBSRC) Senior Fellowship, from 1990 to 1995. Suddenly – no lectures, no administrative chores – just research… utopia. I had proposed two topics in the fellowship application: sperm competition models and the evolution of complex life cycles in parasites. In the Fellowship years, I devoted all my time to sperm competition models; the life cycle work was to be delayed for a further decade. In 1990, Liverpool ecologists and evolutionary biologists merged together into what had been the old Botany building; the Population and Evolutionary Biology Research Group was formed.

My own group (Figure 16.8) quickly grew around sperm competition studies. Leigh Simmons was soon joined by other postdocs: Matt Gage, Nina Wedell, Paula Stockley, and more briefly, Mace Hack, working on contest behaviour, and Markus Frischknecht, on ideal free problems. With additional postgraduates such as Tom Tregenza and Penny Cook, we had a wonderfully stimulating research group. We gained several grants for sperm competition studies, and both theoretical and empirical papers resulted. In addition to sperm competition (both theoretical modelling, and empirical work on dung flies) I became preoccupied by stimulating collaborations with Tim Clutton-Brock (on potential reproductive rates, and

Figure 16.8. Lab photo with my research group, mid-1990s. Back, left to right: Drs Nina Wedell, Mace Hack, Matthew Gage, Tom Tregenza, Leigh Simmons, Paula Stockley.

later on punishment and coercion), with Fritz Vollrath on sex ratios in spiders, with Charles Godfray on parent–offspring conflict, and several others. For me the first half of the 1990s was a sort of academic utopia – with the research group and freedom from departmental duties, it was a fantasy come true.

However, if academic life reached a peak at this time, my personal life certainly hit its worst trough. The first blow was my mother's death from lung cancer in 1992 after several months of increasingly terrible illness; she had never smoked. She had continued courageously after my father's death, insisting on remaining independent. But when her condition made this impossible she spent her final weeks with my brother and his wife, who bore the horror of her decline with unfaltering love and devotion. My wife Sue found it very difficult. She could see her own fate unfolding through my mother's; she found it harder and harder to face.

Sue had been for regular check-ups after her surgery in 1985, and although there had been a few scares, all had gone relatively well. I had been lulled into a false sense of security that Sue never shared. She was right; in early 1994 she began to experience a continual cough, but X-rays indicated nothing obviously wrong. The cough persisted, and after we celebrated our fiftieth birthdays that May, a scan revealed metastases in her lungs. The consultant was reassuring... there were drugs that could well delay things for years... We attended Claire's graduation and enjoyed a delightful family holiday in Jersey that summer, but Sue's health declined rapidly. She became easily exhausted, and experienced waves of nausea. Soon after Claire's twenty-first birthday, Sue was admitted to hospital, and died with myself at her bedside only a couple of days later. Her final decline was mercifully swift compared with my mother's; only one week before we had shopped together for Claire's present.

A state of adrenalin-induced shock pushed Claire, Alan and myself through the days to the funeral, after which we left for a brief holiday in the Cotswolds, a place we had never visited before. We walked and walked, in mutual, indescribable misery, through the old English villages, amid the stone cottages and lanes, beside the rivers and into the woodlands, and in our grief experienced a closeness that words cannot describe. A year later the three of us returned; by then, life was much changed for us all, but I believe we relived the closeness. Looking back, I found it a comfort to be an atheist, and hence unembittered by such questions as why a deity had ordained the tragedy (see Ruse 1999). The deaths of my father, mother, and wife, all within a period of five years, were very hard to bear.

In 1995, I felt very honoured to give the *Niko Tinbergen Lecture*, the Annual Distinguished Lecture of the Association for the Study of Animal Behaviour (ASAB); I have so much enjoyed being a member this society. The same year, the newly restructured biology departments at Liverpool, of which one was Environmental and Evolutionary Biology, were merged into a single School of Biological Sciences, and in 1996, I was given the Derby Chair of Zoology, the Chair that had been frozen after Arthur Cain's retirement in 1988. Receiving this named Chair meant a great deal to me.

By the late 1990s, postcopulatory sexual selection and sperm competition had become established as a discipline in behavioural ecology. Bob Smith had edited a text following the 1980 Tucson meeting (Smith 1984); in the 1990s, further texts appeared. Tim Birkhead and Harry Moore began a series of *Biology of Spermatozoa* conferences, which continue to be wonderfully enjoyable and stimulating workshop-style events. Most of my time was spent modelling sperm competition games, initially on my own (Parker 1990a, b), but in the mid-1990s I began a rewarding long-term collaboration with Mike Ball, then Head of Applied Mathematics in Liverpool. Although my mathematical skills are much improved since the early 1970s, Mike can cope with mathematics that I cannot; it has been great to work with him on more advanced models.

Mike and I usually analysed general models on sperm competition, though these often related to empirical problems, such as fertilisation in externally fertilising species; Paula Stockley and Matt Gage were working on fish. We devised models to complement the empirical interests of the research group, and to relate their observations to specific predictions. Our group had research grants on insect and fish sperm competition, and the insect work involved dung flies. In particular, we wished to examine whether an observed negative relation between copula duration and male size had an adaptive basis. Leigh Simmons and I had produced evidence that it may be, for all but small males (Parker & Simmons 1994), based on the assumption that sperm transfer was direct – i.e. the male transfers each unit of sperm directly into the female's sperm stores. Later however, evidence from Paul Ward's group in Zürich was indicating that direct sperm transfer was improbable; sperm were deposited by the male into the female's bursa – only then were the sperm transferred by movements of the female tract to the sperm stores (Hosken & Ward 2000). We remodelled the system for this indirect form of sperm displacement. Fate smiled on us – we obtained a better fit than with the original direct displacement model (Simmons *et al.* 1999), with no mismatch for small males.

Throughout the 1990s, I had maintained my interest in intra-familial conflict, usually through collaborations with Doug Mock and Charles Godfray. Doug and I planned a monograph and worked on it sporadically throughout the 1990s; it was eventually completed much later than we had hoped (Mock & Parker 1997). On the same theme, Ian Hartley and I gained a research grant to support a postdoc, Nick Royle, to work on parental care and intra-familial conflicts in zebra finch. It had been Nick's idea to apply for the grant, and he did the empirical work in Lancaster; I assisted with models. Work with the evolutionary biologist Linda Partridge allowed me to extend an aspect of the 1979 conflict sexual conflict paper (Parker & Partridge 1998), and Ric Charnov became interested in the dung fly copula duration problem. The 1990s thus proved to be a remarkably exciting research time. Though my work may not have been as original as in the 1970s, the 1990s was certainly my most productive decade, and the sperm competition research group and all the other collaborations made it the most stimulating.

In late 1997 I married Carol Emmett, and we successfully and happily merged our two families. Carol's daughter, Maxine, is a few years younger than my own children, who by then were more or less independent. Our meeting is a remarkable and entertaining story of a secret, double blind date, arranged covertly by friends to pre-empt our certain reluctance. I had not expected to remarry, and I still see myself as having been immensely lucky. Due to Carol's selfless altruism, what many would see as the inevitable sociobiological difficulties of such a merger never became manifest, and close family life prevails.

After 2000

The millennium began with the tragic death of Bill Hamilton. This, and the death of John Maynard Smith four years later, marked the end of the golden era for evolutionary biology and behavioural ecology. For some years, we had all known that John was terminally ill. I had written to him the Christmas before his death suggesting that I take a trip out to Brighton to visit him while in London – he had written back cheerily and I expected we would meet at least one last time. When I rang to make the arrangements, his secretary informed me that he had died the day before. I was deeply saddened that the person I had come to respect most in science had gone forever.

Nevertheless, the decade has generated some wonderful family events; these have become even more precious after the dreadful, sudden loss of my nephew. My daughter Claire's wedding, and the birth of her two little sons, have given me immense pleasure; they live only 1.5 miles away, and I am able to see them often. My son Alan completed his Ph.D., and spent a year travelling the world with his partner before beginning postdoctoral research in London. He now has a research fellowship in Glasgow University; he and his partner now have a baby son whom I see all too infrequently. My stepdaughter Maxine gained a Ph.D. at Bristol, and is now a postdoctoral researcher in Liverpool. The hobbies still flourish (Figures 16.9 and 16.10); I have some very decent Wyandotte bantams these days, and currently play regularly in no fewer than three local jazz bands.

Figure 16.9. Playing clarinet, Jazz Ragg Dixieland Jazz Band, in the Royal Oak, Betws y Coed, where we have played (and continue to play) each week since February 1995. Left to right: myself (clarinet), Stan Williams (trumpet), John Restall (base guitar), Morris Jones (trombone). Out of view: Malcolm Hogarth (piano), Brian Dawson (drums). Around 2002.

Figure 16.10. Judging bantams at Hinckley and District Poultry Club Show, January 2004.

The era around the close of the past millennium marked some big changes for me in Liverpool. The sperm competition research group dwindled to its natural end with the departure of the postdocs: Leigh Simmons moved to the University of Western Australia, Matt Gage to the University of East Anglia, Nina Wedell and Tom Tregenza to Leeds and later to Cornwall Campus, Exeter, and Paula Stockley was appointed to an academic post in Veterinary Science at Liverpool. A year or two later, all staff in biology at Liverpool were relocated to a new building fused alongside a modification of the old Life Sciences Building, which itself had been one of many locations around the campus housing biological groups. This has been a major advance for biology at Liverpool.

Mike Ball and I continued work on sperm competition models by refining the concept of roles (Ball & Parker 2000) and developing a model that involves both sperm competition between males and sperm selection by females (Ball & Parker 2003). However, my work underwent a major shift in focus, to the evolution of complex life cycles in helminth parasites. This work was long overdue since it had begun almost 20 years earlier, on the flight to Salt Lake City in 1983, and had formed an equal part of my SERC Senior Fellowship application in 1990. I had begun collaborations earlier with my academic colleague Jimmy Chubb, a parasitologist who joined the staff of Liverpool Zoology Department almost a decade before I did. We gained a grant to support a postdoc, and began to develop ideas. Our first manuscript, a development of my pilot models from the Utah trip, was rejected by *Nature* on the grounds that the mathematical treatment was too simplistic. To do a more rigorous analysis was difficult – we recruited Mike Ball. Although the new analysis did not greatly change the conditions for incorporation of additional hosts, it was a much better treatment and was accepted (Parker *et al.* 2003). Jimmy and I also began an exciting collaboration with Manfred Milinski at the Max Plank Institute in Plön, the aim being to produce models that can be tested empirically by scientists in his group. I have known Manfred since the 1970s and I had greatly enjoyed our collaborations on various topics while he was at the University of Berne.

Some years ago, an exciting collaboration involved models of the evolution of anisogamy with the eminent Oxford mathematician and biometrician Michael Bulmer (Bulmer & Parker 2002), who by then had retired from his Professorship at Rutgers. This had been stimulated by some, in my view unjustified (see Bulmer et al. 2002), criticisms of the early anisogamy model (Parker *et al.* 1972) by Randerson & Hurst (2001). Once again, I was involved with someone in an entirely different mathematical league from myself; the credit for the mathematical analysis is entirely Michael's. We were able to show that anisogamy should evolve from isogamy as the function relating zygote survival to zygote size pulls away from the function relating gamete survival to gamete size. Such a change would be expected during the evolution of multicellularity.

Though collaborative work on complex life cycles occupies most of my time, theoretical modelling of parental care and intra-familial conflict continues with Doug Mock and Trish Schwagmeyer. From time to time I revisit various other topics, including sperm competition games, and empirical research on sperm competition in dung flies is even now supported at Liverpool. Amazingly, I still find research tremendous fun and highly absorbing. These days, I spend most of my time in my office at home, and when after a few hours I cannot bear the computer any longer, I walk up the garden and admire my bantams, or have a coffee and play the saxophone for ten minutes.

Final reflections

Although sperm competition and adaptations to postcopulatory sexual selection are probably the topics I am best known for, I like to hope that my contributions on evolutionary conflicts, the evolution of anisogamy, animal distributions, and phenotypic strategies, have also been significant. I have been amazingly lucky, as a young biologist fascinated by natural history, to have had the remarkable good fortune to have coincided with the explosion of behavioural ecology as a discipline (Parker 2001, 2006, 2007). I have been very fortunate since 2000 to receive a number of medals and other awards for a distinguished career, and though I never feel I deserve such honours and that doing science is enough reward in itself, this has been most gratifying.

Of course, over the years, there have been a few regrets and disappointments. A lesson I learned in 1970 was that an idea is not much use once someone has already published it. I had produced a manuscript, intended for *American Naturalist*, on what I termed 'perception advertisement' – the idea being that when a prey individual sees an approaching predator, the chances that the predator will be successful suddenly reduce, depending on the distance between predator and prey at the moment of perception. If at that moment, the prey ceases to become profitable for the predator, the predator should instantly quit. Thus signals related to escape (e.g. the white tail flash of rabbits) should become specialised over evolutionary time to signal that the prey has perceived the predator. While discussing these ideas, a Liverpool postgradute drew my attention to Smythe's (1970) 'pursuit invitation' paper, which had just appeared, though I had not seen it. His concept was so similar that I decided to suppress my paper, and instead used some of it later in the bird coloration work with Robin Baker (Baker & Parker 1979).

There have been many similar occurrences since then; another was a theory of ageing similar to that of Kirkwood (1977). Such is the lot of a theorist. Perhaps more importantly, I regret not having made more serious attempts to increase my expertise in certain areas, particularly mathematics and molecular biology, and possibly I should have devoted less time to hobbies other than science (Parker 2007). I also much envy John Maynard Smith's gift for naming phenomena: for instance, he would never have called a model "the opponent-independent costs game" (Parker 1979)! But fortunately, I don't have too many regrets; life is short and to be enjoyed whenever possible. I certainly *don't* regret changing from medicine to zoology in my first month as an undergraduate.

It has been immensely exciting to have worked in the same era as all the great field biologists in this volume, and such giants of theory as Bill Hamilton, John Maynard Smith, Bob Trivers, Ric Charnov, and George Williams. To have played a small part in the conceptual development of some areas in the overlapping fields of behavioural and evolutionary ecology, where natural history becomes science (Parker 2006), has indeed been a great privilege. Theoretical work in these areas is maybe less prominent than it was – whether this is because much has been done already, or because fashions have changed, it is hard to say. But questions form the very heart of science: without the right questions we can make no progress. A new and fundamental question can lead to endless breakthroughs; it represents the foundation upon which each field in science is built (Parker 2009).

When I think of that exciting era between 1965 and 1980, encompassing my first marriage, the birth and early childhood of my children, the behavioural ecology revolution and the excitement of my early research, I experience strong emotions that I cannot easily describe. Most lives owe their definition to certain times and eras, and the 1970s was my time, my nirvana. My feelings for that wonderful epoch undoubtedly reflect the routine, nostalgic indulgences of an elderly man approaching retirement, but I would at least like to hope they have a more solid foundation. That is for others to judge. All I can say is that they are akin to the sharply poignant, indefinable mix of joy and sadness I feel when walking alone with my thoughts in the English countryside during the hour before dusk, on a still, sunny evening in late summer.

References

Baker, R. R. (1978). *The Evolutionary Ecology of Animal Migration*. New York: Holmes & Meier Publishers Inc.

Baker, R. R. & Bellis, M. A. (1995). *Human Sperm Competition: Copulation, Masturbation and Infidelity*. London: Chapman and Hall.

Baker, R. R. & Parker, G. A. (1979). The evolution of bird coloration. *Phil. Trans. R. Soc. Lond.* B**278**: 63–130.

Ball, M. A. & Parker, G. A. (2000). Sperm competition games: a comparison of loaded raffle models and their biological implications. *J. Theor. Biol.* **206**: 487–506.

Ball, M. A. & Parker, G. A. (2003). Sperm competition games: sperm selection by females. *J. Theor. Biol.* **224**: 27–42.

Birkhead, T. R. (2000). *Promiscuity: an Evolutionary History of Sperm Competition and Sexual Conflict*. London: Faber & Faber.

Birkhead, T. & Monaghan, P. (2009). Ingenious ideas – the history of behavioural ecology. In *Evolutionary Behavioral Ecology*, ed. D. F. Westneat & C. W. Fox. New York: Oxford University Press.

Bulmer, M. G. & Parker, G. A. (2002). The evolution of anisogamy: a game-theoretic approach. *Proc. R. Soc. Lond.* **B269**: 2381–8.

Bulmer, M. G., Luttikhuizen, P. C. & Parker, G. A. (2002). Survival and anisogamy. *Trends Ecol. Evol.* **17**: 357–8.

Charnov, E. L. (1973). *Optimal Foraging: Some Theoretical Explorations*. Ph.D. thesis, University of Washington.

Charnov, E. L. (1976). Optimal foraging: the marginal value theorem. *Theor. Pop. Biol.* **9**: 129–36.

Charnov, E. L. (1979). Natural selection and sex change in Pandalid shrimp – test of a life history theory. *Am. Nat.* **113**: 715–34.

Clarke, B. C. (2008). Arthur James Cain. *Biog. Mem. Fell. R. Soc. Lond.* **54**: 47–57.

Dawkins, R. & Carlisle, T. R. (1976). Parental investment, mate desertion and a fallacy. *Nature* **161**: 131–3.

Dawkins, R. & Krebs, J. R. (1979). Arms races between and within species. *Proc. R. Soc. Lond.* **B205**: 489–511.

Fisher, R. A. (1930). *The Genetical Theory of Natural Selection*. Oxford: Clarendon Press.

Foster, W. (1967). Co-operation by male protection of ovipositing female in the Diptera. *Nature* **214**: 1035–6.

Fox, C. W. & Westneat, D. F. (2009). Adaptation. In: *Evolutionary Behavioral Ecology*, ed. D. F. Westneat & C. W. Fox. New York: Oxford University Press.

Fretwell, S. D. (1972). *Populations in a Seasonal Environment*. Princeton, NJ: Princeton University Press.

Gross, M. R. (1996). Alternative reproductive tactics: diversity within sexes. *Trends Ecol. Evol.* **11**: 92–8.

Hamilton, W. D. (1967). Extraordinary sex ratios. *Science* **156**: 477–88.

Hammerstein, P. & Parker, G. A. (1982). The asymmetric war of attrition. *J. Theor. Biol.* **96**: 647–82.

Hammerstein, P. & Parker, G. A. (1987). Sexual selection: games between the sexes. In *Sexual Selection: Testing the Alternatives*, ed. J. W. Bradbury & M. Andersson, pp. 119–42. Chichester: John Wiley & Sons Ltd.

Hosken, D. J. & Ward, P. I. (2000). Copula in yellow dung flies (*Scathophaga stercoraria*): investigating sperm competition models by histological observation. *J. Insect Physiol.* **46**: 1355–63.

Kirkwood, T. B. L. (1977). Evolution of aging. *Nature* **270**: 301–4.

Kitcher, P. 1985. *Vaulting Ambition: Sociobiology and the Quest for Human Nature*. Cambridge, MA: MIT Press.

Maynard Smith, J. (1972). *On Evolution*. Edinburgh: Edinburgh University Press.

Maynard Smith, J. (1974). The theory of games and the evolution of animal conflicts. *J. Theor. Biol.* **47**: 209–21.

Maynard Smith, J. (1978). *The Evolution of Sex*. Cambridge: Cambridge University Press.

Maynard Smith, J. (1982). *Evolution and the Theory of Games*. Cambridge: Cambridge University Press.

Maynard Smith, J. & Parker, G. A. (1976). The logic of asymmetric contests. *Anim. Behav.* **24**: 159–75.

Maynard Smith, J. & Price, G. R. (1973). The logic of animal conflicts. *Nature* **246**: 15–18.

Mock, D. W. & Parker, G. A. (1986). Advantages and disadvantages of egret and heron brood reduction. *Evolution* **40**: 459–70.

Mock, D. W. & Parker, G. A. (1997). *The Evolution of Sibling Rivalry*. Oxford: Oxford University Press.

Myers, G. (1990). *Writing Biology: Texts in the Social Construction of Scientific Knowledge*. Wisconsin: University of Wisconsin Press.

Orians, G. H. (1969). On the evolution of mating systems in birds and mammals. *Am. Nat.* **103**: 589–603.

Owens, I. P. F. (2006). Where is behavioural ecology going? *Trends Ecol. Evol.* **21**: 356–61.

Parker, G. A. (1968). The sexual behaviour of the blowfly, *Protophormia terrae-novae* R.-D. *Behaviour* **32**: 291–308.

Parker, G. A. (1970a). The reproductive behaviour and the nature of sexual selection in *Scatophaga stercoraria* L. (Diptera: Scatophagidae). VII. The origin and evolution of the passive phase. *Evolution* **24**: 774–88.

Parker, G. A. (1970b). Sperm competition and its evolutionary consequences in the insects. *Biol. Rev.* **45**: 525–67.

Parker, G. A. (1970c). Sperm competition and its evolutionary effect on copula duration in the fly *Scatophaga stercoraria* L. *J. Insect Physiol.* **16**: 1301–28.

Parker, G. A. (1970d). The reproductive behaviour and the nature of sexual selection in *Scatophaga stercoraria* L. (Diptera: Scatophagidae). II. The fertilization rate and the spatial and temporal relationships of each sex around the site of mating and oviposition. *J. Anim. Ecol.* **39**: 205–28.

Parker, G. A. (1974a). Courtship persistence and female-guarding as male time investment strategies. *Behaviour* **48**: 157–84.

Parker, G. A. (1974b). Assessment strategy and the evolution of fighting behaviour. *J. Theor. Biol.* **47**: 223–43.

Parker, G. A. (1978). Selfish genes, evolutionary games, and the adaptiveness of behaviour. *Nature* **274**: 849–55.

Parker, G. A. (1979). Sexual selection and sexual conflict. In *Sexual Selection and Reproductive Competition in Insects*, ed. M. S. Blum & N. A. Blum, pp. 123–66. New York: Academic Press.

Parker, G. A. (1982a). Why are there so many tiny sperm? Sperm competition and the maintenance of two sexes. *J. Theor. Biol.* **96**: 281–94.

Parker, G. A. (1982b). Phenotype-limited evolutionarily stable strategies. In *Current Problems in Sociobiology*, ed. King's College Sociobiology Group, pp. 173–201. Cambridge: Cambridge University Press.

Parker, G. A. (1983a). Arms races in evolution: an ESS to the opponent-independent costs game. *J. theor. Biol.* **101**: 619–48.

Parker, G. A. (1983b). Mate quality and mating decisions. In *Mate Choice*, ed. P. P. G. Bateson, pp. 141–66. Cambridge: Cambridge University Press.

Parker, G. A. (1990a). Sperm competition games: raffles and roles. *Proc. R. Soc. Lond.* **B242**: 120–6.

Parker, G. A. (1990b). Sperm competition games: sneaks and extra-pair copulations. *Proc. R. Soc. Lond.* B**242**: 127–33.

Parker, G. A. (2001). Golden flies, sunlit meadows: a tribute to the yellow dung fly. In *Model Systems in Behavioural Ecology: Integrating Conceptual, Theoretical, and Empirical Approaches*, ed. L. A. Dugatkin, pp. 3–26. Princeton, NJ: Princeton University Press.

Parker, G. A. (2006). Behavioural ecology: the science of natural history. In *Essays on Animal Behaviour: Celebrating 50 Years of Animal Behaviour*, ed. J. R. Lucas & L. W. Simmons, pp. 23–56. Burlington, MA: Elsevier.

Parker, G. A. (2007). Q & A Geoff A. Parker. *Curr. Biol.* **17**: R111-12.

Parker, G. A. (2009). In celebration of questions, past, present, and future. In *Social Behaviour: Genes, Ecology and Evolution*, ed. T. Szekely, J. Komdeur & A. Moore. Cambridge: Cambridge University Press. In press.

Parker, G. A. & Maynard Smith, J. (1990). Optimality theory in evolutionary biology. *Nature* **348**: 27–33.

Parker, G. A. & Partridge, L. (1998). Sexual conflict and speciation. *Phil. Trans. R. Soc. Lond.* B**353**: 261–74.

Parker, G. A. & Simmons, L. W. (1994). The evolution of phenotypic optima and copula duration in dung flies. *Nature* **370**: 53–6.

Parker, G. A. & Stuart, R. A. (1976). Animal behaviour as a strategy optimizer: evolution of resource assessment strategies and optimal emigration thresholds. *Am. Nat.* **110**: 1055–76.

Parker, G. A. & Sutherland, W. J. (1986). Ideal free distributions when individuals differ in competitive ability: phenotype-limited ideal free models. *Anim. Behav.* **34**: 1222–42.

Parker, G. A., Baker, R. R. & Smith, V. G. F. (1972). The origin and evolution of gamete dimorphism and the male-female phenomenon. *J. Theor. Biol.* **36**: 529–53.

Parker, G. A., Chubb, J. C., Ball, M. A. & Roberts, G. N. (2003). Evolution of complex life-cycles in helminth parasites. *Nature* **425**: 480–4.

Pennycuick, C. J. & Parker, G. A. (1966). Structural limitations on the power output of the pigeon's flight muscles. *J. Exp. Biol.* **45**: 489–98.

Randerson, J. P. & Hurst, L. D. (2001). The uncertain evolution of the sexes. *Trends Ecol. Evol.* **16**: 571–9.

Richards, O. W. (1927). Sexual selection and related problems in the insects. *Biol. Rev.* **2**: 298–364.

Rose, M. R. (1978). Cheating in evolutionary games. *J. Theor. Biol.* **75**: 21–34.

Ruse, M. (1996). *Monad to Man: the Concept of Progress in Evolutionary Biology.* Cambridge, MA: Harvard University Press.

Ruse, M. (1999). *Mystery of Mysteries: is Evolution a Social Construction?* Cambridge, MA: Harvard University Press.

Salt, G. (1978). Howard Everest Hinton. *Biog. Mem. Fell. R. Soc. Lond.* **24**: 151–82.

Schwagmeyer, P. L. & Parker, G. A. (1987). Queuing for mates in thirteen-lined ground squirrels. *Anim. Behav.* **35**: 1015–25.

Segerstråle, U. (2000). *Defenders of the Truth: the Battle for Science in the Sociobiology Debate and Beyond.* Oxford: Oxford University Press.

Shaw, R. F. & Mohler, J. D. (1953). The selective advantage of the sex ratio. *Am. Nat.* **87**: 337–42.

Simmons, L. W., Parker, G. A. & Stockley, P. (1999). Sperm displacement in the yellow dung fly, *Scatophaga stercoraria* : an investigation of male and female processes. *Am. Nat.* **153**: 302–14.

Smith, C. C. & Fretwell, S. D. (1974). The optimal balance between size and number of offspring. *Am. Nat.* **108**: 499–506.

Smith, R. L., ed. (1984). *Sperm Competition and the Evolution of Animal Mating Systems.* London: Academic Press.

Smythe, N. (1970). On the existence of "pursuit invitation" signals in mammals. *Am. Nat.* **104**: 491–4.

Sober, E. (1993). *Philosophy of Biology.* Oxford: Oxford University Press.

Sutherland, W. J. & Parker, G. A. (1985). Distribution of unequal competitors. In *Behavioural Ecology: the Ecological Consequences of Adaptive Behaviour*, ed. R. M. Sibly & R. H. Smith, pp. 255–73. Oxford: Blackwells.

Trivers, R. L. (1971). The evolution of reciprocal altruism. *Q. Rev. Biol.* **46**: 249–64.

Trivers, R. L. (1974). Parent–offspring conflict. *Am. Zool.* **14**: 249–64.

West-Eberhard, M. J. (1979). Sexual selection, social competition and evolution. *Proc. Am. Phil. Soc.* **123**: 222–34.

Williams, G. C. (1966). *Adaptation and Natural Selection.* Princeton, NJ: Princeton University Press.

Wynne-Edwards, V. C. (1962). *Animal Dispersion in Relation to Social Behaviour.* Edinburgh: Oliver & Boyd.

17

An improbable path

MICHAEL J. RYAN

I know who I am and where I am, some idea of how I got here, but no inkling of why. The oldest of 11, the son of a truck driver, I was born in 1953 in the Bronx, born in a place and a time when, as Bruce Springsteen said, "you're brought up to do what your daddy done"; my current station in life thus seems a most improbable outcome.

The purpose of this chapter is not to review my life. This is a scientific autobiography and the task is to review my science and how I came to do it. Of course, this includes the people and the circumstances that shaped what I do and how I do it. For me, this essay was a joy to write because it has helped me explore how I maintain my childlike fascination with basic

Leaders in Animal Behavior: The Second Generation, ed. L. C. Drickamer & D. A. Dewsbury. Published by Cambridge University Press. © Cambridge University Press 2010.

questions of our natural world, and why I am drawn to certain scientific questions. Thus this scientific autobiography is primarily one of interests and ideas.

Asking questions

June 18, 2007, Gamboa, Panama

"Why are the frogs calling so much right now? Why don't the females call? Why are túngara frogs always in puddles? Why don't the red-eyed tree frogs make chucks? Why do they leave their children"? I am being bombarded by these "Why, Daddy?" questions from my daughters Lucy and Emma, 11 and 8, as they are once again enthralled by these little calling machines we call túngara frogs. Although this is their first night in Panama this year, it is hardly their introduction. Their joining me for my research here has become a yearly tradition, but their captivation with nature, in general, and these gnomes of the night, in particular, has not waned, nor have their questions ceased. Their inquiries can bounce around all of Tinbergen's (1963) four questions: causation, survival value, ontogeny, and evolution. The "whys" change to "wows" when they scoop up a foam nest in order to watch the eggs hatch and the tadpoles develop in our apartment. But all of their questions, and mine as well, center around the one larger question Tinbergen proposed in that same paper, "Why do animals behave like they do?" (p. 411).

Starting out

There was no single epiphany sparking my interest in animal behavior, but rather a series of smaller acts of revelation. (Some of this section is taken directly from Berreby (2003) and Ryan (2006b).) I was always interested in nature and in animals. I lived in New York City, in the Bronx, until I was 10. My mother regularly took us to the American Museum of Natural History and my dad often carted us off to the Bronx Zoo; dinosaurs and snakes were the biggest lures at each. I watched Marlin Perkin's *Mutual of Omaha's Wild Kingdom*, and this motivated my friends and me to organize our own "safaris". We would go into the basement of our apartment building and collect empty whisky bottles left there by homeless persons; we ignorantly but affectionately called them hobos. We would become quite familiar with these basements during the air raid drills associated with the Cuban Missile Crisis. We then scavenged the nearby vacant lots hunting grasshoppers. We imprisoned our quarry in the newly acquired 'collection jars', counted and then released them. I am embarrassed to say that it never dawned on us to do a mark–recapture study.

   In the fifth grade my family moved to rural Sussex County in northwestern New Jersey, and I experienced what was akin to "ecological release". We were surrounded by forests, and those forests were inhabited by creatures we never encountered in alleys of the Bronx. My brothers, friends and I almost lived in the forest, spending all day hiking, looking for animals, and sleeping under the stars as we were serenaded by the nocturnal choruses of insects and frogs. Hunting and fishing were big parts of those years. When I first encountered a formal biology course in high school my interests were well primed.

I attended a Catholic high school and had a wonderful biology teacher, a Benedictine monk, Father Patrick Bonner. From there I attended a small state college in Glassboro, New Jersey, to become a high school biology teacher. Reading and criticizing the primary literature in an ecology class taught by Roger Raimist revealed that the scientific process was accessible to mere mortals. I had two other classes that had an important influence on me, herpetology and animal behavior. The professor was Andy Prieto, who became a good friend and who steered me toward graduate school before I had ever even seen a graduate student. While at Glassboro I also joined the 'biology club'. This was shortly after Earth Day was founded, and to many of my friends in this club, biology was synonymous with environmentalism. We spent most of our time outdoors learning the flora and fauna of the serene and somewhat odd environment of the nearby Pine Barrens. I also was one of three students chosen to accompany Prieto on a trip to the Galapagos Islands. It was my first time out of the country, my first time on an airplane since I was two, and my first time in the tropics. That trip left a lasting impression; how could it not?

My senior semester of student teaching high school biology could not have been more rewarding. I loved teaching and I had an excellent rapport with the students. I had the good fortune of being assigned to a high school near the campus, within walking distance of my house and, critically, where discipline was not a problem. This made it more difficult to choose between graduate school and high school teaching. What made it more even more tempting was an offer from my old high school, Pope John XXIII, to teach biology and coach baseball. When I turned down the teaching job they offered me part-time of just coaching baseball. That was even more difficult to walk away from.

Trying out graduate school

I decided to enter a Master's program in graduate school. If I decided that graduate school was not for me, the MS degree would still contribute to my teaching credentials and ensure a slightly higher salary. I entered Rutgers University, Newark (NJ), in 1975 under the mentorship of James Anderson.

I did not receive any financial support when I began. Newark was a commuter's campus and still devastated from the race riots of the 1960s. It was an awful place to live and I had little luck finding affordable housing anywhere near by. So I lived with a friend in the peace and serenity of the forests of Sussex County, which I have always loved. It meant a 40 mile commute to campus along interstate 80, a main thoroughfare into New York City. But I found that if I left my house by 5 AM, eating my breakfast as I drove, I could avoid much of the traffic. Gas prices at that time were much lower, and I have my father's truck driver's genes, so driving was never a challenge.

Together with another graduate student, Clark Keller, I did some contract work cleaning out abandoned houses in Newark to support myself. Soon afterward, however, Anderson received a contract to determine the reptiles and amphibians that should be granted protected status by the State. He hired several of us on this contract and my job was to determine the range and status of the blue spotted salamander, *Ambystoma laterale*, which was known from only one small area in the northwestern part of the state. Any rainy nights that winter and spring were spent 'road running', driving up and down roads to intercept the salamanders as they migrated from the woodlands where they usually reside to the wetlands where they breed.

During that work Jim Anderson and I discovered a salamander previously unknown to the State. *Ambystoma tremblayi* is an all-female gynogenetic species associated with *A. laterale*. These two taxa and another sexual–asexual pair, *A. jeffersonianum – A. platineum*, are sometimes considered a hybrid swarm. Gynogenetic species are usually of hybrid origin and are clonal. But they have an odd "sexual" requirement: they need sperm to trigger embryogenesis. I could not wrap my head around this odd system. Why would these species not become extinct? Why would males of the sexual species waste their time, energy, and sperm on a clonal female? I didn't study these questions in the salamanders, but I pestered my advisor by not letting go of them. It would be 20 years before I did some research on this topic.

When I began at Rutgers it was clear that I wanted to do a thesis that combined herpetology and animal behavior. Lizards seemed the most social of the herps, but Anderson took me to the Great Swamp National Wildlife Refuge, in the area where we found the *Ambystoma* salamanders, to see a chorus of bullfrogs. It was a stunning experience. Large males with bright yellow throats were emitting a near-deafening call that sounded like "jug-a-rum". They were vigorously defending their territories and when another male intruded they would clasp each other face to face and have a wrestle off. Females were smaller and without yellow throats, but a lot of the frogs lingering on the territories that we thought were females were actually mature but younger males adopting an alternative "satellite" mating strategy. I had found my MS thesis topic.

I studied the bullfrogs for two seasons, 1976 and 1977. I received my first grant, from the Theodore Roosevelt Memorial Fund of the American Museum of Natural History, one of the places that nurtured my early interests in biology. Most of the money went to purchasing a more reliable flashlight and a box of batteries; research was simpler then. My main goal was to document the relationship between territoriality, mating strategies and mating success. What really caught my interest, though, was the frog's mating calls.

There are about 5000 species of frogs. Typically, males produce mating calls that females use to identify and evaluate mates and which also serve in male–male interactions. The Modern Synthesis of Evolutionary Biology put a great emphasis on speciation, and the role of mate recognition as a premating isolating mechanism was a prominent contribution of behavior. Frank Blair's research had demonstrated how the species-specific nature of the anuran mating call resulted in reproductive isolation between species, and Robert Capranica's studies had begun to show how the frog's auditory system decoded conspecific calls. All of this being the case, I was impressed with how different the males all sounded from one another. These males are territorial and males tend to be in the same place each night. As I checked the location of the males I had marked with numbered bands I placed around the waists, I realized that I often could identify a male by his voice. I had decided that after my study of territoriality in bullfrogs I would investigate their communication system, and I would specifically ask whether variation in the males' mating calls influenced their attractiveness to females. I was not yet aware of the intriguing notion of sexual selection.

The Zoology Department had an interesting group of faculty for students interested in herpetology. Jim Anderson was a field biologist, ecologist and evolutionary biologist; Dan Wilhoft a physiologist; and Sam McDowell an anatomist and taxonomist who could draw diagrams of anatomy with both hands simultaneously. There was also a new faculty member from Cornell University, Doug Morrison, who brought sociobiology and behavioral ecology to the department and instantiated in me an early fascination with bats that would later reach fruition.

Another great resource for me was the Institute of Animal Behavior (IAB). The Institute had gained a large degree of fame at this time under the leadership of Danny Lehrman, who passed away shortly before I arrived. Lehrman was well known for his studies of the hormonal mechanisms underlying courtship behavior in ring doves, and for his influential critiques of Konrad Lorenz's theories of instinct. At my time Jay Rosenblatt, who studied behavioral development in rats, was the Institute's director, and there were two faculty in field behavior, Monica Impekoven and Colin Beer. The Institute was a different sort of place; it had both the air and the reputation of elitism. There were only a small number of students; all of them received full fellowship support for their entire graduate school career, and they seemed to have little interest in interacting with those in zoology – or so I was warned. The Institute could only be reached through a private elevator to which only IAB members had keys, and there were no classrooms, just a lounge with a kitchen where their informal classes were held.

I took two classes at the Institute, Rosenblatt's course on behavioral development, and Impekoven's course on social behavior. Regretfully, my teaching duties kept me from taking Beer's ethology course, but I was able to use his Kay Sonograph to start quantifying variation in the bullfrog's mating calls. I also attended many of the seminars at the Institute, and spend some time there socially. I was always welcome, and at one point I was, in a sense, invited to the high table – I was given my own elevator key! Many years later, when I returned to the Institute to give a seminar, it truly warmed my heart when Colin Beer introduced me as "one of our own".

Things could not have gone better at Rutgers. I eventually received a TA and I was excited as much about teaching as about research. Jim Anderson and I became very close friends, and when I decided to transfer to the Ph.D. program in my second year he gave me his full and enthusiastic support. That all ended quickly, however. Jim and the students who were working on the endangered species project all attended a town meeting of citizens concerned with planned development in the Great Swamp region. We were hoping that the presence of threatened *Ambystoma* salamanders in the area might cause the State to halt the development. Anderson gave a wonderful presentation of island biogeography theory as an argument against habitat fragmentation. When the meeting ended we all went our separate ways. It was the last time I saw Jim. He died in a car crash on the way home. Ironically, the NY Times reported the next day on conservation easements in California that were being made to allow salamanders to pass under highways during their migrations. They credited Anderson's studies while he was a student at the Museum of Vertebrate Zoology at the University of California, Berkeley, in motivating these improvements.

Anderson's death was devastating to all of us, not just his students but the entire biology community at Rutgers. He was only in his 40s when he died, full of life and energy, and the most respected person in our department. I had never lost someone so personally close to me, and I was at a loss what to do next, refusing even to discuss it for weeks on end. Doug Morrison became my sponsor, my mentor, and also a close friend. He convinced me that I needed to go to Cornell to be at one of the epicenters of the new sociobiology and behavioral ecology.

Starting graduate school, again

After some delay on their part, I was finally accepted at Cornell into the Section of Neurobiology and Behavior (NB&B). My advisor was Kraig Adler, who was studying amphibian orientation. Kraig is a herpetologist and, luckily for me, was willing to take students who used reptiles and amphibians for subjects for a variety of studies. For that I am eternally grateful. Cornell was a wonderful place, and NB&B an exciting and initially intimidating department. In that small faculty of 15 or so, there were three members of the National Academy of Sciences: Tom Eisner, Bill Keeton and Dick O'Brien.

Cornell was a perfect match for my interests. There were two key faculty members whose research complemented my interests in animal communication and mating behavior. Robert Capranica was a neuroethologist with a lab full of grad students and postdocs trying to understand how the frog's brain decoded acoustic signals. Steven Emlen, one of the founding figures of behavioral ecology, is best known for his studies of avian social behavior but he had also just published a seminal paper on lek organization in bullfrogs. Adler, Emlen, Capranica and Bill Brown, who years before with E. O. Wilson published a groundbreaking paper on reproductive character displacement, would come to constitute my doctoral committee. Cornell was a different place than Rutgers and a different species from Glassboro State College.

I had never been around graduate students like those at Cornell. As with most programs, that is where most of your learning takes place. Adler had three other graduate students at the time: Gordon Rodda, who was working on alligator orientation and went on to do conservation biology with US Fish and Wildlife; Bruce Waldman, who initiated an entire research field with his discovery of sibling recognition in tadpoles, a general topic that he continues to study today; and John Phillips, who initiated a quest he still follows, understanding how animals use the earth's magnetic field in orientation. Other graduate students at that time were Eliot Brenowitz and Steve Nowicki, who both study the neurobiology and behavior of bird song, Gary Rose and Harold Zakon, who now both study the neural basis of communication in electric fish, and Pepper Trail and Doug Lank, who were students of bird behavioral ecology. In retrospect, Cornell offered a very competitive but a very cooperative atmosphere. We all knew we were in a special place, at a special time, and surrounded by special people, both faculty and other graduate students. We all wanted to show we deserved to be there but the competition was all directed inward. Students were always ready to help, discuss, and debate research, and most everyone was close.

There were two courses at Cornell that had an immediate and lasting impact on me. Bob Capranica and Ron Hoy taught *Animal Communication*. It was my first experience in having a course taught by some of the researchers who actually defined the field. The emphasis was integrative. It covered the physics of signals, signal production, and the neural decoding of signals as well as a few topics in behavioral ecology. But it was probably most memorable for its emphasis on Fourier analysis and information theory. This course made me aware of numerous research tools in the field of animal communication and for the first time made me aware of how important it was to have an understanding of mechanisms to truly grasp the totality of animal behavior. The animal communication course I teach today is modeled after that one. The other course that first year was *Vertebrate Social Behavior* taught by Steve Emlen. Steve had just published with Lew Oring a seminal paper in *Science* on the evolution of mating systems (Emlen & Oring 1977) and had made a number of other critical contributions to the fledgling field of sociobiology. His enthusiasm for the subject was contagious, he was an engaging lecturer, and, as if any of us needed to be convinced, he hooked us on the excitement of social behavior. There were other important courses for me that year as well. A symposium on animal communication had a series of invited speakers who gave public lectures and then attended a reception and spent a couple of hours in discussion with the students taking this course. Carl Sagan and Jane Goodall were among the stellar line-up of lecturers.

I also took a field course that introduced me to tropical biology. With Ruth Buskirk and Glenn Hausfater, a group of first- and second-year graduate students visited Guanacaste and Monteverde in Costa Rica, and Barro Colorado Island in Panama. Although I did a project on thermoregulation in an iguanid lizard (*Ctenosaura*) for my independent project, my most memorable experience was watching the diurnal mating behavior of the *Atelopus* frogs. Bright green and fearless, these frogs sat atop rocks in small streams in the forests of Monteverde, chirping their hearts out for females. I had never seen a diurnal frog nor one that was so accessible to the researcher; nothing at all like the bullfrogs I had studied. They are no more, however, having been among the early victims of the world-wide amphibian decline. On Barro Colorado Island (BCI) we hiked through the forest and were amazed by the capuchin monkeys who threw sticks at us from the canopy, the morpho butterfiles that glided in front of us but seemingly could never be caught, and the red-capped manakins that flashed colors and clicked their wings as they flitted through their leks. At dusk we were about to hurry down the flight of 200-plus stairs to the dock to board the boat to Frijoles and then the train into Panama City. One of my friends came running from behind the Kodak House. She yelled, "Mike, come here, look, a frog lek!" It was a group of túngara frogs whining and chucking – this was even before they were widely known as túngara frogs. I had no idea what this was to portend (Figure 17.1).

Figure 17.1. A calling male túngara frog, *Physalaemus pustulosus*, with a small *Corethrella* fly. (photo by Santiago Ron).

STRI and túngara frogs

I had a pretty clear idea of what I wanted to study for my research. Robert Trivers' paper on parental investment and sexual selection (Trivers 1972) and Emlen and Oring's paper on the evolution of mating systems were bringing sexual selection into the forefront of socio-biology. Still, at that point there was scant evidence that variation in male display behavior had any influence on female mate choice. Darwin's notion of aesthetic preferences in females received little support from other evolutionary biologists, including his "bulldog" Thomas Huxley. Julian Huxley (1938) and that great herpetologist G. Kingsley Noble (Noble & Bradley 1933) bemoaned the lack of any evidence for female mate choice based on intraspecific variation in display traits. Darwin himself did not seem convinced that one could ever provide evidence of this phenomenon.

But then again, I thought, why did all these bullfrogs sound so different from one another? Variation in signals does not require that there be variation in their meaning, but this signal variation suggested that the males' calls might be perceived by the females as quite different from one another. Bullfrogs, however, did not seem to be the ideal species for such a venture. Their breeding season was too short and they seemed intractable for experimentation. I wanted to work in the tropics where the breeding seasons of many frogs extended throughout the long months of the rainy season, and to work with a smaller frog with whom I could do experiments. I scoured the literature and came across an account of red-eyed tree frogs, *Agalychnis callidryas*, which seemed to exhibit lekking behavior. I submitted an application for a short-term fellowship to the Smithsonian

Figure 17.2. Embarking for a 1990 field trip to the mountains of western Panamá (from left: Debbie Greene, Walt Wilczynski, Stan Rand, Mike Ryan, Kyra Mills, Ulysses [uncertain of last name], and Frederico Bolanos).

Tropical Research Institute in Panama to conduct this study on BCI. This was right before my Cornell trip to the tropics, and before I saw the túngara frogs on BCI.

I received the short-term fellowship. Before traveling to Panama to begin my thesis research I attended the meetings of the Herpetologists' League in Tempe, AZ. I presented a talk about my bullfrog studies, for which I received an award for the best student talk (the first of three I received), and I met Stan Rand, who gave a talk on vocal communication in túngara frogs, *Physalaemus pustulosus*. At that point, Stan had published only one paper on these frogs, on nesting behavior, and Stan was calling them mud puddle frogs (Figure 17.2).

I soon arrived on BCI and it was all I remembered it to be, except magnified: the overpowering green of the forest, the sounds of social interactions of innumerable animals, and the odors of the flowers advertising their goods. All of my senses were assaulted with diversity. It did not take long to find a sizeable population of treefrogs. I chose one in the lab clearing, near the small cement pond behind Kodak House that Stan had built years ago to study túngara frogs and where I had seen my first túngara frog lek. This would allow me to study the behavior of the treefrogs and have access to nearby facilities where I could do experiments. The problem was with the treefrogs. As their name might imply, much of their life is lived in the trees, often above 3 m or more above the ground. The frogs descended to the water to deposit their eggs as pairs already mated, and the act of mate choice was difficult to observe. And these damned cat-eyed snakes kept eating their eggs! (Much later, with no input from me, my graduate student Karen Warkentin made a wonderful career showing how the eggs detect such intrusions and how they fight back.) Red-eyed tree frogs were also a pitiful choice for a study of acoustic communication. Their calls are of fairly low amplitude, and the din of the then bothersome chorusing túngara frogs at my feet made it all but impossible to record the mating calls of the treefrogs.

I decided, with Stan's permission, to switch my study to túngara frogs. Stan had not studied them for many years and had not published his basic work on their calling behavior. At that time he was deeply involved in fascinating studies of iguana and crocodile behavior with Gordon Burghardt.

Stan's earlier unpublished studies provided a critical foundation for my work. He had shown that males produce a call with two components, a whine to which chucks could be added. Males added chucks, up to six of them, in response to calls from other males. Females preferred whines with chucks to whines without chucks.

The studies of the túngara frogs posed other challenges but soon began to proceed fairly smoothly. By summer's end I had data showing that females choose their mates with little interference from males, larger males were more likely to be chosen as mates, and there was a significant negative correlation between male body size and the fundamental frequency of the chuck. This latter effect was known in some frogs, has since been shown in numerous others, and is simply due to larger males having a larger vocal apparatus that vibrates at a lower frequency.

Before leaving Panama at summer's end I talked to Stan about continuing my studies of túngara frogs for my Ph.D. thesis and he was quite supportive. I presented my preliminary data to my Cornell committee and they approved it, and I remember being especially pleased that Capranica and Emlen were enthusiastic. If I could have paused at that point 30 years ago and peered into the future I would have seen that the associations I was to now solidify with STRI, with Stan Rand, and with the túngara frogs would be the center of my scientific life for decades to come.

I spent the next year at Cornell writing grants to return to BCI. I received both a long-term STRI predoctoral fellowship and a dissertation improvement grant from the National Science Foundation. I was on my way.

Túngara frogs, tractability, and technology

Túngara frogs were a near-perfect system for my study of sexual selection and communication. It is a bit easier to quantify male mating success in frogs than in many other animals. Birds were then and still are now the paradigmatic taxon for behavioral ecology. In birds the mating act is a "cloacal kiss". The male mounts the female, their cloacas touch, and males transfer sperm. In some species this takes but a few seconds. In most frogs, the male clasps the female from the top and they usually remain in this state, amplexus, for hours. In túngara frogs, the pair also constructs a foam nest. The process takes about an hour, they usually begin nest construction towards the end of the night when chorusing has begun to subside, and they frequently do so out in the open in conspicuous places.

Túngara frogs are also quite tractable when it comes to marking and monitoring the behavior of males, recording their calls, and conducting preference tests. The frogs are small, about 30 mm in length, and they call in pools of water often in clear view, their calls are loud, and their vocal sac makes them quite conspicuous. After some practice in the dark, it is easy to locate a microphone on a stand within about half a meter of a frog and have it resume calling fairly quickly. Thus recordings of frog calls are very clean; the caller is easily identified compared to others with whom he is chorusing, and the signal:noise ratio is high. The greatest advantage of studying frogs, however, is the ability to test experimentally female preferences for male signals using the phonotaxis paradigm first used by Martof and Thompson (1958). In these experiments, a reproductively primed female is placed between two speakers that are opposite one another, each broadcasting a stimulus antiphonal to the other speaker. The female usually begins the experiment under a cone in which she can hear the calls and is allowed to acclimate. The cone is then lifted and the female approaches one of the speakers. In nature, females only approach a male's call when searching for a mate. Furthermore, frog calls, compared with bird song, are relatively simple and thus can be more easily synthesized by the addition of various sine waves, a type of reverse engineering in which the call is deconstructed into sine waves with a Fourier analysis and then analogous sine waves are synthesized and combined to synthesize the call.

When I returned to Cornell after my first summer on BCI I talked to Capranica about synthesizing calls. I wanted to test the hypothesis that the large-male mating advantage derived from female preferences for lower-pitched calls. He had his research associate Ann Moffat help me. We did the synthesis in a large room filled with sine wave generators, wave shapers, and a variety of other machines I couldn't identify. They were all linked together with cables, and it was hard to fathom at first sight how all these sounds combined through this spider-web-like network could deliver a synthetic whine–chuck call onto the tape recorder. Ann, however, was as expert in explaining call synthesis as she was in doing it. The usefulness of the knowledge I gained was short-lived, however, as the affordable computing age dawned. This was brought home a few years later when I saw Capranica give a talk on animal communication methods. He showed a slide of that same room filled with sound equipment, the walls teeming with wires and blinking oscilloscopes, and a large Kay Sonograph dominating one of the corners of the room. The next slide was of the same room but it was nearly bare. In the center of the room was a small table with an Apple computer. This was the future. All of the signal analysis and synthesis which was the purview of only large and rich labs was now available on a much smaller, cheaper scale to anyone with a few hundred dollars.

The technology, or lack of it, at STRI then is still hard to imagine. STRI had one computer in the entire Institute. It was in Panama City, part of an environmental science monitoring project. STRI generously let me use it for a few hours on Sunday afternoons. There was no software; I had to write programs in BASIC for all the statistics I used. There was no screen, just an LCD panel that displayed one line at a time, so I would have to print out the entire program to debug it. There were no floppy disks; all the data had to be entered on punch cards. So each Sunday morning, I would fill my backpack with punch cards, a BASIC manual, and a statistics book and take the boat across Gatun Lake and wait for the train into Panama City. The train almost never arrived on time, and sometimes didn't arrive at all. But waiting by the lake, in an open field surrounded by tropical forests and everything that came with the forests, made a late train seem like a treat.

Technology has been a blessing in recording signals. At that time, I used a Nagra tape recorder. Even now when I occasionally use one, to digitize some old reel–reel tapes, I am amazed by their incredible precision and smooth operation. But they weigh 17 lbs! I recorded calls of most of the frogs and numerous other animals on BCI those three years; the frog calls constitute much of the CD that accompanies Ibanez *et al.* (1999). More than once in the pursuit of a sloth or monkey vocalization I unintentionally rode down a muddy hill on my back in the rain clutching a Nagra to my chest. Those are rugged machines – more rugged than me, I hate to admit.

Back to BCI

... it is obviously probable that [females] appreciate the beauty of their suitors. It is, however, difficult to obtain direct evidence of their capacity to appreciate beauty.

(Darwin 1871, p. 111).

Female choice and male mating success

I spent the next three years or so on BCI. I continued to monitor female choice and male mating success, this time for most of an entire breeding season. On each of 152 consecutive nights I measured, toe-clipped, and marked with observation tags all new males and females that were present, remarked any frogs that had lost their tags, and determined who mated each night. Over that time I marked and observed 617 males and documented 751 matings. At the same time I began to conduct female preference tests. I tested the hypothesis that the large-male mating advantage, which was even more strongly supported by the second and larger data set, resulted from preference for chucks having lower fundamental frequencies.

In the first series of experiments I gave females a choice between synthetic calls, courtesy of Moffat and Capranica, which had chucks of fundamental frequencies near the two extremes, indicating small and large males. There was a preference for lower calls of larger males, offering strong support for sexual selection by female choice. As far as I knew, this was the first experimental demonstration that female choice could generate sexual selection. I wanted to be sure the manuscript was strong, thus I asked a large number of key people in the field to read it for me: Kraig Adler, Steve Emlen, Carl Gerhardt, Hank Howe, Bert Leigh, Stan Rand, Bob Trivers, Kent Wells, and Mary Jane West-Eberhard. The care was worth the effort. It was accepted by *Science* and published in 1980 (Ryan 1980).

The rest of the time studying túngara frogs concentrated on quantifying variation in mating success during the longer sampling period, conducting more female choice tests to titrate the range over which females discriminate, and doing various experiments, some successful but with many failures, to determine what benefits accrued to females choosing larger males.

Energy costs of calling and the inspiration of George Bartholomew

When túngara frogs add chucks to their whines the calls become more attractive. When they are calling in a chorus pretty much all of them produce chucks, but when a male is by himself he usually only produces a whine. If complex calls are more attractive, why not always make them? I knew there must be a cost, and I began to explore this question.

One advantage of working at STRI is the constant stream of visiting scientists who work at the various field sites and labs. A regular visitor to BCI during my time there in the late 1970s was George Bartholomew. Bart was one of the founders of the field of environmental physiology. He was also one of the most integrative of the biologists I had known, and he defined the terms of gentleman and scholar.

Bart's work on moths had hit some stumbling blocks during his visit. I approached him and his graduate student Terri Bucher and asked whether they would be interested in measuring the energetic costs of calling in túngara frogs, and specifically to test the hypothesis that complex calls incur an additional energetic cost. Stan had planted this bug in Bart's ear previously, suggesting that they measure the cost of visual displays in anolis lizards. We showed that there was a substantial increase in the rate of oxygen consumption, and thus energy expenditure, during calling but it was no more expensive for a male to make a chuck than not. This was the first direct estimate of metabolic cost of a sexual display in a vertebrate. Although we did not pursue this line of research much further, measuring the energetic cost of calling in frogs became a small cottage industry for some time.

Predation costs, **Trachops,** *and Tuttle*

It seemed crucial that we continue our attempts to understand why males do not always make the most attractive call. In 1976 Robert Jaeger published a short note in *Copeia* in which he reported that large marine toads, *Bufo marinus*, were dining on túngara frogs in Kodak Pond on BCI. He suggested that the túngara frogs' calls were luring the toads. Jaeger did not comment on the simple versus complex calls of the túngara frogs, but in light of Peter Marler's paper about the relationship between call structure and function (Marler 1955), Stan and I both thought that chucks made advertisement calls easier to locate – for both females and potential predators. I tried and failed with experiments to show that marine toads were attracted to túngara frog calls. Then an amazing thing happened. Stan Rand received a letter from Merlin Tuttle, a bat researcher. The year before Tuttle had captured a bat, *Trachops cirrhosus*, with a frog in its mouth (Figure 17.3). Merlin was returning to BCI in the dry season (January) to begin a study to determine if the bats were homing in on frog calls. He wanted to know if there was a herpetologist on BCI who might be interested in helping.

Like most frogs, túngara frogs do not breed in the dry season. Instead of returning to Cornell for the dry season, as I had originally planned, I decided to stay and work with Merlin. This was a collaboration made in heaven. Merlin's interest was from the prospect of bat foraging, not surprisingly since I always thought Merlin was part bat anyway. My

Michael J. Ryan

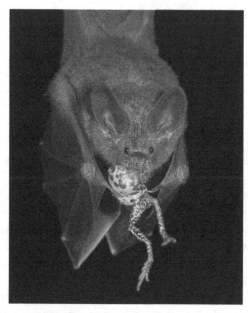

Figure 17.3. A frog-eating bat, *Trachops cirrhosus*, with a túngara frog (photo by Alexander T. Baugh).

interest, of course, was from the frog's point of view, especially the túngara frog's point of view. Merlin is an incredible bat field biologist, and I have rarely met anyone with such a feel for their study animal. Joined with my interest in the frogs, animal communication, and sexual selection, we formed a good team.

In a week or so we had successfully netted some *Trachops* on BCI and had them in a flight cage. Merlin taught me how to net bats and how to train *Trachops* so they would eat from my hand, and introduced me to most of that incredibly diverse chiropteran fauna on BCI. After the *Trachops* became habituated we tried our first experiment. We placed two speakers in opposite corners of the flight cage and played a túngara whine from one side and a whine–chuck from the other. The bat immediately left its perch and landed on the speaker playing the whine–chuck! We were elated. But that was the last bat to respond for weeks! We kept changing details of the experiment and the flight cage but the bats would not leave their perch. Ernst Mayr, who was on BCI at the time, suggested that we might be feeding the bats too much. We are not sure why, but the bats again began to respond and hardly ever stopped. Our first publication, in *Science* in 1981 (Tuttle & Ryan 1981), was on the general phenomenon of bats responding to frog calls and their ability to discriminate between palatable and non-palatable frogs and between small and large ones. We then published a paper in *The American Naturalist* the next year showing that the bats preferred the túngara frogs' complex calls over simple ones (Ryan *et al.* 1982). So it was now clear that when males add chucks to increase their attractiveness to females they do so at the risk of increased predation by what we now refer to as frog-eating bats.

Almost not leaving STRI, and leaving STRI, for a time...

I began to get ready to leave BCI in the summer of 1981after almost three years on the island. That was a heady time to be on BCI. The number of scientists working there was impressive enough, but STRI would often bring other visitors solely for the purpose of enhancing the intellectual climate there. Mary Jane West-Eberhard and Bill Eberhard were doing pioneering work in sexual selection. Although STRI staff they lived in Costa Rica but regularly visited Panama. Ernst Mayr, John Maynard Smith, Amotz Zahavi, and Robert Trivers all visited the island for various lengths of time. As I said, Mayr was interested in our *Trachops* work and offered some good advice on keeping the bats motivated. Maynard Smith wrote his Game Theory book on BCI. I was in charge of the beer machine, so needless to say he and I became quite close. Zahavi was quite clever in how he could explain much of the behavior of túngara frogs in terms of the handicap principle, and he had me chasing one of those ideas with experiments for some time. Trivers was at STRI for a year to write his social evolution book and was always quite encouraging. If these intellectual giants were to visit a university, graduate students might be able to join them for a lunch and ask a few questions. But being on BCI they were a captive audience.

For many who work on BCI the dream is to never leave. That dream almost came to fruition. Ira Rubinoff, the Director at STRI, was away on sabbatical and Mike Robinson, who later became the Director of the National Zoo, was the Acting Director. I received a message that he wanted to see me in his office in the city. I had no idea what it was about. After asking me how I liked it at STRI he offered me a job there. The idea was that I would be the staff scientist on BCI once I received my Ph.D at Cornell. It was a total shock, and even Stan Rand seemed to have no inkling of the job offer. I immediately accepted. But due to a variety of bureaucratic issues and hiring freezes the offer had to be withdrawn. So off I went back to Cornell.

Finishing a thesis and starting a book

While on BCI I received a letter from an editor at the University of Chicago Press, Susan Abrams. She said that she had heard about my work with túngara frogs and suggested that I write a book. I had given this some thought. I was doing some very different kinds of studies with túngara frogs and I thought there might be some advantage to putting all of this work together in one place. Not everyone thought this was a good idea at this stage in my career. Some members on my thesis committee thought that I should be concentrating on primary research and publishing papers, not rehashing research in a book form. Steve Emlen was one of the few who recommended that I go ahead with the project.

Another goal that year was my research on neurophysiology. My plan was to characterize the auditory sensitivity of the túngara frog by recording from the eighth auditory nerve. I had become interested in what I was calling a "neuro-neutral" theory of sexual selection, which I will return to later, and these data would be helpful in exploring that concept. Capranica was on leave that year and my work in his lab just didn't go well. But as I learned years later from

Walt Wilczynski, these frogs are small and it is a real chore to record from their auditory nerves. Even an expert such as Walt retreated from the eighth nerve to the bigger target of the mid-brain to get some insight into what these frogs were hearing.

In addition to writing my thesis and working in Capranica's lab, I needed to find employment for the next year. I had three potential avenues. One was to do a postdoctoral fellowship with Eugene Morton at the National Zoo. Gene had been one of the founders of the field of habitat acoustics in animal communication. He had shown that the songs of many birds evolved to match the acoustic features of the habitat to enhance signal transmission over long distances. I wanted to ask similar questions about frogs. I didn't receive that postdoc; years later Gene told me that as the ornithologist at the zoo he was expected to sponsor students working on birds, not frogs. My second option grew from my interest in a fellowship from the Miller Institute for Basic Research in Science at the University of California, Berkeley. These fellowships were quite prestigious and well salaried and provided the opportunity to be mentored by one of the science faculty at Berkeley. The fellowships also came with great independence and freedom. Dr David Wake is a well known evolutionary biologist at Berkeley who had done classical work on evolution with salamanders, mostly from a morphological and developmental perspective, and he agreed to sponsor my application.

The third avenue for employment was to apply for faculty positions. In January of 1982 (I was 28), I interviewed at the University of Texas in Austin. They were filling five jobs that year and I was the first of many to be interviewed. There was an ice storm while I was there. While I was in Austin, Dave Wake called to let me know that I would be awarded a Miller Fellowship. So at least I had employment for the next two years, and that removed some of the pressure of the interview. I think it also upped my ante there. With the hiring of five new faculty in one year, they seemed to welcome a potential hire who would not come for another two years; they told me if I were hired I could first spend two years at Berkeley. Eventually I was told that I had the job. When I gave that news to George Bartholomew, he told me not to make a decision right yet as I was going to be interviewed at UCLA. I was then offered that job as well. The chance to have Bart as a colleague was a huge temptation, but I thought that the quality of non-academic life for me, given my life style, would be better in Austin than Los Angeles. I might have been right about that choice, but I'll never know. I have been very happy in Austin and in the course of my career, and I have also been honored to have offers from the University of Chicago, the University of California at Davis, and another offer at STRI. All of those institutions are of very high quality and peopled with outstanding researchers in my field. Whether each decision was the best one doesn't really matter; what matters is that life has been good for me in Austin.

As soon as my thesis was defended I started to work on my book, and made substantial progress during the summer. The main purpose of the book was to weave together the various studies on mating systems, communication, and costs of calling, much of this already published in specialty journals and some not yet published. A new idea in the book was what I awkwardly termed the "neuro-neutral model". The idea was fairly simple. Females preferred calls with lower fundamental frequency chucks. The lower the fundamental the closer in frequency the harmonics, thus these calls might squeeze more energy into the frequency window to which the inner ear is sensitive. Greater acoustic stimulation should lead to greater neural stimulation and perhaps increase the attractiveness of the call to females. The general notion was that males are under selection to find ways to increase the sensory stimulation of females. If mutations arise that increase call attractiveness, this could lead to the evolution of more effective displays that might not be related to other components of male fitness, such as overall survivorship. That is, attractive displays need not be reliable indicators of good genes or other attributes of survivorship. The females mating with these males would produce "sexy sons", and it is possible that this process could initiate runaway selection. But it also seemed clear that runaway selection would not be required either. This idea is a derivative of West-Eberhard's sensory trap and was the beginning of the idea I later called "sensory exploitation".

I sent the first draft of my book to Peter Marler and asked him to write a foreword. I didn't know him well, having only talked to him at any length only once when as a grad student I gave a seminar at the Rockefeller University field station in Millbrook. But I considered him the preeminent animal behaviorist of his time. Needless to say, I was thrilled when he agreed. Peter wrote a wonderful essay on how recent changes in the field of animal behavior towards evolutionary biology were critically important, but unfortunately this paradigm shift also led to the abandonment of studies of mechanisms at the field's own peril. He thought the túngara frog system was an example of how one could integrate evolutionary and mechanistic topics.

I finished my book while I was at Berkeley and *The Túngara Frog, A Study in Sexual Selection and Communication* was published by the University of Chicago Press in 1985 (Ryan 1985), right after I arrived at Texas. The elation of having the book published was followed by substantial trepidation waiting for the first review. All scientists receive peer reviews of the manuscripts they submit to journals, but these are private and not published, which is not true of book reviews. That first review appeared in *Science*, and gave high praise: "this is a careful, creative, and informative study that is likely to become a classic of its kind" (Andersson 1986). Other laudatory reviews followed.

As with many things, writing a book took longer and was more work than I had imagined. But it was a smart decision to do this. It brought together otherwise disparate studies to make clearer the links between them. It highlighted the usefulness of integrative studies, and made the túngara frog system well known for studies of sexual selection and animal communication. Because of policies of The University of Chicago Press, I was not allowed to acknowledge the support and friendship of their science editor, Susan Abrams. She was a superb editor, interested in everything, and had wonderful maternal instincts. Until she died a number of years ago she was a fixture at many meetings and seemed to be everyone's friend. She certainly was mine.

It's Miller time!

It is hard to imagine a better postdoctoral situation for me than my time at Berkeley. My sponsor, Dave Wake, believed that Miller Fellows should have total intellectual freedom and not tied to the research or lab of the sponsor. And what a place to be free! The Museum of Vertebrate Zoology (MVZ), where Dave was the director and I had my office, was a wonderful community where most everyone met for coffee each morning and had their own seminar series once a week. I became friends with all of the scientific faculty, including Dave Wake, Harry Greene, Jim Patton, Bill Lidicker, and Ned Johnson, and many of the graduate students, especially Claudia Luke. The Department of Zoology, housed in the same Life Sciences Building, was packed with outstanding scientists. George Barlow was a well-known behaviorist and I attended behavior seminars that included him, Thelma Rowell, Frank Pitelka, and Roy Caldwell.

The intellectual emphasis at Berkeley, and especially in the MVZ, could not have been more different than Cornell. Sociobiology had just been birthed and Cornell was one of its birthing places. Most of the graduate students and faculty in behavior at Cornell were focused on understanding the adaptive significance of behavior. As Emlen often pointed out, the field was young and theory outstripped data. Many of the faculty at Berkeley, especially Dave Wake, were interested in another new field, evolutionary development, or "evo-devo," and were extremely cautious about accepting hypotheses of adaptation that did not consider the importance of constraints. Rather than being on the bandwagon of sociobiology with Wilson, Trivers, and Hamilton, they seemed more impressed with the criticisms of sociobiology being sounded primarily by Stephen Jay Gould and Richard Lewontin. For me, this led to no intellectual discomfort but a healthy dialectic. Although at Cornell I was schooled by some of the luminaries in sociobiology, such as Steve Emlen and Paul Sherman, I was also exposed to neural mechanisms of behavior by folks like Bob Capranica and Ron Hoy. So I appreciated the need to consider constraints on adaptation.

Another development that was paralleling the field of sociobiology was the rebirth of phylogenetics. Publication of Hennig's Systematics book in English gave rise to the push for cladistics on the North American continent. There were some students at Cornell (John Rawlins and Jim Carpenter) who were doing this on their own, but they were in entomology and our intellectual circles intersected only briefly. In NB&B we knew little of any of this. At the MVZ, however, phylogenetics reigned supreme. I began my Miller with a firm foundation in behavioral ecology and the mechanisms of animal behavior. I left with an appreciation for phylogenetics and its techniques, and I began to understand how I would apply this approach to studies of behavior. Although I didn't realize it at the time, the last link to what would become my interest in integrating studies of brain, behavior, and evolution was in place.

I made good use of my freedom at Berkeley. I finished up my túngara frog book, I took a course in quantitative genetics, read some books on phylogenetics, attended numerous lectures, and taught a sexual selection seminar with Frank Pitelka. I returned to bullfrogs briefly with Paul Licht and one of his graduate students, Mary Mendonca, studying the hormonal basis of alternative mating behaviors. I did some work in Panama studying habitat acoustics of a large number of species of frogs, and with Eliot Brenowitz reanalyzed

Morton's classical study of habitat acoustics in birds, emphasizing, perhaps not surprisingly, the importance of morphological constraints and phylogenetic history.

I also spent nearly half a year in Kenya with Merlin Tuttle on a memorable trip studying *Cardioderma cors*, a bat that was reported to eat frogs. This is a microchiropteran bat in the family Megadermatidae, a different family from *Trachops*. We thought that, like *Trachops*, they might be using frog calls as localization cues; if so, it would be a wonderful example of convergence and allow for a series of interesting comparative studies. We did most of our work on the coast, south of Mombasa, where these bats were numerous.

We had a fantastic field assistant in Kenya, Paul Koboche. He was a well-rounded naturalist, could speak most of the languages there, and was incredibly educated about African history. I learned much from him and we became very close friends. I met some of his children and we talked about how when I had my own children I would return and we would all go on safari. We had some adventures in Africa. One was a very close call in the Tsavo when we were charged by an elephant. Paul was driving our Land Rover while Merlin and I were standing in the back as the elephant charged. When the elephant got close, too close it ended up, we yelled for Paul to take off. He popped the clutch and stalled the truck. The elephant pushed the truck and Paul thought that its "nudge" push-started the stalled vehicle, allowing him to speed away just in time to avert disaster. Paul later went on to become a well known conservationist in Kenya.

Paul and I had often talked about my returning to Kenya in the future and we would revisit many of the places we worked with our families. Tragically, a number of years ago he was killed by an elephant in Tsavo.

Merlin moved to Austin, Texas, in 1985, a year after I started at the University of Texas. He became the full-time director of an organization he had founded, Bat Conservation International, which also goes by the acronym of BCI. So one BCI brought us together and another BCI kept us together. Unfortunately, Merlin and I have become so busy with our respective jobs that our paths do not cross as much as either of us would prefer.

At Berkeley I also met Ted Lewis in the engineering school. I knew his work on neuroanatomy of one of the frog's inner ear organs, the amphibian papilla. This organ had a frequency-place map, and its size varied drastically from a simple patch of tissue in the most primitive frogs to a much larger horn-shaped structure in the more modern anurans. If the sensory epithelium is larger then there should be more room to sense sounds of different frequencies. Thus these frogs should be able to hear a greater range of frequencies or there should be more frequency resolution. Capranica had some data suggesting it was the former. With these data in mind, I wrote a paper for *Proceedings of the National Academy of Sciences* suggesting that neuroanatomy can influence speciation. The idea is simple: the range of sensitivity of a receiver can limit the amount of meaningful mutations in a signal. Since the divergence of recognition systems usually occurs with speciation, then speciation should be more likely when divergence of recognition systems is more likely. The paper was accepted in *PNAS* (Ryan 1986).

Beginning at Texas

During my time at Berkeley I was already looking forward to interacting with scientists at the University of Texas, in part because of Mark Kirkpatrick's arrival as a Miller Fellow at Berkeley. He was working on theoretical models of sexual selection, and has always had a knack for simplifying mathematical explanations, so he was a great help in furthering my understanding of the population genetics of sexual selection. He arrived during my second year as a Miller Fellow and was immediately interviewed for a job in the Zoology Department at the University of Texas. He was offered it and accepted. So for the year we overlapped at Berkeley we knew we would both be at Texas. We are both still here.

I was hired with a group of four others in the Zoology Department at the University of Texas. Jim Bull, an evolutionary biologist, and Harold Zakon, a fellow graduate student from Cornell and a neurobiologist who works on electric fish, are still there. Matt Winkler is a molecular biologist who eventually left to found the company Ambion. John Rawlins, also a fellow graduate student from Cornell, who eventually left to join the Carnegie Museum, introduced systematics to much of the department. To me, having just come from Berkeley, he was preaching to the choir.

In general, our teaching was one course a semester, an undergraduate course and a graduate course. My undergraduate course was *Vertebrate Natural History*, which was a wonderful way to be introduced to Texas and to spend lots of time in the field with students. Eventually I traded that course for *Animal Behavior*. For my graduate course I developed one in *Animal Communication* with Harold Zakon. The course was modeled on the one that both Harold and I took with Capranica and Hoy at Cornell. Eventually Walt Wilczynski and I taught that together for a few years. I then changed it into an undergraduate course with a full lab and a requirement for independent projects. This has been a popular course with a full enrollment each time I teach it. For me the teaching has two important values. The first and most important is the interaction with students. The second is the rather clichéd feeling that you don't understand something until you teach it.

Swordtails

At Berkeley, George Barlow had told me about a thesis from Stanford University in which the researcher manipulated the sword length in swordtail fishes (*Xiphophorus*) to examine its influence on mate preference. The basic idea was that male displays seem to be super-normal sexual stimuli, and the swordtails might offer a way to investigate the phenomenon. This rung true in its relation to the neuro-neutral model I had considered for túngara frogs, and I became increasingly interested in incorporating these fish into my research at the University of Texas. Swordtails and platyfish (*Xiphophorus*) are an important model for studies of melanoma. As luck would have it, the UT Cancer Research Center was within an hour's drive and one of its scientists, Don Morizot, was an evolutionary geneticist who worked with these fish. I drove out to meet him and he told me that amazing story of the *P* gene.

Breeding studies by Klaus Kallman showed that in several *Xiphophorus* there were polymorphisms in male body size, and size was genetically passed from father to son. In one species, *X. nigrensis*, there were three body size classes: small, intermediate, and large. In a close relative, *X. pygmaues*, there were only small males. This raised a number of obvious questions for a behavioral ecologist: Did the size classes differ in their mating behavior? Were small males sneakers and large males courters? Were there genetically determined mating strategies? How was this genetic polymorphism maintained, and how did female mating preferences figure into all of this? And what about that species with only small males?

At that point there had been no studies of the behavioral ecology of swordtails. There were some excellent laboratory studies done in Germany on dominance hierarchies. This suggested that there might be some difficulties in studying these fish, but Morizot assured me that was not the case, and he and Klaus Kallman had wondered why this system had evaded the interests of behavioral ecologists. They invited me on a trip to Mexico and introduced me to these fish in the wild.

Although I was now intrigued by the role of *P* genes in the swordtail mating system I still wanted to pursue the role of the sword in mating preferences. A graduate student of mine, Bill Wagner, spent some time attaching false plastic tails to males but we could never show that it influenced female preference in *X. nigrensis*. It seemed we had chosen the wrong species, as another graduate student, Gil Rosenthal, later showed that *X. nigrensis* only had a very weak preference for swords. At the same time, however, there was another graduate student, Alex Basolo (who later married Bill Wagner), studying sex ratio evolution in *X. helleri*. Alex tried these same experiments attaching false tails to species with and without tails and was quite successful (Basolo 1990). In fact, she has made a career of such studies.

In *X. nigrensis* we showed that small males chased females, large males courted them, and females preferred the latter. But what about *X. pygmaeus*, the species with only small males? These males exhibited the "alternative" mating strategy of sneaking even though there are no large courting males. And the females? They prefer large courting males of the closely related but allopatric *X. nigrensis* to their own small non-courting males. Thus they retain the preference for large size and courtship that was lost in this species. We published this work in *Science* in 1987 (Ryan & Wagner 1987), and it made clear to me how important it was to consider the evolutionary history of a species to really understand its current behavior.

Studies of swordtails continue to be a part of my research program, especially as researchers are getting closer to sequencing the *P* gene and understanding its functional significance. The behavioral ecology of swordtails has also become the main focus of a former graduate student and two former postdocs of mine. Gil Rosenthal, a leading expert on the behavioral ecology of these fishes, is now working on olfactory communication and hybridization. He also bought his own field station in prime swordtail habitat. Molly Morris is continuing to work on signaling behavior as well as phylogenetics in swordtails, and Molly Cummings brought her intricate knowledge of visual ecology to bear on these fishes. Another former postdoc, Deborah McLennan, still works occasionally on this system, and always with intriguing results.

Mollies

Since I had worked with gynogenetic salamanders I had been intrigued by their maintenance. Gynogenetic females mate with males of other species. If they cannot do so, they
become extinct. If they are too good at it, they could drive the other species to extinction and
follow them closely into oblivion. A gynogenetic fish that faces the same challenge is the
Amazon molly. They associate with the sailfin molly in the northern part of their range,
which includes Texas.

I met a German undergradate, Ingo Schlupp, at the first International Herpetological
Congress in Canterbury, England, in 1989. I had just started to study mollies; Ingo planned
to do his thesis on mollies and needed lab space in Texas. I offered him space. He later did a
postdoc with me, and he continues to use both my lab and my house as a base for this
research, which he continues as a faculty member at the University of Oklahoma. He is one
of my closest friends.

Ingo and I walked across the UT campus one day in 1994 reviewing the basic assumptions
of the gynogenetic mating system. One was that males of the sexual species did not receive
any benefit from mating with asexual species since there was no fertilization. It then struck
me that there might be a partial solution to this dilemma. Lee Dugatkin had shown mate-
choice copying in guppies: females alter their mating preferences upon observing the mate
choice of other females (Dugatkin & Godin 1992). If this were to happen in mollies, then a
male sailfin molly mating with an Amazon molly might obtain delayed benefits if this
seemingly wasteful act made him more attractive to his own females. We teamed up with
Cathy Marler, one of my postdocs. In a matter of a few weeks we conducted the experiments,
which did indeed show mate-choice copying, and the paper was accepted in *Science*
(Schlupp *et al.* 1994). It was the fastest time from idea to publication in which I had ever
been involved.

Cricket frogs

I was hired at UT to replace Frank Blair. He had made fundamental contributions on the role
of frog mating calls to species isolation, but his health was in decline then and we had little
opportunity to converse. After he passed and I moved into his office, I found a manuscript by
Evitar Nevo and Bob Capranica. I had been meaning to read it when I got a call from Walt
Wilczynski. Walt had been a postdoctoral associate with Capranica while I was finishing up
at Cornell and was hired by the Psycholgy Department at UT while I was at Berkeley. I
thought Walt would be a valuable colleague at UT but I never envisioned how critical a
collaborator and close a friend he would become over the next two decades. Walt was calling
me to say that Capranica sent him a manuscript he wrote with Nevo on geographic variation
in cricket frogs (it was eventually published (Nevo & Capranica 1985)). It was mostly an
analysis of cricket frog calls throughout its large range in North America. There were some
phonotaxis tests and no neurobiology. In that paper they pointed out that there are two
subspecies of cricket frogs in Texas that meet at the border of the grasslands and piney

woods in east Texas. They referred to the subspecies as "incipient species". Usually, studies of species recognition compare well established species. But there has been a lot of evolution that can affect mate recognition once speciation is complete. Differences between species now might have played little or no role in the process of speciation. Studying an ongoing process might add some insights.

In our first study, which appeared in *Science* in 1988 (Ryan & Wilczynski 1988), we showed that there was geographical variation in call dominant frequency, some but not all of this variation was due to variation in body size, and this was correlated with tuning of the frog's inner ear. The correlated variation among populations resulted in local mate preferences which could provide grist for the mill of speciation.

The cricket frog work also led to interesting Ph.D. theses by Bill Wagner and Nicole Kime. Bill showed that males can alter their call's dominant frequency during male-male encounter as a bluff of body size. Nikki showed that the graded series of call complexity in which initial calls of the call bout were shorter and subsequent ones increasing in duration were not, as we had thought, a gradation from male aggressive to more attractive calls. Much still remains to be learned here.

Back to túngara frogs

I had not worked with túngara frogs for a few years since I had finished my Ph.D. thesis, but Stan Rand and I remained close friends. In 1985 I was in Washington, D.C., at the Smithsonian Institution's Natural History Museum and went to visit Stan, who was on a sabbatical leave there. Studies of the acoustic bases of mate preferences were only a small part of my thesis work. I had continued to be interested in what mechanisms of mate choice could tell us about sexual selection: the neuro-neutral model with the túngara frogs, the effects of neuroanatomy of speciation in frogs, the preference for phenotypes long lost in swordtails. I had been thinking of returning to work on túngara frogs and concentrating on understanding the details of how and why females were more attracted to acoustic variation. I told Stan about this and he said he had moved to Gamboa, a small town on the Panama Canal, and suggested we work together there.

We did the first experiments in 1987. This was before STRI acquired the joint lab/living quarters, Building 183, in 1988. We performed phonotaxis experiments in the same small portable chamber that Stan used in the 1960s and that I used for my thesis work. We conducted the experiments on a concrete slab under the apartment I was staying in. It worked most of the time, but whenever a bus rumbled past, the frogs leapt wildly and we had to scratch the experiment.

In 1988 STRI began slowly turning Gamboa into a research site. The first move was to acquire a building that had labs downstairs and apartments upstairs. Stan had an acoustic chamber built in his frog lab where we could conduct our experiments under well controlled conditions, and we repeated all the experiments we had done the year before. The "frog lab", to which I still have access even after Stan's passing thanks to the immense generosity of

STRI, has been upgraded considerably and has been the control center for our túngara project for two decades now.

Sensory exploitation in 1990

When male animals utter sounds in order to please the females, they would naturally employ those which are sweet to the ears of the species; and it appears that the same sounds are often pleasing to widely different animals, owing to the similarity of their nervous systems.

(Darwin 1872, p. 91)

The study we began under a garage in Gamboa in 1987 was published in *Evolution* in 1990 (Ryan & Rand 1990). The gist of that study was that details of the frog's auditory biology can predict acoustic components that are going to be more attractive. Frogs have two inner ear organs that are sensitive to air-borne sound, the amphibian papilla and the basilar papilla. Walt Wilczynski had shown that in túngara frogs the AP is tuned to about 500 Hz and the BP to about 2100 Hz. There is substantial processing of signals by the brain, of course, but in most frogs the inner ears act as filters that initiate call analysis.

With this knowledge in hand, we were able to synthesize calls that were much simpler but just as attractive as the real male call. The first point was that the details of the female's sensory system, through their influence on female choice, generated the selection under which male calls evolved. If males could evolve display components that were more stimulatory to the female, they were perceived as more attractive and thus favored by sexual selection. Rephrased, male display traits can evolve under sexual selection for sensory exploitation. This is what I was getting at with the "neuro-neutral" model in my túngara frog book. Another point we made is that there are alternative pathways to making attractive signals. We were able to synthesize stimuli that were attractive to females but that males did not produce. The particular response of the male to sexual selection should be biased by their evolutionary history, and especially by the types of sounds their larynx can produce.

The tuning of the BP tends to match the chuck's dominant Hz. The match is not perfect. The average tuning of the BP is a few hundred hertz below that of the average chuck's dominant Hz. In my thesis work I had shown that females choose larger males based on their lower-frequency calls. One of Walt's students, Jim Fox, made a computer model of the filtering properties of the BP and showed that chucks with lower frequencies should elicit more stimulation of the BP, and this might be the neural mechanism that ultimately explained preference for larger males.

Not all frogs use both of the inner ear organs in communication; perhaps most do not. The túngara frog and its sister species, *P. petersi*, both add very distinct suffixes, chucks in túngara frogs and squawks in *P. petersi*, to their whine-like advertisement calls. All of the frogs in the *P. pustulosus* species group, nine or so, produce whine-like calls whose dominant frequencies tend to match the tuning of their AP. It is possible in some of these species the BP is recruited in decoding some of the higher harmonics in the whine, but for other species this seems unlikely. Did the tuning of the BP evolve to match the chuck because of benefits accrued to females in choosing larger males or did the male's evolve a chuck that happened to match? In 1989 Stan and I made a short trip to Ecuador to begin studies of the close relatives of túngara frogs. I followed up with a six month stint there in 1990, and we then pursued these frogs in Peru, Venezuela, and Brazil over the next seven years. The neurobiology showed that the BP tuning of these close relatives was quite similar to that of the túngara frog. Thus there was no evidence that females evolved their tuning to favor the frequencies produced by larger males. These data on the tuning of the túngara frogs and *P. coloradorum* was the second paper we published on sensory exploitation in 1990, this one in *Nature* (Ryan *et al.* 1990).

I also did female phonotaxis experiments with *P. coloradorum* and showed that these frogs, which lack chucks, prefer their own calls to which chucks are appended. More recently Santiago Ron, while a graduate student at UT and now a faculty member at Universidad Catolica in Quito, Ecuador, has delved deeper into this class of frogs west of the Andes and has discovered more species and more types of calls than had been previously known. This is a story that is continuing to be developed.

I published a third paper on sensory exploitation in 1990 in *Oxford Surveys In Evolutionary Biology* (Ryan 1990). Earlier, Doug Futuyma visited UT and I told him about these ongoing studies. He urged that I write a more general paper outlining the arguments for sensory exploitation. Here I reviewed data from neurobiology and sexual selection that made a more general argument for this phenomenon.

At about the time these studies were being published, Steffan Ulfstrandt invited me to give a plenary lecture at the third International Conference of Behavioral Ecology in Uppsala, Sweden. The congress was memorable for me, and the talk was well received – I was immediately asked to visit Zurich, Bern, and Basle to talk about this idea in Switzerland. John Maynard Smith later gave a plenary lecture at the same congress. I had come to know John quite well on BCI as a grad student, and like everyone else who knew him, I was struck by his brilliance and his insights, as well as his most generous personality. His talk was memorable because he offered a public apology to Amotz Zahavi, who I happened to be sitting next to at the time, for his criticisms of Zahavi's handicap principle. At one point near the end of his talk John stopped, paused and said, "now about that stuff that Mike Ryan was talking about ... I don't know. I'll have to think about it". (Later at that meeting he tried to arrange for me to meet Magnus Enquist who was working on artificial neural networks because he thought Enquist's studies offered strong support for what I had just talked about. How right he was. I will return to this topic later.)

Brain, behavior, and evolution in túngara frogs

The notion of sensory exploitation made it clear that to have a deep understanding of behavior we needed to integrate an understanding of the neural mechanisms with the past phylogenetic history. Specifically, we wondered how the past history of the receiver influenced how it responded to signals today.

The study published in *Evolution* in 1990 convinced us of something Tinbergen made clear, the sign stimulus. Not all aspects of a signal are salient to the receiver. We are now able to measure many types of signals in excruciating detail, and we sometimes forget that our ability to obtain these quantitative measures does not always inform US about the receiver. That study also showed us that there might be various signal forms that are equally attractive to a receiver. For example, only the fundamental frequency of the whine is needed to elicit phonotaxis from females; the other harmonics do not even increase the attractiveness of the call, and a single tone that excites the basilar papilla is as effective as the full harmonic spectrum of the chuck. The question we began to ask was not whether signals and responses to signals are adaptive, but why is it that signalers use those particular signals and why do receivers show those particular responses? The ideal resolution would be to visualize the evolutionary history of signals and receivers. That wasn't possible. Nor could we compare signals and receiver between daughter (derived) and founder (ancestral) species as in Hawaiian *Drosophila* (Kaneshiro 1980). For most extant species their ancestors are extinct.

Felsenstein wrote an important paper in 1985 in which he introduced the notion of independent contrasts (Felsenstein 1985). He provided a method to estimate quantitative characters at ancestral nodes and control for phylogeny when comparing sister taxa. We used this and some similar methods to estimate the calls of ancestral nodes.

At that time a colleague in my department, David Cannatella, had begun to resolve the phylogenetic relationships of the species in the *Physalaemus pustulosus* species group. He had first worked on this group in 1984 (Cannatella & Duellman 1984). His study provided me and Stan with a phylogenetic background against which to examine the evolution of the communication system.

We measured the average call parameters for the fundamental frequency of the whine of the túngara frog, the other four known species in the species group, and three other congenerics. For each call parameter we estimated the value at each of the ancestral nodes and then took all of the estimated parameters for each node and synthesized the call. We then determined whether females (falsely) recognized each of the heterospecific and ancestral calls as a potential mate and to what degree females preferred their own conspecific call to each of the other calls. The main result was that females made lots of recognition errors, and even some discrimination errors. One might predict that the similarity of the heterospecific/ancestral call to that of the túngara would determine the recognition errors, but the phylogenetic distance, which was not strongly correlated with the call similarity, was even a better predictor of when such errors occurred.

Why would phylogeny predict recognition errors? When an ancestor gives rise to two daughter species the calls and the responses to them diverge. They are not designed from scratch but more likely result from some tweaking of the ancestral signal and receiver systems. A simple prediction was that if one could change the calls that the brains of ancestors had to recognize, this should change how current brains recognize their calls.

This brings us back to Magnus Enquist. I was impressed by the work that he and Antony Arak published on sexual selection and artificial neural networks in *Nature* in 1993 and 1994 (Enquist & Arak 1993, 1994). They trained networks to recognize arbitrary signals and showed that certain well-known effects, such supernormal responses and preference for symmetry, emerged in the absence of selection but as byproducts of network design. That approach seemed to embrace some of the contingency of real brains and how they might evolve that seemed to be missed in population genetic modeling. I was thinking of starting to use networks to approach the historical contingency idea when Steve Phelps, a graduate student, entered the picture. After some discussion and a computer science class, Steve switched to my lab and, with the joint sponsorship of Walt Wilczynski, began his study using networks to simulate brain evolution in túngara frogs.

Steve trained neural networks to recognize túngara frog calls. Once trained, he gave the artificial neural networks other calls with which we had tested real frogs but that he did not use in training the neural networks. The response biases of the networks predicted that of the frogs. Next he varied history. One population of nets were trained with the most ancestral call that we had estimated and then its immediate descendent, etc., on the way to the tungara call. Other populations of neural networks were given histories that differed from the real one in various ways. History did not constrain the nets from being able to evolve to recognize the túngara call. But it did influence how the nets did so. This was revealed by the ability of the nets to predict the response biases of the real frogs. Only if the nets had the real history could they predict the behavior of the frogs (Phelps & Ryan 1998; Phelps *et al.* 2001).

Steve continued some of this work as a postdoc at STRI with me and Stan. He had an important influence on our túngara frog work that continues today, and he remains a sounding board on many ideas, even though he now studies rodents.

Family

This was about the time that my family came to be. I married Marsha Berkman in 1994. My daughter Lucy was born in 1995 and Emma in 1998. Marsha and I divorced in 2002, but the girls remain the center of my life. The joy they bring me each and every day is unmatched, and that joy keeps everything else in life in perspective. Whether we are watching monkeys on BCI in Panama, driving cross-country, or swimming in the creek behind our house, their constant inquisitiveness also reminds me of why I am a scientist (Figure 17.4). I am often asked whether I want my daughters to become biologists. I don't really care. What I do want is that they grow up to maintain their fascination with and respect for the natural world around them. It seems that will be the case.

Large collaborations

This winding path of research experiences eventually led to an incredible opportunity for collaborative research. In 1998 a group of four PIs, David Cannatella, Walt Wilczynski, Stan Rand and I, received a grant from an NSF program, Integrative Challenges in Environmental Biology. Others involved included Cathy Marler at the University of Wisconsin, and Robert Dudley who during the granting period left our department for the University of California

Figure 17.4. Lucy Ryan, Emma Ryan, Rachel Page, a frog-eating bat, and the author.

at Berkeley. The purpose of this grant was to integrate studies of brain, behavior, and evolution in túngara frogs toward a deeper understanding of the function and evolution of socio-sexual behavior.

My initial feeling was that if funded the grant would provide funding for our separate labs to continue our own research. Although we hoped for synergism I was skeptical. I could not have been more wrong. We made remarkable progress. David Cannatella and his students, especially Santiago Ron from Quito, Ecuador, discovered new species doing new things in the *P. pustulosus* species group in western Ecuador. A graduate student of mine, Kathy Boul, discovered two populations of *P. petersi* only 20 km apart in Amazonian Ecuador in which one produced complex calls and the other didn't, and which failed to recognize each other as conspecifics. Together with population genetics work by Chris Funk, another postdoc on the grant, they offered what I think is one of the strongest examples of sexual selection driving speciation (Boul *et al.* 2007.). Two of my other graduate students, Rachel Page and Ximena Bernal, added layers of complication to the communication network of túngara frogs. Rachel documented how the frog-eating bat, *Trachops cirrhosus*, learns to associate the palatability of frogs with their calls and how this information is culturally transmitted among bats (Figure 17.4) (Page & Ryan 2005). Ximena studied blood-sucking flies, *Corethrella* spp., which exhibit phonotaxis to the túngara frog calls and, like the female frogs and frog-eating bats, prefer complex calls to simple calls (Figure 17.1) (Bernal *et al.* 2006). Even the famed insect neurobiologist, Ron Hoy, is stymied as to how these flies can hear the call, although he and Ximena are sure to figure this out in due time.

Another student, Beth Dawson, addressed an issue that had long been an assumption in anuran biology, that frogs don't learn their calls. Although Beth showed that this is basically correct, early acoustic experience does influence some details of the acoustic structure of a túngara frog's call and has a strong influence on the types of sounds that will elicit his calling. We made other advances in neurobiology. Kim Hoke, with the encouragement of Sabrina Burmeister, who was a student of Walt's and now also works on túngara frogs, came to work with me and Walt. Kim uses gene expression to map neural responses in the brain. She has made great progress in showing how the brain analyzes call variation and has also uncovered areas of the brain that seem responsible for differences in how the sexes respond to signal variation (Hoke *et al.* 2004, 2005).

Perhaps the most rewarding aspect of the grant was the forum for interactions it provided. Each year we had a meeting in Austin for everyone associated with the grant and anyone else who cared to listen. Usually about 20–30 attended and we often invited an outside observer and ended with a dinner or a party. The party after the last meeting was at my house and included some local musicians ("Braless" they called themselves) whom Walt and I had followed for some time. It was bittersweet, as the party also served as a *bon voyage* for Walt and his partner Debbie, both of whom I had known since I was in grad school at Cornell. Also, this was the first meeting that Stan missed; up to then he had visited Austin at least once a year for two decades. His health was beginning to decline. We called him and passed the phone around to all in attendance. He knew everyone there, and he could even recognize the music of Braless, whom he had heard on his past visits to Austin.

The official end to this grant was a symposium that I organized for the forty-first Animal Behavior meetings in Oaxaca, Mexico, in 2004 on Sexual Selection and Communication in Túngara Frogs. There were 17 papers by 19 presenters. One of the first talks was by Stan Rand on the natural history of túngara frogs. Stan's health had hardly been worse. The auditorium was packed; I think many realized they were hearing Stan's last talk.

Stan died on November 14, 2005. My last conversation with him was a couple of days before he died. I called him just before I left to give a talk in our engineering department about how túngara frogs produced complex calls. I was talking about Stan's early ideas on this topic while he was undergoing surgery and as I talked I realized Stan would soon be gone. When I returned to my office some of my students were trying to be upbeat, saying that Stan sounded stronger. I told them we had just talked to him for the last time. Stan's influence on tropical biology and the hundreds of students he mentored in more than four decades at STRI will never be forgotten (Ryan, 2006a). He certainly was one of the most important people in my life. That is another story in itself.

The research that I began with Stan continues unabated. The number of students, undergraduate and graduate, male and female, latin and anglo, who cut their scientific teeth on this system continues to grow, as do our insights into why these animals behave as they do.

50 years and going...

A good place to end this rant is at my half-century mark. My dear friend Ingo Schlupp organized a party and symposium at UT celebrating my fiftieth birthday in October, 2003. A logistical nightmare for anyone, Ingo was able to do this from overseas without me having a

Figure 17.5. The author near Boquette, Panama (photo by Ximena Bernal).

clue. He was worried, however, that I might not be in town so he broke the news to me while we were drinking beer with students in an outdoor café in Leiden in The Netherlands on a temperate September night. I was incredulous, and of course, grateful. A number of former postdocs, graduate students, and close friends arrived from various places in North America and a few countries overseas. I was grateful to see all of them. Ingo moderated the symposium and Stan Rand, Peter Marler and Gil Rosenthal all gave presentations that were mixtures of lyrical, insightful, and hilarious. My daughters Lucy and Emma got a big kick out of the event, especially Gil's depiction of me as a bald and bearded toddler roaming the landscape with dinosaurs. We then migrated to another room for a reception and my friend, Andrew "AJ" Johnson, who had come from England for the event, serenaded the crowd from an overhanging balcony with a somewhat inebriated version of Danny Boy.

Yes, a good place to end, to leave the past and get back to the future (Figure 17.5).

Acknowledgements

I thank Karin Akre and Molly Cummings for reading this manuscript.

References

Andersson, M. (1986). Review: mate choice. *Science* **231**: 1317–18.
Basolo, A. L. (1990). Female preference for male sword length in the green swordtail, *Xiphophorus helleri* (Pisces: Poeciliidae). *Anim. Behav.* **40**: 332–8.

Bernal, X., Rand, A. S. & Ryan, M. J. (2006). Acoustic preferences and localization performance of blood-sucking flies (*Corethrella* Coquillett). *Behav. Ecol.* **17**: 709–15.

Berreby, D. (2003). Scientist at work. Michael Ryan: evolving by accident. *New York Times, Science Times* pp. 1, 4. October 14, 2003.

Boul, K. E., Funk, W. C., Darst, C. R., Cannatella, D. C. & Ryan, M. J. (2007). Sexual selection drives speciation in an Amazonian frog. *Proc. R. Soc. Lond.* B**274**: 399–406.

Cannatella, D. C. & Duellman, W. E. (1984). Leptodactylid frogs of the *Physalaemus pustulosus* group. *Copeia* **1984**: 902–21.

Darwin, C. (1871). *The Descent of Man and Selection in Relation to Sex.* London: J. Murray.

Darwin, C. (1872). *The Expression of the Emotions in Man and Animals.* London: J. Murray.

Dugatkin, L. A. & Godin, J. G. J. (1992). Reversal of female mate choice by copying in the guppy (*Poecilia reticulata*). *Proc. R. Soc. Lond.* B**249**: 179–84.

Emlen, S. T. & Oring, L. W. (1977). Ecology, sexual selection, and the evolution of mating strategies. *Science* **197**: 215–23.

Enquist, M. & Arak, A. (1993). Selection of exaggerated male traits by female aesthetic senses. *Nature* **361**: 446–8.

Enquist, M. & Arak, A. (1994). Symmetry, beauty and evolution. *Nature* **372**: 169–70.

Felsenstein, J. (1985). Phylogenies and the comparative method. *Am. Nat.* **125**: 1–15.

Hoke, K. L., Burmeister, S. S., Fernald, R. D. *et al.* (2004). Functional mapping of the auditory midbrain during mate call reception. *J. Neurosci.* **24**: 11264–72.

Hoke, K. L., Ryan, M. J. & Wilczynski, W. (2005). Acoustic social cues shift functional connectivity in the hypothalamus. *Proc. Natn. Acad. Sci. USA* **102**: 10712–17.

Huxley, J. S. (1938). Darwin's theory of sexual selection and the data subsumed by it, in light of recent research. *Am. Nat.* **72**: 416–33.

Ibanez, R., Rand, A. S. & Jaranukko, C. A. (1999). *Los Anfibios del Monumento Natural Barro Colorado, Parque Nacional Soberania y Areas Adyacentes.* Santa Fe de Bogota, Colombia: Impreso por D'Vinni Editorial Ltda.

Jager, R. G. (1976). A possible prey-call window in anuran auditory processing. *Copeia* **1976**: 833–4.

Kaneshiro, K. Y. (1980). Sexual isolation, speciation and the direction of evolution. *Evolution* **34**: 437–44.

Marler, P. (1955). Characteristics of some animal calls. *Nature* **176**: 6–8.

Martof, B. S. & Thompson, E., Jr. (1958). Reproductive behavior of the chorus frog *Pseudacris nigrita*. *Behaviour XIII*, **3–4**: 243–55.

Noble, G. K. & Bradley, H. T. (1933). The mating behavior of lizards: its bearing on the theory of sexual selection. *Annls N.Y. Acad. Sci.* **32**: 25–100.

Nevo, E. & Capranica, R. R. (1985). Evolutionary origin of ethological reproductive isolation in cricket frogs, *Acris*. *Evol. Biol.* **19**: 147–214.

Page, R. & Ryan, M. J. (2005). Flexibility in assessment of prey cues: frog-eating bats and frog calls. *Proc. R. Soc. Lond.* B**272**: 841–7.

Phelps, S. M. & Ryan, M. J. (1998). Neural networks predict response biases in female túngara frogs. *Proc. R. Soc. Lond.* B**265**: 279–85.

Phelps, S. M., Ryan, M. J. & Rand, A. S. (2001). Vestigial preference functions in neural networks and túngara frogs. *Proc. Natn. Acad. Sci. USA* **98**: 13161–6.

Ryan, M. J. (1980). Female mate choice in a Neotropical frog. *Science* **209**: 523–5.

Ryan, M. J. (1985). *The Túngara Frog, A Study in Sexual Selection and Communication.* Chicago, IL: University of Chicago Press.

Ryan, M. J. (1986). Neuroanatomy influences speciation rates among anurans. *Proc. Natn. Acad. Sci. USA* **83**: 1379–82.

Ryan, M. J. (1990). Sensory systems, sexual selection, and sensory exploitation. *Oxford Surv. Evol. Biol.* **7**: 157–95.

Ryan, M. J. (2006a). Profile: A. Stanley Rand (1932–2005). *Iguana* **13**: 43–6.

Ryan, M. J. (2006b). Q&A: Michael Ryan. [Self interview.] *Curr. Biol.* **16**: 1012–13.

Ryan, M. J. & Wagner, W. E. J. (1987). Asymmetries in mating preferences between species: female swordtails prefer heterospecific males. *Science (Wash., D.C.)* **236**: 595–7.

Ryan, M. J. & Wilczynski, W. (1988). Coevolution of sender and receiver effect on local mate preference in cricket frogs. *Science (Wash., D.C.)* **240**: 1786–8.

Ryan, M. J., Tuttle, M. D. & Rand, A. S. (1982). Sexual advertisement and bat predation in a Neotropical frog. *Am. Nat.* **119**: 136–9.

Ryan, M. J. & Rand, A. S. (1990). The sensory basis of sexual selection for complex calls in the túngara frog *Physalaemus pustulosus*: sexual selection for sensory exploitation. *Evolution* **44**: 305–14.

Ryan, M. J., Fox, J. H., Wilczynski, W. & Rand, A. S. (1990). Sexual selection for sensory exploitation in the frog *Physalaemus pustulosus*. *Nature* **343**: 66–7.

Schlupp, I., Marler, C. & Ryan, M. J. (1994). Benefit to male sailfin mollies of mating with heterospecific females. *Science* **263**: 373–4.

Tinbergen, N. (1963). On aims and methods in ethology. *Z. Tierpsychol.* **20**: 410–33.

Trivers, R. L. (1972). Parental investment and sexual selection. In *Sexual Selection and Descent of Man*, ed. B. Campbell, pp. 136–79. Chicago, IL: Aldine Pub. Co.

Tuttle, M. D. & Ryan, M. J. (1981). Bat predation and the evolution of frog vocalizations in the Neotropics. *Science* **214**: 677–8.

18

Birds, babies, and behaving

MEREDITH J. WEST

Childhood schooling

Perhaps because I study development for a living, I think I should be able to explain my own trajectory through developmental mechanisms. We all do so to some extent – harken back to our childhood for clues to the future. But my brand of developmental systems theory does not place as much weight on retrospective explanations: living is for the here and now and what is remembered from the past is often not very predictive unless we make it so through selective memory or, more to the point, selective forgetting. And so I begin my narrative with the caveat that my childhood is perhaps most important because it was happy – with regard to my future, I am not sure my parents or I gave it much thought (or more than likely did not share the thoughts we had)… life would happen but beyond being casually praised

Leaders in Animal Behaviour: The Second Generation, ed. L. C. Drickamer & D. A. Dewsbury. Published by Cambridge University Press. © Cambridge University Press 2010.

for being a good student in school, I don't remember much proscriptive developmental thinking such as whether I should be scientist or a secretary. It is also possible that future-oriented thinking had been blunted by the death of my older brother from cancer; thus good health and care occupied the top rank of concern. But most probably, given the times, thoughts for my future involved a house and family and a career in mothering. It was not until I was a full professor and gave a university-wide lecture that my parents happened to attend that the talking stopped about my settling down to "simply" take care of our sons. There was never tension about my decision (or the one to keep my name), simply loving anxiety that I was doing something untoward by pursuing a career.

I was one of six children and the group of us occupied much of my out of school time. We shared a love for animals and were always after our parents for more pets – chipmunks and snakes are two I remember the most vividly but there was also the stray cat or dog that just showed up to vie for attention with our tenured cats and dogs. My father even contributed to the dream for our zoo by bringing home two baby alligators, Shorty and Pete. They wore out their welcome by literally biting the mailman. We also had a chicken that laid eggs under sofa cushions. As I write this all down, I have to revise my childhood thinking that my parents were not pet-friendly: as a kid it seemed we were often at odds over just how big the extended pet family would be, but I now see it in a better perspective. It also never occurred to me that being surrounded by animals was a pathway to a career, as is the case with so many who enjoy careers in animal behavior and started as amateur naturalists.

My schooling undoubtedly led me closer to a position of scholar… I attended a private Catholic school for girls taught by the Sisters of Mercy, many of whom would leave the order when the Vatican revolution took place in the 1970s. In retrospect, they were great teachers and incipient revolutionaries in ways we students did not appreciate. I know we held them in deep respect. I placed out of several science courses when starting college, another gift from my teachers.

Perhaps one trait that has carried stably into adulthood is reading… I am always reading a book or two. As a child, one of my fears was that the world would run out of books before I died. I would purposely read slowly to make the treasure of books last longer. As another route to reading insurance, I took courses in Russian in high school and got to read some of the classics of that country when in college. I am happy to report my two sons are readers too. Later my years as editor of academic journals served an additional purpose of giving me more to read and extending my reading future.

Graduate school

I graduated from Tufts University, at a time when there was a women's part called Jackson College although we had all of our classes with the Tufts men. Female-oriented things that I remember are the separate dorms, having to wear skirts to class and dinner, and check-in time in the evening. When I entered Tufts, I placed out of introductory chemistry and so I had a space in my freshman schedule that I filled with introductory psychology. One course was enough to determine it as a major and I spent several years in different labs as each psychology course had an experimental component in areas such as perception, personality, motivation, and learning. I did very little animal work and didn't know what I was missing in animal behavior for several more years. Moreover, I somehow missed Kenneth Roeder and his elegant insect work at Tufts. His lab was simply not open to undergraduates or majors in psychology. I liked having to write formal lab reports each week, a tradition now lost to most psychology departments.

The chairman of the psychology department was a woman named Dorothea Crook, who had studied perception at Cornell University. My family doctor, also a woman, had also gone to Cornell. Perhaps for these reasons Cornell was the only school I applied to for graduate training, thinking I would study learning or perception. As I think back, I marvel at my naiveté in choosing only one school but I was lucky enough to have good mentors. I did not even visit Cornell before I attended, relying on the memories of the Cornell alumnae around me. Perhaps I was not desperate to go to school, but as I had always gone to school, I guess I figured I would just keep going.

My education at Cornell started with a series of lectures by the faculty in the summer before the school year started. We were given a stipend ranging in value with how early we came: the earlier the better. I had already been assigned an advisor, Martin E. P. Seligman, and gained experience in his lab at the same time. He worked on learned helplessness in dogs; a phenomenon that had grown out of designing the appropriate controls for studies of Pavlovian conditioning. In such conditioning, an arbitrary stimulus, say, a tone, is paired with an aversive stimulus, in this case, shock to dogs' feet. To show that it was the contingency causing the animal to learn to jump across a gate to avoid the shock, it was necessary to test the condition where the shocks and arbitrary stimulus came in random fashion (procedures stemming from the work of Allan Wagner and Robert Rescorla). Marty had found that such random conditions caused the dogs to give up, to simply take the shock and cease trying to avoid it by jumping, hence the term learned helplessness. Marty thought it might be a good model for human depression and in fact went on to study with clinical psychologists at Penn on this idea, which grew to be very popular as an animal model.

I was not interested in animal models of abnormal behavior, nor was I convinced that the dogs' behavior was due only to the random contingency. If so, it should work with positive or appetitive stimulation as well, but that did not seem to be the case. I also wondered about the effects of context. When we retrieved dogs from the small kennels to participate in the experiment, they did not seem helpless until they reached the experimental room. My memory is that some of the dogs were cooperative and easy-going until they were in the apparatus. Why weren't they helpless all the time? If this was a model of depression, shouldn't we see the effects in all contexts, as depression, to my knowledge, was not context-specific? I was not the only grad student to think this way and a number of us met frequently with Marty to discuss our concerns. He took them seriously but still maintained the dog model as a good analog of human conditions that led to depression, such as poverty, uncontrollable noise, and the draft for the Viet Nam war. But Marty moved to Penn while we were still in our early years at Cornell and has gone on to study human behavior in striking and novel ways, focusing on "positive psychology."

Figure 18.1. Photo of Andrew P. King by Jennifer Miller.

But my interest in learned helplessness, which was never strong, was overshadowed by the summer lecture of William C. Dilger, a joint member of Psychology and Neurobiology and Behavior. He mesmerized me with his description of the field of ethology; he said ethologists looked at animals as a set of answers, and it was our job to figure out the evolutionary questions. He illustrated his thinking by drawing (he was an excellent artist) the beaks of birds and engaging us in thinking up reasons for the different morphologies. He offered a combination of a respect for nature and animals with the instincts of a good experimentalist. He was one of the first to do playbacks in the field and had also done one of the most complete studies of behavioral genetics comparing the behaviors of hybridized members of the Agapornis parrot family. The contrast to the dog work was great: ethologists appreciated the whole nature of the animal they were studying, and did not focus on arbitrary connections among behaviors but on connections set up by natural history.

I did not come to question learning theory solely of my own accord. A fellow graduate student, Andrew King, was also in the Seligman group.

Unlike myself, he had a strong research history in animal learning and had come to Cornell to study classical conditioning with Marty. His first question to me in fact was "What's your theory of learning?" I thought the question indicated perhaps too much of a dose of experimental psychology and that it was rather bold for a first-year graduate student to have so grand a thing as a theory. Something clicked, though, as we eventually would marry and collaborate on theories of development and learning. We did not formally collaborate during graduate school, although we helped each other with our collective research. We both ended up with Dilger as our chairman and shared lab space. Drew's background was similar to mine in that he had raised parakeets and bred fish, but he had more of a sense of direction about his future. His future at that point, though, was compromised by threat of the Viet Nam draft, as was true of all the male American students. E. J. Gibson did her best to make sure they were assigned teaching assignments, which gave some relief. The war seemed to circle overhead during our whole stay at Cornell. The times also produced racial violence when a group of armed black students took over the student union. Cornell cancelled all classes that spring to allow for some chance at reconciliation and change, which slowly came about but probably not to the armed intruders' wishes. Finally, graduates who had received federal funding for postdocs had the money impounded by President Nixon, a swipe at the seemingly left-leaning NIH and widespread antiwar protests at major universities.

Our lab space became available at first through the graciousness of James and Eleanor Gibson, well-known perception experts. We had been encouraged to hang out in their lab by older students, from whom we learned a tremendous amount about science and politics. The Gibsons carried out independent and dependent lines of research ranging from theories of perceptual systems ("JJ" as he was called (Gibson 1966)) and the development of perceptual learning ("EJ" (Gibson 1969)). EJ was also referred to as Mrs. Gibson, which was not meant in any pejorative sense but for practical reasons (although I never heard any one address JJ as "Mr. Gibson"). At one point, I rebelled against what seemed like a sexist tradition and left a note in her box addressed to "Dr. Gibson." A few days, later I saw JJ, who asked me questions about the message in the note, professing confusion as to what to do. Clearly Mrs. G had put the unopened note in Dr. G's box without a second thought. I decided if she could live with "Mrs.," so would I. Although she had had to fight hard to get an independent academic position at Cornell, she never expressed anything but enthusiasm for science to us students. She was a keen mentor for female students and did much to help us find jobs or postdocs.

Due to space constraints at the time, the Gibson's laboratories were located off campus near the Ithaca airport and were known as the "airport labs". The labs had lots of space, enough so that Drew and I could get offices there even though Dilger was our major professor (he had limited space for his many students). EJ was on my research committee. The airport lab was, in the airport lab's opinion, the best place to be in terms of intellectual stimulation and friendship. The lab was large and parking was just outside the door. The Navy funded JJ and some years he did not use up his entire budget and attempted to return the money to the Navy, something the Navy did not know how to do. One year we talked him into buying the first Xerox machine for the lab. JJ was writing a book when we were there and had a weekly seminar discussing his ideas about perceptual systems and things called "affordances"(Gibson 1966). The concept has now slipped into more common usage, at least in psychology, and generally refers to the functions that define an object: for example, a chair's affordance is sit-ability and conversely anything with sit-ability, a log and overturned trash can, have something in common with chairs. Active perceivers explored the environment in search of affordability. For Drew and myself, the concept linked learning and perception; we will talk more about affordances later.

EJ was also finishing a book on perceptual development in animals and humans. The book is still useful today as a complete survey of theories and data on perceptual learning in animals through 1969. She had done work on goats' attachment as well as inventing the visual cliff with Richard Walk. The cliff was a plank of wood in the middle of a glass tabletop with squares attached to the surface of the glass or the floor to simulate depth. Walk and Gibson studied many young animals on the cliff and showed the importance of whiskers in kittens and rats to judging depth, among other findings (E. J. Gibson, 1969). The amount of research they did goes beyond any autobiographical bounds but was an elegant example for me of how a young animal's own activities affected its development, a growing theme in my head for research. Dilger's work on parrots learning nest building also brought up the question of genetic and experiential factors, which was a broad banner for much of psychology and ethology in the 1960s and 1970s (Dilger 1964). Although Dilger taught us much about how to study animals and how to pose questions to animals, it was the Gibson's work on perception as a "performatory" activity taking place in a stimulus-rich world that formed the basis for our later work, even to this day. In both ethology and psychology, the sense was that experiences were imposed on passive perceivers with plastic parts of their nervous system to allow experience to do its trick. We now know that experience creates some of that plasticity and that the processes are interpenetrable. The Gibsons also stressed a different kind of environment than that of the learning theorists: to the Gibsons, the world abounded in information, and the perceiver's job was to discriminate and act upon it. Our later resistance or perhaps anxiety about templates for bird song development and schemas stemmed from the different worldview representing an environment of impoverished information.

I managed to package many of these ideas into my thesis on play and exploration in domestic kittens. My interests were sparked by watching a youngish cat attempt and finally succeed, over a matter of months, in getting an older cat to play. I could not get over the amount of motivation involved – what was so important about pouncing, facing off, and leaping toward one another? The topic included my ethological interest in naturally occur-ring forms of early experience and my psychological interests in perceptual learning, i.e., learning the function of things in the world. These topics also naturally drew me into developmental theories, which I found I loved to read and think about. In some sense the work on play was a Rorschach for portraying experiential effects. I studied a total of 42 kittens from 8 litters from 5 mothers. Three of the mothers were feral cats that lived outside and underneath our rental house. The cats knew well-hidden means of egress and ingress to our house and it took us quite a while to learn how they were entering and why food was disappearing from our kitchen counter. I discovered the scheme one day when I and our dog came home at a time I usually was not in the house and found a huge tom cat sleeping on the bed... he was as surprised as I was but a lot faster and escaped though an opening where the pipes entered the house. I doubt I would have dared catch him as he was not at all friendly when we would encounter him outdoors. The other memorable meeting was when he discovered a litter of kittens in our garage and he gobbled down several of them. We steered clear of "Dad" as he was known.

The feral mothers were more tolerant of my presence as long as I did not try to touch them. Often I watched them through the window to the garage or while sitting in my VW Beetle, with which I could even follow them distances into the woods. The feral mothers were moms and sisters and they shared caretaking and nursing activities, forcing me to rethink the idea of cats as not very social. The feral litters were very small due to poor nutrition, disease, and infanticide. Play seemed especially important in these litters because the mothers could not go off and hunt alone and had to tolerate playful advances that were much less frequent in the non-feral and larger litters where, during social play, mom could easily slip away. Several times I saw feral moms start to head for the fields and forests with a kitten that simply wouldn't stay home, most often because it was singleton. The incipient dangers were realized once when a horse from next door got loose and ran into the same field. Mom scurried away but the kitten froze and was trampled. Marc Bekoff has talked about wild coyote pups and incidental function of play where staying together depended on playing together and I thought the same was true for cats. At some age mom did not protest but usually she brought prey to the kittens rather than the other way around. So although the mothers cooperated for kitten care, hunting appeared to remain a solitary activity.

At the time there was much debate over the definition of play as well as its function. I don't think the same holds true today; I think play is more generally accepted as an important developmental category. But while I was working on it, I remember participating in a seminar on play organized by Bekoff for an ASZ Christmas meeting (West 1974). A comment from an audience member stayed with me. The speaker was Irwin Bernstein and the comment was that we should stop fussing about what we called play and focus on the activities that defined it, such as jumping or running or manipulating. A graph of the amount of "play" meant little without good operational definitions devoid of the motivational confusions the term "play" seemed to attract.

Thus, I began to think of my work as looking at the experience of early experience. What exactly is play doing? My observations of feral cats led to my beliefs about the function of play. I did not think the value was for the future, i.e., practice for hunting for example. I thought the value was for the present to provide the young with a means to explore the properties of the emerging world, as kittens obtained very different reactions when pouncing on a toy versus a real mouse. If anything, play seemed to be a misleading preparation because the object of the play did not respond in ways similar to those of potential prey or peers. For example, a common play move was to rear up on the hind legs with the arms outstretched in a sort of bear hug: what better way to scare off even a non-vigilant mouse or bird? But the experience delivered information for predicting behavioral properties of hunt-able objects. I can recall one of the feral kittens pouncing on Dad and getting swatted to kingdom come – I think the kitten learned rapidly that adults had different properties than other kittens. As they matured, sex differences also became apparent and females began to retreat from males' initiatives because it seemed the males did not distinguish playful from sexual initiatives.

During this time, Drew and I wrote a paper for a course on comparative psychology – we complained that learning psychology did not necessarily emphasize the species typical skills of an animal but looked for abstract capacities to acquire stimulus response contingencies. Much of what was special about different species was deliberately ignored in order to find generalities, generalities we thought to exist probably throughout even simple animal species. To Drew and myself, this approach was like using a dishwasher solely as a storage box but never turning on the electricity to see what the box could do. Thus the major affordances were missed. We sought learning paradigms with more electricity, paradigms focusing on motivated species' typical learning. Such a move distanced us from those seeking laws of learning, or perhaps focused us on different kinds of laws. Laws of function/survival or development became more important. This view was shared by a small number of comparative psychologists such as T. C. Schneirla, Gilbert Gottlieb, Jay Rosenblatt, Daniel Lehrman, and Don Dewsbury. Aside from Don, they represented the "Rutgers" group or the "Museum group" as many had appointments at the Museum of Natural History in New York City. They focused on actual problems animals faced such as the transition to filial, sexual, and parental behavior in rats, cats, or birds. They were among the first to look at multiple levels of analysis, starting with behavior but often extending their thinking to underlying hormonal states.

Dewsbury represented another tradition, looking at a similar behavior, in his case sexual behavior in rodents, among the varied taxa. He compared closely related species and compared reproductive strategies in males and females. Drew and I were inspired by all such comparative work and began to plan a comparative program of our own – we would travel to the Galapagos Islands to study tool use in Darwin's finches, species already studied for differences in beak morphology. We managed to secure fellowships to free up the time, but political events swept through South America that made travel risky with no guarantee of access to the birds (and thus a very risky means to complete dissertations) and so we shelved the idea, always hoping to return to it once we had our degrees. The effort served as a model, however, for the kind of integrative work we wanted to do, and the model would reassert itself when we began to think about the evolution of learning in cowbirds. We owe Cornell a great debt when it comes to stimulating integrative work. Cornell had a strong committee system fostering the creation of cross-disciplinary faculty representation for example between psychology and neurobiology and behavior. Comparative approaches were also fostered. As students, we had nothing to compare to Cornell and so it was not until I got a job at the University of North Carolina that I retrospectively began to appreciate what Cornell had to offer.

Like most graduate programs, Cornell had qualifying exams, which I took with W. Dilger, E. J. Gibson, and R. McLeod, an historian. He asked me the most challenging question: to trace the history of comparative psychology from Aristotle to the present. I also took it as an opportunity to include the growing role of ethology as an alternative way to approach learning and development. His question got me thinking about animals as models or surrogates for humans. At that time, this was the major role for animals in psychology. But I was skeptical that the desired simplicity of animal models would map onto the complexity of human behavior in a linear way. To borrow from computer vernacular of today, learning psychologists (or many of them) saw animals' minds as computer programming versions, with upgrades and updates in "higher" species that flowed linearly from one species to another, and so understanding a bird, rat, or primate bore direct relationships to understanding a human, the most advanced form of the same basic software program. The computer language was the common connection. But too much of the work focused on narrow fields of behavior, primarily conditioning, probably because it was only in such simple environments that rules emerged. While I could see similarities in some mental/neural capacities, one had to ignore too much of the animal to keep the comparative speculation clear. Psychologists such as Hull, Spence, Skinner, or Tolman were articulate visionaries of this view. But what I thought about were all the capacities animals had that humans did not have, in particular, sensory apparatus such as echolocation. Does this ability simply get added into the program or subtracted from the human program, man minus the hard- and software of bat echo locating minus flying minus insectivory, etc.?

The Gibsons' views differed as animals were evolved to fit rich ecological niches. JJ looked carefully and comparatively at animals' eyes and their binocularity to develop his theories of depth perception and the role of motion parallax. Drew and I had a large fish tank in our office and JJ would spend time watching how the fish (cichlids) moved their eyes when swimming forward or backwards. His curiosity about animals encouraged us to think that psychologists could study behavior in "non-traditional" animals and was more in line with the thinking of those in neurobiology and behavior (whose lab was also located at the airport). JJ did not, however, think that studying the neuroanatomical correlates of vision was terribly helpful in creating theories of perception. He felt it was the ecology that held the answer. A review of Gibson's work once noted his theme, as "It's not what inside your head that matters but what your head is inside of"(Mace 1977).

Living in Ithaca brought out the naturalist in everyone, as the environment was stunning: forests, streams, waterfalls, and gorges. Dilger's lab was located at the Laboratory of Ornithology on Sapsuckers Woods Rd., a beautiful drive from campus. Most people we knew watched birds as a hobby. For graduate students, the physical features of Ithaca encouraged lots of outdoor recreation from boating to cross-country skiing; there was an activity for every season. Drew and I lived in the country in a small house bordering acres and acres of New York State Forest. Fortunately, the animals of interest, kittens for me and cowbirds for Drew, could be found in our front yard. We also had a dog, and bred parrots and tried to breed cichlids, a species known as Oscars, who grew to be quite large and were highly aggressive, meaning that we could only keep two in a 75-gallon tank. They were visually quite alert, watching us as much as we watched them. I can remember that I used to read the newspaper at my desk next to the fish tank during the weekdays, but on Sunday, I read the Sunday *Times*, which had color in the magazine section. The fish became quite excited at the color pictures on Sunday but paid little attention to the news in black and white. After a few years, we had assembled a small zoo; something we had both wanted as children but had no idea could become part of a formal education.

A final influence on us during graduate school came from the departmental chairman, Harry Levin, whom we got to know quite well because Drew was hired to assist Harry, whose spine had been injured, a consequence of unsuccessful back surgery. He was a psycholinguist but reached out to students in all fields. Harry treated students with the utmost respect, and in return, students tried hard to live up to that respect. Drew and I learned a lot from Harry about navigating the academic landscape from what turned into a close personal relationship, and he continued to mentor both of us after he became Dean of the Arts College and thereafter up until his death.

Harry, as chair, told all of the students that his vision was that we become skeptics in our respective fields and not follow the mainstream, but pursue overlooked tributaries that might be missed by those following major trends. We took his message to heart and credit our chronic lack of mainstream-ness to his influence. We have never done research that followed a popular trend, even though at times it would have helped us judge whether we were having any impact. Levin, as well as other senior professors, also did not stress publishing much as graduate students because they thought we did not yet know enough to make a substantive contribution (and because you could get a job without many publications). It was a quite different world from the present where students often begin to publish as undergraduates.

In 1970, Drew and I attended a symposium at McMaster University entitled "Can psychiatrists use ethology?" "Everyone" in the psychobiological world was there, headed by Robert Hinde. The Rutgers group was especially well represented and we heard talks by Ernest Hansen on play, Colin Beer on comparative methods, and an overview of developmental studies by Lehrman. Harry Harlow was also there, representing those trying to study affectional systems (the scientific term for mother love) in non-human primates. I had mixed feelings about Harlow. On the one hand, his writing about the comparative psychology of learning was outstanding – he argued that conditioning was probably one of the slowest ways possible to teach something to an animal and he demonstrated the role of cognitive reinforcement (curiosity) showing that rhesus monkeys would "work" for the chance to look out a window at a faster rate than to obtain food (Harlow 1965). But his work on social development bothered me: he did several studies in which he housed primates alone in small, darkened cages (he called them "pits") with no added sensory stimulation for a year (Harlow 1974; Harlow & Harlow 1962). The monkeys developed severe psychopathologies and Harlow's work was cited as an animal model for autism. But I could not understand the parallel. Human infants were not subjected to "pits" in order to show abnormalities and I could think of no ecological parallel to solitary living without any stimulation except that that was self-generated. It is one of those cases where using a scientifically clean method (control of stimulation) was uninterpretable as it was completely bogus, at least in mammals. Young animals did not live alone for years in close confinement with or without added stimulation.

To be completely truthful, I thought the work, although immensely popular, was inhumane and went from person to person explaining my discomfort. I don't think Harlow would have done the work today and I don't think an IACUC would approve it. But setting aside animal welfare concerns, I think Harlow's theoretical point of view would probably still be tolerated given the number of animal studies using isolation as a basic condition. Harlow also indulged in the sometimes tricky business of trying to find humor in some of his work. When he tried to breed motherless monkeys, he found they showed no interest in male

Figure 18.2. Two views of cowbirds. (a) Male cowbirds countersinging to one another, a common social behavior among males; (b) female cowbirds displaying typical poses in one of our aviaries. Photo by Andrew King (APK).

sexual overtures and so he tied the limbs of a female to a metal frame and "allowed" the male to mate with her... he referred to the apparatus as the "rape rack". Such an approach reinforced in me the need for respectful animal conditions and the theoretical uselessness of such draconian methods. Although his work was very influential in showing the role of mothers and peers in development, I was convinced that more ecologically valid methods would have come to better conclusions but with more applicability to humans and monkeys.

Introduction to cowbirds

While Harlow's talk at the conference did not inspire me, those of Hansen, Beer, and Lehrman were exceptionally motivating. Hansen talked about the importance of play, Beer stressed the role of microphyletic comparisons, and Lehrman made intriguing arguments about how to choose animal models (Lehrman 1974). He noted, for example, that scientists tended pick and choose animals to model whatever human behavior was of interest and so birds were often chosen as models of early experience because of the findings of those studying imprinting. But he posed the question: which bird do we choose? He remarked that red-winged blackbirds deprived of parental stimulation showed abnormal species identification and abnormal song learning but brood-parasitic cowbirds, also members of the Icteridae family, did not appear to be influenced by their early experience with host species and lack of contact with "parental" cowbirds. The question, then, was: which species do you choose as a model? Obviously the answer was that no one species would accommodate the role of "the" animal model. His words struck a chord with Drew and myself and we came back excited to look at questions of early experience in cowbirds. There are seven species of cowbirds, one of which is entirely non-parasitic. Resurrecting some of our thinking about the aborted Galapagos project, we envisioned studying all the cowbird species while looking at the role of early experience as the species radiated north and became more and more parasitic. Drew took on the most parasitic species, *Molothrus ater*, the North

Figure 18.3. A copulation solicitation posture to a playback of male song. The posture occurs while the song is in progress. Photo by APK.

American brown-headed cowbird, for his dissertation. Thirty or so years later we are still working on *M. a. ater* although we have studied eight different populations of the three subspecies, but we have found enough to keep us busy still trying to figure out their learning and development.

Ernst Mayr helped to frame the work with his 1974 paper on open and closed behavioral systems (Mayr 1974). He targeted the cowbird as a representative of a completely closed developmental system impervious to postnatal experience, a view based on speculation. This led Drew to test these ideas experimentally.

While I was learning about play, Drew was building his first aviary to study brood parasitism in cowbirds. His deeper interests included social processes such as how cowbirds managed species identification. North American cowbirds were raised by over 200 different species and subspecies until fledging, then fed by their hosts for a short time, and then were found in small flocks with other fledglings. To answer such questions meant having ready access to very young cowbirds. Thus, it came to be that Drew and I began to work as an informal team to try to raise young birds. It was simply more than one person could do. At first we thought we could convince other birds to raise cowbirds and developed a colony of canaries for that purpose. We found we could breed the canaries but that even though canaries would feed young cowbirds, sometimes twice the size as the canaries, something was not right, most likely the diet, leading to the early deaths of the young cowbirds. We also tried zebra finches but were not more successful. Before we knew it, we had colonies of adult cowbirds, canaries, and zebra finches requiring daily care. We sold canary and finch offspring to raise funds to pay for food and supplies. Eventually, we raised one cowbird, #62, that thrived, the last bird of the 1973 season. Unfortunately, or so it seemed at the time, she was a female and we were pretty convinced that vocal communication was key to socializing and females did not sing. We decided to try a playback experiment to her, following in Dilger's footsteps and at the suggestion of Bob Johnston, a new faculty member in psychology. We were hoping that the female would approach the playback speaker but to everybody's astonishment, the female responded with a copulatory posture to cowbird song. Subsequent playbacks revealed the response to be selective to cowbird song and not non-conspecific song.

We rightfully thought we might have the beginnings of a paradigm in which to test quantitatively the functional properties of different birds' songs. But we had to solve the husbandry problems first, which would become the major focus of Drew's dissertation. While this was ongoing, Drew would take data on kitten play bouts in the afternoon and I fed his birds in the morning, giving us both chances to get to campus and do other work. We were spoiled: we had inherited the Howard S. Liddell Animal Behavior Farm when the department moved into a new building and absorbed many off-campus facilities including the airport lab. I had spacious rooms for litters of kittens and Drew had at least five rooms for aviaries. Our only neighbor was someone studying raccoons in one room; they occasionally escaped and trashed our offices but it was a small price to pay for almost unlimited space. Drew also had outdoor aviaries outfitted to study brood parasitism with perches on micro-switches in front of potential host nests to count how often the nests were visited. We learned to make fake eggs (with the help of Joan Johnston) to place in the nests. They were made of frozen chicken yolk dipped in paraffin. The only problem was the heat – nothing smelled worse than overheated egg and wax. But we learned that we could breed cowbirds, and Drew's dissertation revealed a great deal about the egg-laying behavior of females, which, along with the playback response, set the stage to study the developmental cycle: from egg to egg.

North Carolina

But before we could pursue that goal, we had to finish theses and find a place to work where we could have facilities like those at Cornell. We were also hooked on the outdoors after living in Ithaca. Two people drew us toward North Carolina, Harriet L. Rheingold at UNC and Gilbert Gottlieb at the Dorothea Dix Hospital, an almost unique state-funded basic research unit for which minimal clinical duties were expected. We chose to move there after I was offered a one-year visiting assistant professor position, with the expectation that I would work with Rheingold to learn how my thinking about play would fare with human infants.

Drew made arrangements to have contact with Gilbert's lab although a postdoc slot was not available. Instead, he accepted a lectureship at Duke with the support of J. E. R. Staddon. Gilbert also taught a proseminar in the developmental division at UNC where I was located. As a result we also got a good dose of the teachings of Kuo, as well as Gottlieb. Kuo had helped Gottlieb fashion a window in duck eggshell to watch and manipulate the final stages of embryonic development. My duties involved teaching child development, which was a challenge because I had never taken a course in child development, and so I learned along with the students. As a visitor, I was not given any lab space but began to collaborate with Rheingold on studies of toddlers and social pragmatics of communication, which had some overlap with play. I saw enough infants to realize, however, that the much more slowly developing motor system of the human infant did not translate into anything that looked like social play in mammals. As I had no space to study kittens, my attention was changed considerably to look at how active infants socialized those around them. I was especially interested in their vocalizations, as their vocal play seemed closer to mammalian play in its paradoxical repetition and inventiveness. It was also clear that parents seized on infant sounds to carry out proto-conversations and to divulge much information about the environment. I would eventually get an NIH career development award to study both birds and babies. During that time, I worked with a number of students interested in some aspect of parent–child relations. Jim Green and Gwen Gustafson looked longitudinally at mother–child interactiveness, and Gena Emery and Anne Arberg looked at language. Finally, George Holden developed new techniques to look at parental reasoning. I also worked with Dr. Eleanor Leung, who had been a research associate in Rheingold's lab, and we collaborated to look at the properties of maternal speech.

I learned a great deal from Harriet about infants and also academics over the years. She read and commented on every paper I wrote and would discuss the craft of writing at length. Her writing was superb (her dissertation and almost every subsequent publication was published with no revisions). I still see or feel her looking over my shoulder as I write. She had been an undergraduate at Cornell and thus was another alum that helped focus my career. But my interests in cowbirds always overshadowed my interest in infants because the cowbird work seemed theoretically the most important contribution we could make to understanding broader principles of development. We also saw the bird work as well as a way to think up new methodology to use with human mothers and children, much of it modeled from the bird work.

We had started house-hunting immediately, as we needed a place for our animal collection which consisted of something like nine cats (I had given away 42 kittens in pairs as I did my work), one dog, and several dozen birds. We found a rental house with a basement and moved the birds into large flight cages, and we housed the sound attenuation chambers we had had built at Cornell into what should have been the living room. The chambers were 1 m^3 with one-way windows in the front. We could watch bird TV, with the channel set to social development in cowbirds. After much searching we found a small farm with a great barn and a garage and we moved into more permanent quarters. Designing and building outdoor aviaries consumed much time in the first year but not so much that we could not follow up on cowbird #62's response to playback. We raised a new set of females and did playbacks to them of several cowbird songs plus the songs of heterospecifics. They responded with copulatory postures and surprisingly responded more to the song of male cowbirds that sang atypical songs because they had been raised out of earshot of other cowbird males. Why would such song be more potent? We published the results in *Science*, our first collaborative paper, and began to chart a five-year plan of research (King & West 1977). Harriet was somewhat frustrated by the competition that birds gave to watching babies but was enough of a comparative psychologist to know that we had to pursue this anomaly as it might tell us something important about the development of song, and she was for any kind of developmental research. She would be pleased to know that in the last few years both Drew and I have begun doing as much infant as avian research and finding intriguing similarities on early stages of vocal development.

Animal behavior Farm

Beginning in the late 1970s, Drew and I became a formal research team (co-PIs on grants) as it fit both of our interests and needs, i.e., the practicalities of doing the avian research pretty much single-handedly off campus with none of the traditional help a university might provide from animal care to technical assistance. We had decided that an independent lab/farm would keep us out of the space wars that occurred on campus and, more importantly, no one knew more about the care or housing of a wild species than we did and thus we had no one on campus available to help in a meaningful way. We had also learned at Cornell the importance of easy access to one's lab and so living at a farm with the right facilities seemed the optimal course. I believe many of our earlier studies could not have been done any other way and the same is still true today. The best ideas still come from watching the birds. We generally feed and care for the birds ourselves because it keeps us closely tied to the animals' behavior. I think I remember a story about Konrad Lorenz and someone asking him, after he had won his Nobel, why he did not have a technician do the feeding; he quipped that he did not see why someone else should have all the fun.

Fortunately we were successful in grant writing from NIMH and NSF. The grant funds also intersected with the practical question of one vs. two tenure track jobs. We did not see how we could accomplish our research goals and have a family if both of us had the full measure of academic duties. We also saw how hard it was for friends who chose the two-job route to balance raising children with a research career. And so, when a tenure track job became available at Duke, Drew passed on it but remained an adjunct member. It was a risky move, especially given our unusual lab arrangement. But grant money not only facilitated the research but it also helped to legitimize the arrangement: PIs providing their own personal laboratory on private property. The off-campus lab also had the advantage of allowing us to move quickly to construct new aviaries or modify existing structures without the "help" of a University architect or the physical plant.

Our approach might not seem odd to field workers and in many ways they represented a better model than lab workers. We firmly believed that the social ecology in which we kept the birds was playing a major role at a time when birdsong research focused almost entirely on vocal cues. Most birdsong researchers who studied development also used more stringent deprivation paradigms, often raising young males alone to reveal the "isolate" song of the species, which was considered some kind of genetic blueprint for song ontogeny. We did not believe that solitary housing was developmentally legitimate because in nature, although social conditions varied by species, young songbirds did not live in social isolation, thus it seemed evolutionarily suspect to use this as a baseline condition. But we still fell to some extent for the typical avian lingo and referred to our males who were raised with females (who did not sing) or other species as "isolate" males until it became abundantly clear that the term was a misnomer, which we explored in a paper entitled "Enriching cowbird song by social deprivation" (West & King 1980).

During the late 1970s and 1980s we tackled the social question of why cowbirds with "abnormal" songs were preferred by females in playback studies. First, we built aviaries in which to watch flocks of birds as well as using indoor flight cages and more sound-attenuating chambers. We also recorded cowbird song in the various housing contexts. We found the acoustic basis of greater song potency, with the help of J. E. R. Staddon at Duke University, who worked with Drew using a zero-crossings analyzer (ZCA), which allowed the user to see the song in real time without the frequency/time trade-off inherent in sonograms (West *et al.* 1979). Real-time analyzers are now commonplace and inexpensive, but John's invention was very new at the time. John's main area was animal learning but he was very open to ethological considerations and thus was an excellent collaborator.

During our tenure in NC, cowbirds were a common but not an abundant species; the population was a relatively new one from a phylogenetic point of view. We decided, with the help of a student, David Eastzer, to look at other populations of cowbirds and see whether we saw the same pattern of vocal development and function (Eastzer *et al.* 1985). Each summer David would pack up his pickup truck with a portable cage in the bed and drive to Texas or Oklahoma to record more ancestral populations. All in all, he studied 8 populations and found variation in song to be the rule. He also looked at border populations between two cowbird subspecies and found an interesting mix of songs from both populations; when we brought females from that area back to the lab we found they had very broad preferences for the mix of songs present on their natal grounds. The comparative work on geographic variation fascinated us because we saw so much variation within subspecies, as well as across subspecies (King & West 1990). In retrospect, it is not as surprising given the ecological differences between populations from the nature of the habitat to the degree of migration. Cowbirds are found across North America, thus there was much to differ in their life histories (Rothstein *et al.* 1986).

Our home/lab became an even more treasured resource once we had children; it also reinforced our job decisions. At that time, babies at home and work on campus presented us with every parents' major dilemma, "How do I make time for both kids and work?" Even with our caretaking arrangements, I confess that the "kid/job" conundrum posed difficulties, at least at a mental level. When in at the office I thought I should be home, and vice versa. But it was possible to divide the time between our two schedules and care for our sons at home. Raising human infants while raising birds also had much in common in that the adult/caregiver was not really in charge: the demanding youngsters took center stage. Having our lab at home also meant that our biological and academic children met and interacted. When our oldest son was about five, he asked whether David Eastzer was his older brother, a perfectly appropriate question given that David spent so much time at our home/lab. He received the real thing two years later and dove right into the role of sibling.

Figure 18.4. A female wing stroke to a singing male photographed from video monitor. The wing stroke lasts between 200 and 400 msec. Photo by APK.

It was during the early child-rearing years that two very fortunate things happened, albeit of very different natures. First, I received a five-year career development award from NINCDS, freeing me up from any teaching duties. Thus, we could focus full time on the latest interest in the lab, the role of non-singing females in song development in young males. We had raised young males only with females who could not sing. Thus, they could provide social but not vocal stimulation. It took much exploring to discover the signal system of the females, which turned out to be visual gestures such as wing strokes and gapes to song onset. Thus, we discovered a remarkably open system based on social learning (King & West 1983a; West & King 1988). This was the second lucky event, a turn in the research that was unheard of in the field of birdsong. We were led to find a visual gestural system by videotaping the males when singing to females and saw that at times the males became very excited. When we traced back the males' footsteps, we found the female's very rapid wing gestures were hard to see with the naked eye as they lasted only 200–300 ms (West & King 1988).

What was groundbreaking was that the males were not learning by imitation, but through trial and error or operant shaping. The males had to read the female's behavior but not repeat it. Imitation, the thought-to-be core learning mechanism, was not at work. Although the birdsong literature was beginning to find social effects to be important in species where male tutoring took place, this was the first evidence of female tutoring and a role for visual stimulation in vocal development. The nature of this finding was sufficiently surprising to some reviewers as to cause them to propose that the females must be secretly singing. We pursued the effect in a series of papers, and the social role of female behavior in structuring male vocal and social behavior continues to be a theme in our work.

Indiana

While our years in North Carolina were productive, we missed the integrative environment we had known at Cornell. UNC did not offer the transparency between programs that we had grown up in at Cornell. We did not have access to graduate students who came in through biology, and psychology had no program in animal behavior (although the developmental division supported animal work). There was no way to attract students interested in integrative work. Thus began a search for a school offering a more interdisciplinary program, as well as good access to cowbirds, and a nice place to live for our kids. I visited a number of schools in the Midwest, but after searching for a year Indiana University stood out because of the nature of its psychology and biology programs. The IU psychology department could not understand that we were not angling for two jobs and that Drew was perfectly content with a senior scientist position, a decision we have never regretted. We were invited to join Psychology in 1989 and we were very welcome in the Biology Department; in fact our first student at IU, Todd Freeberg, was in the ecology and evolution program, thanks to the efforts of Ellen Ketterson, Val Nolan, and Bill Rowland. I also became a member of the IU Biology Department in 1991.

Our move to Indiana was complex because we wanted to build a bigger lab and we wanted it to be within 10 minutes of campus to facilitate student participation. All moves for senior faculty are difficult because so much must be disassembled and re-created in both personal and professional lives. Many people went far out of their way to help find a property for our lab, with Lloyd Peterson and his wife, Peggy, leading the list, which also included Bill Timberlake, Rod Suthers, and Esther Thelen. Having built a lab at Cornell and then on a bigger scale at UNC, we had in mind some basic principles that had guided us along the way. We had limited resources in both prior settings, first, as grad students, and then as faculty at UNC but with no funds for start-up. At Indiana, the Psychology Department, the Dean's office, and NSF through CISAB all contributed financially to our new laboratory, named the Animal Behavior Farm as we had bought a small farm including a house for student offices and a kitchen to make food, a metal out-building to house equipment, flight cages, and our sound analysis equipment, and finally, large outdoor aviaries with shelters. It took a year to find an appropriate site: the land was too perfect to pass up and so we bought it with the commitment to build a home within a few years. Before that we lived in an IU dorm and then in the renovated old farmhouse at the lab.

We landed at IU at an especially crucial time as they were submitting a grant to NSF to develop interdisciplinary work in animal behavior, by gathering together faculty in biology, psychology, and medical sciences. The co-directors were Bill Timberlake and Ellen Ketterson. The Center for the Integrative Study of Behavior (CISAB) has grown to a faculty of over 40, an expanded mission in teaching and research, and hundreds of students have used CISAB's resources and gone through its program. One the most distinctive features of the Center was its commitment to equipment such as computers, microphones, tape record-ers, and finally a DNA and endocrinology lab run at first by Dr. Amy Poehlman. Dr. Shan Duncan ran the technological side of the center for many years and helped many students and faculty as well creating connections to the Animal Behavior Society. But the most distinctive mark of CISAB was the commitment to vertical and horizontal integrative training through courses, colloquia, and research plans. The nature of our lab and CISAB allowed us to attract graduate students seeking integrative training which we saw as critical to our long term goals to integrate communicative development in birds with converging studies of human communicative development.

Needless to say, the faculty in the developmental program in Psychology was the most compelling factor in our transition. Esther Thelen, Susan Jones, Linda Smith, and Jeff Alberts formed the core of the developmental program, and we were soul mates with them in terms of approaches to developmental questions. The rest of the faculty contained many luminaries, making the new department even more inviting. Indiana was just beginning to support integrative research on a large scale. Our students had more resources to use and more faculty to choose for their committees. I must also add that the then Dean, Mort Lowengrub, never blinked an eye at our request to build an off-campus lab that they would help fund, and used the weight of his office to eliminate bureaucratic obstacles. The administrative web eventually included many people who were supportive, but Mort's enthusiasm and confidence stood out. Every subsequent Dean and system president has been helpful.

During my time at IU, I assumed new duties as Editor of *Animal Behaviour* from 1991 to 1994. Ellen Ketterson was the editor for Short Communications, and Kris Bruner, the managing editor, began her long productive association with the journal on our watch. This was before the journal had decided to go the multi-editor system, which began in a small way in my final year as editor. Now there are four or more American editors and an equal number on the other side of the Atlantic with our sister organization, ASAB. After the editorship, I was elected President of ABS and served during tumultuous times as we wrangled with ASAB and the publisher about profits and managed to strike a new deal giving the American executive editorial office more money, which we dearly needed to keep up with the speed with which papers arrived at our door. We also established a central office for the journal at Indiana, ably staffed by Steve Ramey, Kris Bruner, and Lori Pierce. We could not have found a better or more dedicated manager than Steve, who has spearheaded all of the features of the editorial process that now are handled electronically. Later, in 2000, I became editor of the *Journal of Comparative Psychology*, with Sue Linville as a wonderful managing editor. JCP attracted a somewhat different audience than *Animal Behaviour*, with many more experiments on cognition, especially in non-human primates. I did try to represent other taxa as well, and other topics. So my childhood worry about sufficient reading material was somewhat assuaged. Journal work, as Lee Drickamer, the previous editor, told me, is "relentless" and follows you everywhere. But I truly enjoyed learning about the creative questions investigators thought up to ask and clever ways of answering them.

Aviary work

On the research front, we were shifting our basic experimental designs to meet the affordances offered by flocks in the spacious aviaries. Thus, the major structures, outdoor-indoor aviaries at the Farm were the heart and mind of the lab.

Figure 18.5. A view of one of our largest aviaries also showing one of the shelters between them. Photo by APK.

The experiment that set the tone for the next decade was done with Todd Freeberg (Freeberg *et al.* 1995). We used all our different facilities. First, we individually housed young wild-caught SD male cowbirds with either SD females (FH) or canaries (CH). They lived together through the fall into the spring. The birds were housed in sound attenuating chambers (1 m^3 with a one-way window). Then, in early May, we moved the males to flight cages (2.4 m × 1.8 m × 1.8 m) with the five males in each group housed together. We had never done this before, systematically looking at how birds react to the greater freedom of the flight cage. At first, it seemed there would not be much to see as the birds sat quietly hardly moving – any sound sent then flying in a frenzied fashion with the males seemingly intent on not landing on a perch containing another male. And for some reason, the CH birds seemed to be more affected than the FH birds. But within a week, they were much calmer singing to themselves or to the canaries housed with the CH males. One of the reasons we had housed them in chambers was to control their experience with other cowbirds. So the CH group represented the closest we came to an isolated male. The FH males were also isolated from males but presumably were being socially tutored by female cowbirds. We saw this experiment as a definitive test of whether cowbirds had open or closed systems with respect to species identification. If the cowbird system was innate, as many presumed, all the males should end up courting local females.

We set up a test of social recognition by introducing the males individually into a neutral cage containing local female cowbirds and canaries. The FH males reacted in what appeared to be a species-typical manner: approaching and singing to females and ignoring the canaries. The CH males were different: the males courted the canaries, singing and chasing them. Thus, socially isolated cowbirds do not have a template for species recognition; it is acquired through experience with adult females. We assumed that adult males played a role as well, and our confidence in that statement grew when we placed all of the FH and CH birds in two large (18.3 m × 9.1 m × 3.7 m) aviaries containing many potential mates. Included were female cowbirds from NC and SD and well as IN, canaries, and starlings, a novel species. Each day we recorded their singing, mating, and social behavior. Much to our surprise, the CH males continued their canary pursuit, ignoring other cowbirds most of the time. So even seeing more normally raised males did not induce social learning. However, the FH birds were not very good models, because here in the aviaries they also did not court the solicitous females, nor did they sing much to other males. Typically, male cowbirds exchange songs with one another in a behavior known as countersinging: we saw nothing like it.

As a last gasp effort to extract species-typical behavior, we introduced adult males, who immediately began to court and countersing. But seeing adult models caused no change in the CH and FH males. Obviously for male experience to be effective they must interact much earlier in the year. This set of experiments represented a turning point in the lab: now that we saw the difference in behavior between conventional housing and the aviaries, we knew that work with aviary flocks would continue to be necessary. This work also clearly showed the limitation of assessing song quality by playback, as males needed to learn to use their songs regardless of quality.

Todd Freeberg turned to aviary housing to do a daring dissertation on the cultural transmission of mate preferences, showing that mate preferences were influenced by post-natal experience with adults and that preferences could be passed on to the next generation (Freeberg 1996). The experiment was daring because the birds were outside and could see and hear wild cowbirds and yet appeared to be influenced only by the birds within their aviary. So social interactions appeared to matter greatly. We have since shown that song sharing, commonly seen in the field, can be limited by a transparent aviary wall separating two flocks: we saw no song sharing across flocks, but did see it within flocks. We have repeated this test with eight flocks with the same result each time. Thus, it became clear that social context predicted song learning, not simply exposure to song. This led us to start to investigate the role of larger flock dynamics to gate social stimulation.

Studying birds in flocks brought on many new methodological issues such as how to record flock dynamics. Anne Smith, a student in biology, got us started with a study of a large flock of 74 birds, in which she used paper and pencil measures of near neighbor patterns (Smith *et al.* 2001). These patterns allowed us to see structure in the flock with birds assorting by age and by sex. But Anne also found that juvenile males whose second most frequent neighbor was an adult male fared better in the breeding season. And males for whom females were the second most frequent neighbors had males whose song development progressed faster than that of other juvenile males. This finding fit with earlier IU work on male and female influence using conventional housing and demonstrated the role of social structure in directing different developmental outcomes. Anne also analyzed DNA from the 74-bird flock and found no relationship between kin in terms of social assortment.

We had long wondered what kin relations would look like, since cowbirds can potentially have many siblings due to their brood parasitic habit.

But the problem with more birds was not space but methodology. We had reached the limits of paper and pencil: we needed a way to gather more data from many birds simultaneously. We began to look into alternative methods with the help of Shan Duncan and a new postdoctoral fellow David J. White. We developed the use of voice recognition software so that observers could "speak" the data into a wireless lapel microphone without looking away from the birds; the information was transmitted from the aviaries to the lab building where computers received the codes and a database organized them into summary tables (White *et al*. 2002a). We found we could take many times more data and could for the first time record detailed reactions to songs in real time, not just the songs themselves. For example, in our first large flock study over an 8-month period four observers collected approximately 32,000 data points and entered the data manually into a database. By contrast, using voice recognition and programmable databases, four observers can collect and analyze a comparable data set in about 20 days. With Dave as a collaborator, we went on to study social development of males housed since fledging in flocks ranging between 20 and 30 birds. Dave spearheaded an effort to look at juvenile male song and social development as a function of social context, i.e., the presence or absence of adult males (White *et al*. 2002b). We found that males without adults differed on many dimensions from juveniles with males (all had females present). A behavior that emerged as important was counter singing (CS), the behavior that we had found to be absent when males were housed with just females. CS consists of rapid exchanges of song between two or more males (Figure 18.2). We found that the males without adult males did not show CS but those housed with adult males did. Although females were not the targets of CS, they did appear to notice the behavior, as we found that CS correlated with the number of eggs laid (King *et al*. 2003a,b). Further studies revealed the importance of social learning to CS and we found that CS could be transmitted from one generation to another. But we also found the absence of CS could be transmitted to a new generation of males (White *et al*. 2007). So, competence and incompetence were under cultural control.

The efforts are still ongoing, but one enormous surprise was the finding that the males who seemed most dominant and generally sang the "best" songs did not sire the most eggs. This finding was particularly apparent when individuals were followed over several years, interacting with different social companions from year to year. Thus, the ability to measure cumulative reproductive success over several years is beginning to present us with a very different picture. In particular, the role of singing performance and the ability to adjust behavior from year to year may well turn out to be critical to understanding reproductive success. Many ideas about the functional consequences of different male phenotypes can now be addressed. These findings have led to investigations of the importance of communicative pragmatics. This work extended the Freeberg *et al*. (1995) findings with CH males to show the ability of exogenetic stimulation not only to alter development but to connect to cumulative reproductive success, bringing our measures to the threshold of fitness.

We did not ignore the females in our expanding work with flocks. We had done a series of developmental studies using restricted housing in the early 1980s and had concluded that females, unlike the males, had a closed program of species recognition, a finding we replicated with several geographic populations (King & West 1983b). All of those studies housed females in small groups in sound-attenuating chambers with or without a male and found no evidence that female preference for male song could be modified. Now in the flock setting we discovered a completely different picture of the development of female preferences. We raised one flock of only females. The females could of course hear male cowbirds outside the aviaries. We then tested the females' playback preferences for local and distant South Dakota and Texas song (King *et al.* 2003a). The results from the all-female flock were surprising, as adult females showed no preference for local songs over those of other subspecies. Neither did juvenile females. The finding was especially dramatic with respect to Texas song in that those songs were lexically very different from Indiana song. This was astounding because all females we had ever tested for local vs. distant population preferences preferred local songs. Thus housing females in a flock without males had erased preferences for the adult females at the macro-geographic level. We went on to show that contact with males mattered, even if it came from tape recordings of male song. Thus, by tape or live tutoring outdoors where wild cowbirds were present, we could induce specific preferences in the female cowbirds when they were housed in flocks (West *et al.* 2006). We could now see that the restricted housing work of the 1980s had the effect of freezing female preferences while the flock studies revealed that when females can observe groups of other females react to male behavior, their song preferences are easily changed. This work dispelled the notion of a genetically predetermined safety net for reproductive behavior. Thus, the flock studies on the development of male singing performance as well as female preferences showed the presence of an exogenetic mechanism.

Largely through the work of Grace Freed-Brown, we were also beginning to learn something about the development of female social dynamics in flocks and found important differences in the interaction patterns of juveniles versus adults, with the juvenile females being much more active and less discriminating in their choice of social partners (Freed-Brown *et al.* 2006). With Jennifer Miller, we also found that flock-housed young females (reared from the egg) showed no evidence of same-sex sociality as juveniles (Miller *et al.* 2006). Thus, like males, females did have to engage in early learning, presumably from adults, in order to show species-typical behavior of strong assortment by sex. Taken as a whole, these studies begin to show us the social mechanisms actually responsible for female preferences and we believe that these types of studies will increasingly use social networking statistics to describe the developmental environment that is the safety net for the acquisition of appropriate reproductive behaviors in both males and females. Thus, we see network statistics replacing the ghosts of the innate safety net.

Flock studies are ongoing, with an emphasis on vocal improvisation and its relation to pragmatic ability in young males, as Jennifer Miller has found that males raised with adult females improvised more than juvenile males with juvenile females and that the adult-housed males showed superior courtship skills in their first breeding season. Thus, it appears

Figure 18.6. A starling listening to a human interacting with him. Photo by APK.

that the adult females were socially shaping the song content as well as how to deliver the song. Improvisation has not been studied in anywhere near the detail of song copying but may well prove to be an engine of pragmatic skills. We believe that there is at present a "missing link" in connecting birdsong capacity to birdsong competence, e.g., the use of song as an effective elicitor of reproductive behavior in females. The common assumption is that competence simply flows from some possibly innate by-product of birdsong development. Understanding the dimensions of competence, or what we call "communicative pragmatics," can be summed up as answering the "wh" questions, the "who," "what," "where," "when," and "why" of singing performance. We presently are investigating how young birds (a) learn how to use their signals though social modeling and social operant learning and (b) learn to lengthen their attention span so as to be able to acquire critical feedback from social companions. Indeed, we believe that acquisition of attention span, also critical for human communicative development, becomes the root mechanism for learning and using birdsong in myriad ways, including territorial encounters, song sharing and mating.

During the years of cowbird work, two other lines of research were leading us to the same conclusion about the importance of social shaping. First, we worked for several years with starlings, many of which were hand-reared and kept in human homes under different social circumstances. We found that starlings engaged in vocal sonar and their repertoires showed evidence of human shaping, as they included human words if they had lived in interactive contact with a human (West *et al.* 1983).

Interactive contact simply meant that birds had much more social freedom than the control birds and routinely socialized with humans. Marianne Engle completed a dissertation of the social circumstances for mimicry and created a scale of interactivity. Starlings are an intimidating species to work with because their song is so much more complex than that of cowbirds, but they adapt marvelously to captivity and there is now a niche in the bird lover's world devoted to "pet" starlings, with terrific material on the web.

We also explored the life of one starling in depth, a starling that had been owned by Wolfgang Amadeus Mozart for three years. We were curious about the relationship, which we pictured as comical and affectionate, based on reports from our 1980s starlings. We also thought that Mozart left a requiem for his pet, the piece known as the Musical Joke K522, so-called because it has fragmented parts that do not quite come together and much repetition, including a nine-trill note that sounded to us like the contact call of our twentieth century starlings: they were idiosyncratic calls to say the least but their function seemed clear. Thus, we pictured Mozart's incorporating starling phraseology into a simple folk song. We wrote a paper about the whole historical adventure and its relevance to social shaping for the *American Scientist* (West & King 1990). It is safe to say that it is the article most requested of anything we have ever written. I talked about the experience at an AAAS convention and the press got hold of it in 1991, the 200th anniversary of Mozart's death. Our story became a popular part of Mozartiana and the report was in many science and popular magazines, newspapers, and radio…we still get requests for interviews 17 years later. I also get e-mails from starling owners wanting advice about health and food. Each story is a variation on the theme of "I found this ugly looking little bird cheeping near my home and I fed it for a few days (usually dog food, which was not that bad a choice) and before I knew it I was in love with the most imperious young bird I have ever met." I still hear from owners with updates, as the birds can live into their teens in captivity.

But the starling work was also important because it showed another improvising species that was affected by social contact. We had had a second group of starlings in our first experiment that lived as caged pets in human homes and received good but not extra-ordinary care, such as playing in the kitchen sink. They were raised as most caged birds are raised. They mimicked household sounds but no human voices. Thus, the interactive contact, as in cowbirds, mattered. Merely hearing and seeing humans was not enough. These findings served to deepen our interest in social shaping, especially what we saw as similarities to the development of human speech. This interest began with the work of Michael Goldstein, who decided to follow up on the wing-stroking work in female cowbirds by looking at prelinguistic communication in human infants. As in the cowbird work, he set out to capture the nature of social maternal responses to the activities of their children. We first found that mothers perceived different infants' communicative signals in a consistent manner, ascribing the same function to the different sights and sounds (Goldstein & West 1999). This was important because it meant there was something in the prelinguistic sounds to hear, i.e., communicative meaning. But Mike's most important achievement occurred when looking directly at social shaping. He compared infants whose mothers delivered either contingent or non-contingent non-vocal feedback in an ABA design. He found that infants whose mothers were contingent responders had infants who showed advanced phonology in the second A trial, whereas infants whose mothers were non-contingent did not. Thus, we had strong evidence that social shaping of vocal structure also occurred in humans, at a point in development comparable to the time we saw it in birds (Goldstein *et al.*

2003). The *PNAS* paper garnered much media attention, second only to Mozart's starling. I think it was of interest because of the comparative theme and the fact that the effect did not depend on imitation. Imitation is the supposed mechanism in both species for vocal learning; however, neither the babies nor the male cowbirds were copying their partner's behavior but extracting from it information about what sounds were effective in obtaining a response.

Julie Gros-Louis, a postdoctoral fellow with a background in primatology and communication, brought a fresh eye to both the bird and the baby work. Julie analyzed the content of mothers' behavior during free play and found the mothers responded differentially to infant behaviors and sounds, affording the infants contingent feedback that varied with the infant's vocal or behavioral action (Gros-Louis *et al.* 2006). Because the avian work points to a fundamental role for the development of attention to predict vocal and pragmatic skills, we are beginning to study the development of attention in prelinguistic infants. Specifically we are investigating the role of caregiver contingent stimulation in lengthening or shortening infants' attention. Erin Ables and Jennifer Miller are just completing a series of studies that demonstrate that infants show differences in attention span depending on the nature of caregiver contingencies. We are doing similar studies in the cowbirds, looking at directed or undirected song as proxies for attention or inattention. Juvenile males housed with juvenile females show shorter attention spans, for example, than juvenile males housed with adults. We should note that Jennifer Miller is the first student we have had who simultaneously has initiated research in both the bird and the baby labs and used observation and theory from both preparations to guide a parallel research program. This has been a long-term goal of our research, to have both labs ask essentially similar questions of both birds and babies but to do so in the same time frame, thus driving a comparative synergy.

Summary

I have used the word "surprised" several times in the text to capture our emotions at experimental outcomes. We stand by this attribution and it is our fondest wish that there are more surprises to frame our future. But it is especially important to remember the first "flashbulb" surprise: the deprived female's copulatory response to song. We knew we were being given a chance to ask very different questions about mate choice. Her greater response to the songs of deprived males suggested we had an adaptive "story": cowbirds, with no experience with one another, appeared to have the ultimate safety net to insure mate selection. The male side of the story was just as exciting as in most species isolate songs were less effective than normal songs in eliciting copulatory responses, but in cowbirds, atypical songs seemed to be more potent than normal songs. Thus, without experience on either side, mating mechanisms seemed to be in place, suggesting that females, through sexual selection, mated with males with the "best" song. What a great story for a brood parasite who would seem to need especially closed programs to engage in species and mate recognition.

The "closed program" story began to unravel when we used more naturalistic settings and stimulation. First, we found that females living in flocks showed no song preferences. We surmised that competition among the females induced plasticity. Second, we discovered that we could manipulate preferences through social housing, something we had not seen in more conventional housing. We also had to come to grips with the finding that silent females affected song content and quality. These data suggested that sexual selection might not be focused on song quality, but song use. If so, females may be selecting on the basis of male learning and attentional processes, especially by males watching females. Thus, non-imitative song learning was as basic to song development as imitation. We think, but do not know for sure yet, that improvising males may be the product of more attention to female responsiveness. Thus, females are using song use (directed song, countersinging) when it is available as a cue. The missing link was connecting male and female development. Ironically, we had always had access to female influence, as we never raised birds alone but thus used females as a control for housing with males. Even the males deprived of male company, say in the very first experiment with naïve males, had either other species or females with them. Gottlieb had frequently talked about non-obvious influences on development in his ducklings (such as prenatal vocalizations) and we considered the females' effect to be analogous.

Thus, the "adaptive story" of the lock and key female response to song turned out to be deceptive. Until we understood the sensitivity of the development of male song and social behavior to social context along with modifiable female preferences also responsive to context, we did not see that the actual safety net was an exogenetic mechanism afforded by a stable social ecology. In the early 1970s when we started our research program we were profoundly influenced by the developmental work of Marler, Gottlieb, Lehrman, and Tinbergen, among others. At the time an understanding of development seemed to be central to the investigation of the evolution and function of behavior. We believe that the single most significant contribution of our work is to remind others of that fact. At the present time, it is our perception that many investigators do not consider it necessary to know the development of a behavior to understand its function or what has been selected by evolution, as the mature behavior is assumed to be the endpoint (West *et al.* 2003). This convenient illusion is typically supported by a failure to understand individual and geographic variation of behavior that can be supposed to be genetic rather than ecologically driven. Even in research that focuses on the neural or hormonal basis of behavior, the present reliance on the lock and key connection of male song production and female reaction to song in a restricted housing preparation is likely to be misleading about the systems responsible for reproductive behavior. Consider that female cowbird song preferences are "frozen" by the restricted housing setting.

Thus as we enter the late innings of our career, we are saddened by the present trend, which seems to de-emphasize an appreciation of the essential importance of the development of behavior. Oddly enough, the fact that we were so taken in by the power of innate behavior in the early innings to provide evolutionary answers makes it easy for us to understand the attraction of this perspective, but it also provides us with the motivation for our continued research efforts. We owe a lot to the illusion of an innate answer to provide opportunities for surprises which inform about the stability of exogenetic mechanisms that evolution has chosen to trust for the constructive transmission of critical reproductive behaviors.

Conclusion

As I look to the future, I see more birds and babies but I do not see South American cowbirds or the Galapagos, goals that motivated us during the earliest years. We have found our wheelhouse looking at the myriad ways in which social experience modifies developmental trajectories. To reduce the work to the simplest terms, our research argues that nature and nurture exist in inherited niches: genes inherit environments, in other words, species-typical surroundings that shape the contribution of nurture (West & King 1987). To paraphrase Mace, what we believe is that it's not "what's inside your genes that matters but what your genes are inside of." Elsewhere we have referred to our exogenetic theme as a message in a bottle hugging the shore of seas roiled by the Human Genome Project (West & King 2001). We are up against formidable theories and habits contained in the biological frame of "genes for x". But, in the end, as has been shown to be true for genetic explanations, one has to confront an environment and trace its direct and indirect effects. This gives us hope that our words and deeds will be seen for what they are: descriptions of developmental contingencies that require the "right" environment in order for ontogenetic principles to appear.

References

Dilger, W. C. (1964). The interaction between genetic and experiential factors in the development of species-typical behavior. *Am. Zool.* **4**: 155–60.

Eastzer, D. H., King, A. P. & West, M. J. (1985). Patterns of courtship between cowbird subspecies: evidence for positive assortment. *Anim. Behav.* **33**: 30–9.

Freeberg, T. M. (1996). Assortative mating in captive cowbirds is predicted by social experience. *Anim. Behav.* **52**: 1129–42.

Freeberg, T. M., King, A. P. & West, M. J. (1995). Social malleability in cowbirds (*Molothrus ater artemisiae*): species and mate recognition in the first 2 years of life. *J. Comp. Psychol.* **109**: 357–67.

Freed-Brown, S. G., King, A. P., Miller, J. L. & West, M. J. (2006). Uncovering sources of variation in female sociality: implications for the development of social preferences in female cowbirds (*Molothrus ater*). *Behaviour* **143**: 1293–315.

Gibson, E. J. (1969). *Principles of Perceptual Learning and Development*. New York: Appleton-Century-Crofts.

Gibson, J. J. (1966). *The Senses Considered as Perceptual Systems*. Boston: Houghton-Mifflin.

Goldstein, M. H. & West, M. J. (1999). Consistent responses of human mothers to prelinguistic infants: The effect of prelinguistic repertoire size. *J. Comp. Psychol.* **113**(1), 52–58.

Goldstein, M. H., King, A. P. & West, M. J. (2003). Social interaction shapes babbling: Testing parallels between birdsong and speech. *Proc. Nat. Acad. Sci. USA* **100**: 8030–5.

Gros-Louis, J., West, M. J., Goldstein, M. H. & King, A. P. (2006). Mothers provide differential feedback to infants' prelinguistic sounds. *Int. J. Behav. Devel.* **30**(6): 509–16.

Harlow, H. F. (1965). Mice, monkeys, men, and motives. In *Curiosity and Exploratory Behavior*, ed. H. Fowler, pp. 91–103. New York: MacMillan.

Harlow, H. F. (1974). Induction and alleviation of depressive states in monkeys. In *Ethology and Psychiatry*, ed. N. F. White, pp. 197–208. Toronto: University of Toronto Press.

Harlow, H. F. & Harlow, M. K. (1962). Social deprivation in monkeys. *Scient. Am.* **207**: 136–46.

King, A. P. & West, M. J. (1977). Species identification in the North American cowbird: appropriate responses to abnormal song. *Science* **195**: 1002–4.

King, A. P. & West, M. J. (1983a). Epigenesis of cowbird song – a joint endeavour of males and females. *Nature* **305**: 704–5.

King, A. P. & West, M. J. (1983b). Female perception of cowbird song: a closed developmental program. *Devel. Psychobiol.* **16**: 335–42.

King, A. P. & West, M. J. (1990). Variation in species-typical behavior: a contemporary theme for comparative psychology. In *Contemporary Issues in Comparative Psychology,* ed. D. A. Dewsbury, pp. 331–9. Sunderland, MA: Sinauer.

King, A. P., West, M. J. & White, D. J. (2003a). Female cowbird song perception: evidence for plasticity of preference. *Ethology* **109**: 865–77.

King, A. P., White, D. J. & West, M. J. (2003b). Female proximity stimulates development of male competition in juvenile brown-headed cowbirds, *Molothrus ater. Anim. Behav.* **66**: 817–28.

Lehrman, D. S. (1974). Can psychiatrists use ethology? In *Ethology and Psychiatry,* ed. N. F. White, pp. 187–96. Toronto: University of Toronto Press.

Mace, W. M. (1977). James J. Gibson's strategy for perceiving: Ask not what's inside your head, but what your head's inside of. In *Perceiving, Acting, and Knowing*, ed. R. Shaw & J. Bransford, pp. 43–66. Hillsdale, NJ: Laurence Erlbaum Associates.

Mayr, E. (1974). Behavior programs and evolutionary strategies. *Am. Scient.* **62**: 650–9.

Miller, J. L., Freed-Brown, S. G., White, D. J., King, A. P. & West, M. J. (2006). Developmental origins of sociality in brown-headed cowbirds (*Molothrus ater*). *J. Comp. Psychol.* **120**: 229–38.

Rothstein, S. I., Yokel, D. A. & Fleischer, R. C. (1986). Social dominance, mating, and spacing systems, female fecundity and vocal dialects in captive and free-ranging brown-headed cowbirds. *Curr. Ornithol.* **3**: 127–85.

Smith, V. A., King, A. P. & West, M. J. (2001). The context of social learning: association patterns in a captive flock of brown-headed cowbirds (*Molothrus ater*). *Anim. Behav.* **62**: 23–35.

West, M. J. (1974). Social play in the domestic cat. *Am. Zool.* **14**: 427–36.

West, M. J. & King, A. P. (1980). Enriching cowbird song by social deprivation. *J. Comp. Physiol. Psychol.* **94**: 263–70.

West, M. J. & King, A. P. (1987). Settling nature and nurture into an ontogenetic niche. *Devel. Psychobiol.* **20**: 549–62.

West, M. J. & King, A. P. (1988). Female visual displays affect the development of male song in the cowbird. *Nature* **334**: 244–6.

West, M. J. & King, A. P. (1990). Mozart's Starling. *Am. Scient.* **78**: 106–14.

West, M. J. & King, A. P. (2001). Science lies its way to the truth… really. In *Developmental Psychobiology*, Vol. 13, ed. E. M. Blass, pp. 587–614. New York: Kluwer Academic/ Plenum Publishers.

West, M. J., King, A. P., Eastzer, D. H. & Staddon, J. E. R. (1979). A bioassay of isolate cowbird song. *J. Comp. Physiol. Psychol.* **93**: 124–33.

West, M. J., Stroud, A. N. & King, A. P. (1983). Mimicry of the human voice by European starlings: the role of social interaction. *Wilson Bull.* **95**: 635–40.

West, M. J., King, A. P. & White, D. J. (2003). The case for developmental ecology. *Anim. Behav.* **66**: 617–22.

West, M. J., King, A. P., White, D. J., Gros-Louis, J. & Freed-Brown, S. G. (2006). The development of local song preferences in female cowbirds (*Molothrus ater*): flock living stimulates learning. *Ethology* **112**: 1095–107.

White, D. J., King, A. P. & Duncan, S. D. (2002). Voice recognition technology as a tool for behavioral research. *Behav. Res. Meth. Instrument. Comput.* **34**(1): 1–5.

White, D. J., King, A. P. & West, M. J. (2002). Facultative development of courtship and communication in juvenile male cowbirds (*Molothrus ater*). *Behav. Ecol.* **13**: 487–96.

White, D. J., Gros-Louis, J., King, A. P., Papakhian, M. & West, M. J. (2007). Constructing culture in cowbirds (*Molothrus ater*). *J. Comp. Psychol.* **121**: 113–23.

19

A brief just-so story of my life (a few of the reminiscences that are fit to print)

MARY JANE WEST-EBERHARD

Prologue

Autobiographies often start with an apology to counter the awkward implication that the writer believes there is an interested audience. Even Darwin, when invited by a publisher to write his autobiography, began modestly by saying he thought it might be of interest for his children. But there are always a few readers who, like me, enjoy peering into the lives of others. This may especially be true of ethologists, who are professional voyeurs.

Unfortunately for voyeurs, autobiography is not the most candid medium. Even an honest attempt to find the truth of one's own life is doomed to fall short, for "no man ever understands quite his own artful dodges to escape from the grim shadow of self knowledge" [Joseph Conrad, in *Lord Jim*]. Because of this, the deepest human truths are in novels or in proper scientific studies of behavior, not in autobiographies.

In my generation in the United States, where I was born in 1941, it was still quite unusual for a woman to become a scientist. Women in science still face certain challenges not faced by men, while having at the same time certain definite advantages. I will try to think on paper, here, about what might have been different about my life compared with that of others who did not pursue what was seen as an odd career, or any career, with the same zeal, and about some of the exceptional, largely unplanned events that ended up making a full scientific career possible without giving up a full family life.

Early instars

I was born in Pontiac, Michigan, the second of four children but the firstborn of my parents, for my older sister Charlotte, six years my senior, was adopted after her mother (a younger sister of my mother) died when Charlotte was very small. We have a younger sister, Ann (b. 1943) and a brother, Richard (b. 1944). My father, Earl West, along with four of his brothers, was a small businessman who sold farm machinery, then appliances and cars in Plymouth, Michigan, a town with a population of about 8000 at the time I was in high school in the 1950s. His father, the son of Irish immigrants, ran a general store in Cherry Hill, a rural crossroad in southern Michigan near Ypsilanti. His English mother had immigrated to the US as a young woman.

I think that my father was a competent, but not a shrewd, businessman. He was too idealistic about service to customers, a man who cared more about amiable relationships than extraordinary sums of money. He attended the University of Michigan briefly, but the Great Depression of the early 1930s made it necessary for him to seek a full-time job. I identified closely with my father. Both he and my mother encouraged their children to be interested in things and to ask questions, qualities that are usual in small children but not always rewarded in little girls if they include curiosity about water pumps, radios, and trucks.

My mother, Chloe Losey, grew up on a farm in the Cherry Hill area that was commercially dominated by the West General Store. Her father was a proud, but not wealthy, landed farmer with fine draft horses and one of the first Ford cars. I know that my mother "married up" because the mantel clock she inherited from her family is a humble 24-hour clock, whereas the one inherited by my father is a more aristocratic eight-day clock [George Eliot uses clocks as indicators of wealth and status in *Adam Bede*]. I know little of my mother's ancestors, except that they were of Scottish, German, and English descent, and (like my father's family) Protestants, but for both of my parents church attendance was focused largely on weddings and funerals. Of the five children (three girls and two boys) in my mother's family, she was the only one who finished college and, much later, graduate school (she had bachelors and masters degrees in elementary education). Both of my parents put a high value on hard work and doing even small things well. My mother was less logical in argument than my father, but I think more ambitious and more of a perfectionist in every-thing she did.

My childhood was quite ordinary in many respects, but at the same time it was conducive to developing academic interests. My mother had done what many academically inclined young women of her era and rural middle class did, which was to become a teacher. I think that her career as a teacher was important in giving me early self-confidence in school, something that was reinforced consistently by my own teachers and so carried over into later life. Like her own mother (who was a strident, tough-minded farm woman who had worked in the fields like a man and could cuss like one too), my more refined mother nonetheless did not adopt the traditional wifely submissiveness that was still expected of her generation. She did not hesitate to outspokenly disagree with my father on matters large and small. But they agreed on how to manage children (mainly the domain of my mother) and in being conservative Republicans with the typical mid-western-U.S. Republican values of that time – staunch individualism and the belief that a person can mold his or her own destiny through honest hard work. This made them somewhat blind to the fact that our own lives, if successful, were not entirely due to our own independent efforts, but were full of advantages of birth in solidly middle class white families in communities with reasonably good schools, advantages that made upward mobility possible. At the same time my mother had a strong sympathy for children with special problems, especially those who were socially handi-capped by disabilities or family background. This sympathy extended to neighbors and other members of the community, such as foreigners or recent arrivals, who were marginal in the eyes of others. I mention this because I think that the open-minded attitudes of my parents toward people of diverse backgrounds helped make it easy for me to move out of an essentially provincial small Midwestern town, first to the expanded world of Ann Arbor, with its cosmopolitan university life, and then to an adulthood spent abroad.

That same open-mindeddness perhaps also made it easier for my parents to tolerate and encourage an odd daughter like me. An avid reader of pioneer stories, I wore a coonskin hat and a boyish (and incongruous) P jacket to primary school for more than a year. This was important for the walk to Smith Elementary school, when I tracked Indians by following the dislodged stones, broken twigs, and footprints that Daniel Boone would have found in the dirt trail. I made pioneer huts and hideouts of boards and grass in the vacant field across the street from our house in Plymouth Township, and would hunker inside, enjoying the solitary secrecy of my own little space. I also dug some Indian caches in our back yard, where I kept emergency provisions – a few coins and potatoes, and, once, some moldy cigarettes that I smoked secretly with a friend, perhaps under the influence of Huck Finn, another of my favorite book characters. My mother had to endure the nearly impossible task of shopping with me for dresses that I hated, the worst crises being in springtime, when girls were obliged by fashion to wear pastel colors, hats and white gloves to church on Easter Sunday. Luckily for me, my mother had an intuitive understanding of small children. She probably realized what I now believe about the value of a vivid imagined life, and about playing alone. A child who is always alone is unfortunate, but one who is never alone may be more unfortunate, for playing alone helps to foster the independent imagination that may contribute to thinking in new ways as an adult. My mother saved my coonskin cap. It now hangs in my study, an encouragement to exploration of new frontiers, and a monument to understanding parents.

My interest in animal behavior may trace back to some other childhood experiences. Until I was eight years old we lived on Maceday lake in the rural vacation area of Waterford Township, in southern Michigan. We had a big red canoe that could glide silently through lily pads in a swampy area adjoining the lake, and could get close to turtles before they would plop, startled, out of view. I liked trying to discover the turtles before they discovered us, and then to try to find them again when, after a child's eternity, they would poke their snakelike heads from the surface, in some unexpected place, to breathe. And I would concentrate on scanning the water in a solitary contest of discovery, to find the heads of snakes and turtles among the false heads that were just the tips of the branches of submerged trees, and to be the first to announce them in a careful whisper to the others in the canoe. Mother ducks with ducklings patrolled our waterfront, where they would scramble and race each other to snatch scraps of bread. You learned to find worms for fishing bait, and how to tell a bluegill from a perch from a sunfish. Pole fishing with worms, you learn about the rewards of patient, silent observation. You watch for small signs of movement in the water, and wait quietly for a nibble or a tug, deducing the behavior of an invisible fish by its tentative bites so that you can hook it with a well-timed jerk of your pole.

How much of my later love of patiently watching behavior came from these early experiences is difficult to know. But I have a definite preference for watching animals undisturbed, out of doors, in the places they have found for themselves, as if sneaking up to them in a canoe. And I can watch the behavior of wasps for hours, patiently waiting for something to happen, as if fishing in a still lake with a pole, knowing that my peculiar kind of patience will be rewarded by teaching me something about their lives that is new and real.

When I search my memory of early childhood for intimations of a career in science, one other memory stands out. When I was about seven I wanted a radio of my own. My father said that he could sell me one from his appliance store for the wholesale price of ten dollars, and my parents gave me the chance to earn it by collecting small sticks for fireplace kindling at the rate of ten cents per 24"–28"×12"cardboard box full. That meant I had to gather 100 boxes full of kindling, which I did, leaving them with a cellar full of small sticks when we moved a year later. Clearly my parents knew how to reward goal-oriented persistence. They nurtured the illusion that I could do anything I set my mind to if only I worked hard enough. The rhythmic refrain "I think I can, I think I can…" from *The Little Engine That Could*, one of my favorite children's books, echoed in the unconscious background at thesis writing time, and later, when writing long papers on kin selection, sexual selection, and, most monstrous of all, a 794-page book on developmental plasticity and evolution (a book that would have been shorter had I been willing to spend another ten years paring it down). Each of these projects required years of sustained effort, and all depended on scouring the literature for scattered sticks of formerly unrelated facts. Sheer dogged persistence may be as important for sustaining a career in science, especially for women, as any other personal trait.

It seems worthwhile to mention these early experiences because childhood events are so important in setting the expectations of self and others regarding what a person can do. From an early age I resisted the things (ballet, piano or flute, sewing, knitting) that girls are supposed to like and found the things that boys do (baseball, trumpet, playing cowboys and Indians, chasing insects) more interesting and impressive. If there were things that girls were not supposed to be *able* to do I would simply figure that it was a rule for sissies, not for me. So when a prominent professor at Michigan said that the zoology department needed to accept women graduate students because it was important to train capable lab assistants, I figured that applied to other woman students – like the one I saw knitting in departmental seminars – not to me or to the other women I knew who were serious about their work. It did not occur to me to lead a feminist revolt because I never felt part of an oppressed group. A famous meeting at the Massachusetts Institute of Technology on problems of women in science and engineering, where I represented the University of Michigan as a first-year graduate student (1964) (see Mattfeld & Van Aken 1965) made little impression on me because what other women saw as barriers I saw as inapplicable to those sufficiently willing to push ahead in a career. Just as I had not been sympathetic with sissies, I was not sympathetic – in retrospect, not sufficiently sympathetic – with girls who gave up because they were treated like girls.

Given these attitudes, beginning in childhood, my problem was not so much getting ahead in the man's world of science, as it was to adjust to getting along in the female side of life. I easily visualized myself in a career, but did not think I would make a very good wife or mother even though it was something I vaguely wanted to be. These worries ended up being resolved by a fortunate marriage and some nice little kids.

Primed by playing with a toy microscope, and later by 4-H projects in entomology and wild-flowers, I looked forward to the high-school biology course, where I thought there would be real microscopes and animals and plants to observe and dissect. Instead, the ninth grade biology consisted entirely of filling in the blanks of a dull workbook. The courses, in physics, chemistry, and mathematics did little to revive an interest in science. High-school science meant memorizing large numbers of unrelated facts, something I was not good at, and the students who excelled in those courses seemed too often to be socially awkward slide-rule-toting "squares." I did enjoy doing "reports" in those courses (like one on alchemy as a precursor of modern chemistry), and was inspired by reading *Woman Doctor* (a biography of Elizabeth Blackwell, the first woman to earn a medical degree in the US), and a biography of Marie Curie. My adolescent notions of an exciting intellectual life came to be identified with people interested in the great ideas of philosophy, literature, and history. Professors in tweed jackets. The ability to read existentialist novels in the original French. Literary circles in shabby cafes. Classroom science seemed neither interesting nor glamorous.

Many of the extra-curricular activities in a US high school reward competitiveness and leadership and I threw myself into many of those – girls' athletics, band, and student government – with what now seems to me to have been an exaggerated sense of importance. But this did encourage a leadership psychology, and offer some practice in public speaking, solo music performances under pressure, and politics, things that transfer helpfully into almost any field. In my high school most of the good students – the ones I most admired and wanted to emulate – went to the University of Michigan. I followed their example without hesitation.

A metamorphosis in Ann Arbor

I did not anticipate the many ways in which the four years in college would define and change my life. Ann Arbor was only a 20-minute drive from my home town, but it was a very different world: there were large numbers of Democrats, Jews, and people from other countries, and even a likeable and intelligent girl whose mother was a labor leader in Detroit. It was the early 1960s, and Ann Arbor was a center for the student movement. Having been a leader in high school activities, and being idealistic about trying to do something positive for humanity, I was active in several campus groups. But I soon became disillusioned with student politics and also with the brand of academics I had thought I admired. At the same time I was being forced, by the need to choose a major, to decide what I thought was important in life, and what direction my future should take.

On the academic side, several important things had happened. My interest in biology had been revived by a freshman zoology course, in an honors section taught by Richard Alexander, a new faculty member who (I learned later) had decided to teach a freshman lab section in order to learn fundamentals of general zoology that his training as an entomologist had left out. I finally had a chance to use a real microscope, and also learned that science could be full of interesting concepts: Alexander emphasized evolutionary ideas, and raised questions to stimulate discussions. And he loaned me copies of articles he thought I should read, such as a *Scientific American* article on the origin of life, by George Wald. From that article I most remember Wald's point that given a very large volume of interacting molecules and a very long period of time it was unlikely that the seemingly improbable origin of life would *not* have happened.

Richard Alexander's office was in the Museum of Zoology, in the Insect Division. This was a pleasing coincidence because entomology was the one area of science I thought I knew something about, through my projects in 4-H. This helped give me the courage to knock on his door and inquire about a job. The result was a part-time position as an assistant in the Insect Division. In the friendly setting of the Museum I got to know some of the graduate students and the professors, and heard them discuss their research and the big questions of evolutionary biology and ecology. As part of my job I helped with the library research for a major review of mating behavior in insects, and so became deeply familiar with the museum library and at home in the specialized literature of insect natural history and behavior. And I learned a lot about the rearing and behavior of live crickets and other Orthoptera, Alexander's specialty.

One summer, when I was working full time in the museum, a female short-tailed cricket made her burrow in full view in a big glass jar. Dick Alexander did not have time to observe her behavior, so he passed that opportunity to me. I ended up doing many hours of continuous observations, on some days recruiting my younger brother Dick to take my place while I went to the bathroom or ate lunch. The result was my first publication on insect behavior (West & Alexander 1963).

Because of that little cricket I was hooked on studying behavior. You could watch every detail of her life, how she groomed the eggs, fed the nymphs, and actually defecated meticulously in a special region of her burrow away from the brood. Most exciting was to see her lay tiny, undersized eggs which she seized in her mouthparts and fed to the tiny hatchling nymphs – trophic eggs! This was a phenomenon previously unknown in crickets, but described in ants by E. O. Wilson, and so I wrote him a letter asking some questions and requesting some of his reprints – my first scientific correspondence. He replied kindly, even asking about my plans for graduate school.

The cricket behavior study was also the subject of my first scientific lecture, a contributed paper delivered during the summer of 1962 at a meeting of the Michigan Entomological Society, which met concurrently with the Michigan Academy of Sciences. I didn't know how to make slides and so I drew titles and illustrations with felt-pen magic markers on big sheets of paper, which I clipped to an easel and folded back one at a time as I went through the talk. The Michigan Academy of Sciences awarded a prize for the best student talk. I was sure that the cricket and I would win. But the prize went to George Eickwort, who was a senior undergraduate at Michigan State University and used a slide projector to present the results of his undergraduate research in entomology. I could not believe that they had awarded the prize for what seemed to me to be such a dull talk. The slides were real, but they were almost incomprehensible, crowded with drawings of the tibial spurs of sweat bees – morphological details that, in my innocence, I considered patently uninteresting when compared to the discovery of maternal care and trophic eggs in a cricket. Of course later I realized that the judges had recognized the extraordinary promise of a young insect morphologist, who went on to become a professor of entomology at Cornell, and a leader in the study of the systematics and behavior of bees and mites, a valued friend and colleague until his untimely accidental death in 1994.

I felt very much at home in the museum, where people seemed down-to-earth and enthusiastic about their work, and friendly with each other even while engaging in lively debate. People answered naïve questions kindly. Indeed, simple questions could turn out to be important questions. Meanwhile I had a very different experience in the humanities, where there seemed to be a premium on sophistication. Naïve questions were sometimes treated as stupid questions, reflecting shameful ignorance rather than interest. This was especially true in my freshman English class, where the professor seemed to purposefully intimidate students. His snobbish and cruel attitude turned me away from subjects I was predisposed to like, especially since I had discovered biology to be full of exciting ideas about evolution and behavior, as well as unpretentious people who seemed to consider earnest interest more an advantage than a defect.

Despite the happy experience of Zoology 101 and the museum, I still was not sure I wanted to go into science. So I agonized over the choice of a major during sophomore year, making elaborate lists of possibilities and pros and cons. I liked studying behavior, and liked working with people, so thought that a major in one of the social sciences might be a good choice. But courses in psychology seemed obsessed with methods rather than concepts, and a biologically oriented anthropology course did not appeal to me, for it did not consider the functions or evolution of behavior, and was devoid of the interesting observations of different cultures I had hoped for. The comparative psychology course taught me that I did not like the idea of working on the behavior of animals in cages. So I decided to major in zoology after all, even though I had never intended to become a scientist.

This decision doomed me to take organic chemistry, the great memorizing contest that served as a cruel filter for entry to medical school. I began well, but gradually fell behind. I did not understand anything and so could not remember anything. On the second exam I got a 64% and the professor wrote "sinking fast" next to the grade. True to his prediction, I sank all the way to a 12, the lowest grade I had ever seen, and failed the course with a D. This had the happy result of obliging me to take a make-up course in summer school at Eastern Michigan University that turned out to have an inspired visiting instructor, who started by explaining the logic and history of the periodic table. He then showed how organic molecules are built and how that explains why they behave as they do. Because this made organic chemistry as interesting as animal behavior, rather than turning it into a cramming contest, I enjoyed the subject and easily got an A.

The excellent Honors Program at Michigan especially suited me because it emphasized concepts, readings, essays, and term papers rather than short-answer exams. Later, as an undergraduate in a seminar course on insect behavior taught by Richard Alexander – his first independent course, with only one other student (Bruce Goldman, now an evolutionary endocrinologist) – I wrote a paper on the evolution of the recruitment dances of honeybees with the same excitement I had felt about papers in English and History. But in Alexander's course signs of nascent scholarship were rewarded with discussion and interest. The publication-like lab reports for physiology courses, demanding and carefully evaluated, were incomparable training for scientific writing, as were the short essays on readings in an honors zoology seminar taught by Donald Maynard. These honors-program courses, and the debating-society atmosphere of the Insect-Division lunchtime gatherings at the museum, left me better prepared for a scientific career than were the students I later knew from "better" colleges like Harvard, where students were not engaged and trained with the same dedicated intensity by the professors themselves.

Maynard's graduate course in "The neuromuscular basis of animal behavior," which I took as a senior undergraduate and later helped teach as a graduate student assistant, was the toughest in the zoology department. The text was a medical school tome on neurophysiology that was so technical and dense as to be virtually unreadable. Since "animal behavior" was in the course title, Maynard started the course with "King Solomon's Ring," the famous popular book by Konrad Lorenz, and the only reading assignment that treated the behavior of whole animals. I saw that as an insult designed to portray ethology as light fare compared to neurophysiology. This kind of attitude challenged me to look for the larger significance of my "simple" observations on wasps, and to emphasize general concepts in talks and written reports. I wanted to show the importance of fieldwork on natural history and behavior, not only to the already convinced, but to somewhat arrogant and condescending lab scientists as well.

That mental habit – of always looking toward the larger pattern – was greatly encouraged by Richard Alexander, as my undergraduate thesis advisor and, later, as my doctoral chairman. It would be difficult to overstate my debt to him, and I have tried to pinpoint why it was that his teaching was so effective. He put a premium on thinking, especially critical thinking and was mercilessly critical himself, even while encouraging to students. He handed out reams of mimeographed handouts for us to read. Often, during his complicated lectures, I could not figure out what I was supposed to write down, and when I looked at some of my old notes before writing this essay I found that I had parallel sheets of paper, some of them on the lectures themselves, and some of them labeled "my thoughts" (stimulated by listening and sometimes distillations of what had sunk in from what I had heard). We were hammered with particular examples of sloppiness and wrong-headed thinking, especially when they might have been improved by an evolutionary approach. Dick's own rapid and creative thinking was tangible because he literally thought out loud, needling us into debates (some students, who probably hadn't been needled and teased by their uncles the way I had, felt threatened by this and considered it too aggressive). He was generous with his ideas, not among those who hide their original ideas until after they are published. As he once remarked, the best remedy for paranoid worries about getting scooped is to publish; and in addition to publishing he was producing interesting thoughts at such a fast rate that there were plenty to give away. Perhaps most important, he took his teaching seriously and dedicated great energy to it, something that inspires students to respond in kind. Of course, because of this brand of teaching I still am not sure where Dick Alexander's originality left off and my own began on many topics.

Enter: the wasps

When it came time to start an undergraduate honors thesis I knew that I wanted to work on the behavior of an insect, but I was not sure which one except that I wanted "my own" insect, not a cricket or a katydid like others in the insect division of the museum. I liked the Hymenoptera and liked watching insects interact with each other. So I consulted with Henry Townes, the local hymenopterist, and he suggested the common and little-studied social wasp, *Polistes*. This was in springtime near the end of my penultimate undergraduate year (1962). On sunny days *Polistes* females were just beginning to fly. I caught some of them, and put them in a terrarium to see what they would do. They immediately began to interact aggressively and to engage in the mouth-to-mouth transfers of liquid called "trophallaxis," so it was clear that there would be plenty of behavior to observe. I decided to do my undergraduate thesis on the common local species, *Polistes fuscatus*.

That summer, after a few observations of a *Polistes* colony nesting under the eaves of our family house in Plymouth (Figure 19.1), I embarked on my first academic adventure outside of the state of Michigan – a summer course in Invertebrate Zoology at the Marine Biological Laboratory in Woods Hole, Massachusetts. The summer at Woods Hole was a heady

Figure 19.1. A nest of *Polistes fuscatus*. My first observation colony, photographed with a small plastic Kodak Brownie camera (Plymouth, Michigan, August, 1962). The nests of this cosmopolitan genus, commonly found (as here) beneath the eaves of houses, are familiar in most parts of the world inhabited by ethologists. Behavior of adults and brood is easily observed on the exposed (unenveloped) single comb, and individuals are easily marked for individual identification. From West (1963).

experience for someone already excited about research and now serious about a career in biology. Nobel laureates and students mingled in the friendly atmosphere of the cafeteria, and conversations were easy and interesting, for people were on working holidays, running courses and research in the excellent laboratories and libraries at Woods Hole. I loved the atmosphere and the sense of tradition of the place, where many illustrious zoologists had worked in the past. The students themselves were a select group from universities across the US who worked hard all day in the labs, and played hard in the evenings. In the lab we dissected fresh lobsters one day and then ate them with beer and salad at a lab party in the evening. And when the daytime work held no culinary promise, after a day sketching bryozoa or worms you could buy live lobsters in town for a dollar a pound. I even played blues trumpet with a piano player in a dark corner of the Captain Kidd bar. My twenty-first birthday was an excuse for an unforgettable surprise party in the marine invertebrate lab.

At Woods Hole I learned for the first time that I had a midwestern accent, and had the magic experience of seeing many scientists whom I knew by name from their articles and books. George Wald was there. I pretended not to be shocked to see that he and his wife allowed their small children to run around naked on the beach, something that would never have occurred at a Michigan lake in the 1950s. James Watson, already a Nobel Laureate for his work on DNA and still a bachelor, ogled girls with a round-eyed lizardlike stare. The famed physiologist Albert Szent Giorgi, also a Nobel Laureate, gave a wonderful lecture in which he drew the progress of science as a circuitous wandering line rather than a straight line to discovery. Theodosius Dobzhansky, for my generation one of the great heroes of evolutionary biology, and a man with a charismatic personality, gave a lecture.

One day at Woods Hole someone pointed out Libbie Hyman, a legendary figure of classical zoology, the author of the standard comparative anatomy text, and a world-renowned authority on invertebrate zoology. I was disappointed to see that she was a dour and unhappy-looking woman at age 74, with a homely countenance, walking alone, clad in a severe grey suit. She did not look at all like the kind of person I wanted to be. Her autobiography (Hyman & Hutchinson 1991) showed that she had a sad personal life except for a lifelong passion for her work. The following year the neurobiologist Rita Levi-Montalcini spoke in Ann Arbor, in one of the large auditoriums on campus, about her work on the nerve growth factor, and I was thrilled to see a brilliant lecture delivered by a vivacious woman scientist. I kept the poster advertising her lecture on the wall of my dormitory room for the rest of the year, and so did not forget her name. The next time I saw it was many years later, in a magazine article that mentioned that she had been awarded the Nobel Prize in Physiology or Medicine. Many years later I met her on two different occasions, in Rome, and found her to be dignified, gracious and kind.

Eventually I decided to stay at Michigan for graduate school as a student of Dick Alexander. One of his criteria for accepting a new graduate student (I was his first grad student, an earlier one, Ken Shaw, being an advanced student inherited from someone else), was to be willing to do a thesis on something different from his own research. I especially liked that attitude because the last thing I wanted to be was a member of some professor's adoring gaggle of graduate students, for I had actually witnessed such a little group at a scientific meeting, tagging after a noted ethologist whom I had considered as a potential advisor. Dick did not start publishing on insect, mole-rat, and human sociality until after I had fledged and published my thesis on the social biology of *Polistes*.

One of the unforgettable benefits of staying on at Michigan for graduate school was the chance to participate in a year-long graduate seminar (Fall 1965–Spring 1966) dedicated to a detailed critique of one book: Ernst Mayr's 1963 classic *Animal Species and Evolution*. Richard Alexander, and my fellow graduate students Douglas Futuyma, Dale Hoyt, Joseph Jehl, Bert Murray, Dan Otte, Ann Pace, Bob Vinopal, and I took turns leading a sentence-by-sentence scrutiny of the book, and we sent a 141-page critique to Mayr. As a result we all got imprinted with speciation theory, and I have used that book as an encyclopedia of data and evolutionary thought from its era ever since.

By opting for graduate school at Michigan, where I knew the system and had taken most of the required courses, I was able to finish most of the requirements for a doctorate – teaching, language exams, thesis plan, and the formidable 12-hour-long written comprehensive exam – in the first year. Then I did a busy summer of fieldwork on *Polistes fuscatus* in the countryside near Ann Arbor. I resolved to become an expert on everything about the genus *Polistes*, from taxonomy to endocrinology.

A serendipitous landing in paradise

During that first summer of fieldwork I had a fateful conversation on the campus in Ann Arbor. I happened to see one of the students from the section of the accelerated introduction to zoology (Zoo 105) that I had taught the year before. His name was Fabio Heredia, a Latin American student on a Ford Foundation fellowship. I knew him better than the other students because I had spent so much time helping him with English and the organization of his lab reports. He asked me what I was planning to do the next semester, after my summer of fieldwork on wasps. I had just received an NIH traineeship that had a US$300 stipend for travel, so I told him in rather vague terms that I might use my fellowship to do some fieldwork in the tropics. "Ah," he said, "you should go to Colombia. I can arrange everything!" And he gave me his home phone number so that I could call him later.

That casual conversation changed the course of my life. I looked up Colombia on a map, not sure whether he had said Colombia or Costa Rica since they both start with C, and not really knowing much more than that about either of them. I asked Henry Townes if he thought it would be easy to find a *Polistes* species there, and he said that it would not be a matter of finding just one, but how many of the several likely species there might be. So I began to plan to actually go. It turned out that Fabio Heredia had been the chairman of the biology department at his university, the Universidad del Valle in Cali, and was on leave to learn English and then attend graduate school. He arranged everything for me, including space at the university and a place to stay – a room in the home of a biology professor who lived near the university. Later that summer I had a fairly serious allergic reaction to a sting while alone in the field and drove myself to the emergency room of the University Hospital, where they told me I should stop wearing flowered blouses and sandals and stay away from wasps. But there was no way I was going to give up *Polistes* and a trip to South America, so an allergy kit and a series of desensitizing injections were among the supplies I took along.

The trip to Cali was unforgettable. I was extremely excited when, after weeks of careful planning and packing, I was actually on my way. It was my first trip in an airplane, and I felt as if I was leading a triumphant parade when at last I strode through the Miami airport, followed by a porter with a cart bearing everything I thought I would need: ten pieces of luggage, including a very heavy instrument for recording temperature and humidity called a "thermohumidograph," (now replaceable by a digital instrument the size of a small clock), and my trumpet in its case, which I took everywhere in anticipation of a jam session (something that never occurred in Colombia, which seemed to be devoid of trumpet-playing women).

The economical Ecuatoriana flight from Miami to Cali was on an old prop plane that droned on and on, seemingly endlessly. I wondered how it could have enough fuel for a journey that took so long. The plane landed in the Calipuerto, antecedent of today's modern international airport, having flown low over large expanses of sugar cane and scattered roofs of tile or thatch, then settlements near the Rio Cauca, a tropical scene as I had imagined it would be, with palms and banana trees, mud houses with thatched roofs, and dugout canoes. I fully expected to live in a thatched hut with lizards crawling in the roof above a little wooden table where I would work. So I was surprised to find that the university was located in an upscale section of the city, and that my rented room was in a modern house with running water, electricity, and a tiled bath, and that there was a maid who prepared all of the meals. Such was my innocence about big South American cities, and the romantic hardships and cultural adjustments I was prepared to endure. Still, this psychological overshoot meant that my attitude toward everything I encountered was overwhelmingly positive. I expected to be a social outcast for a time, as a person with an infant vocabulary and a fondness for wasps is likely to be. But Cali is famous as a beautiful city nestled against the western Andes, with a perfect climate and open, hospitable people. It was another situation where, like the zoology department in Michigan, being a somewhat naïve but earnest and friendly girl from Michigan was probably an advantage rather than a liability.

My five months of research in Cali went extremely well but it got off to a slow start because I could not find *Polistes* colonies within walking distance of where I lived. Back in Ann Arbor I had met a student from Cali who said that if I ever needed help I should call his brother, Tulio Jaramillo, an agronomy student in nearby Palmira, and I had written down the name and phone number to be polite. The Jaramillos had a large sugar cane farm and "trapiche" in the valley, where they produced "panela," the crude sugar cakes that gave *Polistes* its local common name – "paneleras." There I found several colonies of *Polistes* that formed the core of my long-term observations.

Several times a week I would go to the Jaramillo residence by bus, wearing a skirt or a dress because it was not considered proper for women to wear slacks in public. Each one-hour trip was a Spanish lesson, for Efraim Jaramillo had an uncanny ability to use my minuscule vocabulary and build upon it to explain complicated things. After several weeks of this, and with the help and moral support of my bilingual friend Maria Eugenia Cobo, plus tolerant coaching by friends at the university, I learned enough Spanish to converse and to find my way to the other places, such as the hydroelectric plant at Anchicayá (now controlled by a guerilla group), that were key to my research. During the five months of fieldwork in Colombia I was extraordinarily dependent on the help of other people. It was there, especially, that I think it was actually an advantage to have the protective extra help accorded a woman.

Early kin-selection research (1964–65)

While I was in Cali, Hamilton's 1964 papers on kin selection appeared and were immediately sent to me by Dick Alexander. I still have the copies he sent, now dogeared from use, the brown and brittle pre-xerox thermal paper showing its age. Because I got the papers so quickly, and because Hamilton referred extensively to *Polistes*, which he had observed in Brazil, I immediately started to think about the relevance of kin selection to my research, and was able to attempt the first quantitative tests of the idea carried out in the field (West 1966, 1967; West-Eberhard 1969).

I use the word "attempt" for a reason. I had direct observations, on both *P. fuscatus* in Michigan, and *P. erythrocephalus* in Colombia, indicating that nestmates co-founded new nests. And in both species I had observed that a single female could be the exclusive egg-layer for a period of time long enough to make all of the workers her daughters. But I had no way of knowing how many times she had mated, and no assurance that I had seen every oviposition event. So some of the workers could have been only half-sisters, rather than full sisters as required by Hamilton's idea regarding the unusual ¾ relatedness of hymenopteran workers and brood.

The first molecular measures of genetic relatedness were more than ten years in the future. But the Colombian wasps were remarkably variable in color, and I had noted that members of the same colony seemed more similar to each other in color than to those from other groups. So I decided to quantify this using a laborious "index of similarity," recording the coloration of 12 body regions for hundreds of individual wasps from a large number of colonies. Then, back in Ann Arbor, I began a multiple analysis of variance with the help of a fellow graduate (Jim [Steve] Ferris) assigned to help others with programming to do statistics on the big mainframe university computer housed in a building across the street from the Museum. The idea was to test, first, the independence of the variation in the 12 color measures, then to see to what degree colony members were more similar to each other than to the population at large.

In those days, long before the invention of desk-top computers, each graduate student was allowed several hours of time on the university computer, and this required the tedious punching of binary data on 8.3 cm × 19 cm cards that were then fed into the computer in ordered stacks. This was by far the most time-consuming aspect of data collection and analysis for my thesis, and the one that caused me the most misery.

Some time during the data analysis I explained to Dick Alexander what I was trying to do. "Well," he asked me "did you do it blind?" At first I didn't even understand what he was trying to say: that for the data to be valid, it would be necessary to conduct the color analysis without knowing the colony membership of the individuals concerned. Otherwise, it would be easy to bias the results. I had not done the measurements blind, and furthermore I realized that there had been a few occasions when I had judged a borderline color marking to be more like the rest of the colony because that was more likely given the pattern I thought I had observed. Either I had to redo the entire set of measurements, or I had to discard this part of my thesis. I was already behind in writing my thesis and in danger of losing a post-doctoral fellowship as a result. So I decided to abandon the phenotypic index of relatedness and rely instead on purely behavioral evidence.

In retrospect this event was as fortunate as it was painful, because the later genetic estimates were so much better. My discarded, primitive phenotypic estimate was seriously flawed. Even if it had been done blind, there was the possibility of environmental effects on coloration that could have been shared, along with kinship, by members of the same colony. But because of this costly lesson I have little tolerance for behavior studies that should be done blind but are not. I feel no hesitation in recommending that they be thrown out no matter how much effort they represent. Because so much subsequent research on kin recognition was carefully done blind, my ill-conceived index of similarity would not have stood the test of time.

An eventful post-doctoral period

My thesis was finished in February, 1967, barely in time to start a post-doctoral fellowship with Howard Evans at the Harvard Museum of Comparative Zoology (MCZ) that should have started in September of the previous year. I spent most of the first days in Cambridge recovering from exhaustion, therapeutically sleeping and reading the *New York Times*.

Those were exciting times at the MCZ for a young biologist interested in behavior and evolution, especially of social insects. Howard Evans was an expert on the solitary precursors of the social insects, and Ed Wilson was writing his classic book *The Insect Societies*. Ernst Mayr was an outspoken member of the audience at seminars, especially if he sniffed some reference to sympatric speciation. He sat head tilted upward as if alert, but with eyes closed as if sleeping. But then he would render some incisive comment at the end. George Gaylord Simpson worked reclusively in his office. I talked with him only twice, once at a lunch at the faculty club at the invitation of the kindly herpetologist Ernest Williams. Simpson and his wife Anne Roe, a noted psychologist, entered the room like royalty. Ernst Mayr was also present and he and Simpson talked about what to do with their gold medals. Simpson suggested that they could be melted to retrieve the gold, and I thought he was kidding but he apparently actually did this later, first having replicas made for display. Ed Wilson's good-natured and entertaining friend and mentor William "ant" Brown spent the next summer in Cambridge working at the Museum, and we became long-time friends with him and his lively wife Doris. Ed Wilson was always busy, but always cordial and encouraging, and continued to be so as years went on. He regularly attended the meetings of the Cambridge Entomological Club, a lively and informal forum for students, amateurs, and professional entomologists alike, as well as "Carpenter's Teas," weekly gatherings sponsored by Frank Carpenter that were a friendly institution for entomologists at Harvard for many years. Robert Jeanne was starting his thesis research on the behavior of social wasps of the genus *Mischocyttarus*, and we had much in common as he planned and carried out fieldwork in Brazil and have remained close colleagues and friends ever since.

But by far the most important event of my two and a half years at Harvard was meeting a student of the Harvard arachnologist Herb Levi, named William (Bill) Eberhard, soon after arrival. Part of the post-doctoral job was to help advise graduate students, and Bill made an appointment with me ("Dr. West") to discuss our common interest in building behavior, of spiders' webs (Bill's thesis topic) and wasps' nests (a favorite topic of mine). Our first date, complete with a bottle of wine, was a trip to the elegant Lexington, Massachusetts, dump, where the quite respectable discards of that prosperous community could be recycled as furnishings of student apartments – one of many items of local folklore that I learned from Bill, who had been both an undergraduate and a graduate student at Harvard. A few months later we were married, and while Bill finished his thesis we had our first child (Jessica, born in 1968).

We first met Bill Hamilton, whose writings on kin selection were already mentioned, while at Harvard. Unbeknownst to me, while I was in Colombia he had been in Brazil, continuing his work on social wasps. He had submitted his *Journal of Theoretical Biology* manuscript and then headed into the field. Our first correspondence was in September of 1967, during my first postdoctoral year at Harvard. Ed Wilson had met Hamilton two years before, at a symposium. Wilson had referred to Hamilton's ideas in a positive light during a symposium in London, where Hamilton was present but did not speak, though he partici-pated in the (published) discussions. [J. S. Kennedy's (1966) brilliant final talk and the discussion it provoked are still worth reading.] Ed told me that Hamilton had indicated in a letter that he was planning another trip to South America to work on the swarm-founding social wasps, so I finally wrote to Hamilton, telling him of my interest in his ideas and enclosing a summary of my thesis and a copy of a manuscript of mine that was in press in *Science* (1967).

Hamilton answered with one of those magnificent long letters full of interesting com-ments that became familiar to his correspondents – this one was 7 single-spaced typed pages long, somewhat less crammed with words and easier to read than the tiny scrawl of his handwritten letters and postcards. He had already seen the published abstract (1966) of a short talk I had given at a AAAS meeting, where there was one sentence saying that "The findings support the ideas of Hamilton (1964) regarding degree of relationship and the evolution of social behavior in insects."[1] He also had some comments on my *Science* paper that made me wish I had sent it to him before it was in press. And, like me, he had tried to invent ways of estimating relatedness by using phenotypic measurements. He also invited me to join him and his recent bride Christine for the planned fieldwork in Brazil, an invitation extended to both Bill and me later, and a subject of much wishful thinking on my part. Even though we were expecting our first child at about the time the expedition would get under way, I still thought, in my innocence regarding motherhood and wifehood, that the trip might be possible.

I urged Hamilton to visit Harvard, and this finally became possible on his first trip to the US in May of 1969, after he spoke at the Smithsonian shortly before we were to move to Colombia. Bob Trivers, then a graduate student, was very anxious to meet him, and so we invited Bob to dinner with Hamilton at our apartment in Cambridge, where that historic first meeting took place.

Despite extensive discussion, including working out a detailed plan to coauthor a book on tropical wasps, Hamilton and I always emphasized different aspects of Bill's own kin-selection formulation, with him preferring to look at the genetic (relatedness, *r*) side and me

---

[1] As small as it was, this was probably the first report of research designed to examine the kin-selection idea, and purporting to confirm it. An earlier publication by Haskins and Whelden (1965) referred to Hamilton's ideas on kin selection in an analysis of data collected before the Hamilton (1964a,b) papers appeared. It cited some findings of potential relevance to the degree of genetic relatedness among colony members in various ponerine ant species and suggested further research that might test whether their findings could be reconciled with the theory, e.g. by looking for behaviors that would raise relatedness above the levels suggested by their preliminary research. Thus, their paper neither supported nor contradicted the theory.

being especially interested in the cost-benefit side (the K in Hamilton's rule, $K > 1/r$). I, along with other students of behavior, used the expression as a behavioral and developmental decision rule, and Bill accepted that usage. But in correspondence he would always put the word "decision" in quotation marks. When some phenomenon, like the large and possibly genetically heterogeneous foundress groups of tropical *Polistes*, seemed difficult to explain, he would look for a genetic explanation such as inbreeding, and I would look for (and find) individual phenotypic differences, such as ovarian development, size or aggressiveness – indicators of differences in reproductive capacity that could be environmentally influenced and would affect the benefit/cost side of Hamilton's Rule. I could never understand why Bill gave so little attention to the conditional, cost–benefit side of his own formulation.[2]

The Cambridge days – getting married, having our first child, and seeing the impossibility of a trip to Brazil – were days of adjustment. It took me by surprise when Bill [Eberhard] objected to my plan to continue using my maiden name on publications. Later this became quite common, but it was not common then, and he thought that a refusal to use my married name would indicate a lack of commitment to marriage. By that time I already had a few publications under the name "West" and I did not want to drop it, so I decided to arm myself for further discussion by talking with other married women scientists who were publishing independently of their husbands. I knew of only one such woman at Harvard, so I talked with her. She said that I should definitely stick with my maiden name because it would be a real mess if, like her, I were to be repeatedly married and divorced! This was not exactly the argument Bill wanted to hear. The question was finally settled by his mother, who advised me to use both names, because it would be convenient for legal purposes. Eventually I had to add a hyphen in order to keep the two names together in reference lists, even though it became awkwardly long.

The decision regarding names was urgent because I had just finished the laborious pre-computer re-formatting and revising my thesis for submission to the journal *Behaviour*, edited at that time by G. P. Baerends, who had done a classic study of the behavior of the wasp *Ammophila*. But Baerends turned down the manuscript without review, as not suitable for the journal. This did not particularly surprise me because it was partly natural history, not pure behavior; and discussions of kin selection were controversial and not in the usual style of the journal. But I could not face reformatting it repeatedly, and did not want to break it into several small papers, as would certainly be encouraged today. So I took advantage of the willingness of Francis Evans, editor of the *Miscellaneous Publications of the University of Michigan Museum of Zoology*, to publish it there with little change.

Even though this was a relatively "obscure" journal, the monograph was not ignored because it was immediately picked up and used by leaders in the field: Charles Michener, O. W. Richards, Leo Pardi and others in addition to Dick Alexander, Bill Hamilton, and

---

[2] Bill Hamilton reveals the background for his brand of genetic determinism in some autobiographical notes in *Narrow Roads of Gene Land*, Vol. I (Hamilton 1996). The most extensive treatment of condition-dependent gene expression by Hamilton known to me is a single paragraph in an endnote to a book chapter (Hamilton 1987), where he wrote (p. 433): "The worst pathologies of the kin-selection criterion arise when genes for social behavior are unconditionally expressed – i.e. expressed by every individual of a given genotype. ... Conditionality, although mentioned, was insufficiently emphasized in my previous work (Hamilton, 1964a,b)."

Ed Wilson, who featured it in the chapter on social wasps in his widely read classic *The Insect Societies* (1971). Many years later I met Baerends at a meeting and he briefly complimented me on my work. That meant a lot to me because I so admired his ethological research and had worried that he didn't approve of mine.

After our first child (Jessica) was born we drove directly from the maternity clinic in Boston to the Huyck Preserve near Albany, NY. We had a joint appointment as summer fellows for that year (1968). I had given seminar lectures at Harvard, Cornell and Yale while pregnant and thought that a mere baby would not interfere with a summer of research. I think that Jessica must have sensed this because she screamed for ten days almost without a break and managed to get my attention for the rest of the summer. The wasp boxes Bill built to help me out attracted few *Polistes*. But Bill, propelled outdoors by the ruckus inside the house, got lots done in that wonderful place, and salvaged our scientific reputations.

There may be a physiological inability to concentrate on other kinds of work while caring for an infant, for even back in Cambridge I had difficulty working despite good intentions. Howard Evans and I were writing our semi-popular book on *The Wasps*. Thanks to his sympathetic patience, probably encouraged by his wife Mary Alice, who maintained a career along with a family, Bill's prodding, and the fact that we both had projects that could be done largely at home (he was writing his thesis), my chapters of the book finally got done. Eventually I went into my office one day a week while a friend cared for Jessica. Then, when Jessica was nearly a year old, we moved to Colombia.[3]

Return to paradise

The move to Colombia in 1969 was not carefully calculated to be the perfect career move it turned out to be. Both of us liked the idea of research in the tropics, and my time there as a graduate student had been extraordinarily productive of new data and new friendships, as well as an exciting adventure in a fascinating place. But the opportunity to return was another serendipitous event: Aníbal Patiño visited Boston briefly just as Bill was finishing his thesis, and, remembering his kindness to me in Cali, we invited him to dinner. He asked Bill if he would consider living in Colombia, and Bill said yes. Within a few weeks a job for Bill was arranged in the biology department of the Universidad del Valle in Cali, and a few months later we were there. We knew that if we did not spend some time in the tropics before settling into conventional academic jobs it might become difficult to ever do so. And the offer looked very good because we thought that the dollar sign on the salary meant US dollars per year, whereas it meant Colombian pesos per month! But we decided that if Colombian professors could live on that salary, even without furniture from the Lexington dump, we could too.

We intended to stay in Colombia for one or two years. But Bill proved to be an unusually gifted teacher of the enthusiastic biology students there, and the opportunities for research in Colombia were wonderful for both of us, especially after we moved to a rented house

---

[3] For more on Howard Evans see West-Eberhard (2004, 2005).

("El Guamo") on the outskirts of town. By that time (1970) we had our second child, Anna, and four years later we adopted our third, Andres, at 8 months of age. Eventually we purchased the house to avoid having to move out. There I was able to follow the social history of a *Metapolybia* colony for eighteen months in a protected place under our patio roof, while simultaneously monitoring the behavior of our children in the patio below, a scene immortalized in Sarah Hrdy's masterful *Mother Nature* (1999). The *Metapolybia* observations provided the first description of social organization – the cycle of swarming, dominance dynamics, and worker participation in queen determination – in a swarm-founding multi-queen tropical social wasp.

One morning, at the nursery school attended by our children, I saw a eumenine wasp fly into a bougainvillea vine. It led me to a nest of *Zethus miniatus*, a rarely encountered group-living species of a primarily solitary-nesting subfamily of wasps and the subject of a classic study by Ducke in the Brazilian Amazon in the early 1900s. Scarcely any discovery more exciting could be imagined for someone interested in the origins of social life in wasps, for the species had never been subject to a long-term study of the behavior of marked individuals. So I dropped other projects in order to concentrate on that.

Those two species – the readily observed and common *Metapolybia aztecoides*, and the rare but fascinating *Zethus miniatus* – changed my perceptions of social evolution as much as marriage and the move to Colombia changed the course of my personal life. The *Metapolybia* colony beneath our roof (see West-Eberhard, 1978; Figure 19.2) showed me how to reconcile the multiple queens of swarm-founding tropical species with kin selection theory, and how to understand what Charles Michener had been talking about when he reasoned that mutualism could explain worker sterility without kin selection – an argument I had not believed until I saw it with my own eyes in *Metapolybia*. *Metapolybia* females never nested alone. They would cooperate while the group essential to their reproduction was small, then compete so strongly when the colony became large and more secure that some – the losers in social competition – were sterilized as a result, even though they had contributed to the success of the winners by laying worker-producing eggs. The predictions of kin-selection theory were confirmed because genetic relatedness within the colonies was kept high by the cyclic reduction of queen number to one, with the effect of cyclic restoration of relatedness to the level of a mother and daughter workers before reproductive swarms were produced. I think it is quite common for biologists to have their understanding directed and limited by what they can see in their favorite organisms, and I think that is why Michener, who saw such a process in bees, was a pioneer in understanding mutualism, and why Ed Wilson, who works with highly evolved social insects (ants), emphasizes the superorganismic aspects of social insects, minimizing the role of competition within groups and emphasizing group selection rather than the positive contributions of kin selection during the transition to eusociality.

I had to visit the *Zethus* nest on afternoons and weekends, when there were no children at the school. Eventually, after a thorough long-term study of marked individuals, I moved the nest to a similar location on our own property. It was actively inhabited for more than four years, and it was a society at the very brink of full sociality, with a few subfertile cooperating

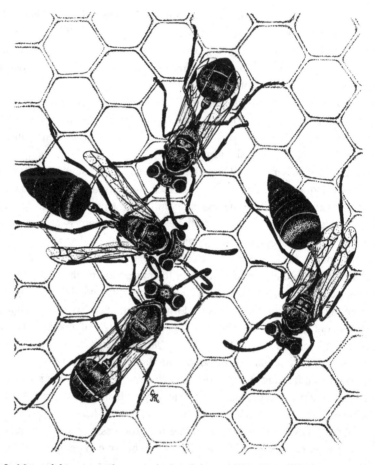

Figure 19.2. *Metapolybia aztecoides*, a tropical social wasp (Vespidae) that initiates nests in swarms. Queens with enlarged abdomens (due to developed ovaries) extend them and adopt a characteristic "bending display" when they encounter other queens on the nest. Workers, whose abdomens are not enlarged, perform a rapid jerking movement near queens. Queens and workers of this species, easily distinguished by their behavior, cannot be reliably distinguished by size or morphological differences. Drawing by Gerardo Ravassa.

females that were just short of being sterile workers. It was clear (see West-Eberhard, 1987) that there need be no "gene for altruism" to push them over the brink, only selection for losers in social competition to remain in the group (I had seen that they instead left and sometimes founded new nests on their own). I surmised that relatedness in the long-lived groups would decline so severely that aid to nestmates would not consistently satisfy the conditions for profitable aid to kin. I also realized, observing *Zethus*, that individual females alternated between workerlike and queenlike behavior, as do many solitary wasps and bees that alternately build and provision nest cells, then lay eggs in them (West-Eberhard, 1987),

Figure 19.3 A nest of *Zethus miniatus* observed on Barro Colorado Island, Panama (July, 1994). Some open cells contain females ensconced, heads facing outward, each guarding an egg or larva. On the lower part of the nest a female, whose thorax bears a paint mark for individual identification, inspects an open cell. Photo by Thomas Eltz.

and so was reminded that the roots of both social-insect castes undoubtedly arise with little phenotypic or genetic change from the maternal behavior of solitary species – an old idea in the literature on social insects that had been lost from view and not elaborated in a modern physiological and theoretical-genetic context. Later I studied a colony of *Zethus miniatus* on Barro Colorado Island in Panama (Figure 19.3).

We soon found that household help was well within the reach of a university professor's salary in Cali. But having two cars or a chauffer definitely was not. So in Cali we economized on driving time by using horses: Bill would ride a few kilometers to within walking distance of a bus service, leaving the horse tethered on the vacant and fenced property of a missionary school, where it trimmed the grass until picked up for the ride home. The good nursery school with the *Zethus* nest was conveniently located en route. Then later some generous neighbor friends allowed our children to pile into their car for rides to the more distant primary school, greatly increasing work time for me, and incidentally providing a nice coffee break at their home when it came time for me to pick them up. Kids were the perfect distraction, an insistent contact with humanity, a delightful and necessary diversion from work. At the same time I like to think that children actually benefit from having a mother who is not completely devoted to them, but has her own life in full view, and so does not depend entirely on them for her chances of success in life, something that puts an undue burden on offspring.

The first months in Cali were occupied by writing a paper on kin selection, showing how the K of Hamilton's rule is as important as the relatedness (r) side for social evolution, something I could do while at home with two small children. This paper, a long time in gestation and revisions, started as a joint effort by Dick Alexander and me to write a general article on factors other than relatedness in the evolution of social behavior. But we did not agree on what the emphasis should be: Dick was most interested in the idea of parental manipulation, whereas I wanted to examine the importance of the cost–benefit (K) side of Hamilton's idea. He thought that parents, especially mother's, interests would always win over those of offspring because of the powerful ability of parents to influence the development of their young, but I thought it necessary to define precisely when the offspring might win. Because I was unable to think of a precise solution for this problem (later solved by Robert Trivers (1974)), and wanted to focus on something else, I decided it was better to go our separate ways, despite the strong wish we both had to write something jointly on the evolution of social behavior. I sent my manuscript to the *American Naturalist*, where the editor, George Williams, decided to "send it upstairs" to the *Quarterly Review of Biology* (QRB) (where he was also editor) because of its length, and it was eventually published in QRB (West-Eberhard 1975). Soon after it was accepted, Dick Alexander submitted his manuscript to the same journal, only to find that they were unwilling to consider it because of having just accepted a very similar manuscript by me! Of course this was not a happy event, but it had a happy ending, for Dick found that his paper could be published almost immediately in *Annual Review of Ecology and Systematics* because a promised manuscript had not arrived. As a result his paper (Alexander 1974) came out earlier than mine and had the publicity advantage of being cited in Wilson's widely read book *Sociobiology* (Wilson 1975).

Wasp research and the paper on kin selection led to various invitations to speak, one of the most memorable being the infamous 1978 AAAS Symposium on Sociobiology in Washington DC, where radical anti-sociobiology demonstrators dumped water on Ed Wilson's head,[4] thereby allowing the always gentlemanly and poised Wilson to win the

---

[4] Wilson (1985) gives his own account of this, and capturing the flavor of controversy at that time, in an autobiographical sketch written soon afterward and published in the earlier volume on leaders in ethology.

day. Seldom in the US, I was unprepared for all the publicity, and appalled by how otherwise good scientists allowed their presentations to be cheapened by it. David Hull had anticipated this in his impressively astute talk, and I sent him a brief outburst giving my reaction, which he included in the published version of his talk (Hull 1980). I could not work up enthusiasm for finishing the manuscript version of my talk, which had emphasized common errors in thinking about kin selection. Richard Dawkins made that the topic of his paper although he had spoken on Evolutionarily Stable Strategies.

Wasp behavior and beyond: sexual and social selection

Even though I decided to study wasps as a naïve undergraduate, attracted simply by their lively behavior and the promise of a common insect that is easy to observe, I could not have chosen a more overtly Darwinian group of organisms. Wasps compete fiercely as individuals, even within highly specialized groups of relatives. Because I could see this so clearly in *Polistes*, where females compete by combat and display in very obvious ways, I was able to appreciate the competitive nature of the more derived and subtle ritualized aggressive displays in the more complex societies of swarm-founding species. *Polistes* taught me how a competitive female could use behavioral cues to decide when to profitably adopt a subordinate position and help a superior reproductive among her relatives (West 1967; West-Eberhard 1969), and how that decision could lead to a division of labor and a coherent society (West-Eberhard, 1979).

Moving from the social displays of wasps to an interest in sexual selection and the idea of social selection in general was triggered by a fateful conversation with Clark University ethologist Nicholas Thompson. He waited patiently to speak with me at the back of a crowded auditorium following my talk in a symposium, and began our conversation by asking whether I thought that the evolution of social behavior is like evolution by sexual selection. I did not immediately see the connection, and asked him to explain. He said, "Both of them lead to something like 'fads.'" Trying to be polite, and not knowing what to say about such an odd-sounding idea, I suggested that we continue the conversation at lunch.

Not until later did I understand the profound significance of what Nick Thompson was trying to tell me: both the competitive displays of social organisms like wasps, and those of sexually competing males like peacocks, are like human fashions in being extravagant and diverse, with no obvious relation to adaptive function. They can evolve as purely social signals that need only affect status and get the attention of others. Literature on classical ethology showed that the signals used in non-sexual social competition among females, sibling nestmates, and territorial animals follow the same patterns of extravagance and diversity as those under sexual selection. I first presented this and related ideas in an invited lecture in 1978, in a symposium at the American Philosophical Society organized by Caryl Haskins. The audience included distinguished social scientists and one of them cautioned me, in conversation, against using the term "social selection" because of dire connotations it would have in the human sciences. So I removed that term from the title and much of the text

(1979). In retrospect, this was a mistake, for the evolutionary biologists who were my primary audience would have had no problem with the term.

I was especially interested in the possible relationship between sexual selection and speciation, and found evidence for this in taxonomic monographs that commonly show sexual-signal morphology to be species-specific even when other kinds of characters are not (West-Eberhard 1983, 1984). My training during the heyday of the Modern Synthesis taught me that courtship behavior evolves under selection for "reinforcement" of barriers to hybridization – as reproductive isolating mechanisms, and this interpretation was consistent with a coherent genetic theory of geographic speciation and reproductive isolation. My advisor Dick Alexander was a pioneer in the use of courtship behavior of insects in studies of speciation; and most (perhaps all) of Mayr's examples of geographic speciation in birds and other animals involved sexually selected traits; yet sexual selection was scarcely mentioned during that era of intensive attention to speciation. Indeed, I did not think of the possible connection between sexual selection and speciation until I happened to read part of a book manuscript sent to me by Randy Thornhill, on insect courtship, speciation, and isolating mechanisms, just as I was beginning to write about sexual and social selection as agents of diversification (see West-Eberhard 1979), something that was not in the ms (of Thornhill and Alcock 1983) I read. It was a timely exchange of ideas for both them and me.

The strong, exclusive emphasis on sexual behavior as isolating mechanisms shows how a whole field of ethology can be led astray by the lure of a coherent or attractive theory, even when alternative explanations – as conspicuous as Darwin's theory of sexual selection – are at hand. A similar distraction occurred later, when ESS theories based on models of genetic polymorphism predicted equal payoffs for alternative behaviors in discussions of behavioral tactics and strategies, even though the conditional alternatives that characterize animal behavior are not expected to satisfy those predictions. But in both eras there was productive excitement about ideas that generated new directions of research and made evolutionary ethology interesting and important.

Old passions in new bottles

The effort to expand an idea beyond what is known about wasps has capitalized on the love of library research that I experienced as an undergraduate. It has also meant a slow rhythm of publication punctuated by large synthetic papers rather than a string of smaller reports on the details of fieldwork, and I have many exciting field observations yet to be published. In fieldwork on wasps I never set out to test a particular evolutionary hypothesis. Rather, my intention has always been to learn everything I can about behavior and natural history of unmanipulated individuals in the circumstances where they are found, with simple experiments (such as removals of dominant individuals) that mimic natural events and therefore illuminate their consequences.

Of course observational research is not devoid of hypothesis testing. On the contrary, it employs the multiple hypotheses involved in childlike *wondering* – wondering, for example, about why an individual behaves in a particular, distinctive way. Productive wondering is the essence of observational ethology. It is a demanding technique because it requires patience and imagination – a head full of hypotheses and ideas about the kinds of observations that would test them. But it is almost certain to tell you something real about behavior, more likely to yield clues about function and evolution than manipulative testing that begins with pure theory. Plain observations of animal behavior are so productive of new insights and testable explanations with a high probability of being correct that I have often wondered why more people do not watch the natural behavior of organisms. Even many people who write about the evolution of behavior do not actually watch it.

One good result of my early infatuation with the humanities was to have passionately studied French, which, combined with high-school Latin and later immersion in Spanish meant I could easily read scientific papers in French, Portuguese, and Italian. The plain ability to read in these languages opened the way to study much ethological literature that was, and still is, historically important even though often ignored, especially by English-speaking biologists. This led indirectly to some of the most rewarding friendships of my professional life, especially in France (among others, with Bernadette and Roger Darchen, who had studied with the termite expert and giant of French zoology, P.-P. Grasse, a powerful critic of Darwinism), and others in Italy and Brazil.

During the 1940s brilliant, path-breaking research on *Polistes* had been done by the Italian ethologist Leo Pardi. Pardi was not only a keen observer of wasp behavior but he had written a brilliant doctoral dissertation on the insect ovary in the 1930s and was a master of classical zoology. He was the first to discover correlations between the ovarian status of females and their dominance behavior, and he made quantitative descriptions of the ovarian and behavioral ontogeny of adult females that are still unsurpassed. It was Pardi who discovered that dominance relations determine caste in *Polistes*. This led him into terrible conflict in scientific meetings and in print with some French biologists who claimed that caste was controlled entirely by other mechanisms they had observed, especially differential oophagy, the selective eating of the eggs of competitors. I had observed all of these things and so in my thesis publication (West-Eberhard 1969) could point out that there was really no conflict between these mechanisms, since all were activities of dominant females that could act together to suppress reproduction of subordinates.

Pardi had been so affected by the vehemence of the public attacks on his research that he had abandoned work on wasps to become a distinguished student of migration in crustaceans. But I sent him my 1969 monograph and he read it closely. Unknown to me at the time, my resolution of the controversy and extensive use of his publications meant a great deal to Pardi. He returned briefly to research on social wasps alongside his student, Stefano Turillazzi, who subsequently founded a vigorous modern school of research on the behavior and biology of *Polistes* and other social vespids. Thanks to Stefano Turillazzi I met Pardi in Florence and visited his beautiful home in the Tuscan hills near Florence. That meeting with Pardi, a fine scholar who was a critical perfectionist in his research and who had long been one of my heroes, was wonderful for me, as have been my friendship and waspish collaborations with Stefano Turillazzi, and my acquaintance with other Italian evolutionary ethologists such as Laura Beani, Francesco Dessi, and Floriano Papi. The worldwide network of colleagues and friends, and the enlightenment one experiences when there is a chance to visit them, is one of the greatest benefits of a career in science. Most of my journeys abroad as a scientist were for research or meetings, but some were connected with many years of work on the Committee on Human Rights of the US National Academy of Sciences, Academy of Engineers, and Institute of Medicine, which uses the prestige of members of these organizations to try to help students and professionals in these fields who are prisoners of conscience. While the committee is small and has a limited budget it has many supporters and a wonderful staff, and has confirmed my conviction that effective humanitarian work can be done quietly, through individual contacts, in ways that complement and reinforce the immense efforts of larger organizations like Amnesty International.

The Smithsonian connection

While in Cali I worked mainly at home for about three years, then began to spend mornings in a small university office, generously provided by the friendly biology department even though I did not have a job there. I worked the equivalent of full time with no salary or research support, but didn't really need any. I could do exciting fieldwork without expensive travel or equipment. For dissections, taxonomy, and morphometrics we used an antique binocular microscope with old but excellent Zeiss optics that Howard Evans had allowed us to "buy" from the MCZ for ten dollars before we left the US. We never knew what he did with the money and it was too good a deal to ask. In lieu of a proper microscope light we used an ordinary table lamp focused through an old magnifying glass mounted on a little stand, another antique that was left by a visiting biologist who had stayed at our house. Bill turned out to be a gifted teacher of the local students, and research was so exciting for both of us that our temporary adventure of tropical research turned into a whole unplanned decade of living in Cali.

Eventually that became a lifetime of research in the tropics, thanks to events that were equally unforeseen and fortunate. After we had been for three years in Cali, Martin Moynihan, a student of Niko Tinbergen and a protegé of Ernst Mayr who had become director of the Smithsonian Tropical Research Institute (STRI) in Panama, called on us and offered us a small research stipend (US$2000), to help support our research and to enlarge the Smithsonian presence in Cali, where STRI had a house for visiting researchers. We think he knew of our presence there through Michael Robinson, another Tinbergen-trained ethologist and a student of spider behavior whom we had met while at Harvard. A year or so later we were invited to Panama to give seminars without being told that it was a kind of job interview, and I was offered a position as a part-time STRI scientist (Bill was still full time at the Universidad del Valle), with the understanding that we were both associates of the Institute. This appointment was a tremendous boon to our research, for it meant not only more financial security and US health insurance for the whole family, but contact with excellent colleagues at STRI and the services of the Smithsonian libraries and other help from the talented supporting staff at STRI.

Our friend Aníbal Patiño once remarked, after we had lived for several years in Cali, that he was surprised to see us still working so hard at our research. He had thought that we would eventually succumb to the tropical lethargy that seemed to affect everyone else, including his fellow biologists who had returned from graduate study abroad. I think he underestimated the power of the work ethic that had been instilled in us by our families, and of the motivation that came from seeing evolutionary ideas that we had studied at University and graduate school played out literally in our back yard.

Many other things contributed to our happiness and productivity in Cali and, later, Costa Rica. We had many lively and helpful friends, liberating help with chores at home, and opportunities for path-breaking long-term research projects on poorly known tropical species. And we loved the sheer beauty of Cali – the tropical setting, the wonderful climate, the endless new sights Although we missed daily contact with the colleagues we had known in the US, we were not really isolated from them. Many came to visit us and stayed in our home, for there were no organized field stations and we lived within easy reach of a variety of altitudes and habitats. Once during a holiday break we hosted what turned out to be a mini-congress of leading insect ethologists when Howard Evans, Robert and Janice Matthews, and O. W. Richards happened to visit us at the same time. Others who arrived were Dutch hymenopterist Jack van der Vecht and his wife Bep, Cornell entomologist Bill "ant" Brown, W. D. (Bill) Hamilton and his family, Charles and Mary Michener, with Michael Breed and Bill Bell from the University of Kansas and, repeatedly, Martin Cooper, a talented entomologist who had studied with Richards but opted for work in a cider factory rather than an academic post. Martin used our house as a base for remarkable treks into little-explored areas of Colombia. He would appear unannounced to organize his notes and pack his specimens in chocolate-cookie tins while we happily consumed the cookies and compiled information on the wasp specimens (especially, many new species of the social sphecid *Microstigmus*) he found. In addition we hosted a number of graduate students and post-doctoral fellows doing research on insect behavior. Many of these people became our long-term correspondents and friends, often sending us books and reprints they knew would help with our research. My location in South America also put me in a favorable position for invitations to international meetings, for there were few English-speaking scientists publishing on social-insect behavior and evolution in that part of the world.

But the most important colleague for me, obviously, has been Bill. It is a great advantage to have someone trapped under the same roof who will read a draft manuscript with a thorough, informed, ruthlessly critical eye, and who is at the same time writing interesting things that need to be closely read in return. Although we have always worked independently our interests overlap broadly, and they have often been complementary in helpful ways. It is no accident, for example, that my research on alternative phenotypes gained momentum during a period when Bill was studying the horn dimorphisms of male beetles, or that we both wrote extensively on sexual selection and on parental care, though we were studying different organisms and emphasizing different points. Inbred thinking might have been the result had both of us not been pig-headedly independent-minded and argumentative – something we have to moderate if discussing science in the presence of dinner guests so that they will not think (as some have) that we are on the brink of divorce.

It would be dishonest to give the impression that I never experienced lapses in motivation. During graduate school there were several months of a spectacular inability to work, sleeping until noon in a period of personal unhappiness, when I was supposed to be finishing my thesis, something which nearly caused the loss of my Harvard post-doctoral appointment. There were also moments of feeling isolated. When *The Wasps* appeared in print in the US in 1970, one year after we moved to Colombia, it happened so far away, geographically and culturally, that there was nothing tangible to celebrate – it was a locally invisible event. Manuscripts seemed to disappear into a black void. Then I would be surprised, at some meeting abroad, that someone had read them and wanted to talk with me as a result. But the growing connection with STRI and the other factors already mentioned helped to alleviate any ill effects of geographic isolation.

After nearly ten years in Colombia we decided to move because of worries about the security of our children in a place where kidnapping was on the rise, and because our idyllic country home had been suddenly surrounded by a slum.

Tropical paradise, part II

We ended up, in late 1978, with a three-way decision to make. Just after realizing that a move was imperative we took one of our rare family trips to the US, where we learned of a job opening at the University of Utah. Interviews were arranged, and we were offered two half time positions with tenure on that very same trip. At the same time, an opportunity for Bill opened in Costa Rica, where Bill had recently taught a field course. Some energetic students encouraged him to try for a job there, where there was a vacancy in evolutionary biology. And we had our much valued Smithsonian connection, which led to an offer of two full-time positions at STRI in Panama. After much soul-searching and making of lists, it was arranged that we would be part-time STRI scientists stationed in Costa Rica, where Bill would teach half time at the Universidad de Costa Rica while working half time for STRI, and I would begin with a half-time appointment at STRI to be gradually increased toward full time if I remained productive. True to this agreement, my appointment increased gradually from half to ¾ to 7/8 and finally, in 1987, to full time. The flexibility of STRI and especially Ira Rubinoff in making an unusual arrangement enabled us to continue in tropical research of the back-yard kind that enables close long-term monitoring of *in situ* natural behavior at any time of day or night, while combined with university teaching by Bill.

Soon after arrival in Costa Rica we attended a workshop on social behavior in Mahabaleshwar, India, and I did some "field" work in Bangalore's Cubbon Park with Raghavendra Gadagkar, then a student at the Indian Institute of Science. Raghavendra was just starting his research on *Ropalidia marginata*, a *Polistes*-like wasp that is now probably the best-studied species of social wasp thanks to the insightful work of Raghavendra and his students (Gadagkar 2001). Then, thanks to my friend since graduate school, Priyamabada Mohanty Hejmadi, I visited Utkal University, Bhubaneshwar, Orissa, where she was head of the same zoology department where J. B. S. Haldane had spent many years. We discovered nests of *Zenorhynchium nitidulum*, one of the few known group-living eumenine wasps, and I was able to observe their behavior for comparison with that of the eumenine (*Zethus miniatus*) I had studied in Cali. The experience of being introduced to the fascinating societies of Indians and Indian wasps in the company of such fine people was one of the most memorable experiences of my life. In the mid-1980s Raghavendra, Bill Hamilton, and I had the privilege of seeing another *Ropalidia* species and another complex human society, in Japan (Okinawa), as guests of Yosiaki Itô, an energetic promoter of research on social wasps and of world pacifism whom I admire for both reasons.

During the period of decision about our move from Colombia a friend suggested that we should take advantage of the opportunity to go back to the US "where the action is."[5] It seemed to us that the action most important for us was happening in our own tropical back yard. But it did raise the question of whether our peculiar life-style made us in fact more isolated from the world of science than perhaps we realized. So we accepted a generous invitation to be visiting professors at the University of Michigan for a semester, in 1981. We decided that this would be my chance to teach, so Bill dedicated his stay to library research while I devoted the semester to presenting a graduate seminar on sexual selection and alternative phenotypes, with special reference to divergence and speciation, using drafts of two manuscripts (see West-Eberhard 1983, 1986) as a basis for discussion. The seminar participants were outstanding, since there was a lot of interest in behavior and evolution among students and associates of Dick Alexander, Bill Hamilton, and Peter and Rosemary Grant and others based at the Museum at that time. Many of "my" students in that seminar are now leaders in ethology and evolutionary biology.

The pleasure of working with lively students in that seminar caused a slight twinge of regret at not having spent a lifetime working with students, something I had dreamed of doing while a student myself. But it also showed that we were not behind in our thinking. On the contrary, one benefit of our geographic isolation from the great centers of academia seemed to us to have been isolation from the distractions of trendy topics in science, which meant that we followed our own ideas as they grew out of observations of living organisms in the field. My Smithsonian job made it possible to do long-term studies of behavior and natural history, while equipping me with fine libraries and necessary equipment, and congenial colleagues. And it has permitted the luxury of long periods devoted to thought and writing, something that is rare for university scientists. As for regrets about not teaching, if there is anything I have learned it is that you cannot do everything in life that you would like to do. I do sometimes regret that I have nothing impressive to show people who want to visit "my lab." Should I show them the kitchen, where I often sort specimens and pour alcohol into jars? The back yard and the bushes along the driveway? The collection of glass jars in the pantry? The microscopes in Bill's crowded home office? Or my "study" – the

---

[5] Whenever I remember this advice I think of a comment made by an elderly American woman who lived happily near a pristine creek in a rural part of Costa Rica. When asked why she had chosen to live so far away she said that it was her friends in the city who seemed to her to live far away (from her, and from what she thought was important).

4 m × 6 m cabin full of books, filing cabinets, and piles of papers where I read, ruminate, and write?

Much of the Smithsonian – Costa Rican period has been spent writing, although I have accumulated many observations on the local wasps and the behavior of other group-living insects here, and some in Panama. The longest writing project, started in the mid-1980s and finally hand-carried to the New York office of Oxford University Press in two carry-on suitcases in August of 2001, was a book on *Developmental Plasticity and Evolution*. (West-Eberhard 2003) It was initially intended to be a compendium of alternative phenotypes in behavior and morphology and their significance for evolution. This book reflects my background as a museum-bred evolutionary ethologist, so it is quite a different amalgam of evolution and development than the "evo-devo" that, to my profit, happened to emerge from molecular embryology during the same period. Like everything else I have written, it grew out of fieldwork on the behavior and natural history of wasps. The *Zethus* in the pre-school garden had taught me how the flexible behavior of a solitary female occurs in modular (switched) phases that could be decoupled to form the behaviorally and physiologically alternatives of social insects, the contrasting worker and queen castes. From the queen–worker alternatives of social insects I moved to thinking about alternative phenotypes – polymorphisms and polyphenisms – in general (1989), and from there to thinking about the parallels of alternative phenotypes with the modular, switch controlled (and therefore modular and easily reorganized) development of all phenotypes. This eventually led to writing the fat book on developmental plasticity and evolution in all organisms (West-Eberhard 2003).

While writing that book various things happened that kept it exciting. Costa Rica is an attractive *venue* for international meetings, and in 1994 Gabriel Macaya, a Costa Rican molecular biologist friend who had studied abroad, helped his ex-professor, Giorgio Bernardi, to organize a very select and well-funded meeting on the evolution of the genome. The organizing committee enlisted my support as a local evolutionary biologist, so I decided to attend, carefully hiding the fact that I hadn't thought about DNA since cramming for a graduate exam more than 30 years before. Wally Gilbert, a Nobel Laureate, talked about the origin of introns and exon shuffling, running to a fax machine to get the latest data from his lab. I had especially interesting conversations with Richard Keynes, a descendent of Darwin and historian of the Darwin family, and with Emile Zuckerkandl, who had worked with Linus Pauling and had many original ideas about molecular evolution. He sent me a remarkable paper of his (Zuckerkandl & Villet 1988) that, although it had been overlooked by evolutionary biologists, is of fundamental importance for anyone (including ethologists) interested in the relationship between environmental and genetic factors in development, showing the biochemical *equivalence* of the two kinds of input. As a result of that meeting I realized that the evolution of the genome resembles that of behavior (and all other aspects of the phenotype) in being reorganizational and subject to conditional induction of novel alternative phenotypes (e.g. through alternative splicing). Throughout the writing of that book my STRI colleague Bill Wcislo was perhaps the person who best understood what I was trying to do, and most steadfastly provided references and helpful discussion. The train of thought that led from studies of wasps to a treatise on development and evolution, including some key inputs from other colleagues (Fred Nijhout, John Bonner, Bill Eberhard, and Marc Kirschner outstanding among them), is outlined in the Preface to the book, and in a very brief autobiographical sketch (West-Eberhard 2009).

Probably there are few families of my generation where two full-time scientific careers and family life itself have been so thoroughly, and I think happily, intermingled, without the work of either partner being subordinated to that of the other. A number of the factors that made this possible have already been mentioned, but perhaps most decisive, in retrospect (for this was not consciously planned), was to leave the pressured world of US academia, with its dearth of affordable childcare and domestic help, to live in Colombia and, later, Costa Rica. There, in the countryside near cities with good schools and a university, our homes have been private field stations where we could both do much of our work at home, being on family duty while working, and at the same time having the on-call baby-sitting services of a hired helper who could also take the drudgery out of routine cooking, washing, and housekeeping chores.

Epilogue

While writing this essay I have been following the development and behavior of some gregarious caterpillars (*Papilio anchysiades*) in view of our kitchen window in Costa Rica. At each larval molt they literally grow out of their skins, bursting into new freedom from the constrictions of an earlier stage even while preserving a basic unity of form. I think my life has been like a series of larval instars, with several distinct sheddings of old ideas and illusions, first with the move from my home town to Ann Arbor, then with the discovery of concept-driven science and two summers at Woods Hole, then with work in Colombia, and finally, following a pupation of thesis writing and social quiescence, a metamorphosis to academic and personal adulthood – a postdoctoral appointment at Harvard, and, eventually, a family and a marvelous permanent position with the Smithsonian Tropical Research Institute. Reproduction and winged flight. By some combination of luck and adaptive behavior I have escaped things that might have cut this trajectory short. There is absolutely no doubt that gregariousness helped. At each phase there were key people and protective associations of them. I watch these caterpillars and hope that my interpretation is correct, that the associations have been *mutually* rewarding and beneficial.

Looking ahead at all there is still to be done I sometimes worry that I could be happy just talking with friends, seeing our kids and our grandchildren, reading good novels, getting an occasional backrub, and watching the water surge through our ditches after a heavy rain.

References

Alexander, R. D. (1974). The evolution of social behavior. *Annual Review of Ecology and Systematics* **5**: 325–81.

Evans, H. E. & West-Eberhard, M. J. (1970). *The Wasps*. Ann Arbor, MI: University of Michigan Press.

Gadagkar, R. (2001). *The Social Biology of* Ropalidia marginata: *Toward an Understanding of the Evolution of Eusociality.* Cambridge, MA: Harvard University Press.

Hamilton, W. D. (1964a,b). The genetical evolution of social behaviour I, II. *Journal of Theoretical Biology* **7**: 1–16; 17–52.

Hamilton, W. D. (1987). Discriminating nepotism: expectable, common, overlooked. In *Kin Recognition in Animals*, ed. D. J. C. Fletches & C. D. Michener, pp. 417–37. New York: John Wiley and sons.

Hamilton, W. D. (1996). *Narrow Roads of Gene Land*, Vol.1, *Evolution of Social Behaviour.* New York: W. H. Freeman.

Haskins, C. P. & Wheldon, R. M. (1965). "Queenlessness," worker sibship, and colony versus population structure in the formicid genus *Rhytidoponera. Psyche* **72**: 87–112.

Hrdy, S. B. (1999). *Mother Nature.* New York: Pantheon.

Hull, D. L. (1980). Sociobiology: another new synthesis. In *Sociobiology: Beyond Nature/Nurture?*, ed. G. W. Barlow & J. Silverberg (eds), pp. 77–89. Boulder, CO: Westview Press.

Hyman, L. H. & Hutchinson, G. E. (1991). *Libbie Henrietta Hyman. Biographical Memoirs*, Vol. 60, pp. 103–13. Washington, D. C.: National Academy Press.

Kennedy, J. S. (1966). Some outstanding questions in insect behaviour. In *Insect Behaviour*, ed. P. T. Haskell *Symposium of the Royal Entomological Society of London* **3**: 97–112.

Mattfeld, J. A. & Van Aken, C. G. (1965). *Women and the Scientific Professions.* Cambridge, MA: M.I.T. Press.

Mayr, E. (1963). *Animal Species and Evolution.* Cambridge, MA: Harvard University Press.

Thornhill, R. & Alcock, J. (1983). *The Evolution of Insect Mating Systems.* Cambridge, MA: Harvard University Press.

Trivers, R. L. (1974). Parent-offspring conflict. *American Zoologist* **14**: 249–64.

West, M. J. (1963). The behavior and natural history of a social wasp, *Polistes fuscatus* (Fabricius), in fall and winter. Unpublished Zoology Honors Thesis, University of Michigan, Ann Arbor.

West, M. J. (1966). The nature and determination of castes in polistine wasps. *American Zoologist* **6**(4): 346.

West, M. J. (1967). Foundress associations in polistine wasps: dominance hierarchies and the evolution of social behavior. *Science* **157**(3796): 1584–5.

West, M. J. and R. D. Alexander. 1963. Subsocial behavior in a burrowing cricket, *Anurogryllus muticus. Ohio Journal of Science* **63**(1): 19–24.

West-Eberhard, M. J. (1969). The social biology of polistine wasps. *Miscellaneous Publications of the Museum of Zoology of the University of Michigan*, no. **140**, pp. 1–101.

West-Eberhard, M. J. (1975). The evolution of social behavior by kin selection. *Quarterly Review of Biology* **50**(1): 1–33.

West-Eberhard, M. J. (1978). Temporary queens in *Metapolybia* wasps: non reproductive helpers without altruism? *Science* **200**: 441–3.

West-Eberhard, M. J. (1979). Sexual selection, social competition, and evolution. *Proceedings of the American Philosophical Society* **51**: 222–34.

West-Eberhard, M. J. (1983). Sexual selection, social competition, and speciation. *Quarterly Review of Biology* **58**: 155–83.

West-Eberhard, M. J. (1984). Sexual selection, competitive communication and species specific signals in insects. In *Insect Communication*, ed. T. Lewis, pp. 283–324. London: Academic Press.

West-Eberhard, M. J. (1986). Alternative adaptations, speciation and phylogeny. *Proceedings of the National Academy of Sciences, USA* **83**: 1388–92.

West-Eberhard, M. J. (1987). Observations of *Xenorhynchium nitidulum* (Fabricius) (Hymenoptera, Eumeninae), a primitively social wasp. *Psyche* **94**: 317–23.

West-Eberhard, M. J. (1989). Phenotypic plasticity and the origins of diversity. *Annual Review of Ecology and Systematics* **20**: 249–78.

West-Eberhard, M. J. (2003). *Developmental Plasticity and Evolution.* New York: Oxford University Press.

West-Eberhard, M. J. (2004). Howard E. Evans: known and little known aspects of his life on the planet. *Journal of the Kansas Entomological Society* **77**(4): 296–322.

West-Eberhard, M. J. (2005). Howard E. Evans 1919–2002. *Biographical Memoirs, Vol. 86*, pp. 1–19. Washington, D.C.: National Academies Press.

West-Eberhard, M. J. (2009). Bio: Mary Jane West-Eberhard. *Evolution and Development*, **11**: 8–10.

Wilson, E. O. (1971). *The Insect Societies.* Cambridge, MA: Harvard University Press.

Wilson, E. O. (1975). *Sociobiology.* Cambridge, MA: Harvard University Press.

Wilson, E. O. (1985). In the queendom of the ants: a brief autobiography. In *Leaders in the Study of Animal Behavior*, ed. D. Dewsbury. New York: Bucknell University Press. [Reprinted, 1989, Chicago, IL: University of Chicago Press.]

Zuckerkandl, E. and Villet, R. (1988). Concentration-affinity equivalence in gene regulation: convergence of genetic and environmental effects. *Proceedings of the National Academy of Sciences, USA* **85**: 4784–8.

20

A bird in the hand

JOHN C. WINGFIELD

Whenever I give a talk at a meeting, or a seminar at another institution, one question I am almost always asked is "how come you work with birds in the field and have a name like that?" Just coincidence but it is, perhaps, worth a few tales of fate and serendipity that led me to blend fieldwork with laboratory investigations in animal behavior.

I was born in 1948 in Ashbourne, Derbyshire, a small market town in rural England that has little claim to fame except for being one of the most beautiful locations in the country. My younger brother, Andrew, and I grew up in this tranquil, rural, setting. There is absolutely nothing in my background that would have predicted an academic career. My father's family were coal miners and servants, and my mother's family were farmers.

Leaders in Animal Behavior: The Second Generation, ed. L. C. Drickamer & D. A. Dewsbury. Published by Cambridge University Press. © Cambridge University Press 2010.

Although I was the first in the extended family to go to college, my parents and grandparents remain some of the brightest people I have known and they would have thrived in college. My father and mother grew up in an era when they were essentially unable to take advantage of higher education but my father did, none the less, manage a highly successful business career. Perhaps this prepared them to raise a son who had non-traditional ambitions. I am sure they wondered from time to time why I did not want a more conventional career, but it is true to say I only wanted to work as a biologist, particularly with birds.

The importance of natural history

Throughout my career I have always regarded knowledge of natural history as the foundation of all my research and teaching. Gaining such a base is possible only through time spent in the field and learning from others who know how to observe nature. My earliest memories are evenings with my grandmother, usually around a glowing coal fire. She told me about the old traditions and how they used natural resources as well as farm animals and plants. She had a complete practical knowledge of nature without any formal training whatsoever, although some of the basic concepts were clearly present. She knew exactly when certain plants would flower and fruit. Some of them were used for medicines and others for spices and food. When making soap she used local plants for fragrance. She knew when to collect certain local mushrooms and preserve them for the winter and when to collect lapwing eggs (considered a delicacy). My grandmother was a resource for the whole community, especially during the trying times of the post-World War II years. In the 1950s she was acutely aware that farming practices and rural lifestyles were changing throughout the country. Many natural products she had used were now becoming rare, making me aware of conservation issues at a very early age. These images and stories have always stuck with me and it was during these very early years of my childhood that a life-long obsession with nature began.

My grandmother was also keenly aware of behavioral traits and cycles of animals in general. As I grew into early teenage years I would frequently relate my observations of behavioral phenomena to her. She was always ready to discuss all I had seen walking out in the woods and farmland around Ashbourne, particularly in Bradley Wood and Dovedale (Figures 20.1 and 20.2). One spring I hand raised two magpie nestlings – an education in itself, especially the way in which these two birds took advantage of the entire family to gain access to food. However, their penchant for harassing pets and plaguing the chickens (particularly the roosters) did not endear them to the rest of my family.

I have many vivid memories of animal observations that strengthened my interests in biology. Examples include a stoat performing gyrations to mesmerize a rabbit and attack and capture it. During approaching storms I repeatedly observed mistle thrushes singing from the tops of trees into gusting winds. These and many other observations were familiar to my grandmother, and her explanations and knowledge of them turned out to be accurate when I later read about them in natural history books. I had biology teachers too, particularly my grade school teacher, Mr. Seals, who was delighted that some of his pupils clearly had great interest in

Figure 20.1. Bradley Wood, Ashbourne, Derbyshire. A mixed woodland of oak, beech, ash and Scots pine. It was here and in the surrounding farmland that I spent many days observing wildlife and how they coped with a changing environment.

Figure 20.2. Dovedale, Derbyshire/Staffordshire border. Within the gorge are remnants of primeval forest – some of the last in England. Here too, I gained early experience observing animals in their environment. It is still an inspiring place to hike and reflect.

nature and the outdoors. He was not a trained biologist either, but none the less he took us out on field trips to impart some of his knowledge of nature. This certainly helped to intensify my interest and avid thirst for information about plants and animals in their environment. I consider myself very privileged to have been part of such an "education" at such an early age.

After a very happy childhood in the very secure environment of a close-knit family in a small rural town, I was then confronted with the notorious "11-plus" examination. This was an early I.Q. test that determined which school you went on to. I was able to pass and attended Queen Elizabeth 1st Grammar School, founded in 1583. Two teachers in particular were instrumental for me – Mr. Regnauld, the biology teacher, and Mr. Harris, who taught geography and geology courses that turned out to be very important in my later career. My schooling gave me a very strong foundation in biology, but one thing I found frustrating was that all the interesting things I saw and heard animals doing in the field were never mentioned in textbooks. There seemed to be little emphasis on what animals actually did in their natural habitat. There were other books such as Gilbert White's *Natural History of Selborne* (White 1860), written in the eighteenth century, that drew my attention. White's detailed accounts and anecdotes of what animals in his neighborhood were doing year-round fascinated and inspired me to observe more. My field observations at that time were anecdotal but seemed to make sense in my naïve way. I was intrigued as to why black-headed and herring gulls would fly off in one direction in the morning and the reverse direction at night. I figured out, correctly, that they were going from a roost site to a feeding area during the day. Not much of an observation in retrospect, but for an 11-year-old in a rural town in England with, as yet, no formally trained naturalists to interact with, it was a revelation. I soon became aware of seasonal migrations, nesting cycles, winter strategies, etc. around Bradley Wood and the adjacent farmlands. Some species appeared in the winter and I would look for them eagerly each autumn. Why were some species common in certain months (such as Bohemian waxwings) and would not be seen again for years? Some were abundant in winter every year, yet others passed through on their way to wintering areas to the south. All of these things made me ponder – why and how did animals do things so differently? I found some bird books in our school and local libraries and these gave me some information about migratory patterns, but much of what I was seeing was not mentioned. My parents had subscribed to several nature magazines in a wonderful effort to try to nurture what obviously was a passion for me. These helped too, but I was seeing so much more that was not explained.

One thing that fascinated me above all else was the suite of coping behaviors animals showed in the face of adversity. There were two extremely cold winters in 1959–60 and 1962–63. Snow cover was deep and temperatures remained well below freezing for weeks. I watched as the animals changed their behavior dramatically. Foxes became much more bold, venturing into towns where at that time they were not tolerated at all. Fox hunting was very widespread and very popular, and gamekeepers also shot them on sight. One morning after a particularly cold week I found a dead little owl by a drain at our house. Steam rising from the drain may have attracted the owl, but was it trying to get at water or was it seeking warmth? I also found crows, and robins that had starved to death. Populations of certain species such as long-tailed tits and other small songbirds declined precipitously, but I could not be sure whether they had left the area or had succumbed to the cold. Other birds were able to leave; for example, lapwings and other open field birds such as thrushes departed as soon as it snowed. One evening wrens entered a nest box in our garden to roost. At dawn the next morning I watched as they left – no fewer than 25 wrens flew out. Inspection of the box revealed 4 dead – all apparently starved to death. I wondered with great curiosity about these behavioral responses to severe conditions. Clearly they were switched on only when necessary, and were different from the regular daily behavioral routines and seasonal changes. How did they do this? There was nothing in the books that even attempted to explain such phenomena. Little did I know that I was preparing myself for one of my life-long research interests – the hormonal bases of stress and coping mechanisms in response to environmental perturbations such as bad weather.

I also performed crude autopsies on birds I found dead (having been introduced to anatomy, etc. in my grammar school biology classes). I still have some of the notes documenting body condition of these birds. What was the reason for their demise – starvation, injury, hypothermia? Again, I was unaware that I was preparing myself for an integrated approach combining morphology, physiology and behavioral observations of birds in the field.

When I was eleven I learned about the Derbyshire Ornithological Society (DOS) through one of my teachers. This was the first time I had contact with people who had similar interests to me and, importantly, had formal training in ornithology. Mr Appleby, the General Secretary of the DOS, endured many long phone calls (I was oblivious of time) as I related to him birds that I had seen and things they were doing. He was an extremely patient man even though he must have been very busy, and I am deeply indebted for his patience and calm mentoring of a young budding ornithologist/naturalist. Mr Appleby passed away a few years ago and I wish I could have had the chance to tell him how much I appreciated his patience, and if it is any consolation, that I have certainly paid my debt in mentoring, counseling and listening patiently to many students during my career. On occasion, when frustrated by difficult or struggling students, I always remember Mr Appleby and how he put up with me.

Mr Barry Potter I remember particularly well. He took me on as an apprentice bird bander, taught me how to mist-net birds and collect other data such as body mass, how to keep meticulous records on just about everything – molt, body condition, parasites, etc. I was so inspired by being able to participate and learn so many new basic field skills. One experience I had with Barry Potter will live with me forever. One April he asked if I would be interested in getting up very early to see a black grouse lek near Warslow and Alstonefield (Figure 20.3) only a few kilometers from where I lived. I was thrilled! I did not fully appreciate that this involved getting up at three o'clock in the morning, a concept that was about as alien to a teenage boy as going to the moon. But, none the less, that first night I could not sleep with excitement and anticipation, as well as anxiety of oversleeping and missing everything! So at 3 am I was up, full of energy and expectation, and went with Barry to this magical lek.

At first light we saw these magnificent birds calling and dancing with the females wandering through totally unconcerned, preening, or just sitting. I will also never forget the other sounds of the birds on the moors nearby (Figure 20.3). The combination of sounds from bubbling songs of curlews, the continuous, breathless songs of skylarks, to the whir of

Figure 20.3. The moors near Warslow and Alstonefield, Derbyshire. August 2005. Note the transition from millstone grit and heather (foreground) to limestone woodland/grassland (background). This was an extremely rich area for watching birds and other animals and how species distribution varied from one habitat to another.

diving wings and tails of lapwings and snipe was like music to my ears. I got up three mornings in a row and I could see Barry was just as enthralled by the spectacle as I was.

It was through the DOS that I also learned of other areas to see wildlife. I went on every field trip my patient parents would allow – they had often to drive me to pick up points, etc. Other kids would go to their soccer and cricket games (which I also did most weekends), but what I really wanted to do was go to, for example, the nearest sewage farm and search for migratory shorebirds that favored such places as feeding stopovers. Looking back it amazes me how tolerant my parents were of their aberrant son.

At this time I met Miss Kitty Hollick, an avian enthusiast and daughter of one of the most revered physicians in the town of Ashbourne. She maintained a wonderful old home and a very famous garden that she opened to the public once a year for a popular garden party. She rode about town on her bicycle with a wicker basket on the front. For me, as young boy of coal mining and farming families, Miss Hollick seemed like an unreachable and professional person that I had no business talking to. Then one day I spotted a northern shrike – a bird I had never seen before and that was not even pictured in local wildlife books. I excitedly announced this to Mr Appleby and Barry Potter and they encouraged me to call Kitty Hollick and tell her where the bird was. I plucked up the courage and called her. She was absolutely delighted and accompanied me to see the shrike. To my astonishment she announced that the bird was probably of the southern sub-species (from the Mediterranean area) and not from the north (its major range). From that time on she gave me all her copies of a national publication, *British Birds*, a journal that publishes papers and notes for birders as well as amateur and professional analytical papers on the natural history, behavior and ecology. Each monthly issue I would read avidly from cover to cover. For the rest of her life she would very religiously bring the latest copy of *British Birds* and when I eventually went off to college, she left them with my parents. I still have all those copies, going back to 1964, and I now subscribe to that journal myself – the first I ever read.

This was a turning point for me. I had a source of serious natural history and I learned how professional scientists and amateurs could combine to contribute to the advancement of biology. Those three people, Barry Potter, Mr Appleby and Kitty Hollick, clearly had major influences on my ambitions and my drive to be a biologist. This was what I wanted to do for the rest of my life, but I needed laboratory training as well. It was clear from what I had learned in Grammar School that a field and laboratory approach might help to answer both the why and the how questions about animals in their natural environment.

The transition to science

When I was 14 it was very clear that I must go to university and then go on to do a Ph.D. It was here that I ran into one of those lifetime roadblocks. I announced early on that I wanted to continue to the sixth form (equivalent to eleventh and twelfth grades in the USA) and take my A-level exams. These were required for entry into a university. Although I had wonderful biology and geography teachers who were very supportive, the Headmaster was much less enthusiastic. I remember his words clearly, asking why I was interested in applying to university. He said very forcefully "Wingfield, you are not university material". My reaction was one of shock and then defiance, although I probably did not show it at the time. The Headmaster's words motivated me far more than I could have imagined.

At that time students applied for a scholarship to go to university because most people could not afford to pay for a college education. The competition was intense. The University of Sheffield made an offer that required I attain specific A-level grades for acceptance. I felt empowered and committed, knowing exactly what I needed to do. Later I learned from my Professors at the University of Sheffield that they had read the letter from my Headmaster and concluded that it was a judgment call on social status and not academic potential. My other Grammar School teachers wrote glowing letters that bolstered the admission committee's decision to take a chance on me. Otherwise my career could have ended right there. The A-level exams were the most difficult I have ever taken but somehow I pulled through and received the required grades.

My university undergraduate career began in 1967. The first three weeks at Sheffield were a shock: physical chemistry, 8 am until 5 pm every day, practicals in the afternoons and more on Saturday from 9 until 12 noon. The first three weeks of winter term focused on inorganic chemistry and in spring term, organic chemistry. If you failed any exams on these topics you were not allowed to continue! It was a grind but very useful because the courses were designed for biologists and I still use that information today. The integrative biology course was also intensive and I went the entire first year in college without going in the field to watch birds and other organisms as I had always done (although I must admit that being a first-year student and partying was another reason). The next summer I took two field courses, one in botany and another in marine biology. Then I realized how much I missed being out in the field and I went back to it with renewed energy. By now I had a strong grounding in basic chemistry, biochemistry and physics that turned out to be essential for my future integration of endocrinology with behavior and ecology.

The Department of Zoology at the University of Sheffield had a number of faculty who were endocrinologists. I had opportunities to conduct undergraduate research with founders of this field such as John F. Ball that gave me a unique combination of skills and background knowledge that was perfect for my developing interests. What continued to be frustrating was that animal behavior was almost completely lacking. This was one of my first interests as a child but it was clear that no formal course in behavior was in sight. There was some mention in endocrinology classes of the behavioral effects of hormones, but the extensive work of Frank Beach, Robert Hinde (1965) and Daniel Lehrman (1965) rarely came to the surface.

At this time I discovered books by Konrad Lorenz: *"King Solomon's Ring"* and *"On Aggression"* (Lorenz 1952, 1963). These fascinated me and piqued my imagination. I read Niko Tinbergen's *"Herring Gull's World"* (Tinbergen 1953) and Thorpe's *"Bird Song"* (Thorpe 1961). Because animal behavior interested me so much, at one of my tutorials in my second year I suggested we cover the book *"On Aggression"*. Our tutorial professor just bristled, stiffened in his seat and looked at me with staring eyes and a contemptuous look on his face. To my dismay he then haughtily told me that he had just returned from a meeting at which this book and the ideas on aggression had been discussed and profoundly criticized. I was utterly dismayed and felt greatly humiliated. How could he summarily dismiss a whole body of work that seemed to me so fascinating and full of interesting questions to pursue? I had clearly stepped over some boundary and the subject of behavior was the line. Despite this traumatic academic experience, my interest in behavior continued undaunted; I felt that if behavior researchers such as Lorenz and Tinbergen induced such intense responses from other researchers and professors, then something very exciting, even revolutionary, was going on.

Graduate school

As graduation approached, I knew I had a chance of a decent grade, and I expressed a desire to go on to graduate school. Of course my first interest was to research in behavior and ecology with a strong field component. On the other hand, I had developed a deep interest in endocrinology because of the role it played in regulating responses to environment during development and day-to-day life. It was clear that this control and integration system had so many interfaces with other branches of biology and biomedicine. These included behavior and ecology, although at the time I had no concept of how enormous and expansive that could, and would, be. I saw endocrinology as a unifying system that brought together morphological, physiological and behavioral changes in natural settings as well as in medicine and agriculture. I had absolutely no desire to go into medicine or agriculture; I was just fascinated by basic biology.

On the advice of my undergraduate research advisor, John Ball, I approached the Chairman of the Department, Professor Ian Chester-Jones, one of the founders of modern comparative endocrinology. He was delighted that I was interested in graduate programs and advised that, because I had done so well in physiology and endocrinology, I had the potential to excel in that area. John Ball was equally enthusiastic. I went off to interviews and at the University College of North Wales (UCNW), where there was a student research fellowship available for the Ph.D. degree. I met the Chairman, Jimmy Dodd, another founder of comparative endocrinology, and a young emerging faculty member, Brian K. Follett. Brian worked with birds (Japanese quail) on photoperiodic control of reproduction – clearly very important for coping with environment change and regulating life cycles. He was very encouraging but unfortunately his laboratory was full at that time and he did not feel he could squeeze another student in.

Jimmy Dodd told me about what he and others were doing with local marine fish in relation to reproductive cycles, osmoregulation and stress. I met another young faculty member, Andrew Grimm, who had just finished a postdoctoral fellowship with the steroid biochemist and endocrinologist David Idler in Halifax, Canada. Andrew was establishing his research program on steroids and reproduction and also in relation to environmental stress. Specifically, the project focused on identifying what steroids were present in the blood of local marine fish, then to use newly developed antibodies for steroid hormones to develop assays for measuring circulating levels. I could see immediately that this would allow us to follow the natural history and life cycles of fish as well as provide some application because of the potential to understand how reproductive processes were controlled and how stress (such as from commercial fishing) could affect these animals. Andrew talked about going out on trawlers to collect fish for the marine station (at Menai Bridge on the island of Anglesey) and sampling fish on board to assess trawling stress. So here was a field component too! The focal species was a commercially important (and delicious) flatfish called the plaice. It is a handsome animal with a green olive color and red spots on the right (upper) side. After expressing great interest in this fellowship and the opportunities to learn new skills and expand my knowledge of endocrinology, to my delight I was accepted.

Up to 1970 many had tried to measure circulating levels of hormones in blood. However, plasma concentrations were so low that it was very difficult, in most cases impossible, to get accurate measures. Development of such techniques would open up a host of opportunities to understand developmental changes, responses to the environment, stress effects, etc. What excited me particularly was the potential to incorporate behavioral components, including coping mechanisms in response to stresses of environmental perturbations. This triggered memories of watching animals around Ashbourne deal with severe weather, and now I was on the verge of being able to start looking at this from a very different angle – experimental and biochemical.

While at UCNW in Bangor I was steeped again in comparative endocrinology. None the less it was possible for me to gain further field experience while accompanying population ecologists (particularly Richard Arnold and Geoff Wood) to islands around the remote and rugged coast of North Wales. They investigated populations of seabirds, rodents, etc., and at the same time I was going out to sea on trawlers, working with fish. I learned that there was great variation from year to year at seabird colonies. In some years virtually no individuals of some species were breeding, yet in other years all were breeding. It seemed logical to me that reproductive hormone studies using newly emerging techniques could be applied to these populations to answer the question of how reproductive function, including behavior, varied so much among individuals and from year to year.

While catching fish at sea I watched as migrant songbirds were attacked by gulls after being exposed at dawn over the ocean. Some of the survivors alighted on our boat and remained close to the deck hands, possibly as protection from the predatory gulls. Did the stress response, and the hormone secretions that accompany it, change the behavior of these migrants to immediate survival strategies?

So many questions were forming in my mind and yet the topic of behavior was still not a major emphasis in the graduate program. I wondered again – why? It would be wonderful to apply some of the techniques I was learning and helping to develop to understand how behavior varied so much from year to year and even moment to moment. Some colleagues actively discouraged my interest in behavior. They worried that it is difficult to quantify, there is far too much variation and how do you control for anything, especially in field studies? I was never convinced by those arguments because I had spent so much time watching animals already. The reality of social relationships among individuals in the natural world must have a profound influence on the integration of morphology and physiology in response to environment change. Such things I pondered daily, but in isolation.

Working with fish endocrinology was extremely productive. I learned a great deal of steroid biochemistry while developing extraction and chromatography systems to separate steroid hormone fractions from the same sample. We had been very successful in using antibodies and binding proteins of steroid hormones to develop radioimmuno- and radio-binding assays to measure steroid concentrations up to three orders of magnitude lower than had ever been possible before. We observed "stress responses" in which glucocorticoid levels in the blood of fish would rise within minutes of applying a stressor such as capture and handling. This simple procedure applied in the field became known as the "stress-series" and has been applied to over 200 species over the years. It enabled us to apply a standardized stress to individuals so the effects of gender, age, season, habitat, etc. on glucocorticoid responses could be investigated.

Dramatic seasonal changes in reproductive hormone levels emerged associated with gonadal development and migration. We could experimentally remove components of the endocrine system such as the anterior pituitary gland and see dramatic effects at the level of circulating hormones – a direct link with the physiological and morphological changes going on. Furthermore, Follett's laboratory had developed a chicken luteinizing hormone (LH) radioimmunoassay to measure responses of non-reproductive quail to spring-like long days that stimulated gonadal development. Research tools were now available that could, for the first time, assess the dynamic nature of the endocrine system and its responses to changing environments. Truly a time of wonder and dreams of what was to come. To me a burning question remained: the relationships of morphological and physiological changes with behavior.

Brian Follett had been collaborating with Robert Hinde's group at Cambridge measuring LH in relation to behavioral interactions and transitions in canaries (Hinde *et al.* 1974). It was here that I discovered and read with great interest this work on behavior as well as the work of Lehrman (e.g. 1965), Beach (1948) and others on hormone–behavior interactions. So many opportunities for behavioral endocrinology presented themselves. My only reservation was that these investigators were working with captive and largely domesticated species of birds (and other animals). Few were thinking about the field except for some very early work (Emlen & Lorenz 1942; Collias 1950; Watson 1970; Trobec & Oring 1972). These papers showed that such research could be done in the field, but many still did not think it was worth while.

Some practical issues

As my postgraduate studies progressed, it was beginning to dawn on me that some critical issues presented considerable problems of interpretation. For example, we went out on trawlers, put down a large pocket-like net and dragged it across the seabed for up to four hours. Then the net was hauled to the surface with fish, seaweed, invertebrates, debris, etc. that were then dumped on the deck. Desperate, flopping fish were grabbed and put in shallow tanks of seawater where they slopped around with the pitching of the boat on the waves. Not only was this stressful enough, but the fish became seasick and vomited so profusely that frequent changes of seawater were required. I had never imagined that fish would get seasick like us! After a day or so of this trauma we transported fish to the marine station and transferred them to large land-based tanks. Here we allowed them to acclimate to the aquaria (if indeed they ever did) and finally began investigations and experiments. It was quite clear to me that by this time the fish were so chronically stressed that our data on seasonal changes of circulating hormone titers must be interpreted with caution. The data were still useful, indeed unique at that time, but I became more and more convinced that we needed to sample animals in the wild in natural settings in addition to investigations on captives.

Investigations of domesticated, or semi-domesticated animals (mice, quail, chickens, etc.) are clearly important, but such creatures no longer live in natural settings and probably would not survive if they were put back into the real world. Investigations of wild animals in their natural habitat were the key. However, working with fish had other problems: they do not live in the same medium as we do. The only way to obtain data from the plaice in its natural settings was to dive to the bottom of the murky Irish Sea and somehow sample them quickly. Birds, on the other hand, are mostly terrestrial, diurnal, and easy to observe. Ecologists and behavioral biologists had already compiled a mass of natural history and experimental data on them. Having said this, I should quickly point out that since that time several investigators have been able to sample fish underwater in their natural habitat. Several years later I had an opportunity to accompany Doug Shapiro and his team to collect blood samples from coral reef fish in Puerto Rico. They captured from and brought them to the surface for sampling within a very few minutes. These underwater field endocrinologists have my lasting admiration.

Becoming more and more convinced that "field endocrinology" was viable and that birds were the ideal group to study, I asked Brian Follett where I could go as a postdoctoral fellow to do this. He advised me to go to the United States of America, suggesting several laboratories, including that of Donald S. Farner at the University of Washington in Seattle. Don Farner's laboratory piqued my interest particularly because he had done much fieldwork on birds both in captivity and in the field, and he was clearly interested in the environmental control of life cycles, including photoperiodism, behavior and migration. Don offered me a postdoctoral position, and I suspect my skills with steroids and radio-immunoassay in general were major reasons why, because they were now in great demand as endocrine research expanded rapidly.

After graduating with a Ph.D. in 1973 I learned that an American friend, Jan Smith, had been accepted into graduate school in Farner's laboratory, so I would actually know some-one when I got there. Now all I had to do was to get to Seattle. My parents probably would have come up with the money for my airfare, but I was 24 and travel costs were high because of the 1973 oil crisis. I worked in a Derbyshire limestone quarry bagging crushed stone for three months to pay for the airfare and then have a little money to get by until my first paycheck in Seattle. I got on the plane on December 28, 1973, and on arrival in Seattle the first hurdle was to get through immigration. Although I had an exchange visitor visa for three years, the immigration official would not let me enter the USA because I did not have enough money. I managed to convince him that I did have a job and everything would be fine, but it was close!

The next day I was introduced to the laboratory in the Department of Zoology, after a brief tour of the spectacular campus of the University of Washington. Don Farner was eager for me to establish the multiple steroid assays as soon as possible. One of his students, Florence Lam from Hong Kong, introduced me to a partition chromatography system using diatoma-ceous earth as column support. This had great potential in separating steroid fractions from small blood samples and had more potential than the thin-layer silica gel plates I had been using, enabling us to develop several protocols for measuring multiple steroids simulta-neously. This left enough plasma to measure LH, the assay Brian Follett had introduced and validated for use in sparrows earlier that year.

Field endocrinology

After the first month or two, I broached the possibility of doing fieldwork that spring. In particular I wanted to collect blood samples from the local Puget Sound white-crowned sparrow (Figure 20.4) that Don and his students had also studied (e.g. Lewis, 1975). The idea was to catch, sample, then band and take many other measures from each bird before releasing it to follow behavior through the breeding season and beyond. Blood would be drawn within a very few minutes to avoid stress effects (that I now knew a lot about), and the

Figure 20.4. White-crowned sparrow, *Zonotrichia leucophrys*. Seattle, May, 2006. The species I first investigated in field endocrinology.

samples would be small enough not to be debilitating. I imagined these marked birds could then be recaptured at intervals to obtain additional samples so we would have profiles of hormone levels for individuals in natural settings. Don listened to me in his polite way with one eyebrow raised and puffing on his ever-present pipe. He said they had tried this kind of thing before by collecting individuals (with a gun), and it was very difficult to get all the information one needed. Obviously it is not possible to follow up once a bird has been collected, but I was proposing a very different approach by catching birds in nets and traps and then releasing them. Don could see the potential, and to his credit, and after my infernal pestering, he gave me a chance, for which I will be eternally grateful.

In reflection it seemed that a number of serendipitous events had presaged this. All the skills I had learned trapping birds and banding them (from my youth in Derbyshire), and observing them in relation to environmental events, kept my mind open to possibilities of studying them in natural settings. Barry Potter, and Richard Arnold from Bangor, who had taught me how to capture and band birds, also showed me how to recognize a brood patch, an indicator of breeding condition, fat scores visible through the skin in the front of the neck and abdomen, and molt. People in Farner's laboratory taught me how to color-band birds to identify individuals in the field, determine whether they were territorial, mated, and what sub-stage of breeding they were in, and to assess social status. Size of the cloacal protub-erance (a copulatory and sperm storage organ) in males allowed additional assessment of breeding condition. I also learned a remarkable technique of unilateral laparotomy that revealed not only sex, but also direct assessment of reproductive condition (e.g. pre-laying, laying or post-laying). I was astonished how tolerant the birds were of this procedure; they flew off strongly when released. Males would sing within minutes, females returned to the nest and raised broods normally (Wingfield & Farner 1976). These skills, coupled with my more formal training in endocrinology, allowed us to measure far more than we could have ever imagined.

That first season I managed to obtain an extensive set of samples from both male and female white-crowned sparrows. The spring of 1974 was particularly cool and rainy and birds were not breeding. I worried that this was because of my attentions. However, by mid-May conditions improved and the birds began nesting. In 1975, spring was much warmer and the birds started breeding a month earlier. I had my first effect of weather (Wingfield & Farner 1978a). I felt there was no limit to the information we could obtain! Dire predictions that birds would desert and nests fail never materialized. Behavioral observations told us much about who was mated to whom and what sub-stage of breeding they were in. Behavioral ecologists looked askance – what was the big deal, they had been doing this for decades. Very true, but the concept of correlating field behavior with hormone titer was absolutely new. Others had collected birds in the field by shooting or by trapping and then sacrificing the bird before collecting trunk blood. In most cases it was not known whether the individual birds were breeding or not. As a result changes in hormones were correlated with date of collection and not in relation to what the bird was doing.

One other benefit of our field-endocrinology approach was that even if the state of the bird could not be determined immediately when the first sample was collected, releasing it allowed us to make further observations (e.g. feeding young, fledglings, etc.). We could then extrapolate back to what the bird was actually doing at the time of first capture.

With the first set of plasma samples in hand I returned to the laboratory and analyzed them for steroid hormones as well as LH (Wingfield & Farner 1975). We were extremely efficient in analysis using one small plasma sample of about 100 μl volume or even less. It took about 5 weeks doing nothing else to complete the analysis. To avoid correlation of hormones with just date of sampling I had to come up with a way to plot the data according to what each bird was doing. From field notes I knew whether birds had just arrived, whether they were territorial and mated, when egg laying began, when they were incubating, feeding young, fledglings, etc. We could also pinpoint when they began the next brood or if they re-nested after a clutch or brood was lost to a predator (e.g. Wingfield and Farner, 1979). In these ways sub-stage in the breeding cycle was used as the unit to plot means and standard errors and then space them in time according to when they were collected and the normal duration of each sub-stage.

Many previous studies in which birds were collected in relation to calendar date showed that gonads, not surprisingly, grew in spring and then collapsed at the end of the breeding season. Plasma levels of hormones followed that simple cycle: increase in spring, high when breeding and then decline post-breeding. This was affirmed by captive studies in which birds were exposed to long, spring-like day lengths. Our data on gonad development and regression matched the other field and laboratory data perfectly, but the hormone data were dramatically different (see Wingfield & Farner 1993; Wingfield & Silverin 2002 for detailed reviews). Although LH and sex steroids did rise as the gonads developed, the subsequent complex changes during the rest of the breeding season were completely unexpected. Sex steroid levels declined markedly once females were incubating and throughout the parental phase. These hormones increased again when they progressed into the sexual phase of a second brood, or when re-nesting after loss of the nest to a predator. This was different from anything we had seen in the laboratory, or from field data paying attention only to date. I can remember that September day in 1974 when I spread the graphs out on the desk for Don Farner to examine. He stared at them, puffing on his pipe, and after several minutes he looked up and said "Handsome data John, handsome data". We talked about the endless possibilities this kind of approach offered, how we could explain it, what new hypotheses immediately sprang to mind. He then asked whether I was willing to try this again next season. I stood there grinning and he smiled. He went back toward his office, turned and said "Keep those data safe, you will be presenting them a lot".

The next year's data showed the power of the system even more because it was becoming clear that we could look at year-to-year variation, age classes, weather, effects of habitat change (by humans), etc. Comparisons among populations were also of interest. For example, most of the laboratory work on white-crowned sparrows had been on the northern long-distance migrant, Gambel's sparrow. Field endocrinology work in Alaska would allow us to test hypotheses of how northern birds, with short breeding seasons, might integrate and synchronize reproductive effort compared with the mid-latitude sub-species near Seattle. Don was able to persuade the National Science Foundation to fund a field season in central Alaska (Wingfield & Farner 1978b). The results showed unequivocally that hormonal patterns in the field are very different from those in captivity or just compiled by date of collection. More importantly, they showed that individual patterns vary according to social status and unique experience of the physical environment. We could relate glucocorticoid levels to natural stressors, energetic demands of breeding, etc. The hormonal status of juveniles and dispersal could be investigated in natural settings. Laboratory data alone would be unlikely to provide the same insight into these processes. Field data, however, generated hypotheses and predictions that could be tested in the laboratory. This combination of field and laboratory investigations proved to be a powerful combination for the future.

After those first publications we had no shortage of collaborators working with other species. One of the first was Ebo Gwinner at the Max-Planck-Institut für Verhaltensphysiologie, and his student Hubert Schwabl. We began a long and very productive collaboration that was cut short by the tragic early passing of Ebo in 2004. He was an inspiration and pursued many aspects of field endocrinology. Other collaborations developed with behavioral ecologists who had spent many years working with a species in the field and for whom the time was ripe for the field-endocrinology to follow. This was a very rich source of new material and publications.

Next there followed a time of reflection for me. A combination of events and skills learned had put me in this position to explore field endocrinology. How might we explore the hormonal bases of mating system, degree of parental care, perturbations of environment, etc., and how should I go about tackling all of this using an experimental approach? It was clear that I still was sadly lacking in formal training in behavior. E. O. Wilson's book on sociobiology was revolutionary (Wilson 1975) and behavioral ecology was developing rapidly. There were animal behavior courses at UW and lots of expertise, but unfortunately the mid-1970s was a time when the rift between proximate mechanisms and ultimate causation was as wide as it has probably ever been (Tinbergen's four questions in behavioral biology notwithstanding!). There was no precedent to bring these two areas of biology together. Once again I was left frustrated, but there were seminars, discussions with students and faculty, and textbooks. Slowly I developed a foundation in behavior, but a formal course eluded me. At this time cell and molecular biology started to grow dramatically, and the rift with organismal biology gaped even wider. Some academics actively preached that you did one or other and not both. So here I was, an upstart standing on a rickety bridge trying to span the proximate and ultimate banks of this conceptual rift. Why I did not fall into the abyss I do not know. All I could do was focus on field and laboratory studies and ignore the resistance of others. The corpus of data would speak for itself.

The enormous intellectual chasm between organismal and cell and molecular biology will not be bridged by one person alone. Biological science is so vast that it will require teams of collaborators with suites of expertise. How many more years will it be before students are encouraged to close this gap, rather than to focus so finely that they gain no appreciation for the larger picture? For me, at that time in the 1970s, it was intimidating to be in the middle, but exciting as well. Others on both sides of the chasm were taking fleeting glimpses at the other side, but the effort needed to bridge it, and vulnerability to withering criticism (almost always unfounded) proved to be too much. None the less, I felt convinced that field endocrinology was the foundation of at least one bridge that could unite the two sides.

Mating systems and breeding strategies

In 1976 Don Farner was approached by George Hunt at the University of California, Irvine, concerning a project he and his wife Molly had initiated on Santa Barbara Island off southern California. They had been finding supernumerary eggs in nests of the western gull (Figure 20.5). Normally gulls lay from 1 to 3 eggs in a clutch, never more. But the Hunts were finding nests with 4, 5 and even 6 eggs. Molly showed that nests with large clutches were attended by two females and both contributed eggs to the nest. Female–female pairing raised many behavioral and evolutionary questions and we could all see how a field endocrine approach might provide new insight. The National Science Foundation liked it too, and funded the project.

Just prior to onset of the study, Fry and Toone (1981) showed that if western gull eggs were treated with DDT (the infamous and now banned pesticide with estrogenic activity), embryos became feminized. This was a potential mechanism by which sex ratio could be skewed toward females that in turn resulted in these female–female pairs. This was an exciting opportunity to expand my field and laboratory expertise in a completely different system with both why and how questions to answer. First, we set out to catch enough birds to determine whether there was indeed a skew in the sex ratio, or whether these females in homosexual pairs might be feminized males. To do this we captured gulls by using cannon nets and cooked popcorn as bait (which the gulls loved to eat). It was critical to set the

Figure 20.5. Western Gull, *Larus occidentalis*. California coast, July 2004. A key study species revealing that patterns of hormone secretions matter.

cannons at the right angle, check links and fold the nets carefully. Bait must be spread far enough away so that birds would not be hit by the leading edge of the net, but not so far that the net would not cover them. Gulls quickly came to the bait and once settled, not hovering above, I completed a circuit to an automobile battery. Boom, the cannons fired, pulling the net over the gulls, and we had up to 70 extremely irate birds to process.

The results revealed a skewed sex ratio of three adult females for each adult male (Hunt *et al.* 1980). Laparotomy also indicated that females of homosexual pairs (identified by color band combination) showed no evidence of feminized gonads. Their ovaries looked perfectly normal, at least from gross inspection. We also conducted many hours of observations from blinds on the colony showing little sexual dimorphism in behavior. Females had similar vocalizations and both members of a pair were territorial (Hunt *et al.* 1984). I had not previously appreciated how some avian species show so little sexual dimorphism in plumage and behavior. Some females even mounted males. For me this raised many fascinating questions about control mechanisms.

The skewed sex ratio was consistent with feminization of embryos by pesticides (Fry & Toone 1981), but the sex ratio of newly hatched chicks was 50 : 50 (Hunt *et al.* 1980). On this island the skewed sex ratio of gulls apparently was not a result of feminized embryos, although such a mechanism could be present in other populations. We went on to show that sex ratio remained equal through sexual maturation at 4–5 years of age and then became skewed in adults. This suggested that adult males either failed to return to the island or did not breed, or that there was a differential mortality of males. This issue remains unresolved.

For me the most fascinating aspect of this work concerned behavior. Females in homo-sexual pairs did everything males did with the exception of courtship feeding. It is possible, then, that these females were behaviorally masculinized. However, a detailed behavioral analysis showed that they were not different from females in heterosexual pairs. Given the lack of sexual dimorphism in plumage, territorial and parental behavior, and low dimorphism in sexual behavior too, it was not difficult to imagine how in a population with more females than males such female–female pairs might form. Furthermore, the analysis of hormone profiles in blood showed that testosterone levels remained very low (Wingfield *et al.* 1982) and there were no differences between males and females. Only when males were establishing territories for the first time (when 4–5 years old) did they show higher testosterone concentrations in plasma.

Comparing these data with our previous studies of white-crowned sparrows indicated how variable patterns of testosterone secretion were. The possibility arose that mating system and breeding strategy were important. This opened the door for many more collaborations, and field studies began in many other vertebrates as well. A huge database began to develop involving investigations of free-living organisms and captive animals in semi-natural conditions. The database for birds alone is enormous and perhaps unmatched by that for any other vertebrate group.

Next, an important collaboration developed with Bengt Silverin at the University of Göteborg in Sweden. He was working on the pied flycatcher, one of the most extensively studied birds in the world in relation to its behavioral ecology. The Swedish populations have two types of male, one socially monogamous and the other sequentially polygynous. Profiles of plasma testosterone in monogamous males showed a brief peak and earlier decline during the parental phase than in the polygynous males (Silverin & Wingfield 1982). Fieldwork on European blackbirds supported the pattern for socially monogamous species (Schwabl *et al.* 1980), and collaborations with Gordon Orians and Les Beletsky confirmed patterns for polygynous icterids (Beletsky *et al.* 1995). Patterns of hormones did match mating system consistently.

Settling down

It was during the gull project that another life-transforming event occurred. I met a beginning graduate student, Marilyn Ramenofsky, who was working with a pioneer of comparative endocrinology, Aubrey Gorbman. I was immediately attracted to her and delighted to discover that she too had an interest in behavior and birds. She had become aware of me through her sister Ann, whom I had met on a field trip while she was conducting a project. I had also heard that Marilyn was a famous swimmer, breaking three world records in freestyle swimming. Capping that, she won a silver medal at the Tokyo Olympics. I could see no reason why anyone with such an illustrious background would show any interest in a country hick from Derbyshire, whose only real claim to anything was being able to run steroid assays and catch birds in the field. It turned out that she had an interest in applying endocrine techniques to her research on Japanese quail to explore the confusing and complex interrelationships of testosterone and aggression. Regardless of this academic relationship, romance prevailed and we were married in 1979. Marilyn became not only my partner over the years since, and mother of our children, but also my best colleague, staunchest supporter and severest critic.

Another urgent and critical issue remained – I needed a permanent position. I could not be a postdoctoral fellow forever, no matter how much I enjoyed it. I applied for some jobs that looked appropriate given what I wanted to do. I got two interviews, but these were mismatches of colleagues and my work had tenuous links to their missions as departments. But then, thanks to a former graduate student at UW and collaborator, Bill Searcy, who was at the time a postdoctoral fellow at Rockefeller University (now at the University of Miami) with Peter Marler, I was invited to give a talk at the RU Field Research Center in Millbrook, New York. Many things clicked during this visit. They could see the enormous potential of combining endocrinology with fieldwork and behavior. I had inspiring discussions with Peter Marler, Fernando Nottebohm, Don Griffin and their co-workers. Everyone seemed totally steeped in animal behavior, and they showed incredible breadth that allowed them to bridge fields and see how integration would be the way of the future. I had found a new home. Later on I visited the main campus in Manhattan and met two other endocrinologists, Bruce McEwen and Don Pfaff, both of whom showed interest across fields although their main research was in the biomedical realm. The Field Research Center and Peter Marler's idea of establishing an endocrinology laboratory right there in a rural setting where field-work could also be done, was most attractive. I was offered a position as Assistant Professor within Peter's larger group. Being Rockefeller University, this position was not tenure-track (which Don Farner was concerned about and thought I was making a risky choice), but I had the opportunity to build my own research group and develop field endocrinology in what was the most supportive atmosphere I could imagine.

In early 1981, Marilyn and I drove from Seattle, through the southern USA, up the east coast to the Field Research Center in Millbrook, New York. The first year was difficult because Marilyn was finishing her Ph.D. and returned to Seattle. We got through the better part of a year commuting across country whenever I could and accruing huge telephone bills. My first focus was to build the laboratory from literally nothing; it was still being painted and no equipment was in place. We made trips to the main campus in Manhattan and scrounged what used equipment we could to set up a laboratory in Millbrook. Soon we became fully operational and independent. Meanwhile I was able to do extensive local fieldwork on song sparrows (Figure 20.6). It was a delight being able to walk from the laboratory to my field sites in minutes. I could do fieldwork in the morning and laboratory work in the afternoon – very efficient. It was a challenge to get everything going at a rural site, but also very exciting. In the six years we spent there I had nothing but lively encouragement, intellectual stimulation and generous support.

Soon after arrival in Milbrook, potential postdoctoral researchers contacted me who were very interested in field endocrinology. During the first two years, three of them received postdoctoral fellowships (NRSAs) from the National Institutes of Health to work in my laboratory. This was a wonderful boost for my developing independent research program. The first was Bob Hegner from Cornell, then Greg Ball from Rutgers University Institute of Animal Behavior, and Al Dufty from SUNY Binghamton. All had backgrounds in

Figure 20.6. Song sparrow, *Melospiza melodia*, Seattle, December 2004. A key study animal that set the stage for the "challenge hypothesis" and subsequent studies of hormone–behavior interactions.

behavioral ecology; working together with them and others at RU with interests from ethology to evolutionary biology, endocrinology and cell mechanisms was incredibly stimulating. We discussed and speculated, mixed and matched different concepts and asked why and how questions. There was little or no debate on whether one should stay on one's own side of that intellectual rift between ultimate and proximate approaches.

Synthesis and theory

Marilyn's thesis work on testosterone measurements and manipulations in relation to expression of aggression in captive quail helped to explain some important hormone–behavior interrelationships and set the stage for theoretical approaches (Ramenofsky 1984). Marilyn and I pored over the contradictory literature on testosterone and aggression in which some found a close relationship and others none. Why such variation? It appeared that the correlation of testosterone and aggression was most clear when social instability was prevalent and least clear in socially stable groups where aggressive relationships and status had already been established. This realization allowed us to then address a burning issue of field endocrinology: the need for a strong experimental framework.

Some still doubted whether controlled experiments in endocrinology could be done in natural settings. However, earlier studies by Trobec and Oring (1972) showed that implants of testosterone in free-living male sharp-tailed grouse had marked effects on aggression, particularly of sub-dominant males at a lek. Changing testosterone profiles experimentally from a socially monogamous type to a polygynous type resulted in a shift in mating system in several species (Watson 1970; Wingfield 1984a). In others this manipulation resulted in dramatic behavioral changes but not a change in mating system (e.g. Ketterson *et al.* 1996).

It was clear that controlled field experiments could indeed be done, and that patterns of testosterone levels mattered, having dramatic effects on mating systems and reproductive success.

Michael Moore, while he was a student in Don Farner's laboratory, and I discussed at length how we could also take experimental aspects of behavioral endocrinology to the field. We worked well together and planned field and laboratory experiments to test whether social interactions would influence the patterns of testosterone we saw from the fieldwork. He gave female white-crowned sparrows subcutaneous implants of estradiol to lock them into the sexual phase of breeding. The prediction was that interactions with sexually receptive females would increase LH and testosterone secretion in males, thus explaining high levels of testosterone in sexual phases of the breeding cycle and lower levels in the parental phase. Males associated with estrogen-implanted females had significantly higher plasma levels of LH and testosterone in both field and laboratory experiments. Mike flourished, going on to finish a fabulous thesis of experimental field endocrinology (Moore 1982, 1984). He is now a Professor at Arizona State University, where he pioneered more field endocrinology working on reptiles.

These subcutaneous implants were easy to make and simple to insert under the skin, and furthermore, our assay systems allowed us to check hormone patterns generated in blood. We could now tailor the size of the implants to give physiological doses of steroids, avoiding possible confounds of high pharmacological levels. Hormone implants could also be used to manipulate the behavior of one individual and determine the consequences on mates, or nearby individuals. An unprecedented suite of experimental techniques now presented itself. These hormonal and other manipulations were later referred to as "phenotypic engineering" by Ketterson *et al.* (1996) and by Sinervo and Svennson (1998).

Cheryl Harding, in collaboration with Brian Follett, challenged free-living and territorial male red-winged blackbirds with a decoy and song, and then captured them to measure changes in LH and testosterone (Harding & Follett 1979). Although she found no change in mean circulating testosterone levels (despite frequent citations of this article to the contrary!), she found an increase in the variance of LH. This paper set the stage for experiments on social interactions in the field with endocrine measures in addition to behavior. The results were interesting because of the lack of change in testosterone, but this was later to become a key observation.

If manipulation of testosterone patterns had the potential to change mating system and potentially increase reproductive success, why were not more male birds polygynous? In a key experiment, Bengt Silverin in Sweden had implanted free-living male pied flycatchers with testosterone and showed several behavioral effects including reduced parental care in testosterone-treated males (Silverin 1980). Bob Hegner conducted an elegant study in which he collected detailed observations of changes in sexual, aggressive and parental behavior of male and female house sparrows near Millbrook (Hegner & Wingfield 1986a,b). He also collected blood samples to correlate behavioral changes with fluctuations of LH and sex steroids. This was one of the most extensive and detailed field investigations to date and confirmed the negative relation between high circulating testosterone and male parental care. Hegner then implanted males with testosterone, with an empty implant as control, or with flutamide (an anti-androgen). His detailed observations confirmed that testosterone-treated males showed reduced parental care, so chicks grew less rapidly and fewer survived to fledging, resulting in reduced reproductive success (Hegner & Wingfield 1987), consistent with the findings of Silverin (1980). It also confirmed that the temporal patterns of testosterone concentrations in blood can have important ramifications for fitness.

Hegner's work emphasized the classic trade-off of high testosterone promoting aggression, larger territories and in some cases more than one mate that would tend to increase reproductive success versus the need for lower testosterone concentrations if males show parental care. Greg Ball, in a landmark study of European starlings, showed patterns of testosterone in males typical of socially monogamous mating systems, and that density of breeding birds, with presumably more social interactions, was correlated with elevated testosterone (Ball & Wingfield 1986). Al Dufty was working with the brood parasitic brown-headed cowbird in which neither males nor females show any parental care. He made a key observation that testosterone patterns were very different with no rapid declines or socially modulated surges of testosterone (Dufty & Wingfield 1986a,b). This was a wonderfully exciting and creative time with almost daily discussions and a limitless stream of ideas.

What remained was to manipulate the behavior of monogamous species (as Cheryl Harding had done for a polygynous species) to determine whether social interactions affected patterns of testosterone secretion. We developed simulated territorial intrusions for free-living male song sparrows to show experimentally the effects of male–male interactions on LH and testosterone levels. Follow-up experiments in which males were removed indicated that replacement males taking over the vacant territory, and their neighbors, had increased testosterone levels in blood. Furthermore, testosterone implants in males (and controls) resulted in neighbors increasing testosterone secretion, but not males at least one territory removed (Wingfield 1984a). Taken together these field experiments led to the "challenge hypothesis", which states that there is a breeding baseline of testosterone with higher levels occurring on a facultative basis depending on social instability when males compete for territories and access to sexually receptive females. At other times testosterone levels must be low, at the breeding baseline, for paternal care to be expressed. These interactions may drive the patterns of testosterone secretion in birds (Wingfield *et al.* 1990) and vertebrates in general, including humans (Sikkel 1993; Sapolsky 2002; Hirschenhauser *et al.* 2003; Archer 2005; Hirschenhauser & Oliveira 2006).

In the mid-1980s Ellen Ketterson and Val Nolan spent time in my laboratory at RU while they were on sabbatical from Indiana University. They learned endocrine techniques in the field as well as the assay systems, applying them to their long-term investigations on dark-eyed juncos at Mountain Lake, Virginia. Their project is now a classic study in field endocrinology from an evolutionary perspective where phenotypic engineering using testosterone implants in males (and now females) indicated many benefits and costs that may have led to evolution of hormone–behavior interactions. Many others also showed experimentally that high levels of testosterone for prolonged periods had potentially deleterious effects. For example, Al Dufty's work at RU on cowbirds revealed that prolonged high testosterone led to increased injuries, with possible fitness consequences (Dufty 1989). Many of these studies have been included in a summary of nine years of data on dark-eyed juncos by Reed *et al.* (2006).

Integrating ecology and endocrinology

By 1986, Marilyn and I were expecting our first child and it was clear that I needed a permanent position with tenure. I was fortunate indeed to land a job as Associate Professor at the University of Washington in Seattle, returning in August 1986. A new laboratory was set up back in our old haunts where all the field endocrinology began, and we had lots of new questions to pursue. One of the first addressed how similar territorial behavior was regulated in spring and autumn. Was testosterone, high in spring, also involved in the non-breeding season? A new graduate student in the laboratory, Tom Hahn (now at the University of California, Davis), helped me conduct field experiments on song sparrows, which were non-migratory and territorial year-round. We compared this species with the local white-crowned sparrow, which was migratory and showed no territorial behavior in autumn. Both species were abundant at sites in western Washington (Figures 20.7 and 20.8) allowing efficient parallel field studies. We showed that autumnal aggression increased, after a quiescent period during molt, with no change in LH or testosterone (Wingfield & Hahn 1993). There was no social modulation of testosterone in autumn (unlike in spring), and no response to sexually receptive females. Moreover, castration did not affect autumnal aggression. Our conclusion was that gonadal steroids were not required for territorial aggression in the non-breeding season.

Figure 20.7. Typical sparrow habitat at False Bay, San Juan Islands, in Puget Sound, western Washington. Song sparrows favor wet thickets, such as along the San Juan River, and adjoining open areas. White-crowned sparrows prefer drier grassy areas nearby.

Figure 20.8. Typical habitat for breeding sparrows along the outer coast of the Pacific Ocean north of Ocean Shores, Washington State. Note the grassy sand dunes in the foreground – preferred by white-crowned sparrows, and coastal rain forest favored by song sparrows. Olympic Mountains in the background.

Another graduate student, Kiran Soma, pointed out that Barney Schlinger at the University of California Los Angeles had shown that birds synthesize steroids in the brain (e.g. Schlinger 1994). Kiran conducted field and laboratory experiments confirming that conversion of androgens to estrogens in the brain was essential for expression of autumnal territorial aggression (e.g. Soma *et al.* 2000; Soma 2006). In the non-breeding season, sex steroids activate aggression locally within the brain thus avoiding peripheral effects of testosterone on reproductive traits that would be inappropriate in autumn. This mechanistic approach is consistent with the evolutionary constraints hypothesis (e.g. Ketterson *et al.* 1992, 1996). Furthermore, multiple mechanisms may exist by which hormones act on the same process and while this may seem redundant, diverse mechanisms may be important at particular times of year or specific social situations (the evolutionary potential hypothesis of Hau 2007) thus avoiding the "costs" of prolonged high circulating hormones (Wingfield *et al.* 2001). This concept may prove widespread for supposed redundant mechanisms in endocrinology.

Another example of potential "costs" of testosterone is the effect on the immune system, now a huge field (ecological immunology) for behavioral ecologists and endocrinologists. It is clear that testosterone does suppress the immune system in some species but not in others. Why this is so remains unclear. Behavior is central to these questions; Noah Owen-Ashley showed that sickness behavior is a potentially important paradigm by which to examine the relationship of testosterone and immunocompetence (e.g. Owen-Ashley & Wingfield 2006; Owen-Ashley *et al.* 2006). He showed how sickness behavior as a variable was less ambiguous than the difficult to interpret results on other aspects of the immune system. Immunologists are now taking new interest in this area and great progress is expected in the years ahead.

Recently, immediate early gene expression was investigated in the song control systems of song sparrows sampled in the field (Jarvis *et al.* 1997) opening up a new era of field endocrinology. Simone Meddle, followed up showing how environmental cues such as photoperiod first trigger immediate early gene expression leading to reproductive development (Meddle & Follett 1995). Later, as a postdoctoral fellow in my laboratory she extended this work further to address how social cues influence reproductive function (e.g. Meddle *et al.* 1997, 1999). Field endocrinology now interfaces with many facets of biology from behavioral ecology to cell and molecular mechanisms (see also Bentley *et al.* 2006). Neurogenomics studies developed from the zebra finch genome project led by David Clayton at the University of Illinois are revolutionizing field endocrinology. Using the spring/autumnal aggression paradigm in song sparrows as a model system, a postdoctoral fellow in my laboratory, Motoko Mukai, is using micro-arrays to explore possible genes involved in social modulation of testosterone. Thus the level of integration has increased and intensified, and new opportunities are exciting indeed.

Finite state machines and seasonality

A former student in my laboratory, Jerry Jacobs, developed finite state machine theory, enabling us to integrate life history strategies with hormone mechanisms (Jacobs & Wingfield 2000). McNamara and Houston's work (e.g. 1996) on annual routine models also allow integration of mathematics and physiological mechanisms. Finite state machine theory (or something like it) is important to generate testable hypotheses addressing mechanisms of hormone regulation of morphological, physiological and behavioral aspects of the life cycle.

This fostered research on seasonality and social interactions from the Arctic to the tropics. Work by Gwinner's group and also by Rachel Levin, Michaela Hau, Martin Wikelski, Wolfgang Goymann, Ignacio Moore and others (reviewed in Wingfield & Silverin 2002; Hau 2007) showed that tropical species are similar in some respects to temperate zone species whereas others are very different. Tom Hahn's ideas on flexibility in seasonality (Hahn *et al.* 1997; Hahn & MacDougall-Shackleton 2008) may provide a theoretical base from which to approach these issues. Mathematical models for dealing with natural history data also show great promise for generating hypotheses on mechanisms.

Much remains to be done to promote the integration of ecology, behavior and endocrinology. The formation of a research coordination network by Marcel Visser and Marcel Lambrechts (Europe), Tony Williams (Canada) and myself was formed to address such issues directly by bringing together ecologists, behavioral and evolutionary biologists, and endocrinologists focused on a single topic, the reproduction of birds. Workshops and technical meetings supported by the network resulted in a compendium of papers that begin spanning the rift between proximate and ultimate causation (Wingfield *et al.* 2008). We wait with anticipation the research that may follow.

Unpredictability

Observations of birds in the field during periods of bad weather revealed several behavioral and physiological correlates that were confirmed in laboratory experiments. Accumulating data suggested an emergency life history stage (ELHS; Wingfield 2003) that can be triggered by any perturbation of the environment, allowing individuals to cope. Postdoctoral fellow Michael Romero was key in formalizing the endocrine bases of these concepts, and Alexander Kitaysky conducted some remarkable field investigations on Alaskan seabirds that provide population and evolutionary perspectives. One important problem remains: how to define the transition from daily routines to a stress response. This is partly a problem with the word stress itself, which has so many meanings. The emerging concept of allostasis in relation to animals in their environment (e.g. McEwen & Wingfield 2003) for me was a breakthrough because one key term, allostatic load (the cumulative costs of daily routines plus stress of injury, infection, weather, predators, etc.) allowed us to conceive a continuum of daily energetic requirements that varies from individual to individual without worrying about where the boundary of daily routines and stress may lie. Allostatic overload, the point when the ELHS is triggered, may be different for each individual but is measurable. The potential is there to provide a new level of integration of ecology and endocrinology with many potential applications to conservation biology as well as biomedicine.

Our future and global change

Many other burgeoning issues have arisen in the past ten years because of the attempts of researchers from diverse fields to integrate ecology, behavior and endocrinology. These areas of "environmental endocrinology" deserve mention, but this is a rapidly developing area and will be a major focus of our research in the future. Marilyn and I recently moved to take advantage of new opportunities at the University of California in Davis after 22 years at the University of Washington. For example, we are all very concerned about global change and loss of biodiversity. The integration of theory, evolutionary, ecological and behavioral biology with physiology – neurobiological and endocrinological – has revealed a biodiversity of mechanisms by which organisms cope with a changing environment. This diversity occurs among populations and within individuals of those populations. The concept has been termed "redundancy" in the system that I feel has ignored and obfuscated the interesting point. Apparently "redundant" mechanisms actually play crucial roles in different environmental and social contexts. Diversity of mechanisms has a bright future.

Our work in the Arctic revealed several hitherto unknown mechanisms associated with coping with a severe environment. These include integration of migration with breeding (arrival biology) under inclement weather of the Arctic spring (e.g. Wingfield & Hunt, 2002), development of behavioral insensitivity to testosterone once breeding is under way, and modulation of sensitivity to acute stress. All of these new mechanisms have ecological bases focusing down to mechanisms at cellular and molecular levels. How many more fascinating mechanisms remain to be discovered in deserts, forests, tropical rain forests, mountains, marine and freshwater environments? This great diversity of mechanisms exists despite a very well conserved endocrine system in which there is about 80% or more homology of hormone types from fishes to humans (Wingfield 2006). This portends a spectrum of investigations on receptor, metabolic and transport mechanisms where potential for even greater diversity lies. Behavior and endocrinology form the glue that holds all this together. Field endocrinology is the link between the animal in its environment and laboratory studies down to cell and molecular levels. It melds why and how questions that feed off each other. Keeping them apart, or focusing on one to the complete exclusion of the other is, to me, a sterile pursuit. Laboratory experiments often reveal surprising results that generate why questions that we can take into the field. Field investigations likewise indicate mechanisms in real situations that we can take back into the laboratory to explore in detail.

Application of field endocrinology techniques and concepts goes beyond basic biology and has proved useful for conservation biology. My colleagues Professor Dee Boersma, students Gene Fowler, Lisa Hayward, Ellen Boyd, Shallin Busch, Brian Walker and I have collaborated over the years showing how stress responses, sex hormone patterns, etc. in the field can provide critical information in relation to pollution, habitat destruction, global warming and ecotourism. Another colleague, Sam Wasser at the University of Washington, now directs a Center for Conservation Endocrinology that is pioneering new advances in applying basic research to urgent practical problems on a global scale. Many further applications will follow.

As a final note, a new cadre of researchers is emerging. These talented young people are very broadly trained and also facile with molecular biology and bioinformatics techniques. They are intellectually fearless – tackling organismal questions in biology with molecular and cell techniques. I have no doubt that this new wave will bridge that chasm between organismal and cell/molecular biology. Future years will be just as wonderful as the past.

Acknowledgments

My greatest gratitude goes to my family, my wife Marilyn and our children, my parents, Don and Nancy Wingfield, and my grandparents. Not only did they show remarkable understanding and patience, they also helped me along in so many ways. Raising two children had a major influence on my research even though I spent much more time at home with our family and less time in the laboratory and office. Watching them grow, playing various games, building things, and observing how they interacted with their physical and social environments fascinated me. Many things they obviously had never encountered before, yet I was always impressed with how logical, given their prior experience, their reactions were. This was very unlike their parents, whose responses to the world around them were influenced by cultural indoctrination and bias. I can state with conviction that our children, Emma and Anna, taught me to see the world again through eyes much less fettered by dogma. This in turn allowed me to attain a greater degree of creativity in my research and develop a healthy skepticism of accepted dogma.

My career in animal behavior was greatly enhanced by the people who worked with me either as students and postdoctoral fellows or as collaborators, Some I have mentioned in the text but there are many others who contributed in so many ways. Integration of ecology, behavior, physiology and endocrinology from the organism in its natural habitat to cellular and molecular levels cannot be achieved by one person alone. I have always tried to run my laboratory as teams by including people with expertise in specific parts of the research project as a whole. I often wonder at my fortune in having so many exceptional people work in my laboratory over the past thirty years. They all built field endocrinology, not me alone. Those not already mentioned include Lee Astheimer, William Buttemer, Kent Dunlap, Tim Boswell, Nicole Perfito, Matt Richardson, Z. Morgan Benowitz-Fredercks, Sharon Lynn, Meta Landys, Lisa Belden, Todd Sperry, Steve Schoech, Kathleen Hunt, Katie O'Reilly, Donna Maney, Creagh Breuner, Rebecca Holberton, Sammrah Raouf, Nigella Hillgarth, Danny Afik, Mark Stanback, Kaoru Kubokawa, Cheryl Wotus, Fran Bonier, Doug Wacker and Alex Coverdill. Current members of my laboratory not mentioned (Jason Davis, Sara O'Brien, Elizabeth Addis, Gang Wang, Patrick Kelly, Aaron Clark and Karen Lizars) continue to challenge the gap between organisms and molecules. We are grateful to my long-time technician and laboratory manager, Lynn Erckmann, without whose help we would have been smothered by rules and regulations.

No research gets done without funding, and the National Science Foundation has been the backbone of support for my research for over thirty years. Additional funding from the National Institutes of Health, the US Fish and Wildlife Service, the National Geographic Society, and a Guggenheim Fellowship helped tremendously. However, it has been the NSF that has provided ongoing core support for all that I, and my teams, have been able to achieve.

References

Archer, J. (2005). Testosterone and human aggression: an evaluation of the challenge hypothesis. *Neurosci. Biobehav. Rev.* **30**: 319–45.

Ball, G. F. & Wingfield, J. C. (1986). Changes in plasma levels of sex steroids in relation to multiple broodedness and nest site density in male starlings. *Physiol. Zool.* **60**: 191–9.

Beach, F. A. (1948). *Hormones and Behavior: A Survey of Interrelationships between Endocrine Secretions and Patterns of Behavior.* New York: Paul B. Hoeber.

Beletsky, L. D., Gori, D. F., Freeman, S. & Wingfield, J. C. (1995). Testosterone and polygyny in birds. *Current Ornithol.* **12**: 1–41.

Bentley, G. E., Kreigsfeld, L. J., Osugi, T. *et al.* (2006). Interactions of gonadotropin-releasing hormone (GnRH) and gonadotropin-inhibitory hormone (GnIH) in birds and mammals. *J. Exp. Zool.* **305A**: 807–14.

Collias, N. E. (1950). Hormones and behavior with special reference to birds and mechanisms of hormone action. In *A Symposium on Steroid Hormones*, ed. E. Gordon, pp. 277–329. Madison, WI: University of Wisconsin Press.

Dufty, A. M. (1989). Testosterone and survival. *Horm. Behav.* **23**: 185–93.

Dufty, A. M. Jr. & Wingfield, J. C. (1986a). Temporal patterns of circulating LH and steroid hormones in a brood parasite, the brown-headed cowbird, *Molothrus ater.* I. Males. *J. Zool. (Lond.)* **208**: 191–203.

Dufty, A. M. Jr. & Wingfield, J. C. (1986b). Temporal patterns of circulating LH and steroid hormones in a brood parasite, the brown-headed cowbird, *Molothrus ater*. II. Females. *J. Zool. (Lond.)* **208**: 205–14.

Emlen, J. T. & Lorenz, F. W. (1942). Pairing responses of free-living valley quail to sex hormone pellet implants. *Auk* **59**: 369–78.

Fry, D. M. & Toone, C. K. (1981). DDT-induced feminization of gull embryos. *Science* **213**: 922–4.

Hahn, T. P. & MacDougall-Shackleton, S. A. (2008). Adaptive specialization, conditional plasticity and phylogenetic history in the reproductive cue response systems of birds. *Phil. Trans. R. Soc. Lond.* B**363**: 267–86.

Hahn, T. P., Boswell, T., Wingfield, J. C. & Ball, G. F. (1997). Temporal flexibility in avian reproduction: patterns and mechanisms. *Curr. Ornithol.* **14**: 39–80.

Harding, C. F. & Follett, B. K. (1979). Hormone changes triggered by aggression in a natural population of blackbirds. *Science* **203**: 918–20.

Hau, M. (2007). Regulation of male traits by testosterone: implications for the evolution of vertebrate life histories. *BioEssays* **29**: 133–44.

Hegner, R. E. & Wingfield, J. C. (1986a). Behavioral and endocrine correlates of multiple brooding in the semi-colonial house sparrow, *Passer domesticus*. I. Males. *Horm. Behav.* **20**: 294–312.

Hegner, R. E. & Wingfield, J. C. (1986b). Behavioral and endocrine correlates of multiple brooding in the semi-colonial house sparrow, *Passer domesticus*. II. Females. *Horm. Behav.* **20**: 313–26.

Hegner, R. E. & Wingfield, J. C. (1987). Effects of experimental manipulation of testosterone levels on parental investment and breeding success in male house sparrows. *Auk* **104**: 462–9.

Hinde, R. A. (1965). Interactions of internal and external factors in integration of canary reproduction. In *Sex and Behaviour*, ed. F. Beach, pp. 381–415. New York: Wiley.

Hinde, R. A., Steel, E. & Follett, B. K. (1974). Non-ovary mediated effect of photoperiod on oestrogen-induced nest building. *J. Reprod. Fertil.* **40**: 383–99.

Hirschenhauser, K. & Oliveira, R. (2006). Social modulation of androgens in male vertebrates: meta-analyses of the challenge hypothesis. *Anim. Behav.* **71**: 265–77.

Hirschenhauser, K., Winkler, H. & Oliveira, R. F. (2003). Comparative analysis of male androgen responsiveness to social environment in birds: the effects of mating system and paternal incubation. *Horm. Behav.* **43**: 508–19.

Hunt, G. L. Jr., Wingfield, J. C., Newman, A. & Farner, D. S. (1980). Sex ratio of western gulls on Santa Barbara Island. *Auk* **97**: 473–9.

Hunt, G. L. Jr., Newman, A., Warner, M. H., Wingfield, J. C. & Kaiwi, J. (1984). The reproductive behavior of the western gull (*Larus occidentalis wymani*), prior to egg-laying. *Condor* **86**: 157–62.

Jacobs, J. D. & Wingfield, J. C. (2000). Endocrine control of life-cycle stages: a constraint on response to the environment? *Condor* **102**: 35–51.

Jarvis, E. D., Schwabl, H., Ribeiro, S. & Mello, C. V. (1997). Brain gene regulation by territorial singing behavior in freely ranging songbirds. *Neuroreports* **8**: 2073–7.

Ketterson, E. D., Nolan, V. Jr., Wolf, L. & Ziegenfus, C. (1992). Testosterone and avian life histories: effects of experimentally elevated testosterone on behavior and correlates of fitness in the dark-eyed junco (*Junco hyemalis*). *Am. Nat.* **140**: 980–99.

Ketterson, E. D., Nolan, V. Jr., Cawthorn, M. J., Parker, P. G. & Ziegenfus, C. (1996). Phenotypic engineering: using hormones to explore the mechanistic and functional bases of phenotypic variation in nature. *Ibis* **138**: 70–86.

Lehrman, D. S. (1965). Interaction between internal and external environments in the regulation of the reproductive cycle of the ring dove. In *Sex and Behavior*, ed. F. A. Beach, pp. 355–380. New York: Wiley.

Lewis, R. A. (1975). Reproductive biology of the white-crowned sparrow (*Zonotrichia leucophrys pugetensis* Grinnell). I. Temporal organization of reproductive and associated cycles. *Condor* **77**: 46–59.

Lorenz, K. Z. (1952). *King Solomon's Ring*. New York: Crowell.

Lorenz, K. Z. (1963). *On Aggression*. Vienna: Verlag Dr. Borotha-Schoeler.

McEwen, B. S. & Wingfield, J. C. (2003). The concept of allostasis in biology and biomedicine. *Horm. Behav.* **43**: 2–15.

McNamara, J. M. & Houston, A. I. (1996). State-dependent life histories. *Nature* **380**: 215–21.

Meddle S. L. & Follett B. K. (1995). Photoperiodic activation of fos-like immunoreactive protein in neurons within the tuberal hypothalamus of Japanese quail. *J. Comp. Physiol.* **176**: 79–89.

Meddle, S. L., King, V. M., Follett, B. K. *et al.* (1997). Copulation activates Fos-like immunoreactivity in the male quail forebrain. *Behav. Brain Res.* **85**: 143–59.

Meddle, S. L., Foidart, A., Wingfield, J. C., Ramenofsky, M. & Balthazart, J. (1999). Effects of sexual interactions with a male on fos-like immunoreactivity in the female quail brain. *J. Neuroendocrinol.* **11**: 771–84.

Moore, M. C. (1982). Hormonal responses of free-living male white-crowned sparrows to experimental manipulation of female sexual behavior. *Horm. Behav.* **16**: 323–9.

Moore, M. C. (1984). Changes in territorial defense produced by changes in circulating levels of testosterone. A possible hormonal basis for mate-guarding behavior in white-crowned sparrows. *Behaviour* **88**: 215–26.

Owen-Ashley, N. T. & Wing field, J. C. (2006). Seasonal modulation of sickness behavior in free-living northwestern song sparrows (*Melospiza melodia Morphna*). *J. Exp. Biol.* **209**: 3062–70.

Owen-Ashley, N. T., Turner, M., Hahn, T. P. & Wingfield, J. C. (2006). Hormonal, behavioral, and thermoregulatory responses to bacterial lipopolysaccharide in captive and free-living white-crowned sparrows (*Zonotrichia leucophrys gambelii*). *Horm. Behav.* **49**: 15–29.

Ramenofsky, M. (1984). Agonistic behavior and endogenous plasma hormones in male Japanese quail. *Anim. Behav.* **32**: 698–708.

Reed, W. L., Clark, M. E., Parker, P. G. *et al.* (2006). Physiological effects on demography: a long term experimental study of testosterone's effects on fitness. *Am. Nat.* **167**: 667–83.

Sapolsky, R. M. (2002). Endocrinology of the stress response. In *Behavioral Endocrinology*, ed. J. B. Becker, S. M. Breedlove, D. Crews & M. M. McCarthy, 2nd edn, pp. 409–450. Cambridge, MA: MIT Press.

Schlinger, B. A. (1994). Estrogens to song: picograms to sonograms. *Horm. Behav.* **28**: 191–8.

Schwabl, H., Wingfield, J. C. & Farner, D. S. (1980). Seasonal variation in plasma levels of luteinizing hormone and steroid hormones in the European blackbird *Turdus merula*. *Vogelwarte* **30**: 283–94.

Sikkel, P. C. (1993). Changes in plasma androgen levels associated with changes in male reproductive behavior in a brood cycling marine fish. *Gen. Comp. Endocrinol.* **89**: 229–37.

Silverin, B. (1980). Effects of long-acting testosterone treatment on free-living pied flycatchers, *Ficedula hypoleuca*, during the breeding period. *Anim. Behav.* **28**: 906–12.

Silverin, B. & Wingfield, J. C. (1982). Patterns of breeding behavior and plasma levels of hormones in a free-living population of pied flycatchers, *Ficedula hypoleuca*. *J. Zool. Lond.* **198**: 117–29.

Sinervo, B. & Svensson, E. (1998). Mechanistic and selective causes of life history trade-offs and plasticity. *Oikos* **83**: 432–42.

Soma, K. K. (2006). Testosterone and aggression: Berthold, birds and beyond. *J. Neuroendocrinol.* **18**: 543–51.

Soma, K. K., Tramontin, A. D. & Wingfield, J. C. (2000). Estrogen regulates male aggression in the non-breeding season. *Proc. R. Soc. Lond.* B**267**: 1089–96.

Thorpe, W. H. (1961). *Bird Song*. London: Cambridge University Press.

Tinbergen, N. (1953). *The Herring Gull's World*. London: Collins.

Trobec, R. J. & Oring, L. W. (1972). Effects of testosterone propionate implantation on lek behavior of sharp-tailed grouse. *Am. Midl. Nat.* **87**: 531–6.

Watson, A. (1970). Territorial and reproductive behavior of red grouse. *J. Reprod. Fertil. Suppl.* **11**: 3–14.

White, G. (1860). *The Natural History of Selborne*. London: Gresham Books.

Wilson, E. O. (1975). *Sociobiology: the New Synthesis*. Cambridge, MA: Harvard University Press.

Wingfield, J. C. (1984a). Androgens and mating systems: testosterone-induced polygyny in normally monogamous birds. *Auk* **101**: 665–71.

Wingfield, J. C. (1984b). Environmental and endocrine control of reproduction in the song sparrow, *Melospiza melodia*. II. Agonistic interactions as environmental information stimulating secretion of testosterone. *Gen. Comp. Endocrinol.* **56**: 417–24.

Wingfield, J. C. (1985). Short-term changes in plasma levels of hormones during establishment and defense of a breeding territory in male song sparrows, *Melospiza melodia*. *Horm. Behav.* **19**: 174–87.

Wingfield, J. C. (2003). Control of behavioural strategies for capricious environments. *Anim. Behav.* **66**: 807–16.

Wingfield, J. C. (2006). Communicative behaviors, hormone-behavior interactions, and reproduction in vertebrates. In *Physiology of Reproduction*, ed. J. D. Neill, pp. 1995–2040. New York: Academic Press.

Wingfield, J. C. & Farner, D. S. (1975). The determination of five steroids in avian plasma by radioimmunoassay and competitive protein binding. *Steroids* **26**: 311–27.

Wingfield, J. C. & Farner, D. S. (1976). Avian endocrinology – field investigations and methods. *Condor* **78**: 570–3.

Wingfield, J. C. & Farner, D. S. (1978a). The endocrinology of a naturally breeding population of the white-crowned sparrow (*Zonotrichia leucophrys pugetensis*). *Physiol. Zool.* **51**: 188–205.

Wingfield, J. C. & Farner, D. S. (1978b). The annual cycle of plasma irLH and steroid hormones in feral populations of the white-crowned sparrow, *Zonotrichia leucophrys gambelii*. *Biol. Reprod.* **19**: 1046–56.

Wingfield, J. C. & Farner, D. S. (1979). Some endocrine correlates of renesting after loss of clutch or brood in the white-crowned sparrow, *Zonotrichia leucophrys gambelii*. *Gen. Comp. Endocrinol.* **38**: 322–31.

Wingfield, J. C. & Farner, D. S. (1993). Endocrinology of reproduction in wild species. In *Avian Biology*, ed. D. S. Farner, J. R. King & K. C. Parkes, vol. IX, pp. 164–327. New York: Academic Press.

Wingfield, J. C. & Hahn, T. P. (1993). Testosterone and territorial behavior in sedentary and migratory sparrows. *Anim. Behav.* **47**: 77–89.

Wingfield, J. C. & Hunt, K. (2002). Arctic spring: hormone-behavior interactions in a severe environment. *Comp. Biochem. Physiol.* **B132**: 275–86.

Wingfield, J. C. & Silverin, B. (2002). Ecophysiological studies of hormone-behavior relations in birds. In *Hormones, Brain and Behavior*, ed. D. W. Pfaff, A. P. Arnold, A. M. Etgen, S. E. Fahrbach & R. T. Rubin, Vol. 2, pp. 587–647. Amsterdam: Elsevier Science.

Wingfield, J. C., Newman, A., Hunt, G. L. Jr. & Farner, D. S. (1982). Endocrine aspects of female-female pairing in the western gull (*Larus occidentalis wymani*). *Anim. Behav.* **30**: 9–22.

Wingfield, J. C., Hegner, R. E., Dufty, A. M.Jr. & Ball, G. F. (1990). The "challenge hypothesis": theoretical implications for patterns of testosterone secretion, mating systems, and breeding strategies. *Am. Nat.* **136**: 829–46.

Wingfield, J. C., Lynn, S. E. & Soma, K. K. (2001). Avoiding the "costs" of testosterone: ecological bases of hormone-behavior interactions. *Brain Behav. Evol.* **57**: 239–51.

Wingfield, J. C., Visser, M. E. & Williams, T. D. (eds). (2008). Integration of ecology and endocrinology in avian reproduction: a new synthesis. *Phil. Trans. R. Soc. Lond.* **B363**.

21

Living with birds and conservation

AMOTZ ZAHAVI

Begininmgs

I don't remember myself without birds. My mother used to say that I was watching them before I could even walk or talk. I was born August 14 1928, in Petach-Tiqva, then a small town near Tel-Aviv, Israel, then Palestine, under the British mandate. Petach-Tiqva was the first Jewish village built in Israel by Jews in the nineteenth century, and my grandfather was among the early settlers there. When I was growing up there were many open fields and orchards among the houses, and I used to set out in the early mornings to watch birds before going to school. I did not know anybody who knew birds other than the most common ones, and there were no illustrated guidebooks available to me to help in their identification. I enjoyed the birds, found their nests and knew their songs, so I had my own names for them.

At the age of 12, in 1940, I met Dr Heinrich Mendelssohn. He was the director of a small zoo at the Biological-Pedagogical Institute in Tel Aviv. From then on, I could identify birds and learn their proper names at the collection of stuffed birds that was exhibited at the zoo,

Leaders in Animal Behavior: The Second Generation, ed. L. C. Drickamer & D. A. Dewsbury. Published by Cambridge University Press. © Cambridge University Press 2010.

and there was someone who was interested in my observations. I had the privilege of accompanying Mendelssohn in the field and I attended many of his talks.

My only sib sister died of pneumonia at the age of 15, in 1941, a few months before the sulfa drugs that could have saved her life were first available. Following her death my family left Petach-Tiqva. I studied at the Pardess Hana Agricultural Secondary School.

In 1941the German army was advancing in North Africa towards the Middle East. The British were preparing to move their army to India. The Jewish community in Israel was facing the danger of extermination. Plans were made to make the Carmel mountain range into a stronghold and fight to the end. I remember how we, the high-school students, participated in these preparations. Soon, however, the British defeated the Nazis in the battle of El Alamein and we were saved.

I was a boy scout. At that time our goal was to establish new farming settlements. I planned to be a member of a kibbutz (a communal settlement) and a farmer. The school taught agriculture and science along with practical work. We worked on the farm two days a week in the first three years, and up to three days a week during the last two years. I got special permission to use these days to watch birds in the spring in exchange for working for a whole month during the summer vacations.

Mendelssohn persuaded me, in 1947, to study biology rather than agriculture at the Hebrew University of Jerusalem. His argument was that I could be a farmer without a university diploma, but that it would be more difficult to go into biology if I studied agriculture. He also argued that, since at that time there were hardly any other young ornithologists in Israel, being a professional ornithologist would be a greater service than adding one more farmer to the many who were already there.

1947/8 was to be my first year at the university, but within a month the War of Independence interrupted my studies for a year. I was not involved in actual fighting (I was exempt because the recent death of my sister left me as an only child). Instead, I drilled young people at the besieged city of Jerusalem. Later I was the commander of a company that was stationed in several kibbutzim in the Jordan valley, near the Sea of Galilee.

In 1949, after my military service, I attended some geology courses besides biology. One of the lecturers, Professor Ben-Tor, was conducting a geological survey of the Israeli Negev, a desert that extends over more than half of the country. Before 1948, the British authorities of Palestine did not allow Jews to travel freely in that part of the country. My knowledge of life in the desert was therefore very restricted. Ben-Tor needed helpers to carry stones and guns, cook, and guard at night. I was among the few students who felt themselves lucky to accompany the survey. During 1949–54 we spent a few hundred days in the field. I missed many lectures and much lab work, but I learned much about the desert and got to know its geology, its flora and its fauna (mainly the vertebrates). The biologists among the student-carriers included Eviatar Nevo, now a famous evolutionary biologist at Haifa University, and the late Amiram Shkolnik, who became a professor of Eco-physiology at Tel-Aviv University; we collected plants, trapped rodents and watched the birds. It was at that time that I fell in love with the open desert.

Figure 21.1. With friends in the Hula swamp 1952.

In Jerusalem I studied with Yaakov Wahrman the chromosomes of gerbils and spiny mice that we collected in the desert (Zahavi & Wahrman 1957). The study revised the systematic relationship of several of the rodents, and found polymorphism in the number of chromosomes of certain populations of one of these species, *Gerbillus pyramidum* (Wahrman & Zahavi 1958). It was good evidence for an active speciation process going on in the coastal dunes of the eastern Mediterranean.

I did not plan to be a professional ornithologist. My idea of ornithology, at that time, consisted of identifying birds, finding nests, and surveying bird populations. Hence, at the end of 1952, I started a project in biochemistry for my master's degree under Professor Reich (there was no B.Sc. degree at the Hebrew university at that time). I was planning to watch birds in my spare time. But in February the birds began to sing, and I could not resist the temptation to go out to the field. I excused myself to Professor Reich and took up a study of the birds of the Huleh swamp as my thesis, with Mendelssohn as my adviser (Zahavi 1957) (Figure 21.1). I was sure that by making this decision I was giving up all intellectual challenges, but it was a daily occupation that provided much enjoyment. Tinbergen's book *The Study of Instinct* (Tinbergen 1951) influenced my decision to learn more about the scientific study of animal behaviour. I applied to work with Niko Tinbergen for a year. A recommendation by Col. R. Meinertzhagen, whom I accompanied for a few days in the Israeli desert, helped me receive a scholarship from the British Council, and Tinbergen agreed to accept me into his group.

In spring 1954 I married Avishag Kadman, a fellow student majoring in botany, whom I courted in the Huleh swamp. Our first daughter, Naama, was born before I left for Oxford.

Tinbergen and Oxford

This was my first visit abroad. In Israel people were dressing very informally. My parents insisted that in Europe I should present myself as a respectable person, so they fitted me with a suit and tie. When Professor Tinbergen invited me to dinner at his home, I put on the suit and struggled with the tie, only to find Niko in his shorts, inviting me to join him in peeling potatoes. I remembered that moment when, ten years later, Rami, my future technician, appeared dressed up in suit and tie for his work-interview with me.

I spent January – March 1955 with Tinbergen's group at Oxford. I attended seminars with, among others, Desmond Morris and Mike and Esthy Cullen, and became acquainted with the jargon and the way of thinking of ethologists. I visited Cambridge and met with Robert Hinde and Peter Marler. In the spring I accompanied Uli and Rita Weidman to study the black-headed gull colony at Ravenglass, and spent most of the breeding season there. I performed an experiment of my own, putting additional nests near the original ones in the territory. When I placed a partition between the two nests, both members of the pair incubated simultaneously for long periods. I concluded that they were competing over incubation, and suggested that the extra urge to incubate was a mechanism that ensured that the nest would be incubated constantly. Like all ethologists at that time, I used group-selection arguments when there was no obvious way to suggest how a certain behaviour benefited the individual directly. Years later, with the understanding of the Handicap Principle, I reinterpreted the behavior of the gulls to suggest that prolonged incubation was used by each bird as means to increase its prestige vis-à-vis its partner.

Later that year I had the privilege of attending the third meeting of the European ethologists, at Groningen (the ethologists tried to keep their conferences small and intimate by restricting attendance). At the conference I became acquainted with Konrad Lorenz, as well as the Barrends and other Dutch ethologists. I fledged from all these meetings as an ethologist.

My stay in Europe was interrupted in August 1955 by the death of my mother.

I did not have the time to write down my observations on the black-headed gulls: back in Israel I was committed to work for conservation.

My involvement with conservation

In 1948, when Israel became an independent state, the Jewish population numbered only 600,000. By 1951, immigration had increased the population to two million. The newly formed state of Israel was concerned with settling new immigrants and with defense. Conservation was the last thing on the mind of the authorities. The Botanical and Zoological Societies formed a joint committee to do something about conservation, mainly by advising the government. Mendelssohn, my mentor, was the very active chairman of that committee.

In 1950 the government of Israel decided to reclaim and drain the Huleh Swamp and lake. I was anxious to study the bird life of the area before it vanished. Once more I asked my teachers at the university to excuse my absence from formal lectures and laboratories, and spent many days observing the birds, observations that turned into my M.Sc. thesis in 1954. One day, I believe in 1952, Mendelssohn asked me to accompany a committee of the national planning department that came to study the feasibility of creating a small nature reserve in the area. I guided them in the swamp and lake in a boat, and realized how little they knew of the area. Together with several senior scientists (all of them my teachers), I attended the final meeting of the governmental committee that was to decide the size of the future nature reserve. Comments of members of the committee and its chairman made me realize that they were not impressed by the scientific arguments. I stood up and described the nest of the white-tailed eagle that I had found earlier that year in the area, the impressive size of the nest, the large and beautiful bird flying down daily to fish in the lake. I insisted that it needed at least 1000 acres of lake and swamp for its survival; otherwise that majestic bird would disappear from the region. Obviously this was not a fully researched scientific argument. A few weeks later I met the chairman of the committee on the street. "Come here, my boy", he called to me (he was over seventy years old); "because of you we decided to allocate 1000 acres to the nature reserve." Suddenly I realized that I could influence decisions of where a proposed nature reserve would be and what its size would be, and could save some of the area and its birds.

In 1952 my friend Azaria Alon, a biology teacher at kibbutz Beith Hashita, another of Mendelssohn's students, suggested that we form a non-governmental organization (NGO) that would act rather than just advise. We suggested that the committee for conservation should itself become that NGO: they agreed, and the Society for the Protection of Nature in Israel (SPNI) was created. Azaria and I were the only young members of the board. All the others were professors or middle-aged biology teachers. Azaria was busy teaching at Beith Hashita. Hence, I happened to be the only person free to handle the office work of the new NGO. I used to walk around with a book of receipts, asking my friends whether they happened to have on them two pounds. When they gave me the money, believing that I needed some personal help, I would give them the receipt and thank them for joining the SPNI. Most of the early members were biology teachers who were personally connected with Mendelssohn in one way or another. Other members of the SPNI were mostly kibbutz members interested in natural history. Soon after, when I went to Oxford, Avraham Toren (Bumi), another biology teacher from kibbutz Maabarot, took over the responsibility of being the secretary of the SPNI. However, as there was no money, he had to teach most of the time, in order to earn his salary.

When I returned from Oxford in August 1955, the SPNI was in a difficult state. There was no money, and membership was not increasing. Something had to be done. Although I had the opportunity to start my Ph.D. studies as an assistant at the department of zoology at the Hebrew University (the only university in Israel at that time), I decided to serve for a few years as a full-time secretary of the SPNI, trying to establish it firmly. My plan was to return to the world of academia after three years, but these stretched to be 14 years.

Tel-Aviv University was created at around that time, by the merger of several independent semi-academic institutions. The Biological Pedagogical Institute, with Mendelssohn as its head, became the biology faculty of the new university. I suggested to Mendelssohn that he should incorporate my activities in conservation with those of the faculty. He had the courage to put me on the payroll of the department of zoology, as a demonstrator, although he knew that I spent most of my time organizing the SPNI. He also helped the SPNI by providing office space and office services within the department. I am sure that without his support, the fledgling SPNI would not have survived.

The limited funds of the SPNI were spent on employing wardens in several key areas of special importance for conservation. The SPNI also organized public events, lectures, and yearly conventions with guided tours for its members and the general public. The events provided some net income and were therefore extended to become weekly tours guided by the local wardens of the SPNI. The income from these events became a major source of funding. It was the immediate need for funds, rather than the long-term benefits of educa-tion, that started the SPNI's education system. We soon realized, though, that our work as guides and our educational efforts actually created a public that backed our endeavors for conservation.

Independently of the SPNI, Yosi Feldman was appointed as the director of a youth hostel in Ein Gedi (an oasis on the shores of the Dead Sea), where he attempted to create a Field School to guide tours of youngsters in the back country of the Israeli desert. The SPNI already had a warden at Ein Gedi, so the SPNI joined Yosi Feldman and together created the first Field Study Center (Field School). The staff consisted of several guides and volunteers, plus a technical staff of 2–3 persons. The staff provided services to the tourists that visited Ein-Gedi and kept the area clean. The ministry of education agreed to let the SPNI provide guides for school outings. This agreement provided financial support for the system of field study centers and facilitated direct contact with a large population of school children. It did not take long to enlarge the staff of the SPNI in other regions from single wardens to small groups, creating additional Field Study Centers that ran tour services to schools. The ministry of housing joined in to help and built hostels and classrooms for the centers.

Our second daughter Tirtza was born in May 1957. Avishag took care of the family besides her own research in plant physiology, and I was busy with the SPNI. In the 1960s the girls were already old enough to travel with us. We often traveled together to the various Field Study Centers. Our daughters enjoyed these travels, sitting together with our dog at the back of our old station-wagon.

In the early 1960s the SPNI played a central role in passing a bill of conservation in the Israeli Knesset (the parliament). We managed to convince the Knesset to establish two separate authorities, one for the national parks (mainly archeological and historical sites for tourism) and another for nature conservation. We said that the creation and sustaining of national parks inherently involves development and change, whereas conservation should strive to leave nature undisturbed as much as possible. We did not realize at the time that the urge to develop is inherent in any authority, no matter what its stated goal is.

I myself served for thirty years on the board of the National Parks authority, trying to take care of nature conservation within the national parks.

Once the governmental authority was established, a retired general, Avraham Yofe, was appointed as its chairman, and Uzi Paz, a former warden of the SPNI, as its director. Most of the burden of fighting for the creation of nature reserves and their management was no longer upon the SPNI. The Nature Reserve Authority increased the number and area of the nature reserves and, together with the SPNI, did a great deal to preserve the wild flowers and wildlife around the country.

The SPNI focused on enlarging the field study center system and its educational activities. It still retained a conservation department, lest the politicians, ruling over the governmental agencies, force the conservation authorities to agree with policies that conflict with conservation. There were several cases in which the SPNI indeed intervened in order to change governmental policies, something that a governmental authority cannot do. One of these cases was the conservation of the spring of the river Dan at Tel-El Kadi, one of the sources of the Jordan River. The authorities decided to capture the water of the spring at the source. The nature reserve authority decided not to fight against this decision. I also thought that there was no chance to win the fight, because of the high economic value of the project. However, our local warden, Yossi Lev-Ari, decided to fight for the spring, and I agreed to provide the support of the SPNI. He succeeded in convincing the local villagers, the ones who were to benefit from the project, to abandon the plan. The spring is still there, within a nature reserve, and a great many people enjoy its beauty every year. This fight convinced me that it was time to let younger people take the lead in fighting for conservation.

Around 1970 my relations with the Nature Reserve Authority (NRA) and General Yoffe became strained over the transfer of an overpopulation of gazelles from the lower Galilee, where they were becoming pests due to strict enforcement of the hunting law and the lack of predators, to the Golan Heights. Together with other Israeli scientists, I objected to the transfer, because the Golan had its own local population of gazelles, and we were afraid that mixing the populations would result in the loss of local adaptations. However, the NRA did not heed our objections. General Yoffe assured me that he could always find a scientist who would endorse his policy.

The Institute for Nature Conservation Research

Now that the NRA had taken over the responsibility for the management of reserves and the SPNI focused on education, there was an obvious need for a scientific foundation for conservation. It seemed reasonable to me that its place should be at a university. Once more, Mendelssohn provided the solution: at that time he was the Dean of Science at Tel-Aviv University, and he agreed to form the Institute for Nature Conservation Research in 1965. I became its head, until 1982. Unfortunately, the institute did not have the hoped-for impact on nature conservation policy. The governmental authority did not like outside advice, and hired its own scientists. So did the SPNI after I left. The Institute did have some impact on the conservation of freshwater habitats (due to the work of Dr Avital Gazit), the monitoring of pesticides in the environment (Professor Al Perry), on landscape surveys (conducted by Dr Zeev Meshel) and on the development of biological control of pests (Dr Baruch Sneh).

When I started my babbler research at Hazeva, the farmers, trying to save their crops, were killing and poisoning everything. They did not know which species of birds or mammals were doing the damage. Some of the habituated babbler groups were poisoned. I recruited Zohar Zuk-Rimon, my assistant from the institute, to monitor and take care of the individual birds and mammals that caused damage to the fields. Over a few years he shot and trapped some porcupines, hares, crested larks and bulbuls that were doing damage, but the rest of the wildlife survived. After that the local municipality and the Nature Reserve Authority took upon themselves the job of controlling pests in the region, and the farmers stopped the widespread use of poisons against birds and mammals.

The Institute for Nature Conservation Research was disbanded in 2005 when the School for Environmental Sciences was established at Tel-Aviv University. The president of the university and the donors were more interested in the study of the physical environment: pollution, energy, planning and environmental laws. As a retired professor, I failed to convince the university authorities of the importance of research for the conservation of the fauna and flora and the natural environment.

At the end of 1969 Azaria Alon replaced me as the general secretary of the SPNI. By that time the SPNI had around 10,000 members and five field study centers. Some conservation research was going on at the centers and provided stimulation and relief for the guides. The society encouraged and helped its guides to study and obtain university degrees.

In 1980 the SPNI and its three former secretaries – myself, Azaria Alon and Yoav Sagi – received the Israel Prize (the highest governmental prize in Israel) in appreciation for the services the SPNI and its directors provided to the state of Israel.

Back to science

Running the SPNI was a very tiring job. It was time to go back to research. In the late 1960s most of my colleagues in the academy were already professors, heads of departments or scientific institutes. However, I was prepared to start from the bottom. During my last few years as the secretary of the SPNI I was already doing research for my Ph.D., on the wintering behaviour of the white wagtail, watching the birds in the mornings before going to the office (Zahavi 1971a). Together with my assistant, Rami Dudai, I caught and colour-ringed a large number of wagtails. Many of them were found on municipal garbage dumps near Tel Aviv. We used to visit these dumps early in the mornings to follow the birds and do some experiments. I got to know a man who was scavenging the garbage, collecting items for his living. One day he asked me what I was doing there. When I told him that I was catching birds for the university, he nodded his head in sympathy, wondering about the strange jobs people were willing to undertake to make a living.

I found that the wintering wagtails could be found in three different social organizations: some males held permanent territories, and many of the ones that did so accepted a female into the territory, whereas the rest of the population wandered in flocks. I found that I could experimentally change the behavior of the wagtails from flocking to territorial, and vice versa, by changing the distribution of food from an even distribution, where the birds fed in flocks, to distinct piles, around which certain individuals formed boundaries and defended territories. Females often paired with the owner of such a territory. Such pairs could be quite stable over several weeks. Although the pair-forming displays at the wintering territory were similar to those of sexual pairing, it had nothing to do with reproduction, but rather allowed the females to obtain access to food that was otherwise only available to the dominant males. The males gained by having a subordinate helper to defend the territory. Both territorial and flocking birds met at night in communal roosts. I suggested that in this way a territorial bird that lost its feeding territory might follow a flock to a good feeding ground, as suggested by Ward for *Quelea* (Ward 1965).

I returned to Oxford in 1970, this time to the Edward Grey Institute. David Lack, the head of the Institute, kindly accepted me as a visiting scientist. Our family lived in Reading, where Avishag was able to pursue her plant physiology research with Dr Daphne Vince-Prue, and I traveled daily to Oxford.

I spent much time at the excellent library of the Institute, writing my Ph.D. dissertation as well as a paper on the function of pre-roost gathering and night-roosts as information centers (Zahavi 1971b). I became interested in the general question of the relationship between a bird's ecology and its social adaptations, and was introduced to the studies of Ian Newton and the long-term tit study at Wytham Wood, supervised by Chris Perrins.

When I returned to Oxford in 1970 I was completely unaware of the big controversy between Wynne-Edwards and Lack and Maynard Smith, about group selection versus individual selection and the then new theory of kin selection. When I worked for conservation I had little time for reading. I kept some personal connection with Tinbergen's group, but it seemed that the ethologists were not involved in that controversy. The year with Lack convinced me that individual selection was the only stable selection mechanism.

At the ornithological congress in the Hague (September 1970) I met Peter Ward, a meeting that resulted in my travelling with him to East Africa and a paper we wrote together on the gatherings of birds as information centers (Ward & Zahavi 1973). That paper was well received, and was later supported by field experiments (Parker-Rabenold 1987; Heinrich, 1988).

In spite of my conviction that behavioural phenomena should be interpreted only according to individual selection, I made the mistake of interpreting the communal displays of the pre-roost gatherings as a means to increase the size of the roost and the amount of information contained in it. This is a classical case of an argument of group selection that does not explain why an individual bird spends energy and time on participating in the group's display. It is easy to fall into the trap of explaining adaptations by their contribution to the group.

Ten years after the paper was published I received a letter from *Current Contents* addressed to Peter Ward, who had passed away a few years earlier, informing us that "This paper has been cited in over 125 publications over the last ten years, making it the 3rd most cited paper ever published in this journal." I was asked to comment on developments in the field in the intervening years. I used the opportunity to correct my mistake: I suggested that participating in the communal displays enables each individual to assess and compare itself to its neighbours and to attach itself the following day to a subgroup of ability comparable to its own, with which it can match (Zahavi 1982, 1983, 1996; Zahavi & Zahavi 1997).

At Reading I cooperated with Don Broom and some other bird ringers; we ringed wagtails at the sewage farm, and learned that British wagtails also form winter pairs on their winter territories and spend the nights in communal roosts, with all the other wagtails. My family was greatly disappointed. They often helped me watch the wagtails on the rubbish dumps in Israel; when we visited abroad, our hosts often took us first to the sewage farms and rubbish dumps, where many birds could be found. Before going to Britain we told our girls that we would spend most of our free time in forests and parks. However, on our first day at Reading, we already followed the British wagtails, which were leaving the beautiful park to roost at the sewage farm.

My stay in Europe was interrupted this time by the sudden death of my father. I became an orphan at the age of 42.

When I returned to Israel I was offered the position of a senior lecturer at the Department of Zoology of Tel-Aviv University. I think I owe this also to Mendelssohn.

The start of the babbler study

My interest in the babblers had already started at Oxford. The study of the wagtails clarified for me that ecological conditions can shape the social organization of organisms: why birds flock, keep territories or pair to defend the territory. In the library of the EGI I read the papers of Skutch on helpers at the nest (Skutch 1935) and others. These papers drew my attention to group-territorial birds and the problem of helping at the nest. I knew that in Israel babblers have helpers, and decided to try to understand that system.

With the help of the late Rami Dudai, my excellent technical assistant, who worked with me in my wagtail study, we ringed a population of some 20 groups of babblers around the Hazeva field study center. Observing the babblers became more and more difficult, however. They became wary of us because we approached their nests and caught members of their group. They considered us to be super-predators, to such an extent that we could watch them only briefly from great distances with telescopes, or from hides. I was desperate and about to quit, when I was lucky enough to visit Glen Woolfenden and his tame scrub jays in Florida in 1973. Glen invited me to come and watch his birds. He carried his binoculars and a pocket full of peanuts. I wondered whether he had left his telescopes in the hides in the field. Once outside the station he whistled, and several jays came flying, one landing on his hand. He told me that jays in the neighborhood were used to people feeding them peanuts, and he managed easily to tame the population he was studying. I hoped that the babblers would be as smart as the jays, and would learn that human presence could be beneficial. We started to provide the babblers with tidbits of bread whenever we met them. It took the whole of 1974 to tame the population. My daughter Tirtza was the first person to hand-feed a free-living babbler, by the name of LZTM. By the end of the year we could stand among babblers without interfering with their behaviour, sit next to a nest without a hide, and watch their intimate copulations, which they do away from the group. We have continued to follow the

Figure 21.2. Introducing our grandchildren Oren and Kinneret Ely to the Babblers, 1998.

Figure 21.3. Babblers play like puppies.

same population ever since (Figure 21.2). In the last few years we tamed a few desert larks, bulbuls, shrikes and blackstarts. One can observe so many details of behaviour from watching a tame wild bird in its natural surroundings. We use these birds as objects for teaching students and as a means to encourage eco-tourism (Figure 21.3).

Throughout the years we were lucky enough to have with us at Hazeva a large number of volunteers and students, many of them from abroad: Kenya, Japan, New Zealand, Australia, the USA, Canada, Britain, Germany and others – so many indeed that it is not possible to name them all here.

The Handicap Principle

The course I taught at the University was called "Socio-Ecology". I did not know at the time that I was a sociobiologist; Wilson's book had not been published yet. In 1972, when I was trying to explain Fisher's model about the evolution of the peacock's tail to my students, Yoav Sagi raised doubts about the validity of Fisher's model. He wondered why the females that, according to Fisher, started the process by choosing males with the longest tail because tail length was correlated with quality, should continue to chose the males by the length of their tails once, according to Fisher, the extra length of the tails reduced the quality of the males, and the correlation between tail-length and quality was lost. It took me several months to come up with the idea of the Handicap Principle (I started to use this term only in September 1973). I suggested that the burden imposed by signals of mate choice serves to test and advertise the quality of the males. Only high-quality males can carry a cumbersome tail. Therefore, females that select males with the most elaborate and cumbersome characters can be sure that they select from among the best male phenotypes available to them. I suggested that these burdens are similar to "handicaps" used in sports, where a more experienced player is disadvantaged in order to make it possible for a less experienced player to participate in the game. For example, a handicap in a horse-race means varying amounts of weight that are added to the saddles of the highest-rated horses. Winning with a handicap confers higher status on the winner. My family accepted the theory wholeheartedly without any reservations. Naama later made use of it in her human history seminars. Not so with my colleagues. Some of the problems might have been semantic: when I picked the term "the Handicap Principle" I was not aware that the term "handicapped" had become the politically correct synonym for "invalid". In his book *The Selfish Gene* Dawkins recalls asking me whether a male should cut off a hand or a leg in order to be attractive. An American researcher was sure that the term handicap meant only "invalid" or "disabled," and, when I tried to explain why I used the term handicap, she was offended and asked me whether I was trying to teach her English.

The following account of my attempt to introduce the Handicap Principle to my colleagues abroad is based on my letters home from that period.

In January 1973 I visited Oxford again. I discussed the idea of the Handicap Principle with Tinbergen, Ian Newton, Mike Cullen, and others. They were interested, but not convinced. As for myself, I realized that the Handicap Principle was not restricted to sexually selected signals, but explained the evolution of other signaling systems as well. It explained why the Egyptians built the Pyramids, and the purpose of the mysterious big statues of Easter Island. Many phenomena that were earlier explained by kin selection could be better explained by the Handicap Principle, which thus made kin selection redundant. I realized that I had to write a book in order to disseminate these ideas. It took us more than twenty years to write that book.

Unlike the papers on information centers, the idea of the Handicap Principle raised much opposition. During August and September 1973 I had the opportunity to discuss the Handicap Principle with many colleagues. I discussed it with Maynard Smith and Robert Selander at the International Congress of Systematic and Evolutionary Biology in Boulder, Colorado. Maynard Smith was interested but not convinced. He promised, however, to consider my paper for publication once I had written it. Selander was more agreeable. We had lengthy discussions, which continued later when I visited him in Austin, Texas. There I used the term "Handicap Principle" for the first time, in my letter home. On that same visit to the USA I presented my ideas at the Ethological congress in Washington, D.C. People were interested, but, as usual, not convinced. I think that the problem was that the idea was too basic and too simple to be "scientific". It was received as "legitimate" only years later, when Allen Grafen made formal models of the Handicap Principle (Grafen 1990a,b). Interestingly, I found later that economists, anthropologists, and members of other disciplines that deal with humans did not have any problem in accepting the Handicap Principle. At the Smithsonian Museum in Washington I was especially impressed by the handicap represented by the huge stone slabs that were used as money in the Yap Islands in the Pacific.

During that summer I collected information about a number of studies of different cooperative breeders. Following the conference we, a group of Israeli ethologists including Mendelssohn, traveled to Costa Rica, where I spent several days with Sandy Vehrencamp in her study area and observed the anis. From Jack Bradbury I learned about social bats, and was impressed by studies of the polyandric jacanas. From there we went to Barro Colorado Island, Panama, where I learned from Yael Lubin about the social spiders. I also visited Glen Woolfenden in Florida and his tame jays.

In Austin I had lengthy conversations with Selander. We discussed how the ecological conditions shape the social and breeding organization of different organisms, including the social insects, suggesting that breeding systems should be interpreted as production lines. In particular we discussed how the ecology can explain why one sex rather than the other is dominant. (Chapter 14 in Zahavi & Zahavi 1997). Selander tried to get me interested in some papers with formal models – but I found that, although the mathematics might be quite elaborate, the biological assumptions that were at the base of the models were usually too simplified, and often completely wrong. I still cannot understand why formal models are considered superior to a verbal logical analysis in the discussion of the evolution of social interactions. Social interactions are too complex for formal mathematical models. However, the logic of the evolution of social interactions can be discussed and resolved verbally. In Austin I also realized that the Handicap Principle can explain altruism: altruism could be interpreted as a signal that displays the social status of the altruist, and the investment in the altruistic act could be interpreted as the "handicap" that attests to the honesty of that signal (Zahavi 1977a). Altruism was thus explained by individual selection. I also became aware of the weakness in models of kin selection and reciprocity, a weakness shared by all models of indirect selection: all of them are vulnerable to social parasites, which can destroy any system based on indirect selection (Zahavi 2003).

Robert Trivers told me that when I visited him at Harvard later that month, he was tempted to throw me out of the room because of that heresy – but remembered in time that Selander, in a telephone call, told him that he should try and listen to me, because although I was not using the conventional terminology, I was saying something important. At Harvard I spent most of my time trying to convince Trivers, to no avail. I also met with E. Mayr, E. O.Wilson and others. I returned home on the eve of Yom-Kippur, 1973. The next morning the war broke out.

The Yom-Kippur war

In October 1973 Israel was attacked by Egypt and Syria. During the war I volunteered to serve as an ambulance driver. Our regiment was used as a reserve and we were not involved in fighting. I had ample time to read Darwin's book on sexual selection. During the long dark evenings in the desert I was listening to the jokes and chatter of the young soldiers and realized that, although I knew the lexical meaning of all the words, I was often at a loss to understand the jokes, the meaning of sentences and other subtleties of their conversations. There was a gap of only 20 years between us, and there were already so many changes in the use of idioms.

In the summer of 1974 I visited Britain again, on my way to the 16th International Ornithological congress at Canberra, Australia. I often stayed with friends and colleagues, learning about their research and discussing new ideas. I apologize for not being able to thank each of them individually here.

In Oxford I gave a seminar arguing against the idea of kin selection and tried to explain the logic of the Handicap Principle. Several years later I had a lengthy discussion with Allen Grafen. It took ten more years and two more lectures to convince him, and subsequently the sociobiology community, about the validity and importance of the Handicap Principle.

On my way back from Australia I visited John Maynard Smith at his home in Sussex. We discussed the idea of the Handicap Principle while walking in the open fields around the house. He admitted that the Handicap Principle was an attractive idea. However, he insisted that unless a formal model was made to test the idea, he was not convinced that it could work. My argument was that the logic of the handicap can be assessed with a verbal model. I am grateful to Maynard-Smith for accepting my paper in the *Journal of Theoretical Biology* (of which he was the editor) without a formal model. Moreover, his paper opposing the Handicap Principle (Maynard-Smith 1976) stimulated others to pay attention to my paper. I am much in debt to him for advertising the principle that otherwise might have been ignored, like some of my later papers dealing with basic problems in evolution such as "The Testing of the Bond" (Zahavi 1977c).

My paper "Mate Selection: a Selection for a Handicap" was finally published in 1975. Right away, I found myself debating the logic of the Handicap Principle with theoreticians (Davis & O'Donald 1976; Kirkpatrick 1986). Like Maynard Smith, they rejected the Handicap Principle using genetic models, even though I explicitly discussed its use in phenotypic interactions, especially after 1977 (Zahavi 1977a,b). The simple argument of the

Handicap Principle was considered by theoreticians to be "intuitive", despite the fact that it was based on a sound logical argument. For some reason that I cannot understand, logical models expressed verbally are often rejected as being "intuitive" and only mathematical models are considered by theoreticians even when their basic assumptions are unrealistically simplistic.

However, there were exceptions. Hamilton and Zuk (1982) used the Handicap Principle to suggest that bright-coloured plumage can advertise health and the absence of parasite infections. Diamond (1990) used the Handicap Principle to explain why people drink alcohol and use drugs. When I asked him how it was that he used the Principle that everyone else rejected, he answered that it "simply made sense."

In a talk at the Ethological Conference in Parma, Italy, in September 1975, I presented the Handicap Principle and pointed out that it can explain the evolution of altruism. Lorenz was greatly impressed. I think that he felt relieved, because he did not like the use of formal modeling that seemed to be taking over sociobiology.

At about that time I was greatly impressed by Schelling's book *The Strategy of Conflict* (Schelling 1960), especially with his suggestion that often conflicts are not just a matter of interaction between two parties: third parties are frequently involved as witnesses. I used his ideas to suggest that in addition to a signaler and a receiver, third parties are involved, and they play a role in shaping the pattern of some signals (Zahavi & Zahavi 1997). In my short paper "Why Shouting" (Zahavi 1978a) I suggested that the loud begging calls of fledglings of babblers are also directed to the attention of predators, in order to force their parents to take care of them.

In autumn 1976, when I visited Harvard again, I met with T. Schelling and his student, Michael Spence, who suggested an idea similar to the Handicap Principle, explaining the importance of the burden imposed on students to obtain university degrees as a test for their quality (Spence 1973). I found I could easily explain the Handicap Principle to them (both received the Nobel Prize several years later).

Hamilton was among the visitors to the babbler study area. I am not sure whether this visit convinced him of the validity of the Handicap Principle, but I certainly failed to draw his attention to the weakness of kin selection theory. Upon my suggestion, Hamilton kindly invited me in 1989 to present a series of talks in Oxford. I was hoping that by a series of consecutive talks I would be able to convert at least a few Oxfordians to the Handicap Principle. This time I succeeded. At the end of my second talk, Grafen declared that I was right, and that he believed he found a way to create a formal model of the Handicap Principle. Consequently he published two formal models (Grafen 1990a,b) and thus vindicated the Handicap Principle to people who insisted on mathematical models. In his papers Grafen also stated that the "main biological conclusions" of his models were "the same as those of Zahavi's original papers on the Handicap Principle" and that "the Handicap Principle is a strategic principle, properly elucidated by game theory, but actually simple enough that no formal elucidation is really required." At last the Sociobiology community accepted it. Although many prefer the terms "costly signaling," "honest advertisement" or similar synonyms rather than the term "Handicap," the main idea that signals require a special investment to ensure their reliability is now generally accepted. At a conference in August 1990, Maynard Smith strongly endorsed the Handicap Principle. Still, both Maynard Smith and Grafen, as well as many other sociobiologists, did not accept my opinion that all signals are loaded with handicaps (Maynard Smith & Harper 2003).

Many of my colleagues claim that I present new ideas because of an urge "to be different." In fact, the logic of the Handicap Principle, together with a very stubborn belief that selection works only directly on the individual, made it necessary for me to reconsider many of the interpretations of social behaviour presently accepted by sociobiologists. In the second edition of *The Selfish Gene* (p. 313), Dawkins remarks that if the Handicap Principle is correct "it might necessitate a radical change in our entire outlook". Unfortunately these implications of the Handicap Principle have not been discussed by the scientific community. Many of these interpretations are presented and discussed in our joint book *The Handicap Principle* (Zahavi & Zahavi 1997). Nearly every one of the 18 chapters of the book reconsiders some social phenomena in a new way. In a recent paper "Indirect Selection and Individual Selection in Sociobiology: My Personal Views on Theories of Social Behaviour" (Zahavi 2003), I comment that

being on the periphery has its benefits: if I were dependent on my colleagues for the advancement of my scientific career or my social status, I would not have been able to continue developing the Handicap Principle over the many years when it was nearly unanimously rejected. Luckily I was living in a distant corner of the world, and usually interacted with other sociobiologists only once a year, at conferences. At home, my social status and my scientific career were well secured because of my previous "altruistic" work in conservation.

Some implications of the Handicap Principle

Although much of the following is described in our book and my papers, I believe it may interest readers to follow the lines of thought that led to the development of some of the implications of the Handicap Principle. My wife Avishag helped me define and describe ideas in better form in my book and in many papers, and often had a significant part in the shaping and the clarification of ideas.

The inflation of signals as a test for the Handicap Principle

For many years I was troubled by a nagging question: how can I verify my hypothesis that in signals, investment is obligatory, rather than being a side effect as is the case with any other trait? In order to answer this question, I had to define the difference between the selection of signals and that of other traits. The distinction between the selection mechanism of signals and that of all other traits became clearer to me when my daughter Naama studied the history of the use of lace. Naama, by that time a student of history at the Hebrew University of Jerusalem, became interested in the use of lace for decorations in Europe in the sixteenth and seventeenth centuries. Lace was very expensive because of the large amount of skilled labor that was needed to produce it. Its value was above that of its weight in gold, and as such it served to advertise people's wealth. Soon after lace-making machines were invented, the price of lace dropped drastically (Pond 1973). At first everyone used large amounts of lace – but very soon it went out of fashion, as it could no longer serve to distinguish the very rich from the moderately well-to-do.

I realized that Naama's description of the use of lace could be the solution to my problem. Signals can function to differentiate between signalers only when the investment in the signal is differential: easier to high-quality signalers and more difficult to low-quality signalers. If the investment is reduced to the extent that all individuals can signal alike, the signal can no longer function to highlight differences between signalers, and is selected out by a process similar to inflation in money – unlike non-signal traits, the benefit of which increases when the investment required to develop them is reduced. I therefore suggested that natural selection includes two selection mechanisms: a selection for efficiency, by which most traits are selected; and signal-selection, which encompasses the signaling components of Darwin's sexual selection, by which all signals are selected. The process of inflation can test and support the theory that signals are selected by a different mechanism from that of all other traits (Zahavi 1981a,b.; Zahavi & Zahavi, 1997). However, the evolutionary process of inflation cannot be tested in multicellular organisms by short-time experiments. I therefore became interested in the social behaviour of microorganisms, thinking that their brief generation-period might enable one to test the idea experimentally (see below). I have not yet managed to find a laboratory that would conduct the experiment.

Providing clear and precise information as a handicap

When I tried to define the borderline dividing signals with and without handicaps, I focused on a set of markers that were considered as markers of belonging to a certain set, such as species, gender or age group. Markers that are considered set-specific, such as a dot, or a line across the tail, are often very small and seem to require only minute investment. The question was whether these signals entailed handicaps. I resolved this question by pointing out that these patterns do not carry messages in their own right, but rather they are standards by which information on body shape, size, movements, etc. is better shown. A dot on the forehead can help to evaluate the shape of the forehead. Other ornaments, such as lines, stripes, and colour patches, can serve as reference points to the size or shape of the individual or its body parts, or to the quality of its movements (Zahavi 1978b). Like all other signals, these set specific signals have evolved through competition among individuals to advertise their advantages over other members of their own set, rather than to advertise their belonging to the set (Zahavi 1978b; Zahavi & Zahavi 1997). Obviously, once these markings were present, they could be used to infer that an individual belonged to a certain set. However, this was not the primary reason these signals were selected for and maintained by natural selection.

Hasson (1991) and Maynard Smith and Harper (2003) suggest that such set-specific characters provide information without imposing a cost on the signaler. They term them

"Amplifiers" and "Indices", respectively. I suggest that providing clear and honest informa-tion about the exact quality of a signaler is in itself a handicap. The gains and losses of such displays are differential: the better individual usually gains, because its superiority becomes more obvious. The lesser one loses, because its defects become more noticeable, and it can no longer exploit the margin of doubt as to its qualities. However, even the best individual in a group may lose – when a still better one joins the competition; and the lowest-quality one may gain when someone of a still lower quality joins in.

Ritualization

This is the process by which signals evolve out of traits that were not signals to begin with (Huxley 1914). Huxley suggested that the process of ritualization evolves due to the common interest of partners to communicate clearly. However, the idea that standard patterns evolve through competition enabled me to suggest that ritualization also evolves as the result of competition among individuals to advertise their qualities. Observers can judge small differences between competitors better if the competitors display their abilities in a standard way. This is the reason that strict standards are imposed at the highest levels of human contests in areas such as sport, beauty, or music. Ritualization is the process by which such competitive standards evolve (Zahavi 1980; Zahavi & Zahavi, 1997).

Cheating

I use the following definition for a signal: a trait that has evolved, in the signaler, in order to transfer information to receivers to affect the behaviour of the receivers in a manner that is beneficial to the signaler. The receiver should respond only to reliable signals, to ensure that the change in its behaviour suits its own interests as well. Hence receivers select signals to be reliable. With the HP I was able to reinterpret, as honest signals, many interactions that were considered as cheating or manipulations (e.g. Dawkins & Krebs 1978).

Threat signals

The phenomenon most often interpreted as cheating is threat displays, in which the signaler is described as trying to display its size as bigger than it really is, by growing manes, extending crests and spreading fins. For example, the manes of monkeys and lions are supposed to increase the apparent size of the head. However, if that is the reason, why are the manes coloured differently from the rest of the head? I suggested that the coloration of the mane displays it clearly as an appendix. In fact, manes impose additional handicaps, since by forming a frame around the head they reduce somewhat the apparent size of the head. This is a handicap that a mature lion or monkey can afford, but a young one that wishes to display its skull as big as it is cannot afford (Zahavi & Zahavi 1997; Zahavi 2006).

Cuckoos

Cheating is also supposed to explain the relationship of some cuckoos and their hosts. In 1975 I stayed with Karl Vernon at his study site of the helmet shrikes in Zimbabwe. In the evenings, sitting outside the bungalow, I was listening to his stories about his studies of social parasites among the South African birds. One of the stories was about a cuckoo that pushed the host out of its nest, laying its egg while the host perched on the rim of the nest. After the cuckoo departed, the host resumed its incubation, not bothering to evict the parasite's egg, which was very different in colour and size from its own. This story did not fit the prevalent idea that a social parasite cheats its host by mimicking the host's egg. Also, it was not reasonable to assume that natural selection has "not yet" had the time to select for a cleverer host. During one of these long nights I figured out the "Mafia model." In my paper (Zahavi 1978c) I did not use the term "Mafia," but I described a strategy similar to the one used by the Mafia to explain the logic of the model: I suggested that the cuckoo should revisit the nest and predate on the offspring of a host that rejected its eggs, but not predate if a cuckoo nestling was present in the nest. It is important to note that the benefit to the cuckoo is not in punishing of the host, but in the food it gets, or in forcing the host to re-lay in a territory governed by the individual parasite. The term "Mafia model" was coined by Soler in his paper on the great crested cuckoo (Soler *et al.* 1995). Avishag and I visited Soler and stayed with his team in their residential caves, and enjoyed watching the interactions of the magpies and the crested cuckoos. (Zahavi & Zahavi 1997, p. 189).

The story of the European cuckoo and its reed warbler host is quite different. The European cuckoo is a successful nest parasite. When its egg is accepted, the parasite's nestling kills all of the host's offspring. Still, in the same host population, some individuals evict the cuckoo's egg, whereas others do not evict it. Such mixed populations are found across Eurasia from Britain to Japan. I could not believe that over all that area selection has not "yet" eliminated the reed warblers that cannot recognize a cuckoo egg. I speculated that the coexistence of rejecters and acceptors was the result of some equilibrium. Obviously I did not know what that equilibrium was. My student Arnon Lotem decided to study why some reed warblers do not reject the eggs of the European cuckoo. Nick Davies could not accommodate him at Oxford, so Arnon decided to study the interactions of cuckoos and reed warblers in Japan, in collaboration with Hiroshi Nakamura. We met Hiroshi and his family a year earlier, on our visit to Japan, and enjoyed their hospitality, including a visit to a potential study site and to the mountains. The conclusion from Arnon's study was that the difference between acceptors and rejecters was not a consequence of genetic polymorphism, but rather of age. Arnon found that the acceptors were first-year breeders, whereas older birds usually rejected the cuckoo eggs. He suggested that at its first breeding, the warbler was not yet familiar with the coloration of its own eggs – there was therefore a danger that it would eject its own egg instead of that of the cuckoo (Lotem *et al.* 1995). It is important to note that, although selection can develop cuckoos that mimic the eggs of reed warblers in general, they cannot match the pattern of the eggs of the particular individual they parasitize.

Arnon's study did not provide an answer to the question of why hosts continue to feed their parasite's nestling and fledgling, even though the shape and size of the cuckoo nestling is so different from that of their own offspring. My present hypothesis is that, once the host loses the chance to breed that year, tending to an alien nestling provides the host with higher social prestige than the alternative of not having anyone to feed at all. Such prestige helps it breed successfully in the following years. This answer is similar to my interpretation of helpers at the nest in cooperative breeders: they feed offspring that are not their own and are often not related to them.

Babblers and altruism

When I started to study the babblers in 1971, I did not believe in interpretations based on group selection, but I was willing to explain social adaptations by kin selection or reciprocity, as most people still do today. However, data collected during the first years of the study revealed that large groups of babblers did not reproduce better than groups of average size (Zahavi 1974, 1989). Hence, I was faced with the problem of why all individuals in large groups invest in helping when their help is not needed. Furthermore, detailed observations revealed that individuals compete with one another to help, and sometimes are even aggressive towards other group members that try to help. The first indication that babblers compete to act as altruists came from the observations of my daughter Tirtza, who collected data on sentinel activity. The data revealed that dominants replaced subordinates as sentinels, but subordinates did not replace dominants directly. Allofeeding in adult birds is also nearly always unidirectional: in almost all cases, higher-ranking individuals feed lower-ranking ones. Subordinates that are not ready to give up their post as sentinels, or ones that do not accept a feeding, are sometimes attacked by the dominant. Most allofeedings are carried out in the open, and the donors usually advertise their donations by loud purrs, calling the attention of the rest of the group: we often see group members raising their heads and watching allofeeding interactions. Sometimes they come to intervene (Kalishov *et al.* 2005) (figure 21.4).

Since I developed the Handicap Principle at the same time that I started my study of the babblers, I was able to suggest that in babblers, and in many other group-living animals, altruism could replace overt aggression as means of advertising claims for social prestige. Furthermore, the investment in altruism could be interpreted as a handicap that attests to the reliability of the claim (Zahavi 1977a,b.). Data collected by my students and collaborators further support the theory that altruism in babblers can best be interpreted as an investment in claiming social prestige. (A partial list of these studies is given in our book, several of these are in Hebrew with brief English summaries only, as some of these students chose not to pursue an academic career, and thus were not pressed to publish their work in English.)

In 1981 Tamsie Carlisle joined our team. Tamsie came to Hazeva after conducting a B.Sc. study with Dawkins and a Ph. D. with Trivers – that is, she came from the hard-core of kin selection and reciprocity theories. She studied the helping at the nest by yearling

Figure 21.4. A dominant babbler feeds and exchanges a subordinate sentinel.

babblers. Her results convinced her that the only way to interpret her data was to assume that the babblers compete to help, an assumption that could not be accounted for by kin selection or reciprocal altruism (Carlisle & Zahavi 1986). Also, babblers often help non-relatives that join their group, although they know that they are not related (as shown by the patterns of avoidance of incest). Competitive altruism in babblers represents a constant quest for prestige among these group-living birds (Zahavi 1990). The idea that social prestige has an important role in all social interactions, including those of the mating pair, resolved my observations at Ravenglass years ago, when I noticed the competition between the male and female black-headed gulls. Instead of trying to exploit one another, letting their partner invest in incubation, they compete to incubate. I was now able to suggest that each one invested to increase its prestige in the eyes of its partner.

Not everyone who studied our population of babblers at Hazeva agreed with my con-clusions. John Wright, who spent several seasons studying the feeding at the nest and collected data on sentinel activity, did not notice evidence for competition to serve the group, and interpreted his results with a model of group-augmentation, which is in fact a model of group selection (Wright 1997). I believe he underestimated or probably missed behaviours that we interpret as competition.

Signals among microorganisms

I became interested in studying signaling and social behaviour of microorganisms for two reasons: I could not believe that even an individual bacterium or amoeba would kill itself for the sake of other individuals, and I hoped that the study of the signaling of microorganisms would suggest an experimental model by which I could simulate the inflation process of signals (Zahavi & Ralt 1984).

We have not yet found a laboratory that would be able and willing to conduct an experiment in evolution to study the process of inflation in signaling among microorganisms. However, using the Handicap Principle, we suggested that traits that used to be interpreted as altruism by models of group selection (Shapiro 1998) could be interpreted better by models of individual selection. Perhaps the most important of these studies was our interpretation of the social life of slime molds, explaining the evolution of the active cell death performed by the stalk-forming cells as a selfish trait (Atzmoni *et al* 1997; Zahavi 2005). We suggested that, for low-quality amoebas under stress, active cell death may be the best option to have a chance to pass at least some of their genes to the next generation by transfection. Small as that chance may be, it is still better than the alternative of death by lysis. I believe that a similar interpretation may apply to bio-films of bacteria that are presently interpreted according to models of group selection.

The production of antibiotics as a selfish trait

Bruce Levin, then at Amherst, directed my attention to the problem of explaining the evolution of antibiotics by a model of individual selection. Microbiologists interpret the production of antibiotics by bacteria as weapons by which populations of bacteria fight each other. This interpretation begs the question: why should every individual bacterium invest in producing the antibiotics? There are millions of bacteria in a population. Any individual that refrains from investing in the production of antibiotics would have an advantage over those that invest for the sake of the population. Trivers (1971) already pointed out that when an individual has the chance to exploit another and does not do so, that individual is an altruist. Bruce told me about his studies of bacteria in sewage. He expected that the fierce competition among the many bacteria species would result in a high secretion of antibiotics in sewage. However, he found the opposite: the amount of antibiotic secretion was very low, too low to eliminate stocks of susceptible bacteria. This suggested that in sewage bacteria were not using antibiotics as a weapon against other populations. At about that time, I saw in Jim Shapiro's lab in Chicago how individual bacteria space themselves: each is surrounded by a tiny space, like the individual space around a bird within a flock. I suggest that the antibiotic serves to protect the tiny space surrounding an individual bacterium against encroachment by its neighbors. This suggestion is supported by the well-known fact that the antibiotic industry has to select special bacterial lines in order to produce large quantities of antibiotics. The evolution of antibiotics can be compared to the development of the longbow in Britain. It is reasonable to assume that the longbow was used by British people to fight their fellow British. King Henry V used the British bowmen to fight the French. The battle of Agincourt was won by the longbow – however, the longbow was not developed for that purpose.

Our dog Namir and his contribution to the theory of testing the bond

My first and only dog was Namir, a beautiful and lovely German pointer (Figure 21.5). I intended to use him to locate the nests of the red-legged partridge that I was planning to study once I finished with the babblers. All through his ten years with us Namir was frustrated, because he was not allowed to run after the partridges in the study area. He was not interested in babblers, however, and readily stayed in the jeep when we were in the field. As soon as I left the jeep, he would move to the front to take over the driver's seat, which he probably considered to be the most prestigious. Later he would relax and lie down across the back, completely ignoring the babblers that would often hop around him, looking for tidbits. But although I was able to command him not to run after partridges, I could not stop him when a hare ran in front of the car. He kept pursuing them, although he never managed to overtake any of them.

Namir's main contribution to science was his habit of jumping on me when I came home tired from work, just as my little daughters used to do years earlier. His favourite place at home was the passage between the kitchen and the dining room, where we had to push him to pass by, or on the doorstep when he realized that I was getting ready to leave the house. At that time I was observing daily the clumping and allopreening of babblers, and especially their morning dances, in which they jump over one another, push one another, and interfere

Figure 21.5. Our dog Namir.

Figure 21.6. Allopreening (photo by Oded Keinan).

Figure 21.7. "Water dance". When these desert birds find water they often cannot resist getting wet – followed by the water dance.

with the movements of their group members (Figures 21.6 – Figures 21.8). I could not avoid noticing the common elements in the behaviour of the dog, the babblers, and my daughters when they were young. I concluded that by imposing a burden, individuals are testing the social bond between them. The reaction to the interference provides a clue and a measuring

Figure 21.8. Babblers in a night roost.

stick to the motivation of the social partner to bond. Indeed, why would one who is not interested in a social bond accept the burden of an interfering partner? In my short, but I believe very important paper: "The testing of the bond" (Zahavi 1977c), I acknowledged and expressed my thanks to Namir.

Faithful bees

During our long drives, whether abroad or during the weekly shuttle between our study area at Hazeva and our home in Tel-Aviv, we discuss (and often argue over) interpretations of our own observations or those of colleagues. One such endeavor was to understand why bees stay faithful to the flowers they pollinate. The stimulus to ask the question came one day when we were standing with our friend and colleague Professor. Danny Cohen in front of the botany department at the Hebrew University of Jerusalem. The flowers in the bed resembled those of a legume. However, we learned from Danny that they were in fact not a legume. Danny suggested off-hand that these flowers resemble the legumes in order to cheat the bees and lure them to visit them. On our way to Hazeva, we started to ponder about the system of plant–bee interactions. We knew that when collecting pollen, honey bees and some other bees are faithful to a single plant species on each collecting trip. Obviously it is in the interest of the plant to be pollinated by, and send its pollen to, flowers of its own species only. However, we could not see why the bee should care to be faithful, instead of collecting pollen from the nearest flower of whatever kind. We also knew that when honey bees collect nectar, they are not necessarily faithful to the species, often moving between flowers of very different structures. We therefore decided that somehow the plants must be manipulating bees to stay faithful to them when collecting pollen. The logical conclusion was to look for the answer in the grains of pollen.

We knew that pollen grains vary greatly in shape, size, and structure even among closely related species. We hypothesized that perhaps the pollen of a single species sticks together to form a neat pollen pack, but that trying to pack together pollen of different species could cause the bee to lose its load of pollen. If pollen from a different species would cause the bee to lose the load, it would be worth while for the bee to spend time on searching for flowers of a single species. In cooperation with Dr Avner Cohen from the Volcani center and Professor Dini Eisikowitch from Tel-Aviv University, an expert on pollination, we examined loads of pollen under light electron microscopes. Indeed, loads composed of pollen of a single species were packed neatly, the grains fitted tightly to each other. However, even a single foreign grain disrupted the packing and often caused the loads to crack (Zahavi *et al.*1984). Subsequently Dini and his students extended and verified these observations.

Colour

Wherever we traveled – in the deserts of the Namib or the Kalahari, in Peru or in Australia, in the tropical forests of the Old and the New World, in the swamps or on the sea shore – we tried to observe and make sense of the choice of colour for display in birds' plumage. Avishag's expertise in the effects of light spectra on plant development was of help. Our general conclusions about the factors that select for the use of a particular colour in advertising are described in our book (Zahavi and Zahavi 1997).

In our travels we met many fellow researchers. We visited their research areas, discussed their observations with them, and often enjoyed their hospitality at their homes. I apologize for not being able to thank most of them by name in this chapter.

The Hazeva Field Study Center

The center was established in 1969 by the SPNI when I was still its general secretary. From 1971 it was the base for our babbler study, housing our volunteers and students. I usually spend 2–3 days there every week. The center also forms a base for conservation, education and research for the Rift Valley and the central Negev. The activities of the center facilitated the formation of two small but beautiful Nature reserves, the Shezaf and the Gidron, adjacent to the Jordanian border. The babbler research in these areas played a role in convincing the authorities to establish the reserves.

In 1997 the SPNI decided to close its activities at the Hazeva Field Study Center. The Center's buildings were derelict houses of a former army camp and a set of badly designed buildings that presented difficult maintenance problems. Maintenance was very costly, and it was not balanced by the income provided by visitors. Closing the center was also in line with a new policy of the SPNI, to concentrate on activities in city communities rather than in the field. I opposed that trend. Hence, I suggested to the SPNI that I should take upon myself the responsibility for the operations and financing of the center. The burden was much greater than I had expected. Much of my time and efforts over the following ten years were invested in maintenance, looking for financial support, finding the staff to operate the center, and renovating the derelict hostel, dining room, kitchen, and classrooms. Still, I succeeded. I did not do it alone: several guides and technical staff, and a good number of young volunteers who came to help for one year or more, carried out the day-to-day operations. I would not have been able to succeed without their help. Not least, I am obliged to several donors including my family, who provided financial support. In 2006 the center was maintaining itself financially and I handed it back to the SPNI, whose new director promised to keep it going.

Obviously, I was doing less scientific research during these years, but I continued to observe the babbler with my students, and had the satisfaction of having around me young people who felt responsible for conservation and education with the same spirit that was found in the SPNI when I headed it. I am especially grateful to my former student Roni Ostreicher, who has been living at the center with his family since 1989 and continues to participate in the babbler study.

The evolution of pheromones and hormones

Over the past few years, and especially since 2006, my main scientific challenge has been trying to understand the evolution of chemical signals. I suggest that, like other signals, pheromones are shaped by the Handicap Principle – that is, there is a relationship between their chemical structure and the messages encoded in them. The similarity between the structure of many pheromones and hormones led me to conclude that the Handicap Principle is effective even among cells of the multicellular organism. In the multicellular organism, handicaps prevent hormone production by cell phenotypes that have not developed properly. This means that the patterns of hormones – their chemical properties – should also bear a relationship to the messages encoded in them. (Zahavi 1993, 2006, 2008). With my students, we read literature on microbiology, chemistry, endocrinology and histology, and discuss what we find with professional colleagues. Preliminary results of the study of the sex steroid hormones already helped to suggest what is the message in testosterone and in estrogen and why testosterone is secreted by one type of cells and estrogen by another (Zahavi & Fuks, in preparation.)

Group selection vs. individual selection

I am convinced that natural selection works only through the individual. It troubles me very much to find that models of group selection are again often used to explain evolutionary phenomena (Wilson & Holldobler 2005; Wilson & Wilson 2007). I am also concerned about the use of models of indirect selection such as kin selection and reciprocity by mainstream researchers. I am concerned because these theoretical approaches affect the definition of objectives, the planning of research, the set of facts to be collected by observations or experiments, and, of course, the interpretation of results. When a certain trait helps the group but seems to harm individuals possessing it, researchers often follow the easy way out to explain the observations with a model of indirect selection. They stop searching for the possibility that the individual within the group is doing the best it could do to serve its own interests. Interpretations based on individual selection may require more data and may not be easy to develop. In the past I was able to suggest interpretations based on individual selection for data that were interpreted by indirect selection by other researchers. What I consider as major successes in that effort are my re-interpretations of altruism in babblers and in slime molds.

It took 15 years to "vindicate" the Handicap Principle as a means by which signals evolve to be reliable. It may take longer to appreciate its use in signaling within the body and its importance in explaining signaling in microorganisms and in the social insects with models of individual selection. This is one of the reasons why I continue to teach undergraduates, in the hope that sooner or later they will look for the facts and find the means for such interpretations.

Epilogue

I am still deeply involved with conservation, trying to preserve as much as possible of the natural landscape and wildlife of Israel, especially the desert. I advise the SPNI and governmental agencies, and educate young volunteers and visitors at the field study center at Hazeva to appreciate nature and to fight for its conservation.

Figure 21.9. From right to left: Naama, Kinneret, Tirtza, Amotz. Avishag is behind the camera. (2007).

Figure 21.10. Avishag, 2008.

With my wife Avishag, both of us now retired, we continue to watch the babblers in the early mornings several days a week. We know each babbler personally, by a name coined from the initials of its four rings. We have followed most of them since they fledged from the nest. We watch how they compete to establish their prestige in their own group. We follow their fate when they disperse to join other groups, or become refugees. A few individuals eventually succeed in breeding for a short, or sometimes even a very long period. We mourn when they die. We collect their stories and hope that eventually we shall find the time to write them down (Figures 21.9 and 21.10).

References

Atzmoni, D., Zahavi, A. & Nanjundiah, V. (1997). Altruistic behaviour in *Dictyostelium discoideum* explained on the basis of individual selection. *Curr. Sci.* **72**: 142–5.

Carlisle, T. R. & Zahavi, A. (1986). Helping at the nest, allofeeding and social status in immature Arabian babblers. *Behav. Ecol. Sociobiol.* **18**: 339–51.

Davis, G. W. F. & O'Donald, P. (1976). Sexual selection for a handicap. A critical analysis of Zahavi's model. *J. Theor. Biol.* **57**: 345–54.

Dawkins, R. (1989). *The Selfish Gene*, 2nd edn. Oxford: Oxford University Press

Dawkins, R. & Krebs, J. R. (1978). Animal signals: information or manipulation. In *Behavioral Ecology*, ed. N. Davis & Krebs, J. R., pp. 282–309. Blackwell. London:

Diamond, J. (1990). Kung Fu kerosene drinking. *Nat. Hist.* **99**: 20–4.

Grafen, A. (1990a). Biological signals as handicaps. *J. Theor. Biol.* **144**: 517–46.

Grafen, A. (1990b). Sexual selection unhandicapped by the Fisher process. *J. Theor. Biol.* **144**: 473–516.

Hamilton, W. D. & Zuk, M. (1982). Heritable true fitness and bright birds: a role for parasites? *Science* **218**: 384–7.

Hasson, O. (1991). Sexual displays as amplifiers. Practical examples with an emphasis on feather decorations. *Behav. Ecol.* **2**: 189–97.

Heinrich, B. (1988). Winter foraging at carcasses by the three sympatric corvids, with emphasis on recruitment by the raven *Corvus corax*. *Behav. Ecol. Sociobiol.* **23**: 141–56.

Huxley, J. S. (1914). The courtship habits of the great crested grebe (*Podiceps cristatus*) with an addition to the theory of sexual selection. *Proc. Zool. Soc. Lond.* **35**: 491–562.

Kalishov, A., Zahavi, A. & Zahavi, A. (2005). Allofeeding in Arabian Babblers (*Turdoides squamiceps*). *J. Ornithol.* **146**: 141–50.

Kirkpatrick, M. (1986). The handicap mechanism of sexual selection does not work. *Am. Nat.* **127**: 222–40.

Lotem, A., Nakamura, H. & Zahavi, A. (1995). Constraints on egg discrimination and cuckoo-host co-evolution. *Anim. Behav.* **49**: 1185–209.

Maynard Smith, J. (1976). Sexual selection and the handicap principle. *J. Theor. Biol.* **57**: 239–42.

Maynard Smith, J. & Harper, D. (2003). *Animal Signals*. Oxford: Oxford University Press.

Parker-Rabenold, P. (1987). Recruitment to food in black vultures; evidence for following from communal roosts. *Anim. Behav.* **35**: 1775–85.

Pond, G. (1973). *An Introduction to Lace*. London: Garnstone Press.

Schelling, T. C. (1960). *The Strategy of Conflict*. Cambridge, MA: Harvard University Press.

Shapiro, J. A. (1998). Thinking about bacterial populations as multicellular organisms. *A. Rev. Microbiol.* **52**: 81–104.

Skutch, A. F. (1935). Helpers at the nest. *Auk* **52**: 257–73.

Soler, M., Soler, J., Martinez, J. G. & Moller, A. P. (1995). Magpie host manipulation by great spotted cuckoos: evidence for an avian mafia? *Evolution* **49**: 770–75.

Spence, M. (1973). Job market signaling. *Q. J. Econom.*, **87**: 355–74.

Tinbergen, N. (1951). *The Study of Instinct*. London: Oxford University Press.

Trivers, R. L. (1971). The evolution of reciprocal altruism. *Q. Rev. Biol.* **46**: 35–57.

Wahrman, J. & Zahavi, A. (1958). Cytogenetic analysis of mammalian sibling species by means of hybridization. *Proc. Xth Int. Congr. Genetics* **11**: 304–5.

Ward, P. (1965). Feeding ecology of the black-faced dioch *(Quelea quelea)* in Nigeria. *Ibis* **107**: 173–214.

Ward, P. & Zahavi, A. (1973). The importance of certain assemblages of birds as "information-centers" for food-finding. *Ibis* **115**: 517–34.

Wilson, D. A. & Wilson, E. O. (2007). Rethinking the theoretical foundation of sociobiology. *Q. Rev. Biol.* **82**: 327–48.

Wilson, E. O. & Holldobler, B. (2005). Eusociality: origin and consequences. *Proc. Natl Acad. Sci. USA* **102**: 13367–71.

Wright, J. (1997). Helping at the nest in Arabian babblers: signaling social status or sensible investment in chicks? *Anim. Behav.*, **45**: 1439–48.

Zahavi, A. (1957). The breeding birds of the Huleh Swamp and Lake (Northern Israel). *Ibis* **99**: 600–7.

Zahavi, A. (1971a). The social behaviour of the white wagtail, *Motacilla alba*, wintering in Israel. *Ibis* **113**: 203–11.

Zahavi, A. (1971b). The function of pre-roost gatherings and communal roosts. *Ibis* **113**: 106–9.

Zahavi, A. (1974). Communal nesting by the Arabian Babbler, a case of individual selection. *Ibis* **116**: 84–7.

Zahavi, A. (1975). Mate selection: a selection for a handicap. *J. Theor. Biol.* **53**: 205–14.

Zahavi, A. (1977a). Reliability in communication systems and the evolution of altruism. In *Evolutionary Ecology*, ed. B. Stonehouse, & C. M. Perrins, pp. 253–9. London: Macmillan Press.

Zahavi, A. (1977b). The cost of honesty (further remarks on the Handicap Principle). *J. Theor. Biol.* **67**: 603–5.

Zahavi, A. (1977c). The testing of the bond. *Anim. Behav.* **25**: 246–7.

Zahavi, A. (1978a). Why shouting. *Am. Nat.* **113**: 155–6.

Zahavi, A. (1978b). Decorative patterns and the evolution of art. *New Scient.* **80**: 182–4.

Zahavi, A. (1978c). Parasitism and nest predation in parasitic cuckoos. *Am. Nat.* **113**: 157–9.

Zahavi, A. (1980). Ritualization and the evolution of movement signals. *Behaviour* **72**: 77–81.

Zahavi, A. (1981a). Natural selection, sexual selection and the selection of signals. In *Evolution Today*, ed. G. G. E. Scudder & J. L. Reveal, pp. 133–8. Pittsburgh, PA: Carnegie-Mellon University Press.

Zahavi, A. (1981b). Some comments on sociobiology. *Auk* **98**: 412–14.

Zahavi, A. (1981c). The lateral display of fishes: bluff or honesty in signaling? *Behav. Anal. Lett.* **1**: 233–5.

Zahavi, A. (1982). Some further comments on the gatherings of birds. *Acta 18th Congr. Int. Ornithol.*, ed. V. D. Ilychev & V. M. Gavrilov, vol. 2, pp. 919–20. Moscow: Academy of Sciences of the USSR, 1985.

Zahavi, A. (1983). This week's citation classic: The importance of certain assemblages of birds as "information centers" for food finding. *Curr. Cont.* **15**: 26.

Zahavi, A. (1989). Arabian Babbler. In *Lifetime Reproduction in Birds*, ed. I. Newton, pp. 253–76. London: Academic Press.

Zahavi, A. (1990). Arabian Babblers: The quest for social status in a cooperative breeder. In *Cooperative Breeding in Birds. Long-term Studies of Ecology and Behavior*, ed. P. B Stacey & W. D. Koenig, pp. 103–30. Cambridge: Cambridge University Press.

Zahavi, A. (1993). The fallacy of conventional signaling. *Phil. Trans. R. Soc. Lond.* **B338**: 227–30.

Zahavi, A. (1996). The evolution of communal roosts as information centers and the pitfall of group-selection: a rejoinder to Richner and Heeb. *Behav. Ecol.* **7**: 118–19.

Zahavi, A. (2003). Indirect selection and individual selection in sociobiology: my personal views on theories of social behaviour. *Anim. Behav.*, **65**: 859–63.

Zahavi, A. (2005). Is group selection necessary to explain social adaptations in microorganisms? *Heredity* **94**: 143–4.

Zahavi, A. (2006). Sexual selection, signal selection and the handicap principle. In *Reproductive Biology and Phylogeny of Birds*, Part B, ed. B. G. M. Jamieson, pp. 143–59. Plymouth: Science Publishers.

Zahavi. A. (2008). The handicap principle and signaling in collaborative systems. In *Sociobiology of Communication, an Interdisciplinary Perspective*, ed. P. D'Ettorre & D. P. Hughes, pp. 1–10. Oxford: Oxford University Press.

Zahavi, A. & Ralt, D. (1984). Social Adaptations in Myxobacteria. In *Myxobacteria: Development and Cell Interactions*, ed. E. Rosenberg, pp. 215–20. New York: Springer Verlag.

Zahavi, A. & Wahrman, J. (1957). The cytotaxonomy, ecology and evolution of the gerbils and jirds of Israel (*Rodentia: Gerbillinae*). *Mammalia* **21**: 341–80.

Zahavi, A. & Zahavi, A. (1997). *The Handicap Principle*. New York: Oxford University Press.

Zahavi, A., Eisikowitch, D., Kadman-Zahavi, A. & Cohen, A. (1984). A new approach to flower constancy in honey bees. In *21 Symposium International sur la Pollinisation* (Les Colloques de l'INRA No 21) INRA Publ.(ed.), pp. 89–95. Versailles: INRA.

Printed in the United States
by Baker & Taylor Publisher Services